Plant Response and Tolerance to Abiotic Oxidative Stress: Antioxidant Machinery as a Paradigm of Defense

Plant Response and Tolerance to Abiotic Oxidative Stress: Antioxidant Machinery as a Paradigm of Defense

Editor

Nafees A. Khan

MDPI • Basel • Beijing • Wuhan • Barcelona • Belgrade • Manchester • Tokyo • Cluj • Tianjin

Editor
Nafees A. Khan
Aligarh Muslim University
India

Editorial Office
MDPI
St. Alban-Anlage 66
4052 Basel, Switzerland

This is a reprint of articles from the Special Issue published online in the open access journal *Antioxidants* (ISSN 2076-3921) (available at: https://www.mdpi.com/journal/antioxidants/special_issues/Plant_Tolerance).

For citation purposes, cite each article independently as indicated on the article page online and as indicated below:

LastName, A.A.; LastName, B.B.; LastName, C.C. Article Title. *Journal Name* **Year**, *Volume Number*, Page Range.

ISBN 978-3-0365-7782-1 (Hbk)
ISBN 978-3-0365-7783-8 (PDF)

Contents

About the Editor

Nafees A. Khan

Professor Nafees A. Khan, Department of Botany, Aligarh Muslim University, Aligarh, India, researches the mechanisms of the hormonal and nutritional regulation of plant development, with an emphasis on plants' resilience against stressful environments via the use of nutriomics and metabolomics. He has published research papers in crucial high-impact-factor journals and collaborated internationally on research to unravel the strategies adopted by plants to protect and strengthen antioxidant machinery via optimizing nutrient use efficiency through the judicious modulation of plant hormones/signaling molecules for the better survival of crop plants under changing climatic conditions. Prof. Khan has been noted as a Highly Cited Researcher in the area of Plant Science by Elsevier every year from 2019 to 2022. He has edited nineteen books published by Elsevier, Springer-Nature, Frontiers, NOVA, Alpha Science, and others. He has served as the Editor/Guest Editor of leading plant science journals published by Elsevier, Frontiers, Springer-Nature, and MDPI. He is a Fellow of the National Academy of Sciences, India (NASI), as well as being a Fellow of the Linnean Society, the Indian Botanical Society, and the Indian Society for Plant Physiology. Dr. Khan has received several accolades, among them being the UGC Research Award, Scientist of the Year Award, Distinguished Scientist Award, UGC Mid-Career Award, Teaching Excellence Award, and Innovation and Research Excellence Award. He has a high number of global citations (Google Scholar cumulative citation 18530; h-index 70 and i-10 index 192).

Preface to "Plant Response and Tolerance to Abiotic Oxidative Stress: Antioxidant Machinery as a Paradigm of Defense"

Plants are continuously exposed to a large number of abiotic stress factors during their life cycles. These influence their cellular metabolisms, physiological processes, and growth as well as development. Thus, plants have evolved inherent strategies with which to acclimatize to changing abiotic stress conditions and adapt through in-built tolerance mechanisms; however, drastic changes in climatic conditions further aggravate the impact of stress on plants, which exceeds plants' tolerance capacities. Factors such as low or high temperatures, drought or waterlogging, high salinity, ultraviolet radiation, heavy metals, nutrient stress, etc., are grouped under abiotic stress, which generates the excessive production of reactive oxygen species (ROS), causing oxidative stress. One of the essential ways through which to improve plants' tolerance and relieve the pressure of abiotic-stress-induced oxidative changes for the survival of plants is to improve the capacity of antioxidant machinery. The research/review articles included in this book provide collective knowledge on the antioxidant machinery for abiotic stress tolerance mechanisms.

Nafees A. Khan
Editor

antioxidants

Article

Effect of Thallium(I) on Growth, Nutrient Absorption, Photosynthetic Pigments, and Antioxidant Response of *Dittrichia* Plants

Francisco Espinosa *, Alfonso Ortega, Francisco L. Espinosa-Vellarino and Inmaculada Garrido

Research Group FBCMP(BBB015), Faculty of Sciences, Campus Avenida de Elvas s/n, University of Extremadura, 06006 Badajoz, Spain
* Correspondence: espinosa@unex.es

Abstract: *Dittrichia* plants were exposed to thallium (Tl) stress (10, 50, and 100 µM) for 7 days. The Tl toxicity altered the absorption and accumulation of other nutrients. In both the roots and the leaves, there was a decline in K, Mg, and Fe content, but an increase in Ca, Mn, and Zn. Chlorophylls decreased, as did the photosynthetic efficiency, while carotenoids increased. Oxidative stress in the roots was reflected in increased lipid peroxidation. There was more production of superoxide ($O_2^{\cdot-}$), hydrogen peroxide (H_2O_2), and nitric oxide (NO) in the roots than in the leaves, with increases in both organs in response to Tl toxicity, except for $O_2^{\cdot-}$ production in the roots, which fluctuated. There was increased hydrogen sulfide (H_2S) production, especially in the leaves. Superoxide dismutase (SOD), ascorbate peroxidase (APX), dehydroascorbate reductase (DHAR), monodehydroascorbate reductase (MDHAR), and glutathione reductase (GR) showed increased activities, except for APX and MDHAR in the roots and GR in the leaves. The components of the ascorbate–glutathione cycle were affected. Thus, ascorbate (AsA) increased, while dehydroascorbate (DHA), reduced glutathione (GSH), and oxidized glutathione (GSSG) decreased, except for in the roots at 100 µM Tl, which showed increased GSH. These Tl toxicity-induced alterations modify the AsA/DHA and GSH/GSSG redox status. The NO and H_2S interaction may act by activating the antioxidant system. The effects of Tl could be related to its strong affinity for binding with -SH groups, thus altering the functionality of proteins and the cellular redox state.

Keywords: antioxidant defense system; ascorbate; *Dittrichia*; glutathione; hydrogen sulfide; nitric oxide; photosynthesis; reactive oxygen species; thallium toxicity

Citation: Espinosa, F.; Ortega, A.; Espinosa-Vellarino, F.L.; Garrido, I. Effect of Thallium(I) on Growth, Nutrient Absorption, Photosynthetic Pigments, and Antioxidant Response of *Dittrichia* Plants. *Antioxidants* **2023**, *12*, 678. https://doi.org/10.3390/antiox12030678

Academic Editor: Stanley Omaye

Received: 30 January 2023
Revised: 2 March 2023
Accepted: 7 March 2023
Published: 9 March 2023

1. Introduction

Plants' response to stress includes a sharp increase in reactive oxygen, nitrogen, and sulfur species (ROS, RNS, and RSS, respectively) production, which alters the cellular redox balance [1–4]. The damage that is induced by this alteration, and by the excess ROS, is dealt with by the enzymatic and non-enzymatic antioxidant systems [5–7]. The former includes superoxide dismutase (SOD, EC 1.15.1.1), peroxidases (POD, EC 1.11.1.7), catalases (CAT, EC 1.11.1.6), ascorbate peroxidase (APX, EC 1.11.1.11), glutathione reductase (GR, EC 1.6.4.2), monodehydroascorbate reductase (MDHAR, EC 1.6.5.4), and dehydroascorbate reductase (DHAR, EC 1.6.4.2), and the latter includes ascorbate (AsA), reduced glutathione (GSH), phenolics, alkaloids, non-protein amino acids, α-tocopherols, and carotenoids [2,8]. The ascorbate–glutathione (AsA–GSH) cycle removes hydrogen peroxide (H_2O_2), maintaining redox homeostasis [2]. The production of nitric oxide (NO), the main RNS, is also related to plants' response to stress [9–11]. It acts at the level of the expression of the defense genes that are involved in eliminating ROS [12–14] and plays a key role in the defense mechanisms against different stressors [15], including heavy metals, such as Cd and As [9,16–19]. Heavy metals increase the synthesis of NO [9,16,20], which is involved in activating the antioxidant defense system and in eliminating the excess ROS that is produced, thus contributing to the maintenance of redox homeostasis [9,21–23]. Hydrogen sulfide (H_2S) is involved in numerous metabolic processes in plants [24–26], including the response to heavy metal toxicity [20,25,27,28]. The exogenous application of H_2S enhances the antioxidant defense system's response, allowing the oxidative stress that is induced by heavy metals to be reduced [29,30]. H_2S modulates the activation of the antioxidant system and the gene expression of its components [31,32]. Both NO and H_2S act by controlling the AsA–GSH cycle components and the ROS levels, efficaciously removing H_2O_2 [33]. The levels of GSH are key in the processes of defense against heavy metals, being able to bind to them and also contribute to the biosynthesis of phytochelatins [27,34]. Through the induction of the antioxidant system, the interaction between H_2S and GSH lowers ROS production under conditions of heavy metal toxicity [29,35,36]. The increase in NO and H_2S, and their interaction, may act on the antioxidant systems that are involved in the stress response, enhancing their activity and the expression of the genes that are involved in these antioxidant defense systems, thus contributing to the elimination of ROS and the maintenance of redox homeostasis [26,32,37].

Thallium (Tl) is a very toxic element for living beings. It belongs to group IIIA of the periodic table, occurring as Tl(I) and Tl(III). The former is the most stable form, and the latter is the most toxic [38]. Tl forms minerals, including silicates and sulfates [39]. It is widely distributed in the environment in low concentrations of between 0.3 and 0.5 $\mu g\ g^{-1}$ [40]. It is used as a catalyst in alloys, optical lenses, jewelry, low-temperature thermometers, semiconductors, dyes, etc., and its salts as rodenticides and insecticides; although, the WHO recommends against its use due to its great toxicity [40]. It is found in ionic form in drainage waters [41–43], being released from the aqueous waste and soils of disused Pb–Zn and As mines.

Since Tl and K have similar ionic radii, they can be absorbed by plants through the same mechanisms, depending on the plant species [38]. Tl-induced toxicity is commonly a symptom of the replacement of K with Tl [44]. While in *Arabidopsis thaliana* Tl and K have antagonistic effects in that they mutually interfere with each other's absorption [45], in *Biscutella laevigata,* increased K does not inhibit the uptake of Tl and, therefore, it seems that the absorption of Tl would not be carried out through the same systems as K, but it will be carried out through specific transporters [46].

Tl hyperaccumulator plants have been identified in the Brassicaceae family [44]. They accumulate Tl in their leaves and roots, and, to a lesser extent, in their stems and fruits, with a dependence on species and soil type. Thus, in *Brassica juncacea,* the maximum Tl accumulation in the aerial part is in the leaves and the minimum in the flowers [39], whereas in other species, Tl accumulation occurs mainly in the roots, then the leaves, and, to a lesser extent, in the stem [47–49]. Some ferns and aquatic plants of the genus *Lemna* are reported to have a great capacity for Tl accumulation, which would make them interesting for use in Tl phytoremediation processes [42,43].

Various hydroponic culture studies have been carried out on the absorption, accumulation, and toxicity of Tl [46,50,51]. In *Sinapis alba* under these conditions, most of the Tl is transported to the leaves [50]. In *Silene latifolia* and *Biscutella laevigata,* low Tl concentrations induce a hormetic increase in biomass in response to slight toxic stress [46,51–53]. This contrasts with the results of the in vitro culture of *Arabidopsis,* which show a decline in biomass [45]. Although Tl(I) is not a redox metal, it can alter photosynthetic electron transport, leading to increased ROS and MDA [49]. The oxidation of pigments, the alteration of complexes, and the disappearance of the grana occur in the discolored foliar areas [44]. Chang et al. [45] describe the inhibition of gene expression of LCH II subunits. Tl toxicity induces oxidative stress [44], with increases in H_2O_2 and in SOD, APX, and POX activities [49,54]. These enzymes are also positively regulated at the gene expression level under Tl toxicity [44]. The effect differs between the ionic forms, with Tl(III) inducing greater ROS production than Tl(I) [55]. Pu et al. [48] describe a decrease in the SOD activity and an increase in POX, while Liu et al. [56] describe increases in both of these activities. This activity in response to heavy metal toxicity can vary with exposure time and can also inhibit the synthesis of these enzymes and alter their assembly [57]. Tl has a strong affinity for the amino imino and the sulfhydryl groups of proteins and other biological macromolecules, an example being glutathione, which decreases in its reduced form (GSH) due to oxidation or to the formation of complexes with Tl ($Tl(SG)_3$) [27,34]. There has yet to be evidence for the participation of NO and H_2S in these defense responses. *Dittrichia viscosa* (L.) Greuter is a species belonging to the Asteraceae family that develops in a wide range of soils and climatic conditions in the Mediterranean basin. *Dittrichia* can colonize poor soils and also those that are degraded due to anthropic activities, such as mining, with high concentrations of heavy metals and metalloids [58]. *Dittrichia* are capable of absorbing and accumulating large amounts of Cd, Cu, Fe, Ni, Pb, Zn, As, and Sb [11,58–64]. The participation of the components of the AsA–GSH cycle, as well as NO and H_2S, in the Sb accumulation process has recently been reported [11,63]. However, this capacity varies depending on the different genotypes or populations, which shows great genetic plasticity [60–62]. In accordance with this, *Dittrichia* can be considered a good candidate to carry out phytoremediation processes for heavy metals and metalloids, despite not being a hyperaccumulator plant [52]. The present study was therefore aimed at determining the involvement of ROS, NO, H_2S, and the antioxidant systems in *Dittrichia* plants' response to Tl, and the morphophysiological alterations that areinduced by the toxicity of this element.

2. Materials and Methods

2.1. Plant Materials, Growth Conditions, and Treatments

The seeds of *Dittrichia viscosa* (L.) Greuter obtained from "Semillas Silvestres, SL" were surface sterilized for 15 min in 10% sodium hypochlorite solution (40 g L^{-1}), rinsed several times with distilled water, and, before their germination, were imbibed in distilled water, aerated, and agitated for 2 h at room temperature. After imbibition, the seeds were germinated in a plastic container (30 × 20 × 10 cm) filled with a sterilized perlite mixture substrate wetted with Hoagland solution and were kept at 27 °C in the dark for 48 h. After germination, the seedlings were cultivated for 5 days at 27 °C and 85% relative humidity, with a light/dark photoperiod of 16 h/8 h, and a constant illumination under photosynthetic photon flux density of 350 μmol m^{-2} s^{-1} during the day.

After 7 days, the plants were grown in hydroponic culture in lightweight polypropylene trays (20 × 15 × 10 cm; 4 plants per container) and the same environmental conditions (except for relative humidity, 50%). The plants were cultivated in a basal nutrient solution composed of the following: 4 mM KNO_3, 3 mM $Ca(NO_3)_2$ $4H_2O$, 2 mM $MgSO_4$ $7H_2O$, 6 mM KH_2PO_4, 1 mM NaH_2PO_4 $2H_2O$, 10 μM $ZnSO_4$ $7H_2O$, 2 μM $MnCl_2$ $4H_2O$, 0.25 μM $CuSO_4$ $5H_2O$, 0.1 μM Na_2MoO_4 $2H_2O$, 10 μM H_3BO_3, and 20 μM NaFeIII-EDTA. After 10 days in hydroponic culture, we started our assay with Tl. For the Tl treatment, the basal solution was supplemented with Tl (I) sulfate (Tl_2SO_4) in final concentrations of 0 μM (control), 10 μM, 50 μM, and 100 μM Tl. Each cultivation solution was adjusted to pH 5.8, continuously aerated, and changed every 5 days. The plants were exposed to the Tl treatments for 7 days.

The plants of each treatment were divided into roots and shoots, which were washed with distilled water, dried on filter paper, and weighed to obtain the fresh weight (FW). Half of the roots and shoots from each Tl treatment were dried in a forced-air oven at 70 °C for 24 h to obtain the dry weight (DW) and the subsequent analysis of the concentration of Tl. The other half of the fresh roots and leaves were used for the biochemical analyses. The relative water content (RWC) of the leaves was determined at the time of harvest from fresh material in accordance with the method described by Smart and Bingham [65]. Leaf disks were collected from the different treatments, and their FWs were determined. They were then immersed in distilled water for 1 h, dried externally with filter paper, and weighed again to obtain the turgid weight (TW). Finally, they were oven-dried at 70 °C for 24 h and weighed to obtain the DW. The RWC was calculated as RWC% = (FW − DW)/(TW − DW) × 100.

2.2. Determination of Tl and Mineral Content

The plant material (roots and leaves) of the control and Tl treatments was harvested and rinsed with distilled water. After 24 h of drying at 70 °C, the root and leaf material was crushed in a marble ceramic mill. The Tl, K, Mg, Ca, Fe, Mn, Cu, and Zn content was measured by inductively coupled plasma-mass spectrometry (ICP-MS, model NexION 300, PerkinElmer) in accordance with Lehotai et al. [66]. The bioaccumulation factor (BF) was calculated from the ratio between the concentration of the element in the roots or leaves and that present in the hydroponic solution, and the de translocation factor (TF) was calculated from the ratio between the concentration of the element in the leaves and in the roots.

2.3. Determination of the Soluble Amino Acid and Protein Contents

The fresh plant material (roots and leaves) of the control and Tl treatments was homogenized (0.2 g mL^{-1}) in 50 mM sodium phosphate buffer, pH 7.0, then filtered through muslin, centrifuged at 12,360× *g* for 15 min, and the supernatant was used for the protein and amino acid assays. For the protein assay, Bradford's method was used [67]. The results are expressed as mg g^{-1} FW against an albumin standard curve. For the amino acid assay, the method of Yemm and Cocking [68] was used, with ninhydrin reagent. The results are expressed as mg g^{-1} FW against a glycine standard curve.

2.4. Determination of Photosynthetic Pigment Contents and Photosynthetic Efficiency

The chlorophyll and carotenoid contents of the leaves were determined at the end of each trial. About 0.125 g of fresh leaves were incubated in 10 mL methanol for 24 h in the dark. The concentrations of chlorophylls and carotenoids were measured spectrophotometrically (Shimadzu UV1603) at A_{666}, A_{653}, and A_{470}. The total chlorophyll and carotenoid content was calculated as described by Wellburn [69] and expressed as µg g^{-1} FW.

For the determination of the photosynthetic parameters, the middle region of the fully expanded upper leaves at the end of each Tl treatment were adapted in the dark for 10 min, and then the minimal fluorescence (F_0), the maximal chlorophyll fluorescence (F_m), and the maximum photosynthetic efficiency (F_v/F_m) were recorded with a handheld fluorometer (Chlorophyll Fluorometer, OS-30p, Opti-Sciences). The variable fluorescence (F_v) and the rate constants of photochemical and nonphotochemical deactivation of excited Chl molecules (F_v/F_0) was calculated.

2.5. Determination of Lipid Peroxidation and Reactive Oxygen Species ($O_2^{\cdot-}$ and H_2O_2), NO, and H_2S Contents, and SOD Activity

To analyze oxidative stress, the formation of malondialdehyde (MDA) was determined using thiobarbituric acid (TBA). Briefly, 0.25 g of plant material (roots or leaves) was homogenized with 2.5 mL of solution containing 0.25% TBA and 10% trichloroacetic acid (TCA). The mixture was incubated at 95 °C for 30 min. The reaction was stopped by immersing the tubes in ice, then filtering, and centrifuging at 8800× *g* for 10 min. The MDA was determined spectrophotometrically in the supernatant at $A_{532}-A_{600}$ with ε = 155 mM^{-1} cm^{-1} and expressed as µmol MDA g^{-1} FW [70]. The H_2O_2 content was analyzed in accordance with Velikova et al. [71]. Fresh material (roots or leaves) was homogenized (0.2 g mL^{-1}) in 0.1% TCA. The homogenate was centrifuged at 12,000× *g* for 15 min. The reaction mixture contained 0.5 mL of supernatant, 0.5 mL 10 mM of potassium phosphate buffer (pH 7.0), and 1 mL of 1 M KI solution. The H_2O_2 concentration was estimated based on the reaction mixture absorbance at A_{390} using a standard curve of H_2O_2. The $O_2^{\cdot-}$ generating and SOD activities were measured in an extract obtained from roots or leaves that was homogenized (0.5 g mL^{-1}) at 4 °C in 50 mM phosphate buffer, pH 6.0, 0.5 mM phenylmethylsulfonyl fluoride (PMSF), 1 mM β-mercaptoethanol, and 1 g L^{-1} plyvinylpolypyrrolidone (PVPP). The homogenate was filtered and centrifuged at 39,000× *g* for 30 min at 4 °C, and the supernatant was collected as an enzyme extract. The $O_2^{\cdot-}$ generation was assayed spectrophotometrically by measuring the oxidation of epinephrine to adrenochrome at A_{480} (ε = 4.020 mM^{-1} cm^{-1}) [72]. The reaction mixture contained 1 mM epinephrine in acetate buffer 25 mM, pH 5.0. The SOD activity was determined in 50 mM phosphate buffer, pH 7.8, 0.1 mM ethylene diamine tetra-acetic acid (EDTA), 1.3 μM riboflavin, 13 mM methionine, and 63 μM nitro blue tetrazolium (NBT) [73]. The reaction mixture was maintained in the dark at 25 °C, and the reaction was started by the addition of the riboflavin and enzyme extract and was illuminated for 2 min. The changes in absorbance A_{560} were measured. A unit of SOD is defined as the amount of enzyme required to cause 50% inhibition of NBT reduction. The NO content was determined using the method described by Zhou et al. [74]. Roots or leaves (0.5 g) were homogenized in 3 mL of 50 mM cool acetic acid buffer (pH 3.6, containing 4% zinc diacetate). The homogenates were centrifuged at 10,000× *g* for 15 min at 4 °C. The supernatant was collected. The pellet was washed in 1 mL of acetic acid buffer and centrifuged as before. The two supernatants were combined, and charcoal was added, followed by vortexing and filtration. A mixture of filtrate and the Griess reagent (1:1) was incubated at room temperature for 30 min. The absorbance was determined at A_{540}. The NO content was calculated by comparison against a standard curve of $NaNO_2$ and expressed as nmol $NaNO_2$ g^{-1} FW. The H_2S concentration was determined using Li's method [75]. Roots or leaves were ground in liquid nitrogen, homogenized (0.5 g mL^{-1}) in 20 mM Tris-HCl buffer, pH 8.0, containing 10 mM EDTA and 20 mM $Zn(OAc)_2$, and centrifuged at 15,000× *g* for 15 min at 4 °C. The supernatant was combined with 30 mM $FeCl_3$ (in 1.2 M HCl) and 20 mM DMPD (in 7.2 M HCl) (1:1:1). The mixture was incubated at room temperature for 15 min and the A_{670} was determined. The H_2S content was calculated by comparison against a standard curve of NaHS and expressed as nmol H_2S g^{-1} FW.

2.6. Determination of the Components and Antioxidant Enzymes of the AsA–GSH Cycle

To determine the total ascorbate and glutathione, fresh roots or leaves (1 g mL^{-1}) were homogenized at 4 °C in 5% metaphosphoric acid. The homogenate was centrifuged at 20,000× *g* for 20 min at 4 °C, and the supernatant was collected for the determination of ascorbate and glutathione. The total ascorbate pool (AsA + DHA) and total glutathione pool (GSH + GSSG) were determined in accordance with De Pinto et al. [76]. Total ascorbate was determined by the reduction of DHA to AsA, and the concentration of DHA was estimated from the difference between the total ascorbate pool and the AsA. The ascorbate pool was determined at A_{525}. The glutathione pool was determined from the change in absorbance at A_{412} over 1 min. GSH was estimated as the difference between the amount of total glutathione pool and that of GSSG.

To determine the activities of the enzymes involved in the AsA–GSH cycle—APX, MDHAR, DHAR, and GR—the roots or leaves (0.5 g L^{-1}) were homogenized at 4 °C in 50 mM phosphate buffer, pH 7.5, 0.5 mM PMSF, 1 mM β-mercaptoethanol, 1 g L^{-1} PVPP, and 5 mM AsA for the APX activity. The homogenate was filtered and centrifuged at 39,000× *g* for 30 min at 4 °C, and the supernatant was used for the enzyme determinations. The APX activity was determined spectrophotometrically by measuring the oxidation of ascorbate at A_{290} for 2 min (ε = 2.8 mM^{-1} cm^{-1}) [77]. The reaction mixture contained 0.5 mM ascorbate, 0.2 mM H_2O_2, and the enzyme extract, at 25 °C, in 0.1 M phosphate buffer, pH 7.5, and EDTA 0.5 mM, expressing the result as μmol ascorbate min^{-1} mg^{-1} protein. The DHAR activity was determined from the oxidation of DHA at A_{265} for 1 min (ε = 14 mM^{-1} cm^{-1}) [78] in a medium containing 0.1 M phosphate buffer (pH 6.5), 0.5 mM EDTA, 2.5 mM GSH, 0.5 mM DHA, and the enzyme extract. The DHAR activity is expressed as nmol ascorbate min^{-1} mg^{-1} protein. The MDHAR activity was determined from the oxidation of NADH at A_{340} for 1 min (ε = 6.22 mM^{-1} cm^{-1}) [79] in a medium containing 50 mM Tris-HCl buffer (pH 7.8), 10 mM AsA, 0.2 mM NADPH, 0.5 units of ascorbate oxidase, and the enzyme extract, expressing the result as μmol NADH min^{-1} mg^{-1} protein. The GR activity was determined at A_{340} from the oxidation of NADPH for 3 min (ε = 6.22 mM^{-1} cm^{-1}) [78] in a medium containing 0.1 M phosphate buffer (pH 7.5), 0.5 mM EDTA, 0.5 mM GSSG, 0.2 mM NADPH, and the enzyme extract, expressing the result as nmol NADPH min^{-1} mg^{-1} protein.

2.7. Statistical Analyses

The data to be presented are the means ± SE of at least 10 replicates obtained from three independent experiments. For each measurement, a Shapiro–Wilk normality test was performed (since $n < 50$) to verify if they had a normal distribution. Later, we applied a parametric test of one-way ANOVA. Those values where there are significant differences are marked with different letters, that is, where $p \leq 0.05$. All statistical analyses were performed with Microsoft Excel 365 (Microsoft Inc., Alburquerque, NM, USA) and the SPSS v. 24 package (SPSS Inc., Chicago, IL, USA).

3. Results

3.1. Effect of Tl on the Growth, Accumulation of Mineral Elements, and Soluble Amino Acid and Protein Content of Dittrichia

The growth of *Dittrichia* is very similar at all of the Tl concentrations, with little difference between them (Figure 1A, Table 1). On the contrary, the growth that can be observed in the control plants, with 0 µM Tl, is much greater. The plants that were grown under Tl toxicity show a significantly smaller size than the control plants in both the shoot and the root parts (Table 1); furthermore, the leaves appear chlorotic, especially in the venal area and, to a lesser extent, in the interveinal area. The length of the shoots decreases drastically with the Tl concentration to values of 38% and 32% for 10 µM and 100 µM, respectively. The effect of the three concentrations of Tl on the growth of the roots is similar, with an inhibition of 48% with respect to the length of the control roots.

Table 1. Effect of Tl on the length, fresh weight (FW), dry weight (DW), and fresh weight percentage (%DW) of the *D. viscosa* plants. The data are the means ± SE from ten independent experiments ± SE, each carried out in triplicate. The different letters indicate significant differences at $p < 0.05$.

Treatments	Length (cm)		FW (g)		DW (g)		% DW	
	Roots	Shoots	Roots	Shoots	Roots	Shoots	Roots	Shoots
Control	46.29 ± 2.33a	17.21 ± 1.89a	2.135 ± 0.332a	4.603 ± 0.495a	0.181 ± 0.010a	0.690 ± 0.047a	6.47 ± 0.15b	11.78 ± 0.29c
10 µM Tl	18.10 ± 2.08b	6.55 ± 0.81b	0.532 ± 0.077b	1.267 ± 0.132b	0.079 ± 0.004b	0.237 ± 0.032b	10.05 ± 0.38a	18.03 ± 0.50b
50 µM Tl	19.90 ± 1.32b	6.27 ± 0.50b	0.520 ± 0.075b	1.094 ± 0.113c	0.072 ± 0.008b	0.230 ± 0.031b	9.81 ± 0.73a	17.62 ± 0.87b
100 µM Tl	19.25 ± 1.68b	5.67 ± 0.43b	0.500 ± 0.053b	0.845 ± 0.055d	0.046 ± 0.009c	0.183 ± 0.036c	7.20 ± 0.97b	22.16 ± 1.07a

The fresh weight of both the shoots and the roots decreases in response to the Tl dose. This weight decreases significantly in the shoots as the Tl concentration increases (73%, 77%, and 82% for 10 µM, 50 µM, and 100 µM Tl, respectively). On the contrary, in the roots, the fresh weight is similar in the different concentrations of Tl that were used, being 75% of the control values. The dry weight decreases in both the shoots and the roots, depending on the concentration of Tl in the medium. The DW/FW percentage increases in the plants that were subjected to Tl toxicity, with the greatest value in the shoots being observed at 100 µM Tl. In the roots, an increase in this percentage can be observed due to the effect of Tl, although this increase is less when increasing the Tl. The RWC (Table S1) shows no alteration in the roots, while it decreases very slightly in the shoots (≈9%).

The absorption and the accumulation of Tl increases with the concentration of Tl in the medium (Table 2). In the roots, the Tl content is positively correlated with that of the medium, but not in leaves, where the accumulation of Tl is similar for 50 µM and 100 µM. The accumulation observed here is much greater in the roots than in the shoots. Thus, in the treatment with 10 µM Tl, the accumulation in the roots is 6× that produced in the leaves, with this accumulation being of greater magnitude with the higher Tl concentrations. This greater capacity for accumulation in the roots is reflected in a higher BF (i.e., $BF_{roots} > 350$, $BF_{shoots} < 100$). On the other hand, the value of TF_{Tl} is low, decreasing progressively with the increase in Tl in the medium. As the concentration of Tl in the roots increases, its transport capacity decreases (from 0.163 to 0.043 for 10 µM and 100 µM Tl, respectively).

Table 2. Tl content in roots and leaves, bioaccumulation factor (BF), and translocation factor (TF) in *D. viscosa* plants. The data are the means ± SE from 10 independent experiments, each carried out in triplicate. The different letters indicate significant differences at $p < 0.05$. nd = not detect.

Treatments	Tl (mg kg^{-1} DW)		BF		TF
	Roots	Shoots	Roots	Shoots	
Control	1.9d	nd	nd	nd	nd
10 µM Tl	2286.0 ± 378.2c	374.7 ± 48.9b	571a	93a	0.163a
50 µM Tl	7991.2 ± 786.0b	623.9 ± 89.5a	399b	31b	0.078b
100 µM Tl	14,035.7 ± 634.9a	601.1 ± 121.0a	351b	15c	0.043c

Table 3 shows the effect of Tl toxicity on the absorption and the accumulation of different macroelements (K, Mg, and Ca) and microelements (Fe, Mn, Cu, and Zn). The K content in the root and shoot samples shows a decreasing trend as the Tl concentration increases, and this decrease is greater in the roots than in the leaves. However, there is a wide variability in the measured data for this microelement. With regard to the value of TF_K, it is similar to the control value, except with 10 µM Tl, in which a slight decrease can be observed (Table S2). Similarly to K, the Mg content (Table 3) decreases due to Tl independently of the concentration, and this is more pronounced in the roots than in the shoots. The TF_{Mg} values increase (×2) with respect to the control, although they are similar for the three Tl concentrations. The lowest Ca content in the roots is observed in the plants that were grown in 10 µM of Tl, with a decrease of 52%, relative to the control. However, in the roots of the plants that were grown with 50 µM Tl, the decrease is 20%, and with 100 µM Tl, the amount of this element increases (by 28%, relative to the control). In the shoots, the Ca content is not significantly modified.

Table 3. Effect of Tl on the K, Mg, Ca, Fe, Mn, Cu, and Zn contents in roots and shoots of *D. viscosa* plants. The data are the means ± SE from ten independent experiments, each carried out in triplicate. The different letters indicate significant differences at $p < 0.05$.

Treatments	K (mg kg^{-1} DW)		Mg (mg kg^{-1} DW)		Ca (mg kg^{-1} DW)	
	Roots	Shoots	Roots	Shoots	Roots	Shoots
Control	84,014.1 ± 27,699.2a	96,366.0 ± 31,295.4a	34,884.6 ± 2213.3a	6093.3 ± 1696.1a	44,796.5 ± 10,631.2ab	14,465.0 ± 399[illegible].8a
10 µM Tl	75,916.4 ± 29,047.9a	61,058.8 ± 17,123.4a	11,205.2 ± 4113.9b	4863.9 ± 782.5b	21,418.2 ± 5794.7c	15,267.4 ± 313[illegible].7a
50 µM Tl	58,217.0 ± 20,738.0b	61,336.0 ± 21,882.6a	12,748.7 ± 5783.5b	4729.2 ± 918.6b	35,963.2 ± 10626.0b	14,249.8 ± 279[illegible].5a
100 µM Tl	50,594.4 ± 21,353.5b	72,777.5 ± 32,277.6a	13,252.0 ± 4551.0b	5813.6 ± 1731.1a	57413.1 ± 10,935.5a	16,019.9 ± 593[illegible].9a

Treatments	Fe (mg kg^{-1} DW)		Mn (mg kg^{-1} DW)		Cu (mg kg^{-1} DW)		Zn (mg kg^{-1} DW)	
	Roots	Shoots	Roots	Shoots	Roots	Shoots	Roots	Shoots
Control	5466.5 ± 3559.0a	134.9 ± 68.2a	882.2 ± 131.3c	55.3 ± 5.0c	45.7 ± 12.5	12.1 ± 2.8a	52.7 ± 20.7b	38.7 ± 7.8
10 µM Tl	2530.1 ± 1131.4b	151.2 ± 75.8a	1450.3 ± 594.1a	225.9 ± 93.9a	41.9 ± 12.5	8.4 ± 2.6b	83.2 ± 35.8ab	51.2 ± 17.5a
50 µM Tl	3031.6 ± 1149.0b	111.6 ± 69.0b	1083.1 ± 128.5b	144.6 ± 69.8b	51.2 ± 17.3	8.3 ± 4.0b	103.0 ± 32.6a	47.0 ± 23.9ab
100 µM Tl	4134.8 ± 1034.3a	84.1 ± 39.5b	1345.4 ± 17.3a	172.6 ± 72.3a	74.4 ± 15.5a	8.3 ± 2.6b	161.7 ± 56.5a	53.8 ± 23.4a

With regard to the absorption and the accumulation of Fe, Mn, Cu, and Zn (Table 3), in the roots there is an increase in the content of Mn, Cu, and Zn, while the content of Fe decreases. While Cu and Zn increase with an increasing Tl dose, the increase in Mn is similar for all of them. The decrease in the Fe content is smaller in response to an increase in the concentration of Tl in the medium (54%, 45%, and 24% for 10 µM, 50 µM, and 100 µM Tl, respectively). In the leaves, Fe decreases with the increase in Tl in the medium and Cu decreases by 30% (independently of the Tl concentration), while Mn and Zn increase. With regard to the values of TF (Table S2), the TF_{Fe} increases at 10 µM Tl, but is similar to the control value for the other concentrations. The TF_{Mn} increases, while TF_{Cu} and TF_{Zn} decrease.

Tl induces an increase in the amino acid content, which is greater in the leaves (≈×4.3) than in the roots (≈×1.8) (Table 4). This strong increase does not depend on the dose of Tl that is used. With regard to the protein content (Table 4), in the roots that were subjected to 10 µM of Tl toxicity, there is little difference from the control, with a slight increase being observed; however, with 50 µM and 100 µM, a progressive decrease can be observed, being by 16% and 41%, respectively. In the leaves, there is a decrease, which is greatest with 10 μM Tl (by 43%, relative to the control).

Table 4. Effect of Tl on the soluble amino acid and protein contents in *D. viscosa* plants. The data are the means ± SE from ten independent experiments, each carried out in triplicate. The different letters indicate significant differences at $p < 0.05$.

Treatments	Total Soluble Amino Acids (mg g^{-1} FW)		Total Proteins (mg g^{-1} FW)	
	Roots	Leaves	Roots	Leaves
Control	2.375 ± 0.517b	2.540 ± 0.236b	2.12 ± 0.18a	2.82 ± 0.43a
10 μM Tl	4.176 ± 0.461a	10.816 ± 1.628a	2.39 ± 0.07a	1.60 ± 0.17c
50 μM Tl	4.439 ± 0.665a	10.171 ± 0.920a	1.77 ± 0.16b	1.90 ± 0.12b
100 μM Tl	4.459 ± 0.759a	11.192 ± 1.015a	1.25 ± 0.28c	1.82 ± 0.21b

3.2. Effect of Tl on Photosynthetic Pigment Content and Photosynthetic Efficiency

Tl toxicity induces a decrease in both chlorophyll a and b, and, consequently, in total chlorophyll (Figure 1A–C). The decrease in these contents is smaller and very similar in 10 μM and 50 μM, with values of 54% and 63% of the control, respectively. For 100 μM Tl, this content is 45% of the control. The chlorophyll a/b ratio (Figure 1D) increases in response to Tl, going from 1.60 to 1.92 for the control leaves and the leaves that were subjected to Tl, respectively. In contrast, the carotenoid content (Figure 1E) shows a clear increase (by around 140%) in the plants that were subjected to Tl, although with no difference between the different concentrations that were used. The decrease in chlorophylls and the increase in carotenoids caused by Tl translate into an increase in the carotenoids/chlorophylls ratio (Figure 2F).

In the Tl treatments, the F_0 value increases as the Tl concentration increases (Figure 1G). Thus, the F_0 increases by 10% for 10 μM Tl, while with 50 μM and 100 μM Tl the increases are much greater, at 65% and 71%, respectively. The efficiency of photosystem II (F_v/F_m) (Figure 1H) is also strongly affected by the Tl doses, with the decreases being dependent on the increase in Tl, ranging from 10% to 36% for 10 μm and 100 μM Tl, respectively. With regard to the F_v/F_0 ratio (Figure 1H), there is a marked decrease with the increase in the Tl concentration that was used. Thus, while in the control plants the F_v/F_0 ratio is 3.604, in those that were treated with 100 μM this ratio is 1.026.

3.3. Lipid Peroxidation, Production of $O_2^{\cdot-}$ and SOD Activity, and H_2O_2, NO, and H_2S Content

Figure 2 shows the response to the treatments with Tl at the level of lipid peroxidation production of $O_2^{\cdot-}$ and SOD activity, and H_2O_2, NO, and H_2S content. In the roots, lipid peroxidation increases in response to Tl (Figure 2A). On the contrary, in the leaves, the levels of lipid peroxidation are lower, with similar values for the three concentrations of Tl that were used. The production of $O_2^{\cdot-}$ (Figure 2B) also differs between the roots and the leaves, being greater in the former. In the roots, the differences between the treatments are slight and varying, with decreases of 18% and 23% for 10 μM and 100 μM Tl, respectively, but with no significant change for 50 μM Tl. In the leaves however, the $O_2^{\cdot-}$ production increases with increasing Tl concentrations in the medium. The SOD activity, in response to Tl (Figure 2C), increases in both the roots and the leaves, with the latter presenting the greatest values. In both of the organs, the SOD activity increases for all of the concentrations of Tl. For the treatment with 10 μM Tl, the SOD activity increases by ×1.8 and ×1.4 for the roots and the leaves, respectively, and for 50 μM Tl, by ×2.3 and ×1.8, respectively. With 100 μM of Tl, the SOD activity also increases, although at lower values (×1.6 and ×1.4, in the roots and the leaves, respectively). In both of these organs, the lowest and highest Tl concentrations cause similar increases in SOD activity. The H_2O_2 content (Figure 2D) increases in both the roots and the leaves in response to Tl-induced stress. The H_2O_2 content in the roots increases from 125 μmol g^{-1} FW in the controls to values that reach 844 μmol g^{-1} FW for 100 μM Tl and shows a strong positive correlation with the Tl concentration. The relative increases are ×2.5, ×4.8, and ×6.7 for 10 μM, 50 μM, and 100 μM Tl, respectively. A similar response is seen in the leaves, although in this case the levels of H_2O_2 stabilize at 50 μM Tl, with values for 50 μM and 100 μM being similar. The production of NO (Figure 2E) is much greater in the roots than in the leaves, and it increases in response to the increase in Tl treatments. Thus, 10 μM of Tl induces a ×1.8 increase, with the increases for the two greater concentrations being very similar to each other (×2.5 and ×2.6, respectively). In the leaves, there are also increases in response to Tl toxicity, by very similar factors (≈×1.25) for all of the concentrations. With regard to the production of H_2S (Figure 2F), the levels are higher in the leaves than in the roots. In the roots, the greatest increase in H_2S content corresponds to the treatment with 10 μM Tl (×1.8). The greater Tl concentrations induce only slight changes in the H_2S content, with values that are very similar to the control. In the leaves, there is increased H_2S content for all of the concentrations, with the maximum value corresponding to the 100 μM Tl treatment (×3.7, relative to the control).

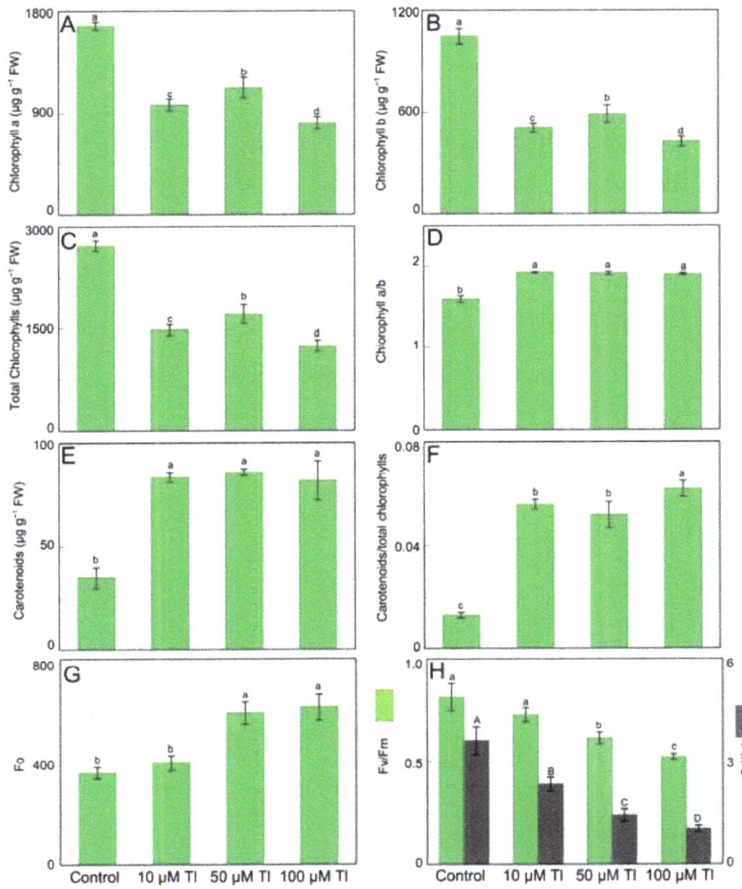

Figure 1. Effect of Tl on photosynthetic pigments and parameters in leaves of *D. viscosa*. Chlorophyll a content (**A**), chlorophyll b content (**B**), total chlorophylls (**C**), chlorophyll a/b (**D**), carotenoids content (**E**), carotenoids/total chlorophylls (**F**), F_o (**G**), and F_v/F_m and F_v/F_o (**H**). The data are the means ± SE from ten independent experiments. The different letters indicate significant differences at $p < 0.05$.

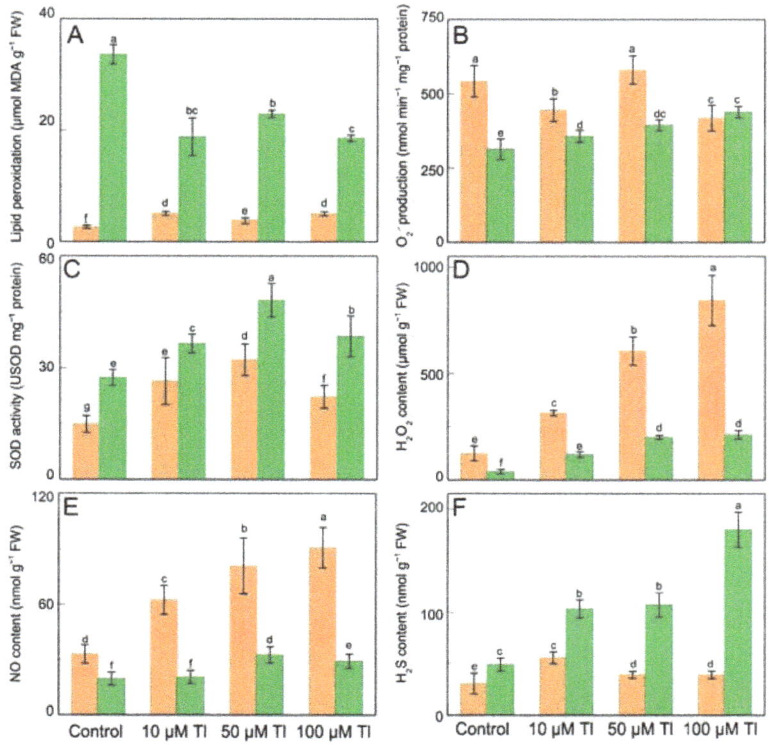

Figure 2. Effect of Tl on the lipid peroxidation (**A**), $O_2{\cdot}^-$ production (**B**), SOD activity (**C**), and H_2O_2 (**D**), NO (**E**), and H_2S (**F**) content in roots (orange) and leaves (green) of *D. viscosa*. The data are the means ± SE from ten independent experiments. The different letters indicate significant differences at $p < 0.05$.

3.4. APX, DHAR, MDHAR, and GR, and Ascorbate and Glutathione Pool

The effect that Tl has on the enzymes of the AsA–GSH cycle was determined (Figure 3). APX, DHAR, MDHAR, and GR all present behavior that differs between the roots and the leaves. In the roots, the APX activity (Figure 3A) decreases after the stress that is induced by 10 µM Tl but is unaltered by 50 µM or 100 µM Tl. On the contrary, in the leaves, the APX activity increases at the higher Tl concentrations, but is unaffected by 10 µM Tl. The DHAR activity in the roots increases strongly with 10 µM Tl, but the greater concentrations show less activation (Figure 3B). In the leaves, DHAR increases by very similar factors (≈×2.0, relative to the control) in response to all of the Tl concentrations. The MDHAR activity (Figure 3C) also differs in behavior between the roots and the leaves; while in the roots there are slight decreases in this activity at 10 µM, in the leaves there are increases (by between ×2.8 and ×3.7) for all of the Tl concentrations, as was the case for the DHAR activity. The GR activity (Figure 3D) increases in the roots, but this increase is less the greater the Tl concentration (1.8- and 5.4-fold for 10 µM and 100 µM Tl, respectively). In the leaves, the GR activity decreases with 10 µM and 50 µM Tl but is not significantly altered with 100 µM Tl.

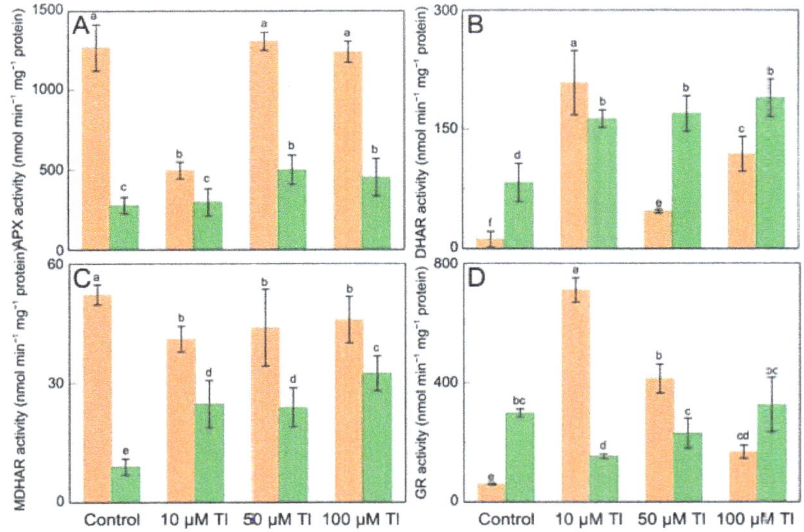

Figure 3. Effect of Tl on the APX (**A**), DHAR (**B**), MDHAR (**C**), and GR (**D**) activities in roots (orange) and leaves (green) of *D. viscosa*. The data are the means ± SE from ten independent experiments. The different letters indicate significant differences at $p < 0.05$.

The AsA and DHA contents are altered by Tl (Figure 4A,B). Thus, in the roots, the control AsA content is 121 nmol g $^{-1}$ FW, and Tl duplicates (×2.0) this content for 10 μM and 50 μM Tl, while 100 μM does not alter from the control value. On the contrary, the DHA content decreases by 14–48%, relative to the control, depending on the Tl concentration. In the leaves, the result is very similar, although the contents of both AsA and DHA are greater than those in the roots. The increase in AsA content is 2.23-fold in the presence of 10 μM Tl, and 1.63- and 1.49-fold for 50 μM and 100 μM Tl, respectively. The DHA content decreases for all of the concentrations. The total ascorbate pool (AsA + DHA) (Figure 4C) is unaltered, except with 100 μM Tl, which reduces this pool in both of the organs. The redox state (AsA/DHA) (Figure 4D) is altered, increasing in the roots and the leaves, in response to the Tl concentrations. With regard to the GSH content (Figure 4E), in the roots there is an increase only in those that were subjected to 100 μM Tl, but not for the other concentrations. The GSSG content (Figure 4F) decreases for the two lower concentrations, but not for 100 μM Tl, which produces levels that are similar those of to the control. In the leaves, both the GSH and the GSSG content (Figure 4G) decrease with the Tl concentration, with very similar values for 50 μM and 100 μM Tl. With regard to the glutathione pool, in the roots that were subjected to 10 μM and 50 μM Tl, there is a reduction of 35% of the total content, relative to the control. On the contrary, 100 μM causes a 23% increase (1.23-fold) in the AsA + DHA pool. Whereas, in the leaves, this pool decreases for all of the concentrations of Tl that were used. The GSH/GSSG ratio (Figure 4H) increases in the roots from 1.32 in the control to 4.06, 6.02, and 2.14 for 10 μM, 50 μM, and 100 μM, respectively. In the leaves, this ratio presents a different behavior, decreasing with Tl.

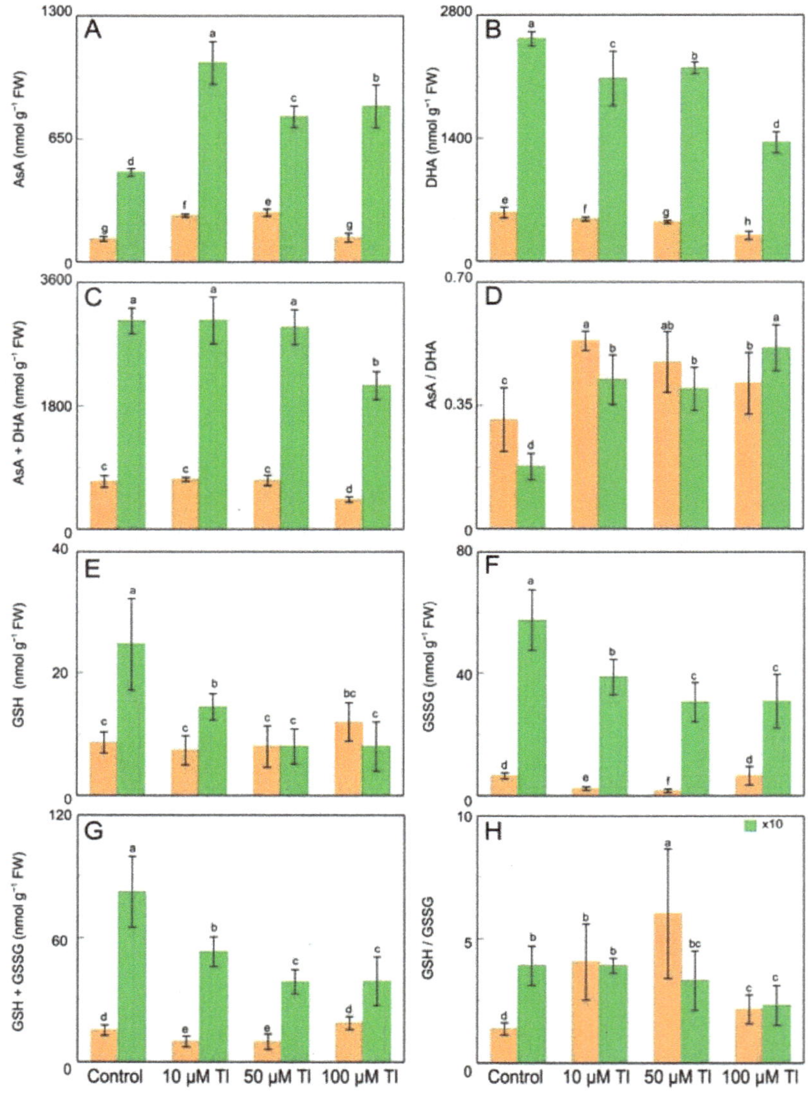

Figure 4. Effect of Tl on the AsA (**A**), DHA (**B**), and ascorbate pool (AsA + DHA) (**C**) contents; the AsA/DHA ratio (**D**), the GSH (**E**), GSSG (**F**), and glutathione pool (GSH + GSSG) (**G**) contents; and the GSH/GSSG ratio (H), in *D. viscosa*. The data are the means ± SE from ten independent experiments. The different letters indicate significant differences at $p < 0.05$.

4. Discussion

In *Dittrichia viscosa*, the concentrations of Tl that were used here cause a decrease in growth and biomass production (Figure S1, Tables 1 and S1), with these results being similar to those that have been observed in other plants [45,47,48,80–82]. The effect of Tl on the dry matter content and RWC is not significant. Only in the shoots is a slight decrease in RWC observed, a result that is very similar to that described by Radić et al. [49]. This contrasts with the decreases in RWC that has been observed in plants stressed by Cd [83–85].

The Tl that is absorbed accumulates mostly in the roots [48,49,86,87], with high values of BF_{Tl}, and decreases in TF_{Tl} with an increasing dose of Tl in the medium (Table 2). This preferential accumulation in the roots may obviate more serious damage to photosynthetic pigments [49].

Tl toxicity produces alterations in the absorption and accumulation of other elements (Table 3). There is a tendency to decrease the K content, although this is not significant. This variability may be due to the great genetic variability that is evidenced in *Dittrichia* [61,62]. Thus, the decline in K content is similar to that described by Buendia-Valverde et al. [88] for *Capsicum annuum* leaves, with decreases in the content at a low Tl concentration and a recovery of control values at a greater concentration. The Tl affects K uptake by the roots, but not its mobility within the plant, with TF_K (Table S2) being unaltered and the lower shoot content being due to the decrease in the root concentration rather than any reduced translocation. This effect of Tl on K uptake is a result of their mutual antagonism, since Tl uses K transport systems [34]. Chang et al. [45] also describe the antagonistic effect of Tl and K in *Arabidopsis*, in which Tl interferes with the K uptake. In *Sinapis alba* that are grown in hydroponic culture, the effect of Tl is to reduce the K content in the roots, as well as in the stems and leaves [50]. With regard to Mg, the Tl inhibits its absorption in *Dittrichia*, decreasing the content in the roots. In the leaves, the decrease in the Mg content is less, since there is a significant increase in TF_{Mg}. However, despite the increase in TF_{Mg} and the lower effect that is observed on the Mg content in the leaves, this does not prevent the decrease in the chlorophyll content. In contrast, in *S. alba*, Holubick et al. [50] describe low Tl concentrations leading to a decrease in Mg in shoots but not in roots, whereas, in *C. annuum*, the changes in the Mg content depend on the variety [88]. With regard to the Ca content, the data show that there is a decrease in its accumulation in the roots, except for those with the highest concentration that was used. The content of this element in the leaves is unaffected, however. Once again, the effects of Tl on the absorption and accumulation of Ca are diverse, depending, as in the case of Mg, on the variety [88] and the accumulating organ [50]. The microelements Fe, Mn, Cu, and Zn behave in different ways. While Fe decreases in both the roots and the leaves [50], Mn and Zn increase in both of these organs, and Cu only increases in the roots but decreases in the leaves. The inhibited growth that is induced by Tl stress may be due to this alteration in the absorption of essential mineral nutrients [89]. Our results show decreases in the absorption of K, Mg, and Fe, and alterations in the absorption and the accumulation of other essential elements, such as Ca, Mn, Cu, and Zn, which have increased presence in both the roots and the leaves, except for Cu, which is below the control levels in the leaves (Tables 3 and S2).

The increase in amino acid content in response to Tl toxicity could be associated with the observed decrease in the protein content, as well as with an accumulation of amino acids in response to this toxicity [90–93]. The increase in the soluble amino acid content and the decrease in the protein content show a clear effect of Tl toxicity (Table 4). Similar results have been described in different plants that were subjected to toxicity, not only of Tl [81], but also of other heavy metals, such as Cu, Cd, and Ni [94–96]. The decrease in protein content could be due to the effect of oxidative damage on protein synthesis, to alterations at the DNA level that would affect protein synthesis, or to the inhibition of the absorption of other elements, such as Mg and K [94]. It could also be due to increases in protease activity [97]. The increase in amino acid accumulation could be due to protein turnover increasing proteolysis, as well as due to the de novo synthesis of osmoprotectant amino acids, which is usually the case for stress-induced proline accumulation [90,91,93]. The amino acids can act on osmoregulation, heavy metal detoxification, and the formation of polyamines and glutathione, secondary metabolites, or the organic nitrogen reserve [92].

The decrease in the chlorophyll content that was observed in *Dittrichia* in response to stress that was induced by Tl toxicity (Figure 1A–D) coincides with results that have been described by other authors, but not the increase that has been observed in the carotenoid content (Figure 1E,F). Thus, Naumann et al. [80] in *Lemna minor* and Mazur et al. [44] in *Sinapis alba* find Tl to induce decreases in both the chlorophyll and the carotenoid content.

In *Arundo donax* and *Coix lacryma-jobi*, Pu et al. [48,86] find 50 $\mu g\ L^{-1}$ Tl to cause a decrease in the chlorophyll content, while lower concentrations have no effect. The increase in the carotenoid content and the carotenoid/chlorophyll ratio that has been observed in *Dittrichia* could be due to the role of carotenoids as ROS scavengers, thus protecting PSII functionality from oxidative damage [11,98–101]. The chlorophyll a/chlorophyll b ratio indicates the degree of appression of the thylakoid membranes in the chloroplast, being inversely proportional to the degree of appression, which, in turn, causes the "light harvesting complexes II" (LHCII) to be more closely connected, increasing the light-gathering capacity, and, therefore, the efficiency of energy transmission. In addition, the increase in the thylakoid appression enhances the electron transfer from photosystem II to the cytochrome b_6f complex, thus increasing non-cyclic photosynthetic electron transport [102]. That ratio is lower in control plants than in those that were grown in the presence of Tl, which would mean that the control plants capture light more efficiently. The interaction between Tl and the -SH groups affects the functionality of the enzyme systems that are involved in the biosynthesis and/or degradation of chlorophylls and the stability of the chloroplasts, as described for Sb toxicity [11,102].

The photosynthetic efficiency of control plants and those that were treated with 10 μM Tl are very close to the values that are described as normal by Murchie and Lawson [103]. The higher concentration Tl treatments show greater decreases, indicating an impairment of the photochemical activity. The stress that is induced by Tl toxicity causes an increase in the minimum fluorescence value (F_0) (Figure 1G), reflecting an alteration in the photosynthetic apparatus. This reveals that there is excess $O_2{}^{\cdot -}$ production in the chloroplasts, which negatively affects the PSII activity. The F_0 fluorescence originates from the chlorophylls that are associated with the antenna complexes. Its increase thus implies a reduced energy transfer from LHCII to PSII, i.e., there is a disconnection of the antennas in response to stress [104,105]. The F_v/F_m ratio represents the maximum PSII quantum yield, which correlates with the net photosynthesis quantum yield. It is indicative of the level of photoinhibition affecting the PSII complexes [106]. Tl toxicity reduces F_v/F_m (Figure 1H) to different degrees depending on the Tl concentration, these results being similar to those that have been described in other plants [44,47,48,86]. Changes in the PSII activity reduce photosynthesis levels [105,107]. In addition, the ratio F_v/F_0, which indicates the maximum ratio of the quantum yields of the photochemical and non-photochemical PSII processes, allows the photosynthesis capacity to be estimated. *Dittrichia viscosa* under Tl toxicity conditions show large decreases in this ratio (Figure 1H), which may indicate an interruption of donor-side PSII photosynthesis [108]. These results indicate that Tl toxicity leads to a strong alteration of photosynthetic activity, affecting PSII, as indicated by the increases in F_0 and the decreases in the F_v/F_m and F_v/F_0 ratios. Recently, Chang et al. [45] have described the negative effect of Tl on the photosynthetic process by decreasing the expression levels of the LCHII subunit genes, thus reducing the aggregation of the LCHII complexes, and, with them, the levels of photosynthesis.

With regard to ROS, by accumulating the Tl preferentially in the roots (Table 2), there is an increase in ROS production, which can cause the lipid peroxidation that is observed in the roots (Figure 2A). This behavior is similar to that described in *Dittrichia* plants in response to stress induced by Sb toxicity [11]. The increase in lipid peroxidation has also been described in response to Tl-induced stress [49,81]. On the contrary, Sirova and Vaculik [109] describe an increase in lipid peroxidation in leaves due to Cd toxicity, but not in roots, since a large part of the Cd is translocated to the leaves, provoking oxidative stress that would damage the membranes. Yao et al. [55] describe a decrease in lipid peroxidation in rice roots that were subjected to Tl(I) toxicity. In *Dittrichia*, TF_{Tl} decreases progressively with increasing Tl (Table 2), and this, together with the increase in SOD activity, may prevent increases in lipid peroxidation in leaves [110]. In the roots, the accumulation of Tl is very high, inducing strong increases in the production of $O_2{}^{\cdot -}$ and H_2O_2, and, although the SOD activity increases, damage by lipid peroxidation occurs (Figure 2). The stress induced in *Dittrichia* by Tl leads to increases in the content of $O_2{}^{\cdot -}$ and H_2O_2 and in the SOD activity

(Figure 2B–D), which is a response that is similar to that described in other plants [49,81]. On the contrary, while decreases in $O_2^{\cdot-}$ and H_2O_2 content have also been observed, the SOD activity increases [55]. However, the toxicity that is induced by Cd and As induces increases in the production of $O_2^{\cdot-}$ and H_2O_2, as well as in SOD activity [9,23,85,111]. NO and H_2S perform signaling functions, both under physiological conditions and under stress conditions, intervening at the level of nitration, nitrosation, and persulfidation [112,113]. Tl induces increases in the NO content in *Dittrichia* (Figure 2E), similar to those that have been described in response to toxicity by heavy metals such as As and Sb [10,11,16,114]. However, a decrease in the NO content in As toxicity has also been described [85,115,116]. These contradictory results could be explained by the magnitude of the stress, the plant age, the organ, or the treatment duration [116,117]. Increases in the H_2S content (Figure 2F) have also been described in response to toxicity by Sb [11], Cd [9,85], and Cr [118]. The amount of H_2S is greater in leaves than in roots, possibly because of the greater production of H_2S in chloroplasts than in cytoplasm [119].

AsA and GSH are involved in the activation of the antioxidant systems of plants against stress [120]. In addition, they maintain cellular redox homeostasis [121]. Our results show that Tl toxicity induces alterations in APX, DHAR, MDHAR, and GR activities (Figure 3), as well as increases in AsA content and decreases in DHA, GSH, and GSSG (Figure 4). The AsA content in the roots increases, and that of DHA decreases, thus raising the AsA/DHA ratio (Figure 4A–D). The increase in AsA may be due to the increase in DHAR and GR and a decrease in the APX activity (Figure 3A,B,D), with an increase in H_2O_2 content. However, the MDHAR activity decreases (Figure 3C) despite it being the main enzyme responsible for AsA levels. In roots, although there is an increase in NO, the H_2S levels only increase significantly with 10 μM of Tl, whereas the greater concentrations barely raise this content. In leaves, as in roots, the AsA content increases and the DHA content decreases, with consequent increases in the AsA/DHA ratio. Both DHAR and MDHAR increase, which raises the AsA content, since APX, although increasing in activity, does so to a lesser extent. The increases that are observed in the levels of NO and H_2S in leaves may increase the APX activity, which is modulated by both S-nitrosation and by persulfidation [122,123]. Espinosa-Vellarino et al. [11] describe Sb toxicity in *Dittrichia* leading to increased AsA and decreased DHA in leaves, with a decrease in both in the roots. On the contrary, decreases in AsA content have been described due to the toxicity of Tl [74], Cd, and As [85,124]. Hasanuzzaman et al. [125], also with Cd toxicity, describe decreases in AsA and increases in DHA.

In the roots, the GSH content does not change with 10 or 50 μM of Tl, but increases with 100 μM Tl, while that of GSSG decreases with 10 and 50 μM and remains at the control levels with 100 μM of Tl. The GSH/GSSG ratio increases with Tl (Figure 4H), and, although there is greater GR activity, the GSH content does not increase in absolute terms, although it does so relative to GSSG. The increased DHAR might consume the GSH that is produced by the GR, or this GSH might be used for other reactions, such as the direct elimination of ROS, phytochelatin synthesis, or direct binding to Tl, which has a high affinity for -SH groups [27,34]. In leaves, GSH and GSSG decrease, as does the GSH/GSSG ratio, which could be due to increased DHAR activity and decreased GR activity. These alterations are similar to those that are described by Hasanuzzaman et al. [125], with a decrease in the content of GSH and redox pool, an increase in APX activity, decreases in DHAR and MDHAR, and variable behavior of the GR. In *Dittrichia* under Sb toxicity [11], there are similar alterations in the leaves, although in the roots, the GSH content also decreases, as well as that of GSSG. The decrease in the GSH content in the leaves contrasts with the increase that has been described in response to Zn toxicity [126]. Alterations in the AsA/DHA and GSH/GSSG ratios are fundamental as a protective mechanism of the cell against ROS [127]. In our case, Tl produces an increase in the AsA/DHA ratio (Figure 4D) in both the roots and the leaves, and in the GSH/GSSG ratio (Figure 4H) in the roots, but with this ratio remaining constant, or slightly lowered, in the leaves. Tl induces an alteration of the cellular redox state. These results show the great capacity of *Dittrichia* to activate antioxidant defense systems, from amino acids, carotenoids, antioxidant enzymes, and AsA, maintaining a reduced state of the main antioxidants, especially in the leaves, in response to Tl stress.

5. Conclusions

In *Dittrichia*, the stress induced by Tl led to a major reduction in plant biomass, with reductions in the size and FW, altered the DW/FW ratio, but hardly affected RWC, strongly increased the amount of soluble amino acids, and decreased that of proteins. At the level of nutrient absorption and accumulation, it induced a major decrease in K, Mg, and Fe contents in both the roots and the leaves. On the contrary, Ca, Mn, and Zn accumulated to a greater extent in both of the organs. While chlorophylls decreased, carotenoids increased, possibly to act as a protection system by reducing the photosynthetic efficiency. With regard to stress markers, an increase in lipid peroxidation was observed, especially in the roots. The production of $O_2^{\cdot -}$, H_2O_2, and NO was greater in the roots than in the leaves, increasing in both of the organs in response to Tl toxicity, except for $O_2^{\cdot -}$ production in roots, which presented fluctuations. There was also greater H_2S production, especially in the leaves. The activities of SOD, APX, DHAR, MDHAR, and GR increased, except for APX in the roots and GR in the leaves. The components of the AsA–GSH cycle were also affected, with AsA increasing while DHA, GSH, and GSSG decreased. These Tl toxicity-induced alterations modified the AsA/DHA and GSH/GSSG redox status. The effects of Tl could be related to its high -SH group binding affinity, altering the functionality of proteins and the cellular redox state.

Supplementary Materials: The following are available online at https://www.mdpi.com/article/10.3390/antiox12030678/s1, Figure S1: *Dittrichia viscosa* plants subjected to Tl toxicity for 7 days; Table S1: Effect of Tl toxicity on the relative water content (RWC) in roots and leaves of *D. viscosa* plants; Table S2: Effect of Tl toxicity on the translocation factor of K (TF_K), Mg (TF_{Mg}), Ca (TF_{Ca}), Fe (TF_{Fe}), Mn (TF_{Mn}), Cu (TF_{Cu}), and (TF_{Zn}) of *D. viscosa* plants.

Author Contributions: All authors contributed to the manuscript. Conceptualization: F.E. and I.G. Funding acquisition: F.E. and I.G. Data curation, investigation, validation, and writing—original draft: F.E., A.O., F.L.E.-V. and I.G. Writing—review and editing: F.E. and I.G. All authors have read and agreed to the published version of the manuscript.

Funding: This study was made possible thanks to the Junta de Extremadura for the support given to the Research Group FBCMP (GR21112) and for the research project IB18074.

Institutional Review Board Statement: Not applicable.

Informed Consent Statement: Not applicable.

Data Availability Statement: The data presented in this article will be made available without any reservation.

Conflicts of Interest: The authors declare no conflict of interest.

References

1. Corpas, F.J.; Leterrier, M.; Valderrama, R.; Airaki, M.; Chaki, M.; Palma, J.M.; Barroso, J.B. Nitric oxide imbalance provokes a nitrosative response in plants under abiotic stress. *Plant Sci.* **2011**, *181*, 604–611. [CrossRef]
2. Foyer, C.H.; Noctor, G. Ascorbate and glutathione: The heart of the redox hub. *Plant Physiol.* **2011**, *155*, 2–18. [CrossRef] [PubMed]
3. Das, K.; Roychoudhury, A. Reactive oxygen species (ROS) and response of antioxidants as ROS-scavengers during environmental stress in plants. *Front. Environ. Sci.* **2014**, *2*, 53. [CrossRef]
4. Liu, H.; Wang, J.; Liu, J.; Liu, T.; Xue, S. Hydrogen sulfide (H2S) signaling in plant development and stress responses. *Abiotech* **2021**, *2*, 32–63. [CrossRef]
5. Mittler, R. Oxidative stress, antioxidants and stress tolerance. *Trends Plant Sci.* **2002**, *7*, 405–410. [CrossRef] [PubMed]
6. Zhang, F.-Q.; Wang, Y.-S.; Lou, Z.-P.; Dong, J.-D. Effect of heavy metal stress on antioxidative enzymes and lipid peroxidation in leaves and roots of two mangrove plant seedlings (*Kandelia candel* and *Bruguiera gymnorrhiza*). *Chemosphere* **2007**, *67*, 44–50. [CrossRef] [PubMed]

7. Gill, S.S.; Tuteja, N. Reactive oxygen species and antioxidant machinery in abiotic stress tolerance in crop plants. *Plant Physiol. Biochem.* **2010**, *48*, 909–930. [CrossRef]
8. Anjum, N.A.; Gill, S.S.; Gill, R.; Hasanuzzaman, M.; Duarte, A.C.; Pereira, E.; Ahmad, I.; Tuteja, R.; Tuteja, N. Metal/metalloid stress tolerance in plants: Role of ascorbate, its redox couple, and associated enzymes. *Protoplasma* **2014**, *251*, 1265–1283. [CrossRef]
9. Khan, M.N.; Siddiqui, M.H.; AlSolami, M.A.; Alamri, S.; Hu, Y.; Ali, H.M.; Al-Amri, A.A.; Alsubaie, Q.D.; Al-Munqedhi, B.M.A.; Al-Ghamdi, A. Crosstalk of hydrogen sulfide and nitric oxide requires calcium to mitigate impaired photosynthesis under cadmium stress by activating defense mechanisms in *Vigna radiata*. *Plant Physiol. Biochem.* **2020**, *156*, 278–290. [CrossRef]
10. Espinosa-Vellarino, F.L.; Garrido, I.; Ortega, A.; Casimiro, I.; Espinosa, F. Effects of Antimony on Reactive Oxygen and Nitrogen Species (ROS and RNS) and Antioxidant Mechanisms in Tomato Plants. *Front. Plant Sci.* **2020**, *11*, 674. [CrossRef]
11. Espinosa-Vellarino, F.L.; Garrido, I.; Ortega, A.; Casimiro, I.; Espinosa, F. Response to Antimony Toxicity in *Dittrichia* viscosa Plants: ROS, NO, H_2S, and the Antioxidant System. *Antioxidants* **2021**, *27*, 1698. [CrossRef] [PubMed]
12. Qiao, W.; Li, C.; Fan, L.M. Cross-talk between nitric oxide and hydrogen peroxide in plant responses to abiotic stresses. *Environ. Exp. Bot.* **2014**, *100*, 84–93. [CrossRef]
13. Correa-Aragunde, N.; Foresi, N.; Lamattina, L. Nitric oxide is a ubiquitous signal for maintaining redox balance in plant cells: Regulation of ascorbate peroxidase as a case study. *J. Exp. Bot.* **2015**, *66*, 2913–2921. [CrossRef]
14. Rizwan, M.; Mostofa, M.G.; Ahmad, M.Z.; Imtiaz, M.; Mehmood, S.; Adeel, M.; Dai, Z.; Li, Z.; Aziz, O.; Zhang, Y.; et al. Nitric oxide induces rice tolerance to excessive nickel by regulating nickel uptake, reactive oxygen species detoxification and defense-related gene expression. *Chemosphere* **2018**, *191*, 23–35. [CrossRef]
15. Gibbs, D.J.; Conde, J.V.; Berckhan, S.; Prasad, G.; Mendiondo, G.M.; Holdsworth, M.J. Group VII ethylene response factors coordinate oxygen and nitric oxide signal transduction and stress responses in plants. *Plant Physiol.* **2015**, *169*, 23–31. [CrossRef] [PubMed]
16. Leterrier, M.; Airaki, M.; Palma, J.M.; Chaki, M.; Barroso, J.B.; Corpas, F.J. Arsenic triggers the nitric oxide (NO) and S-nitrosoglutathione (GSNO) metabolism in Arabidopsis. *Environ. Pollut.* **2012**, *166*, 136–143. [CrossRef]
17. Singh, P.K.; Indoliya, Y.; Chauhan, A.S.; Singh, S.P.; Singh, A.P.; Dwivedi, S.; Chakrabarty, D. Nitric oxide mediated transcriptional modulation enhances plant adaptive responses to arsenic stress. *Sci. Rep.* **2017**, *7*, 3592. [CrossRef]
18. Yordanova, R.; Baydanova, V.; Peeva, V. Nitric oxide mediates the stress response induced by cadmium in maize plants. *Genet. Plant Physiol.* **2017**, *7*, 121–134.
19. Sharma, A.; Soares, C.; Sousa, B.; Martins, M.; Kumar, V.; Shahzad, B.; Sidhu, G.P.; Bali, A.S.; Asgher, M.; Bhardwaj, R.; et al. Nitric oxide-mediated regulation of oxidative stress in plants under metal stress: A review on molecular and biochemical aspects. *Physiol. Plant.* **2019**, *168*, 318–344. [CrossRef]
20. Khan, M.N.; Mobin, M.; Abbas, Z.K.; Siddiqui, M.H. Nitric oxide-induced synthesis of hydrogen sulfide alleviates osmotic stress in wheat seedlings through sustaining antioxidant enzymes, osmolyte accumulation and cysteine homeostasis. *Nitric Oxide* **2017**, *68*, 91–102. [CrossRef]
21. Khan, M.N.; Mohammad, F.; Mobin, M.; Saqib, M.A. Tolerance of plants to abiotic stress: A role of nitric oxide and calcium. In *Nitric Oxide in Plants: Metabolism and Role in Stress Physiology*; Khan, M.N., Mohammad, F., Mobin, M., Corpas, F.J., Eds.; Springer International Publishing: Cham, Switzerland, 2014; pp. 225–242. [CrossRef]
22. Khan, M.N.; Mobin, M.; Khorshid Abbas, Z. Nitric oxide and high temperature stress: A physiological perspective. In *Nitric Oxide Action in Abiotic Stress Responses in Plants*; Khan, M.N., Mohammad, F., Mobin, M., Corpas, F.J., Eds.; Springer International Publishing: Cham, Switzerland, 2015; pp. 77–92. [CrossRef]
23. Siddiqui, M.H.; Alamri, S.; Nasir Khan, M.; Corpas, F.J.; Al-Amri, A.A.; Alsubaie, Q.D.; Ali, H.M.; Kalaji, H.M.; Ahmad, P. Melatonin and calcium function synergistically to promote the resilience through ROS metabolism under arsenic-induced stress. *J. Hazard. Mater.* **2020**, *398*, 122882. [CrossRef]
24. Aroca, A.; Gotor, C.; Romero, L.C. Hydrogen Sulfide Signaling in Plants: Emerging Roles of Protein Persulfidation. *Front Plant Sci.* **2018**, *9*, 1369. [CrossRef]
25. Corpas, F.J.; Palma, J.M. H_2S signaling in plants and applications in agriculture. *J. Adv. Res.* **2020**, *24*, 131–137. [CrossRef] [PubMed]
26. Corpas, F.J.; González-Gordo, S.; Palma, J.M. Nitric oxide and hydrogen sulfide modulate the NADPH-generating enzymatic system in higher plants. *J. Exp. Bot.* **2021**, *72*, 830–847. [CrossRef]
27. Khan, M.J.; Mukhtiar, M.; Qureshi, M.M.; Jan, S.U.; Ullah, I.; Hussain, A.; Khan, M.F.; Gul, R.; Shahwani, N.A.; Rabbani, I. Spectrophotometric investigation of Glutathione modulation by Thallium chloride in aqueous médium. *Pak. J. Pharm. Sci.* **2018**, *31*, 1463–1467. [PubMed]
28. Wang, H.R.; Che, Y.H.; Huang, D.; Ao, H. Hydrogen sulfide mediated alleviation of cadmium toxicity in *Phlox paniculata* L. and establishment of a comprehensive evaluation model for corresponding strategy. *Int. J. Phytoremed.* **2020**, *22*, 1085–1095 [CrossRef] [PubMed]
29. Mostofa, M.G.; Rahman, A.; Ansary, M.U.; Watanabe, A.; Fujita, M.; Tran, L.-S. Hydrogen sulfide modulates cadmium-induced physiological and biochemical responses to alleviate cadmium toxicity in rice. *Sci. Rep.* **2015**, *5*, 14078. [CrossRef] [PubMed]
30. Jia, H.; Wang, X.; Dou, Y.; Liu, D.; Si, W.; Fang, H.; Zhao, C.; Chen, S.; Xi, J.; Li, J. Hydrogen sulfide—cysteine cycle system enhances cadmium tolerance through alleviating cadmium-induced oxidative stress and ion toxicity in Arabidopsis roots. *Sci. Rep.* **2016**, *6*, 39702. [CrossRef]

31. Kushwaha, B.K.; Singh, V.P. Glutathione and hydrogen sulfide are required for sulfur-mediated mitigation of Cr(VI) toxicity in tomato, pea and brinjal seedlings. *Physiol. Plant.* **2019**, *168*, 406–421. [CrossRef]
32. Pandey, A.K.; Gautam, A. Stress responsive gene regulation in relation to hydrogen sulfide in plants under abiotic stress. *Physiol. Plant.* **2020**, *168*, 511–525. [CrossRef]
33. Da-Silva, C.; Fontes, E.; Modolo, L. Salinity-induced accumulation of endogenous H2S and NO is associated with modulation of the antioxidant and redox defense systems in Nicotiana tabacum L. cv. Havana. *Plant Sci.* **2017**, *256*, 148–159. [CrossRef] [PubMed]
34. Mulkey, J.P.; Oehme, F.W. A review of thallium toxicity. *Vet. Hum. Toxicol.* **1993**, *35*, 445–453. [PubMed]
35. Kaya, C.; Ashraf, M.; Alyemeni, M.N.; Ahmad, P. Responses of nitric oxide and hydrogen sulfide in regulating oxidative defence system in wheat plants grown under cadmium stress. *Physiol. Plant.* **2019**, *168*, 345–360. [CrossRef] [PubMed]
36. Jia, H.; Wang, X.; Shi, C.; Guo, J.; Ma, P.; Ren, X.; Wei, T.; Liu, H.; Li, J. Hydrogen sulfide decreases Cd translocation from root to shoot through increasing Cd accumulation in cell wall and decreasing Cd2+ influx in Isatis indigotica. *Plant Physiol. Biochem.* **2020**, *155*, 605–612. [CrossRef]
37. Sahay, S.; Gupta, M. An update on nitric oxide and its benign role in plant responses under metal stress. *Nitric Oxide* **2017**, *67*, 39–52. [CrossRef]
38. Yao, Y.; Yang, R.; Liao, W.; Wang, Y.; Liu, W.; Huang, X.; Wang, X.; Zhang, P. Is Oxalic Acid Secretion A Detoxification Strategy for Rice Exposed to Tl(I) or Tl(III)? *Bull. Environ. Contam. Toxicol.* **2022**, *109*, 920–926. [CrossRef]
39. Rader, S.T.; Maier, R.M.; Barton, M.D.; Mazdab, F.K. Uptake and Fractionation of Thallium by Brassica juncea in a Geogenic Thallium-Amended Substrate. *Environ. Sci. Technol.* **2019**, *53*, 2441–2449. [CrossRef]
40. Galván-Arzate, S.; Santamaría, A. Thallium toxicity. *Toxicol. Letters.* **1998**, *99*, 1–13. [CrossRef]
41. Pavoni, E.; Covellia, S.; Adamib, G.; Baracchinib, E.; Cattelanc, R.; Croserab, M.; Higuerasd, P.; Lenaza, D.; Petranicha, E. Mobility and fate of Thallium and other potentially harmful elements in drainage waters from a decommissioned Zn-Pb mine (North-Eastern Italian Alps). *J. Geochem. Explor.* **2018**, *188*, 1–10. [CrossRef]
42. Sasmaz, M.; Öbekb, E.; Sasmaz, A. Bioaccumulation of cadmium and thallium in Pb-Zn tailing waste water by *Lemna minor* and *Lemna gibba*. *Appl. Geochem.* **2019**, *100*, 287–292. [CrossRef]
43. Wei, X.; Zhoua, Y.; Tsangb, D.C.W.; Songc, L.; Zhangd, C.; Yina, M.; Liua, J.; Xiaoa, T.; Zhanga, G.; Wang, J. Hyperaccumulation and transport mechanism of thallium and arsenic in brake ferns (*Pteris vittata* L.): A case study from mining área. *J. Hazard. Mate.* **2020**, *388*, 121756. [CrossRef]
44. Mazur, R.; Sadowska, M.; Kowalewska, L.; Abratowska, A.; Kalaji, H.M.; Mostowska, A.; Garstka, M.; Krasnodębska-Ostręga, B. Overlapping toxic effect of long term thallium exposure on white mustard (*Sinapis alba* L.) photosynthetic activity. *BMC Plant Biol.* **2016**, *16*, 191. [CrossRef]
45. Chang, H.F.; Tseng, S.C.; Tang, M.T.; Hsiao, S.S.Y.; Lee, D.C.; Wang, S.L.; Yeh, K.C. Physiology and Molecular Basis of Thallium Toxicity and Accumulation in *Arabidopsis thaliana*. 2022. Available online: https://ssrn.com/abstract=4196841 (accessed on 8 November 2022).
46. Corzo-Remigio, A.; Pošćić, F.; Nkrumah, P.N.; Edraki, M.; Spiers, K.M.; Brueckner, D.; Van der Ent, A. Comprehensive insights in thallium ecophysiology in the hyperaccumulator Biscutella laevigata. *Sci. Total Environ.* **2022**, *838*, 155899. [CrossRef]
47. Wu, Q.; Leung, J.Y.; Huang, X.; Yao, B.; Yuan, X.; Ma, J.; Guo, S. Evaluation of the ability of black nightshade *Solanum nigrum* L. for phytoremediation of thallium-contaminated soil. *Environ. Sci. Pollut. Res. Int.* **2015**, *22*, 11478–11487. [CrossRef]
48. Pu, G.; Zhang, D.; Zeng, D.; Xu, G.; Huang, Y. Physiological response of *Arundo donax* L. to thallium accumulation in a simulated wetland. *Mar. Freshw. Res.* **2017**, *69*, 714. [CrossRef]
49. Radic, S.; Cvjetko, P.; Glavas, K.; Roje, V.; Pevalek-Zozlina, B.; Pavlica, M. Oxidative stress and DNA damage in broad vean (*Vicia faba* L.) seedlings induced by thallium. *Environ. Toxicol. Chem.* **2009**, *28*, 189–196. [CrossRef]
50. Holubík, O.; Vaněk, A.; Mihaljevic, M.; Vejvodová, K. Tallium uptake/tolerance in a model (hyper)accumulating plant: Effect of extreme contaminant loads. *Soil Water Res.* **2021**, *16*, 129–135. [CrossRef]
51. Corzo-Remigio, A.; Nkrumah, P.; Pošćić, F.; Edraki, M.; Baker, A.; Ent, A. Thallium accumulation and distribution in Silene latifolia (Caryophyllaceae) grown in hydroponics. *Plant Soil.* **2022**, *480*, 213–226. [CrossRef]
52. Poschenrieder, C.; Cabot, C.; Martos, S.; Gallego, B.; Barceló, J. Do toxic ions induce hormesis in plants? *Plant Sci.* **2013**, *212*, 15–25. [CrossRef] [PubMed]
53. Salinitro, M.; Mattarello, G.; Guardigli, G.; Odajiu, M.; Tassoni, A. Induction of hormesis in plants by urban trace metal pollution. *Sci. Rep.* **2021**, *11*, 20329. [CrossRef] [PubMed]
54. Pu, G.; Zhang, D.; Zeng, D.; Xu, G.; Huang, Y. Effects of Thallium Stress on Photosynthesis, Chlorophyll Fluorescence Parameters and Antioxidant Enzymes Activities of *Coix lacryma-jobi*. *J. Environ. Sci. Eng. A* **2017**, *6*, 15–21. [CrossRef]
55. Yao, Y.; Wang, M.; Zhang, P.; Wang, X.; Huang, X.; Liu, W.; Wang, Z.; Yang, R. Different responses in metallothionein gene expression and antioxidative enzyme activity lead to more ROS accumulation and toxicity in rice exposed to Tl(III) than to Tl(I). *Chemosphere* **2020**, *259*, 127258. [CrossRef]
56. Liu, T.; Deng, H.M.; Wang, Y.L.; Chen, Y.H.; Chen, M.C.; Wu, G.M.; Zeng, D.M.; Du, G.Y.; Gu, Y.S.; Xu, X.C. Growth Inhibition of Thallium(I)-nitrate on Duckweed (*Lemna gibba* L.) with Relation to Its Oxidative Stress. *Asian J. Ecotoxicol.* **2014**, *6*, 1112–1117. [CrossRef]

57. Deng, H.M.; Chen, Y.H.; Liu, T.; Wu, C.Q.; Qiu, L.; Wu, G.M.; Zeng, D.M. Study on the translocation and accumulation of Tl in soil-plant system. *Enviro. Chem.* **2013**, *32*, 1749–1757.
58. Parolin, P.; Ion Scotta, M.; Bresch, C. Biology of *Dittrichia viscosa*, a mediterranean ruderal plant: A review. *Phyton Int. J. Exp. Bot.* **2014**, *83*, 251–262.
59. Murciego, A.M.; Sánchez, A.G.; González, M.R.; Gil, E.F.; Gordillo, C.T.; Fernández, J.C.; Triguero, T.B. Antimony distribution and mobility in topsoils and plants (*Cytisus striatus, Cistus ladanifer* and *Dittrichia viscosa*) from polluted Sb-mining areas in Extremadura (Spain). *Environ. Pollut.* **2007**, *145*, 15–21. [CrossRef]
60. Pérez-Sirvent, C.; Martínez-Sánchez, M.J.; SMartínez-López, S.; Bech, J.; Bolan, N. Distribution and bioaccumulation of arsenic and antimony in *Dittrichia viscosa* growing in mining-affected semiarid soils in southeast Spain. *J. Geochem. Explor.* **2012**, *123*, 128–135. [CrossRef]
61. Guarino, F.; Conte, B.; Improta, G.; Sciarrillo, R.; Castiglione, S.; Cicatelli, A.; Guarino, C. Genetic characterization, micropropagation, and potential use for arsenic phytoremediation of *Dittrichia viscosa* (L.) Greuter. *Ecotoxicol. Environ. Saf.* **2018**, *148*, 675–683. [CrossRef]
62. Papadia, P.; Barozzi, F.; Angilé, F.; Migoni, D.; Piro, G.; Fanizzi, F.P.; Di Sansebastiano, G.-P. Evaluation of *Dittrichia viscosa* performance in substrates with moderately low levels of As and Cd contamination. *Plant Biosyst. Int. J. Deal. All Asp. Plant Biol.* **2020**, *154*, 983–989. [CrossRef]
63. Garrido, I.; Ortega, A.; Hernández, M.; Fernández-Pozo, L.; Cabezas, J.; Espinosa, F. Effect of antimony in soils of an Sb mine on the photosynthetic pigments and antioxidant system of *Dittrichia viscosa* leaves. *Environ. Geochem. Health* **2021**, *43*, 1367–1383. [CrossRef]
64. Jiménez, M.N.; Bacchetta, G.; Navarro, F.B.; Casti, M.; Fernández-Ondoño, E. Native Plant Capacity for Gentle Remediation in Heavily Polluted Mines. *Appl. Sci.* **2021**, *11*, 1769. [CrossRef]
65. Smart, R.E.; Bingham, G.E. Rapid estimates of relative water content. *Plant Physiol.* **1974**, *53*, 258–260. [CrossRef] [PubMed]
66. Lehotai, N.; Kolbert, Z.; Peto, A.; Feigl, G.; Ordog, O.; Kumar, D.; Tari, I.; Erdei, L. Selenite-induced hormonal and signalling mechanisms during root growth of *Arabidopsis thaliana* L. *J. Exp. Bot.* **2012**, *65*, 5677–5687. [CrossRef]
67. Bradford, M.M. A rapid and sensitive method for the quantitation of microgram quantities of protein utiliz- ing the principle of protein-dye binding. *Anal. Biochem.* **1976**, *72* 248–254. [CrossRef] [PubMed]
68. Yemm, E.W.; Cooking, E.C. The determination of amino acids with ninhidrin. *Analyst* **1955**, *80*, 209–213. [CrossRef]
69. Wellburn, A.R. The spectral determination of chlorophyll a and chlorophyll b, as well as total carotenoids, using various solvents with spectrophotometers of different resolution. *J. Plant Physiol.* **1994**, *144*, 307–313. [CrossRef]
70. Fu, J.; Huang, B. Involvement of antioxidants and lipid peroxidation in the adaptation of two cool-season grasses to localized drought stress. *Environ. Exp. Bot.* **2001**, *45*, 105–114. [CrossRef] [PubMed]
71. Velikova, V.; Yordanov, I.; Edreva, A. Oxidative stress and some antioxidant system in acid rain-treated bean plants: Protective role of exogenous polyamines. *Plant Sci.* **2000**, *151*, 59–66. [CrossRef]
72. Misra, H.P.; Fridovich, I. The role of superoxide anion in the autoxidation of epinephrine and a simple assay for superoxide dismutase. *J. Biol. Chem.* **1972**, *247*, 3170–3175. [CrossRef]
73. Beauchamp, C.; Fridovich, I. Superoxide dismutase: Improved assays and an assay applicable to acryl-amide gels. *Anal. Biochem.* **1971**, *44*, 276–287. [CrossRef]
74. Zhou, B.; Guo, Z.; Xing, J.; Huang, B. Nitric oxide is involved in abscisic acid-induced antioxidant activities in *Stylosanthes guianensis*. *J. Exp. Bot.* **2005**, *56*, 3223–3228. [CrossRef]
75. Li, Z.G. Quantification of hydrogen sulfide concentration using methylene blue and 5,5′-dithiobis(2-nitrobenzoic acid) methods in plants. *Methods Enzymol.* **2015**, *554*, 101–110. [CrossRef]
76. De Pinto, M.C.; Francis, D.; De Gara, L. The redox state of ascorbate-dehydroascorbate pair as a specific sensor of cell division in tobacco TBY-2 cells. *Protoplasma.* **1999**, *209*, 90–97. [CrossRef] [PubMed]
77. Rao, M.V.; Paliyath, G.; Ormrod, D.P. Ultraviolet-B- and ozone-induced biochemical changes in antioxidant enzymes of *Arabidopsis thaliana*. *Plant Physiol.* **1996**, *110*, 125–136. [CrossRef]
78. De Gara, L.; De Pinto, M.C.; Moliterni, V.M.C.; D'Égidio, M.G. Redox regulation and storage processes during maturation in kernels of *Triticum durum*. *J. Exp. Bot.* **2003**, *54*, 249–258. [CrossRef] [PubMed]
79. Hossain, M.A.; Nakano, Y.; Asada, K. Monodehydroascorbate Reductase in Spinach Chloroplasts and Its Participation in Regeneration of Ascorbate for Scavenging Hydrogen Peroxide. *Plant Cell Physiol.* **1984**, *25*, 385–395. [CrossRef]
80. Naumann, B.; Eberius, M.; Appenroth, K.J. Growth rate based dose-response relationships and EC-values of ten heavy metals using the duckweed growth inhibition test (ISO 20079) with *Lemna minor* L. clone St. *J. Plant Physiol.* **2007**, *164*, 1656–1664. [CrossRef]
81. Babić, M.; Radić, S.; Cvjetko, P.; Roje, V.; Pevalek-Kozlina, B.; Pavlica, M. Antioxidative response of *Lemna minor* plants exposed to thallium(I)-acetate. *Aquat. Bot.* **2009**, *91*, 166–172. [CrossRef]
82. Vanek, A.; Komárek, M.; Chrastny, V.; Becka, D.; Mihaljevic, M.; Sebek, O.; Panusková, G.; Schusterová, Z. Thallium uptake by white mustard (*Sinapis alba* L.) grown on moderately contaminated soils-Agro-environmental implications. *J. Hazard. Mater.* **2010**, *182*, 303–308. [CrossRef]
83. He, S.; Yang, X.; He, Z.; Baligar, V.C. Morphological and physiological responses of plants to cadmium toxicity: A review. *Pedosphere* **2017**, *27*, 421–438. [CrossRef]

84. Naeem, A.; Zafar, M.; Khalid, H.; Zia-ur-Rehman, M.; Ahmad, Z.; Ashar Ayub, M.; Qayyum, M.F. Chapter 12–Cadmium-Induced Imbalance in Nutrient and Water Uptake by Plants. In *Cadmium Toxicity and Tolerance in Plants*; Hasanuzzaman, M., Prasad, M.N.V., Fujita, M., Eds.; Academic Press: Cambridge, MA, USA, 2019; pp. 299–326. [CrossRef]
85. Kaya, C.; Akram, N.A.; Ashraf, M.; Alyemeni, M.N.; Ahmad, P. Exogenously supplied silicon (Si) improves cadmium tolerance in pepper (*Capsicum annuum* L.) by up-regulating the synthesis of nitric oxide and hydrogen sulfide. *J. Biotechnol.* **2020**, *10*, 35–45. [CrossRef] [PubMed]
86. Pu, G.; Zhang, D.; Zeng, D.; Xu, G.; Huang, Y. Accumulation of cadmium and its effects on physiological characteristics in *Coix lacryma-jobi* l. *Bangladesh J. Bot.* **2017**, *46*, 1071–1077.
87. Kim, D.J.; Kang, B.S.; Le, J.H. Comparative Uptake and Translocation of Thallium in Selected Plant Species Grown in Artificially Contaminated Soils. *Korean Soc. Soil Sci. Fert.* **2021**, *54*, 347–358. [CrossRef]
88. Buendía-Valverde, M.; Gómez-Merino, F.C.; Corona-Torres, T.; Cruz-Izquierdo, S.; Mateo-Nava, R.A.; Trejo-Tellez, L.I. Thallium Differentially Affects Macronutrients Concentration and Stoichiometric Ratios with Nitrogen in the Leaves of Chili Pepper Varieties. *Water Air Soil Pollut.* **2022**, *233*, 201. [CrossRef]
89. Rehman, S.; Abbas, G.; Shahid, M.; Saqib, M.; Umer Farooq, A.B.; Hussain, M.; Murtaza, B.; Amjad, M.; Naeem, M.A.; Farooq, A. Effect of salinity on cadmium toler- ance, ionic homeostasis and oxidative stress responses in conocarpus exposed to cadmium stress: Implications for phytoremediation. *Ecotoxicol. Environ. Saf.* **2019**, *171*, 146–153. [CrossRef] [PubMed]
90. Rai, V. Role of Amino Acids in Plant Responses to Stresses. *Biol. Plant.* **2002**, *45*, 481–487. [CrossRef]
91. Sharma, S.S.; Dietz, K.J. The significance of amino acids and amino acid-derived molecules in plant responses and adaptation to heavy metal stress. *J. Exp. Bot.* **2006**, *57*, 711–726. [CrossRef]
92. Hildebrandt, T.M. Synthesis versus degradation: Directions of amino acid metabolism during Arabidopsis abiotic stress response. *Plant Mol. Biol.* **2018**, *98*, 121–135. [CrossRef] [PubMed]
93. Khan, N.; Ali, S.; Zandi, P.; Mehmood, A.; Ullah, S.; Ikram, M.; Ismail, M.A.S.; Babar, M.A. Role of sugars, amino acids and organic acids in improving plant abiotic stress tolerance. *Pakistan J. Bot.* **2020**, *52*, 355–363. [CrossRef]
94. Hou, W.; Chen, X.; Song, G.; Wang, Q.; Chi Chang, C. Effects of copper and cadmium on heavy metal polluted waterbody restoration by duckweed (*Lemna minor*). *Plant Physiol. Biochem.* **2007**, *45*, 62–69. [CrossRef]
95. El-Amier, Y.; Elhindi, K.; El-Hendawy, S.; Al-Rashed, S.; Abd-ElGawad, A. Antioxidant System and Biomolecules Alteration in *Pisum sativum* under Heavy Metal Stress and Possible Alleviation by 5-Aminolevulinic Acid. *Molecules* **2019**, *24*, 4194. [CrossRef]
96. Bari, M.A.; Prity, S.A.; Das, U.; Akther, M.S.; Sajib, S.A.; Reza, M.A.; Kabir, A.H. Silicon induces phytochelatin and ROS scavengers facilitating cadmium detoxification in rice. *Plant Biol.* **2020**, *22*, 472–479. [CrossRef]
97. Umer Chattha, M.; Arif, W.; Khan, I.; Soufan, W.; Bilal Chattha, M.; Hassan, M.U.; Ullah, N.; Sabagh, A.E.; Qari, S.H. Mitigation of Cadmium Induced Oxidative Stress by Using Organic Amendments to Improve the Growth and Yield of Mash Beans [*Vigna mungo* (L.)]. *Agronomy* **2021**, *11*, 2152. [CrossRef]
98. Simkin, A.J.; Moreau, H.; Kuntz, M.; Pagny, G.; Lin, C.; Tanksley, S.; McCarthy, J. An investigation of carotenoid biosynthesis in *Coffea canephora* and *Coffea arabica*. *J. Plant Physiol.* **2008**, *165*, 1087–1106. [CrossRef] [PubMed]
99. Di Toppi, L.S.; Castagna, A.; Andreozzi, E.; Careri, M.; Predieri, G.; Vurro, E.; Ranieri, A. Occurrence of differentinter-varietal and inter-organ defence strategies towards supra-optimal zinc concentrations in two cultivars of *Triticum aestivum* L. *Environ. Exp. Bot.* **2009**, *66*, 220–229. [CrossRef]
100. Chai, L.Y.; Wang, Y.; Yang, Z.H.; Mubarak, H.; Mirza, N. Physiological characteristics of *Ficus tikoua* under antimony stress. *T. Nonferr. Metal Soc.* **2017**, *27*, 939–945. [CrossRef]
101. Palm, E.; Nissim, W.G.; Giordano, C.; Mancuso, S.; Azzarello, E. Root potassium and hydrogen flux rates as potential indicators of plant response to zinc, copper and nickel stress. *Environ. Exp. Bot.* **2017**, *143*, 38–50. [CrossRef]
102. Zhou, X.; Sun, C.; Zhu, P.; Liu, F. Effects of antimony stress on photosynthesis and growth of Acorus calamus. *Front. Plant Sci.* **2018**, *9*, 579. [CrossRef]
103. Murchie, E.H.; Lawson, T. Chlorophyll fluorescence analysis: A guide to good practice and understanding some new applications. *J. Exp. Bot.* **2013**, *64*, 3983–3998. [CrossRef]
104. Yamada, M.; Hidaka, T.; Fukamachi, H. Heat tolerance in leaves of tropical fruit crops as measured by chlorophyll fluorescence. *Sci. Hortic.* **1996**, *67*, 39–48. [CrossRef]
105. Puteh, A.; Saragih, A.A.; Ismail, M.R.; Mondal, M.M.A. Chlorophyll fluorescence parameters of cultivated (*Oryza sativa* L. ssp. *indica*) and weedy rice (*Oryza sativa* L. var. *nivara*) genotypes under water stress. *Aust. J. Crop Sci.* **2013**, *7*, 1277–1283.
106. Roháček, K. Chlorophyll fluorescence parameters: The definitions, photosynthetic meaning, and mutual relationships. *Photosynthetica* **2002**, *40*, 13–29. [CrossRef]
107. Tezera, W.; Mitchell, V.; Driscoll, S.P.; Lawlor, D.W. Effects of water deficit and its interaction with CO_2 supply on the biochemistry and physiology of photosynthesis in sunflower. *J. Exp. Bot.* **2002**, *53*, 1781–1791. [CrossRef] [PubMed]
108. Borkowska, B. Growth and photosynthetic activity of micropropagated strawberry plants inoculated with endomycorrhizal fungi (AMF) and growing under drought stress. *Acta. Physiol. Plant.* **2002**, *24*, 365–370. [CrossRef]
109. Šírová, K.; Vaculík, M. Toxic effects of cadmium on growth of *Aloe ferox* Mill. *S. Afr. J. Bot.* **2022**, *147*, 1181–1187. [CrossRef]
110. Zhao, H.; Guan, J.; Liang, Q.; Zhang, X.; Hu, H.; Zhang, J. Effects of cadmium stress on growth and physiological characteristics of sassafras seedlings. *Sci. Rep.* **2021**, *11*, 1–11. [CrossRef]

111. Yeboah, A.; Lu, J.; Gu, S.; Liu, H.; Shi, Y.; Amoanimaa-Dede, H.; Agyenim-Boateng, K.G.; Payne, J.; Yin, X. Evaluation of two wild castor (*Ricinus communis* L.) accessions for cadmium tolerance in relation to antioxidant systems and lipid peroxidation. *Environ. Sc.i Pollut. Res.* **2021**, *28*, 55634–55642. [CrossRef]
112. Corpas, F.J.; González-Gordo, S.; Cañas, A.; Palma, J.M. Nitric oxide and hydrogen sulfide in plants: Which comes first? *J. Exp. Bot.* **2019**, *70*, 4391–4404. [CrossRef] [PubMed]
113. Muñoz-Vargas, M.A.; González-Gordo, S.; Palma, J.M.; Corpas, F.J. Inhibition of NADP-malic enzyme activity by H2S and NO in sweet pepper (*Capsicum annuum* L.) fruits. *Physiol. Plant.* **2020**, *168*, 278–288. [CrossRef]
114. Ortega, A.; Garrido, I.; Casimiro, I.; Espinosa, F. Effects of antimony on redox activities and antioxidant defence systems in sunflower (Helianthus annuus L.) plants. *PLoS ONE* **2017**, *12*, e0183991. [CrossRef]
115. Rodríguez-Serrano, M.; Romero-Puertas, M.C.; Pazmiño, D.M.; Testillano, P.S.; Risueño, M.C.; Del Río, L.A.; Sandalio, L.M. Cellular response of pea plants to cadmium toxicity: Cross talk between reactive oxygen species, nitric oxide, and calcium. *Plant Physiol.* **2009**, *150*, 229–243. [CrossRef]
116. Rodríguez-Ruiz, M.; Aparicio-Chacón, M.V.; Palma, J.M.; Corpas, F.J. Arsenate disrupts ion balance, sulfur and nitric oxide metabolisms in roots and leaves of pea (*Pisum sativum* L.) plants. *Environ. Exp. Bot.* **2019**, *161*, 143–156. [CrossRef]
117. Wei, L.; Zhang, M.; Wei, S.; Zhang, J.; Wang, C.; Liao, W. Roles of nitric oxide in heavy metal stress in plants: Cross-talk with phytohormones and protein S-nitrosylation. *Environ. Pollut.* **2020**, *259*, 113943. [CrossRef]
118. Fang, H.; Liu, Z.; Long, Y.; Liang, Y.; Jin, Z.; Zhang, L.; Liu, D.; Li, H.; Zhai, J.; Pei, Y. The Ca^{2+} /calmodulin2-binding transcription factor TGA3 elevates LCD expression and H_2 S production to bolster Cr^{6+} tolerance in Arabidopsis. *Plant J.* **2017**, *91*, 1038–1050. [CrossRef] [PubMed]
119. Arif, M.S.; Yasmeen, T.; Abbas, Z.; Ali, S.; Rizwan, M.; Aljarba, N.H.; Alkahtani, S.; Abdel-Daim, M.M. Role of Exogenous and Endogenous Hydrogen Sulfide (H_2S) on Functional Traits of Plants Under Heavy Metal Stresses: A Recent Perspective. *Front Plant Sci.* **2021**, *11*, 545453. [CrossRef] [PubMed]
120. Nanda, R.; Agrawal, V. Elucidation of zinc and copper induced oxidative stress, DNA damage and activation of defence system during seed germination in *Cassia angustifolia* Vahl. *Environ. Exp. Bot.* **2016**, *125*, 31–41. [CrossRef]
121. Noctor, G.; Reichheld, J.P.; Foyer, C.H. ROS-related redox regulation and signaling in plants. *Semin. Cell Dev. Biol.* **2018**, *80*, 3–12. [CrossRef] [PubMed]
122. Begara-Morales, J.C.; Sánchez-Calvo, B.; Chaki, M.; Valderrama, R.; Mata-Pérez, C.; López-Jaramillo, J.; Padilla, M.N.; Carreras, A.; Corpas, F.J.; Barroso, J.B. Dual regulation of cytosolic ascorbate peroxidase (APX) by tyrosine nitration and S-nitrosylation. *J. Exp. Bot.* **2014**, *65*, 527–538. [CrossRef]
123. Palma, J.M.; Mateos, R.M.; López-Jaramillo, J.; Rodríguez-Ruiz, M.; González-Gordo, S.; Lechuga-Sancho, A.; Corpas, F.J. Plant catalases as NO and H_2S targets. *Redox Biol.* **2020**, *34*, 101525. [CrossRef]
124. Kaya, C.; Ashraf, M.; Alyemeni, M.N.; Corpas, F.J.; Ahmad, P. Salicylic acid-induced nitric oxide enhances arsenic toxicity tolerance in maize plants by upregulating the ascorbate-glutathione cycle and glyoxalase system. *J. Hazard. Mater.* **2020**, *399*, 123020. [CrossRef]
125. Hasanuzzaman, M.; Nahar, K.; Anee, T.I.; Fujita, M. Exogenous Silicon Attenuates Cadmium-Induced Oxidative Stress in Brassica napus L. by Modulating AsA-GSH Pathway and Glyoxalase System. *Front. Plant Sci.* **2017**, *8*, 1061. [CrossRef] [PubMed]
126. Prasad, K.V.S.K.; Paradha Saradhi, P.; Sharmila, P. Concerted action of antioxidant enzymes and curtailed growth under zinc toxicity in *Brassica juncea*. *Environ. Exp. Bot.* **1999**, *42*, 1–10. [CrossRef]
127. Kocsy, G.; Brunold, C. Role of glutathione in adaptation and signalling during chilling and cold acclimation in plants. *Physiol. Plant.* **2001**, *113*, 158–164. [CrossRef] [PubMed]

antioxidants

MDPI

Article

Leucine Contributes to Copper Stress Tolerance in Peach (*Prunus persica*) Seedlings by Enhancing Photosynthesis and the Antioxidant Defense System

Maoxiang Sun [1], Suhong Li [1], Qingtao Gong [2], Yuansong Xiao [1,*] and Futian Peng [1,*]

[1] State Key Laboratory of Crop Biology, College of Horticulture Science and Engineering, Shandong Agricultural University, Tai'an 271018, China

[2] Shandong Institute of Pomology, Tai'an 271000, China

* Correspondence: ysxiao@sdau.edu.cn (Y.X.); pft@sdau.edu.cn (F.P.); Tel.: +86-15163873786 (Y.X.); +86-13563821651 (F.P.)

Abstract: Heavy metal contamination has a severe impact on ecological health and plant growth and is becoming increasingly serious globally. Copper (Cu) is a heavy metal that is essential for the growth and development of plants, including peach (*Prunus persica* L. Batsch); however, an excess is toxic. In plants, amino acids are involved in responses to abiotic and biotic stresses, such as water deficit, extreme temperatures, high salinity, and heavy metal stress. However, the role of leucine in the regulation of heavy metal stress is currently unclear. Therefore, we investigated the effects of exogenous leucine on the growth of peach seedlings under Cu stress. Exogenous leucine improved the leaf ultrastructure and ionic balance and increased the chlorophyll content, the net photosynthetic rate, and the maximum photochemical efficiency. Furthermore, it attenuated Cu-stress-induced oxidative damage via a decrease in reactive oxygen species (ROS) and the regulation of the antioxidant and osmotic systems. These effects, in turn, ameliorated the reductions in cell viability, cellular activity, and biomass under Cu stress. Moreover, exogenous leucine increased the activities of nitrate reductase (NR), glutamine synthetase (GS), and glutamic acid synthetase (GOGAT) and thus improved the nitrogen metabolism efficiency of plants. In conclusion, leucine significantly improved the photosynthetic performance and antioxidant capacity, reduced Cu accumulation, and promoted nitrogen metabolism, which in turn improved the resistance of peach seedlings to Cu stress.

Keywords: antioxidant systems; Cu stress; leucine; nitrogen metabolism; peach

Citation: Sun, M.; Li, S.; Gong, Q.; Xiao, Y.; Peng, F. Leucine Contributes to Copper Stress Tolerance in Peach (*Prunus persica*) Seedlings by Enhancing Photosynthesis and the Antioxidant Defense System. *Antioxidants* **2022**, *11*, 2455. https://doi.org/10.3390/antiox11122455

Academic Editor: Nafees A. Khan

Received: 14 November 2022
Accepted: 9 December 2022
Published: 13 December 2022

1. Introduction

Heavy metal pollution of farmland soil is one of the most severe environmental problems in the world [1]. Globally, there are 5 million sites of soil pollution, covering 500 million ha of land in which the soils are contaminated by different heavy metals or metalloids [2]. Heavy metal pollution in soil has a combined worldwide economic impact that is estimated to be in excess of USD 10 billion per year [3]. Heavy metal pollution not only decreases crop yield and quality but also affects human health via the food chain [4]. However, certain heavy metals, such as Cu at lower concentrations, are critical for plant growth and development. Cu is an essential trace element involved in many physiological processes of plants. It acts as a cofactor of enzymes such as the Cu/Zn-superoxide dismutase (Cu/Zn SOD) enzyme and polyphenol oxidase. It is involved in physiological processes such as photosynthesis, cell wall metabolism, and ethylene perception [5]. Cu is a component of the plastocyanin in plant chloroplasts, which is involved in the photosynthetic electron transport process and is also an activator of certain enzymes during chlorophyll formation [6]. Cu deficiency in plants can hinder normal growth, while excess Cu can be toxic. At higher concentrations, Cu ions (Cu^{2+}) combine with oxygen to produce free radicals and other reactive oxygen species (ROS), which induce oxidative stress and lead to cell death. Moreover, the accumulation of Cu in the food chain can lead to poisoning in humans due to excessive consumption [7]. In recent years, Cu has become a major environmental pollutant due to overuse in the manufacturing and agricultural industries [8]. Therefore, in order to reduce plant damage due to Cu stress in soil, it is necessary to improve plant tolerance to Cu stress and reduce Cu ion enrichment under Cu stress.

Under normal conditions, the production and removal of reactive oxygen species (ROS) in plants are in dynamic balance. Oxidative stress occurs when the accumulated concentration of ROS exceeds the threshold of plant defense mechanisms under heavy metal stress [9]. Cu stress can induce cells to produce a large amount of ROS, leading to lipid peroxidation of the cell membrane and resulting in an increase in malondialdehyde (MDA) content, a reduction in the selective permeability of the plasma membrane, the leaching of intracellular materials, and then damage to photosynthetic organelles, thereby affecting the normal operation of physiological metabolic processes such as material exchange and photosynthesis [10]. ROS can change the distribution of ions and initiate the expression of nuclear genes in the process of transmitting and amplifying signals so that plants can tolerate various stresses [11]. Amino acids enhance the adaptive responses in plants to various stresses by directly/indirectly influencing the physiological processes or regulating stress-related gene expression and enzyme activities in plants [12]. In plants, leucine is not only a nutrient matrix for cell metabolism but also a signal molecule that can mediate protein metabolism, lipid decomposition, and other biological reactions. Leucine spray at different stages increased biomass and nitrogen content, promoted nitrogen absorption, and improved grain yield in rice [13].

Peach (*Prunus persica* L. Batsch) is one of the most widely cultivated fruit trees in China [14]. Previous studies have found that excessive copper in peach orchard soil can reduce the dry matter content of peach trees, reduce the photosynthetic rate of peach leaves, and reduce the chlorophyll content of leaves, which are not conducive to fruit yield and quality [15]. Therefore, it is crucial to reduce the Cu content in plant parts and alleviate the impact of Cu stress to improve the yield and quality of peach. Several studies have reported the beneficial effects of amino acids in plants exposed to metal stress [16]. In general, amino acid molecules have three major functions in response to heavy metal stress, namely metal binding, antioxidant defense, and signaling [17]. Due to their ability to bind metals, amino acids and their by-products can be used to respond to metal toxicity. Amino acids facilitate the chelation of heavy metal ions in cells and xylem sap, thereby resulting in the detoxification of heavy metals and an increase in plant resistance to toxic metal ions by activating antioxidant systems [18]. However, the role of leucine in the regulation of heavy metal stress is currently unclear. Therefore, in this study, peach seedlings were used as experimental materials to explore the mechanism of leucine alleviating Cu stress in peach seedlings.

2. Materials and Methods

2.1. Experimental Design

Experiments were performed in the experiment center of Shandong Agricultural University (Tai'an, China) in April 2020 (117°13′ E and 36°16′ N). First, 60-day-old *'lu xing'* peach plants, *Prunus persica* (L.) Batsch., were planted in pots. Seedlings (10 cm tall) grown from peach seeds with the same growth trend and without diseases or insect pests were selected and planted. The pots were cylindrical with an inner diameter of 20 cm and a height of 30 cm. Approximately 2.5 kg of garden soil was placed in each pot, and each treatment was repeated in 60 pots (replicates). The basic physical and chemical properties of the tested soil were as follows: the pH value was 6.68, the alkaline hydrolyzable nitrogen content was 45.65 $mg \cdot kg^{-1}$, the organic matter content was 12.57 $g \cdot kg^{-1}$, the available phosphorus content was 35.33 $mg \cdot kg^{-1}$, and the available potassium content was 83.72 $mg \cdot kg^{-1}$. The growing conditions for the peach trees were day and night temperatures of 29 °C/23 °C, a natural photoperiod of around 12.5 h, and a constant relative humidity of 30%. The seedlings were maintained following conventional management practices. In the preliminary Cu stress screening test (Supplementary File, Figure S1), we found that Cu-induced damage to the peach seedlings began to appear after their treatment with 5 mmol·L^{-1} $CuCl_2 \cdot 2H_2O$ for 6 days, and leucine (10 mmol·L^{-1}) significantly alleviated Cu stress for 6 days, which met the test requirements. Therefore, in this experiment, the concentrations of 5 mmol·L^{-1} for $CuCl_2 \cdot 2H_2O$ and 10 mmol·L^{-1} for leucine were chosen. The seedlings were treated as follows: water (control), 10 mmol·L^{-1} leucine (Leu, Shanghai Yuan ye Biotechnology Co., Ltd., Shanghai, China), 5 mmol·L^{-1} $CuCl_2 \cdot 2H_2O$ (Cu), and 5 mmol·L^{-1} $CuCl_2 \cdot 2H_2O$ + 10 mmol·L^{-1} leucine (Cu+Leu). In this experiment, $CuCl_2 \cdot 2H_2O$ and leucine were uniformly applied to the soil one time at a reagent dosage of 200 mL per tree. All treatments were applied in soil pots at 9 am to avoid excessively high temperatures. On the sixth day after treatment, image indicators were tested with fresh samples, and for the other indicators samples were frozen in liquid nitrogen and placed in an ultra low temperature refrigerator at −80 °C for further testing. Three biological repetitions were included per treatment.

2.2. Determination of Endogenous Leucine Content

On the sixth day after treatment, the leaves, stems, and roots were dried and sifted through a 100-mesh sieve. Approximately 0.5 g of the powdered sample was accurately weighed, placed in a 10 mL centrifuge tube, and mixed with 10 mL of ultrapure water. This mixture was ultrasonicated for 30 min and centrifuged at 10,000 rpm for 2 min. The pH of the supernatant was adjusted from 5.3 to 2.2 with 1 mol/L hydrochloric acid. The resultant sample was used to determine the leucine content with an automatic amino acid analyzer (Biochrom, Cambridge, UK) [19].

2.3. Determination of SPAD Value, Photosynthetic Rate, and Maximum Photochemical Efficiency

In order to observe the dynamic changes in the physiological indexes of peach seedlings after the treatments, we chose to measure them on the 2nd, 4th, and 6th days after treatment. Chlorophyll fluorescence was measured using an Imaging-PAM chlorophyll fluorometer (HeinzWalz GmbH, Effeltrich, Germany). Before the determination of Fo, the plants were dark-adapted for 30 min. Then, the leaves were adapted to actinic light (250 mmol/m^2/s), and the maximum fluorescence (Fm) and steady-state fluorescence (Fs) under actinic light were measured with a saturated pulse (3000 mmol/m^2/s). The variable fluorescence/fluorescence maximum (Fv/Fm), the quantum efficiency of PSII (ϕPSII), the photochemical quenching coefficient (qP), and the non-photochemical quenching coefficient (NPQ) were automatically calculated by the software.

The fifth and sixth uppermost leaves from the top of the plant were selected and fully unfolded. The SPAD values were measured using a chlorophyll tester (SPAD-502 plus, Tokyo, Japan), and the photosynthetic rate was measured using a ciras-3 portable photosynthetic instrument (PP Systems, Hitchin, UK). On the sixth day after treatment, the maximum photochemical efficiency (Fv/Fm) was measured using a portable pulse adjustment fluorometer (Handy PEA, Hansatech, UK). Two leaves per plant and three plants per treatment were used.

2.4. Determination of Biomass

Thirty days after the treatment, the peach seedlings were uprooted, washed, and dried off with an absorbent paper. Then, the fresh weights of the aboveground and belowground parts were measured. The dry weights of the aboveground and belowground parts were measured after drying the samples in an oven at 80 °C for 30 min and at 60 °C to a constant weight.

2.5. Analysis of Leaf Ultrastructure

On the sixth day after treatment, mesophyll tissue (2 mm × 2 mm) was cut with a blade (avoiding the vein), immersed in 2.5% glutaraldehyde, vacuum-dried, and stored at 4 °C for 24 h. The samples were dehydrated with alcohol, soaked in resin, and heated. The sections were stained and observed under an electron microscope (JEOLTEM-1230EX, Tokyo, Japan) [20].

2.6. Determination of Cu Element Contents

On the sixth day after treatment, the powdered peach seedling sample was digested with a combination of nitric acid and perchloric acid, boiled to a constant volume, and filtered through a filter membrane (0.22 μm). The Cu element content in the final sample was analyzed with an atomic absorption spectrometer [21].

2.7. Determination of Evans Blue Assay and Reactive ROS

On the sixth day after treatment, the leaf blades and roots were cleaned with ultrapure water, dried with an absorbent paper, and soaked in an Evans blue dye solution (0.25%) for 24 h. The stained leaf blades were washed with ultrapure water, dried, and boiled in a mixture of anhydrous ethanol and glycerin (9:1) to remove the chlorophyll entirely. Images were captured. Then, the cell viability of the roots was quantified with Evans blue using a standard curve. Fluorescence was measured at 600 nm [22].

The production rate of superoxide radicals ($O_2^{\bullet-}$) and the content of hydrogen peroxide (H_2O_2) were determined as described previously [23]. The production rate of $O_2^{\bullet-}$ was determined by analyzing the formation of nitrite (NO_2-) from hydroxylamine in the presence of $O_2^{\bullet-}$. The leaf sample was homogenized and centrifuged. The supernatant was mixed and incubated. The absorbance of the supernatant was read at a wavelength of 530 nm, and a standard curve with NO_2- was used to calculate the production rate of $O_2^{\bullet-}$. The content of H_2O_2 was determined by monitoring the formation of the hydrogen peroxide–titanium complex. The sample was homogenized with cold acetone. Then, a titanium reagent (15% $Ti(SO_4)_2$) was added to a final concentration of 4%. To precipitate the peroxide–titanium complex, 0.2 mL of concentrated NH_4OH was added per 1 mL of the reaction mixture. After centrifugation (5 min at 10,000× *g*), the pellet was washed twice with acetone and solubilized in 2 mL of 2 N H_2SO_4. The absorbance of the solution was measured at 410 nm.

2.8. Determination of Proline, Malondialdehyde, and Antioxidant Enzyme Activities on the Sixth Day after Treatment

The determination of the free proline content was performed according to Bates et al. [24]. Leaf samples (0.5 g) from each group were homogenized in 3% (w/v) sulfosalicylic acid, and the homogenate was filtered through filter paper. After the addition of acid ninhydrin and glacial acetic acid, the resulting mixture was heated at 100 °C for 1 h in a water bath. The reaction was then stopped using ice bath. The mixture was extracted with toluene, and the absorbance of the fraction with toluene aspired from the liquid phase was read at 520 nm. The proline concentration was determined using a calibration curve and expressed as μmol proline g^{-1} FW.

Lipid peroxidation can be measured by determining the malondialdehyde (MDA) content. The MDA determination followed the method described by Zhao [25] with minor modifications. Briefly, we took 1 mL of the sample supernatant extracted above and added 2 mL of 0.67% thiobarbituric acid (TBA). We used a sample containing only 1 mL of water as a negative control. Next, we placed the samples in a boiling water bath for 15 min, rapidly cooled them by immersion in cold water, and poured them into 10 mL centrifuge tubes. We then centrifuged all tubes at 4000 rpm for 20 min and determined the absorbance of all samples at 600 nm, 532 nm, and 450 nm using a spectrophotometer. The MDA content was determined using the following formula: MDA (μmol·g^{-1}) = [6.452 (A_{532} − A_{600}) − 0.56 × A450] × VT/(V0 × W), where VT = the total volume of the extract; V0 = the assay volume; and W = plant fresh weight.

All enzymatic activities were determined at 25 °C and expressed as U g^{-1} protein. The CAT activity was measured by monitoring the decrease in H_2O_2 at 240 nm for 1 min at 25 °C. The 3 mL reaction mixture contained 100 μL of enzyme extract and 2.9 mL of sodium phosphate buffer (50 mM, pH 6.0) containing 10 mM H_2O_2. The CAT activity was calculated as the amount of enzyme that caused a reduction in the absorbance at 240 nm of 0.01 per minute [25]. The peroxidase (POD) activity was determined by a colorimetric method [25] in a reaction mixture containing guaiacol as the substrate. The POD activity was determined based on the change in absorbance at 470 nm due to the oxidation of guaiacol to tetraguaiacol. The POD activity was defined as the amount of enzyme that caused an increase in absorbance at 470 nm of 0.001 per minute. The superoxide dismutase (SOD) activity was determined based on the ability to inhibit the photochemical reduction in nitroblue tetrazolium (NBT) [25]. The 3 mL reaction mixture was initiated by illumination for 2 min at 25 °C, and the absorbance of blue formazan was measured with a spectrophotometer at 560 nm. One unit of SOD activity (U) was defined as the amount of enzyme that caused a 50% inhibition of the NBT reduction. The APX activity (ascorbate peroxidase) was determined by measuring the oxidation rate of ascorbate at 290 nm according to Zhao [25]. The decrease in the AsA concentration was followed as a decline in the optical density at 290 nm, and the activity was calculated using the extinction coefficient (2.8 mM^{-1} cm^{-1} at 290 nm) for AsA. One unit of APX was defined as the amount of enzyme that breaks down 1 μmol AsA min^{-1}. The guaiacol peroxidase (GPX) activity was measured using a modification of the procedure by Zhao [25]. The reaction mixture in a total volume of 2 mL contained 25 mM (pH 7.0) sodium phosphate buffer, 0.1 mM EDTA, 0.05% guaiacol (2-ethoxyphenol), 1.0 mM H_2O_2, and 100 μL of enzyme extract. The increase in absorbance due to the oxidation of guaiacol was measured at 470 nm (E = 26.6 mM^{-1} cm^{-1}). The dehydroascorbate reductase (DHAR) activity was determined according to Zhao [25]. The reaction mixture in a total volume of 2 mL contained 25 mM (pH 7.0) sodium phosphate buffer, 0.1 mM EDTA, 3.5 mM GSH, 0.4 mM dehydroascorbate (DHAR), and 100 μL of enzyme extract. The DHAR activity was measured by the formation of ascorbate at 265 nm (E = 14 mM^{-1} cm^{-1}). The enzyme activity was expressed as unit g^{-1} min^{-1} FW.

2.9. Total RNA Extraction and Quantitative PCR Analysis on the Sixth Day after Treatment

Genes of the antioxidant system were identified from the NCBI database (https://www.ncbi.nlm.nih.gov/, accessed on 14 November 2022). RNA of leaves was extracted using the RNA Extraction Kit (Kangwei Century Technology Co., Ltd, Beijing, China) and reverse-transcribed to cDNA (Takara). The primers used in this study are shown in Supplementary Table S1. Quantitative PCR was performed to determine the expression of related genes. The reactions were performed using an Ultra SYBR Mixture (Kangwei, Beijing, China) Kit. Three biological replicates and three technical repeats were used in all qPCR reactions. The relative gene expression levels were calculated using the $2^{-\Delta\Delta CT}$ method [26].

2.10. Determination of TTC Assay and Electrolyte Leakage on the Sixth Day after Treatment

The root viability was determined by the 2,3,5-triphenyltetrazolium chloride (TTC) reduction method and expressed as the amount of TTC reduced by per gram of root [25].

The root electrolyte leakage was determined according to a previously described method [27]. Approximately 0.5 g of a root was taken in a test tube, mixed with 20 mL of distilled water, and the vacuum was evacuated three times for 20 min each time. The initial conductivity (S1) of the mixture was measured with a conductivity meter (DDS-11A, Shanghai Kanglu Instrument Equipment Co., Ltd., Shanghai, China). The mixture was then sealed and incubated in a boiling water bath for 10 min. The conductivity was measured after cooling (S2). The conductivity of the distilled water was also measured (S0). Then, the percentage of electrolyte leakage was calculated as follows: electrolyte leakage (%) = [(S1 − S0)/(S2 − S0)] × 100.

2.11. Determination of Nitrogen Content and Nitrogen-Metabolism-Related Enzyme Activities in Leaves

On the sixth day after the treatment, the nitrate–nitrogen (NO_3-N) content of the leaves was determined by salicylic acid colorimetry [25]. Under an acidic condition, the leaf extract was mixed with salicylic acid to form nitrosalicylic acid, whose absorbance was read at 410 nm.

The ammonium–nitrogen (NH_4^+-N) content was determined by Nessler's spectrophotometric method [28]. The leaf extract was mixed with hypochlorite and phenol under a strong alkaline condition to form a water-soluble blue indophenol whose absorbance was measured at 625 nm.

The nitrate reductase (NR) activity in the leaves was determined by sulfanilamide colorimetry [25]. The leaf extract was mixed with sulfanilamide in hydrochloric acid and N-1-naphtyl-ethylenediamine to form a red azo dye whose absorbance was measured at 520 nm.

We determined the glutamine synthetase (GS) activity using the plant GS enzyme activity assay kit and determined the glutamic acid synthetase (GOGAT) activity using the plant GOGAT enzyme activity assay kit (Jiangsu enzyme label Biotechnology Co., Ltd., Jiangsu, China), following the manufacturer's instructions.

2.12. Data Processing and Analysis

We collected three biological replicates for each treatment. Origin version 9.8 was used to conduct all statistical analyses. Duncan multiple range tests, which are included in SPSS version 20.0, were performed to detect any statistically significant differences in the mean values (IBM SPSS, Chicago, IL, USA). The threshold of statistical significance used for all tests was $p < 0.05$.

3. Results

3.1. Effects of Exogenous Leucine on Growth, Chlorophyll Fluorescence Imaging, Net Photosynthetic Rate, SPAD Value, and Maximum Photochemical Efficiency of Peach Seedlings under Cu Stress

It can be seen from Figure 1a that exogenous Leu can enhance the tolerance of peach seedlings under copper stress. Chlorophyll fluorescence can reflect the photosynthetic efficiency and the degree of stress of plants. We measured the changes in chlorophyll fluorescence (Figure 1b), the net photosynthetic rate (Figure 1c), the SPAD value (Figure 1d), and the maximum photochemical efficiency (Fv/Fm) (Figure 1e) of peach seedlings under Cu stress after leucine treatment. Figure 1a shows that the photosynthetic efficiency of leaves decreased under Cu stress, while the application of Leu improved the photosynthetic efficiency under Cu stress. The net photosynthetic rate, SPAD value, and maximum photochemical efficiency of peach seedlings decreased with an increase in time under Cu stress. However, Leu could alleviate the damage of photosynthetic system under Cu stress. Cu stress reduced the net photosynthetic rate, SPAD value, and maximum photochemical efficiency by 53.8%, 28.2%, and 14.5% compared with control, respectively, in peach seedlings within 6 days. Exogenous leucine increased the net photosynthetic rate, SPAD value, and maximum photochemical efficiency of seedlings under Cu stress by 83.35%, 28.6%, and 8.5% compared with the Cu treatment, respectively, within 6 days. The difference in the effect of Leu on the photosynthetic system was not significant under normal conditions compared to the control.

3.2. Effects of Exogenous Leucine on Leaf Ultrastructure of Peach Seedlings under Cu Stress

The ultrastructural changes in the cells of peach leaves under Cu stress and with exogenous leucine are shown in Figure 2a–c. In the control seedlings, mesophyll cells and organelles were visible, and cell walls were intact and smooth. The chloroplasts appeared full and fusiform, and the inner grana lamellae were stacked closely and arranged in order. Few starch granules and plastid globules were observed. Under Cu stress, compared with the control, the starch granule volume increased significantly; starch grains occupied almost half of the chloroplast space. In addition, the number of plastid globules increased, and the chloroplast structure was deformed. With exogenous leucine, the volume of starch granules and the number of plastid globules decreased, and the chloroplast structure was similar to control, even under Cu stress.

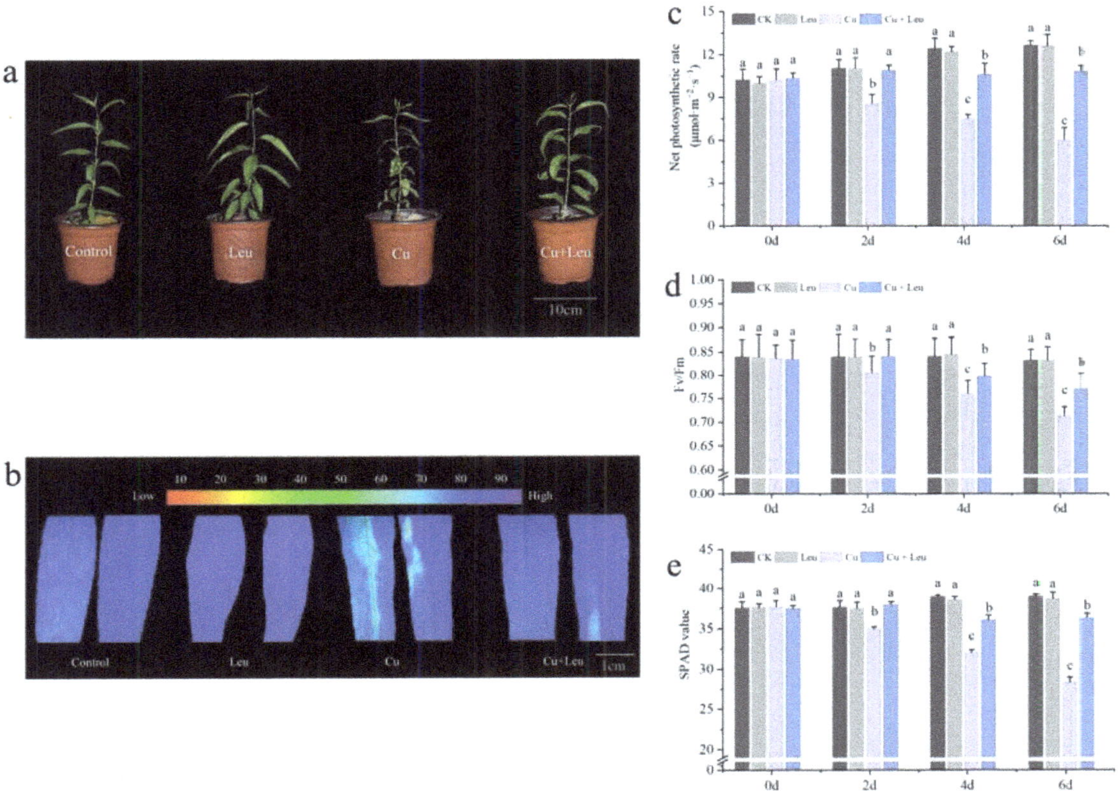

Figure 1. Effects of different treatments on growth, chlorophyll fluorescence imaging, net photosynthetic rate, SPAD value, and maximum photochemical efficiency of peach seedlings. (**a**) Peach seedlings under different treatments; (**b**) chlorophyll fluorescence imaging; (**c**) net photosynthetic rate; (**d**) SPAD value; (**e**) maximum photochemical efficiency. Each data point represents the mean (±SD) of three replicates. Error bars represent standard deviations of the means (n = 3). Different lowercase letters indicate significant differences among different treatments (Duncan test, $p < 0.05$).

3.3. *Effects of Exogenous Leucine on Antioxidant Enzyme Gene Expression and Enzyme Activities in Peach Seedlings under Cu Stress*

We took the logarithm of the relative expression of antioxidant protective enzyme genes based on $\log_2$, so that the final value > 0 was positive regulation and otherwise it was negative regulation. We found that the relative gene expression of antioxidant protective enzymes in the control and Leu treatments was negative, indicating that the control and Leu treatments did not activate the antioxidant protective enzyme system. However, the Cu treatment and the Cu+Leu treatment could activate the antioxidant protective enzyme system in order to remove ROS in the plant. The relative expression of antioxidant protective enzymes in the Cu treatment with Leu addition was higher than that in the Cu treatment, indicating that Leu could reduce the production of ROS in the plant under Cu stress (Figure 3a). The activities of CAT, SOD, GPx, APX, POD, and DHAR increased by 67.6%, 148.9%, 50.3%, 152.1%, 187.1%, and 181.4% compared with control, respectively, under the Cu+Leu treatment. However, the activities of CAT, SOD, GPx, APX, POD, and DHAR decreased by 29.5%, 28.9%, 14.7%, 6.8%, 41.3%, and 51.3%, respectively, under Cu stress compared with the Cu+Leu treatment, indicating that the production of ROS could be effectively reduced by Leu (Figures 3b–g and 4e,f).

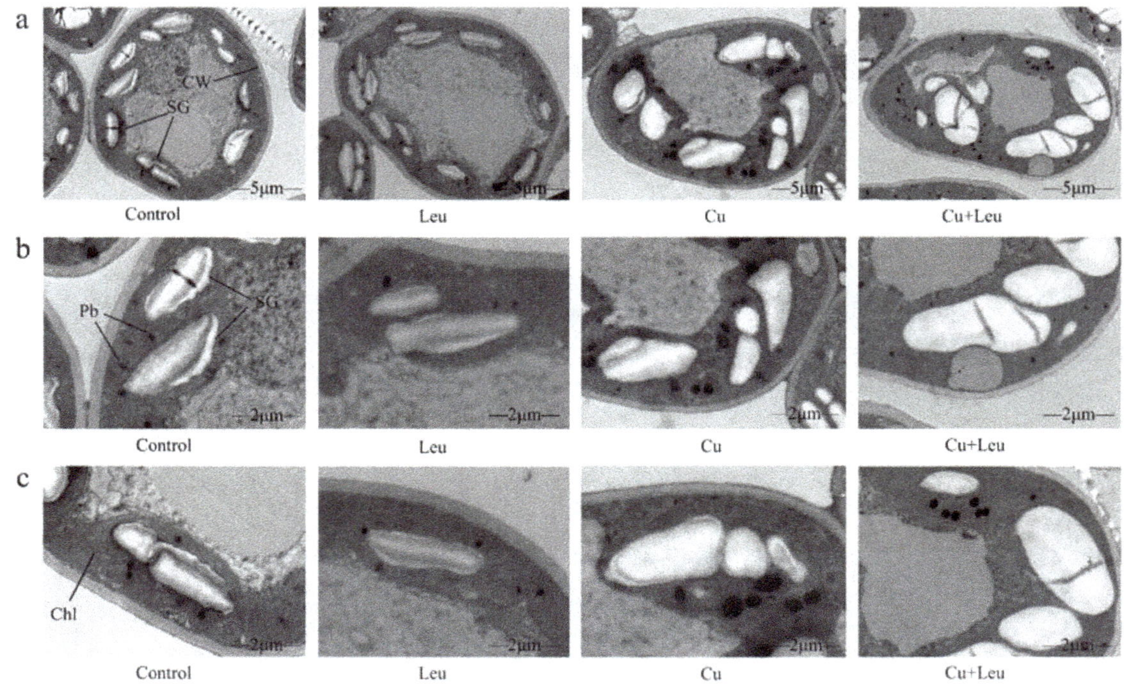

Figure 2. Electron micrographs of leaf mesophyll cells of peach seedlings under different treatments. (**a**) Overall structure of cells with different treatments (bars correspond to 5 μm); (**b**,**c**) Organelle architecture with different treatments (bars correspond to 2 μm). CW: Cell wall; Chl: Chloroplast; SG: Starch granule; Pb: Plastid globule.

3.4. Effects of Exogenous Leucine on Biomass, Leucine Content, Cell Viability, and ROS Accumulation in Leaves of Peach Seedlings under Cu Stress

The fresh weights of the aboveground and belowground plant parts of seedlings under Cu stress were significantly lower than those of the control seedlings after 20 days. However, the fresh weights of the aboveground and belowground parts of seedlings under Cu stress with exogenous leucine were higher than those without exogenous leucine (Figure 4a). These findings indicate that Cu stress inhibited plant growth and reduced the fresh and dry weights of peach seedlings; exogenous leucine alleviated these effects and helped plants retain growth under Cu stress. The endogenous leucine contents in the leaves, stems, and roots of peach seedlings under Cu stress were significantly increased compared to those in the control seedlings. The exogenous leucine further significantly increased the endogenous leucine contents in the leaves, shoots, and roots (Figure 4b).

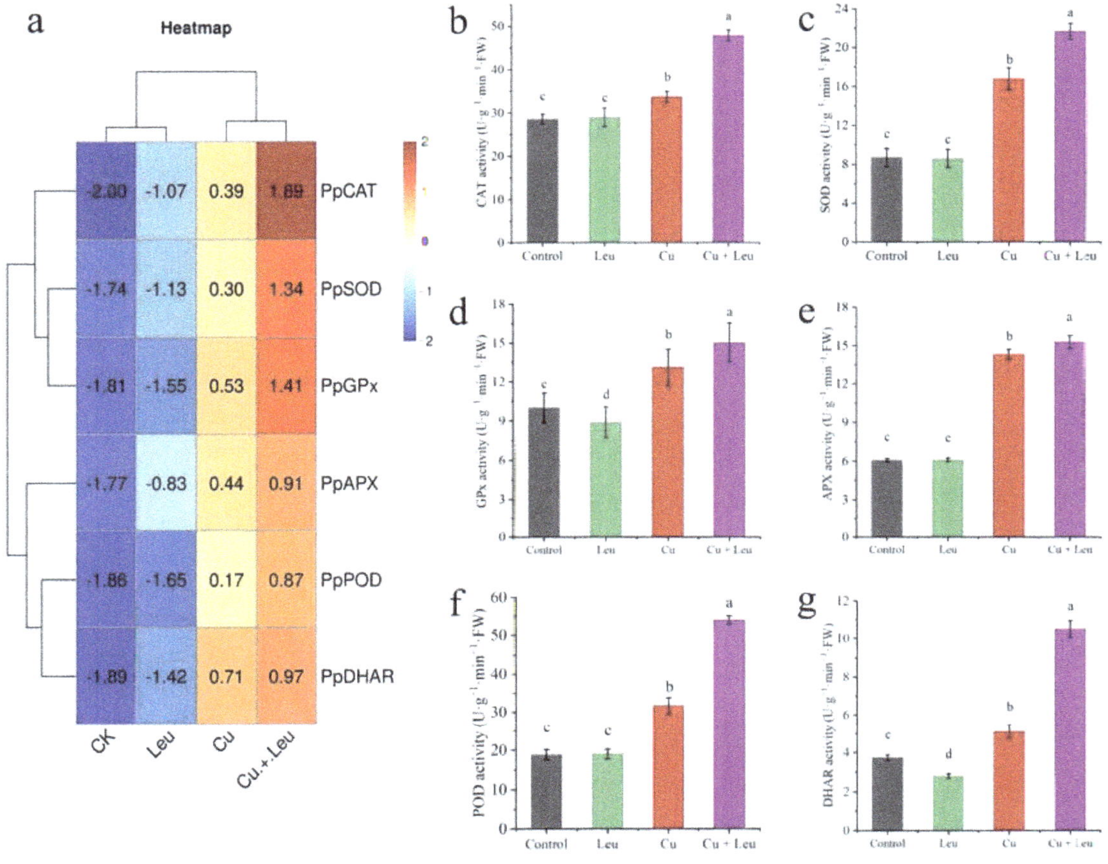

Figure 3. Effects of different treatments on antioxidant protective enzyme gene expression and enzyme activities in peach seedlings. (**a**) The expression of antioxidant enzyme genes; (**b**) CAT activity; (**c**) SOD activity; (**d**) GPx activity; (**e**) APX activity; (**f**) POD activity, (**g**) DHAR activity). Error bars represent standard deviations of the means ($n = 3$). Different lowercase letters indicate significant differences among different treatments (Duncan test, $p < 0.05$).

Evans blue staining was used to analyze the extent of damage in leaf cells after 6 days under Cu stress. The leaves of peach seedlings under Cu stress were darker, with a large percentage of the area stained. Meanwhile, the leaves of peach seedlings under Cu stress with exogenous leucine were lighter, with a smaller percentage of the area stained compared with that of the leaves under Cu stress alone (Figure 4c). These findings indicate that cell viability decreased under Cu stress and exogenous leucine mitigated Cu-stress-induced cell death (Figure 4d). High levels of ROS induced under heavy metal stress disrupt membrane stability and hinder plant growth. Cu stress significantly increased the production rate of $O_2^{\bullet -}$ and the content of H_2O_2. However, exogenous leucine reduced the Cu-stress-induced $O_2^{\bullet -}$ production rate and the H_2O_2 content (Figure 4e,f).

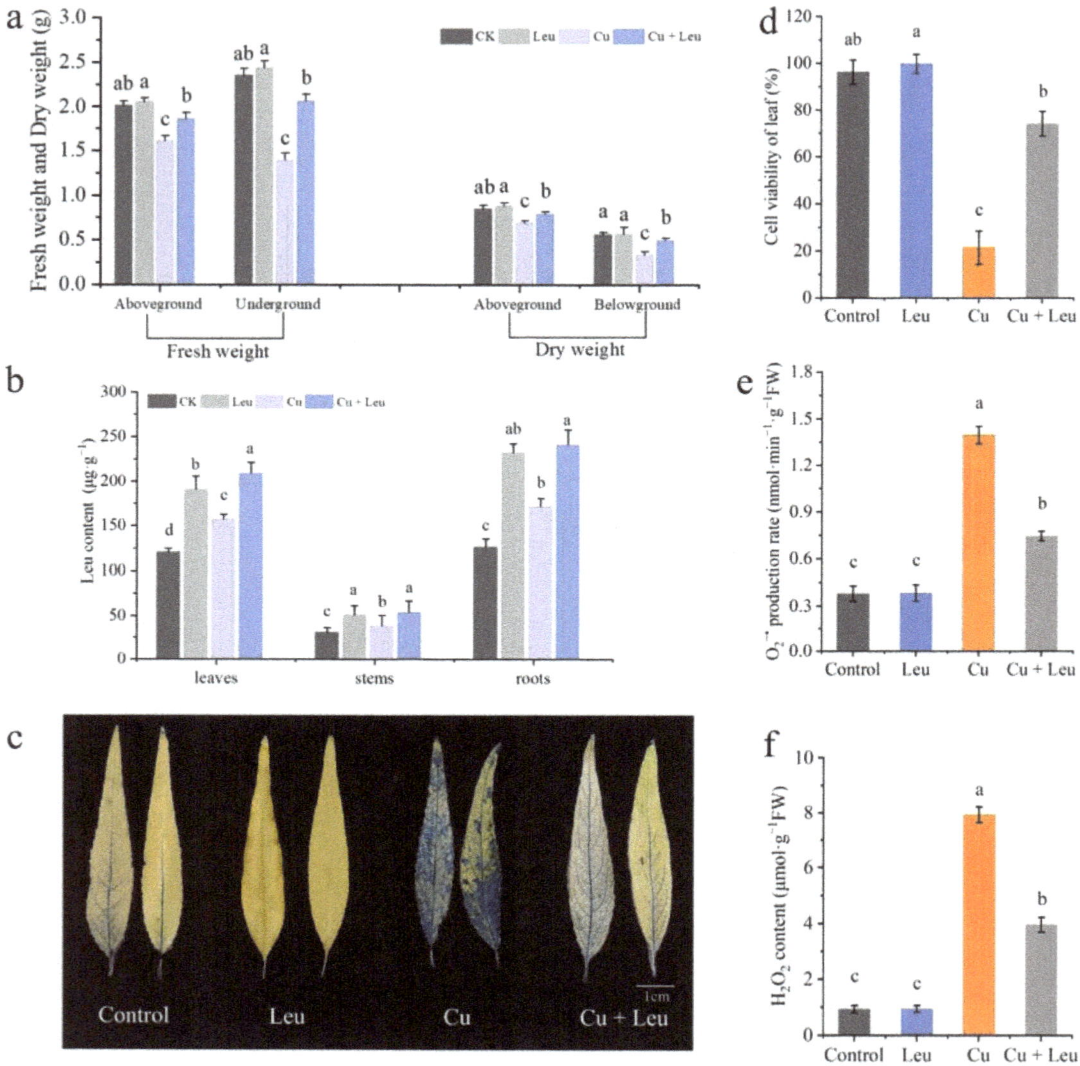

Figure 4. Effects of different treatments on biomass, leucine content, cell viability, and ROS accumulation in leaves of peach seedlings. (**a**) Fresh weight and dry weight of peach seedlings thirty days after the treatment; (**b**) leucine content; (**c**) Evans blue staining of leaves; (**d**) cell viability of leaves; (**e**) leaf $O_2^{\bullet-}$ production rate; (**f**) H_2O_2 content. Error bars represent standard deviations of the means ($n = 3$). Different lowercase letters indicate significant differences among different treatments (Duncan test, $p < 0.05$).

3.5. Effects of Exogenous Leucine on Root Biomass, Root Cell Viability, Root Viability, Proline, Malondialdehyde Contents, and Electrolyte Leakage of Peach Seedlings under Cu Stress

Figure 5a and Supplementary Table S2 show that Leu had little effect on the roots of peach seedlings under normal conditions; however, the root biomass was significantly reduced under Cu stress, and Leu could effectively reduce the reduction in root biomass of peach seedlings. The plant root system is the organ that is directly exposed to heavy metals in soil. Evans blue staining was used to analyze the viability of root cells under Cu stress. The roots were darker under Cu stress, and the color became lighter with exogenous leucine (Figure 5b). In addition, the Cu+Leu treatment significantly increased root cell viability and root viability relative to the Cu treatment (Figure 5c,d). The contents of proline in the peach seedlings under Cu stress were significantly higher than those in the control seedlings (1.3 times higher, Figure 5e). Exogenous leucine significantly reduced the Cu-stress-induced increase in proline contents in peach seedlings. The results indicate that the leucine treatment significantly alleviated the osmotic stress caused by high concentrations of Cu in peach seedlings. The malondialdehyde (MDA) content of seedlings under Cu stress was significantly higher than that of the control seedlings. Exogenous leucine reduced the stress-induced increase in the MDA content; however, the content was significantly higher than that of the control. The results indicate that leucine reduced the MDA content and that electrolyte leakage improved membrane stability in peach seedlings under Cu stress (Figure 5f,g). As can be seen from Supplementary Table S3, Cu ions were mainly enriched in the roots of peach seedlings, and the Cu+Leu treatment could significantly reduce the Cu ion content in the roots compared with the Cu treatment.

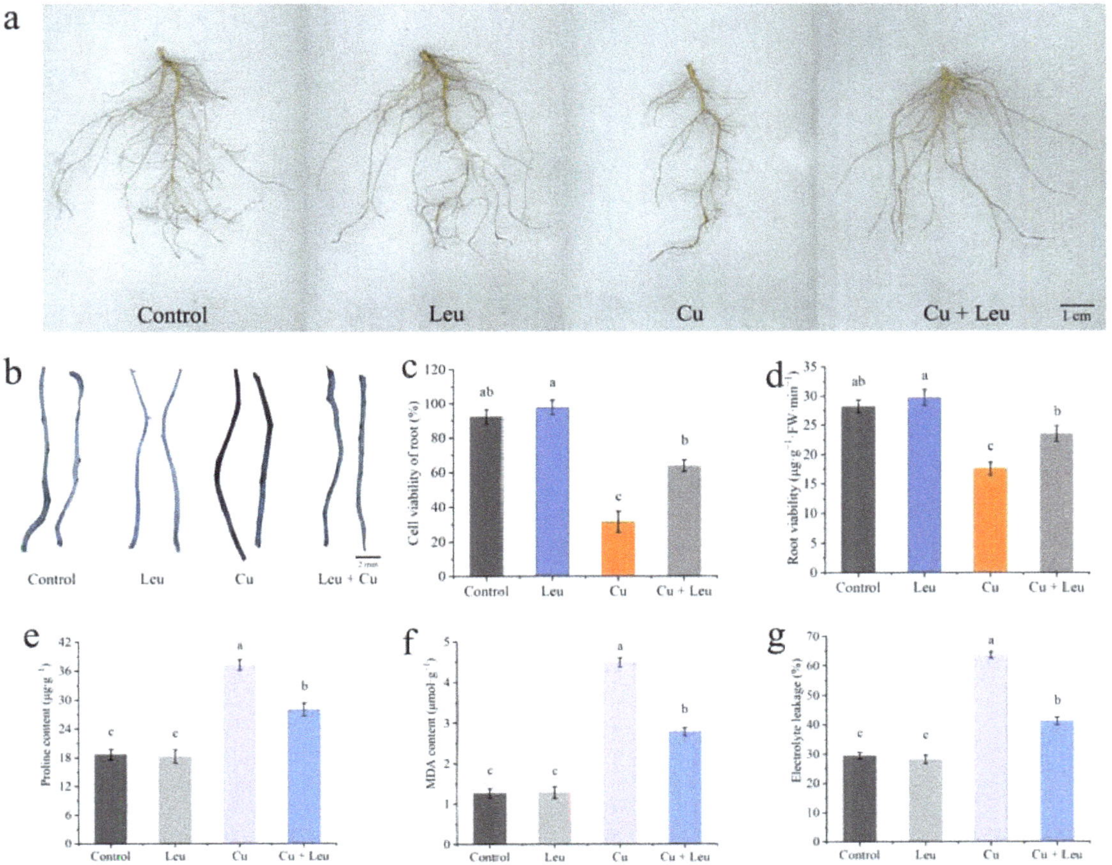

Figure 5. Effects of different treatments on root biomass, root cell viability, root viability, proline, malondialdehyde content, and electrolyte leakage of peach seedlings. (**a**) Root biomass; (**b**) Evans blue staining of roots; (**c**) root cell viability; (**d**) root viability; (**e**) proline content; (**f**) MDA content; (**g**) electrolyte leakage. Error bars represent standard deviations of the means (n = 3). Different lowercase letters indicate significant differences among different treatments (Duncan test, $p < 0.05$).

3.6. Effects of Exogenous Leucine on Nitrogen Metabolism in Peach Seedlings under Cu Stress

Under normal conditions, the activities of NR, GS, and GOGAT treated with control and Leu were at high levels from 0 to 6 days. The activities of NR, GS, and GOGAT decreased significantly with time under Cu stress. Although the enzyme activity of leucine-treated seedlings under Cu stress was significantly lower than that of control seedlings, the enzyme activity decreased slightly after 4 days, which was significantly higher than that of pure Cu stress seedlings (Figure 6a–c). Cu stress decreased the nitrate–nitrogen content and increased the ammonium–nitrogen and free amino acid contents in the peach seedlings (Figure 6d–f). The application of leucine to the seedlings under Cu stress increased the nitrate–nitrogen content; however, it was lower than that in the control group. Meanwhile, with exogenous leucine, the ammonium–nitrogen content was the same as that in the control group, while the total free amino acid content was significantly higher than that in control group.

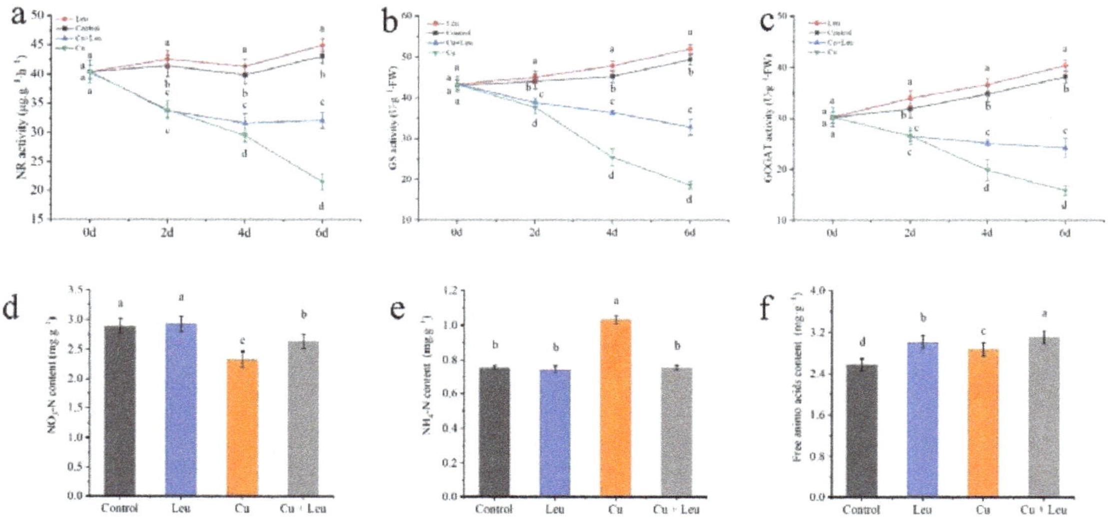

Figure 6. Effects of different treatments on nitrogen metabolism in peach seedlings. (**a**) Nitrate reductase (NR) activity; (**b**) glutamine synthase (GS) activity; (**c**) glutamic acid synthetase (GOGAT) activity; (**d**) nitrate–nitrogen (NO_3-N) content; (**e**) ammonium–nitrogen (NH_4+-N) content; (**f**) free amino acid content. Error bars represent standard deviations of the means ($n = 3$). Different lowercase letters indicate significant differences among different treatments (Duncan test, $p < 0.05$).

3.7. Correlation Analysis of Leucine Content with Osmotic System and Antioxidant System in Peach Seedlings

Figure 7 exhibits a heatmap correlation matrix among different physiological and molecular traits. For the correlation study, the leucine content of the peach seedlings as well as osmotic regulatory system and antioxidant system activity (SOD, POD, CAT, APX, GPx, and DHAR) were evaluated. As can be seen from Figure 7, the leucine content in peach seedlings was positively correlated with plant antioxidant enzyme activities, and the correlation was significantly different and was negatively correlated with the hydrogen peroxide content and the MDA content. The MDA content was negatively correlated with the proline content, the SPAD value, the cell viability of roots, the cell viability of leaves, the APX activity, and the net photosynthetic rate. The hydrogen peroxide content was negatively correlated with the SOD activity, POD activity, DHAR activity, CAT activity, and GPx activity.

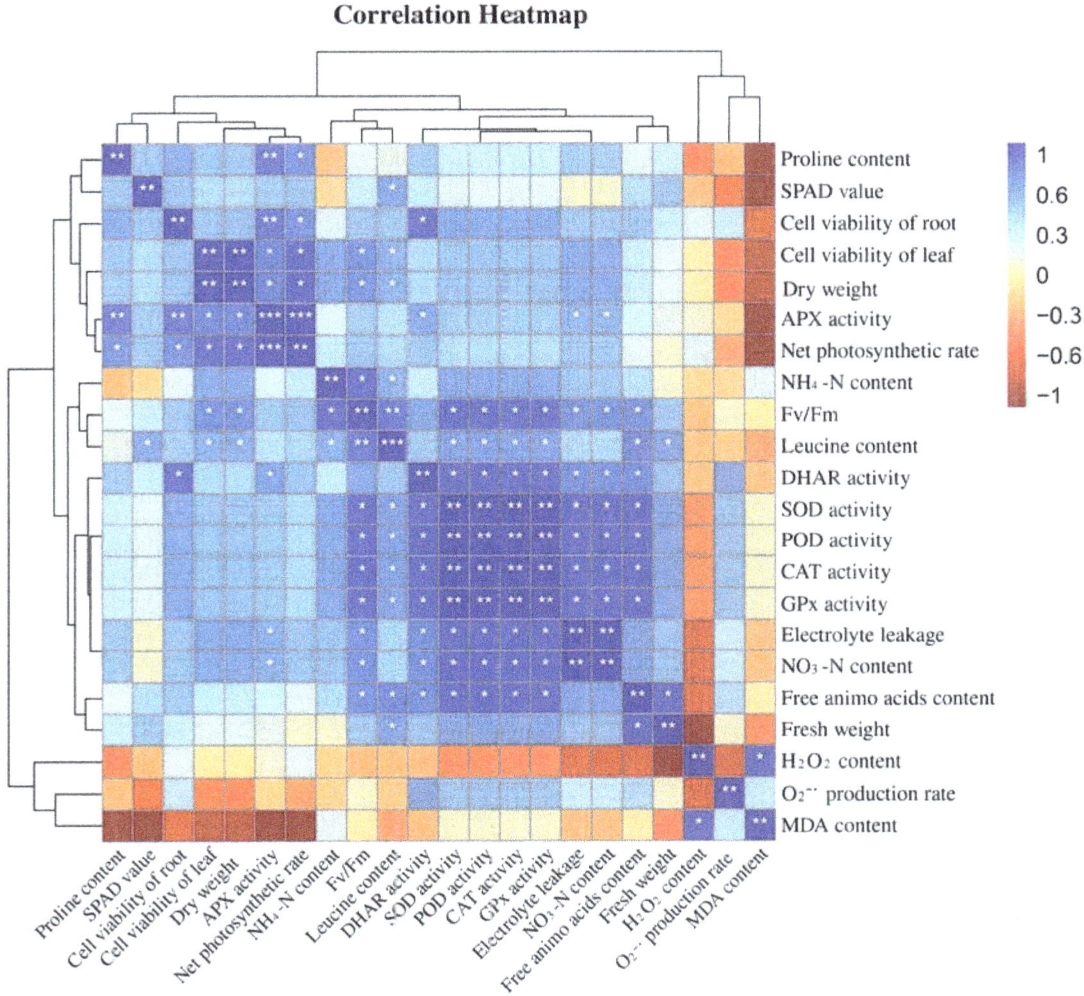

Figure 7. Correlation analysis of leucine content with the osmotic system and the antioxidant system in peach seedlings (Duncan test, * $p \leq 0.05$, ** $p \leq 0.01$, *** $p \leq 0.001$).

4. Discussion

Plants are inevitably challenged by various environmental stresses, in particular salt, heat, intense irradiance, and heavy metal stress. Abiotic stress can reduce crop growth, plant leaf area, and photosynthesis rates [29]. Leaves, the central organs of photosynthesis in plants, respond to changes in the external environment or internal metabolism via changes in the ultrastructure, opening angle, aspect ratio, or photosynthesis [30]. It has been reported that a low concentration of Cu^{2+} (<300 mg·kg^{-1}) could increase the chlorophyll content of *'Hanfu'* apple seedlings and keep the leaves in a healthy state. However, when the Cu^{2+} concentration exceeded 300 mg·kg^{-1}, the plants were stressed, and the chlorophyll content and photosynthetic efficiency of seedlings decreased [31]. Under Cu stress, the net photosynthetic rate of peach seedlings decreased, probably due to the change in chloroplast composition [32]. Chlorophyll a/chlorophyll b were significantly decreased after the application of 600 mg·kg^{-1} Cu^{2+}, indicating that a high Cu concentration seriously inhibited the synthesis process of chlorophyll and thus reduced plant photosynthetic efficiency [33].

The maximum photochemical efficiency is a useful indicator of photosystem function and efficiency. The maximum photochemical efficiency is inversely proportional to stress, and any significant change in this value reflects the influence of stress on plants [34]. In this study, the SPAD values of leaves decreased significantly under Cu stress. However, exogenous leucine significantly increased the chlorophyll fluorescence, net photosynthetic rate, SPAD value, and maximum photochemical efficiency (Figure 1b–e).

The morphology and ultrastructure of chloroplasts also directly affect the photosynthetic performance of plants. Several studies have investigated the effect of Cu on the chlorophyll ultrastructure [35,36]. Lin et al. (2008) found that higher concentrations of Cu ruptured the chloroplasts completely with an expanded thylakoid matrix [37]. Ji et al. (2007) also found expanded chloroplasts, distorted grana lamella, and ruptured membranes in *Potamogeton malaianus* under Cu stress [38]. An appropriate concentration of Cu^{2+} can promote plant photosynthesis, but an excessive concentration of Cu^{2+} will inactivate chlorophyll proteins, change the chloroplast ultrastructure, destroy the structure and function of thylakoids, ultimately inhibit photosynthesis, and seriously affect nutrient accumulation [39]. In this study, under Cu stress the cell ultrastructure changed, the chloroplasts were distorted, and the number and volume of starch granules and plastid globules increased. In addition, the cell walls were ruptured, and starch granules occupied almost half of the chloroplasts. After the application of leucine, the number of starch granules and plastid globules decreased significantly, and the chloroplast morphology returned to normal (Figure 2). The possible reason is that exogenous leucine inhibited pigment oxidative decomposition, increased protochlorophyllide reductase activity, promoted chlorophyll synthesis, increased the SPAD value, and maintained the shape of chloroplasts, which in turn improved the photosynthetic system of leaves.

Plants suffering from adverse heavy metal stress, including Cu stress, iron stress, and arsenic stress, produce excess internal reactive ROS, such as superoxide (O_2-), hydrogen peroxide (H_2O_2), and hydroxyl radicals (·OH-), which negatively elicit oxidative stress on cellular structures and metabolism [40–43]. The antioxidant system is divided into the enzymatic system and the non-enzymatic antioxidant system. These enzymes include SOD, POD, CAT, APX, GPx, and other antioxidant enzymes as well as non-enzymatic antioxidant substances such as GSH, ascorbic acid (ASA), MTs, and proline [44]. Previous studies found that the activities of POD, SOD, and CAT in grape roots increased first and then decreased under different Cu concentrations (0.5, 1, 1.5, and 2 mmol/L) [45]. In this study, we found that the expression levels of antioxidant genes were significantly upregulated when seedlings were exposed to Cu stress, among which the expression levels of *PpCAT*, *PpSOD*, and *PpGPx* were more upregulated. Under Cu stress, the expression of antioxidant genes and the activity of antioxidant enzymes were increased after adding Leu, indicating that Leu could reduce the peroxidation damage under Cu stress (Figure 3). According to Figure 4e,f, compared with the Cu treatment, Cu+Leu can significantly reduce the content of H_2O_2 and the production of O_2- in plants. Therefore, it is speculated that Leu has no obvious effect on the antioxidation system under normal conditions, but under Cu stress Leu can activate the antioxidant system and reduce the oxidative damage caused by ROS.

The AsA-GSH cycle is an important way for plants to respond to stress, mainly through the joint action of multiple enzymatic reactions involving the reduced antioxidant AsA and GSH as well as APX and DHAR so as to realize the process of H_2O_2 removal and the regeneration of ASA and GSH and maintain the REDOX homeostasis of cells, improving the stress resistance of plants under stress conditions [46]. Previous studies have shown that the ASA and GSH contents in wheat roots were significantly higher under the high concentration of Cu^{2+} at 1 mmol/L than in the control treatment [47]. Wu et al. found that Taxus chinensis var. chinensis synthesized a large amount of ASA, which was used to remove a large amount of H_2O_2 accumulated in the root system to weaken high Cu toxicity [48]. In addition, plants can regulate the osmotic balance by increasing the contents of proline and other osmotic regulators, thus maintaining normal cell metabolism and improving plant stress resistance under heavy metal stress. Our results showed that the relative expression and enzyme activities of *PpAPX*, *PpGPx*, and *PpDHAR* were enhanced by Leu supplementation under Cu stress, which partially explained the enhanced ROS scavenging capacity and the decreased proline and MDA contents in Leu-treated peach seedlings under Cu stress (Figures 3–5). As can be seen from the Figure 7, the leucine content in peach seedlings was positively correlated with plant antioxidant enzyme activities, and the correlation was significantly different and was negatively correlated with the hydrogen peroxide content and the MDA content. These results indicate that exogenous leucine reduced the accumulation of ROS in peach seedlings under Cu stress, thus reducing the oxidative damage.

Excessive Cu^{2+} is preferentially accumulated in plant roots, and root length will decrease with an increase in the Cu^{2+} concentration, which will affect the absorption of water and mineral elements by roots and inhibit the growth of the aboveground parts [49]. Michaud et al. found that the root length of wheat decreased by 10%, 25%, and 50%, respectively, when the Cu ion concentration in the root reached 100, 150, and 250–300 mg/kg, indicating that in a certain range, the higher the Cu ion concentration, the more obvious the inhibitory effect on root growth [50]. In this study, the Cu content was the highest in the roots and the lowest in the stems (Supplementary Table S3). In the seedlings treated with leucine under Cu stress, the Cu contents in the roots, stems, and leaves were significantly lower than those in the seedlings under Cu stress alone. In addition, the Cu content in the roots decreased to a greater extent with exogenous leucine. The decrease may be due to the chelating function of amino acids such as leucine and their derivatives [51]. As a result, Cu ions may not be transported through the roots to the aboveground parts. Thus, exogenous leucine also increased the root viability under Cu stress by reducing the accumulation of Cu in the roots and leaves, which in turn improved the root absorption ability, nutrient transportation, and photosynthetic performance, which finally promoted plant growth. Root damage can directly affect the growth and development of plant aboveground parts, in which biomass is usually used as a physiological index to detect the degree of the metal toxicity of plants. Huang et al. found that the biomass of white pomelo did not change significantly under the Cu concentration of 0.5–300 μmol/L, but when the Cu concentration reached 400 μmol/L, the biomass decreased significantly. Studies have shown that Cu stress inhibits plant growth [52]. In Abutilon theophrasti, Cu inhibited root tip cells [53]. Meanwhile, sublethal levels of Cu resulted in lipid peroxidation, which destroyed the membrane structure and affected the root physiological function of beans [54]. In this study, Cu stress significantly reduced the length, area, and volume of roots, the number of root tips, and the number of bifurcations and fibrous roots (Figure 5a, Supplementary Table S2). In addition, the fresh and dry weights of the aboveground and belowground plant parts were significantly reduced (Figure 4a). These changes are consistent with a reduction in root viability under Cu stress (Figure 5b–d).

Nitrogen metabolism is the primary source of protein and amino acids in plants [55]. Plants use a series of enzymes, such as NR, GS, and GOGAT, that are involved in nitrogen metabolism to absorb and efficiently utilize nitrogen [56]. NR is a rate-limiting enzyme in the nitrate assimilation pathway, and its activity is sensitive to H_2O_2 [57]. In this study, Cu stress significantly increased the H_2O_2 content of plants, which led to a decrease in NR enzyme activity. GS and GOGAT are key enzymes that convert inorganic nitrogen to organic nitrogen. The higher content of ammonium–nitrogen in peach seedlings under Cu stress may be related to the decrease in GS/GOGAT activities. The inhibition of the GS/GOGAT pathway hindered the assimilation of inorganic nitrogen into soluble protein [58]. However, a high ammonium–nitrogen content was harmful to peach growth. Cu stress also induced an increase in free amino acids, which may be related to autophagy and accelerated protein degradation [59]. In seedlings with exogenous leucine under Cu stress, the free leucine content increased significantly, probably from exogenous leucine. To maintain an amino acid balance, plants absorb and synthesize more of other amino acids, which further increases the total amount of free amino acids in the peach seedlings. Thus, in this study, leucine retained nitrogen metabolism efficiency in peach seedlings under Cu stress (Figure 6).

5. Conclusions

Exogenous leucine improved the leaf ultrastructure, ionic balance, and photosynthetic parameters. Furthermore, it improved the nitrogen metabolism efficiency of plants and attenuated Cu-stress-induced oxidative damage via a decrease in reactive oxygen species (ROS) and the regulation of the antioxidant and osmotic systems. These effects in turn ameliorated cell viability and biomass accumulation with improved resistance of peach seedlings to Cu stress. This study illustrates that leucine alleviates the damage caused by Cu stress in peach seedlings, which may provide more reference data for environmental risk assessments of Cu and make it possible to reuse soils with excessive copper contents caused by heavy metals.

Supplementary Materials: The following are available online at https://www.mdpi.com/article/10.3390/antiox11122455/s1, Table S1, Primer sequences, Figure S1, growth of 66-day-old peach seedlings under different amino acid treatments, Table S2, Effect of exogenous leucine on root architecture of peach seedlings under copper stress, Table S3, Effect of exogenous leucine on the copper ions content of peach seedlings under copper stress.

Author Contributions: M.S.: Conceptualization, Methodology, Software, Data curation, and Writing—original draft; S.L.: Visualization, Software, Validation, and Investigation; Q.G.: Software and Investigation; Y.X.: Methodology, Supervision, and Funding acquisition; F.P.: Conceptualization, Methodology, Software, Supervision, and Writing—review and editing. All authors have read and agreed to the published version of the manuscript.

Funding: This work was supported by the Shandong Province Rural Revitalization Science and Technology Innovation Boost Action Plan Project (2021TZXD013) and the National Modern Agroindustry Technology Research System Fund (No. CARS-30-2-02).

Institutional Review Board Statement: Not applicable.

Informed Consent Statement: Not applicable.

Data Availability Statement: The data are contained within the article.

Conflicts of Interest: The authors declare that they have no known competing financial interests or personal relationships that could have appeared to influence the work reported in this paper.

References

1. Yang, Q.; Li, Z.; Lu, X.; Duan, Q.; Huang, L.; Bi, J. A review of soil heavy metal pollution from industrial and agricultural regions in China: Pollution and risk assessment. *Sci. Total Environ.* **2018**, *642*, 690–700. [CrossRef] [PubMed]
2. Liu, L.; Li, W.; Song, W.; Guo, M. Remediation techniques for heavy metal-contaminated soils: Principles and applicability. *Sci. Total Environ.* **2018**, *633*, 206–219. [CrossRef] [PubMed]
3. He, Z.; Shentu, J.; Yang, X.; Baligar, V.C.; Zhang, T.; Stoffella, P.J. Heavy metal contamination of soils: Sources, indicators and assessment. *J. Environ. Indic.* **2015**, *9*, 17–18.
4. Ahammed, G.J.; Li, C.X.; Li, X.; Liu, A.; Chen, S.; Zhou, J. Overexpression of tomato RING E3 ubiquitin ligase gene SlRING1 confers cadmium tolerance by attenuating cadmium accumulation and oxidative stress. *Physiol. Plant.* **2021**, *173*, 449–459. [CrossRef] [PubMed]
5. Yruela, I.; Pueyo, J.J.; Alonso, P.J.; Picorel, R. Photoinhibition of photosystem II from higher plants: Effect of copper inhibition. *J. Biol. Chem.* **1996**, *271*, 27408–27415. [CrossRef]
6. Gong, Q.; Wang, L.; Dai, T.; Zhou, J.; Kang, Q.; Chen, H.; Li, K.; Li, Z. Effects of copper on the growth, antioxidant enzymes and photosynthesis of spinach seedlings. *Ecotoxicol. Environ. Saf.* **2019**, *171*, 771–780. [CrossRef]
7. Chary, N.S.; Kamala, C.; Raj, D.S.S. Assessing risk of heavy metals from consuming food grown on sewage irrigated soils and food chain transfer. *Ecotoxicol. Environ. Saf.* **2008**, *69*, 513–524. [CrossRef]
8. Bouazizi, H.; Jouili, H.; Geitmann, A.; El Ferjani, E. Cell wall accumulation of Cu ions and modulation of lignifying enzymes in primary leaves of bean seedlings exposed to excess copper. *Biol. Trace Elem. Res.* **2011**, *139*, 97–107. [CrossRef]
9. Sharma, P.; Jha, A.B.; Dubey, R.S.; Pessarakli, M. Reactive oxygen species, oxidative damage, and antioxidative defense mechanism in plants under stressful conditions. *J. Bot.* **2012**, *2012*, 217037. [CrossRef]

10. Li, L.; Zhang, K.; Gill, R.A.; Islam, F.; Farooq, M.A.; Wang, J.; Zhou, W. Ecotoxicological and interactive effects of copper and chromium on physiochemical, ultrastructural, and molecular profiling in *Brassica napus* L. *BioMed Res. Int.* **2018**, *2018*, 9248123. [CrossRef]
11. Gill, S.S.; Tuteja, N. Reactive oxygen species and antioxidant machinery in abiotic stress tolerance in crop plants. *Plant Physiol. Biochem.* **2010**, *48*, 909–930. [CrossRef] [PubMed]
12. Sanders, A.; Collier, R.; Trethewy, A.; Gould, G.; Sieker, R.; Tegeder, M. AAP1 regulates import of amino acids into developing Arabidopsis embryos. *Plant J.* **2009**, *59*, 540–552. [CrossRef] [PubMed]
13. Zhaohui, Z.; Zhiyuan, Z.; Gangqiao, D.; Xuanming, P.; Jianzhou, T.; Hongke, X.; Xian, L.; Leping, Z.; Jing, L. Eeffects of Exogenous to Rice Dry Weight and Nitrogen Content. *J. Nucl. Agric. Sci.* **2016**, *30*, 1435.
14. Wang, X.; Gao, Y.; Wang, Q.; Chen, M.; Ye, X.; Li, D.; Chen, X.; Li, L.; Gao, D. 24-Epibrassinolide-alleviated drought stress damage influences antioxidant enzymes and autophagy changes in peach (*Prunus persicae* L.) leaves. *Plant Physiol. Biochem.* **2019**, *135*, 30–40. [CrossRef]
15. Hammerschmitt, R.K.; Tiecher, T.L.; Facco, D.B.; Silva, L.O.; Schwalbert, R.; Drescher, G.L.; Trentin, E.; Somavilla, L.M.; Kulmann, M.S.; Silva, I.C. Copper and zinc distribution and toxicity in 'Jade'/'Genovesa'young peach tree. *Sci. Hortic.* **2020**, *259*, 108763. [CrossRef]
16. Noreen, S.; Fatima, Z.; Ahmad, S.; Athar, H.-u.-R.; Ashraf, M. Foliar application of micronutrients in mitigating abiotic stress in crop plants. In *Plant Nutrients and Abiotic Stress Tolerance*; Springer: Berlin/Heidelberg, Germany, 2018; pp. 95–117. [CrossRef]
17. Sharma, S.S.; Dietz, K.-J. The significance of amino acids and amino acid-derived molecules in plant responses and adaptation to heavy metal stress. *J. Exp. Bot.* **2006**, *57*, 711–726. [CrossRef]
18. Kocaman, A. Combined interactions of amino acids and organic acids in heavy metal binding in plants. *Plant Signal. Behav.* **2022**, 2064072. [CrossRef]
19. Li, H.; Liang, Z.; Ding, G.; Shi, L.; Xu, F.; Cai, H. A natural light/dark cycle regulation of carbon-nitrogen metabolism and gene expression in rice shoots. *Front. Plant Sci.* **2016**, *7*, 1318. [CrossRef]
20. Li, L. Physio-Biochemical and Molecular Mechanism of Exogenous Brassinosteroids in Regulating Growth of *Brassica napus* under Copper and Chromium Stress. Ph.D. Thesis, Zhejiang University, Hangzhou, China, 2019.
21. Lu, R. *Analysis Method of Agricultural Chemistry in Soil*; Agricultural Science and Technology Press: Beijing, China, 2000.
22. Liu, C.; Zhao, L.; Yu, G. The dominant glutamic acid metabolic flux to produce γ-amino butyric acid over proline in *Nicotiana tabacum* leaves under water stress relates to its significant role in antioxidant activity. *J. Integr. Plant Biol.* **2011**, *53*, 608–618. [CrossRef]
23. Wanying, Y.; Shasha, S.; Biao, G.; Xiaotong, L.; Yue, L.; Qinghua, S. Effects of Overexpressing SlSAMS1 on Tomato Tolerance to Cadmium Toxicity and Antioxidant System. *J. Nucl. Agric. Sci.* **2020**, *34*, 487.
24. Bates, L.S.; Waldren, R.P.; Teare, I.D. Rapid determination of free proline for water-stress studies. *Plant Soil* **1973**, *39*, 205–207. [CrossRef]
25. Zhao, S.J.; Cang, J. *Experimental Instruction in Plant Physiology*; Agricultural Science and Technology Press: Beijing, China, 2015.
26. Zhang, X.; Zhang, H.; Lou, X.; Tang, M. Mycorrhizal and non-mycorrhizal *Medicago truncatula* roots exhibit differentially regulated NADPH oxidase and antioxidant response under Pb stress. *Environ. Exp. Bot.* **2019**, *164*, 10–19. [CrossRef]
27. Li, H.S. *Principles and Techniques of Plant Physiological Biochemical 560 Experiment*; Higher Education Press: Beijing, China, 1999.
28. Molins-Legua, C.; Meseguer-Lloret, S.; Moliner-Martinez, Y.; Campíns-Falcó, P. A guide for selecting the most appropriate method for ammonium determination in water analysis. *TrAC Trends Anal. Chem.* **2006**, *25*, 282–290. [CrossRef]
29. Ma, D.; Sun, D.; Wang, C.; Qin, H.; Ding, H.; Li, Y.; Guo, T. Silicon application alleviates drought stress in wheat through transcriptional regulation of multiple antioxidant defense pathways. *J. Plant Growth Regul.* **2016**, *35*, 1–10. [CrossRef]
30. Xiao, Y.; Wu, X.; Sun, M.; Peng, F. Hydrogen sulfide alleviates waterlogging-induced damage in peach seedlings via enhancing antioxidative system and inhibiting ethylene synthesis *Front. Plant Sci.* **2020**, *11*, 696. [CrossRef] [PubMed]
31. Bu, F. Physiological and Biochemical Responses of 'Hanfu' Appleseedlings to Heavy Metal Copper Stress. Master's Thesis, Shenyang Agricultural University, Shenyang, China, 2019.
32. Calatayud, A.; Roca, D.; Martínez, P.F. Spatial-temporal variations in rose leaves under water stress conditions studied by chlorophyll fluorescence imaging. *Plant Physiol. Biochem.* **2006**, *44*, 564–573. [CrossRef] [PubMed]
33. Chen, J.; Shafi, M.; Li, S.; Wang, Y.; Wu, J.; Ye, Z.; Peng, D.; Yan, W.; Liu, D. Copper induced oxidative stresses, antioxidant responses and phytoremediation potential of Moso bamboo (*Phyllostachys pubescens*). *Sci. Rep.* **2015**, *5*, 13554. [CrossRef]
34. Zhang, S.; Wang, H.; Wang, W.; Wu, X.; Xiao, Y.; Peng, F. Effects of sucrose on seedling growth and development and SnRK1 activity in Prunus persica. *Chin. Bull. Bot.* **2019**, *54*, 744.
35. Saleem, M.H.; Ali, S.; Kamran, M.; Iqbal, N.; Azeem, M.; Tariq Javed, M.; Ali, Q.; Zulqurnain Haider, M.; Irshad, S.; Rizwan, M. Ethylenediaminetetraacetic acid (EDTA) mitigates the toxic effect of excessive copper concentrations on growth, gaseous exchange and chloroplast ultrastructure of *Corchorus capsularis* L. and improves copper accumulation capabilities. *Plants* **2020**, *9*, 756. [CrossRef]
36. Parveen, A.; Saleem, M.H.; Kamran, M.; Haider, M.Z.; Chen, J.-T.; Malik, Z.; Rana, M.S.; Hassan, A.; Hur, G.; Javed, M.T. Effect of citric acid on growth, ecophysiology, chloroplast ultrastructure, and phytoremediation potential of jute (*Corchorus capsularis* L.) seedlings exposed to copper stress. *Biomolecules* **2020**, *10*, 592. [CrossRef]

37. Lin, Y.-Z.; Zhang, S.-Y.; Zhu, H.-S. Effect of copper stress on ultra-structure of mesophyll cells in Chinese cabbage. *Chin. J. Eco-Agric.* **2008**, *16*, 948–951. [CrossRef]
38. Ji, W.-D.; Shi, G.-X.; Yang, H.-Y.; Xu, Q.-S.; Xu, Y.; Zhang, H. Effects of Cu2+ stress on leaf physiological indice and ultrastructure of *Potamogeton malaianus*. *Ying Yong Sheng Tai Xue Bao J. Appl. Ecol.* **2007**, *18*, 2727–2732.
39. Rehman, M.; Liu, L.; Bashir, S.; Saleem, M.H.; Chen, C.; Peng, D.; Siddique, K.H. Influence of rice straw biochar on growth, antioxidant capacity and copper uptake in ramie (*Boehmeria nivea* L.) grown as forage in aged copper-contaminated soil. *Plant Physiol. Biochem.* **2019**, *138*, 121–129. [CrossRef] [PubMed]
40. Foyer, C.H.; Noctor, G. Oxidant and antioxidant signalling in plants: A re-evaluation of the concept of oxidative stress in a physiological context. *Plant Cell Environ.* **2005**, *28*, 1056–1071. [CrossRef]
41. Turchi, A.; Tamantini, I.; Camussi, A.M.; Racchi, M.L. Expression of a metallothionein A1 gene of *Pisum sativum* in white poplar enhances tolerance and accumulation of zinc and copper. *Plant Sci.* **2012**, *183*, 50–56. [CrossRef] [PubMed]
42. Ahammed, G.J.; Wu, M.; Wang, Y.; Yan, Y.; Mao, Q.; Ren, J.; Ma, R.; Liu, A.; Chen, S. Melatonin alleviates iron stress by improving iron homeostasis, antioxidant defense and secondary metabolism in cucumber. *Sci. Hortic.* **2020**, *265*, 109205. [CrossRef]
43. Ahammed, G.J.; Yang, Y. Anthocyanin-mediated arsenic tolerance in plants. *Environ. Pollut.* **2022**, *292*, 118475. [CrossRef]
44. Geng, A.; Wang, X.; Wu, L.; Wang, F.; Wu, Z.; Yang, H.; Chen, Y.; Wen, D.; Liu, X. Silicon improves growth and alleviates oxidative stress in rice seedlings (*Oryza sativa* L.) by strengthening antioxidant defense and enhancing protein metabolism under arsanilic acid exposure. *Ecotoxicol. Environ. Saf.* **2018**, *158*, 266–273. [CrossRef]
45. Guan, L.P. Study on the Grape Root Physiology and Cytological Response to Copper and Cadmium Stress. Master's Thesis, Shandong Agricultural University, Shandong, China, 2015.
46. Qin, S.; Liu, H.; Nie, Z.; Gao, W.; Li, C.; Lin, Y.; Zhao, P. AsA–GSH cycle and antioxidant enzymes play important roles in Cd tolerance of wheat. *Bull. Environ. Contam. Toxicol.* **2018**, *101*, 684–690. [CrossRef]
47. Paradiso, A.; Berardino, R.; de Pinto, M.C.; Sanità di Toppi, L.; Storelli, M.M.; Tommasi, F.; de Gara, L. Increase in ascorbate–glutathione metabolism as local and precocious systemic responses induced by cadmium in durum wheat plants. *Plant Cell Physiol.* **2008**, *49*, 362–374. [CrossRef]
48. Wu, J.; Hu, J.; Wang, L.; Zhao, L.; Ma, F. Responses of Phragmites australis to copper stress: A combined analysis of plant morphology, physiology and proteomics. *Plant Biol.* **2021**, *23*, 351–362. [CrossRef] [PubMed]
49. Yuan, M. Study on Excess Cu and Mn Tolerance and Its Physiological Response of Four Citrus Rootstocks. Master's Thesis, Southwest University, Chongqing, China, 2018.
50. Michaud, A.M.; Chappellaz, C.; Hinsinger, P. Copper phytotoxicity affects root elongation and iron nutrition in durum wheat (*Triticum turgidum durum* L.). *Plant Soil* **2008**, *310*, 151–165. [CrossRef]
51. Jin, C.; Sun, Y.; Shi, Y.; Zhang, Y.; Chen, K.; Li, Y.; Liu, G.; Yao, F.; Cheng, D.; Li, J. Branched-chain amino acids regulate plant growth by affecting the homeostasis of mineral elements in rice. *Sci. China Life Sci.* **2019**, *62*, 1107–1110. [CrossRef] [PubMed]
52. Lombardi, L.; Sebastiani, L. Copper toxicity in *Prunus cerasifera*: Growth and antioxidant enzymes responses of in vitro grown plants. *Plant Sci.* **2005**, *168*, 797–802. [CrossRef]
53. Pingping, W.; Yong, T.; Kuo, W. Effects of Cu2+ Stress on the Antioxidant Enzymes' Systems and Ultrastructure in the Root Tip Cells of *Abutilon theophrasti*. *J. Shenyang Agric. Univ.* **2007**, *38*, 362.
54. Shainberg, O.; Rubin, B.; Rabinowitch, H.D.; Tel-Or, E. Loading beans with sublethal levels of copper enhances conditioning to oxidative stress. *J. Plant Physiol.* **2001**, *158*, 1415–1421. [CrossRef]
55. Below, F.E. Nitrogen metabolism and crop productivity. In *Handbook of Plant and Crop Physiology*; CRC Press: Boca Raton, FL, USA, 2001; pp. 407–428. [CrossRef]
56. Pathak, R.R.; Ahmad, A.; Lochab, S.; Raghuram, N. Molecular physiology of plant nitrogen use efficiency and biotechnological options for its enhancement. *Curr. Sci.* **2008**, *94*, 1394–1403.
57. Yamasaki, H.; Sakihama, Y. *Nitrate Reductase as a Producer of Nitric Oxide in Plants: Temperature-Dependence of the Enzymatic Active Nitrogen Formation*; CSIRO Publishing: Clayton, Australia, 2001.
58. Liu, X.; Hu, B.; Chu, C. Nitrogen assimilation in plants: Current status and future prospects. *J. Genet. Genom.* **2022**, *49*, 394–404. [CrossRef]
59. Hirota, T.; Izumi, M.; Wada, S.; Makino, A.; Ishida, H. Vacuolar protein degradation via autophagy provides substrates to amino acid catabolic pathways as an adaptive response to sugar starvation in *Arabidopsis thaliana*. *Plant Cell Physiol.* **2018**, *59*, 1363–1376. [CrossRef]

antioxidants

Article

Vanadium Stress Alters Sweet Potato (*Ipomoea batatas* L.) Growth, ROS Accumulation, Antioxidant Defense System, Stomatal Traits, and Vanadium Uptake

Sunjeet Kumar [1,2], Mengzhao Wang [1,2,*], Yonghua Liu [1,2], Zhixin Zhu [1,2], Shah Fahad [3], Abdul Qayyum [4] and Guopeng Zhu [1,2,*]

1 Key Laboratory for Quality Regulation of Tropical Horticultural Crops of Hainan Province, School of Horticulture, Hainan University, Haikou 570228, China
2 Sanya Nanfan Research Institute, Hainan University, Sanya 572025, China
3 Department of Agronomy, Abdul Wali Khan University Mardan, Mardan 23200, Pakistan
4 Department of Agronomy, The University of Haripur, Haripur 22620, Pakistan
* Correspondence: zhuguopeng@hainanu.edu.cn (G.Z.); mzwang@hainanu.edu.cn (M.W.)

Abstract: Vanadium (V) is a heavy metal found in trace amounts in many plants and widely distributed in the soil. This study investigated the effects of vanadium concentrations on sweet potato growth, biomass, root morphology, photosynthesis, photosynthetic assimilation, antioxidant defense system, stomatal traits, and V accumulation. Sweet potato plants were grown hydroponically and treated with five levels of V (0, 10, 25, 50, and 75 mg L^{-1}). After 7 days of treatment, V content at low concentration (10 mg L^{-1}) enhanced the plant growth and biomass; in contrast, drastic effects were observed at 25, 50, and 75 mg L^{-1}. Higher V concentrations negatively affect the relative water content, photosynthetic assimilation, photosynthesis, and root growth and reduce tolerance indices. The stomatal traits of sweet potato, such as stomatal length, width, pore length, and pore width were also decreased under higher V application. Furthermore, V concentration and uptake in the roots were higher than in the shoots. In the same way, reactive oxygen species (ROS) production (hydrogen peroxide), lipid peroxidation (malondialdehyde), osmolytes, glutathione, and enzymes (catalase and superoxide dismutase) activities were increased significantly under V stress. In conclusion, V at a low level (10 mg L^{-1}) enhanced sweet potato growth, and a higher level of V treatment (25, 50, and 75 mg L^{-1}) had a deleterious impact on the growth, physiology, and biochemical mechanisms, as well as stomatal traits of sweet potato.

Keywords: vanadium stress; sweet potato; oxidative damage; antioxidant defense system; photosynthesis; stomatal traits

Citation: Kumar, S.; Wang, M.; Liu, Y.; Zhu, Z.; Fahad, S.; Qayyum, A.; Zhu, G. Vanadium Stress Alters Sweet Potato (*Ipomoea batatas* L.) Growth, ROS Accumulation, Antioxidant Defense System, Stomatal Traits, and Vanadium Uptake. *Antioxidants* **2022**, *11*, 2407. https://doi.org/10.3390/antiox11122407

Academic Editor: Nafees A. Khan

Received: 25 October 2022
Accepted: 2 December 2022
Published: 5 December 2022

1. Introduction

Vanadium (V) is world's fifth most abundant transition element and deposit, mainly in China, the USA, Russia, and South Africa [1,2]. China is the leading producer and consumer of V, with 57% of V production globally. Around 26.5% of V-contaminated soil is present in southwest China [3,4]. V is widely distributed and mobilized in the surrounding environments by several natural events and anthropogenic activities, including weathering of parental rocks, redox processes, leaching, fertilizers usage, combustion, and industrial wastes, which as a result, contaminates the water, soil, and atmosphere [4,5]. V is deposited naturally in the soil in different mineral forms, and the average concentration of V ranges from 3 to 310 mg kg^{-1} in soil. The average V concentration in fresh, ground, and drinkable water is 0.5 µg L^{-1}, with peak concentrations in volcanic areas reaching 127.4 µg L^{-1} [6]. Accumulating V in the natural habitat increases soil and water contamination, ultimately affecting human health by causing nausea, vomiting, dizziness, and more seriously, leading to kidney damage [7]. Different studies have reported that V has detrimental effects on the growth and development of plants [6,8]. V toxicity, bioaccumulation, and bioavailability rely on the oxidation state of vanadium. Vanadate is the toxic chemical compound of V, and at high concentrations hinders phenotypic, physiological, and biochemical processes of plants, eventually obstructing plant growth and yield [8–10].

V can be easily uptake by the plants from the soil; however, the effect is dependent on the V content in the soil. A low level of V treatment enhances plant growth, photosynthesis, and gas exchange elements [11,12]. Furthermore, Altaf et al. reported that 35 mg L^{-1} V treatment caused a drastic decline in the biomass of rice plants [13]. Similarly, Chen et al. exposed *Ipomoea aquatica* to 0–2.50 mg L^{-1} V treatments, and they concluded that concentration-dependent effects of V can be observed for the physiological properties and the plants may adapt to the toxicity of V [12]. Another study demonstrated that tobacco has good vanadium tolerance at < 2.0 mg L^{-1} [10]. A high level of V treatment negatively influences plant growth and yield by increasing reactive oxygen species (ROS) production, hindering lipid membranes, antioxidant enzyme activity, the metabolic process, and gene expression [14,15]. The plants overcome oxidative stress and scavenge ROS by enhancing the production of osmolytes, antioxidants, and stimulating the activities of the antioxidant enzymes [16,17].

Previous studies highlighted the V effect on plant physiology by reducing or altering the photosynthetic rate, shoot and root survival, and leaf chlorosis [8,18]. Photosynthesis is a vital plant metabolic process that increases carbon absorption and yield production; however, its production and fixation can be strongly hindered under heavy metal stress [13,19]. Different studies reported a significant reduction in photosynthetic pigments and gas exchange elements in pepper and watermelon under heavy metals, including vanadium, nickel, and selenium stress [11,20]. Photosynthetic pigments can be impaired by obstructing electron transport and hindering the membrane integrity of the chloroplast. Similarly, plant roots are the essential organ that interacts with and absorbs metal and other components of the soil. Roots are considered the first line of defense and give structural support to the plant against heavy metal toxicity. It helps to defend itself by minimizing the absorption of unnecessary metal [21]. Previous studies also depicted that a higher level of V supplementation significantly reduced root architecture in several plants [11,13,22]. Different plants react differently to V stress, so more research is needed to determine the best concentration of V as a biostimulant in different cultivars of the crops.

Sweet potato (*Ipomoea batatas* L.) has been used as a major source of carbohydrate in many countries around the word, especially Asia and Africa [23]. Additionally, sweet potato roots are used for biofortification and bioethanol production. Thus, the production of sweet potato can be an integral part of food security in the future [24,25]. Moreover, sweet potato stems and leaves can be used as a vegetable for humans and animal feed [26,27]. In addition, sweet potato leaves are rich in protein, iron, calcium, fiber, carotenoids, vitamins, and total polyphenols and possess medicinal properties [28,29]. Sweet potato can severely affect productivity and quality under heavy metal stresses [23,30]. Several studies found that trace amounts of certain elements may stimulate the growth and production of horticultural crops [11,31]. Conversely, V application showed deleterious effects on the growth of different plant species by obstructing their antioxidant defense system [13,20]. However, the physiological and biochemical response of the sweet potato plant under V stress has not been well-studied and requires further research. Therefore, it is essential to explore the influence of V in sweet potato plants, primarily to determine the toxicity level of V for sweet potato. We designed this study to investigate V's effects on sweet potato plants and identify the toxic level of V. Our objective was to determine the effects of V on sweet potato growth, biomass, root morphology, photosynthesis, photosynthetic assimilation, antioxidant defense, stomatal traits, and V uptake.

2. Methodology

2.1. Seedling Collection, Growth Conditions, and Experimental Design

In this experiment, we used the "Haida HD7791" sweet potato cultivar. For the disinfection of sweet potato cuttings, 1 g L^{-1} carbendazim was used for 5–8 min. Afterwards, the sweet potato cuttings were kept in Ro water until the roots appeared. For the acclimation, the cuttings of sweet potato were grown in half Hoagland media (pH 5.8 ± 0.2). A hydroponic experiment in a controlled environment (25–27 °C for 16 h of photoperiod) was conducted to assess V's effect on sweet potato plants. For proper nutrient availability, the nutrient media were replaced after every 5 to 6 days. Subsequently, healthy and uniform seedlings were distributed among the five treatment groups. V (0, 10, 25, 50, and 75 mg/L) was applied as ammonium metavanadate (NH_4VO_3). For the analysis of morphological and physiological measurements, samples were collected after 7 days of treatment with V or normal growth conditions and immediately transferred into a liquid nitrogen tank.

2.2. Growth Variables

Three independent seedlings were used to measure plant height (PH), the leaf area (LA), number of leaves (LN), and fresh and dry weights of the roots and shoots. A portable laser leaf area meter (CI-202) was used to measure the LA (topmost leaves). A ruler was used to measure the height of the seedlings. The seedlings were cut, and the fresh weight of the shoot and root was recorded. For recording the dry weight (DW), the samples were first dried at 105 °C for 30 min and then kept for drying at 70 °C for 3 days [32]. The plant's shoot DW susceptibility index (SDSI) was calculated as follows:

$$\text{SDSI} = \frac{\text{Shoot DW (stressed plants)}}{\text{Shoot DW (controlled plants)}} \times 100 \tag{1}$$

Similarly, the following formula was used to calculate the plant's RDSI;

$$\text{RDSI} = \frac{\text{Root DW (stressed plants)}}{\text{Root DW (controlled plants)}} \times 100 \tag{2}$$

2.3. Relative Water Content Analysis

After recording the FW, the sweet potato leaves were immersed in ddH_2O for four hours. After 4 h the leaves were weighed to determine the turgor weight (TW). To determine the dry weight (DW), the leaves were oven-dried for one day at 70 °C [32]. Finally, the RWC was measured using the following formula:

$$\text{RWC\%} = [(\text{FW} - \text{DW})/(\text{TW} - \text{DW})] \times 100 \tag{3}$$

2.4. Root Morphology

After surface rinsing, the roots were washed with ddH_2O and scanned with the Imagery Screen (Epson Expression 11000XL, Regent Instruments, Chemin Sainte-Foy, QC, Canada) to observe different root traits. The images obtained from the root scanner were analyzed with the WinRHIZO 2003a software program [13].

2.5. Gas Exchange Parameters

Gas exchange parameters of sweet potato leaves were determined for completely matured leaves utilizing a portable photosynthesis system (CIRAS-3, Hansatech Co., Amesbury, MA, USA) [13].

2.6. V Determination, Uptake, and Translocation

A super microwave-assessed digestion system (Anton Paar, Multiwave 7000, Styria, Garz, Austria) was used to digest plant samples (100 mg dry weight) with 2 mL HNO_3, 0.5 mL H_2O_2, and 1 mL deionized water. We used the standard reference material (GBW10015) in triplicate for quality control and assurance (obtained from the Chinese

Academy of Geological Sciences, Langfang, China). The reference CRM standard value was $V(10^{-6})$ 0.87 ± 0.23. The working standards of V (0–200 µg/L) were made using a standard stock solution (GSB04-1759-2004, Beijing, China) containing 1000 mg/L of V. The inductively coupled plasma mass spectrometer (ICP-MS) (Perkin Elmer, NexION 5000, Waltham, MA, USA) was used to measure the V content [32]. The detection limit (DL) of standard V in solution was 0.01 µg/L (0.001522 ppb), the correlation coefficient was 0.999977, and the recovery of CRM standard V ranged from 86.5% to 93%. The subsequent formula to measure V uptake and translocation was used:

$$\text{V uptake (mg)} = \text{V concentrations in the tissues} \times \text{dry weight of the tissues} \quad (4)$$

$$\text{Translocation} = \frac{\text{V concentration in the plant shoots}}{\text{V concentration in the plant roots}} \quad (5)$$

2.7. Measurement of Photosynthetic Pigments

Leaf samples (0.1 g FW) were mixed with 80% acetone, followed by centrifugation for 15 min at 8000× *g*. A microplate reader (Infinite M200 PRO, TECAN, Männedorf, Swiss) was used for measuring the absorbance of chlorophyll a (chl a), chl b, and carotenoids (Car) at 663, 646, and 470 nm, respectively. Finally, the concentration of chlorophyll was calculated with the subsequent formula reported by Kumar et al. [16].

2.8. Determination of Malondialdehyde (MDA)

A kit (A003-1-1) was used to quantify MDA following the thiobarbituric acid (TBA) method. A glass homogenizer was used to properly homogenize 100 mg of fresh leaves in 900 µL of extraction buffer provided by the company. The homogenate was centrifuged at 5000× *g* for 15 min, followed by three centrifugations of 15 s each at 4000× *g*, with a 30 s interval between each centrifugation, with the final centrifugation of 3500× *g* for 10 min. After that, the supernatant was collected and mixed with the working fluid (combination of R1: clarifying agent, R2: buffering agent, and R3: color developer in the ratio of 0.1:3:1) provided by the company, then the mixture was boiled at 95 °C for 20 min. After cooling, the absorbance at 530 nm was determined using a full-wavelength microplate reader (Infinite M200 PRO, TECAN, Männedorf, Swiss) [32].

2.9. Determination of Hydrogen Peroxide, Proteins, GSH, and Antioxidant Enzymes

A homogenized sample of 500 mg of fresh leaves was centrifuged at 10,000× *g* for 15 min with 4.5 mL of 0.1 M PBS. The hydrogen peroxide (H_2O_2), total proteins, reduced glutathione (GSH), and antioxidant enzymes, catalase (CAT), peroxidase (POD), superoxide dismutase (SOD), and ascorbate peroxidase (APX) activities were measured using commercially available test kits purchased from Nanjing Jiancheng Bioengineering Institute, Nanjing, China and their absorbance were determined using a full-wavelength microplate reader (Infinite M200 PRO, TECAN, Männedorf, Swiss) [17,32,33].

2.10. Determination of Proline and Soluble Sugars

Proline was calculated by using an assay kit (A107-1-1). Fresh leaf samples were homogenized with buffer available in the kit and tested at 520 nm following the company's protocol. Approximately 50 mg of fresh leaves were homogenized in 0.45 mL ddH_2O for the analysis of soluble sugars. The homogenate was boiled at 95 °C for 15 min and then centrifuged at 7500× *g* for 15 min. After that, the supernatant was collected and diluted with ddH_2O at 1:9. Using a test kit (A145-1-1), the soluble sugar content of the diluted extracts was determined at 620 nm [17,32].

2.11. Determination of Total Polyphenols and Flavonoid Content

A fresh leaf sample of 1 g was homogenized with 60% ethanol. After that, 1.25 mL of 10% Folin–Ciocalteu reagent was added to 250 µL of extract and 1 mL of sodium carbonate (0.75 g/mL). After incubating for 15 min at 45 °C, the mixture was allowed to remain at room temperature for 30 min. As a final step, the absorbance was recorded at 765 nm, and the results were presented as Gallic acid equivalents per gram (GAE/g) to quantify total polyphenols [16,34].

Approximately 0.25 mL of $NaNO_2$ solution (0.5 g/mL) and 2 mL ddH_2O were mixed with 0.5 mL of extract to measure flavonoids. After being retained at 25–28 °C for 5 min, 150 µL of aluminium chloride (1 g/mL), 1 mL of NaOH (1 M), and 1.2 mL of ddH_2O were added simultaneously. As a final step, its absorbance was measured at 510 nm with Catechin (CAE) used as a standard, and its results were presented as CAE/g [16,34].

2.12. Scanning Electron Microscopy (SEM)

To observe the stomatal morphology, we used a published protocol [30]. To remove any debris, leaves were acetylated in 80% ethanol for two to three min. The tiny sections of leaf were prepared using s-cutting, and after that platinum was used to fix the abaxial and adaxial surfaces and sputtered using Leica Mikrosystem GmbH (ACE600) for 25 min, and finally examined under a SEM (Thermo Scientific, Verios G4 UC, Waltham, MA, USA).

2.13. Statistical Analysis

Three individual replications were used to obtain phenotypic, physiological, and biochemical indices. Significant differences ($p \leq 0.05$) between means were determined using SPSS 25.0 software, and Duncan tests were applied for the means comparison, while ± represents a standard error (S.E). Figures were plotted with GraphPad Prism 7. The "ggplot2" package in R (version 3.3.4, https://CRAN.R-project.org/package=ggplot2 (accessed on 24 August 2022)) was used for principal component analysis (PCA) and Pearson correlation analysis.

3. Results

3.1. Growth Parameters, RWC, and Tolerance Index (TI)

After 7 days of V treatment, the growth traits of sweet potato were significantly influenced by the treatment of V stress (Figure 1). The phenotypic parameters, such as PH, LA, LN, shoot and root FW, DW, SDSI, and RDSI were increased at 10 mg L^{-1} V treatment. Conversely, a significant reduction in the mentioned parameters was observed at 25, 50, and 75 mg L^{-1} V treatment compared to the control ($p < 0.05$; Tables 1 and 2). The PH (8.6%), LN (11.7%), LA (7.6%), SFW (14%), RFW (16.4%), SDW (25.5%), RDW (28.9%), root-to-shoot ratio (3.1%), SDSI (25.5%), and RDSI (30.2%) were increased at 10 mg L^{-1} V treatment than the control. In contrast, at higher concentrations, V treatment showed a negative correlation with plant growth. Furthermore, a drastic reduction in the PH (40.6%), LN (58.8%), LA (61.1%), SFW (57.8%), RFW (68.6%), SDW (62.2%), RDW (77.2%), root-to-shoot ratio (38.8%), SDSI (62.1%), and RDSI (76.3%) were observed at 75 mg L^{-1} V treatment (Tables 1 and 2). Moreover, RWC in the leaves of sweet potato reduced significantly as the level of V treatment increased, and the utmost reduction was noted at 75 mg L^{-1} V treatment (Table 1).

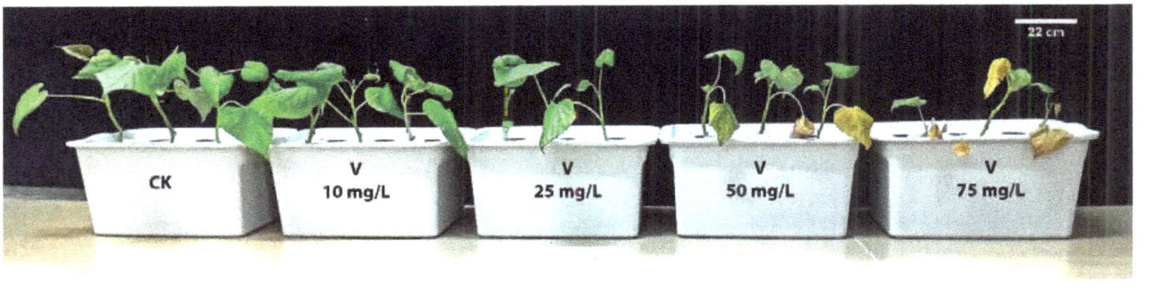

Figure 1. The influence of various V treatments on the growth of sweet potato.

Table 1. The influence of various V treatments on growth parameters of sweet potato.

Vanadium (mg L^{-1})	Height (cm)	Leaf Area (cm^2)	Number of Leaf	Shoot FW (g)	Root FW (g)	RWC (%)
Ck	42.67 ± 4.0^{cd}	69.55 ± 8.0^{c}	5.67 ± 0.6^{c}	5.18 ± 0.4^{d}	2.55 ± 0.5^{cd}	92.98 ± 0.88^{e}
10	46.33 ± 3.1^{d}	74.86 ± 5.8^{c}	6.33 ± 0.6^{c}	5.90 ± 0.5^{e}	2.97 ± 0.4^{d}	90.18 ± 1.66^{d}
25	38.00 ± 4.6^{bc}	52.27 ± 4.3^{b}	4.33 ± 0.6^{b}	4.09 ± 0.4^{c}	2.03 ± 0.3^{bc}	83.89 ± 0.68^{c}
50	31.67 ± 3.5^{ab}	44.24 ± 5.6^{b}	3.67 ± 0.6^{b}	3.08 ± 0.3^{b}	1.49 ± 0.2^{b}	76.52 ± 1.33^{b}
75	25.33 ± 4.2^{a}	27.03 ± 4.7^{a}	2.33 ± 0.6^{a}	2.19 ± 0.2^{a}	0.80 ± 0.1^{a}	70.97 ± 1.12^{a}

Duncan's test indicates a significant difference ($p < 0.05$) between the means of the five treatments indicated by different alphabetical letters.

Table 2. The influence of various V treatments on root and shoot dry weight and root and shoot dry weight susceptibility index of sweet potato.

Vanadium (mg L^{-1})	Shoot DW (g)	Root DW (g)	Root-Shoot Ratio	SDSI	RDSI
Ck	0.518 ± 0.05^{d}	0.324 ± 0.07^{d}	0.623 ± 0.10^{bc}	100	100
10	0.651 ± 0.09^{e}	0.418 ± 0.05^{e}	0.643 ± 0.01^{c}	125.491 ± 12.3^{d}	130.297 ± 10.5^{d}
25	0.387 ± 0.03^{c}	0.225 ± 0.02^{c}	0.582 ± 0.01^{bc}	74.819 ± 5.5^{c}	70.663 ± 8.8^{c}
50	0.288 ± 0.02^{b}	0.148 ± 0.03^{b}	0.511 ± 0.05^{b}	55.725 ± 4.6^{b}	45.967 ± 3.6^{b}
75	0.196 ± 0.02^{a}	0.074 ± 0.01^{a}	0.382 ± 0.07^{a}	37.881 ± 2.8^{a}	23.727 ± 6.7^{a}

Duncan's test indicates a significant difference ($p < 0.05$) between the means of the five treatments indicated by different alphabetical letters.

3.2. Root Morphology

The root growth of sweet potato was induced significantly at 10 mg L^{-1} V treatment; however, 25, 50, and 75 mg L^{-1} V treatments presented a significant reduction in the root characteristics (Figure 2). The root length (13.1%), volume (48.6%), average diameter (13.3%), surface area (8.7%), projected area (32.5%), length per volume (15%), and crossing (17.1%) were more enhanced under 10 mg L^{-1} V than the control. The tips and forks of the roots reduced by 2.6% and 24.2%, respectively, at 10 mg L^{-1} V (Figure 2A–I). Conversely, a substantial reduction was detected from 25 to 75 mg L^{-1} V treatments, and a highest decrease in root characteristics was observed at 75 mg L^{-1} V treatment. The maximum reduction in the root length was 81%; likewise, root volume exhibited 41%, surface area 84.6%, average diameter 39.3%, projected area 80.9%, tips 82.6%, forks 89.2%, crossing 86.8%, and length per volume 81.5% reduction at 10 mg L^{-1} V treatment (Figure 2A–I).

3.3. Leaf Gas Exchange Elements

This study showed that gas exchange elements were enhanced by the application of 10 mg L^{-1} V treatment; in contrast, a higher application of V (25, 50, and 75 mg L^{-1}) showed a negative effect on the gas exchange elements of sweet potato leaves in comparison to the control plant (Figure 3). Comparing the 10 mg L^{-1} V-treated group to the control group, the transpiration rate (Tr), photosynthesis rate (Pn), stomatal conductance (Gs), and intercellular CO_2 (Ci) increased by 10.6%, 23.9%, 33.2%, and 0.8%, respectively (Figure 3). Conversely, maximum reductions of 65.4%, 76.5%, 61.8%, and 62.8% were noticed in the Tr, Pn, Gs, and Ci, respectively when compared to the control (Figure 3).

3.4. Concentration, Uptake, and Translocation of Vanadium

This study depicted that V treatment significantly raised the V concentration in both shoots and roots, and maximum concentrations were recorded in 75 mg L^{-1} treated plants; 32.03 mg kg^{-1} DW in the shoots of sweet potato and 52.68 mg kg^{-1} DW in the roots of sweet potato (Table 3). Similarly, the accumulation of V was found to be higher in the roots than in the shoots. Moreover, V uptake by shoots and roots of sweet potato significantly augmented as the level of V increased. In the same way, the translocation of V from root to shoot was also raised significantly by the increment of V concentration (Table 3).

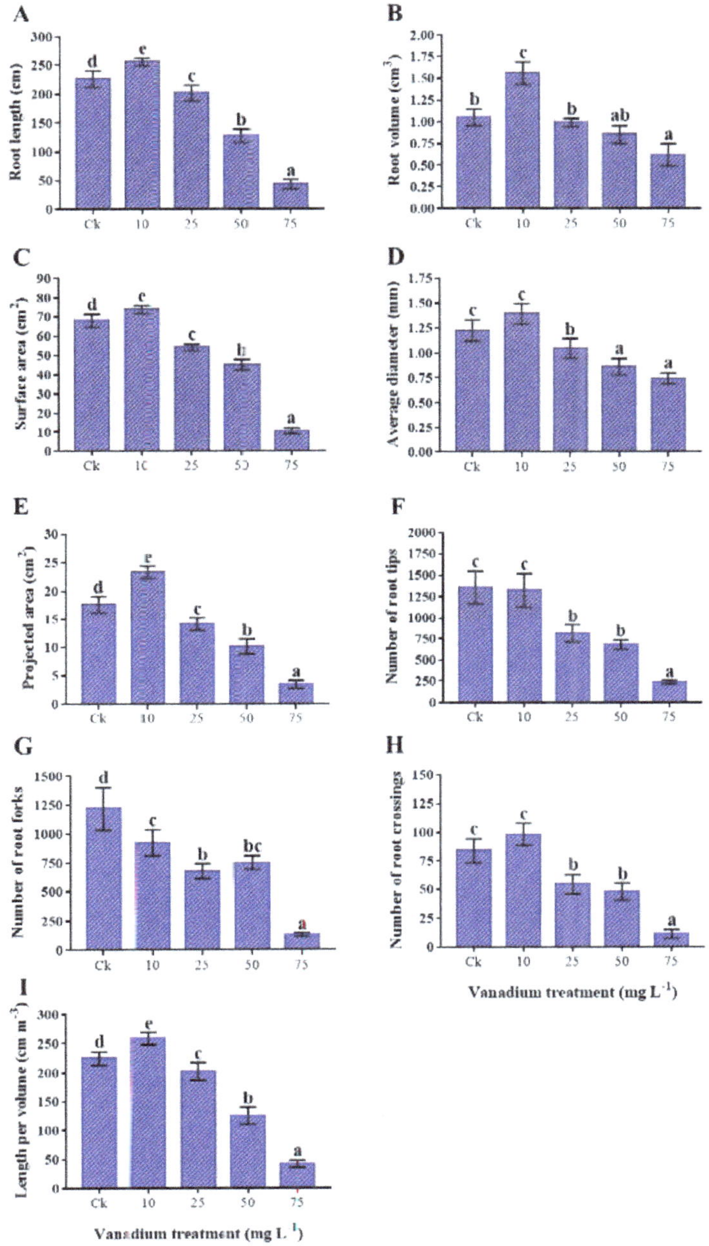

Figure 2. The influence of various V treatments on root morphological traits; (**A**) root length, (**B**) root volume, (**C**) surface area, (**D**) average diameter, (**E**) projected area, (**F**) number of root tips per plant, (**G**) number of root forks per plant, (**H**) number of root crossings per plant, and (**I**) root length per volume. Duncan's test specifies a significant difference ($p < 0.05$) between the means of the five treatments indicated by different letters.

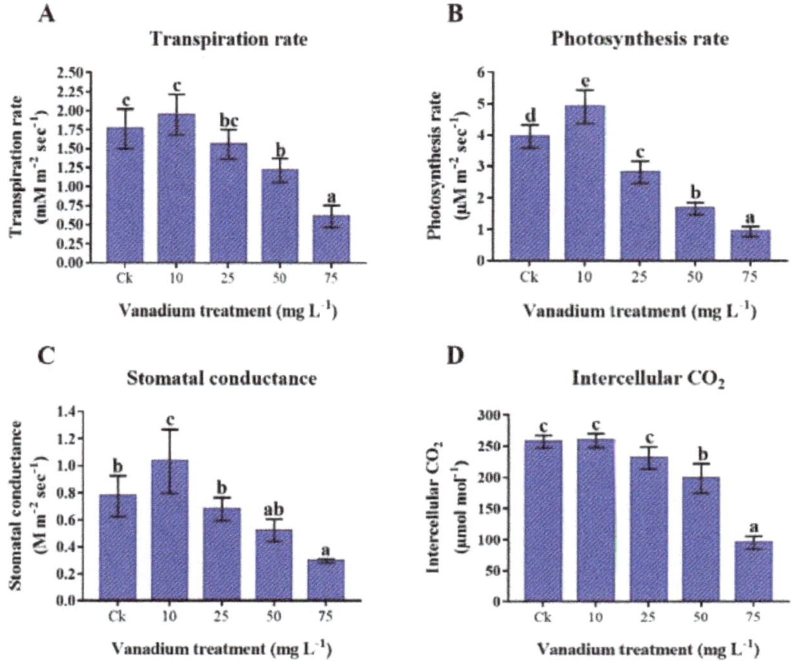

Figure 3. The influence of various V treatments on leaf gas exchange elements; (**A**) transpiration rate, (**B**) photosynthetic rate, (**C**) stomatal conductance, and (**D**) intercellular CO_2. Duncan's test indicates a significant difference ($p < 0.05$) between the means of the five treatments indicated by different alphabets.

Table 3. The influence of various V treatments on the concentrations, uptake, and translocation (root to shoot) of V.

Vanadium (mg L^{-1})	Concentration (mg kg^{-1} DW)		Uptake (mg kg^{-1} DW)		Translocation (Root to Shoot)
	Shoot	Root	Shoot	Root	
Ck	0.07 ± 0.01 ^{a}	0.89 ± 0.30 ^{a}	0.01 ± 0.00 ^{a}	0.09 ± 0.02 ^{a}	0.08 ± 0.01 ^{a}
10	3.53 ± 0.30 ^{b}	7.91 ± 0.72 ^{b}	0.37 ± 0.03 ^{b}	0.82 ± 0.06 ^{b}	0.45 ± 0.00 ^{b}
25	9.50 ± 0.79 ^{c}	19.19 ± 1.62 ^{c}	0.99 ± 0.09 ^{c}	1.99 ± 0.19 ^{c}	0.50 ± 0.01 ^{bc}
50	21.22 ± 1.49 ^{d}	37.41 ± 3.02 ^{d}	2.20 ± 0.17 ^{d}	3.88 ± 0.26 ^{d}	0.57 ± 0.08 ^{cd}
75	32.03 ± 2.41 ^{e}	52.68 ± 3.27 ^{e}	3.32 ± 0.27 ^{e}	5.46 ± 0.39 ^{e}	0.61 ± 0.08 ^{d}

Duncan's test indicates a significant difference ($p < 0.05$) between the means of the five treatments indicated by different alphabets.

3.5. Photosynthetic Pigments

The chlorophyll (Chl) was significantly influenced by the application of V stress. Compared to the control, adding 10 mg L^{-1} V did not significantly raise the concentration of chlorophyll and carotenoid (Car) (Figure 4). On the other hand, higher treatment of V (25, 50, and 75 mg L^{-1}) showed a significant negative impact on the photosynthetic pigments, and utmost reduction was detected at 75 mg L^{-1}. Compared to the control, 77% reduction in the content of total chl, 65.9% in chl a, 70.5% in chl b, and 50.3% in Car were detected (Figure 4).

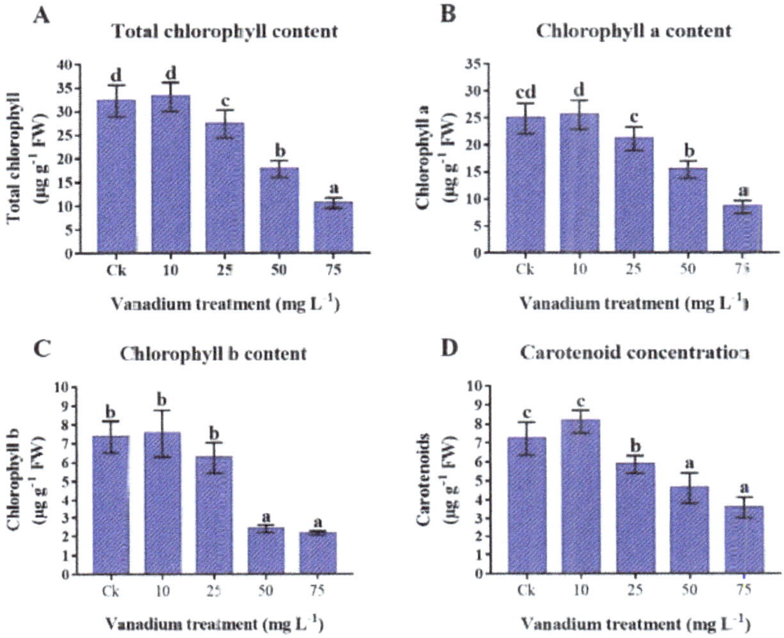

Figure 4. Influence of various V treatments on the photosynthetic pigments in the leaves of sweet potato. (**A**) Total chlorophyll content (T. Chl), (**B**) chlorophyll a (chl a), (**C**) chlorophyll b (chl b), and (**D**) carotenoid (Car) content. Duncan's test indicates a significant difference ($p < 0.05$) between the means of the five treatments indicated by different alphabets.

3.6. Lipid Peroxidation (MDA) and Reactive Oxygen Species (H_2O_2) Content

The V treatment considerably provoked MDA and H_2O_2 levels in the leaves ($p < 0.05$; Figure 5A,B). The rise of V treatment caused an increase in MDA and H_2O_2 content; the highest MDA and H_2O_2 content was present in 75 mg L^{-1} V treatment compared to the control. The MDA content in 75 mg L^{-1} V was 928.9% higher than the control (Figure 5A), where the H_2O_2 content was 665% higher (Figure 5B).

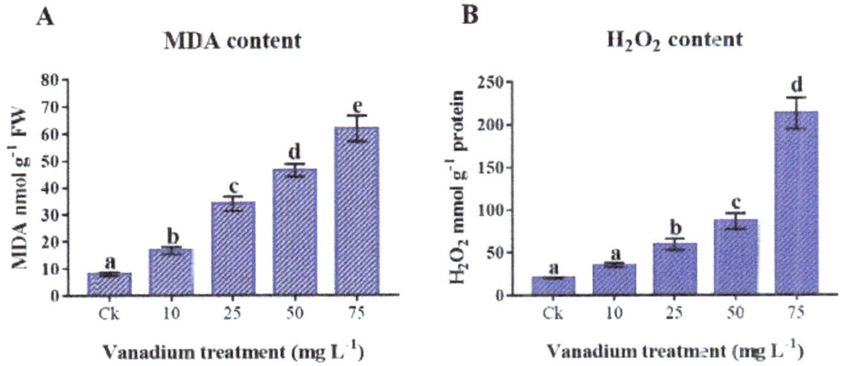

Figure 5. The influence of various V treatments on the production of reactive oxygen species (ROS) and lipid peroxidation in the leaves of sweet potato. (**A**) Malondialdehyde (MDA) and (**B**) hydrogen peroxide (H_2O_2) content. Duncan's test specifies a significant difference ($p < 0.05$) between the means of the five treatments indicated by different alphabets.

3.7. Osmolytes Production

The proline content was significantly increased as the level of V increased ($p < 0.05$). The maximum rise of 448.3% in proline content was detected at 75 mg L^{-1} V treatment (Figure 6A). Soluble sugars were also significantly higher in V-treated plants ($p < 0.05$). The content of soluble sugars in the leaves of sweet potato was increased to 50 mg L^{-1}; however, at 75 mg L^{-1} V, the content of soluble sugars decreased but was still significantly higher (97.6%) than the control plants (Figure 6B). Furthermore, the results exhibited that the total proteins were significantly increased at 10 mg L^{-1} V treatment; later, a significant decrease was observed at a higher level of V treatment ($p < 0.05$). The lowest protein content (0.029 mg g^{-1}) was observed at 75 mg L^{-1} V treatment compared to the control (Figure 6C).

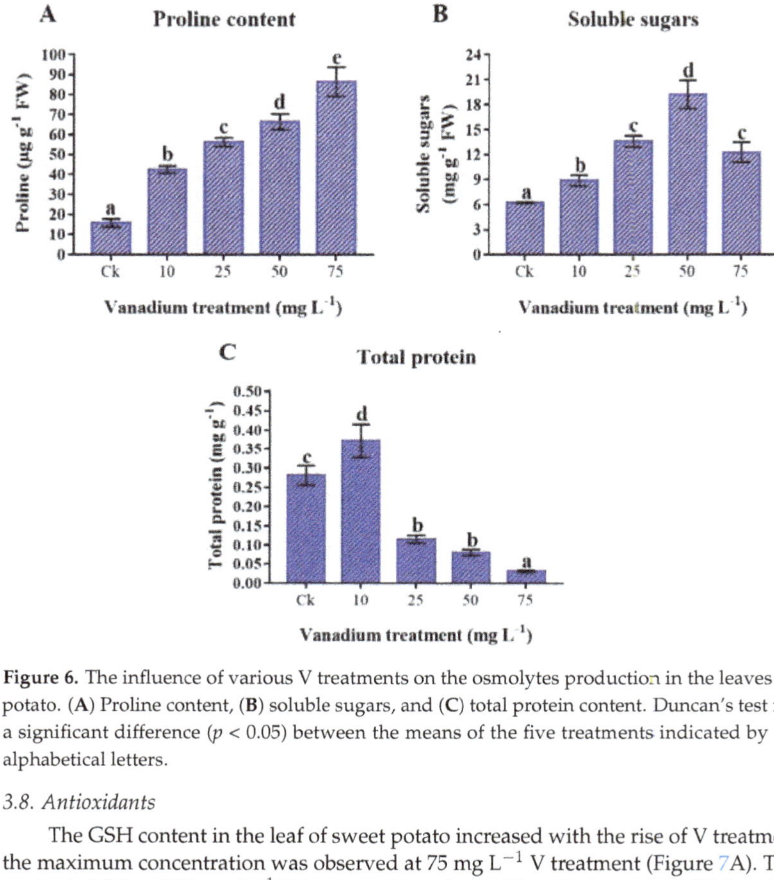

Figure 6. The influence of various V treatments on the osmolytes production in the leaves of sweet potato. (**A**) Proline content, (**B**) soluble sugars, and (**C**) total protein content. Duncan's test indicates a significant difference ($p < 0.05$) between the means of the five treatments indicated by different alphabetical letters.

3.8. Antioxidants

The GSH content in the leaf of sweet potato increased with the rise of V treatment, and the maximum concentration was observed at 75 mg L^{-1} V treatment (Figure 7A). The GSH content at 10 and 25 mg L^{-1} V treatment was insignificantly increased (16.4 and 28.5%); however, a significant increase was observed at 50 and 75 mg L^{-1} V treatment, which were 134% and 324% higher than the control, respectively. Total polyphenols and flavonoid concentrations decreased significantly with the rise of V treatment ($p < 0.05$). Interestingly, total polyphenols and flavonoids drastically reduced at 10 and 25 mg L^{-1} V treatment, then again increased at 50 and 75 mg L^{-1} V treatment, however still significantly lower than in the control plants (Figure 7B,C).

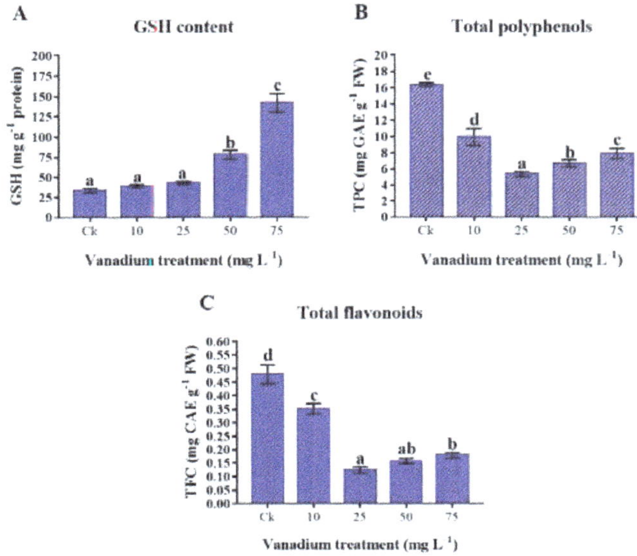

Figure 7. The influence of various V treatments on the antioxidants in the leaves of sweet potato. (**A**) GSH content, (**B**) total polyphenols (TPC), and (**C**) total flavonoid (TFC) content. Duncan's test indicates a significant difference ($p < 0.05$) between the means of the five treatments indicated by different alphabetical letters.

3.9. Antioxidant Enzymes

Antioxidant enzyme activities were significantly influenced by V treatment ($p < 0.05$). We found a significant decrease in APX and POD activities with the rise of V concentration ($p < 0.05$; Figure 8A,D), and maximum reduction was detected at 75 mg L^{-1} of V treatment. As compared to the control, a 62.1% reduction in APX and a 57.5% in POD was detected at 75 mg L^{-1} V treatment. Conversely, the CAT and SOD were positively influenced by V treatment, and a significant increase was observed with the rise in V concentration ($p < 0.05$; Figure 8B,C). Furthermore, the highest activities of CAT and SOD were observed at the 75 mg L^{-1} V treatment, and a 1085% increase in CAT and a 164.7% increase in SOD were observed compared with the control.

3.10. Effects on Leaf Morphology

The present work also attempted to study the effects of V treatment on leaf morphology under SEM. The results showed that leaf morphology under 10 mg L^{-1} V treatment was statistically the same as the control plants (Figure 9). Compared to the control plants, stomatal length, width, pore length, and pore width were increased by 4.7%, 2.6%, 10.1%, and 10.1%, respectively, under 10 mg L^{-1} V treatment (Table 4). In contrast, stomata size was significantly affected under a high level of V treatment (25, 50, and 75 mg L^{-1}), and the maximum deleterious effects were observed at 75 mg L^{-1} V treatment (Figure 9). Compared to the control leaf, stomata length under 75 mg L^{-1} V treatment was reduced by 47.2%; likewise, the width of the stomata was decreased by 80.1%, pore length by 71%, and pore width by 87.1%. Closed and small stomata were observed due to stress conditions, which showed small stomatal openings. The result indicated that V induced stomatal closing and reduced its size.

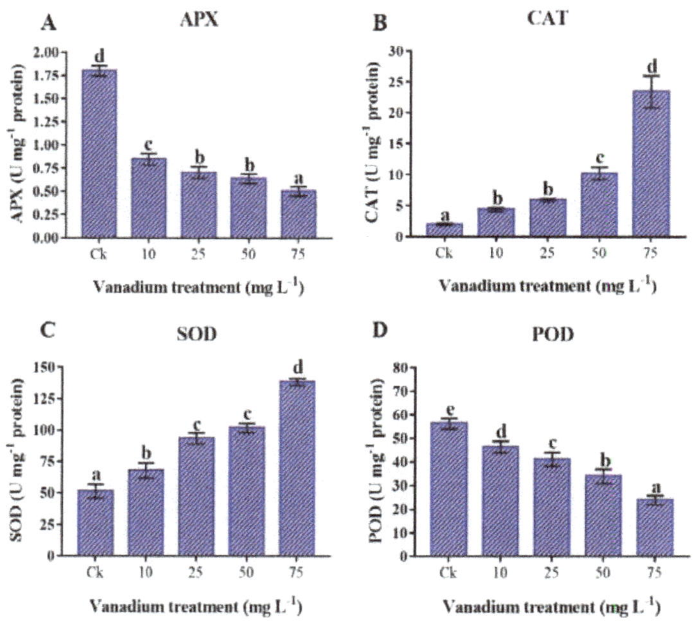

Figure 8. Influence of various V treatments on the antioxidant enzymes activities in the leaves of sweet potato. (**A**) APX, (**B**) CAT, (**C**) SOD, and (**D**) POD. Duncan's test indicates a significant difference ($p < 0.05$) between the means of the five treatments indicated by different alphabetical letters.

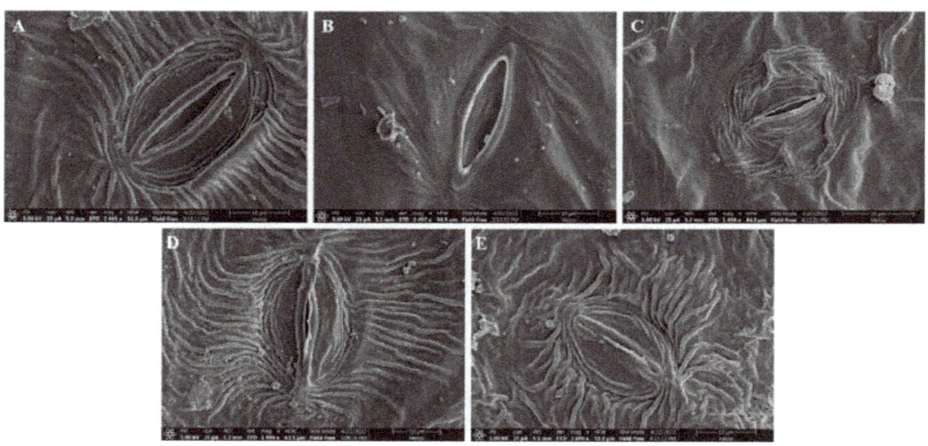

Figure 9. Effect of different V treatments on the stomatal traits of sweet potato leaf. (**A**) Ck, (**B**) 10 mg L^{-1}, (**C**) 25 mg L^{-1}, (**D**) 50 mg L^{-1}, and (**E**) 75 mg L^{-1}.

3.11. Pearson's Correlation and Heat-map Analysis

The negative correlation between physiological parameters and osmolytes, GSH, CAT, and SOD in sweet potato plants treated with different concentrations of V demonstrates the significant reduction in the plant's phenotypic and physiological traits (Figure 10). All plant phenotypic parameters and root and shoot concentrations of V were negatively correlated, showing that V treatment adversely affected plant growth and development. Similarly, osmolytes, GSH, CAT, SOD, and V concentrations and uptake in sweet potato showed a negative correlation with photosynthetic pigments, assimilation, and stomatal traits (Figure 10). However, the phenotypic parameters showed a positive correlation with photosynthetic pigments and assimilation, indicating that plants can grow larger and produce more biomass at a high rate of photosynthetic pigments.

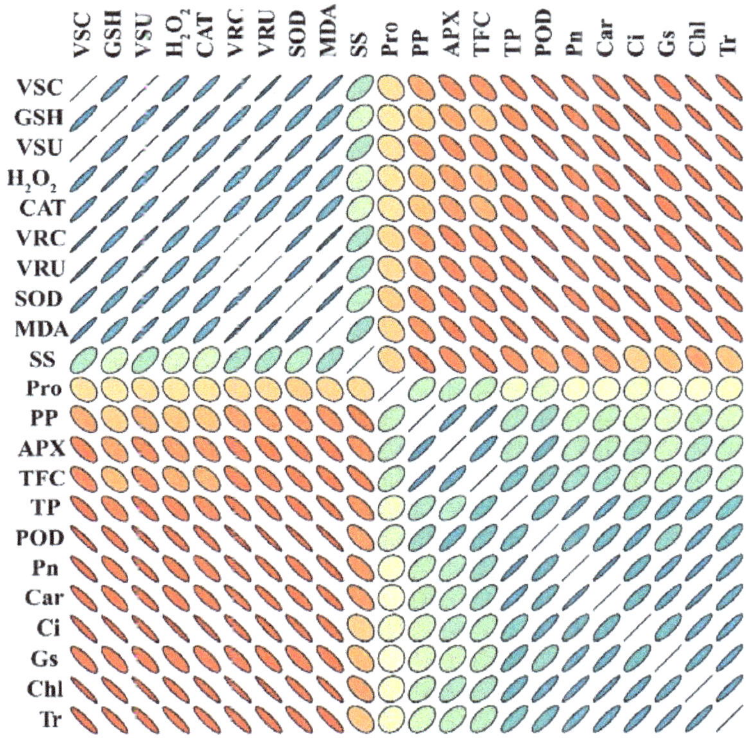

Figure 10. Pearson's correlation analysis (PCA) ($p < 0.05$) was calculated among different traits of V-treated sweet potato seedling. VSC (V shoot concentration), GSH (reduced glutathione), VSU (V shoot uptake), H_2O_2 (hydrogen peroxide), CAT (catalase), VRC (V root concentration), VRU (V root uptake in root), SOD (superoxide dismutase), MDA (malondialdehyde), SS (soluble sugars), Pro (proline), PP (total polyphenols), APX (ascorbate peroxidase), TF (total flavonoids), TP (total proteins), POD (peroxidase), Pn (photosynthetic assimilation), Car (carotenoids), Ci (Intercellular CO_2), Gs (stomatal conductance), Chl (total chlorophyll), and Tr (transpiration rate). Measured determinants and strength correlation matrix. The color code for Pearson correlation coefficients (r), with r = 2, r = 0, and r = −2 denoting red, orange, green, and gray, respectively. A stronger association is denoted by better anisotropy, and the slope of the corresponding line or ellipse represents the trend of that association (positive or negative). Additionally, the ellipses' direction and anisotropy indicate the slope and strength of the relationship. According to the first principal component order, variables were arranged

A heatmap-histogram analysis of different growth traits of sweet potato under different levels of V treatment was also constructed (Figure 11). A significant difference was observed with different colors in the different V treatments and responses of different physiological and biochemical parameters, as well as V concentration and uptake. However, traits with red color indicate insignificant differences within the V treatments. This heatmap-histogram showed a noticeable difference among the growth traits and uptake of V in the sweet potato plant.

Table 4. Effect of different V treatments on the stomatal traits of sweet potato leaf.

Vanadium ($mg\ L^{-1}$)	Stomata Length (μm)	Stomata Width (μm)	Stomatal Pore Length (μm)	Stomatal Pore Width (μm)
Ck	25.52 ± 2.45 [c]	15.52 ± 1.92 [c]	16.30 ± 2.11 [c]	3.92 ± 1.08 [b]
10	26.71 ± 2.97 [c]	15.93 ± 1.78 [c]	18.16 ± 1.31 [c]	4.32 ± 0.92 [b]
25	18.62 ± 2.31 [b]	11.04 ± 1.51 [b]	12.26 ± 1.63 [b]	1.69 ± 0.47 [a]
50	16.22 ± 1.73 [ab]	5.70 ± 1.01 [a]	6.19 ± 1.49 [a]	0.95 ± 0.36 [a]
75	13.47 ± 1.69 [a]	3.08 ± 0.90 [a]	4.72 ± 1.01 [a]	0.51 ± 0.16 [a]

Duncan's test specifies a significant difference ($p < 0.05$) between the means of the five treatments indicated by different alphabets.

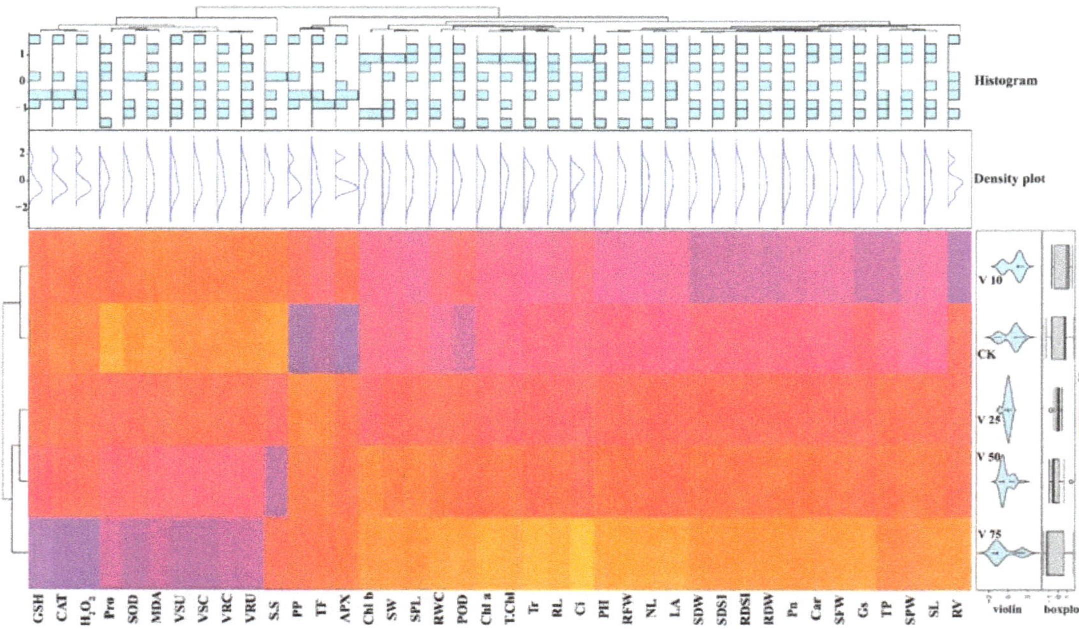

Figure 11. Heatmap-histogram correlation between studied physiological and biochemical parameters under various V treatments in sweet potato.

4. Discussion

Globally, agricultural soil has been polluted with several kinds of soil pollutants with anthropogenic activities. Several studies have indicated that heavy metals, such as Cd, Ni, Pb, and V, are the primary cause of soil pollution. V used in the steel industries and accumulation in the agricultural land and water gained consideration by researchers in recent years [35–37]. V accumulation showed a deleterious effect on living organisms, including plants, animals, and humans. In order to better understand the mechanisms underlying V toxicity in sweet potato, we investigated the phenotypic, physiological, and biochemical processes under various levels of V treatments. Plant growth and biomass are not only used to study different types of heavy metal stress, but also used to evaluate the tolerance level of plants against them. Heavy metals cause plant toxicity and negatively affect the plant length and fresh and dry biomass [38]. However, its toxicity varies with the plant species, chemical structure and formula, concentration, and recurrence of use [39]. However, at higher concentrations, V severely inhibits plant growth and development [40]. In the present study, we observed an improvement in the growth of sweet potato under 10 $mg\ L^{-1}$ V treatment compared to the control plant (Figure 1). According to some studies, plants treated with V had increased height, growth, and fresh and dry biomass [11,12]. Aihemaiti et al. stated that plants generate more biomass at low levels of V due to enhanced chlorophyll biosynthesis [22]. However, many studies reported a decrease in the plant height, growth, and fresh biomass under a higher level of V stress [38,41], and the impact varies from organ to organ of the plants [8]. Similarly, the current study depicted that 25, 50, and 75 mg L^{-1} V treatment considerably reduced the growth of sweet potato, and maximum reduction was observed at 75 mg L^{-1} (Tables 1 and 2). The SDSI and RDSI increased at 10 mg L^{-1}; in contrast, a negative correlation was observed with a further increase of the V level (25, 50, and 75 mg L^{-1}) in the growth medium. A higher level of V stress initiated an ionic imbalance and interrupted their function in metabolic pathways, which eventually affected the process of growth and development of the plant. RWC is a simple and reliable parameter for calculating plant stress. Under various V treatments, we found a decrease in the RWC, indicating that the sweet potato plants were under stress. Osmotic adjustment is impaired in sensitive plant species. A previous study on lettuce also showed a decrease in water content with an increase of V concentration [42]. Under heavy metal stress, various studies also reported a reduction in RWC in the leaves of barley and maize [43,44]. The present study showed that a low level of V treatment (10 mg L^{-1}) induced the root traits of sweet potato (Figure 2). Variations in root size and morphology can improve ionic uptake and translocation, ultimately enhancing plant growth and development [45]. A previous study also reported the expansion of root traits under low V treatment (10 mg L^{-1}) [11]. In contrast, a higher level of V (25, 50, and 75 mg L^{-1}) caused a decrease in these root traits (Figure 2). At a higher level of V treatment, the plant produced comparatively shorter, less lateral, and coralloid structural roots. The decline in root traits under high V levels might have been due to the disruption in mitotic cell division, possibly hampering root tip development [46,47]. Previous studies also reported the decrease of root traits under a higher level of V treatment in watermelon, pepper, and rice [11,13,20]. In addition, V application significantly reduced root morphological traits in many plants [22]; these studies are in agreement with the findings of the present study.

Photosynthetic assimilation is the plant's ability to use CO_2 and perform many metabolic activities [48]. Heavy metal stress drastically affects photosynthesis and significantly affects carbohydrate synthesis. In the current study, we found an increase in these gas exchange elements under 10 mg L^{-1} treatment; however, a significant decrease in these gas exchange elements was observed at a higher level of V stress (Figure 3). A previous study reported a minor increase in the photosynthetic rate and intercellular CO_2 at a low level of V (10 mg L^{-1}) treatment; however, a higher level of V showed a significant reduction in these gas exchange elements in pepper plants [11]. Similarly, another study revealed that V treatment in rice plants caused a reduction in gas exchange elements, which agrees with the current study [13]. At high concentrations, V imposes adverse effects on the physiological processes and photosynthetic assimilation, limiting energy production, and impairing biomass and development of plants [49,50]. V stress possibly reduces photosynthetic activities by affecting the chloroplast and ultrastructures and also disturbs their electron transport mechanisms.

The leaf stomata are responsible for regulating transpiration and CO_2 transport under adverse environmental conditions [51]. The plant experiences cytotoxicity due to the increased concentration of heavy metals, which affects ionic absorption, cell cycle arrest, lipid peroxidation, and ultimately results in cell death [52,53]. Heavy metal interaction with guard cells triggers stomatal closure [54]. V toxicity mediates stomatal abnormalities, such as a reduction in the stomatal length, width, pore length, and width of guard cells (Figure 9 and Table 4). A study reported that the reduction in the stomata size is linked with more distorted stomata under an elevated level of heavy metals in the leaves [30,55]. Photosynthesis, transpiration, and gas exchange can be adversely affected by reductions in stomatal size and closure frequency. From the present study, it has been elucidated that increasing V level decreases stomatal size and closing stomata, which reflects a decrease in photosynthetic assimilation, transpiration rate, and gas exchange.

In the present study, the V concentration, uptake, and translocation had a positive correlation with the rise of V application, and the roots showed higher V content than the shoots (Table 3). The concentration, uptake, and translocation of V to the aerial part are reduced by chelation and fixation of V with a polar compound, stimulation of calcium with a stable compound, and root and stem compartmentalization [56]. A previous study reported that tomato plants treated with V have more V in the roots than in the leaves [36]. In the same way, lettuce, tobacco, alfalfa, milkvetch root, and swamp morning glory treated with V also showed the same trend, and they have more V accumulation in roots than the leaves [1,10,35,42]. Generally, the increase in chlorophyll level characterizes plant photosynthesis assimilation and growth. The reduction in the chlorophyll content was found under different environmental stresses. A higher level of V treatment (25, 50, and 75 mg L^{-1}) reduced the T.Chl, Chl a, b, and carotenoid content (Figure 4). Similarly, previous studies revealed a significant reduction in the chlorophyll content under exposure to V stress in watermelon and pepper [11,13,20]. The reduction in photosynthetic pigments under V stress might be due to the variations in the membrane permeability and the devastation of the elements prompted by oxidative stress [57]. This increase in ROS production causes a reduction in chlorophyll content [15].

Lipids peroxidation is an indication of oxidative stress that is induced by different abiotic stress, and the higher MDA level also indicates cell membrane injury; it is a well-established sign for assessing plants introduced to metal stress [32]. The current study described that the V application considerably boosted the MDA content of sweet potato (Figure 5A). Similarly, previous studies described that V-treated rice and pepper seedlings exhibited an increase in the MDA content [11,13]. Plants produce ROS due to the reaction of heavy metals and fatty acids [58]. H_2O_2 is the main component of ROS and its production increases with heavy metal stress. The current study also reported that an increase in V stress in sweet potato causes a significant increment in H_2O_2 level (Figure 5B). Previous studies also observed a significant increment in the H_2O_2 level when exposed to V stress in rice, tomato, and watermelon plants [13,20,36]. A review article by Chen et al. also highlighted that many articles followed the same pattern of increasing MDA and H_2O_2 under V stress [59]. To deal with these circumstances, the plant possesses a defense mechanism, including osmolytes, antioxidants, and enzymes.

The reduction of osmolytes, such as proline, soluble sugars, and proteins are linked with the leaf water content, which could lead to cellular desiccation and osmotic stress in sweet potato (Figure 6). According to the previous literature, plants exhibit a similar pattern by increasing proline and soluble sugars in response to increased cellular desiccation under V stress [6,11,22,42]. These osmolytes increase under abiotic stress conditions [17,32]. Moreover, these osmolytes might not only assist in protecting plant cellular membranes, but also help in maintaining turgor pressure, which minimizes the deleterious effect of vanadium toxicity. The total proteins are sensitive to heavy metal stress, and different studies reported a reduction of proteins with increased metal stress [13]. The present study showed a significant decrease in total protein content at a higher level of V treatment (25, 50, and 75 mg L^{-1}) (Figure 7A). Heavy metal stress reduces the proteins by prompting toxic effects, damaging the ability of many enzymes with functional thiol groups [60]. The results of the present study agree with the reports of the Refs. [36,61], who stated that total protein content was reduced under V stress. The GSH can enhance the tolerance of plants under metal stress. Furthermore, GSH acts as a ROS scavenger, detoxifying the $O_2^{\bullet-}$ and $HO^{\bullet}$ radicals [32]. This study depicted an increase of GSH under V stress (Figure 7B). GSH levels increase under heavy metal stress due to increased activities of γ-glutamylcysteine synthetase and glutathione synthetase [62]. Our findings agree with previous reports, which described the increment in GSH content under heavy metal stress [61,63]. The secondary metabolites, such as polyphenols and flavonoids, not only enhance the enzymatic activities of the plants but also play a vital role as antioxidants in stress environments [17]. Izbiańska et al. and Kisa et al. reported a decrease in phenolics under heavy metal stress [64,65]. This study depicted that V-treated plants have a significant reduced level of polyphenols and flavonoids (Figure 7C,D). This reduction in the polyphenols and flavonoids biosynthesis under V stress could be due to the decreased activity of essential enzymes of phenylpropanoid pathways [65,66]. Antioxidant enzymes are essential in reducing ROS production and oxidative stress under different environmental stress conditions [17,32]. In this study, we found a decrease in the activity of POD and APX with the rise in V level. In contrast, CAT and SOD significantly increased with the increase of V treatment (Figure 8). This increase in antioxidant enzyme activity may be attributed to the effect of the V ion on free-oxygen radical production. Tobacco plants treated with V have depicted an increment of CAT and SOD activities [10]. Likewise, watermelon and pepper plants under V stress have also depicted increased CAT and SOD activities [11,20]. The results of the current study also agree with studies on rice, chickpeas, oilseed, and tomato under V, Se, and Ni stress, respectively. [41,61,67,68]. Similarly, many other plants have also shown increased activities of these antioxidant enzymes under different levels of V treatments [59]. The antioxidant enzyme activities are powerful ROS scavengers and regulate the plant's cellular membranes under abiotic stress conditions.

5. Conclusions

In the present study, we found that 10 mg L^{-1} V treatment improved the growth and biomass of the sweet potato plant. However, V at higher levels (25, 50, and 75 mg L^{-1}) significantly reduced the growth of sweet potato by altering its physiological and biochemical mechanisms. The higher level of V treatments (25, 50, and 75 mg L^{-1}) significantly reduced the RWC, chlorophyll content, gas exchange elements of leaf, and stomatal traits, which eventually affected the growth of sweet potato. The decline in root morphological traits was observed at the higher V treatment, and we also detected maximum V concentration and uptake in the roots than in the shoots of sweet potato. It was also observed that the V application increased ROS production. Overall, this study will help to understand the physiological tolerance mechanism of V in sweet potato plants. In this study, we focused on the effect of V stress on the morphological and physiological aspects of sweet potato. Besides, this study was conducted in a hydroponically controlled environment, and thus, open-field experiments are encouraged to unravel the more precise effects of V on sweet potato, as in open-field conditions the plants are exposed to several other biotic and abiotic stresses. Moreover, further studies are also required to investigate the molecular mechanism underlying cell death induced by V toxicity in the sweet potato plant. Furthermore, we suggest using various chemicals, phytohormones, and nanoparticles to prevent V toxicity and stimulate sweet potato growth and production.

Author Contributions: All authors contributed to the manuscript. Conceptualization: G.Z. and S.K. Funding acquisition: G.Z. Data curation, investigation, validation, and writing—original draft: S.K. Methodology: S.K., Y.L. and M.W. Writing—review and editing: S.K., S.F., A.Q. and Z.Z. All authors have read and agreed to the published version of the manuscript.

Funding: The earmarked fund for CARS-10-Sweetpotato supported this work.

Institutional Review Board Statement: Not applicable.

Informed Consent Statement: Not applicable.

Data Availability Statement: The data presented in this article will be made available without any reservation

Acknowledgments: The authors acknowledge all the staff members of the Laboratory of Analytical and Testing Center, Hainan University, for providing technical support and help during the experimental work.

Conflicts of Interest: The authors declare that they have no conflict of interest.

References

1. Gan, C.; Chen, T.; Yang, J. Growth responses and accumulation of vanadium in alfalfa, milkvetch root, and swamp morning glory and their potential in phytoremediation. *Bull. Environ. Contam. Toxicol.* **2021**, *107*, 559–564. [CrossRef]
2. Amorim, F.A.C.; Welz, B.; Costa, A.C.S.; Lepri, F.G.; Vale, M.G.R.; Ferreira, S.L.C. Determination of vanadium in petroleum and petroleum products using atomic spectrometric techniques. *Talanta* **2007**, *72*, 349–359. [CrossRef] [PubMed]
3. Yang, J.; Teng, Y.; Wu, J.; Chen, H.; Wang, G.; Song, L.; Yue, W.; Zuo, R.; Zhai, Y. Current status and associated human health risk of vanadium in soil in China. *Chemosphere* **2017**, *171*, 635–643. [CrossRef] [PubMed]
4. Chen, L.; Zhu, Y.; Luo, H.; Yang, J. Characteristic of adsorption, desorption, and co-transport of vanadium on humic acid colloid. *Ecotoxicol. Environ. Saf.* **2020**, *190*, 110087. [CrossRef] [PubMed]
5. Tian, L.-Y.; Yang, J.-Y.; Huang, J.-H. Uptake and speciation of vanadium in the rhizosphere soils of rape (*Brassica juncea* L.). *Environ. Sci. Pollut. Res.* **2015**, *22*, 9215–9223. [CrossRef] [PubMed]
6. Wu, Z.; Zhang, Y.; Yang, J.; Zhou, Y.; Wang, C. Effect of vanadium on testa, seed germination, and subsequent seedling growth of alfalfa (*Medicago sativa* L.). *J. Plant Growth Regul.* **2021**, *40*, 1566–1578. [CrossRef]
7. He, W.; Liao, W.; Yang, J.; Jeyakumar, P.; Anderson, C. Removal of vanadium from aquatic environment using phosphoric acid modified rice straw. *Bioremediat. J.* **2020**, *24*, 80–89. [CrossRef]
8. Imtiaz, M.; Rizwan, M.S.; Xiong, S.; Li, H.; Ashraf, M.; Shahzad, S.M.; Shahzad, M.; Rizwan, M.; Tu, S. Vanadium, recent advancements and research prospects: A review. *Environ. Int.* **2015**, *80*, 79–88. [CrossRef]
9. Larsson, M.A.; Baken, S.; Gustafsson, J.P.; Hadialhejazi, G.; Smolders, E. Vanadium bioavailability and toxicity to soil microorganisms and plants. *Environ. Toxicol. Chem.* **2013**, *32*, 2266–2273. [CrossRef]
10. Wu, Z.; Yang, J.; Zhang, Y.; Wang, C.; Guo, S.; Yu, Y. Growth responses, accumulation, translocation and distribution of vanadium in tobacco and its potential in phytoremediation. *Ecotoxicol. Environ. Saf.* **2021**, *207*, 111297. [CrossRef]
11. Altaf, M.A.; Shu, H.; Hao, Y.; Zhou, Y.; Mumtaz, M.A.; Wang, Z. Vanadium toxicity induced changes in growth, antioxidant profiling, and vanadium uptake in pepper (*Capsicum annum* L.) seedlings. *Horticulturae* **2021**, *8*, 28. [CrossRef]
12. Chen, T.; Li, T.-Q.; Yang, J.-Y. Damage suffered by swamp morning glory (*Ipomoea aquatica* Forsk) exposed to vanadium (V). *Environ. Toxicol. Chem.* **2016**, *35*, 695–701. [CrossRef] [PubMed]
13. Altaf, M.M.; Diao, X.P.; Ur Rehman, A.; Imtiaz, M.; Shakoor, A.; Altaf, M.A.; Younis, H.; Fu, P.; Ghani, M.U. Effect of vanadium on growth, photosynthesis, reactive oxygen species, antioxidant enzymes, and cell death of rice. *J. Soil Sci. Plant Nutr.* **2020**, *20*, 2643–2656. [CrossRef]
14. Reiter, R.; Tan, D.-X.; Zhou, Z.; Cruz, M.; Fuentes-Broto, L.; Galano, A. Phytomelatonin: Assisting plants to survive and thrive. *Molecules* **2015**, *20*, 7396–7437. [CrossRef] [PubMed]
15. Sridhara Chary, N.; Kamala, C.T.; Samuel Suman Raj, D. Assessing risk of heavy metals from consuming food grown on sewage irrigated soils and food chain transfer. *Ecotoxicol. Environ. Saf.* **2008**, *69*, 513–524. [CrossRef] [PubMed]
16. Kumar, S.; Li, G.; Huang, X.; Ji, Q.; Zhou, K.; Hou, H.; Ke, W.; Yang, J. Phenotypic, nutritional, and antioxidant characterization of blanched *Oenanthe javanica* for preferable cultivar. *Front. Plant Sci.* **2021**, *12*, 639639. [CrossRef]
17. Kumar, S.; Li, G.; Yang, J.; Huang, X.; Ji, Q.; Zhou, K.; Khan, S.; Ke, W.; Hou, H. Investigation of an antioxidative system for salinity tolerance in *Oenanthe javanica*. *Antioxidants* **2020**, *9*, 940. [CrossRef]
18. Imtiaz, M.; Mushtaq, M.A.; Rizwan, M.S.; Arif, M.S.; Yousaf, B.; Ashraf, M.; Shuanglian, X.; Rizwan, M.; Mehmood, S.; Tu, S. Comparison of antioxidant enzyme activities and DNA damage in chickpea (*Cicer arietinum* L.) genotypes exposed to vanadium. *Environ. Sci. Pollut. Res.* **2016**, *23*, 19787–19796. [CrossRef]
19. Altaf, M.A.; Shahid, R.; Ren, M.-X.; Altaf, M.M.; Jahan, M.S.; Khan, L.U. Melatonin mitigates nickel toxicity by improving nutrient uptake fluxes, root architecture system, photosynthesis, and antioxidant potential in tomato seedling. *J. Soil Sci. Plant Nutr.* **2021**, *21*, 1842–1855. [CrossRef]
20. Nawaz, M.A.; Jiao, Y.; Chen, C.; Shireen, F.; Zheng, Z.; Imtiaz, M.; Bie, Z.; Huang, Y. Melatonin pretreatment improves vanadium stress tolerance of watermelon seedlings by reducing vanadium concentration in the leaves and regulating melatonin biosynthesis and antioxidant-related gene expression. *J. Plant Physiol.* **2018**, *220*, 115–127. [CrossRef]
21. Andresen, E.; Küpper, H. Cadmium Toxicity in Plants. In *Cadmium: From Toxicity to Essentiality*; Sigel, A., Sigel, H., Sigel, R.K.O., Eds.; Springer: Dordrecht, The Netherlands, 2013; Volume 11, pp. 395–413. ISBN 9789400751781.
22. Aihemaiti, A.; Gao, Y.; Meng, Y.; Chen, X.; Liu, J.; Xiang, H.; Xu, Y.; Jiang, J. Review of plant-vanadium physiological interactions, bioaccumulation, and bioremediation of vanadium-contaminated sites. *Sci. Total Environ.* **2020**, *712*, 135637. [CrossRef] [PubMed]
23. Kumar, S.; Wang, M.; Liu, Y.; Fahad, S.; Qayyum, A.; Jadoon, S.A.; Chen, Y.; Zhu, G. Nickel toxicity alters growth patterns and induces oxidative stress response in sweetpotato. *Front. Plant Sci.* **2022**, *13*, 1054924. [CrossRef] [PubMed]
24. Laurie, S.M.; Faber, M.; Claasen, N. Incorporating orange-fleshed sweet potato into the food system as a strategy for improved nutrition: The context of South Africa. *Food Res. Int.* **2018**, *104*, 77–85. [CrossRef] [PubMed]
25. Mussoline, W.A.; Bohac, J.R.; Boman, B.J.; Trupia, S.; Wilkie, A.C. Agronomic productivity, bioethanol potential and postharvest storability of an industrial sweetpotato cultivar. *Ind. Crops Prod.* **2017**, *95*, 96–103. [CrossRef]
26. Shekhar, S.; Mishra, D.; Buragohain, A.K.; Chakraborty, S.; Chakraborty, N. Comparative analysis of phytochemicals and nutrient availability in two contrasting cultivars of sweet potato (*Ipomoea batatas* L.). *Food Chem.* **2015**, *173*, 957–965. [CrossRef]
27. Fu, Z.F.; Tu, Z.C.; Zhang, L.; Wang, H.; Wen, Q.H.; Huang, T. Antioxidant activities and polyphenols of sweet potato (*Ipomoea batatas* L.) leaves extracted with solvents of various polarities. *Food Biosci.* **2016**, *15*, 11–18. [CrossRef]

28. Sun, H.; Mu, T.; Xi, L.; Zhang, M.; Chen, J. Sweet potato (*Ipomoea batatas* L.) leaves as nutritional and functional foods. *Food Chem.* **2014**, *156*, 380–389. [CrossRef]
29. Kurata, R.; Adachi, M.; Yamakawa, O.; Yoshimoto, M. Growth suppression of human cancer cells by polyphenolics from sweetpotato (*Ipomoea batatas* L.) leaves. *J. Agric. Food Chem.* **2007**, *55*, 185–190. [CrossRef]
30. Kumar, S.; Wang, M.; Fahad, S.; Qayyum, A.; Chen, Y.; Zhu, G. Chromium induces toxicity at different phenotypic, physiological, biochemical, and ultrastructural levels in Sweet potato (*Ipomoea batatas* L.) plants. *Int. J. Mol. Sci.* **2022**, *23*, 13496. [CrossRef]
31. García-Jiménez, A.; Trejo-Téllez, L.I.; Guillén-Sánchez, D.; Gómez-Merino, F.C. Vanadium stimulates pepper plant growth and flowering, increases concentrations of amino acids, sugars and chlorophylls, and modifies nutrient concentrations. *PLoS ONE* **2018**, *13*, e0201908. [CrossRef]
32. Kumar, S.; Li, G.; Yang, J.; Huang, X.; Ji, Q.; Liu, Z.; Ke, W.; Hou, H. Effect of salt stress on growth, physiological parameters, and ionic concentration of water dropwort (*Oenanthe javanica*) cultivars. *Front. Plant Sci.* **2021**, *12*, 660409. [CrossRef] [PubMed]
33. Yang, J.; Li, G.; Bishopp, A.; Heenatigala, P.P.M.; Hu, S.; Chen, Y.; Wu, Z.; Kumar, S.; Duan, P.; Yao, L.; et al. A comparison of growth on mercuric chloride for three *Lemnaceae* species reveals differences in growth dynamics that effect their suitability for use in either monitoring or remediating ecosystems contaminated with mercury. *Front. Chem.* **2018**, *6*, 112. [CrossRef] [PubMed]
34. Kumar, S.; Huang, X.; Ji, Q.; Qayyum, A.; Zhou, K.; Ke, W.; Zhu, H.; Zhu, G. Influence of blanching on the gene expression profile of phenylpropanoid, flavonoid and vitamin biosynthesis, and their accumulation in *Oenanthe javanica*. *Antioxidants* **2022**, *11*, 470. [CrossRef] [PubMed]
35. Gan, C.; Chen, T.; Yang, J. Remediation of vanadium contaminated soil by alfalfa (*Medicago sativa* L.) combined with vanadium-resistant bacterial strain. *Environ. Technol. Innov.* **2020**, *20*, 101090. [CrossRef]
36. Altaf, M.A.; Shahid, R.; Ren, M.-X.; Khan, L.U.; Altaf, M.M.; Jahan, M.S.; Nawaz, M.A.; Naz, S.; Shahid, S.; Lal, M.K.; et al. Protective mechanisms of melatonin against vanadium phytotoxicity in tomato seedlings: Insights into nutritional status, photosynthesis, root architecture system, and antioxidant machinery. *J. Plant Growth Regul.* **2021**, *41*, 1–17. [CrossRef]
37. Yu, Y.; Li, J.; Liao, Y.; Yang, J. Effectiveness, stabilization, and potential feasible analysis of a biochar material on simultaneous remediation and quality improvement of vanadium contaminated soil. *J. Clean. Prod.* **2020**, *277*, 123506. [CrossRef]
38. Xuebin, Q.; Yatao, X.; Ahmad, M.I.; Shehzad, M.; Zain, M. Silicon and its application methods improve physiological traits and antioxidants in *Triticum aestivum* (L.) under cadmium stress. *J. Soil Sci. Plant Nutr.* **2020**, *20*, 1110–1121. [CrossRef]
39. Pilon-Smits, E.A.; Quinn, C.F.; Tapken, W.; Malagoli, M.; Schiavon, M. Physiological functions of beneficial elements. *Curr. Opin. Plant Biol.* **2009**, *12*, 267–274. [CrossRef]
40. Saldaña-Sánchez, W.D.; León-Morales, J.M.; López-Bibiano, Y.; Hernández-Hernández, M.; Langarica-Velázquez, E.C.; García-Morales, S. Effect of V, Se, and Ce on growth, photosynthetic pigments, and total phenol content of tomato and pepper seedlings. *J. Soil Sci. Plant Nutr.* **2019**, *19*, 678–688. [CrossRef]
41. Yuan, Y.; Imtiaz, M.; Rizwan, M.; Dong, X.; Tu, S. Effect of vanadium on germination, growth and activities of amylase and antioxidant enzymes in genotypes of rice. *Int. J. Environ. Sci. Technol.* **2020**, *17*, 383–394. [CrossRef]
42. Wu, Z.; Zhang, Y.; Yang, J.; Jia, Z. Effect of vanadium on *Lactuca sativa* L. growth and associated health risk for human due to consumption of the vegetable. *Environ. Sci. Pollut. Res.* **2022**, *29*, 9766–9779. [CrossRef] [PubMed]
43. González, A.; Gil-Díaz, M.; Lobo, M.C. Response of two barley cultivars to increasing concentrations of cadmium or chromium in soil during the growing period. *Biol. Trace Elem. Res.* **2015**, *163*, 235–243. [CrossRef] [PubMed]
44. Bashir, M.A.; Wang, X.; Naveed, M.; Mustafa, A.; Ashraf, S.; Samreen, T.; Nadeem, S.M.; Jamil, M. Biochar mediated-alleviation of chromium stress and growth improvement of different maize cultivars in tannery polluted soils. *Int. J. Environ. Res. Public Health* **2021**, *18*, 4461. [CrossRef] [PubMed]
45. Wissuwa, M.; Kretzschmar, T.; Rose, T.J. From promise to application: Root traits for enhanced nutrient capture in rice breeding. *J. Exp. Bot.* **2016**, *67*, 3605–3615. [CrossRef] [PubMed]
46. Meisch, H.-U.; Benzschawel, H.; Bielig, H.-J. The role of vanadium in green plants. *Arch. Microbiol.* **1977**, *114*, 67–70. [CrossRef] [PubMed]
47. Yang, J.; Wang, M.; Jia, Y.; Gou, M.; Zeyer, J. Toxicity of vanadium in soil on soybean at different growth stages. *Environ. Pollut.* **2017**, *231*, 48–58. [CrossRef] [PubMed]
48. Guo, Y.; Liu, Y.; Wang, R.; Wang, S.; Lu, X.; Wang, B. Effect of mercury stress on photosynthetic characteristics of two kinds of warm season turf grass. *Int. J. Environ. Monit. Anal.* **2015**, *3*, 293–297. [CrossRef]
49. Abedini, M.; Mohammadian, F. Vanadium effects on phenolic content and photosynthetic pigments of sunflower. *South-West. J. Hortic. Biol. Environ.* **2018**, *9*, 77–86.
50. Yang, J.Y.; Tang, Y. Accumulation and biotransformation of vanadium in *Opuntia microdasys*. *Bull. Environ. Contam. Toxicol.* **2015**, *94*, 448–452. [CrossRef]
51. Khan, M.N.; Zhang, J.; Luo, T.; Liu, J.; Rizwan, M.; Fahad, S.; Xu, Z.; Hu, L. Seed priming with melatonin coping drought stress in rapeseed by regulating reactive oxygen species detoxification: Antioxidant defense system, osmotic adjustment, stomatal traits and chloroplast ultrastructure perseveration. *Ind. Crops Prod.* **2019**, *140*, 111597. [CrossRef]
52. Balasaraswathi, K.; Jayaveni, S.; Sridevi, J.; Sujatha, D.; Phebe Aaron, K.; Rose, C. Cr–induced cellular injury and necrosis in *Glycine max* L.: Biochemical mechanism of oxidative damage in chloroplast. *Plant Physiol. Biochem.* **2017**, *118*, 653–666. [CrossRef] [PubMed]

53. Wakeel, A.; Xu, M.; Gan, Y. Chromium-induced reactive oxygen species accumulation by altering the enzymatic antioxidant system and associated cytotoxic, genotoxic, ultrastructural, and photosynthetic changes in plants. *Int. J. Mol. Sci.* **2020**, *21*, 728. [CrossRef] [PubMed]
54. Rucińska-Sobkowiak, R. Water relations in plants subjected to heavy metal stresses. *Acta Physiol. Plant.* **2016**, *38*, 257. [CrossRef]
55. Gautam, V.; Kohli, S.K.; Kapoor, D.; Bakshi, P.; Sharma, P.; Arora, S.; Bhardwaj, R.; Ahmad, P. Stress protective effect of *Rhododendron arboreum* leaves (MEL) on chromium-treated *Vigna radiata* plants. *J. Plant Growth Regul.* **2021**, *40*, 423–435. [CrossRef]
56. Kaplan, D.I.; Adriano, D.C.; Carlson, C.L.; Sajwan, K.S. Vanadium: Toxicity and accumulation by beans. *Water. Air. Soil Pollut.* **1990**, *49*, 81–91. [CrossRef]
57. Aihemaiti, A.; Jiang, J.; Blaney, L.; Zou, Q.; Gao, Y.; Meng, Y.; Yang, M.; Xu, Y. The detoxification effect of liquid digestate on vanadium toxicity to seed germination and seedling growth of dog's tail grass. *J. Hazard. Mater.* **2019**, *369*, 456–464. [CrossRef]
58. Shah, K.; Kumar, R.G.; Verma, S.; Dubey, R. Effect of cadmium on lipid peroxidation, superoxide anion generation and activities of antioxidant enzymes in growing rice seedlings. *Plant Sci.* **2001**, *161*, 1135–1144. [CrossRef]
59. Chen, L.; Liu, J.; Hu, W.; Gao, J.; Yang, J. Vanadium in soil-plant system: Source, fate, toxicity, and bioremediation. *J. Hazard. Mater.* **2021**, *405*, 124200. [CrossRef]
60. Tanyolaç, D.; Ekmekçi, Y.; Ünalan, Ş. Changes in photochemical and antioxidant enzyme activities in maize (*Zea mays* L.) leaves exposed to excess copper. *Chemosphere* **2007**, *67*, 89–98. [CrossRef]
61. Imtiaz, M.; Tu, S.; Xie, Z.; Han, D.; Ashraf, M.; Rizwan, M.S. Growth, V uptake, and antioxidant enzymes responses of chickpea (*Cicer arietinum* L.) genotypes under vanadium stress. *Plant Soil* **2015**, *390*, 17–27. [CrossRef]
62. Mendoza-Cózatl, D.G.; Moreno-Sánchez, R. Control of glutathione and phytochelatin synthesis under cadmium stress. Pathway modeling for plants. *J. Theor. Biol.* **2006**, *238*, 919–936. [CrossRef] [PubMed]
63. Zeng, F.; Qiu, B.; Wu, X.; Niu, S.; Wu, F.; Zhang, G. Glutathione-mediated alleviation of chromium toxicity in rice plants. *Biol. Trace Elem. Res.* **2012**, *148*, 255–263. [CrossRef] [PubMed]
64. Izbiańska, K.; Arasimowicz-Jelonek, M.; Deckert, J. Phenylpropanoid pathway metabolites promote tolerance response of lupine roots to lead stress. *Ecotoxicol. Environ. Saf.* **2014**, *110*, 61–67. [CrossRef] [PubMed]
65. Kisa, D.; Kayır, Ö.; Sağlam, N.; Şahin, S.; Öztürk, L.; Elmastaş, M. Changes of phenolic compounds in tomato associated with the heavy metal stress. *Bartın Univ. Int. J. Nat. Appl. Sci.* **2019**, *2*, 35–43.
66. André, C.M.; Schafleitner, R.; Legay, S.; Lefèvre, I.; Aliaga, C.A.A.; Nomberto, G.; Hoffmann, L.; Hausman, J.; Larondelle, Y.; Evers, D. Gene expression changes related to the production of phenolic compounds in potato tubers grown under drought stress. *Phytochemistry* **2009**, *70*, 1107–1116. [CrossRef]
67. Ulhassan, Z.; Huang, Q.; Gill, R.A.; Ali, S.; Mwamba, T.M.; Ali, B.; Hina, F.; Zhou, W. Protective mechanisms of melatonin against selenium toxicity in *Brassica napus*: Insights into physiological traits, thiol biosynthesis and antioxidant machinery. *BMC Plant Biol.* **2019**, *19*, 507. [CrossRef]
68. Jahan, M.S.; Guo, S.; Baloch, A.R.; Sun, J.; Shu, S.; Wang, Y.; Ahammed, G.J.; Kabir, K.; Roy, R. Melatonin alleviates nickel phytotoxicity by improving photosynthesis, secondary metabolism and oxidative stress tolerance in tomato seedlings. *Ecotoxicol. Environ. Saf.* **2020**, *197*, 110593. [CrossRef]

antioxidants

MDPI

Article

Exogenously Applied Rohitukine Inhibits Photosynthetic Processes, Growth and Induces Antioxidant Defense System in *Arabidopsis thaliana*

Sajad Ahmed [1,2,*], Mohd Asgher [3], Amit Kumar [1,4] and Sumit G. Gandhi [1,4,*]

1 Indian Institute of Integrative Medicine, Canal Road, Jammu 180001, Jammu and Kashmir, India; amitkumar@iiim.ac.in
2 Department of Botanical and Environmental Sciences, Guru Nanak Dev University, Amritsar 143005, Punjab, India
3 Plant Physiology and Biochemistry Laboratory, Department of Botany, School of Biosciences and Biotechnology, Baba Ghulam Shah Badshah University, Rajouri 185234, Jammu and Kashmir, India; asghermohd@gmail.com
4 Academy of Scientific and Innovative Research (AcSIR), Ghaziabad 201002, Uttar Pradesh, India
* Correspondence: choudharysajad@gmail.com (S.A.); sumit@iiim.res.in (S.G.G.)

Abstract: The secondary metabolite rohitukine has been reported in only a few plant species, including *Schumanniophyton magnificum*, *S. problematicum*, *Amoora rohituka*, *Dysoxylum acutangulum* and *D. gotadhora*. It has several biological activities, such as anticancer, anti-inflammatory, antiadipogenic, immunomodulatory, gastroprotective, anti-implantation, antidyslipidemic, anti-arthritic and anti-fertility properties. However, the ecological and physiological roles of rohitukine in parent plants have yet to be explored. Here for the first time, we tried to decipher the physiological effect of rohitukine isolated from *D. gotadhora* on the model system *Arabidopsis thaliana*. Application of 0.25 mM and 0.5 mM rohitukine concentrations moderately affected the growth of *A. thaliana*, whereas a remarkable decrease in growth and the alteration of various morphological, physiological and biochemical mechanisms were observed in plants that received 1.0 mM of rohitukine as compared to the untreated control. *A. thaliana* showed considerable dose-dependent decreases in leaf area, fresh weight and dry weight when sprayed with 0.25 mM, 0.5 mM and 1.0 mM of rohitukine. Rohitukine exposure resulted in the disruption of photosynthesis, photosystem II (PSII) activity and degradation of chlorophyll content in *A. thaliana*. It also triggered oxidative stress in visualized tissues through antioxidant enzyme activity and the expression levels of key genes involved in the antioxidant system, such as superoxide dismutase (SOD), peroxidase (POD) and ascorbate peroxidase (APX). Rohitukine-induced changes in levels of metabolites (amino acids, sugars, organic acids, etc.) were also assessed. In light of these results, we discuss (i) the likely ecological importance of rohitukine in parent plants as well as (ii) the comparison of responses to rohitukine treatment in plants and mammals.

Keywords: antioxidants; *Arabidopsis*; metabolome; ROS; rohitukine

Citation: Ahmed, S.; Asgher, M.; Kumar, A.; Gandhi, S.G. Exogenously Applied Rohitukine Inhibits Photosynthetic Processes, Growth and Induces Antioxidant Defense System in *Arabidopsis thaliana*. *Antioxidants* **2022**, *11*, 1512. https://doi.org/10.3390/antiox11081512

Academic Editors: Stanley Omaye and Filomena Nazzaro

Received: 1 June 2022
Accepted: 19 July 2022
Published: 3 August 2022

1. Introduction

Plants synthesize a large number of small molecules for use in defence against biotic and abiotic stresses [1–3]. These secondary phytochemicals have commercial applications in pharmaceuticals, flavourings and fragrances, insecticides, etc. [4,5]. They are divided into three main groups: phenolic compounds, terpenoids and nitrogen-containing alkaloids [6]. Usually, alkaloids contain basic nitrogen, derived from an amino acid or purine/pyrimidine, while in some pseudoalkaloids the source of nitrogen is a transamination reaction [7]. Among alkaloids, the chromone alkaloids are a unique group with many biological activities, structurally consisting of a nitrogen system (pyridine, piperidine, pyrrolidine) linked to the A-ring of chromone [8].

The chromone alkaloid rohitukine was isolated for the first time from *A. rohituka* [9] and its structure has been shown to be that of a highly polar molecule [9–11]. It is restricted to only two plant families: Rubiaceae (*Schumanniophyton magnificum* and *S. problematicum*) and Meliaceae (*Dysoxylum gotadhora, D. acutangulum* and *Amoora rohituka*) [12]. Both families are mainly distributed in tropical areas of the world [13]. *D. gotadhora* (Indian white cedar) has been in use in Ayurvedic and other traditional systems of indigenous medicine for many years to treat diabetes, jaundice, leucorrhoea, piles, leprosy, osteomyelitis, etc. [14]. *D. gotadhora* is known as the main source of rohitukine. Within the plant body, rohitukine is found to accumulate in leaves, bark, seeds and fruits [15]. The pharmaceutical potential of rohitukine has been deeply assessed in breast, ovarian and lung cancer cell lines, where it inhibits cyclin-dependent kinase (CDK) CDK2/A and CDK9/T1 complexes by blocking their ATP binding sites [10]. In budding yeast, rohitukine induces ROS generation and apoptosis [16]. P-276-00, IIIM-290 and flavopiridol are semisynthetic derivatives of rohitukine.

The rohitukine-inspired molecule flavopiridol has gained considerable attention in the last two decades for its potent cytotoxic activity against a wide range of cancer cell lines [17] and it has now been approved by the European Medicines Agency for the treatment of chronic lymphocytic leukemia (CLL) [18]. Flavopiridol potently inhibits CDKs 1, 2 and 4, causing cell cycle arrest in G1 and G2 phases in mammalian cells [19]. P-276-00 (Piramal Healthcare Limited, Mumbai, India) is another derivative of rohitukine that has advanced into clinical trials for cancer treatment. P-276-00 selectively inhibits CDK4/D1, CDK1/B and CDK9/T1, and its antiproliferative effect has been observed against a wide range of cancer cell lines [20]. IIIM-290, an orally bioavailable anticancer drug that is already being examined in apre-clinical study, is also derived from rohitukine [21,22].

Similar to many other medicinal plants, the presence of rohitukine in *D. gotadhora* may likely exert allelopathic effects on neighbouring plants. It has been observed that plants release chemical compounds into their surrounding environment which influence their growth and also contribute to restricting invasion by exotic plant species [23,24]. These phytotoxic compounds are usually biosynthesized in plants as secondary products and many of them have been explored for their pharmacological activities [25,26]. To address the phytotoxic effects of secondary metabolites, many studies have been published which show that secondary metabolites have an impact mainly through damage to photosynthetic machinery and frequent decomposition of photosynthetic pigments, the decrease in photosynthetic pigments leading to blockage of energy/electron transfer and inhibition of ATP synthesis [27–29]. Several alkaloids isolated from the medicinal plant *Ruta graveolens* have also been investigated for their photosynthetic inhibitory activities in recipient plants [30]. Antidesmone is a plant secondary metabolite that causes disruption to photosynthetic machinery [31]. Additionally, the secondary compounds of *Satureja hortensis* affect seed germination, morphology and bleach out chlorophyll content in *Amaranthus retroflexus* and *Chenopodium album* [32]. It is quite evident that the targets of toxic plant secondary metabolites are achieved through perturbations to PSII activity [33,34]. In plants, photosynthetic damage is also linked to increased levels of reactive oxygen species (ROS),which indicates oxidative stress [35–37]. Increased ROS levels lead to oxidative damage to cells. Similarly, exposure to toxic secondary metabolites also triggers the ROS pathway in recipient plants [38,39]. To resist oxidative stress, plants, upon exposure to such chemicals, alter the activities of antioxidant enzymes, such as SOD, POD, APX and CAT [27,40,41].

The above-mentioned findings indicate that rohitukine as well as its semisynthetic derivatives have been explored for their remarkable biological activities in animal cells, wherein rohitukine has been reported to trigger ROS generation and apoptosis [16,42]. However, the phytotoxic effect of rohitukine on the growth and development of plants has not been explored. Therefore, here we tried to understand the morphological, physiological and biochemical changes in *A. thaliana* (a model system) treated with appropriate rohitukine concentrations isolated from *D. gotadhora*. The main objectives of the study were to understand the interference of rohitukine with the antioxidant system of *A. thaliana* and its impacts on photosynthesis. Furthermore, we sought to gain insights into photosynthetic pigments, the phytotoxicity attributed to ROS generation and changes in levels of metabolites, such as amino acids, sugars and other organic acids, in *A. thaliana*.

2. Material and Methods

2.1. Source Plant Material, Extraction, Fractionation and Isolation of Pure Compounds

Rohitukine was extracted from dry leaf powder of *D. gotadhora*, as described previously by Mahajan et al. (2015) [15]. Briefly, shade-dried leaves were extracted thrice with 50% ethanol by sonication for one hour each at 45 °C, and the extract was dried using a rotary evaporator. The crude extract was then subjected to acid–base fractionation to obtain a fraction enriched with the alkaloid rohitukine and purified by repeated column chromatography over a silica gel mesh. The purity of the isolated rohitukine was confirmed by liquid chromatography–mass spectrometry (LC–MS) and high-performance liquid chromatography (HPLC). The purity of compounds was checked using HPLC by following the protocol of Kumar et al. (2016) [43]. An RP-C18 column was used (Neo Sphere, 250 mm × 4.6 mm, 5 µm). The mobile phase consisted of methanol–water (75:25 *v*/*v*) at flow rate of 1.0 mL/min. The temperature of the column oven was 40 °C and the injection volume was 4 µL. The detector used was a diode array detector, and the detection wavelength was 254 nm. Solutions of pure rohitukine were prepared by dissolving the alkaloid in sterile double-distilled water, followed by filter sterilization using a syringe filtration unit fitted with a 0.22 µm pore size (Millex-GV, Durapore). The filtered stock solution of rohitukine was then diluted appropriately in autoclaved double-distilled water for foliar treatment.

2.2. Plant Material and Growth Conditions of A. thaliana

Wild-type *A. thaliana* seeds (Col-0 background) were used in the experiment. The seeds were surface-sterilized with 70% ethanol, washed with autoclaved double-distilled water and sown on plates containing Murashige and Skoog (MS) basal medium (for media composition, see Table 1), sucrose 1.5% (*w*/*v*) and agar 0.6% (*w*/*v*) (Hi Media Laboratories Pvt. Ltd. Mumbai, India) [44], supplemented with appropriate rohitukine concentrations: 0 (control), 0.25 mM, 0.5 mM and 1.0 mM concentrations. For foliar treatment, plants were grown in plastic pots filled with 300 g autoclaved soil mixture (soil rite–sand–soil) at the ratio of (3:1:1). Seeds were kept at 4 °C in the dark for 48 h to ensure homogenous germination. After 48 h, the plates containing seeds were transferred to a growth chamber with the following conditions: photosynthetically active radiation (PAR): 680 µmole/m^2/s; temperature: 24 °C; photoperiod light/dark cycles: 16/8 h; and relative humidity: 65%, in the Indian Institute of Integrative Medicine, Jammu and Kashmir, India. The plants grown in soil were irrigated with autoclaved distilled water at intervals of 24 h, while 1 mL of quarter-strength nutrient medium was applied at intervals of 48 h up to five weeks.

2.3. Rohitukine Treatment for A. thaliana

Five-week-old plants of *A. thaliana* with uniform growth and maturity were exposed to 0.25 mM, 0.5 mM and 1.0 mM concentrations of pure rohitukine. Ten plants were treated with each concentration, and plants sprayed with autoclaved double-distilled water without rohitukine were used as controls. The plants were sprayed thrice with the appropriate concentrations of rohitukine dissolved in 50 mL autoclaved double-distilled water (1 mL to each plant) at intervals of 24 h, and 1 h post-treatment the samples were collected for analysis. Rohitukine is polar in nature, with a water solubility of >10 mg/mL and lipophilicity (log D) of <1.0 [10].It was found to have pH-dependent solubility, with its highest solubility in simulated gastric fluid (SGF) [45]. The reported value of acidity constant (pKa) is 5.83 [10]. We prepared the spray reagent at around pH 7.0 and it is therefore expected to be present at unionized form in solution, leading to higher absorption/permeation.

Table 1. Composition of MS nutrient media used for the growth of *A. thaliana* plants.

Class/Category	Ingredients	mg/L
Macro elements	Ammonium nitrate	1650.000
	Calcium chloride	332.200
	Magnesium sulphate	180.690
	Potassium nitrate	1900.000
	Potassium phosphate monobasic	170.000
Microelements	Boric acid	6.200
	Cobalt chloride hexahydrate	0.025
	Copper sulphate pentahydrate	0.025
	EDTA disodium salt dihydrate	37.300
	Ferrous sulphate heptahydrate	27.800
	Manganese sulphate monohydrate	16.900
	Molybdic acid (sodium salt)	0.213
	Potassium Iodide	0.830
	Zinc sulphate heptahydrate	8.600
Vitamins	Myo-inositol	100.000
	Nicotinic acid (free acid)	0.500
	Pyridoxine HCl	0.500
	Thiamine hydrochloride	0.100
Amino Acid	Glycine	2.000

2.4. Determination of Rohitukine Content in A. thaliana

The content of rohitukine in treated samples was determined by thoroughly washing an equal quantity (200 mg) of rosette leaves, followed by homogenization in HPLC grade methanol and sonication, after which the homogenate was centrifuged at 12,000 rpm for 10 min at room temperature (RT). The supernatant was filtered through a syringe filter of 0.22 µm pore size and analysed following the protocol of Kumar et al. (2016) [43], using a HPLC system (Shimadzu, UFLC) consisting of a quaternary pump with a vacuum degasser, a thermos tatted column compartment, an autosampler and a PDA detector. A reverse-phase column (Lichrosphere RP C18e, 5 µm, 250 mm × 4 mm) was used and the column temperature was maintained at 40 °C. The HPLC mobile phase consisted of two solutions. Solution A was composed of water with 0.1% formic acid. The solution was filtered through a 0.45 µm membrane filter and degassed in a so nicator for 3 min. Solution B was pure HPLC-grade acetonitrile. The mobile phase was run using gradient elution: 0.01 min, 10% B; in the next 20 min, 50% B; in the next 5 min, 70% B; in the next 5 min, 90% B, and maintained at 80% B for 5 min; in the next 5 min, 10% B; followed by an equilibration period of 5 min. The flow rate was 0.8 mL/min, and the injection volume was 20 µL. The eluents were detected and analysed at 254 nm. A quantity of 1mg/mL of pure rohitukine was taken as standard.

2.5. Measurement of Leaf Area and Plant Weight

Leaf area was measured using the easy leaf area tool described by Easlon et al. (2014) [46]. Briefly, *A. thaliana* rosette leaves treated with rohitukine were photographed by placing them on plain paper along with a red-coloured piece of paper of 4 cm^2, and measurements were made using easy leaf area software by following the instructions provided. The fresh weight of the aerial parts (rosette leaves) of plants was determined immediately after harvesting. For dry weight, the samples were completely dried in an oven until weight was constant, then weighed to obtain the dry weight using a fine weighing balance (Mettler Toledo, Columbus, OH, USA).

2.6. Photosystem II Activity

Chlorophyll fluorescence parameters were assessed for the rohitukine-treated plants of *A. thaliana* using a Junior-PAM chlorophyll fluorometer (Heinz Walz, Effeltrich, Germany), following the manufacturer's instructions. The parameters assessed were: actual photosystem II (PSII) efficiency (Φ PSII), intrinsic PSII efficiency (Fv/Fm), maximum PSII efficiency (Fv/Fm), photochemical quenching (qP), non-photochemical quenching (NPQ) and electron transport rate (ETR). ETR was calculated usingthe formula Φ PSII × photosynthetic photon flux density × 0.5 × 0.84, as given in [47].

2.7. Photosynthetic Pigment Quantification

Total chlorophyll content was determined in *A. thaliana* plants exposed to 0.25 mM, 0.5 mM and 1.0 mM rohitukine by adopting the method of Arnon (1949) [48], with minor modifications. Briefly, 200 mg fresh leaf samples were homogenized in 0.2 mL 80% acetone, followed by centrifugation at 12,000 rpm for 5 min. The supernatants were collected and their absorbances were recorded at 663 and 645 nm for chlorophyll content using a spectrophotometer.

2.8. Histochemical Detection of ROS

The accumulation of H_2O_2 and O_2^- was analysed by the histochemical staining method, as described by Shi et al. (2010) [49], using 3,3-diaminobenzidine (DAB) (Sigma-Aldrich, St. Louis, MO, USA, cat. no. D5637) and nitro blue tetrazolium (NBT) (Hi Media, cat. no. RM578), respectively. For H_2O_2 detection, a solution of 1mg/mL DAB was prepared in 10 mM phosphate buffer, and the pH was adjusted to 3.8 with1N HCL. *A. thaliana* leaves of similar maturity levels were immersed in DAB solution at RT in the dark until brown spots were visible. After six hours of staining, the samples were incubated in a mixture of ethanol, acetic acid and glycerol (3:1:1) at 80 °C for 15 min to bleach out chlorophyll for proper visualization. ForO_2^- localization, the samples were immersed in a 1 mg/mL solution of NBT prepared in 10 mM phosphate buffer (pH 7.8) until blue-coloured spots appeared. The immersed samples were then boiled in ethanol for better visualization, and photographs were taken by placing the stained leaves in a clean place with a white background using a digital camera.

2.9. The Activity of Antioxidant Enzymes

Fresh leaf tissue of rohitukine-treated and untreated (control) samples of *A. thaliana* were homogenized in potassium–phosphate extraction buffer (100 mM, pH 7.0), using a precooled pestle and mortar. The extraction buffer was composed of 0.05% (*v*/*v*) Triton X-100 and 1% (*w*/*v*) polyvinylpyrrolidone (PVP). The activity of ascorbate peroxidase (APX; EC, 1.11.1.11) was calculated following the protocol described by Asgher et al. (2014) [50] Briefly, APX activity was determined by the decrease in absorbance of ascorbate at 290 nm due to its enzymatic breakdown. A total of 1 mL mixture of 50 mM phosphate buffer (pH 7.0), 0.5 mM ascorbate, 0.1 mM EDTA, 0.1 mM H_2O_2 and enzyme extract was prepared. The calculation of APX activity was carried out using the extinction coefficient 2.8 mM^{-1} cm^{-1}, and one unit of enzyme was required to decompose 1 µmol of substrate per min at 25 °C.

2.10. RNA Extraction, cDNA Synthesis and qRT-PCR

Rohitukine-treated leaves of *A. thaliana* were used as source material for RNA isolation, using TRizol reagent (Ambion, Life Technologies, Carlsbad, CA, USA), according to the manufacturer's protocol. RNA quality and concentration were assessed on a 2% agarose gel and nanodrop spectrophotometer (Thermo Fisher Scientific, Waltham, MA, USA) by measuring the absorbance ratio at 260/280 nm. The isolated RNA samples were subjected to DNase (Ambion TURBO DNA-free, Life Technologies, Carlsbad, CA, USA) treatment to remove traces of genomic DNA. DNase-treated RNA samples were reverse-transcribed using a cDNA synthesis kit (Promega, Madison, WI, USA), following the manufacturer's instructions, with oligo (dT) primers and 1 µg of DNase-treated RNA used as a template. Primer pairs were designed from coding sequences (CDSs) of selected antioxidant system genes of *A. thaliana* (Table 2). The qRT-PCR reactions were set in the CFX96^{TM} Real-Time PCR Detection System (Bio-Rad, Hercules, CA, USA), based on SYBR green chemistry. The PCR reaction mixtures (10 µL) comprised SYBR Green Master mix at a volume of 5.0 µL, 1.0 µM of each primer (Integrated DNA Technologies, Coralville, IA, USA), appropriately diluted cDNA as template and MQ water was used to make up the final volume of 10 µL. Thermoprofiles of the reactions for qRT-PCR were included by preincubation at 95 °C for 10 min, followed by 45 cycles of 3-step amplification with melt (95 °C for 10 s, 60 °C for 10 s and 72 °C for 15 s) and melt. The normalization of the reaction was achieved using the primers of the actin gene as a control. The analysis of samples was carried out in triplicate, and the specificity of each primer pair was authenticated by a dissociation curve (a single peak was observed for each primer pair). The generated (threshold cycle) CT values were then transferred to Microsoft Excel and the $2^{-\Delta\Delta CT}$ method of relative quantification was used to determine the quantitative variation between the samples examined [51].

Table 2. Table showing the details of primers used for gene expression analysis.

S. No	Gene Name	Primer Sequence, Forward	Primer Sequence, Reverse	TM (°C)
1	Peroxidase (POD)	GAGTCAATCGAACAACAACATCC	CCTCTGTCTGAAACTCGTGC	60
2	Catalase (CAT)	TGGAAGAAGATGCAATTCGTGTT	CCAGGTCTTGGTCACATCG	60
3	Copper/zinc Superoxide dismutase (CU/Zn-SOD)	GAGATGATGGAACTGCCAC	TGGCTACTGGAAACGCAGG	60
4	Manganese Superoxide dismutase (Mn-SOD)	CAAGCTGTGAACAAGGGAGA	AGTGAGCGT CAATGG CACT	60
5	Ascorbate peroxidase (APX)	GCAGATGGGCTTATCTGAC	AGGCCTTCCTTCTCTCC	60
6	Actin 2	GTTGACTACGAGCAGGAG	CAGCAGCTTCCATTCCC	60

2.11. Metabolite Extraction and Enrichment Analysis

Metabolite quantification for *A. thaliana* samples (treated and control) was carried out using the method described by Lisec et al. (2006) [52], with minor changes. Briefly, 200 mg of fresh leaves from all four samples of *A. thaliana* treated with 0.25 mM, 0.5 mM and 1.0 mM rohitukine concentrations and the untreated control were excised using a fine pair of scissors and crushed in liquid nitrogen to a fine powder. Afterwards, 3 mL precooled HPLC-grade methanol (100%) was added and the homogenate was transferred to a glass vial. Then, 100 µL of ribitol (2 mg/10 mL) (Sigma-Aldrich, St. Louis, MI, USA) was added as an internal standard, followed by incubation at 70 °C for 10 min with continuous shaking. After incubation, the homogenate was centrifuged for 10 min at 15,000 rpm in the cold centrifuge; the supernatant was collected and 1.5 mL chloroform and 3 mL precooled autoclaved double-distilled water was added and slight vortexed. The upper polar and lower nonpolar layers were transferred to new well-labelled glass vials and dried over a rotary evaporator. The dried samples were then derivatized by adding 80 µL of 20 mg/mL Methoxy amine hydrochloride (Sigma-Aldrich USA), prepared in pyridine, to each tube, followed by incubation at 37 °C for 2 h. N, O-Bis (trimethylsilyl) trifluoroacetamide (BSTFA) (derivatization-grade) (Sigma-Aldrich USA) was added and another incubation at 37 °C took place for 1 h. After incubation, the mixture was transferred to suitable GC vials with inserts and analysed using GC–MS.

The sample was analysed using a GC–MS 4000 system (Varian, Crawley, UK) equipped with a Supelco capillary column (15 m × 60.32 mm × 60.25 m). A 1 µL volume of derivatized sample was injected; the injection temperature was 230 °C, the interface was set to

150 °C and the ion source was adjusted to 250 °C. The program of gradient temperature was: initial temperature of 40 °C for 6 min, +10 °C/min up to 300 °C and hold at 300 °C for 6 min. Mass spectrometry was determined by the full-scan method, ranging from 35 to 780 (*m*/*z*). The metabolites were identified by comparison of mass spectra with the NIST17 library using the molecular ion masses (*m*/*z*) of the fragments and retention time indexes (RI). The downstream analysis of each metabolite was carried out with the peak area of each metabolite taken as source data, using the MetaboAnalyst 5.0 online tool [53].

3. Results

3.1. Qualitative and Quantitative Assessment of Isolated Rohitukine

A quantity of 1.6 g of rohitukine was obtained after purification from 200 g shade-dried *D. gotadhora* leaves, i.e., 0.8% dry weight. After purification, the compound was subjected to a purity check by TLC, HPLC and LC–MS, in which the isolated compound was found to have 98% purity compared with the standard (Figure 1).

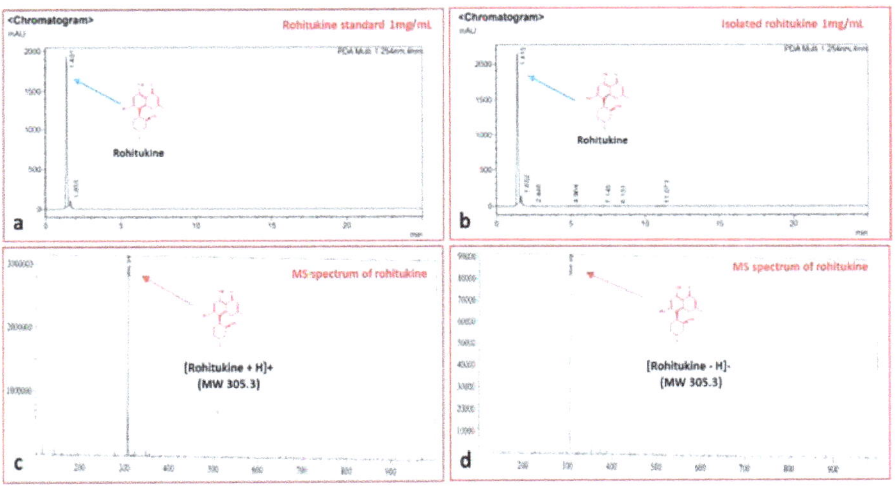

Figure 1. HPLC and MS chromatograms showing the purity of rohitukine isolated from leaves of *D. gotadhora*. (**a**) Chromatogram of standard rohitukine, 1 mg/mL. (**b**) Peak of rohitukine isolated from *D. gotadhora*. (**c**,**d**) Mass spectra showing the presence of a single peak of rohitukine (MW 305.3).

3.2. Determination of Rohitukine Uptake by *A. thaliana*

After 24 h of treatment, the samples were subjected to HPLC for detection and quantification of rohitukine. In leaf tissues treated with 0.25 mM, 0.5 mM and 1.0 mM rohitukine, 0.009%, 0.013% and 0.014% of rohitukine was found to have accumulated, respectively (Figure 2), whereas in the control samples we did not find a peak corresponding to rohitukine, indicating its absence. A standard sample of 1mg/mL rohitukine was used for the calibration curve. However, in plants that received 1.0 mM of rohitukine treatment, the amount of assimilated rohitukine was of a higher quantity as compared to the lesser amount obtained with the 0.25 mM concentration. Therefore, the uptake and stability of rohitukine was confirmed.

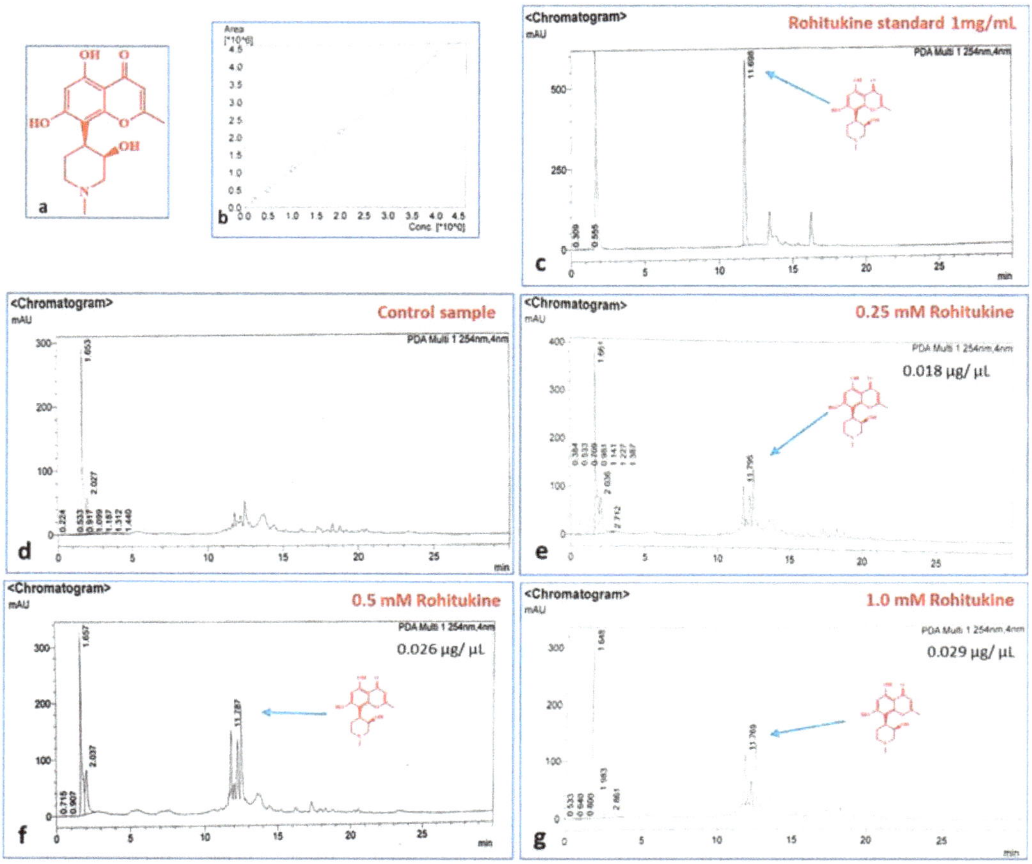

Figure 2. Detection and quantification of rohitukine in treated *A. thaliana* samples. (**a**) Structure of rohitukine. (**b**) Calibration curve of standard rohitukine. (**c**) Rohitukine standard, 1 mg/mL. (**d**) *A. thaliana* control peaks showing no traces of rohitukine. (**e**–**g**) Peaks indicating the presence of rohitukine at 0.25 mM, 0.5 mM and 1.0 mM concentrations.

3.3. Effect of Rohitukine on Plant Morphology

The *A. thaliana* plants treated with rohitukine showed concentration-dependent growth inhibition as compared to the control. A significant decrease in leaf area of 17.7%, 25.1%, and 60.3% with the0.25 mM, 0.5 mM and 1.0 mM rohitukine concentrations, respectively, was observed as compared to the untreated group. The fresh weight of plants treated with 0.25 mM, 0.5 mM and 1.0 mM of rohitukine decreased by 11.6%, 17.4% and 38.7%, respectively. Similarly, dry weight was found to be reduced by 12.5% 19.5% and 40.2% with 0.25 mM, 0.5 mM and 1.0 mM rohitukine, respectively (Figure 3). The maximum growth inhibitory effect of rohitukine was observed in *A. thaliana* plants treated with a 1.0 mM concentration.

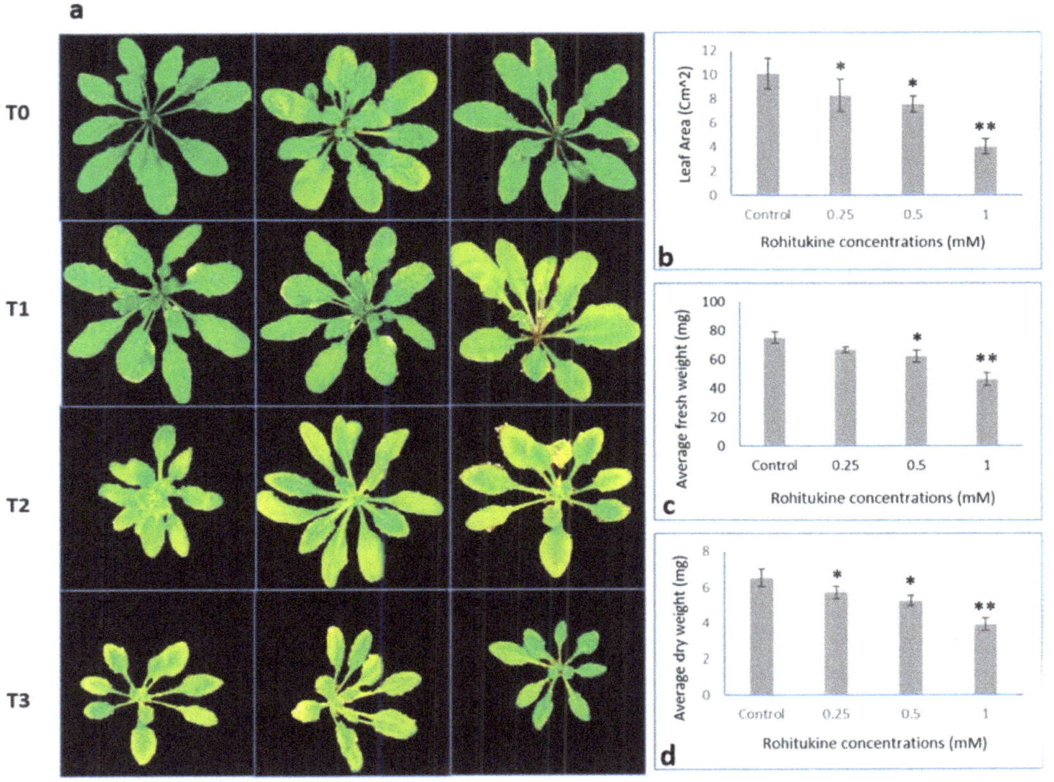

Figure 3. Morphology of *A. thaliana* plants treated with T0 (Control), T1 (0.25 mM), T2 (0.5 mM) and T3 (1.0 mM) rohitukine concentrations. (**a**) Five-week-old *A. thaliana* plants with and without rohitukine treatment. (**b**) Effects of 0.25 mM, 0.5mM and 1mM Rohitukine on leaf area of *A. thaliana* plants. (**c**) Average fresh weight of plants. (**d**) Effect of 0.25 mM, 0.5mM and 1mM Rohitukine on the dry weight of *A. thaliana*. The results are the means of more than three replicates expressed as means ± SD values. Statistical significance was determined by Student's *t*-test. Asterisks * and ** denote the significance level of values at *p*-values < 0.5 and 0.05, respectively.

3.4. Influence of Rohitukine on Chlorophyll Content and PSII Activity

Total chlorophyll content and PSII activity were studied in *A. thaliana* exposed to different concentrations of pure rohitukine by comparing it with untreated control plants. Plants treated with rohitukine showed a dose-dependent decrease in chlorophyll content. With 0.25 mM, 0.5 mM and 1.0 mM rohitukine concentrations, total chlorophyll content decreased by 1.8, 1.7 and 1.6 mg/g fresh weight (FW), respectively, as compared to the untreated control plants having 2.5 mg/gFW (Figure 4g). Actual PSII efficiency (Φ PSII), maximum PSII efficiency (Fv/Fm), intrinsic PSII efficiency, photochemical quenching (qP) and electron transport rate (ETR) decreased by 6.9%, 4.81%, 9.21% and 5.19%, respectively, under 0.25 mM rohitukine. On the other hand, non-photochemical quenching (NPQ) increased by 26.4% compared to the control. Application of 0.5 mM rohitukine decreased Φ PSII, Fv/Fm, Intrinsic PSII efficiency, qP and ETR by 9.7%, 6.02%, 10.52% and 7.79%, respectively, and non-photochemical quenching increased by 32% when compared to the control. Plants that received 1.0 mM rohitukine exhibited maximally decreased Φ PSII, Fv/Fm, Intrinsic PSII efficiency, qP and ETR, by 25%, 21.68%, 18.42% and 18.18%, respectively, and a 54% increase in NPQ as compared to the control (Figure 4).

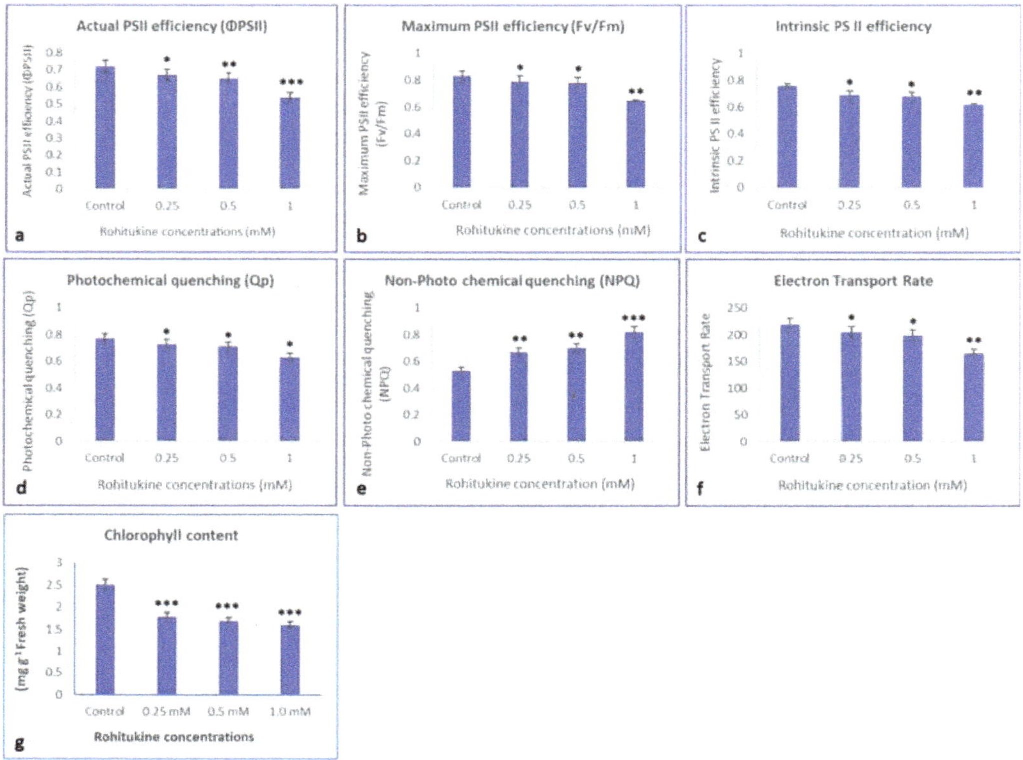

Figure 4. Effects of 0.25 mM, 0.5 mM and 1.0 mM rohitukine on (**a**) actual PSII efficiency, (**b**) maximum PSII efficiency, (**c**) intrinsic PSII efficiency, (**d**) photochemical quenching, (**e**) non-photo chemical quenching, (**f**) electron transport rate and (**g**) total chlorophyll content in five-week-old *A. thaliana* plants treated with rohitukine. Data are presented for the treatments as means ± SDs (n = 3). Statistical significance was determined using Dunnett's multiple comparisons test. Asterisks *, ** and *** denote significance level at p-values < 0.5, 0.05 and 0.01, respectively.

3.5. Accumulation of O_2^- and H_2O_2 in A. thaliana Plants in Response to Rohitukine

Increased levels of O_2^- were noticed in *A. thaliana* plants as dark blue-coloured spots scattered on the leaves treated with rohitukine as compared to the untreated control. Similarly, the *A. thaliana* plants treated with pure rohitukine showed a significant increase in brown-coloured spots when immersed in DAB staining dye. These scattered brown spots were indications of increased levels of H_2O_2 in rohitukine-treated leaves as compared to untreated samples (Figure 5). Asin animal cells, rohitukine induces ROS in plant tissues, with maximum ROS observed for the1.0 mM treatment.

3.6. Effect of Rohitukine on Antioxidant Enzyme Activity

The activity of APX, SOD and POD increased by 2.9%, 2.8% and 6.6%, respectively, with the 0.25 mM rohitukine treatment, whereas with the0.5 mM treatment, the activity of APX, SOD and POD increased by 4.47%, 4.34%, and 17.7%, respectively. The application of the 1.0 mM rohitukine concentration maximally increased APX activity by 10.44%, SOD by 13.04% and POD by 31.1% as compared to the untreated control (Figure 6).

Figure 5. The effects of rohitukine concentrations on T0 (control), T1 (0.25 mM), T2 (0.5 mM) and T3 (1.0 mM) as detected through histochemical detection of ROS in *A. thaliana* leaves. Superoxide ions were detected with NBT staining dye; the H_2O_2 visualization in leaves was performed via DAB staining.

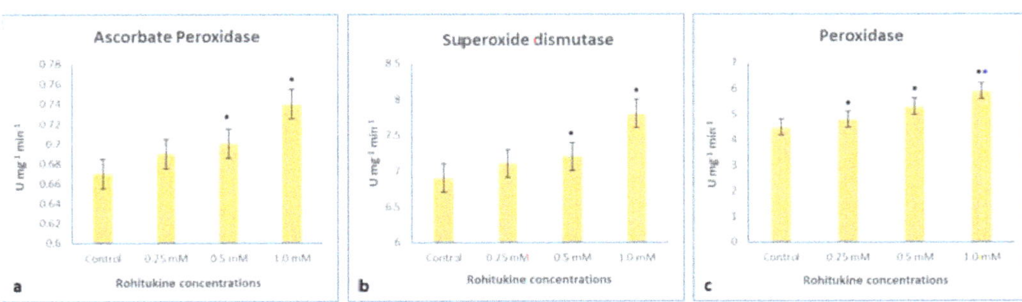

Figure 6. Effects of 0.25 mM, 0.5 mM and 1.0 mM rohitukine on (**a**) APX, (**b**) SOD and (**c**) POD of five-week-old *A. thaliana* plants treated with rohitukine. Data are presented for the treatments as means ± SDs (n = 3). Statistical significance was determined by Dunnett's multiple comparisons test. Asterisks * and ** denote significance level at p-values < 0.5 and 0.05, respectively.

3.7. Effect of Rohitukine on the Expression of Key Genes Involved in the Antioxidant System

The expression patterns of key genes involved in the antioxidant system, such as manganese Mn-SOD, Cu/Zn-SOD, APX and POD, were analysed at transcript level in rohitukine-treated *A. thaliana* samples. The significantly increased expression of these genes was noticed in treated samples as compared to untreated controls. The Mn-SOD gene was found to have a two-fold increased expression in the 0.25 mM and 0.5 mM rohitukine treatments, while in the 1.0 mM treatment we noticed around a three-fold expression of Mn-SOD when compared to controls. Similarly, APX gene transcripts were found to have six-fold upregulated expression in the 0.5 mM and 1.0 mM treatments with rohitukine. In the treatment with the 0.25 mM concentration, the APX was found to be upregulated by 3.5-fold as compared to the control. In the treatment with the 1.0 mM rohitukine concentration, significantly higher transcripts of Cu/Zn-SOD were noticed in *A. thaliana* as compared to the control, whereas with the 0.25 and 0.5 mM rohitukine concentrations, the expression of Cu/Zn-SOD was found to be higher as compared to the control. The expression of the POD gene was also upregulated by rohitukine. The 0.5 mM and 1.0 mM concentrations of rohitukine increased the expression of POD by 2.5 and 4.5-fold, respectively (Figure 7). In all the treatments, overall expression profiles of Mn-SOD, APX, Cu/Zn-SOD and POD were increased when compared to controls. However, plants exposed to 1.0 mM of rohitukine showed the highest transcript levels for all the genes examined.

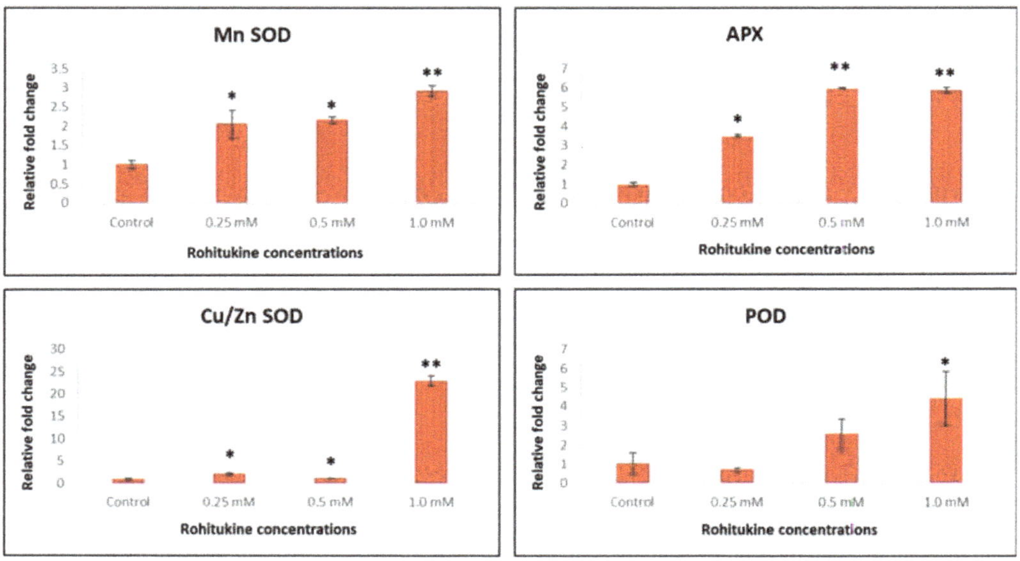

Figure 7. qRT-PCR-based expression analysis of key genes of the antioxidant system of *A. thaliana*. The bar diagrams represent the fold change expression of each gene upon 0.25 mM, 0.5 mM and 1 mM rohitukine exposure as compared to untreated controls. The results are presented as the means of three replicates and as means ± SDs. Statistical significance was determined by Student's *t*-test. Asterisks * and ** denote the significance of fold changes at *p*-values < 0.05 and 0.005, respectively, as compared to untreated controls.

3.8. Metabolite Profiling

With GC–MS analysis of rohitukine-treated *A. thaliana* leaf samples after derivatization and comparison of identified spectra with the NIST17 library, a total of 75, 73, 70 and 71 metabolites were identified with their known structures in the control, 0.25 mM, 0.5 mM and 1.0 mM rohitukine-treated samples, respectively. The metabolites identified were fatty acids, sugars, amino acids, organic acids, polyamines, carbohydrates, etc. To obtain the robust metabolome datasets, only those metabolites which were found to be present in at least three replicates were considered, and their presence in all four samples was ensured before processing. A total number of 37 metabolites were found to be common to all four samples (0.0 mM, 0.25 mM, 0.5 mM and 1.0 mM) and in each replicate. To understand the differences between samples and the similarity between replicates, and to determine the variables that contributed most to these differences, principal component analysis (PCA) was carried out for 37 metabolites, where PCA1 showed 41.8% and PCA2 showed 24.9% of the variation. The accumulation patterns of metabolites in the control as well as in all three treatments are shown by a heatmap generated from 37 metabolites present in all the samples (Figure 8). In all four major clusters, distinct patterns of altered levels of metabolite were reported. Therefore, the patterns of metabolite abundance and clustering indicate the metabolic changes caused by the exposure to rohitukine in *A. thaliana* leaves.

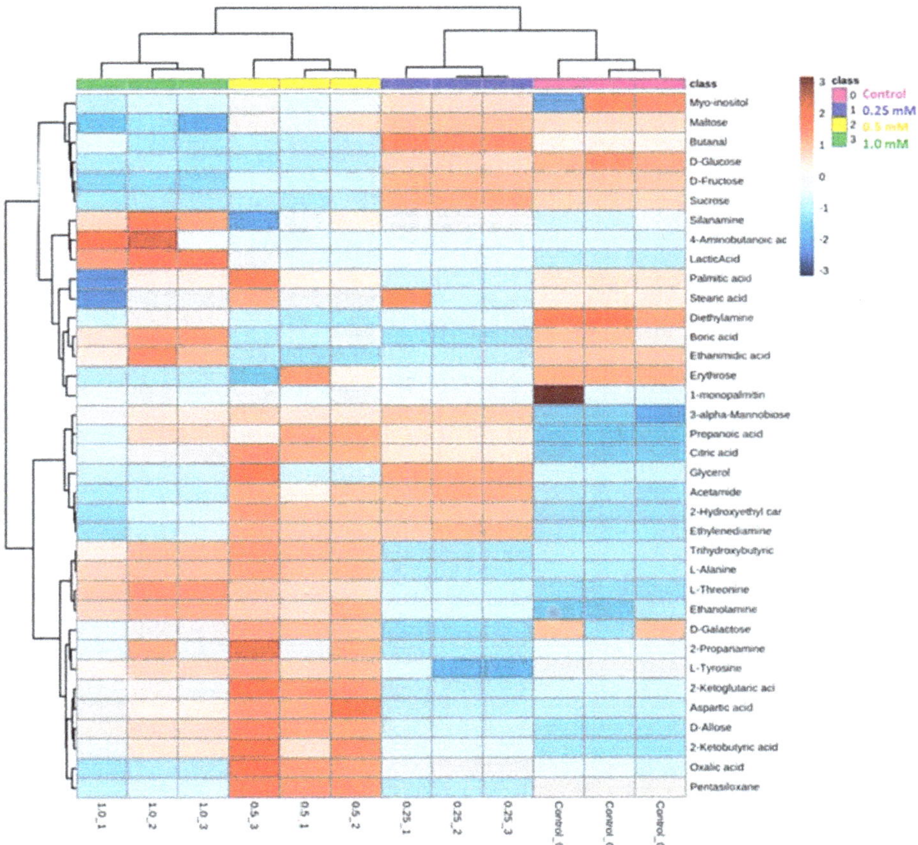

Figure 8. Heatmap illustration of quantities of commonly found metabolites in *A. thaliana* after 0.25 mM, 0.5 mM and 1.0 mM rohitukine treatments as compared to controls. The colour intensities of each box represent the level of each metabolite in each rohitukine-treated group.

Among sugars, the most frequent members found in all the four samples were: erythrose, fructose, galactose, glucose, maltose and sucrose. Dose-dependent decreased levels of fructose, glucose, maltose, sucrose and galactose were noticed in rohitukine-treated samples as compared to controls. Amino acids, L-alanine, aspartic acid, L-threonine and L-tyrosine were found with significantly higher concentrations in rohitukine-treated *A. thaliana* plants when compared to controls (Figure 8). Some other metabolites which were found to be significantly higher after rohitukine exposure were 3-alpha-Mannobiose, ethanolamine and silanamine. Meanwhile, the accumulation of several metabolites was found to be significantly lower after rohitukine exposure, including glycerol, Myo-inositol, acetamide, diethylamine, pentasiloxane, etc. Moreover, we also noticed changes in the levels of several organic acids in rohitukine-treated samples; the upregulated organic acids included: lactic acid, trihydroxy butyric acid and 4-amino butanoic acid. Conversely, several organic acid metabolites, such as palmitic acid, boric acid, stearic acid and oxalic acid, were found to have significantly lower concentrations in the 0.5 mM and 1.0 mM rohitukine treatments.

4. Discussion

Although the biological activities of rohitukine in mammalian as well as in yeast strain cells have been deeply studied, the significance of this molecule in plant systems has not yet been elucidated. It is for the first time that we have tried to understand the physiological and biochemical impacts of rohitukine inside the *A. thaliana* model system. However, the rohitukine biosynthesis pathway in parent plants has not been elucidated yet. As far as the accumulation of rohitukine in source plant *D. gotadhora* is concerned, it was reported to accumulate in leaves, bark, fruits, seeds and twigs, but the highest concentration of rohitukine was reported in seeds (2.42%), followed by leaves (1.06%) [15,54]. Later, Kumar et al. (2016) [10] introduced the chromatography-free protocol of rohitukine isolation, in which they extracted 98% pure compound with a 1% (dry weight) yield. In the present study, we isolated 1.6 g of rohitukine from 200 g dry leaves of *D. gotadhora* with 98% purity (Figure 1). The pure compound of rohitukine was extracted from *D. gotadhora* (source plant) and applied to *A. thaliana*, where rohitukine interfered with plant growth and development. It is possible that rohitukine may leach out from the leaves of the parent plant during rain and it could also be found in soil samples as an allelochemical, which may also be due to the decomposition of plant tissues in soil. The leaching of toxic alkaloids from plant tissues into soil and drainage water has also been evidenced by previous studies [55,56].

For screening of the best inhibitory/modulatory concentration of rohitukine, we applied a range of rohitukine concentrations (0.01 mM to 10 mM) on *A. thaliana* seedlings. Among the concentrations screened so far, 1.0 mM rohitukine maximally inhibited the growth of *A. thaliana*, while 0.25 mM and 0.5 mM concentrations moderately affected growth as scored visually. Therefore, we applied 0.25 mM, 0.5 mM and 1.0 mM concentrations of rohitukine to analyse the dose-dependent effects of the molecule on *A. thaliana*. In previous studies, the dose-dependent growth inhibitory effects of extracts from an important medicinal plant *Hyptissuaveolens* were examined in several plant species [57].The stability and uptake of the compound were also assessed through HPLC, and rohitukine was found to be stable inside the plant tissues and uptake was also confirmed (Figure 3) [10,43]. The question related to the stability of rohitukine at various physiological pH levels is very important for its phytotoxic efficiency. In already published reports, the stability of rohitukine has been assessed by incubating the pure compound of rohitukine in buffers of different pH (1.2, 4.0, 6.8 and 7.4), bio-relevant fluids, such as SGF (pH 1.2) and simulated intestinal fluid (SIF) pH 6.8, and also in rat plasma. Rohitukine was found to be stable in all the tested conditions [10,43].

It has been reported that rohitukine is highly stable in diverse biological fluids. Moreover, it is found to be in the category of high-permeability molecules (log $p_{app} > -5$) based on Parallel Artificial Membrane Permeability Assay (PAMPA) data, which confirm passive transcellular permeation [52]. Computational analysis of rohitukine suggests that it is a substrate of P-gp, which is the key member of the ABC transporter system [45]. In addition, stress situations are reported to regulate transporter expression in the plant system [58]. The above-mentioned evidence in the literature indicates the possible uptake of rohitukine in biological systems. However, further investigations have to be performed to explore the exact mechanism of rohitukine uptake in *A. thaliana* [59]. Additionally, the plasma protein binding efficiency of rohitukine and its semisynthetic derivatives have also been reported [60]. However, in plants, the exact mechanism of membrane transport of rohitukine is unclear. An attempt has been made, using isolated epidermal cells, to understand the mechanism of alkaloid transport through plasma membranes, suggesting the involvement of transporter proteins [61].

The negative effects of plant secondary metabolites, such as L-mimosine, syringaldehyde, juglone, and vanillin, have already been assessed for *A. thaliana* and other crop species [62–64]. In the present study, rohitukine exposure significantly decreased total leaf area and plant biomass. Moreover, a moderate decrease in photosynthesis was observed with 0.25 and 0.5 mM rohitukine concentrations, while photosynthesis decreased maximally in plants that received 1.0 mM of rohitukine when compared to the controls. Similarly, a gradual decrease in total chlorophyll content was observed with increased rohitukine concentration. Similar results have been shown by Hussain and Reigosa (2021) [65], who investigated the influence of two plant secondary metabolites, ferulic acid and p-hydroxybenzoic acid, on the photosynthesis of *Rumex acetosa*, where both the molecules inhibited photosynthetic parameters, such as Fv/Fm, Φ PSII, qP and NPQ. Rutin is another secondary metabolite that is reported to inhibit Fv/Fm and the concentrations of chlorophyll pigments in *A. thaliana* [66]. At increased rohitukine concentrations, we noticed a decrease in ETR, Φ PSII, Fv/Fm, qP and intrinsic PSII efficiency in *A. thaliana*. Consequently, increased NPQ was observed upon rohitukine exposure.

Generally, ROS are produced inside living cells when they encounter any external stress and activate the antioxidant defence system to overcome the such caused by oxidative stress [67]. In plants, ROS are present in $\bullet O_2^-$ ionic states, such as hydroxyl radicals ($\bullet OH$), and molecular states, including H_2O_2 and singlet oxygen ($\bullet O_2$) [68,69]. $\bullet O_2^-$ is reported to increase during external stress and is the precursor of various ROS. The excessive generation of $\bullet O_2^-$ causes an increment in ROS that leads to cell death [70,71]. Rohitukine exposure affected various physiological, biochemical and molecular mechanisms via the excessive production of ROS in *A. thaliana*. In rohitukine-treated *A. thaliana* plants, we detected the accumulation of $\bullet O_2^-$ and H_2O_2 as blue- and dark brown-coloured spots, respectively, through the histochemical staining of leaves. The already published literature suggests that rohitukine induces ROS in yeast strains after 24 h of treatment [16]. Moreover, in cell lines, rohitukine and its semisynthetic derivatives have been reported to induce ROS-mediated apoptosis [21,72,73]. Similarly, two other alkaloids, Graveoline and vitrine, isolated from *Ruta graveolens* and *Evodilitoris*, with immunomodulatory, anti-inflammatory and anticancerous activities are reported to generate ROS in the root coleoptile of wheat [74,75]. We also noted the rohitukine-mediated induced expression of APX, SOD and POD, which are involved in the antioxidant defence system in plants. These genes are exclusively associated with ROS metabolism to combat the stress response. Therefore, it is likely that rohitukine may trigger cell death in plants via ROS generation and affect hormonal transport. Under all defined concentrations of rohitukine, there was a significant increment in APX, SOD and POD enzyme contents in *A. thaliana*. Khan et al. (2011) [76] investigated the SOD-, CAT- and POD-mediated growth inhibitory effects of aqueous and ethanol extracts of *Peganum multisectum* on ryegrass.

The reduction in maximum quantum yield of PSII shows that excitation energy trapping of PSII reaction centers was reduced. A decrease in photosynthesis was suggested to be due to stomatal closure [65]. Reduction in CO_2 passage due to stomatal closure is responsible for the accumulation of ROS, the degradation of xanthophyll pigments and lipids and protein oxidation [77]. The slight decrease in PSII activity observed was due to more ROS accumulation, which increases oxygen production at 1 mM rohitukine. Moreover, the slightly higher ETRs recorded for treatments with 0.25 mM and 0 5 mM of rohitukine compared to the 1 mM treatment were due to less ROS accumulation, as revealed by DAB and NBT staining, which limits ROS production. Our results also show that rohitukine can block the electron acceptor to inhibit photosystem II. These results support the hypothesis that there was a reduction in photosystem II photochemistry and photosynthetic electron transport, which is responsible for ROS accumulation in *A. thaliana*. Similarly, secondary metabolites isolated from several endophytes inhibit PSII electron transport on the water-splitting enzyme and on the acceptor side between P680 and QA. The results of this study were confirmed by chlorophyll-a fluorescence measurements [78]. The reduction in chlorophyll fluorescence under metal stress shows HM antenna pigment disruption due to the hindrance of electron transport flow from PSII to PSI [79–81]. To address the phytotoxic effects of secondary metabolites, many studies have been published which show that secondary metabolites have impacts mainly through damage to photosynthetic machinery and frequent decomposition of photosynthetic pigments. Consequently, decrease in photosynthetic pigments leads to blockage of energy/electron transfer and inhibition of ATP synthesis [28].

The changes caused by any elicitor at the morphological and transcriptomic levels should be reflected in the final end products of gene and protein expression. The metabolome offers better visualization of the changes in the levels of many different metabolites. In recent years, metabolomics has gained considerable attention as a tool for acquiring better insight into the biological processes of organisms [82–85]. In an attempt to analyse the changes in metabolite levels caused by rohitukine in *A. thaliana*, GC–MS-based metabolomics was performed, in which amino acids, carbohydrates, other organic acids, etc., were identified. Among amino acids, L-alanine, aspartic acid, L-threonine and L-tyrosine were detected at higher concentrations in 0.5 mM and 1.0 mM rohitukine-treated samples. Generally, aromatic amino acids, such as tyrosine, play important roles in the synthesis of a wide range of secondary metabolites when plants encounter any external stress. L-tyrosine and other aromatic amino acids act as precursors for the synthesis of phenylpropanoids, a major group of plant secondary metabolites whose function is to protect the plant from abiotic stresses [86–88]. Many essential amino acids, such as valine, methionine, alanine, and leucine, as well as non-essential amino acids, such as histidine, proline cysteine, etc., have been reported to increase under abiotic and biotic stresses [89–91]. The metabolic analysis also identified dose-dependent decreases in concentrations of some sugars, including fructose, glucose, maltose, sucrose and galactose, in rohitukine-treated *A. thaliana* samples as compared to controls. The maximum decreased level of sugars was noticed with the1.0 mM rohitukine treatment. Disruption to photosynthesis and chlorophyll bleaching by rohitukine exposure in *A. thaliana* plants could be the reason for lesser concentrations of sugars in treated samples. There are several studies that have shown the negative impacts of extracts and pure compounds isolated from different plant species on recipient plants in the form of decreased levels of carbohydrates [92–94]. The concentration of myo-inositol was found to be slightly decreased upon rohitukine exposure. It has been reported that Myo-inositol participates in cellular functions and metabolism in plants [95]. It also mediates ROS-induced cell death in the presence of salicylic acid and ethylene towards stress tolerance [96]. Along with Myo-inositol, several other metabolites, such as acetamide, diethylamine, penta siloxane and glycerol, were found to have decreased levels. Indole-3-acetamide triggers stress responses in *A. thaliana* and participates in the crosstalk of auxin and abscisic acid [97].

5. Conclusions

In conclusion, we found that rohitukine not only perturbs physiological mechanisms in mammalian and yeast cells but also affects the growth parameters of *A. thaliana* by triggering ROS generation and metabolic changes and interfering with photosynthetic machinery. The inhibition of PSII activity combined with the upregulation of antioxidant system genes is most likely the basis of its property as an allelochemical. The activity of rohitukine inside *A. thaliana* tissues represents an example of a mechanism whereby active medicinal compounds exert their influence on multiple targets instead of single sites of action. The relative importance and the effectiveness of rohitukine perturbation inside the model plant *A. thaliana* give us an idea of how similarly plant and animal cells respond to the medicinally important molecule rohitukine. Moreover, the phytotoxic potential of rohitukine may help in maintaining the ecological interactions of the parent plant.

Author Contributions: Methodology, formal analysis, writing—original draft preparation, S.A.; writing—review and editing, data curation, M.A.; HPLC, GC–MS, LC–MS, A.K.; supervision, conceptualization, resources, writing—review and editing, funding acquisition, S.G.G. All authors have read and agreed to the published version of the manuscript.

Funding: Support for this study was provided by the Council of Scientific and Industrial Research, India (MLP110006 and HCP38).

Institutional Review Board Statement: Not applicable.

Informed Consent Statement: Not applicable.

Data Availability Statement: Data are contained within the article.

Acknowledgments: S.A. is thankful to the Council of Scientific and Industrial Research, India, for Senior Research Fellowships. The authors also thank Utpal Nandi for his valuable advice during the editing of the manuscript.

Conflicts of Interest: The authors declare no conflict of interest.

References

1. Akula, R.; Ravishankar, G.A. Influence of abiotic stress signals on secondary metabolites in plants. *Plant Signal. Behav.* **2011**, *6*, 1720–1731. [CrossRef] [PubMed]
2. Jan, R.; Asaf, S.; Numan, M.; Kim, K.-M. Plant secondary metabolite biosynthesis and transcriptional regulation in response to biotic and abiotic stress conditions. *Agronomy* **2021**, *11*, 968. [CrossRef]
3. Aftab, T. A review of medicinal and aromatic plants and their secondary metabolites status under abiotic stress. *J. Med. Plant.* **2019**, *7*, 99–106.
4. Varma, V. Advancements in the production of secondary metabolites. *J. Nat. Prod.* **2010**, *3*, 112–123.
5. Jain, C.; Khatana, S.; Vijayvergia, R. Bioactivity of secondary metabolites of various plants: A review. *Int. J. Pharm. Sci. Res.* **2019**, *10*, 494–504.
6. Kabera, J.N.; Semana, E.; Mussa, A.R.; He, X. Plant secondary metabolites: Biosynthesis, classification, function and pharmacological properties. *J. Pharm. Pharmacol.* **2014**, *2*, 377–392.
7. Buckingham, J.; Baggaley, K.H.; Roberts, A.D.; Szabo, L.F. *Dictionary of Alkaloids with CD-ROM.*; CRC Press: Boca Raton, FL, USA, 2010.
8. Houghton, P.; Hairong, Y. Further chromone alkaloids from *Schumanniophyton magnificum*. *Planta Med.* **1987**, *53*, 262–264. [CrossRef]
9. Harmon, A.D.; Weiss, U.; Silverton, J. The structure of rohitukine, the main alkaloid of Amoora rohituka (syn. *Aphanamixis polystachya*) (Meliaceae). *Tetrahedron Lett.* **1979**, *20*, 721–724. [CrossRef]
10. Kumar, V.; Guru, S.K.; Jain, S.K.; Joshi, P.; Gandhi, S.G.; Bharate, S.B.; Bhushan, S.; Bharate, S.S.; Vishwakarma, R.A. A chromatography-free isolation of rohitukine from leaves of *Dysoxylum binectariferum*: Evaluation for in vitro cytotoxicity, Cdk inhibition and physicochemical properties. *Bioorg. Med. Chem. Lett.* **2016**, *26*, 3457–3463. [CrossRef]
11. Jain, S.K.; Meena, S.; Qazi, A.K.; Hussain, A.; Bhola, S.K.; Kshirsagar, R.; Pari, K.; Khajuria, A.; Hamid, A.; Shaanker, R.U. Isolation and biological evaluation of chromone alkaloid dysoline, a new regioisomer of rohitukine from *Dysoxylum binectariferum*. *Tetrahedron Lett.* **2013**, *54*, 7140–7143. [CrossRef]
12. Kumara, P.M.; Soujanya, K.; Ravikanth, G.; Vasudeva, R.; Ganeshaiah, K.; Shaanker, R.U. Rohitukine, a chromone alkaloid and a precursor of flavopiridol, is produced by endophytic fungi isolated from *Dysoxylum binectariferum* Hook. f and *Amoora rohituka* (Roxb). Wight & Arn. *Phytomedicine* **2014**, *21*, 541–546. [CrossRef] [PubMed]
13. Simpson, M.G. Diversity and classification of flowering plants: Eudicots. *Plant Syst.* **2019**, 285–466. [CrossRef]
14. Arya, D.; Goel, S.; Shinde, P.; Joshi, G.; Sharma, O.R.; Sharma, S. Dysoxylum binacteriferum Hook. F.: A promising herbal drug used in folk medicine by Tharu community of Uttarakhand. *World J. Pharmaceut Res* **2017**, *6*, 296–301. [CrossRef]
15. Mahajan, V.; Sharma, N.; Kumar, S.; Bhardwaj, V.; Ali, A.; Khajuria, R.; Bedi, Y.; Vishwakarma, R.A.; Gandhi, S.G. Production of rohitukine in leaves and seeds of *Dysoxylum binectariferum*: An alternate renewable resource. *Pharm. Biol.* **2015**, *53*, 446–450. [CrossRef]
16. Kamil, M.; Jadiya, P.; Sheikh, S.; Haque, E.; Nazir, A.; Lakshmi, V.; Mir, S.S. The chromone alkaloid, rohitukine, affords anti-cancer activity via modulating apoptosis pathways in A549 cell line and yeast mitogen activated protein kinase (MAPK) pathway. *PLoS ONE* **2015**, *10*, e0137991. [CrossRef]
17. Wang, H. Flavopiridol. National Cancer Institute. *Curr. Opin. Investig. Drugs* **2001**, *2*, 1149–1155.
18. Hallek, M.; Pflug, N. State of the art treatment of chronic lymphocytic leukaemia. *Blood Rev.* **2011**, *25*, 1–9. [CrossRef]
19. Carlson, B.A.; Dubay, M.M.; Sausville, E.A.; Brizuela, L.; Worland, P.J. Flavopiridol induces G1 arrest with inhibition of cyclin-dependent kinase (CDK) 2 and CDK4 in human breast carcinoma cells. *Cancer Res.* **1996**, *56*, 2973–2978.
20. Manohar, S.M.; Joshi, K.S. Promising Anticancer Activity of Multitarget Cyclin Dependent Kinase Inhibitors against Human Colorectal Carcinoma Cells. *Curr. Mol. Pharmacol.* **2022**. [CrossRef]
21. Mintoo, M.; Khan, S.; Wani, A.; Malik, S.; Bhurta, D.; Bharate, S.; Malik, F.; Mondhe, D. A rohitukine derivative IIIM-290 induces p53 dependent mitochondrial apoptosis in acute lymphoblastic leukemia cells. *Mol. Carcinog.* **2021**, *60*, 671–683. [CrossRef]

22. Bharate, S.B.; Kumar, V.; Jain, S.K.; Mintoo, M.J.; Guru, S.K.; Nuthakki, V.K.; Sharma, M.; Bharate, S.S.; Gandhi, S.G.; Mondhe, D.M. Discovery and preclinical development of IIIM-290, an orally active potent cyclin-dependent kinase inhibitor. *J. Med. Chem.* **2018**, *61*, 1664–1687. [CrossRef] [PubMed]
23. Inderjit; Nilsen, E.T. Bioassays and field studies for allelopathy in terrestrial plants: Progress and problems. *Crit. Rev. Plant Sci.* **2003**, *22*, 221–238. [CrossRef]
24. Callaway, R.M.; Aschehoug, E.T. Invasive plants versus their new and old neighbors: A mechanism for exotic invasion. *Science* **2000**, *290*, 521–523. [CrossRef] [PubMed]
25. Hussein, R.A.; El-Anssary, A.A. Plants secondary metabolites: The key drivers of the pharmacological actions of medicinal plants. In *Herbal Medicine*; Builders, P.F., Ed.; IntechOpen: London, UK, 2019; Volume 1, p. 13.
26. Velu, G.; Palanichamy, V.; Rajan, A.P. Phytochemical and pharmacological importance of plant secondary metabolites in modern medicine. In *Bioorganic Phase in Natural Food: An Overview*; Springer: Berlin/Heidelberg, Germany, 2018; pp. 135–156. [CrossRef]
27. Yu, J.Q.; Ye, S.F.; Zhang, M.F.; Hu, W.H. Effects of root exudates and aqueous root extracts of cucumber (*Cucumis sativus*) and allelochemicals, on photosynthesis and antioxidant enzymes in cucumber. *Biochem. Syst. Ecol.* **2003**, *31*, 129–139. [CrossRef]
28. Fengzhi, W.; Kai, P.; Fengming, M.; Xuedong, W. Effects of Cinnamic Acid on Photosynthesis and Cell Ultrastructure ofCucumber Seedling. *Acta Hortic. Sin.* **2004**, *31*, 183.
29. Meazza, G.; Scheffler, B.E.; Tellez, M.R.; Rimando, A.M.; Romagni, J.G.; Duke, S.O.; Nanayakkara, D.; Khan, I.A.; Abourashed, E.A.; Dayan, F.E. The inhibitory activity of natural products on plant p-hydroxyphenylpyruvate dioxygenase. *Phytochemistry* **2002**, *60*, 281–288. [CrossRef]
30. Sampaio, O.M.; Vieira, L.C.C.; Bellete, B.S.; King-Diaz, B.; Lotina-Hennsen, B.; Da Silva, M.F.d.G.F.; Veiga, T.A.M. Evaluation of alkaloids isolated from *Ruta graveolens* as photosynthesis inhibitors. *Molecules* **2018**, *23*, 2693. [CrossRef]
31. Sampaio, O.M.; de Castro Lima, M.M.; Veiga, T.A.M.; King-Díaz, B.; Lotina-Hennsen, B. Evaluation of antidesmone alkaloid as a photosynthesis inhibitor. *Pestic. Biochem. Physiol.* **2016**, *134*, 55–62. [CrossRef]
32. Hazrati, H.; Saharkhiz, M.J.; Niakousari, M.; Moein, M. Natural herbicide activity of Satureja hortensis L. essential oil nanoemulsion on the seed germination and morphophysiological features of two important weed species. *Ecotoxicol. Environ. Saf.* **2017**, *142*, 423–430. [CrossRef]
33. Weir, T.L.; Park, S.-W.; Vivanco, J.M. Biochemical and physiological mechanisms mediated by allelochemicals. *Curr. Opin. Plant Biol.* **2004**, *7*, 472–479. [CrossRef]
34. Wang, C.-M.; Chen, H.-T.; Li, T.-C.; Weng, J.-H.; Jhan, Y.-L.; Lin, S.-X.; Chou, C.-H. The role of pentacyclic triterpenoids in the allelopathic effects of Alstonia scholaris. *J. Chem. Ecol.* **2014**, *40*, 90–98. [CrossRef] [PubMed]
35. Das, K.; Roychoudhury, A. Reactive oxygen species (ROS) and response of antioxidants as ROS-scavengers during environmental stress in plants. *Front. Environ. Sci.* **2014**, *2*, 53. [CrossRef]
36. Triantaphylides, C.; Krischke, M.; Hoeberichts, F.A.; Ksas, B.; Gresser, G.; Havaux, M.; Van Breusegem, F.; Mueller, M.J. Singlet oxygen is the major reactive oxygen species involved in photooxidative damage to plants. *Plant Physiol.* **2008**, *148*, 960–968. [CrossRef] [PubMed]
37. Sharma, P.; Jha, A.B.; Dubey, R.S.; Pessarakli, M. Reactive oxygen species, oxidative damage, and antioxidative defense mechanism in plants under stressful conditions. *J. Bot.* **2012**, *2012*, 217037. [CrossRef]
38. Bais, H.P.; Vepachedu, R.; Gilroy, S.; Callaway, R.M.; Vivanco, J.M. Allelopathy and exotic plant invasion: From molecules and genes to species interactions. *Science* **2003**, *301*, 1377–1380. [CrossRef]
39. Ding, J.; Sun, Y.; Xiao, C.L.; Shi, K.; Zhou, Y.H.; Yu, J.Q. Physiological basis of different allelopathic reactions of cucumber and figleaf gourd plants to cinnamic acid. *J. Exp. Bot.* **2007**, *58*, 3765–3773. [CrossRef]
40. Zeng, R.S.; Luo, S.M.; Shi, Y.H.; Shi, M.B.; Tu, C.Y. Physiological and biochemical mechanism of allelopathy of secalonic acid F on higher plants. *Agron. J.* **2001**, *93*, 72–79. [CrossRef]
41. Zuo, S.; Ma, Y.; Ye, L. In vitro assessment of allelopathic effects of wheat on potato. *Allelopath. J* **2012**, *30*, 1–10.
42. Bhattacharya, T.; Dutta, S.; Akter, R.; Rahman, M.; Karthika, C.; Nagaswarupa, H.P.; Murthy, H.C.A.; Fratila, O.; Brata, R.; Bungau, S. Role of Phytonutrients in Nutrigenetics and Nutrigenomic Perspective in Curing Breast Cancer. *Biomolecules* **2021**, *11*, 1176. [CrossRef]
43. Kumar, V.; Bharate, S.S.; Vishwakarma, R.A. Modulating lipophilicity of rohitukine via prodrug approach: Preparation, characterization, and in vitro enzymatic hydrolysis in biorelevant media. *Eur. J. Pharm. Sci.* **2016**, *92*, 203–211. [CrossRef]
44. Lindsey III, B.E.; Rivero, L.; Calhoun, C.S.; Grotewold, E.; Brkljacic, J. Standardized method for high-throughput sterilization of Arabidopsis seeds. *J. Vis. Exp.* **2017**, *128*, e56587. [CrossRef] [PubMed]
45. Chhonker, Y.S.; Chandasana, H.; Kumar, A.; Kumar, D.; Laxman, T.S.; Mishra, S.K.; Balaramnavar, V.M.; Srivastava, S.; Saxena, A.K.; Bhatta, R.S. Pharmacokinetics, tissue distribution and plasma protein binding studies of rohitukine: A potent anti-hyperlipidemic agent. *Drug Res.* **2015**, *65*, 380–387. [CrossRef] [PubMed]
46. Easlon, H.M.; Bloom, A.J. Easy Leaf Area: Automated digital image analysis for rapid and accurate measurement of leaf area. *Appl. Plant Sci.* **2014**, *2*, 1400033. [CrossRef] [PubMed]
47. Krall, J.P.; Edwards, G.E. Relationship between photosystem II activity and CO_2 fixation in leaves. *Physiol. Plant.* **1992**, *86*, 180–187. [CrossRef]
48. Arnon, D.I. Copper enzymes in isolated chloroplasts. Polyphenoloxidase in Beta vulgaris. *Plant Physiol.* **1949**, *24*, 1. [CrossRef] [PubMed]

49. Shi, J.; Fu, X.-Z.; Peng, T.; Huang, X.-S.; Fan, Q.-J.; Liu, J.-H. Spermine pretreatment confers dehydration tolerance of citrus in vitro plants via modulation of antioxidative capacity and stomatal response. *Tree Physiol.* **2010**, *30*, 914–922. [CrossRef]
50. Asgher, M.; Khan, N.A.; Khan, M.I.R.; Fatma, M.; Masood, A. Ethylene production is associated with alleviation of cadmium-induced oxidative stress by sulfur in mustard types differing in ethylene sensitivity. *Ecotoxicol. Environ. Saf.* **2014**, *106*, 54–61. [CrossRef]
51. Livak, K.J.; Schmittgen, T.D. Analysis of relative gene expression data using real-time quantitative PCR and the $2^{-\Delta\Delta CT}$ method. *Methods* **2001**, *25*, 402–408. [CrossRef]
52. Lisec, J.; Schauer, N.; Kopka, J.; Willmitzer, L.; Fernie, A.R. Gas chromatography mass spectrometry–based metabolite profiling in plants. *Nat. Prot.* **2006**, *1*, 387–396. [CrossRef]
53. Pang, Z.; Chong, J.; Zhou, G.; de Lima Morais, D.A.; Chang, L.; Barrette, M.; Gauthier, C.; Jacques, P.-É.; Li, S.; Xia, J. MetaboAnalyst 5.0: Narrowing the gap between raw spectra and functional insights. *Nucleic Acids Res.* **2021**, *49*, W388–W396. [CrossRef]
54. Kumara, P.M.; Srimany, A.; Ravikanth, G.; Shaanker, R.U.; Pradeep, T. Ambient ionization mass spectrometry imaging of rohitukine, a chromone anti-cancer alkaloid, during seed development in *Dysoxylum binectariferum* Hook. f (Meliaceae). *Phytochemistry* **2015**, *116*, 104–110. [CrossRef] [PubMed]
55. Hama, J.R.; Jorgensen, D.B.G.; Diamantopoulos, E.; Bucheli, T.D.; Hansen, H.C.B.; Strobel, B.W. Indole and quinolizidine alkaloids from blue lupin leach to agricultural drainage water. *Sci. Total Environ.* **2022**, *834*, 155283. [CrossRef] [PubMed]
56. Hama, J.R.; Strobel, B.W. Natural alkaloids from narrow-leaf and yellow lupins transfer to soil and soil solution in agricultural fields. *Environ. Sci. Eur.* **2020**, *32*, 126. [CrossRef]
57. Islam, A.M.; Kato-Noguchi, H. Plant growth inhibitory activity of medicinal plant Hyptis suaveolens: Could allelopathy be a cause? *Emir. J. Food Agric.* **2013**, *25*, 692–701. [CrossRef]
58. Isah, T. Stress and defense responses in plant secondary metabolites production. *Biol. Res.* **2019**, *52*, 39. [CrossRef] [PubMed]
59. Singh, A.; Wazir, P.; Chibber, P.; Kapoor, N.; Gautam, A. Re-Validation of New Develop Highly Sensitive, Simple LCMS/MS Method for the Estimation of Rohitukine and its Application in ADME/Pre-Clinical Pharmacokinetics. *Mass Spectrom. Purif. Tech.* **2017**, *3*, 2. [CrossRef]
60. Khalipha, A.B.R.; Bagchi, R.; Hossain, M.S.; Mondal, M.; Biswas, S.; Ray, P.; Smrity, S.Z.; Asha, U.H. A Literature Based Review on Rohitukine and Molecular Docking Studies of Flavopiridol (Rohitukine Derived) and Its Derivatives against Cyclin-Dependent Kinases (CDKs) for Anticancer Activity. *Int. J. Sci. Res.* **2019**, *01*, 15–28.
61. Shitan, N.; Yazaki, K. Accumulation and membrane transport of plant alkaloids. *Curr. Pharm. Biotechnol.* **2007**, *8*, 244–252. [CrossRef]
62. Reigosa, M.; Pazos-Malvido, E. Phytotoxic effects of 21 plant secondary metabolites on *Arabidopsis thaliana* germination and root growth. *J. Chem. Ecol.* **2007**, *33*, 1456–1466. [CrossRef]
63. Jose, S.; Gillespie, A.R. Allelopathy in black walnut (*Juglans nigra* L.) alley cropping. II. Effects of juglone on hydroponically grown corn (*Zea maysl.*) and soybean (*Glycine maxl. Merr.*) growth and physiology. *Plant Soil* **1998**, *203*, 199–206. [CrossRef]
64. Janovicek, K.J.; Vyn, T.J.; Voroney, R.P. No-till corn response to crop rotation and in-row residue placement. *Agron. J.* **1997**, *89*, 588–596. [CrossRef]
65. Hussain, M.I.; Reigosa, M.J. Secondary metabolites, ferulic acid and p-hydroxybenzoic acid induced toxic effects on photosynthetic process in *Rumex acetosa* L. *Biomolecules* **2021**, *11*, 233. [CrossRef] [PubMed]
66. Hussain, M.I.; Reigosa, M.J. Plant secondary metabolite rutin affects the photosynthesis and excitation energy flux responses in *Arabidopsis thaliana*. *Allelopath. J.* **2016**, *38*, 215–228.
67. Yang, Z.; Deng, C.; Wu, Y.; Dai, Z.; Tang, Q.; Cheng, C.; Xu, Y.; Hu, R.; Liu, C.; Chen, X. Insights into the mechanism of multi-walled carbon nanotubes phytotoxicity in Arabidopsis through transcriptome and m6A methylome analysis. *Sci. Total Environ.* **2021**, *787*, 147510. [CrossRef] [PubMed]
68. Blokhina, O.; Virolainen, E.; Fagerstedt, K.V. Antioxidants, oxidative damage and oxygen deprivation stress: A review. *Annals Bot.* **2003**, *91*, 179–194. [CrossRef]
69. Apel, K.; Hirt, H. Reactive oxygen species: Metabolism, oxidative stress, and signal transduction. *Annu. Rev. Plant Biol.* **2004**, *55*, 373–399. [CrossRef]
70. Gill, S.S.; Tuteja, N. Reactive oxygen species and antioxidant machinery in abiotic stress tolerance in crop plants. *Plant Physiol. Biochem.* **2010**, *48*, 909–930. [CrossRef]
71. Zeng, J.; Dong, Z.; Wu, H.; Tian, Z.; Zhao, Z. Redox regulation of plant stem cell fate. *EMBO J.* **2017**, *36*, 2844–2855. [CrossRef]
72. Singh, A.; Chibber, P.; Kolimi, P.; Malik, T.A.; Kapoor, N.; Kumar, A.; Kumar, N.; Gandhi, S.G.; Singh, S.; Abdullah, S.T. Rohitukine inhibits NF-κB activation induced by LPS and other inflammatory agents. *Int. Immunopharmacol.* **2019**, *69*, 34–49. [CrossRef]
73. Mishra, S.K.; Tiwari, S.; Shrivastava, S.; Sonkar, R.; Mishra, V.; Nigam, S.K.; Saxena, A.K.; Bhatia, G.; Mir, S.S. Pharmacological evaluation of the efficacy of *Dysoxylum binectariferum* stem bark and its active constituent rohitukine in regulation of dyslipidemia in rats. *J. Nat. Med.* **2018**, *72*, 837–845. [CrossRef]
74. Ratnayake, W.; Suresh, T.; Abeysekera, A.; Salim, N.; Chandrika, U. Acute anti-inflammatory and anti-nociceptive activities of crude extracts, alkaloid fraction and evolitrine from Acronychia pedunculata leaves. *J. Ethnopharmacol.* **2019**, *238*, 111827. [CrossRef] [PubMed]

75. Ghosh, S.; Bishayee, K.; Khuda-Bukhsh, A.R. Graveoline isolated from ethanolic extract of Ruta graveolens triggers apoptosis and autophagy in skin melanoma cells: A novel apoptosis-independent autophagic signaling pathway. *Phytother. Res.* **2014**, *28*, 1153–1162. [CrossRef] [PubMed]
76. Khan, A.M.; Qureshi, R.A.; Ullah, F.; Gilani, S.A. Phytotoxic effects of selected medicinal plants collected from Margalla Hills, Islamabad Pakistan. *J. Med. Plant Res.* **2011**, *5*, 4671–4675.
77. Souza, R.; Machado, E.; Silva, J.; Lagôa, A.; Silveira, J. Photosynthetic gas exchange, chlorophyll fluorescence and some associated metabolic changes in cowpea (*Vigna unguiculata*) during water stress and recovery. *Environ. Exp. Bot.* **2004**, *51*, 45–56. [CrossRef]
78. Sánchez-Ortiz, B.; Sánchez-Fernández, R.; Duarte, G.; Lappe-Oliveras, P.; Macías-Rubalcava, M. Antifungal, anti-oomycete and phytotoxic effects of volatile organic compounds from the endophytic fungus *Xylaria* sp. strain PB 3f3 isolated from *Haematoxylon brasiletto*. *J. Appl. Microbiol.* **2016**, *120*, 1313–1325. [CrossRef]
79. Liu, W.; Shang, S.; Feng, X.; Zhang, G.; Wu, F. Modulation of exogenous selenium in cadmium-induced changes in antioxidative metabolism, cadmium uptake, and photosynthetic performance in the 2 tobacco genotypes differing in cadmium tolerance. *Environ. Toxicol. Chem.* **2015**, *34*, 92–99. [CrossRef]
80. Parmar, P.; Kumari, N.; Sharma, V. Structural and functional alterations in photosynthetic apparatus of plants under cadmium stress. *Bot. Stud.* **2013**, *54*, 45. [CrossRef]
81. Asgher, M.; Per, T.S.; Verma, S.; Pandith, S.A.; Masood, A.; Khan, N.A. Ethylene supplementation increases PSII efficiency and alleviates chromium-inhibited photosynthesis through increased nitrogen and sulfur assimilation in mustard. *J. Plant Growth Regul.* **2018**, *37*, 1300–1317. [CrossRef]
82. Dao, T.T.H.; Puig, R.C.; Kim, H.K.; Erkelens, C.; Lefeber, A.W.; Linthorst, H.J.; Choi, Y.H.; Verpoorte, R. Effect of benzothiadiazole on the metabolome of *Arabidopsis thaliana*. *Plant Physiol. Biochem.* **2009**, *47*, 146–152. [CrossRef]
83. Beale, M.H.; Sussman, M.R. Metabolomics of *Arabidopsis thaliana*. *Ann. Plant Rev. Online* **2018**, *43*, 157–180. [CrossRef]
84. Rowe, H.C.; Hansen, B.G.; Halkier, B.A.; Kliebenstein, D.J. Biochemical networks and epistasis shape the *Arabidopsis thaliana* metabolome. *Plant Cell* **2008**, *20*, 1199–1216. [CrossRef] [PubMed]
85. Schmidt, H.; Günther, C.; Weber, M.; Spörlein, C.; Loscher, S.; Böttcher, C.; Schobert, R.; Clemens, S. Metabolome analysis of *Arabidopsis thaliana* roots identifies a key metabolic pathway for iron acquisition. *PLoS ONE* **2014**, *9*, e102444. [CrossRef]
86. Dixon, R.; Paiva, N. Stress-induced phenylpropanoid metabolism. *Plant Cell* **1995**, *7*, 1085–1097. [CrossRef]
87. Dixon, R.A. Natural products and plant disease resistance. *Nature* **2001**, *411*, 843–847. [CrossRef] [PubMed]
88. Tzin, V.; Galili, G. The biosynthetic pathways for shikimate and aromatic amino acids in *Arabidopsis thaliana*. In *The Arabidopsis Book*; American Society of Plant Biologists: Rockville, MA, USA, 2010; p. 8. [CrossRef]
89. Tripathi, R.D.; Tripathi, P.; Dwivedi, S.; Dubey, S.; Chatterjee, S.; Chakrabarty, D.; Trivedi, P.K. Arsenomics: Omics of arsenic metabolism in plants. *Front. Physiol.* **2012**, *3*, 275. [CrossRef] [PubMed]
90. Rai, V. Role of amino acids in plant responses to stresses. *Biol. Plant.* **2002**, *45*, 481–487. [CrossRef]
91. Sharma, S.S.; Dietz, K.-J. The significance of amino acids and amino acid-derived molecules in plant responses and adaptation to heavy metal stress. *J. Exp. Bot.* **2006**, *57*, 711–726. [CrossRef]
92. Khalil, R.; Yusuf, M.; Bassuony, F.; Gamal, A.; Madany, M. Phytotoxic effect of Alhagi maurorum on the growth and physiological activities of *Pisum sativum* L. *S. Afr. J. Bot.* **2020**, *131*, 250–258. [CrossRef]
93. Molitor, D.; Liermann, J.C.; Berkelmann-Löhnertz, B.; Buckel, I.; Opatz, T.; Thines, E. Phenguignardic acid and guignardic acid, phytotoxic secondary metabolites from *Guignardia bidwellii*. *J. Nat. Prod.* **2012**, *75*, 1265–1269. [CrossRef]
94. Chen, Z.; Guo, Z.; Niu, J.; Xu, N.; Sui, X.; Kareem, H.A.; Hassan, M.U.; Yan, M.; Zhang, Q.; Wang, Z. Phytotoxic effect and molecular mechanism induced by graphene towards alfalfa (*Medicago sativa L.*) by integrating transcriptomic and metabolomics analysis. *Chemosphere* **2022**, *290*, 133368. [CrossRef]
95. Kaur, H.; Shukla, R.K.; Yadav, G.; Chattopadhyay, D.; Majee, M. Two divergent genes encoding L-myo-inositol 1-phosphate synthase1 (CaMIPS1) and 2 (CaMIPS2) are differentially expressed in chickpea. *Plant Cell Environ.* **2008**, *31*, 1701–1716. [CrossRef]
96. Hu, L.; Zhou, K.; Ren, G.; Yang, S.; Liu, Y.; Zhang, Z.; Li, Y.; Gong, X.; Ma, F. Myo-inositol mediates reactive oxygen species-induced programmed cell death via salicylic acid-dependent and ethylene-dependent pathways in apple. *Hortic. Res.* **2020**, *7*, 138. [CrossRef] [PubMed]
97. Pérez-Alonso, M.-M.; Ortiz-García, P.; Moya-Cuevas, J.; Lehmann, T.; Sánchez-Parra, B.; Björk, R.G.; Karim, S.; Amirjani, M.R.; Aronsson, H.; Wilkinson, M.D. Endogenous indole-3-acetamide levels contribute to the crosstalk between auxin and abscisic acid, and trigger plant stress responses in Arabidopsis. *J. Exp. Bot.* **2021**, *72*, 459–475. [CrossRef] [PubMed]

antioxidants

Article

Genome-Wide Identification and Characterization of the Polyamine Uptake Transporter (Put) Gene Family in Tomatoes and the Role of Put2 in Response to Salt Stress

Min Zhong [1,†], Lingqi Yue [1,†], Wei Liu [1], Hongyi Qin [1], Bingfu Lei [2], Riming Huang [3], Xian Yang [1,*] and Yunyan Kang [1,*]

1 College of Horticulture, South China Agricultural University, Guangzhou 510642, China
2 Key Laboratory for Biobased Materials and Energy of Ministry of Education, Guangdong Provincial Engineering Technology Research Center for Optical Agriculture, College of Materials and Energy, South China Agricultural University, Guangzhou 510642, China
3 College of Food Science, South China Agricultural University, Guangzhou 510642, China
* Correspondence: yangxian@scau.edu.cn (X.Y.); kangyunyan@scau.edu.cn (Y.K.)
† These authors contributed equally to this work.

Abstract: The polyamine uptake transporter (Put), an important polyamines-related protein, is involved in plant cell growth, developmental processes, and abiotic stimuli, but no research on the Put family has been carried out in the tomato. Herein, eight tomato Put were identified and scattered across four chromosomes, which were classified into three primary groups by phylogenetic analysis. Protein domains and gene structural organization also showed a significant degree of similarity, and the *Put* genes were significantly induced by various hormones and polyamines. Tissue-specific expression analysis indicated that *Put* genes were expressed in all tissues of the tomato. The majority of *Put* genes were induced by different abiotic stresses. Furthermore, *Put2* transcription was found to be responsive to salt stress, and overexpression of *Put2* in yeast conferred salinity tolerance and polyamine uptake. Moreover, overexpression of *Put2* in tomatoes promoted salinity tolerance accompanied by a decrease in the Na^+/K^+ ratio, restricting the generation of reactive oxygen and increasing polyamine metabolism and catabolism, antioxidant enzyme activity (SOD, CAT, APX, and POD), and nonenzymatic antioxidant activity (GSH/GSSG and ASA/DHA ratios, GABA, and flavonoid content); loss of function of *put2* produced opposite effects. These findings highlight that Put2 plays a pivotal role in mediating polyamine synthesis and catabolism, and the antioxidant capacity in tomatoes, providing a valuable gene for salinity tolerance in plants.

Keywords: polyamine uptake protein; Put2; antioxidants; reactive oxygen species; salt stress; tomato

Citation: Zhong, M.; Yue, L.; Liu, W.; Qin, H.; Lei, B.; Huang, R.; Yang, X.; Kang, Y. Genome-Wide Identification and Characterization of the Polyamine Uptake Transporter (Put) Gene Family in Tomatoes and the Role of Put2 in Response to Salt Stress. *Antioxidants* **2023**, *12*, 228. https://doi.org/10.3390/antiox12020228

Academic Editor: Nafees A. Khan

Received: 17 December 2022
Revised: 16 January 2023
Accepted: 17 January 2023
Published: 18 January 2023

1. Introduction

Polyamines, one of the organic polycations, are abundant in various plant organisms and are involved in various cellular processes such as cell growth, nucleic acid stability, and protein synthesis [1,2]. The most abundant polyamines in plant cells, diamine putrescine, triamine spermidine, and tetraamine spermine, are strongly associated with plant responses to biotic and abiotic cues [3,4]. Manipulation of putrescine, spermidine, and spermine by chemical or genetic means is essential for many developmental processes [5,6]. The crucial roles of polyamine synthesis and metabolism in response to numerous stresses have been demonstrated by genetic manipulation [7]. Intracellular polyamine pools are critical for the intermediary part of nitrogen metabolism, and also crosstalk with other metabolic pathways, such as hormones, small molecule signals, and stress-response complexes [8,9].

Besides, increasing evidence indicates that the uptake of polyamine plays essential functions in coordinating the response of plants to a variety of environmental stresses. A salt-sensitive cultivar of rice supplied with putrescine in roots exhibited an increased grain yield [10]. The *arginine decarboxylase* (*adc*) mutant in *Arabidopsis* showed hypersensitivity to low-temperature stress, but the tolerance was enhanced after being fed putrescine in *adc* mutants [11]. These studies show that polyamine transport could be an important component of diverse environmental protection. In *Arabidopsis*, methyl viologen 1 (At5g05630, *AtRMV1*), a L-type amino acid transporter, was recently found to be a protein important for paraquat (PQ) and uptake of polyamine [12]. The *Arabidopsis* mutant (AT1G31830, *pqr2/AtPut2*) also encodes a polyamine transporter and negatively responds to ABA signaling [13,14]. The protective influences of polyamine in opposition to PQ toxicity are partly attributed to transport interactions between polyamines and PQ because both have similar uptake characteristics. In addition, the *paraquat resistant1 (pr1)* mutant exhibited inefficient absorption of PQ [13]. The *put5* plants produced fewer flowers, flowered earlier, and had smaller leaves than wild-type (WT) plants, while the *OsPut1* or *OsPut3* over-expression plants showed a buildup of spermidine and conjugated-spermidine in leaves, larger leaves, more flowers, and a delay in the flowering time, which is implicated in polyamine transport [15]. The rice *OsPut1/2/3* mutant was created by CRISPR/Cas9 gene editing, and the *OsPut1/2/3* mutant increases PQ tolerance without significant yield loss [16]. Moreover, *OsPut1-OsPut3* was shown to have a high affinity for spermidine uptake through the substrate assay with a yeast polyamine uptake mutant (*agp2*Δ), and *AtPut1-AtPut3* has similar properties [17]. However, much less attention has been paid to transport proteins in plants. To our knowledge, very little has been unraveled regarding the tomato polyamine uptake protein (Put) family and their functions in abiotic stress.

The tomato is one of the world's most significant cash crops and is sensitive to biotic and abiotic pressures such as salt stress, low and high temperatures, and so on. These unfavorable environmental factors seriously compromise tomato growth and yield. Under stressful conditions, polyamine anabolism and catabolism have been found to have important roles involving a multitude of mechanisms [18]. For example, polyamine oxidase, arginine decarboxylase, *S*-adenosylmethione synthetase, and spermine synthase act as critical mediators in multiple stress conditions [19–23]. To date, it has been widely accepted that polyamine acts as a crucial antioxidant in plants [24]. The increase of endogenous polyamines levels saves cells by eliminating reactive oxygen species (ROS) and boosting the antioxidant capacity in response to oxidative stress [25]. Nevertheless, the implications of Put in response to environmental stresses remains elusive. In particular, the role of Put in polyamine anabolism and catabolism, as well as in antioxidant activity, remains largely unknown. In this study, by identifying and characterizing members of the Put gene family in the tomato, a family known for polyamines uptake, we unraveled that Put2, a candidate of the Put family, had a favorable function in salt tolerance via modulating polyamine metabolism and antioxidants. This study sheds fresh light on the important role of Put-mediated polyamine homeostasis in tomatoes, as well as its significance for plant fitness.

2. Material and Methods

2.1. Indentation and Sequence Analysis of the Tomato Put Family

The protein sequence of tomato Put and *Arabidopsis* AtPut were downloaded from the NCBI (http://www.ncbi.nlm.nih.gov/, accessed on 10 January 2022). The SGN database (https://solgenomics.net/, accessed on 10 January 2022) was used to get the tomato (SL4.0) reference genome sequence and annotations. We identified eight putative Put proteins encoded in the tomato genome based on investigations of the Arabidopsis Put protein. The eight Put genes in tomato were called by their chromosomal locations. Expasy, an online software, was used to gather basic information for all Put proteins, including their molecular weight (Mw) and isoelectric point (pI).

2.2. Alignment of the Protein Sequence and Phylogenetic Tree Construction

Using MEGA X, the protein sequences of Put in tomato, rice, and *Arabidopsis* were aligned [26]. Poor alignment areas from all protein sequences were removed using the trimAl tool, and a phylogenetic tree was created using the maximum-likelihood (ML) technique with the Poisson correction and 1000 bootstrap repetitions in IQ-TREE [27]. The depiction of the phylogenetic tree was constructed by Evolview (www.evolgenius.info/, accessed on 22 January 2022). A phylogenetic tree of the tomato Put protein was also built independently. The Put protein sequences are provided in Supplemental data S1.

2.3. Analyses of Conserved Motifs, Conserved Domains, Cis-Acting Elements in Promoters, and miRNA Prediction

The conserved motifs of the tomato Put proteins were performed by the MEME tool (5.05) (http://meme.nbcr.net/meme/, accessed on 22 January 2022), and Pfam (http://pfam.xfam.org/, accessed on 22 January 2022) was used to predict the conserved domains of the tomato Puts proteins. The promoter regions of Put were created using the 2.0 kb genomic DNA sequence upstream of the translation start codon (ATG). PlantCare took the cis elements from the Put promoter regions. In Supplemental data S2, the cis elements are listed. The relevant data visualization was conducted using TBtools. According to the targeted candidate, as described previously, the Put coding sequences were submitted to the psRNATarget serve to predict the miRNAs (https://www.zhaolab.org/psRNATarget/, accessed on 22 January 2022) [28]. The Puts protein's transmembrane domains were predicted using the TMHMM program [29].

2.4. Plant Material and Treatments

Three-week-old tomato (*Solanum lycopersicum* L. cv. Ailsa Craig) seedlings were treated with different exogenous polyamines, hormones, and oxidative stress, or abiotic stresses. Briefly, 2.0 mM Put, 1.0 mM Spd, or 2.0 mM Spm were sprayed over the seedlings; for hormone treatments and oxidative stress, 100 μM ABA, 2 mM SA, 100 μM GA_3, 40% ethylene (ETH), 100 μM paraquat, and water were also sprayed onto the tomato plants, respectively. For RNA extraction, at 0, 30 min, 1, 3, 6, and 12 h, samples of leaves were taken, accordingly. The control was the water treatment at 0 h.

We then investigated Put genes response to different abiotic stresses. Salt and drought stress were initiated by irrigating the plants with 200 mM NaCl or 20% PEG6000 solution, respectively. Tomato seedlings were subjected to 42 °C (high temperature), and 4 °C (low temperature) for heat and cold stress, respectively. After, the treated samples were respectively collected after 0, 30 min, 1, 3, 6, and 12 h; the samples at 0 h were used as the control. In addition, tissues (root, stem, leaf, bud, flower, and fruit) were harvested for investigation of tissue-specific expression. After each treatment, leaves from different plants (three biological replicates) were quickly frozen in liquid nitrogen and kept at −80 °C for further analysis.

2.5. Yeast Strain and Culture Conditions

The wild type (WT), G19 (Δ*ena1–4*), failure to mediate Na^+ uptake; CY162, the K^+ uptake-deficient, and a yeast strain impaired in spermidine uptake, *agp2*Δ (strains obtained from open biosystems, http://www.openbiosystems.com/GeneExpression/Yeast/ORF/, accessed on 10 May 2022), were used to describe the potential transporter. The yeasts were grown in YPD media at 28 °C. The yeast cells were converted using lithium acetate. Put1-8 coding sequences were respectively cloned into the pYES2 expression vector.

Wild type-empty vector and *agp2*Δ-Put1-8 transformants were grown in yeast extract peptone galactose (YFG) medium. Cell suspensions were serially diluted as OD_{600} of 0.6 for growth tests, and 5 μL aliquots were spotted onto YPG plates containing 25 mM spermidine and 1.5 mM paraquat. After 3–4 days of incubation at 28 °C, the plates were photographed. For polyamine transport assays, the yeast cells were harvested at the mid-logarithmic phase, washed with ddH_2O, and suspended in the YPG media at a dose of 10^8 cells/mL. One-hundred-microliter aliquots cells were transferred to the Eppendorf tubes, and polyamine absorption was activated by the addition of Spermidine or Putrescine at 25 μM concentrations. The absorption was inhibited by adding 1.5 mL of ice-cold uptake buffer with excessive spermidine content at selected times, filtered by a 0.45 μm membrane, and washed with 2 mL ice-cold ddH_2O (three times) to remove the exogenous polyamines. Polyamine determination in vivo was performed by an Agilent high-performance liquid chromatography 1200 series system (HPLC, Agilent Technologies, Santa Clara, CA, USA). Polyamines were obtained from Sigma-Aldrich (St Louis, MO, USA).

For salt tolerance assays, the final pYES2-empty and pYES2-Put1-8 vectors cultured in G19 and CY162 were performed on SD-U (Synthetic Dextrose Minimal Medium without Uracil) medium, and then diluted until the OD_{600} value = 0.6. 5 μL. Aliquots were spotted onto YPG plates containing 100 mM NaCl and 0.1 mM KCl, respectively, and incubated at 28 °C. No treatment was added for the control. After 3–4 days of incubation, the plates were photographed. For Na^+ and K^+ uptake treatment, the empty and positive yeast were incubated to OD_{600} = 1.0, the supporting was discarded, and 50 mL ddH_2O was used to wash the yeast. The yeast cells were obtained by centrifugation, followed by starvation treatment with AP liquid medium without K^+ and Na^+. After starvation treatment, the yeast cells were obtained by centrifugation and treated as follows: inoculating the yeast with an AP liquid medium including 200 μM NaCl and 200 μM KCl, respectively. Subsequently, the liquids were put on a 28 °C shaker (220 r/min), and 4 mL of bacterial solution was taken every 10 min, centrifuged, and the supernatant was collected for analysis of Na^+ and K^+ contents. Three biological replicates were made. The primers for vector construction are listed in Supplemental data S3 (Table S1).

2.6. Plasmid Construction and Plant Transformation

The *Put2*-overexpression vector (full-length coding sequence of *Put2*) was constructed as previously described [30]. Gene loss-function of *put2* lines was generated through gene editing approaches. To generate the CRISPR/Cas9 vector, the two target sequences for *put2* were designed using the online software CRISPR-GE (http://skl.scau.edu.cn/targetdesign/, accessed on 22 January 2022), which were inserted into two single guide RNA (sgRNA) expression cassettes through overlap PCR, followed by cloning into the pYLCRISPR/Cas9Pubi-H vector via Golden Gate ligation method [31]. The confirmed pFGC1008-*Put2*-3HA vector and pYLCRISPR/Cas9Pubi-H-*put2* binary vector were transformed into *Agrobacterium tumefaciens* strain GV3101 by electroporation after transgenic plants were generated with *Agrobacterium*-mediated cotyledon transformation of *Solanum lycopersicum* cv. Ailsa Craig via a method previously described [32]. Two separate homozygous T2 lines from mutation and overexpression lines were confirmed with Sanger sequencing and qRT-PCR. The *put2* mutants and *Put2*-OE plants were used in this study. The primers for vector construction are listed in Supplemental data S3 (Table S1).

2.7. Salt Treatment and Salt Tolerance Assays

The tomato seedlings (WT, *put2* mutants, and *Put2* overexpression lines) were used for salt tolerance experiments. After seed germination and two cotyledons full expansion, the seedlings were cultured in 250 cm^3 plastic pots filled with a peat-vermiculite combination (2:1, *v:v*). The seedlings were placed in a greenhouse at 28 ± 2 °C/20 ± 2 °C (day/night) under a maximum photosynthetic photon flux density (PPFD) of approximately 1200 μmol m^{-2} s^{-1} and a relative humidity of 70–80%. They were watered daily using Hoagland nutrient solution. Three-week-old WT and transgenic tomatoes of uniform size and health growth status were selected and subjected to salt stress treatment. The seedlings were treated by watering the plants with 200 mL of 200 mM NaCl every other day for salt stress. The control treatment was replaced with an equal amount of water. The salt stress treatment lasted for 7 days, and pictures were captured. The maximum quantum yield of PSII (Fv/Fm) was examined with the Imaging-PAM system (IMAG-MAXI; Heinz Walz, Effeltrich, Germany), as previously described by Zhong et al. [30]. The relative electrolyte leakage (REL%), Na^+, and K^+ analysis was performed as described previously by Zhong et al. [30]. The plants were enclosed in envelopes and placed in an oven at 105 °C for 30 min, and then the oven temperature was adjusted to 75 °C to obtain a permanent dry weight (DW).

2.8. Determination of Polyamine Content

The free polyamines content was analyzed by Agilent 1200 High-performance Liquid Chromatography (HPLC, Agilent Technologies, Santa Clara, CA, USA), as previously described in Zhong et al. [33] with slight modifications. Frozen plant tissue (leaves sample, 0.5 mg) was ground with liquid nitrogen, used 10:1 (*v/w*) of extraction buffer (5% cold aqueous perchloric acid (PCA), *w/w*). Samples were incubated for 1 h at 4 °C, then centrifuged at 15,000× *g* for 10 min at 4 °C. Volumes of 200 μL of collected supernatant were mixed with 15 μL benzoyl chloride and incubated for 1 h at 60 °C in darkness conditions. Four milliliters of saturation NaCl solution was used to quench the reaction and diethyl ether was added; 5 mL cold ethyl acetate was then added to extract polyamines. Then, organic layers were evaporated to dryness, redissolved in 100 μL methanol, and filtered with a 0.45 μm pore nylon filter. A volume of 25 μL extraction solution was used to determine the endogenous polyamines levels. The mobile phase was with 64% (*v/v*) methanol and had a flow rate of 0.8 mL min^{-1}. Putrescine, spermidine, spermine, and cadaverine (Sigma, St. Louis, MO 63178, USA) were chosen as standard samples and treated similarly.

2.9. Determination of PAO Enzymatic Activities and H_2O_2 Content

Amine oxidase was determined as previously described by Su et al. [34] and Urra et al. [9]. Leaf samples (0.5 g) were ground with liquid nitrogen, homogenized 2:1 (*v/w*) in 100 mM sodium phosphate buffer (pH 6.5), and centrifuged at 12,000× *g* for 20 min at 4 °C. One-hundred-microliters of the recovered supernatant was mixed with the 3 mL reaction mix, which contained 2.5 mL sodium phosphate buffer (100 mM, pH 6.5), 200 μL 15 mM 4-aminoantipyrine/0.2% (*v/v*) *N*, *N*-dimethylaniline, 100 μL 250 U ml^{-1} peroxidase, and 100 μL 20 mM putrescine as a substrate. The $CuAO_{Put}$, PAO_{Spm}, and PAO_{Spd} were determined using Putrescine, Spermine, and Spermidine as substrates, respectively. A 0.01 value of the changes in absorbance at 555 nm was assayed to one activity unit of the PAO enzyme after being incubated for 30 min at 22 °C. Control samples without polyamines were used to calculate these activities. Leaf H_2O_2 content was determined by specific detection kits according to the manufacturer's instructions (Nanjing Jiancheng Bioengineering Institute, Jiangsu, China).

2.10. Antioxidant Assay

To assess the antioxidant enzyme activity, 0.5 g leaf samples were homogenized with 3 mL ice-cold 50 mM phosphate buffer (pH 7.8), which contained of 2 mM L-ascorbic acid, 2% (*w/v*) PVPP, and 0.2 mM EDTA. Then, the homogenates were centrifuged for 20 min at 12,000× *g*, and supernatants were collected to determine the enzyme activity. The activities of SOD, CAT, APX, and POD were assayed as previously described [30]. AsA/DHA and GSH/GSSG were measured as described by Zhong et al. [35]. The GABA content was determined by the Berthelot reaction with some modifications [36]. Leaf samples (0.5 g) were ground with methanol, centrifuged at 6000× *g* for 15 min, and the supernatant was discarded. The sediment was dissolved in 1.5 mL ddH_2O and heated in a water bath at 50 °C for 2 h, followed by centrifugation for 15 min at 7000× *g*. A volume of 1 mL supernatant was mixed with 100 μL 2 mol L^{-1} $AlCl_3$ and oscillated, and then centrifuged for 10 min at 12,000× *g*. The supernatant was added with 300 μL KOH and incubated for 5 min, then centrifuged at 12,000× *g* for 10 min. The GABA content was measured according to the following procedure: 300 μL supernatant was mixed with the reaction mix, composed of 500 μL 0.1 mol L^{-1} sodium tetraborate (pH 10.0), 400 μL 6% phenol, and 600 μL 5% sodium hypochlorite. The solution was boiled for 10 min and then placed in an ice bath for 5 min. Finally, the absorbance at 645 nm was measured after shaking the solution with 2 mL 60% ethyl alcohol. The assessment of the total flavonoid concentration was determined by Zhishen et al. [37], with some modifications. Dried leaf samples (0.5 g) were homogenized with 2 mL 80% ethanol, and then added 300 μL 20 mol L^{-1} $NaNO_2$; 3 mL 1 mol L^{-1} $AlCl_3$ was added after 5 min, and after 6 min, 2 mL 1 mol L^{-1} NaOH was added and mixed well. Finally, the absorbance was measured at 510 nm.

2.11. Analysis of Gene Expression

Total RNA was extracted with the RNAsimple Total RNA Kit (Tiangen, DP419) and reverse transcribed with the HiScript^{TM} qRT SuperMix for qPCR (+gDNA wiper) kit (Vazyme, Nanjing, China). The qPCR reaction was performed by the ABI VII7 real-time PCR system (Applied Biosystems, Waltham, MA, USA). *Actin* was used as the tomato reference gene. The qRT-PCR primers are listed in Supplemental data S3 (Table S2).

2.12. Statistics

The data are presented as the means ± SDs and were analyzed using SPSS 20 statistical software. The experimental data were analyzed with Duncan's multiple range test at $p < 0.05$.

3. Results

3.1. Identification of Tomato Put Family Genes

To analyze Put proteins, a query search against the tomato genome database was accomplished using *Arabidopsis* and rice Put protein as the control search (Table 1). Eight potential *Put* proteins with high sequence similarity to *AtPut* and *OsPut* were identified and called Put1-8. They had CDs sizes ranging from 1077 bp (Put6) to 1605 bp (Put3), with polypeptides of 359–535 amino acids. The theoretical isoelectric points (pI) of *Put* varied from 5.41 (*Put3*) to 9.37 (*Put8*). The molecular weights (MW) of *Put* ranged from 40.28 (*Put6*) to 58.8 (*Put3*) (Table 1). Additionally, all Puts were predicted to contain transmembrane domains (Figure S1).

Table 1. Put gene identification and characterization in the tomato.

Gene	Gene Code	Chr Number	Location on CHR	CDS_Length	AA_Length	PI	MW(KDa)
Put6	Solyc01g005920	chr1	611660-613377	1077	359	7.96	40.28
Put4	Solyc01g034080	chr1	33621254-33622666	1413	471	6.05	52.67
Put7	Solyc01g111800	chr1	90326196-90327818	1386	462	8.67	51.33
Put2	Solyc08g005540	chr8	411003-415969	1593	531	6.28	58.25
Put3	Solyc08g075710	chr8	57951470-57958170	1605	535	5.41	58.8
Put1	Solyc08g078100	chr8	60059627-60061033	1407	469	8.83	51.71
Put8	Solyc09g092420	chr9	67625291-67628225	1092	364	9.37	41.19
Put5	Solyc10g049640	chr10	45328007-45330412	1473	491	9.28	54.5

3.2. Analysis of the Phylogenetic, Chromosomal Distribution, Gene Structure, and Promoter Sequences of Tomato Family Members

To better study the evolutionary relationships among the tomato Put protein sequences and those of other plants, a phylogenetic tree was performed through the Neighbour-Joining method. By comparing the protein sequences of the Put genes, three main groupings could be distinguished. Two *Put* genes, *Put4* and *Put6,* were included in Group I; Group II only included *Put5*, and the remaining five genes (*Put1*, *Put2*, *Put3*, *Put7*, and *Put8*) were classified into Group III (Figure 1A). According to their location, the chromosomal positions of the Put genes were analyzed. Among these *Put* genes, three (*Put4*, *Put6* and *Put7*) were mapped to chromosome 1, and three (*Put1*, *Put2* and *Put3*) were located on chromosome 8, while *Put5* and *Put8* were located in chromosomes 9 and 10, respectively. These results suggest an uneven distribution of these genes on chromosomes (Figure 1B).

To better analyze the evolutionary relationships in *Put* genes, the phylogenetic tree was also constructed. *Put1*, *Put2*, and *Put3* were observed to have a close evolutionary relationship. A close evolutionary relationship was also found in *Put7* and *Put8*, and *Put4* and *Put6*, respectively. Interestingly, *Put5* has a separate evolutionary branch. In addition, their structural domains are highly conserved and contain the polyamine transport protein PotE structural domain (Figure 1C). The Put family gene structure analysis was carried out to learn more about their intron/exon structures. One intron was present in *Put2* and *Put5*, two in *Put3*, while the other five (*Put1*, *Put7*, *Put8*, *Put4*, and *Put6*) were intron deletion genes (Figure 1D). To investigate the feasible roles of Put family genes in different abiotic stress and developmental steps, we used the PlantCARE database to estimate the presence of *cis*-acting elements in the promoter regions of the Put family. We obtained *cis*-acting elements that are associated with stress, hormone, and light responsiveness. Surprisingly, we only found five typical ethylene responsive motifs (ERE, GCC-box motifs) in the promoter of Put2, but not in that of other Puts (Figure 1E).

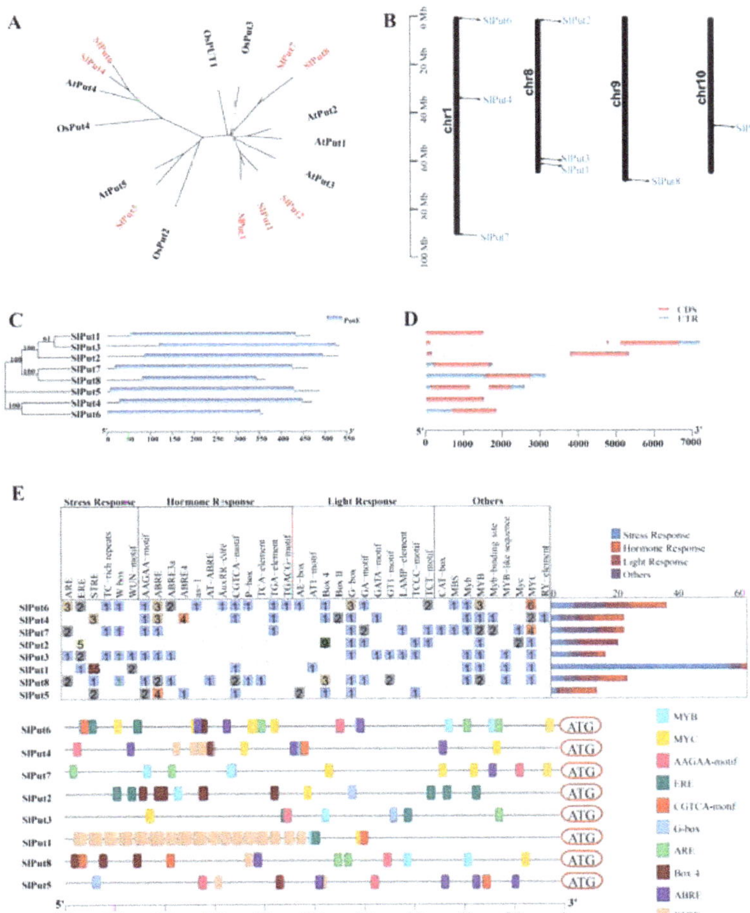

Figure 1. The characteristics of Put genes in the tomato. (**A**) Phylogenetic relationships of Put genes in Arabidopsis, rice, and tomato. (**B**) Gene distribution within the tomato chromosomes. The chromosome numbers are indicated on the left, and the position marked in the chromosome indicates the location of Put genes. (**C**) The protein motifs and phylogenetic trees in the Put famous members. (**D**) Gene structure of the Put family members in tomatoes. (**E**) Cis element distribution of the Put genes.

miRNAs are an important kind of non-coding signal strand RNAs of around 22 nucleotides that are encoded by genes in the organism and are closely involved in the regulation of genes in answer to various life processes and stresses. A total of 29 miRNAs targeting the polyamine transporter genes in the tomato are listed in Table 2, including three sly-miRNAs targeting Put1 (sly-miR390a-5p, sly-miR390b-5p, and sly-miR6022); sly-miR6024 and sly-miR6026 targeting Put2; sly-miR6023 and sly-miR1916 targeting Put3;

six sly-miRNAs targeting Put4 (sly-miR9479-3p, sly-miR6024, sly-miR171c, sly-miR171a, sly-miR9472-5p, and sly-miR9478-3p); sly-miR164a-5p and sly-miR164b-5p targeting Put5; Put6 was targeted by sly-miR1917; six sly-miRNAs targeting Put7 (sly-miR156a, sly-miR156b, sly-miR156c, sly-miR390a-5p, sly-miR396a-3p, and sly-miR6023); and three sly-miRNAs targeting Put8 (sly-miR156a, sly-miR156b, sly-miR156c, sly-miR319b, sly-miR319c-3p, sly-miR396a-3p, and sly-miR6023) (Table 2).

Table 2. Prediction of miRNAs targeting the Put genes in tomato.

miRNA	Target	Expectation	Target_Site	miRNA_Aligned_Fragment	Alignment	Target_Aligned_Fragment	Inhibition
sly-miR1917	Solyc01g005920	4	400–420	AUUAAUAAAGAGUGCUAAAGU	:::::::::.::: ::.:	CCUUUAGCACUUUUUUCUAGU	Cleavage
sly-miR9479-3p	Solyc01g034080	3.5	180–201	GAGAAUGGUAGAGGGUCGGACC	: :: ::.:.:.::::::.	UGGCCCACUUUUUGCCAUUCUU	Cleavage
sly-miR6024	Solyc01g034080	4	721–742	UUUUAGCAAGAGUUGUUUUACC	::.: :::.:: ::::.	CCGAAGAGAACUUUUCCUAAAG	Cleavage
sly-miR171c	Solyc01g034080	4.5	775–795	UAUUGGUGCGGUUCAAUGAGA	:.::: ::.: ::.:::.	GGUUAUUUAAUCCCACUAAUG	Cleavage
sly-miR171a	Solyc01g034080	5	964–984	UGAUUGAGCCGUGCCAAUAUC	: : ::.:.::: :::.:	CUUUUAGGUAUGGCUGAAUUA	Cleavage
sly-miR9472-5p	Solyc01g034080	5	1262–1282	UUUCAGUAGACGUUGUGAAUA	.::.:.::.: :::.:: ::	CGUUUAUAAUGGCUAUUGCAA	Translation
sly-miR9478-3p	Solyc01g034080	5	1125–1145	UUCGAUGACAUAUUUGAGCCU	:.:.:.::::. :.:.:	UAGUUUAGGUAUGUUUUUGGA	Cleavage
sly-miR156a	Solyc01g111800	4.5	1089–1109	UUGACAGAAGAUAGAGAGCAC	:.:::: ::.:::::	GGAAUUUCUAUGUUUUGUCAA	Translation
sly-miR156b	Solyc01g111800	4.5	1089–1109	UUGACAGAAGAUAGAGAGCAC	:.:::: ::.:::::	GGAAUUUCUAUGUUUUGUCAA	Translation
sly-miR156c	Solyc01g111800	4.5	1089–1109	UUGACAGAAGAUAGAGAGCAC	:.:::: ::.:::::	GGAAUUUCUAUGUUUUGUCAA	Translation
sly-miR390a-5p	Solyc01g111800	5	1017–1037	AAGCUCAGGAGGGAUAGCACC	: :::.::.. ::: ::.::	GCUGCUGUCUUGCCUUAGUUU	Translation
sly-miR396a-3p	Solyc01g111800	5	682–702	GUUCAAUAAAGCUGUGGGAAG	::::.: ::::::: ::	AUUCCUAAGGCUUUAUUCUAC	Cleavage
sly-miR6023	Solyc01g111800	5	599–620	UUCCAUGAAAGAGUUUUUGGAU	::.::: :: :::. ::::	AUCUAAAUACACUUUUUUGGAA	Cleavage
sly-miR6024	Solyc08g005540	5	880–901	UUUUAGCAAGAGUUGUUUUACC	::.: ::::: ::::.	CCGAAGAAAACUCUGCCUAAAG	Cleavage
sly-miR6026	Solyc08g005540	5	1181–1202	UUCUUGGCUAGAGUUGUAUUGC	... ::: :::::.:.::	AUGGAACACCUCUAGUCGGGAU	Cleavage
sly-miR6023	Solyc08g075710	4	824–845	UUCCAUGAAAGAGUUUUUGGAU	:::..:: ::::::. ::::	AUCUGAAUACUCUUUUCUGGAA	Cleavage
sly-miR1916	Solyc08g075710	5	526–545	AUUUCACUUAGACACCUCAA	::: :: ::.:::.:.:	AUGAAAUGGCUGAGUGGAGU	Cleavage
sly-miR390a-5p	Solyc08g078100	3.5	778–798	AAGCUCAGGAGGGAUAGCACC	:: :::::.:::.::::: :	GGAGCUAUUCCUCUUGAGCAU	Cleavage
sly-miR390b-5p	Solyc08g078100	3.5	778–798	AAGCUCAGGAGGGAUAGCGCC	:: :::::.:::.::::: :	GGAGCUAUUCCUCUUGAGCAU	Cleavage
sly-miR6022	Solyc08g078100	5	518–538	UGGAAGGGAGAAUAUCCAGGA	: ::: ::::.::::...	UACUGUCAAUUCUUCCUUUUG	Cleavage
sly-miR156a	Solyc09g092420	3	945–965	UUGACAGAAGAUAGAGAGCAC	:.::::::::.:::::	GGAAUUUCUAUCUUUUGUCAA	Cleavage
sly-miR156b	Solyc09g092420	3	945–965	UUGACAGAAGAUAGAGAGCAC	:.::::::::.:::::	GGAAUUUCUAUCUUUUGUCAA	Cleavage
sly-miR156c	Solyc09g092420	3	945–965	UUGACAGAAGAUAGAGAGCAC	:.::::::::.:::::	GGAAUUUCUAUCUUUUGUCAA	Cleavage
sly-miR319b	Solyc09g092420	5	1013–1033	UUGGACUGAAGGGAGCUCCCU	::: :. ::::::.:.	UAGGAAGUUACUUCAGUUCAG	Cleavage
sly-miR319c-3p	Solyc09g092420	5	1013–1033	UUGGACUGAAGGGAGCUCCUU	:::: :. ::::::.:.	UAGGAAGUUACUUCAGUUCAG	Cleavage
sly-miR396a-3p	Solyc09g092420	5	538–558	GUUCAAUAAAGCUGUGGGAAG	::::.: ::::::: ::	AUUCCUAAGGCUUUAUUCUAC	Cleavage
sly-miR6023	Solyc09g092420	5	455–476	UUCCAUGAAAGAGUUUUUGGAU	::.::: :: :::. ::::	AUCUAAAUACACUUUUUUGGAA	Cleavage
sly-miR164a-5p	Solyc10g049640	5	1195–1215	UGGAGAAGCAGGGCACGUGCA	::.:: .:. ::::::..:	GUCAUGUAUCUGGCUUCUUUA	Translation
sly-miR164b-5p	Solyc10g049640	5	1195–1215	UGGAGAAGCAGGGCACGUGCA	::.:: .:. ::::::..:	GUCAUGUAUCUGGCUUCUUUA	Translation

3.3. Responsiveness of Put Gene Expression under Hormone, Polyamine, and PQ Treatment

The hormones in the plant kingdom play crucial roles in plant growth and development. Therefore, four different hormone-induced Put transcript-level changes were analyzed. After ABA treatment, the expression of all Puts was induced at 30 min, while that of *Put2* and *Put5* was induced after 6 h (Figure 2A). The expression of all *Puts* was induced at 30 min after SA treatment, which were also upregulated in the later time points, except for *Put4* and *Put6* (Figure 2A). In contrast, the transcripts of *Put6*, *Put7*, and *Put8* were reduced initially following GA_3 and ETH treatments, and ETH also reduced the *Put4* and *Put5* expression, a GA_3 induced both. *Put2* and *Put3* were reduced after GA_3 treatment, and then induced after 6 h of treatment, Put1 was induced after GA_3 and ETH treatments (Figure 2A).

To examine the possible roles of *Put* genes in the tomato, transcripts of *Puts* after three treatments with polyamines, including Spermidine, Putrescine, and Spermine, were analyzed. Exogenous Spermidine treatment induced all *Put* genes, and *Put2* was the most obvious, followed by *Put5* (Figure 2B). However, the expression of all *Put* genes was reduced by Putrescine, and the reduction magnitude differed among them. After the seedlings were treated with Spermine, all *Put* genes were upregulated, but reduced after 12 h treatment. Additionally, exogenous PQ treatment caused upregulation of all Put genes similar to that of Spermidine treatment (Figure 2B).

3.4. Differential Expression of Put Genes during Abiotic Stress

Upon exposure to drought stress, the mRNA transcript of *Put1–5* was rapidly induced, and *Put2* was the most significant, but other members were reduced (Figure 2C). Salinity stress resulted in the downregulation of *Put3*, *Put4*, *Put6*, *Put7*, and *Put8*, and upregulation of *Put1*, *Put2*, and *Put5*. Cold stress resulted in the most pronounced induction of *Put4*. Furthermore, when the seedlings were submitted to heat stress, sustained and stable upregulation of *Put6* was observed, whereas two genes (*Put7* and *Put8*) were reduced post-treatment (Figure 2C).

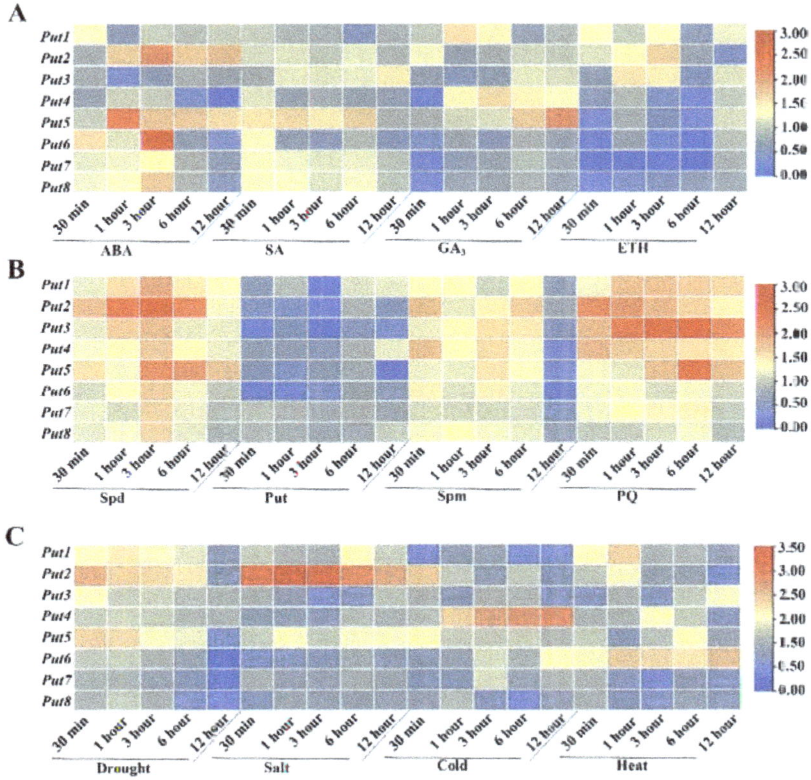

Figure 2. Expression patterns of *Put* genes in tomatoes under hormone, polyamine, paraquat, or abiotic stress. (**A**) Heatmap representation of the responsiveness of *Put* genes after treatment with ABA, SA, GA_3, and ET. The scale (0 to 3) represents the expression level (from low to high). (**B**) Heatmap representation of the responsiveness of *Put* genes after treatment with Spd, Put, Spm, and PQ. The scale (0 to 3) represents the expression level (from low to high). (**C**) Heatmap representation of the responsiveness of *Put* genes after drought, salt, cold, and heat treatment. The scale (0 to 3.5) represents the expression level (from low to high). qRT-PCR was conducted after treatment, according to the methods section. The expression of WT at 0 min in the different treatments was set to 1.

3.5. Expression Analysis of Put Genes in Different Tissues

To research the functions of the Put family of genes in the tomato, the expression patterns of *Put* in various tissues (e.g., root, stem, leaf, bud, flower, and fruit) were performed by qRT-PCR. As shown in Figure 3, the transcripts of *Put2* and *Put5* had high levels in the leaves and roots, *Put3* exhibited high levels of expression in the leaves and flowers, while *Put4* showed a high level of transcript expression in the roots and flowers. In addition, all *Put*, except *Put6*, *Put7* and *Put8*, were highly expressed in flowers. Subsequently, the spatial expression pattern of *Put2* and *Put5* was analyzed by qRT-PCR in vegetative tissues of WT plants with salt stress. *Put2* and *Put5* showed increased expression in organs of plants under salt stress, with the highest level of *Put2* transcripts in the leaves. and the highest level of *Put5* transcripts in the roots. After 7 days of salt treatment, the *Put2* gene is significantly more highly induced than *Put5* in the leaves (Figure 3B,C).

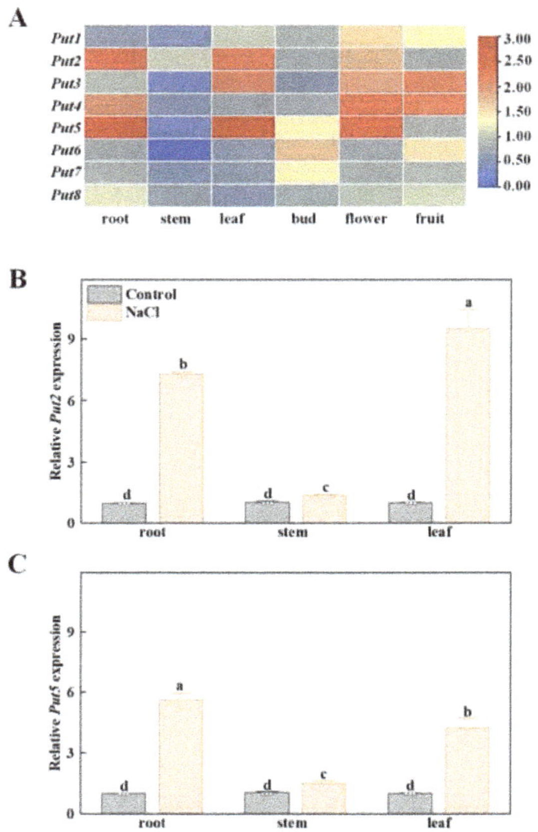

Figure 3. The expression profiles of the tomato Put family of genes in different tissues. (**A**) Heatmap representation of the relative expression of *Put* genes in different tomato tissues. The scale (0 to 3) represents the expression level (from low to high). (**B**,**C**) The spatial expression of *Put2* and *Put5* were performed by qRT-PCR in different organs of WT plants without NaCl (Control) and salt stress (NaCl) (7 days salt treatment). The WT expression in the control condition was set to 1. The data in B and C are presented as mean values ± SD; n = 3. Different letters indicate significant differences between treatments ($p < 0.05$, Duncan's multiple range test). Three independent experiments were performed with similar results.

3.6. Functional Analysis of Put Genes in Yeast

To examine the potential function of these proteins in polyamine transport, cDNA fragments containing the ORF were cloned and introduced into the yeast expression vector pYES2, driving expression under the *GAL1* promoter. High concentrations of polyamines or paraquat are toxic to wild-type yeast strains, whereas mutant *agp2*Δ lacking the polyamine uptake transporter protein impairs the sensitivity to the high concentration of polyamines. The candidate positive polyamine transporter was introduced to the yeast mutant *agp2*Δ. The expression of polyamine transporter proteins (Put1, Put3, Put6, and Put7) did not affect the phenotype of the *agp2*Δ mutant, however, among the eight Put proteins, Put2 and Put5 transformants were more sensitive than other Put transformants under 25 mM Spermidine conditions. The expression of Put2 or Put5 in the *agp2*Δ mutant conferred sensitivity to 25 mM Spermidine (Figure 4A). In addition, transformants appeared sensitive to the polyamine s analog PQ, showing that these proteins are also involved in the uptake of PQ.

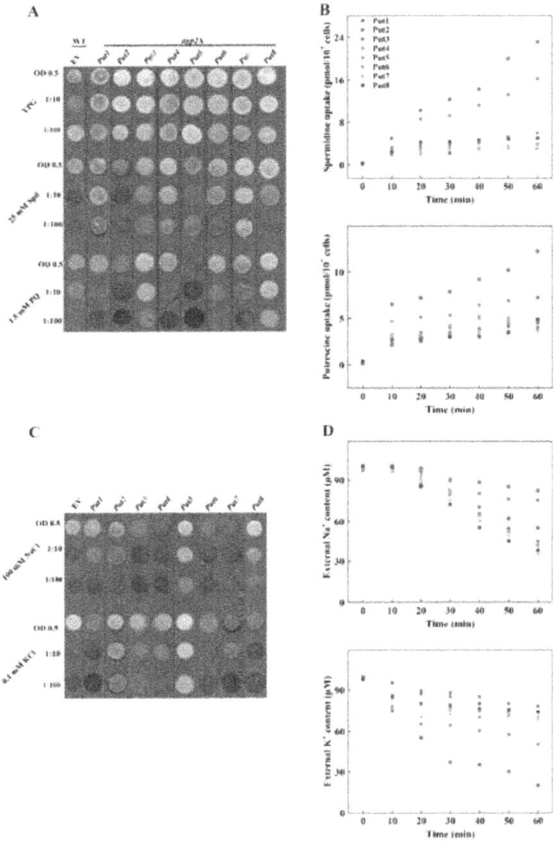

Figure 4. Comparison of polyamines, Na^+, and K^+ uptake among tomato *Puts* in yeast. (**A**) Functional complementation of tomato *Puts* in the yeast mutant *agp2Δ*. *agp2Δ-Puts* vector strains were grown overnight on SC medium supplemented with 2% galactose. Cell suspensions (the starting OD_{600} is 0.5) were serially diluted as indicated and 3 μL of each were spotted onto YP-galactose plates containing 25 mM spermidine or 1.5 mM paraquat. Plates were photographed after 3–4 days of incubation at 30 °C. The data are representative of one of three independent experiments. EV, empty vector. (**B**) Time course uptake of Spermidine and Putrescine for 0, 10, 20, 30, 40, 50, and 60 min to determine the intracellular amount of Spermidine or Putrescine. (**C**) Functional analysis of tomato *Puts* in the yeast mutant G19 (*Δena1–4*) (does not mediate Na^+ uptake) under NaCl treatment, and in the yeast mutant CY162 (a K^+ uptake-deficient mutant strain) under deficiency medium. Cell suspensions (the starting OD600 is 0.5) were serially diluted as indicated and 3 μL of each were spotted onto YP-galactose plates containing 100 mM NaCl or 1.5 mM KCl. Plates were photographed after 3–4 days of incubation at 30 °C. The data are representative of one of three independent experiments. EV, empty vector. (**D**) Time course uptake of Na^+ and K^+ for 0, 10, 20, 30, 40, 50, and 60 min to determine the external levels of Na^+ or K^+.

Further analysis of these genes in light of their competence to transport polyamine was done by incubating with the liquid AP medium supplemented with either 15 μm Spermidine or Putrescine. Transformants of these *Put* genes improved the ability to transport Spermidine or Putrescine than in those of the *agp2Δ* mutant. These transformants possessed higher uptake of Spermidine or Putrescine relative to that mediated by the *agp2Δ* mutant. The Put2- and Put5-transformants were effective for Spermidine and Putrescine, respectively. The uptake ability was more pronounced for Put2 than for Put5 (Figure 4B).

The yeast mutant G19 (*Δena1–4*) (does not mediate Na^+ uptake) and CY162 (K^+ deficient strain) strain were used to further disclose whether Put could transport Na^+ or K^+. The growth status had a significant difference between G19 transformed with Put2 or Put5 than the G19 empty, Put5 had a better growth status, whilst other Put trans-yeast strains were not significant different to the G19-empty vector under 100 mM NaCl treatment (Figure 4C). Similarly, under 0.1 mM KCl, CY162 yeast transformed with Put2 or Put5 had a better-growing status compared to the CY162-empty vector, while other Put trans-yeast strains grew similarly to the CY162-empty vector, indicating that Put2 could transport K^+, and Put5 could transport Na^+ and K^+ (Figure 4C). Furthermore, the ion depletion assay exhibited that K^+ ions decreased significantly faster than Na^+ inoculated into the liquid AP medium with 200 μM NaCl + 200 μM KCl, confirming the role of Put2 in Na^+ and K^+ transport (Figure 4D).

3.7. Put2 Is a Positive Regulator Protein of Tomato Plant Salt Tolerance

We focused our present study on Put2, since *Put2* was highly expressed in leaves after 7 days of salt stress. Furthermore, Put2 had the highest uptake ability of polyamines among the eight Puts and overexpression of *Put2* increased salt tolerance in yeast (Figures 3 and 4). To further determine the role of Put2 in salt stress, as shown in Figure 5, we generated five overexpression lines (named OE#1 to OE#5) and selected two lines (OE#1 and OE#2) for further study after examining Put2 mRNA levels. Meanwhile, two lines of *Put2* mutants (*put2*#1 *and put2*#2) were generated by CRISPR/Cas9 technology, which induced frameshift mutations. Next, we compared the salt tolerance of these tomato plants. In comparison to WT plants, the *Put2*-OE#1 and *Put2*-OE#2 plants exhibited decreased sensitivity to salt stress with less wilted leaves, higher *Fv/Fm* and dry weight, and lower REL after a salt treatment for 7 days. Whereas *put2* mutants displayed salt hypersensitivity with lower *Fv/Fm* and higher REL than WT. These results suggest that Put2 positively regulates salinity tolerance.

To explore whether Put2 responds to salinity stress through mediating the Na^+ balance, we further analyzed the Na^+ and K^+ contents in WT, *put2* mutants, and *Put2*-OE plants subjected to NaCl treatment. K^+ and Na^+ content did not differ significantly among the genotypes examined under normal conditions. However, after treatment with 200 mM NaCl for 7 days, in comparison to WT plants, the levels of Na^+ in shoots parts of the *put2* mutants were much higher, the K^+ content was significantly lower in mutants, and so the ratios of Na^+/K^+ were increased. Conversely, the contents of Na^+ and the Na^+/K^+ ratio in the *Put2*-OE plants were lower, and the levels of K^+ were higher (Figure 5H).

The transcription levels of encoding salt response genes such as *SOS1-3* and *NHX1-3* were examined in the WT, *put2* mutants, and *Put2*-OE plants using qRT-PCR. The mRNA abundance of *SOS1-3* and *NHX1-3* in the mutants was lower than that in WT. On the contrary, the transcript levels of these genes were significantly increased in *Put2*-OE plants. These results demonstrated that Put2, as a positive regulator, is mediated by Na^+/K^+ homeostasis (Figure S2).

3.8. Put2 Improves Salt Tolerance by Facilitating Polyamines Synthesis

Previous studies reported that polyamines uptake proteins can facilitate polyamines synthesis, at least in rosette leaves of *Arabidopsis thaliana* plants (Ahmed et al., 2017), which enlightened us to next investigate the polyamine metabolites. In the polyamine synthesis pathway, key representative molecules (Put, Spd, Spm, and Dap) are derived from the amino acid Arg (Figure 6A). Significantly, there was a considerable increase in the concentrations of Arg (the Put/Spd/Spm precursor), Spd, and Put in the leaves of *Put2*-OE plants, but the content in the *put2* mutants were lower than those of WT plants in the absence of salt stress. When exposed to the salt treatment, the endogenous content of Arg, Spd, and Put increased significantly in *Put2*-OE plants than that of WT and *put2* mutants, whereas the levels of these polyamines in *Put2*-OE plants were still considerably higher than those in the WT plants (Figure 6B–D). Additionally, the Spm and Dap content showed no significant differences among the WT, *put2* mutants, and *Put2*-OE lines (Figure 6E,F).

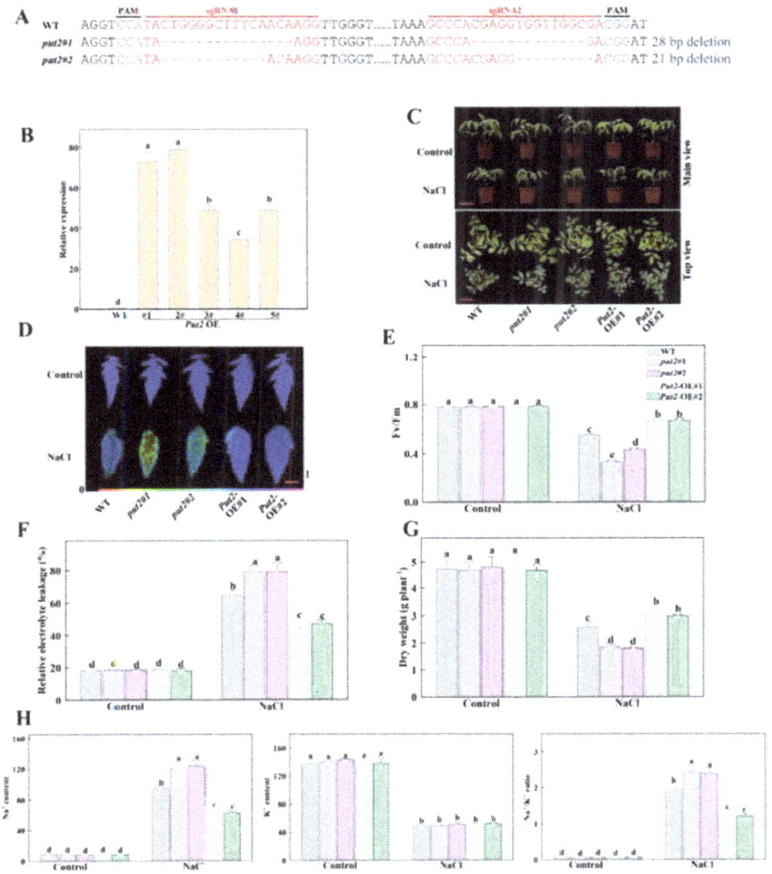

Figure 5. Put2 positively regulates salt tolerance. (**A**) Genotyping of mutations in *put2#1* and *put2#2*. Red letters indicate the target sites, '-' represent sequence deletion, and blue letters represent the protospacer adjacent motif (PAM). (**B**) qRT-PCR analysis of *Put2* transcript levels in WT and Put2 OE (1#, 2#, 3#, 4# and 5#). (**C**) Phenotypes of WT, *put2* mutants, or *Put2* OE lines after exposure with or without salt stress for 7 days. (**D**) *Fv/Fm* in WT, *put2* mutants, or *Put2* OE lines leaves after exposure with or without salt stress for 7 days. The false-color code depicted at the bottom of image range from 0 (black) to 1.0 (purple), showing the level of damage in the leaves. (**E**) Quantitative analysis of *Fv/Fm* as shown in (**D**). (**F**) The relative electrolyte leakage and (**G**) dry weight in WT, *put2* mutants, or *Put2* OE lines after exposure with or without salt stress for 7 days. (**H**) Ion content and Na^+/K^+ ratio in shoots of WT, *put2* mutants, or *Put2* OE lines after exposure with or without salt stress for 7 days. Data are presented as mean values ± SD; n = 3. Different letters indicate significant differences between treatments ($p < 0.05$, Duncan's multiple range test). At least three independent experiments were performed.

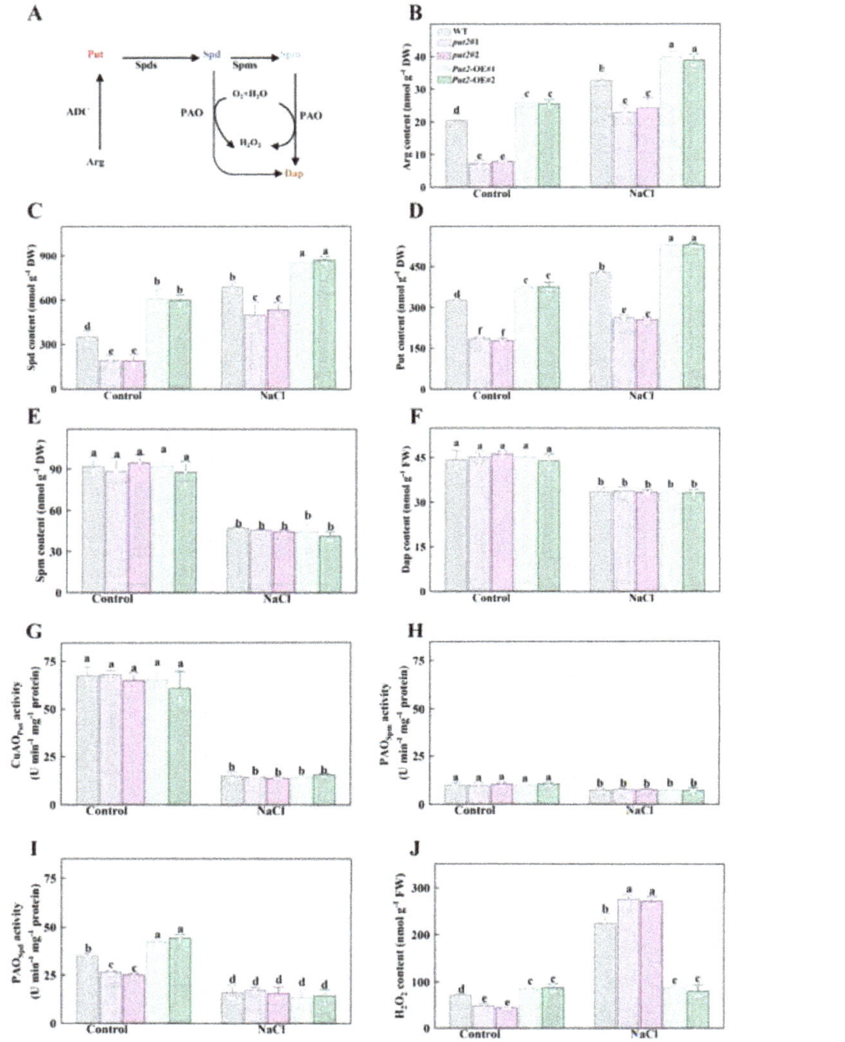

Figure 6. The role of Put2 in the regulation of polyamine metabolism and H_2O_2 content. (**A**) Simplified scheme of polyamine biosynthesis relevant to this study in plants; Arg, arginine, ADC, arginine decarboxylase, Put, putrescine, Spds, spermidine synthase, Spd, spermidine, Spms, spermine synthase, Spm, spermine, PAO, polyamine oxidase, Dap, 1,3-diaminopropane. (**B**) The Arg content in WT, *put2* mutants, or *Put2* OE lines with or without salt stress for 7 days. The content of the polyamines Spd (**C**), Put (**D**), Spm (**E**), and Dap (**F**) in WT, *put2* mutants, or *Put2* OE lines with or without salt stress for 7 days. The Put-dependent CuAO (**G**), Spd-dependent PAO (**H**), and Spm-dependent PAO (**I**) enzymatic activity in WT, *put2* mutants, or *Put2* OE lines with or without salt stress for 7 days. (**J**) The H_2O_2 content in WT, *put2* mutants, or *Put2* OE lines with or without salt stress for 7 days. Data are presented as mean values ± SD; n = 3. Different letters indicate significant differences between treatments ($p < 0.05$, Duncan's multiple range test). At least three independent experiments were performed.

In leaves, knockout or gain of function of Put2 did not influence the Put-dependent PAO and Spm-dependent PAO activity in the absence or presence of salt treatment (Figure 6G,H). By contrast, *Put2*-OE plants showed an increase in Spd-dependent PAO activity, whereas *put2* mutants exhibited a decrease in this activity, relative to WT plants under normal conditions. During salinity stress, there were no significant changes in the activity of Spd-dependent PAO between WT, *put2* mutants, and *Put2*-OE lines (Figure 6I). Since H_2O_2 is one of the PAO reaction products, the levels of H_2O_2 were measured. The leaves of *Put2*-OE plants showed an evident increase in H_2O_2 content in the absence of salt stress, whereas the largest rise was shown in *put2* mutants in the presence of 200 mM NaCl (Figure 6J), proving that *put2* mutants experience more severe oxidative damage from salt stress than WT and *Put2*-OE plants. Thus, it is possible that Put2-induced H_2O_2 production is closely associated with PAOs under normal conditions. On the other hand, Put2 contributes to the decrease of H_2O_2 content under salt conditions, which declined oxidative damage and enhanced salt tolerance. These results demonstrate that Put2 is required for some polyamines metabolism intermediates (including Arg, Spd, and Put), as well as the enzyme (PAO).

3.9. Put2 Decreases ROS Levels under Salinity Stress

The production of ROS is known to be increased under stress conditions, and H_2O_2 is the most stable ROS. To determine whether Put2 reduces ROS accumulation through antioxidant enzyme and non-enzymatic compound regulation under salinity stress, the following relevant indicators were detected. During salt stress, the level of H_2O_2 and MDA, two indicators of oxidative damage during salt stress, were significantly lower in *Put2*-OE plants than in WT; increases in H_2O_2 and MDA content were detected in *put2* mutants (Figures 6J and 7A). Subsequently, antioxidant enzymes, including SOD, CAT, APX, and POD, are in the midst of the primary enzyme defense against ROS. The activities of SOD, CAT, APX, and POD were more pronounced in *Put2*-OE than in WT. In contrast, *put2* mutants showed lower antioxidant enzyme activity than WT (Figure 7), which is consistent with lower levels of H_2O_2 and MDA in *Put2*-OE than in WT and *put2* mutants. These results indicate that Put2 improves salinity tolerance by reducing oxidative damage.

To confirm whether the Put2-mediated ROS decrease is regulated by non-enzyme compounds, we determined the effects of glutathione redox homeostasis, GABA, and flavonoid contents in *put2* mutants and *Put2*-OE plants, which however, revealed no difference from the WT under normal conditions, except for GABA. The GABA content in *put2* mutants was significantly lower than that of WT and *Put2*-OE plants before salt treatment. Overexpression of *Put2* also considerably elevated the AsA/DHA and GSH/GSSG ratios, as well as the content of GABA and flavonoid, while knockout of *put2* compromised the increase in these parameters compared to that in the WT under salinity conditions (Figure 8).

3.10. Put2 Triggers Upregulation of Polyamine Synthesis and Is Related to Detoxification Gene Expression

To investigate the transcriptional regulation of polyamine synthesis and detoxification by Put2, we examined the transcript levels of 27 genes involved in polyamine synthesis, ROS detoxification, and GABA synthesis by qRT-PCR in leaves under normal or salinity conditions. qRT-PCR analysis of polyamine synthesis genes in *put2*#1 mutant and *Put2*-OE#1 plants revealed that the set genes of encoding polyamine synthesis (including *ADC1*, *ADC2*, *SPDS1*, *SPDS2*, *SPMS1*, *SPMS2*, *PAO1*, *PAO3*, and *PAO5*) were upregulated in *Put2*-OE#1 plants; in contrast, they were downregulated in *put2*#1 mutants Furthermore, salinity treatment caused relative increases in the expression of polyamine synthesis genes only in WT and *Put2*-OE#1 plants (column 3 vs. column 1, and column 5 vs. column 3), while they decreased in *put2*#1 mutants (column 4 vs. column 3). Collectively, the analysis of differentially expressed genes also suggests that the modifications in gene expression in response to the stress conditions are more intense in the *Put2*-OE#1 plant. Additionally, we noticed an interesting finding with the greatest upregulation of the *PAO5* gene in the five *PAOs* genes in the *Put2*-OE#1 plant under normal conditions, which may be a mechanism for the relatively higher PAO activity and levels of H_2O_2 in *Put2* OE plants (Figures 6J and 8).

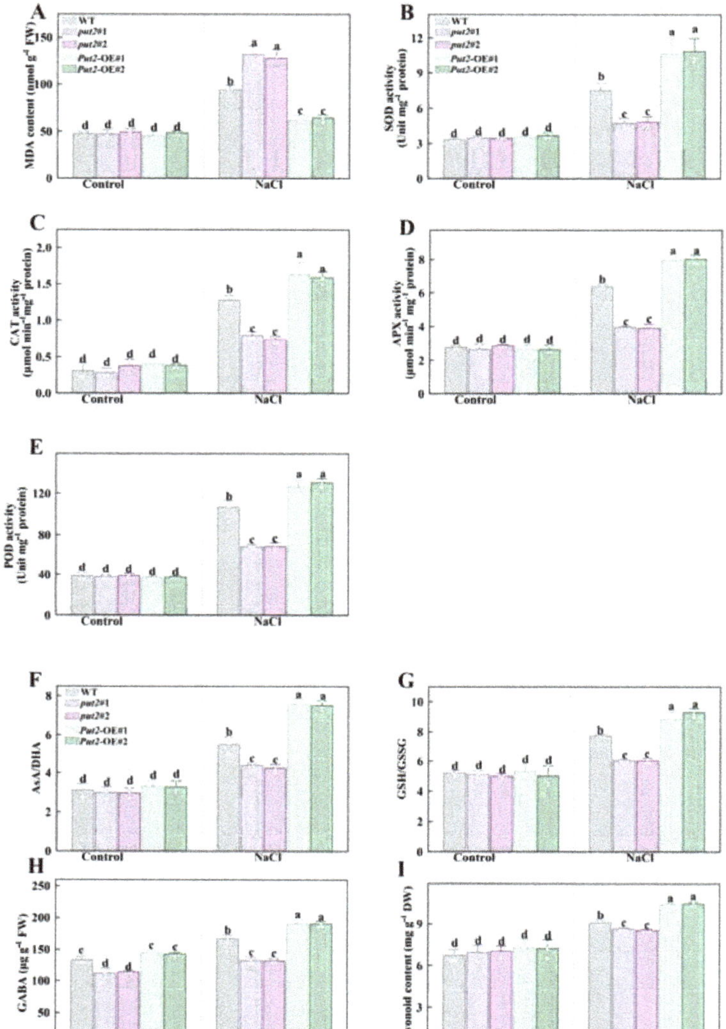

Figure 7. Put2 enhances salt tolerance via positively regulating ROS-scavenging enzyme activity and nonenzymatic antioxidant process. The MDA content (**A**), SOD (**B**), CAT (**C**), APX (**D**), and POD (**E**) activity in WT, *put2* mutants, or *Put2* OE lines with or without salt stress for 7 days. (**F**) The ratio of ascorbic acid (AsA) to dehydroascorbate (DHA), (**G**) the ratio of glutathione (GSH) and glutathione disulfide (GSSG), (**H**) the gamma-aminobutyric acid (GABA), and (**I**) the total flavonoids in WT, *put2* mutants, or *Put2* OE lines with or without salt stress for 7 days. Data are presented as mean values ± SD; n = 3. Different letters indicate significant differences between treatments ($p < 0.05$, Duncan's multiple range test). At least three independent experiments were performed.

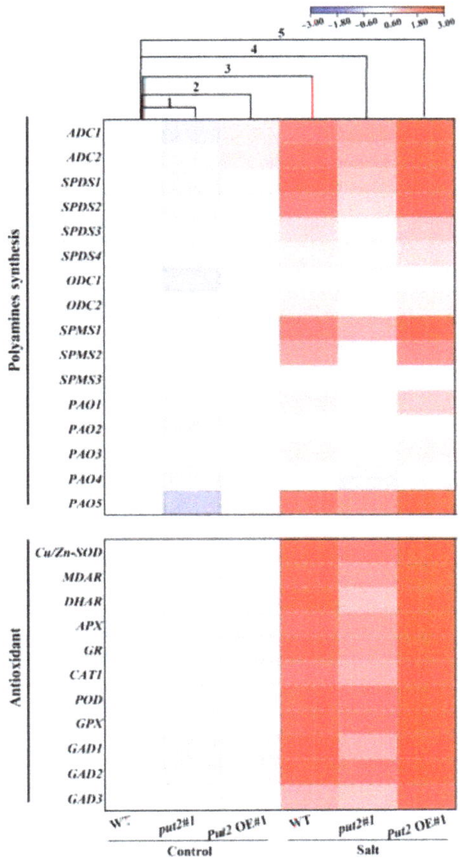

Gene	1	2	3	4	5
	Log₂ fold change				
ADC1	**−0.41**	**0.492**	**1.77**	**1.2**	**2.4**
ADC2	**−0.293**	**0.462**	**1.91**	**1.01**	**2.31**
SPDS1	**−0.374**	**0.29**	**2.19**	**0.835**	**2.33**
SPDS2	**−0.387**	**0.173**	**1.8**	**0.694**	**2.05**
SPDS3	−0.0422	**0.23**	**0.666**	**0.242**	**0.839**
SPDS4	−0.0418	−0.0138	**0.58**	**0.24**	**0.729**
ODC1	**−0.54**	0.00146	**0.303**	−0.168	**0.315**
ODC2	0.0369	0.0442	**0.441**	**0.135**	**0.587**
SPMS1	−0.0435	−0.0144	**1.67**	**1.07**	**2.4**
SPMS2	−0.026	−0.089	**1.2**	**0.259**	**1.54**
SPMS3	−0.0199	0.0572	**0.253**	0.0768	**0.377**
PAO1	**−0.138**	−0.0457	**0.587**	**0.138**	**0.855**
PAO2	**−0.398**	0.0875	**0.123**	0	**0.342**
PAO3	**−0.162**	**0.245**	**0.428**	**0.195**	**0.542**
PAO4	**−0.332**	**0.112**	**0.147**	−0.332	**0.248**
PAO5	**−0.987**	**0.272**	**1.96**	**1.55**	**2.49**
Cu/Zn-SOD	0.0283	−0.0172	**2.4**	**1.67**	**2.86**
MDAR	−0.0303	0.00285	**2.18**	**1.2**	**2.7**
DHAR	−0.0416	−0.00993	**2.25**	**0.947**	**2.78**
APX	−0.0156	−0.0531	**1.96**	**1.2**	**2.66**
GR	−0.0142	−0.04	**2.08**	**1.24**	**2.77**
CAT1	−0.0649	−0.0155	**1.64**	**1.17**	**2.37**
POD	−0.0149	**0.118**	**2.14**	**1.74**	**2.5**
GPX	−0.0244	0.0119	**2.09**	**1.63**	**2.64**
GAD1	**−0.369**	**0.256**	**2.37**	**1.2**	**2.83**
GAD2	**−0.228**	**0.286**	**2.29**	**1.69**	**2.78**
GAD3	0.00142	0.00426	**1.2**	**0.903**	**2.06**

Figure 8. Put2 positively mediates the expression of polyamine synthesis and antioxidant enzyme-encoding genes. Heatmap of polyamine synthesis and antioxidant enzyme-encoding genes. Levels were differentially modified in *put2* vs. WT plants, and *Put2* OE vs. WT plants under normal conditions or salinity stress. Log_2 fold changes (relative to expression levels in sample of WT plants under control treatment) are shown with a color scale. A chart on the right side of the heatmap shows log_2 fold change of expression levels in *put2* (mutant) compared to those in WT plants before salt treatment (column 1), log_2 fold change of expression levels in *Put2* OE (overexpression) compared to those in WT plants before salt treatment (column 2), log_2 fold change caused by salinity treatment in WT plants (column 3), and log_2 fold change caused by salinity treatment in *put2* (column 4) or *Put2* OE (column 5). Statistically significant differences are indicated in bold font ($p < 0.05$).

Similarly, the transcriptional levels of ROS detoxification-related genes, including *Cu/Zn-SOD*, *MDAR*, *DHAR*, *APX*, *GR*, *CAT1*, *POD*, and *GPX*, were further upregulated in the *Put2*-OE#1 plant compared with WT plants after the salinity treatment. By contrast, the induction of ROS detoxification-related genes in response to salt stress was compromised in *put2* mutants; increased ROS-scavenging capacity failed in the *put2*#1 mutant (Figure 8). In addition, the expression of *GAD1*, *GAD2*, and *GAD3*, encoding glutamate decarboxylase, likely the key enzyme for GABA biosynthesis in the tomato, was induced in the *Put2*-OE#1 plant before salt treatment. Accordingly, the transcript levels of *GADs* were further enhanced after salt treatment in the *Put2*-OE#1 plant. Notably, increases in the expression of *GADs* in response to salt stress were suppressed in the *put2*#1 mutant. The enhanced

GADs transcripts and GABA content was observed in the WT and *Put2*-OE plants but not in the *put2* mutants after salt stress (Figures 7 and 8). Thus, the results indicate that Put2 participates in enhanced salt tolerance by mediating the ROS-scavenging capacity.

4. Discussion

Polyamines, which are small polycationic molecular regulators and signaling molecules, not only orchestrate fundamental growth and development in plants, but also induce a series of stress cascades [38]. Polyamines are involved in various pathways in plants and lessen the lethal effects of abiotic stresses by regulating transcription factors, hormonal responses, antioxidant enzymes, and the activation of signaling cascades [39,40]. Polyamine biosynthesis and degradation play important roles in various abiotic stress tolerance pathways and are closely associated with coping with ROS production [41]. The polyamine transporter belongs to the mammalian L-type amino acid transporter family, like polyamine synthesis and metabolism protein, which plays a critical role in plant growth, development, and the stress responses [42]. Though Put has been analyzed in *Arabidopsis*, rice, and the sweet orange, the effects of Put on abiotic stresses in the tomato have not been reported [43,44]. Here, we looked at the physiological and molecular function mechanisms of Put in the resistance to abiotic stress in the tomato. Eight Put proteins in the tomato were identified and characterized at the complete genome level through alignment with *Arabidopsis* Put proteins. Then, extensive analyses on the tomato Put proteins, including phylogenetic development, gene and protein structure, physiochemical characteristics, motifs, miRNAs, and cis elements, and their substrates were performed; and the effects of Puts on abiotic stress tolerance was evaluated in yeast and tomatoes. Furthermore, we demonstrated the importance of the Put2 mediation of polyamine metabolism and antioxidant capacity, and that overexpression of *Put2* increased polyamines to influence the homeostasis of antioxidant capacity, showing that Put2 was a positive regulator for salinity stress in tomatoes. These findings enlarge the comprehension of Put family members, which may be employed in breeding for genetic modification and the development of abiotic stress tolerant crops.

4.1. Identification of Tomato Put Family

The tomato Put proteins were divided into four groups, which have a high similarity to those in rice and *Arabidopsis* (Figure 1). Differences among these proteins are probably due to environmental impacts, and analysis of the structure and motif in these genes indicated that *Put* genes disclosed a close exon-intron and motif structure, showing that a closer evolutionary pattern in these genes and diverse functional relationships also exist among the other group members (Figure 1). Interestingly, protein analysis indicated that the PotE (Putrescine-ornithine antiporter) motif is highly conserved, suggesting that it may be involved in amino acid and polyamine transport (Figure 1C) [45,46]. A series of cis elements is a specific sequence at the promoter region of a given gene, which is involved in the expression of protein-coding transcripts and is mediated by transcriptional regulation and small RNAs. Repression and activation of gene expression by binding with these cis elements are a general means of modulating various life processes [47]. Plant miRNAs have a function in regulating related genes that are involved in response to abiotic stresses [48]. miR390 was strongly induced after exposure to salinity during lateral root formation in poplar and positively regulated auxin signaling subjected to salt stress [49]. miR6024 negatively mediates the resistance genes and defense system, facilitates disease by the necrotrophic pathogen *A. solani*, and perturbs immunity in the tomato [50]. The module of miR164a-NAM3 affords cold resistance by increasing ethylene production in the tomato [51]. Tomato Put was targeted by miRNAs including miR390, miR6024, and miR164a, which might be associated with various stress responses (Table 2). We also discovered several types of conserved cis-regulatory elements in the promoter regions of Put; these cis elements are associated with transcriptional regulation of the core gene network, and of plant growth and development [52]. The *Put* genes of the tomato contain various cis elements including stress, hormone, light, auxin, GA, ABA, and methyl jasmonate (MeJA) responsive elements (Figure 1E). The presence of numerous abiotic stress-specific cis-regulatory motifs and hormonal cis elements implicate these genes stimulating the hormone signaling pathways and providing stress tolerance in the tomato [47]. Of note, the cis elements activate their downstream genes after binding to specific transcription factors, playing an important amplifier role during various abiotic/biotic stresses. Meanwhile, we also speculate that Put is involved in salinity stress since these cis-regulatory elements are also closely implicated in salinity tolerance [53].

4.2. Expression Profiles of the Put Gene Family after Treatment with Various Hormones, Polyamines, and Abiotic Stresses, in Different Tissues

A gene expression profile can provide critical symbols for its biological functions. Here, we examined the expression pattern of the *Put* genes via qRT-PCR under treatment with exogenous hormones and polyamines, as well as abiotic stress conditions. We observed the *Put* genes appeared to be upregulated in ABA treatment. On the other hand, all Put genes, except for *Put7*, were induced by SA. *Put6, Put7,* and *Put8* were inhibited initially after GA_3 treatment. The expression of five *Put* (*Put4*, *Put5*, *Put6*, *Put7*, and *Put8*) genes were downregulated, and the other *Put* (*Put1*, *Put2* and *Put3*) genes were induced in leaves at 3 and 6 h of ET treatment. Similarly, the critical role of growth factors and hormones in increasing polyamine transport rates in mammalian cells has been demonstrated [54]. Together with the presence of hormone-responsive cis elements and altered transcripts levels after hormone treatment, this implies that Put may have a crucial role in the hormone regulatory pathway. Furthermore, the differential expression profiling of *Put* in response to polyamine treatment was evaluated. All *Put* genes are involved in exogenous polyamines treatment; the *Puts* exhibited varied patterns in response to the same polyamine. For instance, *Put2* and *Put5* were dramatically upregulated, while the other genes were slightly induced after spermidine treatment. However, putrescine resulted in suppression of these *Put* genes. Additionally, spermine led to the upregulation of *Puts*, especially *Put2* and *Put5*. On the other hand, the qRT-PCR analysis revealed pronounced effects of the spermidine and spermine-induced *Put* gene expression (Figure 2B). Since polyamines are used as substrates required for the polyamine uptake proteins [43], the different polyamine responses may involve the substrate specificity of polyamine transport and homeostasis. In fact, rice Put1 is a specific and high-affinity spermidine uptake transporter involved in polyamines uptake, leading to the accumulation of polyamines in yeast [55]. The lower affinity of Put may be the reason for its higher proportion in the free state. Therefore, spermidine, spermine, and putrescine, may severe as Put substrates. Interestingly, after salt and drought stress, *Put2* and *Put5* were significantly induced compared with the others. However, their expression was significantly inhibited under cold and heat temperature stress (Figure 2C). Altogether, these results indicate that *Put* genes are potentially involved in hormone and polyamine induction, as well as in response to salinity stress.

The *Put* genes exhibited a divergent expression pattern in different tissues. *Put2* and *Put5* displayed the highest expression in leaves and roots, and all genes had high expression in flowers and fruit (Figure 3A). Similarly, the Put family appears to have a distinct tissue expression profile in *Arabidopsis* and *Citrus sinensis* [43,44], and a divergent pattern of intracellular localization [13], which implied specialization in a spatial manner. Furthermore, *Put2* was more significantly induced by salt stress than *Put5* in leaves and roots, indicating that the functional role of *Put2* related to salt stress may be important. In the tomato, the functional role of *Put2* in abiotic stress tolerance remains largely unknown.

4.3. Put2 Contributes to Polyamins Biosynthesis and Catabolism Associated with Salt Tolerance

Although our results above have shown that Put is involved in abiotic stress, its function is yet to be understood, especially regarding polyamine transport and salt stress. Previous studies have shown that yeast is an excellent heterologous expression system to study the function of genes in polyamine transport and salt stress [43]. Here, we used the yeast model to preliminarily investigate their function in polyamine uptake and salt stress tolerance (Figure 4). The results showed that transformants of *agp2Δ* mutants expressing *Put2* and *Put5* had higher sensitivity to spermidine and paraquat, indicating that both function as an importer. Here, we also showed that the yeast *agp2Δ* cells' capacity to transport paraquat may be compensated by Put2 and Put5 (Figure 4A), since paraquat is transported by the polyamine transport system [12]. A time course absorption experiment directly provided evidence that *Put2* and *Put5* encoded a transporter that can regulate polyamines import, with high activity of polyamine uptake for Put2 (Figure 4B). Furthermore, overexpression of *Put2* increased salt tolerance in yeast, hampered the influx of Na^+, and enhanced K^+ uptake (Figure 4C,D). Indeed, polyamine transporters have recently been linked to the regulation of salt stress through promoting Na^+ efflux and K^+ channels [56]. Thus, combining the previous results in this article (Figures 2–4), we speculate that the induction of *Put2* expression by salt may be regulating the polyamines and Na^+/K^+ homeostasis to alleviate salt damage.

To further validate the function of Put2, the *put2* mutants and overexpression lines were generated (Figure 5A,B). Our results showed that *put2* mutants were more shriveled than WT plants under salinity stress, whereas overexpression elevated salt tolerance (Figure 5C). Likewise, *Put2*-OE plants displayed increased levels of Fv/Fm and dry weight, and reduced levels of relative electrolyte, which agrees with an increase in salt tolerance and a decrease in Na^+ content and the Na^+/K^+ ratio (Figure 5D–H). Similarly, Put3 is critical for Na^+ and K^+ homeostasis by physically interacting with SOS1 and SOS2, forming a complex with SOS2 under stress conditions [56]. As such, the induction of *SOS1-3* and *NHX1-3* in *Put2*-OE plants could also synergistically activate the SOS1 and SOS2 (Figure S2). Thus, subsequent increases in the Put2 activity would enhance salt tolerance by activating the Na^+/H^+ exchange activity.

Polyamines, an important regulator in the plant kingdom, are necessary for plant growth, development, and the stress response. The dynamic balance of polyamines in the plant is stringently regulated by polyamine synthesis, degradation, and transport [57,58]. The latter was previously involved in subcellular polyamine transport through the complementation experiment in yeast [55], and the transport of paraquat in *Arabidopsis* [12]. While the implication of *Put2* in polyamine biosynthesis and catabolism was not noted. Here, under control conditions, overexpression of *Put2* increases the endogenous Arg, Spd, and Put content, which failed to increase in *put2* mutants. Upon salt stress, meanwhile, the polyamine content in WT plants performed much better than *put2* mutants, and *Put2*-OE plants were better than the WT (Figure 6A–D). NaCl supply treatment also increased the activation of polyamine synthesis-related genes more clearly in *Put2*-OE plants (Figure 8), demonstrating the important function of Put2 in the polyamine biosynthesis process. On the other hand, considering polyamine catabolism, the higher activity of PAO_{spd} confirmed the acceleration of polyamine catabolic reactions in *Put2*-OE plants (Figure 6I), and evidenced that Put2 positively regulates PAO activity. In addition, two main sources of ROS were indicated to exist in plants, including NADPH oxidases and polyamines catabolism by PAO activity [8,59]. In *put2* leaves, H_2O_2 was lowered with respect to WT plants under normal conditions; however, an upregulation was observed in *Put2*-OE plants, and likewise PAO activity and *PAO* gene expression were altered (Figures 6J and 8), further demonstrating that Put2 attributes to PAO activity. Accordingly, enhancement of PAO activity has been shown to alleviate salinity damage and increase the polyamines and H_2O_2 content [60,61]. Therefore, Put2 may contribute to governing polyamine biosynthesis and catabolism. However, we cannot completely rule out other possibilities, for example, Put2 may be capable of regulating the long-distance and appropriate tissue distribution of polyamines [62].

4.4. *Put2-Mediated Antioxidant Capacity Establishes Suitable ROS Levels under Salt Stress*

PAO activity has been reported to contribute to an increase in salt tolerance through the production of H_2O_2 [63]. However, *Put2*-OE plants treated with NaCl showed that the increase in H_2O_2 content was lower than in WT plants. Furthermore, the H_2O_2 content increased considerably in *put2* mutants accompanied by serious salt stress injury, which correlated inversely with the activities of PAO (Figure 6J). Thus, PAO-induced H_2O_2 production was stunted under salt conditions in *Put2*-OE plants; therefore, another mechanism to eliminate H_2O_2 must exist. Indeed, plants exposed to salt stress generate a super-excess of ROS, which is highly toxic and can overwhelm the PAO-induced H_2O_2, ultimately damaging cellular activity and leading to plant death [8]. In the current study, *Put2*-OE plants exhibited obviously increased activities of antioxidant enzymes after salt treatment. Additionally, the ASA-GSH cycle was also activated by overexpression of *Put2*, which could be another exploration for Put2 enhanced salt tolerance, as the enzymatic system and scavenging procedure could be activated in *Put2*-OE plants (Figures 7 and 8). These findings indicate that Put2 functions as an important regulator linking polyamines and ROS, and affects both the production and elimination of ROS. The results coincided with a previous study showing that Put2 promoted phyA-mediated germination by sensing seed oxidation and protecting the decaying seed from oxidative damage [64]. Therefore, there is strong evidence that Put2 increased salt tolerance probably by promoting antioxidants in plants. A similar report has been conducted showing that CsPUT4 can protect against cold stress by modulating polyamine homeostasis and turning on the antioxidant enzyme defense system in the sweet orange [44]. Meanwhile, GABA and flavonoids, as free radical scavengers, have been exhibited to alleviate salinity damage and heat damage by inducing polyamine enhancement [65,66], which are increased in *Put2*-OE plants under salinity stress; this further demonstrates positive feedback regulation by Put2.

5. Conclusions

In this work, eight Put family proteins were found in the tomato, and their chromosomal location, structure, phylogenetic tree, and physiochemical properties were investigated. Furthermore, molecular characterization was performed in yeast to understand their involvement in polyamine uptake and salt stress tolerance. Additionally, the cis elements in the promoter, miRNAs targeting Put, and the expression profiles of *Put* genes in different tissues and their responses to exogenous hormones and polyamines, as well as abiotic stress, were analyzed, proving they may play a vital function in abiotic stress, growth, and development. Furthermore, we show the role of Put2, which, to our knowledge, is the first polyamine uptake protein characterized in the tomato shown to play a role in salinity tolerance. Firstly, in yeast, Put2 was highly tolerant to salt stress, as indicated by less Na^+ invasion and K^+ efflux, which also could be attributable to an enhancement in the absorption of polyamines. Importantly, overexpression of *Put2* in the tomato decreased salinity sensitivity, evidenced by enhanced polyamine biosynthesis and catabolism and maintained Na^+/K^+ homeostasis, in addition to activated ROS-scavenging enzyme activities and nonenzymatic antioxidant process. These findings shed light on Put2-regulated salinity tolerance in the tomato. Here, we provide comprehensive deciphering of the mechanisms of Put2 for enhancing salt tolerance and some valuable evidence for interpreting the potential functions of tomato Put genes in abiotic stress tolerance.

Clearly, further studies are required to understand the precise function of Put and the upstream and downstream targets of the individual Puts. Generating loss- and gain-of-function mutations and characterizing their roles will provide useful tools, generating new evidence and new findings.

Supplementary Materials: The following supporting information can be downloaded at: https://www.mdpi.com/article/10.3390/antiox12020228/s1, Figure S1. Cartoon of the predicted topology of tomato Put proteins. Transmembrane domains are indicated as blue rectangle. Figure S2. Relative transcript levels of SOS1, SOS2 and SOS3 (A), NHX1, NHX2 and NHX3 (B) gene expression in WT, put2 mutants, or Put2 OE lines after exposure with or without salt stress for 7 days. Data are presented as mean values ± SD; n = 3. Different letters indicate significant differences between treatments ($p < 0.05$, Duncan's multiple range test). At least three independent experiments were performed. Figure S3. The MDA content in WT, put2 mutants, or Put2 OE lines after exposure with or without salt stress for 7 days. Data are presented as mean values ± SD; n = 3. Different letters indicate significant differences between treatments ($p < 0.05$, Duncan's multiple range test). At least three independent experiments were performed. Supplemental data S1. The protein sequences of Put in Arabidopsis, rice and tomato, respectively. Supplemental data S2. The Put genes Cis-element distribution. Supplemental data S3. Primers used in this study.

Author Contributions: M.Z.: Writing—original draft. L.Y., W.L. and H.Q.: Methodology. B.L. and R.H.: Review and editing. X.Y. and Y.K.: Supervision. All authors have read and agreed to the published version of the manuscript.

Funding: This work was financially supported by Guangdong Provincial Special Fund for Modern Agriculture Industry Technology Innovation Teams: No. 2023KJ122, Guangzhou basic and applied basic research foundation (SL2022A04J00131), GuangDong Basic and Applied Basic Research Foundation (2214050009633).

Institutional Review Board Statement: Not applicable.

Informed Consent Statement: Not applicable.

Data Availability Statement: Data are contained within the article and Supplementary Material.

Conflicts of Interest: The authors declare no conflict of interest.

References

1. Del Duca, S.; Serafini-Fracassini, D.; Cai, G. Senescence and programmed cell death in plants: Polyamine action mediated by transglutaminase. *Front. Plant Sci.* **2014**, *5*, 120. [CrossRef] [PubMed]
2. Aloisi, I.; Cai, G.; Serafini-Fracassini, D.; Del Duca, S. Transglutaminase as polyamine mediator in plant growth and differentiation. *Amino Acids* **2016**, *48*, 2467–2478. [CrossRef] [PubMed]
3. Romero, F.M.; Maiale, S.J.; Rossi, F.R.; Marina, M.; Ruíz, O.A.; Garriz, A. Polyamine metabolism responses to biotic and abiotic stress. *Polyamines* **2018**, *1694*, 37–49.
4. Wang, W.; Paschalidis, K.; Feng, J.-C.; Song, J.; Liu, J.-H. Polyamine catabolism in plants: A universal process with diverse functions. *Front. Plant Sci.* **2019**, *10*, 561. [CrossRef]
5. Sagor, G.; Kusano, T.; Berberich, T. A polyamine oxidase from *Selaginella lepidophylla* (SelPAO5) can replace AtPAO5 in *Arabidopsis* through converting thermospermine to norspermidine instead to spermidine. *Plants* **2019**, *8*, 99. [CrossRef]
6. Zhang, Y.; Wang, Y.; Wen, W.; Shi, Z.; Gu, Q.; Ahammed, G.J.; Cao, K.; Shah Jahan, M.; Shu, S.; Wang, J. Hydrogen peroxide mediates spermidine-induced autophagy to alleviate salt stress in cucumber. *Autophagy* **2021**, *17*, 2876–2890. [CrossRef]
7. Chen, D.; Shao, Q.; Yin, L.; Younis, A.; Zheng, B. Polyamine function in plants: Metabolism, regulation on development, and roles in abiotic stress responses. *Front. Plant Sci.* **2019**, *9*, 1945. [CrossRef]
8. Pottosin, I.; Velarde-Buendía, A.M.; Bose, J.; Zepeda-Jazo, I.; Shabala, S.; Dobrovinskaya, O. Cross-talk between reactive oxygen species and polyamines in regulation of ion transport across the plasma membrane: Implications for plant adaptive responses. *J. Exp. Bot.* **2014**, *65*, 1271–1283. [CrossRef]
9. Urra, M.; Buezo, J.; Royo, B.; Cornejo, A.; López-Gómez, P.; Cerdán, D.; Esteban, R.; Martínez-Merino, V.; Gogorcena, Y.; Tavladoraki, P. The importance of the urea cycle and its relationships to polyamine metabolism during ammonium stress in *Medicago truncatula*. *J. Exp. Bot.* **2022**, *73*, 5581–5595. [CrossRef]
10. Ndayiragije, A.; Lutts, S. Long term exogenous putrescine application improves grain yield of a salt-sensitive rice cultivar exposed to NaCl. *Plant Soil* **2007**, *291*, 225–238. [CrossRef]
11. Cuevas, J.C.; López-Cobollo, R.; Alcázar, R.; Zarza, X.; Koncz, C.; Altabella, T.; Salinas, J.; Tiburcio, A.F.; Ferrando, A. Putrescine is involved in Arabidopsis freezing tolerance and cold acclimation by regulating abscisic acid levels in response to low temperature. *Plant Physiol.* **2008**, *148*, 1094–1105. [CrossRef]
12. Fujita, M.; Fujita, Y.; Iuchi, S.; Yamada, K.; Kobayashi, Y.; Urano, K.; Kobayashi, M.; Yamaguchi-Shinozaki, K.; Shinozaki, K. Natural variation in a polyamine transporter determines paraquat tolerance in *Arabidopsis*. *Proc. Natl. Acad. Sci. USA* **2012**, *109*, 6343–6347. [CrossRef]
13. Li, J.; Mu, J.; Bai, J.; Fu, F.; Zou, T.; An, F.; Zhang, J.; Jing, H.; Wang, Q.; Li, Z. Paraquat Resistant1, a Golgi-localized putative transporter protein, is involved in intracellular transport of paraquat. *Plant Physiol.* **2013**, *162*, 470–483. [CrossRef]
14. Dong, S.; Hu, H.; Wang, Y.; Xu, Z.; Zha, Y.; Cai, X.; Peng, L.; Feng, S. A *pqr2* mutant encodes a defective polyamine transporter and is negatively affected by ABA for paraquat resistance in *Arabidopsis thaliana*. *J. Plant Res.* **2016**, *129*, 899–907. [CrossRef]
15. Ahmed, S.; Ariyaratne, M.; Patel, J.; Howard, A.E.; Kalinoski, A.; Phuntumart, V.; Morris, P.F. Altered expression of polyamine transporters reveals a role for spermidine in the timing of flowering and other developmental response pathways. *Plant Sci.* **2017**, *258*, 146–155. [CrossRef]
16. Lyu, Y.-S.; Cao, L.-M.; Huang, W.-Q.; Liu, J.-X.; Lu, H.-P. Disruption of three polyamine uptake transporter genes in rice by CRISPR/Cas9 gene editing confers tolerance to herbicide paraquat. *aBIOTECH* **2022**, *3*, 140–145. [CrossRef]

17. Aouida, M.; Leduc, A.; Poulin, R.; Ramotar, D. AGP2 encodes the major permease for high affinity polyamine import in *Saccharomyces cerevisiae*. *J. Biol. Chem.* **2005**, *280*, 24267–24276. [CrossRef]
18. Minocha, R.; Majumdar, R.; Minocha, S.C. Polyamines and abiotic stress in plants: A complex relationship. *Front. Plant Sci.* **2014**, *5*, 175. [CrossRef]
19. Tao, Y.; Wang, J.; Miao, J.; Chen, J.; Wu, S.; Zhu, J.; Zhang, D.; Gu, H.; Cui, H.; Shi, S. The spermine synthase OsSPMS1 regulates seed germination, grain size, and yield. *Plant Physiol.* **2018**, *178*, 1522–1536. [CrossRef]
20. Lv, Y.; Shao, G.; Jiao, G.; Sheng, Z.; Xie, L.; Hu, S.; Tang, S.; Wei, X.; Hu, P. Targeted mutagenesis of POLYAMINE OXIDASE 5 that negatively regulates mesocotyl elongation enables the generation of direct-seeding rice with improved grain yield. *Mol. Plant* **2021**, *14*, 344–351. [CrossRef]
21. Liu, X.; Liu, S.; Chen, X.; Prasanna, B.M.; Ni, Z.; Li, X.; He, Y.; Fan, Z.; Zhou, T. Maize miR167-ARF3/30-polyamine oxidase 1 module-regulated H_2O_2 production confers resistance to maize chlorotic mottle virus. *Plant Physiol.* **2022**, *189*, 1065–1082. [CrossRef] [PubMed]
22. Yang, S.; Wang, J.; Tang, Z.; Li, Y.; Zhang, J.; Guo, F.; Meng, J.; Cui, F.; Li, X.; Wan, S. Calcium/calmodulin modulates salt responses by binding a novel interacting protein SAMS1 in peanut (*Arachis hypogaea* L.). *Crop J.* **2022**, *11*, 21–32. [CrossRef]
23. Wang, P.; Xu, Z.; Zhang, Y.; Ma, Y.; Yang, J.; Zhou, F.; Gao, Y.; Li, G.; Hu, X. Over-expression of spermidine synthase 2 (SlSPDS2) in tomato plants improves saline-alkali stress tolerance by increasing endogenous polyamines content to regulate antioxidant enzyme system and ionic homeostasis. *Plant Physiol. Biochem.* **2022**, *192*, 172–185. [CrossRef] [PubMed]
24. Khajuria, A.; Sharma, N.; Bhardwaj, R.; Ohri, P. Emerging role of polyamines in plant stress tolerance. *Curr. Protein Pept. Sci.* **2018**, *19*, 1114–1123. [CrossRef]
25. Gupta, K.; Dey, A.; Gupta, B. Plant polyamines in abiotic stress responses. *Acta Physiol. Plant.* **2013**, *35*, 2015–2036. [CrossRef]
26. Kumar, S.; Stecher, G.; Li, M.; Knyaz, C.; Tamura, K. MEGA X: Molecular evolutionary genetics analysis across computing platforms. *Mol. Biol. Evol.* **2018**, *35*, 1547–1549. [CrossRef]
27. Minh, B.Q.; Schmidt, H.A.; Chernomor, O.; Schrempf, D.; Woodhams, M.D.; Von Haeseler, A.; Lanfear, R. IQ-TREE 2: New models and efficient methods for phylogenetic inference in the genomic era. *Mol. Biol. Evol.* **2020**, *37*, 1530–1534. [CrossRef]
28. Dai, X.; Zhuang, Z.; Zhao, P.X. psRNATarget: A plant small RNA target analysis server (2017 release). *Nucleic Acids Res.* **2018**, *46*, W49–W54. [CrossRef]
29. Möller, S.; Croning MD, R.; Apweiler, R. Evaluation of methods for the prediction of membrane spanning regions. *Bioinformatics* **2001**, *17*, 646–653. [CrossRef]
30. Zhong, M.; Song, R.; Wang, Y.; Shu, S.; Sun, J.; Guo, S. TGase regulates salt stress tolerance through enhancing bound polyamines-mediated antioxidant enzymes activity in tomato. *Environ. Exp. Bot.* **2020**, *179*, 104191. [CrossRef]
31. Xie, X.; Ma, X.; Zhu, Q.; Zeng, D.; Li, G.; Liu, Y.-G. CRISPR-GE: A convenient software toolkit for CRISPR-based genome editing. *Mol. Plant* **2017**, *10*, 1246–1249. [CrossRef] [PubMed]
32. Fillatti, J.J.; Kiser, J.; Rose, R.; Comai, L. Efficient transfer of a glyphosate tolerance gene into tomato using a binary Agrobacterium tumefaciens vector. *Bio/Technol.* **1987**, *5*, 726–730. [CrossRef]
33. Zhong, M.; Wang, Y.; Shu, S.; Sun, J.; Guo, S. Ectopic expression of CsTGase enhances salt tolerance by regulating polyamine biosynthesis, antioxidant activities and Na^+/K^+ homeostasis in transgenic tobacco. *Plant Sci.* **2020**, *296*, 110492. [CrossRef] [PubMed]
34. Su, G.; An, Z.; Zhang, W.; Liu, Y. Light promotes the synthesis of lignin through the production of H_2O_2 mediated by diamine oxidases in soybean hypocotyls. *J. Plant Physiol.* **2005**, *162*, 1297–1303. [CrossRef] [PubMed]
35. Zhong, M.; Wang, Y.; Hou, K.; Shu, S.; Sun, J.; Guo, S. TGase positively regulates photosynthesis via activation of Calvin cycle enzymes in tomato. *Hortic. Res.* **2019**, *6*, 92. [CrossRef]
36. Zhang, G.; Bown, A.W. The rapid determination of γ-aminobutyric acid. *Phytochemistry* **1997**, *44*, 1007–1009. [CrossRef]
37. Zhishen, J.; Mengcheng, T.; Jianming, W. The determination of flavonoid contents in mulberry and their scavenging effects on superoxide radicals. *Food Chem.* **1999**, *64*, 555–559. [CrossRef]
38. Saha, J.; Brauer, E.K.; Sengupta, A.; Popescu, S.C.; Gupta, K.; Gupta, B. Polyamines as redox homeostasis regulators during salt stress in plants. *Front. Environ. Sci.* **2015**, *3*, 21. [CrossRef]
39. Liu, J.-H.; Wang, W.; Wu, H.; Gong, X.; Moriguchi, T. Polyamines function in stress tolerance: From synthesis to regulation. *Front. Plant Sci.* **2015**, *6*, 827. [CrossRef]
40. Song, J.; Sun, P.; Kong, W.; Xie, Z.; Li, C.; Liu, J.H. SnRK2. 4-mediated phosphorylation of ABF2 regulates ARGININE DECARBOXYLASE expression and putrescine accumulation under drought stress. *New Phytol.* **2022**. [CrossRef]
41. Chan, Z.; Yokawa, K.; Kim, W.-Y.; Song, C.-P. ROS regulation during plant abiotic stress responses. *Front. Plant Sci.* **2016**, *7*, 1536. [CrossRef]
42. Fujita, M.; Shinozaki, K. Identification of polyamine transporters in plants: Paraquat transport provides crucial clues. *Plant Cell Physiol.* **2014**, *55*, 855–861. [CrossRef]
43. Mulangi, V.; Chibucos, M.C.; Phuntumart, V.; Morris, P.F. Kinetic and phylogenetic analysis of plant polyamine uptake transporters. *Planta* **2012**, *236*, 1261–1273. [CrossRef]
44. Alhag, A.; Song, J.; Dahro, B.; Wu, H.; Khan, M.; Salih, H.; Liu, J. Genome-wide identification and expression analysis of Polyamine Uptake Transporter gene family in sweet orange (*Citrus sinensis*). *Plant Biol.* **2021**, *23*, 1157–1166. [CrossRef]
45. Kashiwagi, K.; Miyamoto, S.; Suzuki, F.; Kobayashi, H.; Igarashi, K. Excretion of putrescine by the putrescine-ornithine antiporter encoded by the potE gene of *Escherichia coli*. *Proc. Natl. Acad. Sci. USA* **1992**, *89*, 4529–4533. [CrossRef]

46. Tomitori, H.; Kashiwagi, K.; Igarashi, K. Structure and function of polyamine-amino acid antiporters CadB and PotE in *Escherichia coli*. *Amino Acids* **2012**, *42*, 733–740. [CrossRef]
47. Schmitz, R.J.; Grotewold, E.; Stam, M. Cis-regulatory sequences in plants: Their importance, discovery, and future challenges. *Plant Cell* **2022**, *34*, 718–741. [CrossRef]
48. Sunkar, R.; Li, Y.-F.; Jagadeeswaran, G. Functions of microRNAs in plant stress responses. *Trends Plant Sci.* **2012**, *17*, 196–203. [CrossRef]
49. He, F.; Xu, C.; Fu, X.; Shen, Y.; Guo, L.; Leng, M.; Luo, K. The MicroRNA390/TRANS-ACTING SHORT INTERFERING RNA3 module mediates lateral root growth under salt stress via the auxin pathway. *Plant Physiol.* **2018**, *177*, 775–791. [CrossRef]
50. Dey, S.; Sarkar, A.; Chowdhury, S.; Singh, R.; Mukherjee, A.; Ghosh, Z.; Kundu, P. Heightened miR6024-NLR interactions facilitate necrotrophic pathogenesis in tomato. *Plant Mol. Biol.* **2022**, *109*, 717–739. [CrossRef]
51. Dong, Y.; Tang, M.; Huang, Z.; Song, J.; Xu, J.; Ahammed, G.J.; Yu, J.; Zhou, Y. The miR164a-NAM3 module confers cold tolerance by inducing ethylene production in tomato. *Plant J.* **2022**, *111*, 440–456. [CrossRef] [PubMed]
52. Marand, A.P.; Schmitz, R.J. Single-cell analysis of cis-regulatory elements. *Curr. Opin. Plant Biol.* **2022**, *65*, 102094. [CrossRef] [PubMed]
53. Yang, Y.; Guo, Y. Unraveling salt stress signaling in plants. *J. Integr. Plant Biol.* **2018**, *60*, 796–804. [CrossRef] [PubMed]
54. Seiler, N.; Delcros, J.; Moulinoux, J.P. Polyamine transport in mammalian cells. An update. *Int. J. Biochem. Cell Biol.* **1996**, *28*, 843–861. [CrossRef]
55. Mulangi, V.; Phuntumart, V.; Aouida, M.; Ramotar, D.; Morris, P. Functional analysis of OsPUT1, a rice polyamine uptake transporter. *Planta* **2012**, *235*, 1–11. [CrossRef]
56. Chai, H.; Guo, J.; Zhong, Y.; Hsu, C.C.; Zou, C.; Wang, P.; Zhu, J.K.; Shi, H. The plasma-membrane polyamine transporter PUT3 is regulated by the Na^+/H^+ antiporter SOS1 and protein kinase SOS2. *New Phytol.* **2020**, *226*, 785–797. [CrossRef]
57. Shi, H.; Chan, Z. Improvement of plant abiotic stress tolerance through modulation of the polyamine pathway. *J. Integr. Plant Biol.* **2014**, *56*, 114–121. [CrossRef]
58. Pál, M.; Szalai, G.; Gondor, O.K.; Janda, T. Unfinished story of polyamines: Role of conjugation, transport and light-related regulation in the polyamine metabolism in plants. *Plant Sci.* **2021**, *308*, 110923. [CrossRef]
59. Sagi, M.; Fluhr, R. Production of reactive oxygen species by plant NADPH oxidases. *Plant Physiol.* **2006**, *141*, 336–340. [CrossRef]
60. Wu, J.; Liu, W.; Jahan, M.S.; Shu, S.; Sun, J.; Guo, S. Characterization of polyamine oxidase genes in cucumber and roles of CsPAO3 in response to salt stress. *Environ. Exp. Bot.* **2022**, *194*, 104696. [CrossRef]
61. Wang, W.; Liu, J.-H. CsPAO4 of Citrus sinensis functions in polyamine terminal catabolism and inhibits plant growth under salt stress. *Sci. Rep.* **2016**, *6*, 1–15. [CrossRef]
62. Martinis, J.; Gas-Pascual, E.; Szydlowski, N.; Crèvecoeur, M.; Gisler, A.; Bürkle, L.; Fitzpatrick, T.B. Long-distance transport of thiamine (vitamin B1) is concomitant with that of polyamines. *Plant Physiol.* **2016**, *171*, 542–553. [CrossRef]
63. Gémes, K.; Kim, Y.J.; Park, K.Y.; Moschou, P.N.; Andronis, E.; Valassaki, C.; Roussis, A.; Roubelakis-Angelakis, K.A. An NADPH-oxidase/polyamine oxidase feedback loop controls oxidative burst under salinity. *Plant Physiol.* **2016**, *172*, 1418–1431. [CrossRef]
64. Kim, W.; Zeljković, S.Ć.; Piskurewicz, U.; Megies, C.; Tarkowski, P.; Lopez-Molina, L. Polyamine uptake transporter 2 (put2) and decaying seeds enhance phyA-mediated germination by overcoming PIF1 repression of germination. *PLoS Genet.* **2019**, *15*, e1008292. [CrossRef]
65. Xu, J.; Liu, T.; Yang, S.; Jin, X.; Qu, F.; Huang, N.; Hu, X. Polyamines are involved in GABA-regulated salinity-alkalinity stress tolerance in muskmelon. *Environ. Exp. Bot.* **2019**, *164*, 181–189. [CrossRef]
66. Sun, W.; Hao, J.; Fan, S.; Liu, C.; Han, Y. Transcriptome and Metabolome Analysis Revealed That Exogenous Spermidine-Modulated Flavone Enhances the Heat Tolerance of Lettuce. *Antioxidants* **2022**, *11*, 2332. [CrossRef]

antioxidants

Article

Morpho-Physiological and Biochemical Responses of Hydroponically Grown Basil Cultivars to Salt Stress

Michele Ciriello [1], Luigi Formisano [1], Marios C. Kyriacou [2], Petronia Carillo [3], Luca Scognamiglio [1], Stefania De Pascale [1,*] and Youssef Rouphael [1]

1 Department of Agricultural Sciences, University of Naples Federico II, 80055 Portici, Italy
2 Department of Vegetable Crops, Agricultural Research Institute, 1516 Nicosia, Cyprus
3 Department of Environmental, Biological and Pharmaceutical Sciences and Technologies, University of Campania "Luigi Vanvitelli", 81100 Caserta, Italy
* Correspondence: depascal@unina.it

Abstract: Depending on duration and magnitude, abiotic stresses interfere with plant metabolic processes and may severely impact developmental and qualitative attributes. In this study, in addition to characterizing three different cultivars of basil ('Anise', 'Cinnamon', and 'Lemon') grown under hydroponics, we appraised the impact of NaCl salt stress (60 mM) on morphophysiological and nutraceutical properties of the basil crop. Salt stress significantly reduced fresh yield (51.54%, on average) and photosynthetic parameters (ACO_2, E, and gs) in all cultivars by raising tissue concentrations of Na^+ and Cl^-. In addition to reducing the concentration of nitrate (77.21%), NaCl salt stress increased the concentrations of key bioactive molecules, notably carotenoids (lutein and β-carotene), phenolic acids and flavonoid derivatives, thus resulting in a higher antioxidant activity of salt-treated basil plants compared to the untreated ones. Analysis by UHPLC revealed that chicoric acid was the most abundant polyphenolic compound in all basil cultivars, with the highest values recorded in 'Cinnamon'.

Keywords: antioxidant capacity; bioactive molecules; carotenoids; flavonoids; osmotic stress; phenolic acids; secondary metabolites

Citation: Ciriello, M.; Formisano, L.; Kyriacou, M.C.; Carillo, P.; Scognamiglio, L.; De Pascale, S.; Rouphael, Y. Morpho-Physiological and Biochemical Responses of Hydroponically Grown Basil Cultivars to Salt Stress. *Antioxidants* **2022**, *11*, 2207. https://doi.org/10.3390/antiox11112207

Academic Editor: Nafees A. Khan

Received: 9 October 2022
Accepted: 5 November 2022
Published: 8 November 2022

1. Introduction

The world population has peaked at eight billion people, and recent estimates project an increase of more than two billion people by 2050. In the coming decades, the imperative for community policies will be to ensure food security, a daunting challenge when we consider that it is mainly dependent on agricultural production [1]. Currently, the agricultural sector is managing to bridge the gap between demand and production, but demographic growth will put further pressure on the entire sector. As if that were not enough, changing climate scenarios have thrown even more fuel on the fire, endangering the food supply for future generations.

The increase in the frequency and intensity of drought has led to an increase in desertification with consequent soil salinization, a concern that should not be underestimated considering that horticultural crops are widespread in regions with high levels of water salinity [2–4]. In an era of climate change, salinity is undoubtedly a constraining factor in agricultural cultivation [5]. More than 20% of irrigated agricultural soils have high levels of salinity, resulting from the natural erosive process (i.e., primary salinity) but mainly from anthropogenic activities (i.e., secondary salinity) such as intensifying agricultural practices, deforestation, and irrigation, promoting seawater infiltration [6]. If not remedied, negative salinity impacts will affect 50% of the world's agricultural land by 2050. NaCl salinity causes a rapid osmotic effect that has drastic consequences on plant water and nutrient availability, photosynthetic and transpiration rates, stomatal regulation and control mechanisms, and root functional activities [7–11]. If the stress continues, morphological and physiological changes could occur, leading to reduced yields. Taking into account the long-term economic unsustainability of water and soil desalination processes and the low availability of salt-tolerant genotypes, finding an appropriate use for salinized areas is one of the biggest challenges. One possible solution could be to allocate such regions to the cultivation of medicinal plants, appreciated in gastronomic, pharmaceutical, medical, and cosmetic fields due to their richness in secondary metabolites. In fact, although salt stress leads to the production of reactive oxygen species (ROS), which limits production performance, it prompts the plant to activate defensive mechanisms that culminate in the bioaccumulation of phenolic molecules with an antioxidant function. As observed by Valifard et al. [12] on *Salvia mirzayanii* and *Salvia acrosiphon* and by Perin et al. [13] on *Fragaria ananassa*, the increase in polyphenols in response to salt stress is positively related to the up-regulation of phenylalanine ammonia-lyase (PAL) enzyme activity. Although secondary metabolites do not play a specific role in growth processes in medicinal plants, these valuable species-specific molecules increase under stress conditions, improving plant quality traits. The growing interest in products of natural origin that are readily available and have no side effects has increased the demand for medicinal plants because of their beneficial antimicrobial and anti-inflammatory properties, making them inexpensive and renewable "ingredients" for producing natural preservatives, new types of drugs, and cosmetics [14,15].

Currently, in developing countries, about 80% of medicines are of plant origin, while in developed countries, the proportion is only 25%. Among medicinal plants, basil (*Ocimum basilicum* L.), which is exceptionally rich in essential oils, is among the most widespread and well known worldwide, so much so that it has earned the nickname "King of Herbs", with the existence of at least 18 cultivars selected and developed over the years [16]. Its fame is mainly attributable to its gastronomic role as tender and fragrant leaves, critical ingredients of tasty dishes typical of Italian gastronomic tradition [17]. Not surprisingly, as with other plants belonging to the Lamiaceae family, basil, in addition to its recognized culinary aptitude, is cultivated for its secondary metabolites. Most studies have focused primarily on Genovese basil's productive and sensory characteristics without considering the genetic diversity typical of the genus Ocimum. Over time, many basil cultivars that have genetically distinct phytochemical profiles have been selected for their shape, color, aroma, and flavor. Basil contains relatively high concentrations of carotenoids (lutein and β-carotene) and polyphenols, which belong mainly to flavonol-glycoside classes (rutin, quercetin, and di-hydroquercetin) and phenolic acids (rosmarinic, chicoric, caffeic, chlorogenic, kaftaric, salvianic A, salvianolinic A, L and K acids), which, in addition to acting as stress mitigators, are beneficial for human health. However, although Genovese basil cultivated in soil is known to be tolerant of salinity (up to 100 mM NaCl) under, few studies have evaluated the effects of salt stress on the polyphenolic profile of non-Genovese basil for the pharmaceutical and cosmetic industries. The few contributions available in the literature have mainly focused on evaluating the effects of salinity on the yield and quality of Genovese basil while neglecting other non-Genovese types belonging to the genus Ocimum. Our work was aimed at characterizing three types of basil (*Ocimium basilicum* var thyrsiflora, *Ocimum basilicum* cv Cinnamon, and *Ocimum* × *Citriodorum*) under salt stress, both from a yield and phytochemical point of view. We evaluated the yield, morphophysiological response, antioxidant activity [FRAP (ferric ion reducing antioxidant power), DPPH (1,1-diphenyl-2-picrylhydrazyl), and ABTS (2,2′-azinobis-(3-ethylbenzothiazoline-6-sulfonate)], mineral profile (by ion chromatography) and phenolic profile (by ultrahigh performance liquid chromatography).

2. Materials and Methods

2.1. Experimental Site and Plant Material

The experimental trial, aimed at evaluating the effects of NaCl salt stress in basil (*Ocimum basilicum* L.), was carried out in spring–summer 2021 in the greenhouses of the Federico II University of Naples Department of Agriculture (DIA) (Portici, Italy; 40°48′ N, 14°20′ E, 29 m.s.l.). On 5 May 2021, basil plants 'Anise' (*Ocimum basilicum* L. var thyrsiflora;

Blumen, Milan, Italy), 'Cinnamon' (*Ocimum basilicum* L. cv Cinnamon; Blumen, Milan, Italy) and 'Lemon' (*Ocimum* × *citriodorum*; Pagano Domenico & Figli Sementi, Scafati, Italy) were seeded in 54-hole polystyrene trays (52 × 32 × 6 cm, L × W × D; volume: 0.06 L) until two true leaves appeared. At 14 days after sowing (29 May 2021; 1 day after transplanting), seedlings were transplanted into round anti-spiraling pots (0.1 m × 0.1 m × 0.15 m) filled with a mixture (*v/v*) of 1/3 perlite and 2/3 peat (Vigorplant, Fombio, Italy) (Figure 1). The pots were placed in rows with a spacing of 0.27 m × 0.15 m with a density of 25 plants m^{-2}. Nutrient solution (NS) was distributed through drippers with a flow rate of 2 L h^{-1} (1 dripper/plant).

Figure 1. Illustrative pictures of the different types of basil at transplant. From left to right: *Ocimum basilicum* L. var thyrsiflora (Anise), *Ocimum basilicum* L. cv Cinnamon, and *Ocimum* × *citriodorum* (Lemon).

2.2. Experimental Design

Basil seedlings were arranged in the greenhouse according to a bifactorial design in which three basil cultivars ('Anise', 'Cinnamon', and 'Lemon') and two NSs (salt and a non-salt control) were considered as factors. The control NS was a modified Hoagland with the following nutrient element composition: 13.0 mM NO_3-N, 1.0 mM NH_4-N, 1.5 mM P, 5.0 mM K, 1.75 mM S, 4.5 mM Ca, 2 mM Mg, 9 µM Mn, 20 µM Fe, 0.3 µM Cu, 20 µM B, 1.6 µM Zn, and 0.3 µM Mo Each experimental unit was replicated three times and included 15 plants (45 plants per treatment). Saline NS was prepared by adding 60 mM NaCl to the control NS.

2.3. Plant Collection

At harvest (26 days), eight plants per replicate were cut at root collar and sampled to determine biometric parameters and yield. Freshly sampled plants were weighed for total fresh and leaf weight measurements (g $plant^{-1}$). The height (cm) and number of leaves per plant were then determined. The collected samples were dried in a ventilated oven at 60 °C until constant weight (about three days) for the measurement of the dry weight (g $plant^{-1}$) and the percentage of dry matter. The dried plant material was then finely ground using an MF10.1 cutting head mill (IKA®, Staufen im Breisgau, Baden-Württemberg, Germany) for the measurement of the mineral concentration. Four plants per replicate were sampled and immersed in liquid nitrogen, stored at −80 °C, and subjected to a freeze-drying cycle (Christ, Alpha 1–4 (Martin Christ Gefriertrocknungsanlagen GmbH, Osterode am Harz, Germany) for the measurement of antioxidant activity, carotenoids, and phenolic acids, while another part was stored at −20 °C for the measurement of chlorophyll concentration.

2.4. Digital Quantification of the Leaf Area

Digital quantification of the leaf area was performed using photos of the leaves of each plant sampled. Specifically, at 26 days, the leaves of each plant were placed on a white panel perpendicular to the camera lens (Nikon D80, Nikon AF S DX 18-135/3.5-5.6G IF-ED lens; Nikon Corporation, Tokyo, Japan). The captured photos were processed with Adobe® Lightroom Classic and Adobe® Photoshop 2022 software (Adobe Inc., San Jose, CA, USA) for distortion correction and brightness and contrast adjustment. Leaf area was quantified using ImageJ v1.52a software (U.S. National Institutes of Health of the United

States, Bethesda, Rockville, MD, USA). The analyses were performed in triplicate and the leaf area was expressed in cm^2.

2.5. Color Measurement

Just before harvest, colorimetric indices were determined on ten healthy and fully expanded leaves per replicate using a Minolta CR-400 portable colorimeter (Minolta Camera Co. Ltd., Osaka, Japan). Color was converted according to the CIE 1976 L, a^*, b^* (CIELAB) color space, where L indicates brightness, a and b chromaticity (greenness and yellowness, respectively). Chroma and Hue angle (°) were calculated according to the formulas proposed by Kheng [18].

$$Chroma = \sqrt{a^{*2} + b^{*2}} \tag{1}$$

$$Hue = tan^{-1}\left(\frac{a^*}{b^*}\right) \tag{2}$$

where,

$$0^\circ < Hue < 90^\circ \; if \; a^*, b^* > 0$$

$$90^\circ < Hue < 180^\circ \; if \; a^* < 0, b^* > 0$$

$$180^\circ < Hue < 270^\circ \; if \; a^*, b^* < 0$$

$$270^\circ < Hue < 360^\circ \; if \; a^* > 0, b^* < 0$$

2.6. Leaf Gas Exchange and Leaf Fluorescence

At 26 days, six fully expanded and well illuminated leaves per replicate were labeled and used to determine gas exchange and leaf fluorescence. Measurements were taken in the middle of the day (11:00 a.m.–3:00 p.m. solar time).

The net CO_2 assimilation rate (ACO_2; μmol CO_2 m^{-2} s^{-1}), transpiration (E; mmol H_2O m^{-2} s^{-1}), and stomatal conductance (gs; mol H_2O m^{-2} s^{-1}) were measured with the LI-6400 portable gas exchange system (LI-COR Biosciences, Lincoln, NE, USA). The CO_2 in the measurement chamber was set at ambient values (about 400 ppm) and the photosynthetically active radiation at 1000 μmol m^{-2} s^{-1}. The leaves were closed in the measurement chamber until equilibrium was reached (about 15 min).

Chlorophyll fluorescence measurement was performed using a portable F_v/F_m Meter fluorometer (Plant Stress Kit, Opti-Sciences, Hudson, NH, USA) on dark-adapted leaves for 10 min using leaf clips. Maximum chlorophyll fluorescence (F_m) was induced with a saturating light pulse of 3000 μmol photons m^{-2} s^{-1} for 1s, while initial fluorescence (F_o) was recorded with an internal blue LED light of 1–2 μmol photons m^{-2} s^{-1}. F_v/F_m was estimated as $(F_m - F_o)/F_m$.

2.7. Pigment Measurement

The concentrations of leaf chlorophyll and carotenoid (lutein and β-carotene) concentrations were determined by spectrophotometry and high-performance liquid chromatography with diode array detection (HPLC-DAD), respectively. All reagents were purchased from Sigma-Aldrich (Milan, Italy). Analyses were performed in triplicate.

To determine chlorophyll according to Wellburn [19]'s protocol, 0.5 g of fresh frozen sample was extracted in 3 mL of 90% ammonia acetone (v/v) and homogenized by IKA® T10 basic Ultra Turrax® homogenizer (Staufen im Breisgau, Baden-Württemberg, Germany). Chlorophyll a and b concentrations were determined using a UV-Vis spectrophotometer (DR 4000, Hach Co., Loveland, CO, USA) with an absorbance of 647 and 664 nm, respectively. Total chlorophyll was calculated as chlorophyll a + chlorophyll b and expressed as mg g^{-1} of fresh weight (fw).

Lutein and β-carotene concentrations were determined in freeze-dried basil leaves according to the protocol of Salomon et al. [20]. A 0.1 g freeze-dried sample was extracted in a mixture of ultra-pure water (1 mL) and ethanol/n-hexane (5 mL, 60:50, v/v) and subjected to a vacuum centrifugation cycle to separate the pellet from the solvent. The pellet was mixed with methanol and MTBE (methyl-t-butyl ether) in a 1:1 (v/v) ratio and analyzed by HPLC-DAD. The results were expressed as mg kg^{-1} of dry weight (dw).

2.8. Antioxidant Activity

To determine the antioxidant activity, three spectrophotometric methods were compared: DPPH (1,1-diphenyl-2-picrylhydrazyl), ABTS (2,2′-azinobis-(3-ethylbenzothiazoline-6-sulfonate)), and FRAP (ferric ion reducing antioxidant power), according to protocols of Brand-Williams et al. [21], Re et al. [22], and Rajurkar and Hande [23], respectively.

To determine the antioxidant activity of DPPH, 1 mL of DPPH solution (0.1 mM) and 96% methanol were added to 200 μL of the aqueous extract, mixed and incubated at room temperature for 30 min in the dark. The absorbance was recorded against the blank at 517 nm.

To determine ABTS antioxidant activity, a stock solution of $ABTS^{+}$ was prepared by reacting 7 mM aqueous solution of $ABTS^{+}$ with 2.45 mM aqueous solution of potassium persulfate in equal parts. After incubation in the dark (16 h at 23 °C), the stock solution was diluted with ethanol in a ratio of 1:88 until an absorbance of 0.700 ± 0.020 at 734 nm was reached. A 0.1 mL aliquot of each sample, previously filtered and diluted (1:10) with 70% methanol, was mixed with 1 mL of $ABTS^{+}$ solution and stored at room temperature for 2.5 min. The absorbance was read at 734 nm.

For measurement of FRAP antioxidant activity, a FRAP solution was prepared containing 1.25 mL of Fe^{2+}/2,4,6-tris (2-pyridyl)-s-triazine (10 mM) in HCl (40 mM) + 1.25 mL of $FeCl_3$ (20 mM) in water + 12.5 mL of acetate buffer (0.3 M, pH 3.6). An aliquot of 150 μL of the sample was mixed with 2.850 mL of FRAP solution and incubated for 4 min in the dark. The absorbance was read at 593 nm against a blank containing all the reagents.

The absorbances of the DPPH, ABTS, and FRAP essays were recorded with a UV-vis spectrophotometer (Shimadzu, Japan). The results were expressed as Trolox equivalent mmol kg^{-1} dw. All analyses were performed in triplicate.

2.9. Measurement of the Polyphenolic Profile

For the measurement of the phenolic profile, 5 μL samples extracted according to the procedure described by Vallverdu-Queralt et al. [24] were analyzed by ultrahigh performance liquid chromatography (UHPLC; Dionex Ultimate 3000, Thermo Fisher Scientific™, Waltham, MA, USA) and Orbitrap high-resolution mass spectrometry (HRMS; Thermo Fisher Scientific™, Waltham, MA, USA) according to the protocol detailed by El-Nakhel et al. [25]. The chromatographic separation of polyphenols was carried out with a Luna Omega PS column (1.6 μm, 50 × 2.1 mm, Phenomenex, Torrance, CA, USA) set at 25 °C, in which the mobile phases consisted of water (A) and acetonitrile (B), both containing 0.1% formic acid (v/v). The Q-Exactive Orbitrap mass spectrometer was set in a fast negative/positive ion switching mode with two scan events (Full ion MS and all-ion fragmentation, AIF) for all compounds of interest. The calibration and accuracy of the equipment was monitored using a standard reference mixture (Thermo Fisher Scientific™, Waltham, MA, USA). Data processing was performed with Quan/Qual Browser Xcalibur software, v. 3.1.66.10 (Thermo Fisher Scientific™, Waltham, MA, USA). Polyphenols were expressed as mg kg^{-1} dw.

2.10. Mineral Concentration

The measurement of cations (K, Ca, Mg, and Na) and anions (Nitrate, P, and Cl) of basil leaves was carried out by ion chromatography coupled with an electrical conductivity detector (ICS3000, Thermo Scientific™ Dionex™, Sunnyvale, CA, USA) according to the detailed protocol of Formisano et al. [26]. By comparing peak areas, the integration and quantification of mineral concentration was performed using the z Chromeleon™ 6.8 Chromatography Data System software (Thermo Scientific™ Dionex™, Sunnyvale, CA, USA) data of samples with those of certified reference standards.

All treatments were analyzed in triplicate and the concentrations of the concentrations of anion and cations were expressed in g kg^{-1} dw, except for the for the nitrate, expressed in mg kg^{-1} fw.

2.11. Statistics

Statistical analysis was performed with IBM SPSS 20 software (Armonk, NY, USA) for Microsoft Windows 11. A two-way analysis of variance (ANOVA) was applied to assess the significance of the effects and interactions between the cultivar factors (CV) and salt stress (S). The mean effect of CV factor was compared by one-way analysis of variance, and the mean effect of S was compared by Student's t test. Statistical significance of the CV × S interaction and the CV factor was determined using the Tukey–Kramer HSD test at the level of $p < 0.05$. All data were presented as mean ± standard error.

3. Results

3.1. Biometric Parameters

Except for total fresh weight and fresh leaf weight (Figure 2A,B), the biometric parameters shown in Figure 3 and Supplementary Table S1 were significantly affected by the interaction between cultivar (CV) and stress (S) factors. Compared to the Control, salt treatment (Salt) reduced the total fresh and leaf weight by 51.52 and 47.32%, regardless of the cultivar (Figure 2A,B). The highest total fresh weight was found in the cultivar Cinnamon, followed by 'Anise' and 'Lemon'. Relative to fresh leaf weight, no significant differences were observed between 'Cinnamon' and 'Anise', unlike 'Lemon', which showed the lowest value (40.11 g $plant^{-1}$). Compared to the Control, Salt treatment reduced the number of leaves by 39.15, 16.11, and 44.54% in 'Anise', 'Cinnamon', and 'Lemon', respectively (Figure 3B). A similar trend was observed for the leaf area, with the lowest value obtained from the Lemon × Salt interaction (725.29 cm^2; Figure 3C). This latter interaction also resulted in the lowest total dry weight (6.20 g $plant^{-1}$; Figure 3D). Differently, the percentage of dry matter of all cultivars increased significantly after salt treatment.

3.2. Colorimetric Indices

As shown in Table 1, the CIELAB colorimetric parameters a^*, b^*, the chroma and the Hue angle were significantly affected by the interaction of the CV and S factors. Specifically, for 'Lemon', an increase in a^* (+18.43%) and Hue angle (+2.21%) was observed in plants under salt stress, compared to the Control, while b^* (−26.49%) and Chroma (23.79%) decreased. In 'Anise' and 'Cinnamon', Salt treatment did not result in significant differences for a^* and Chroma. An opposite trend was observed for the hue angle between the Anise and Cinnamon cultivars. Specifically, compared with Control, the above colorimetric parameter was not affected by salt treatment in 'Anise', while in 'Cinnamon' there was a 1.90% increase.

3.3. Leaf Physiological Parameters

The CV × S interaction did not result in significant differences for the physiological parameters shown in Figure 4 and Supplementary Table S2. Regardless of the cultivar, compared to the control, salt stress significantly ($p < 0.001$) reduced net CO_2 assimilation (ACO_2), transpiration (E), and stomatal conductance (gs) by 23.72, 30.95, and 29.16%, respectively. The same parameters were influenced by the mean effect of the cultivar. Specifically, 'Lemon' showed the highest E and gs values compared to 'Anise', but the lowest ACO_2 values compared to 'Cinnamon'. On the contrary, the maximum quantum efficiency of photosystem II (F_v/F_m) was solely affected by S. The data in Figure 4 D show that plants under salt stress decreased the F_v/F_m values (3.79%) compared to the control.

Table 1. Effect of basil cultivars (CV) and stress (S) on CIELab colorimetric indices.

Treatment	L	a	b	Chroma	Hue Angle
Cultivar (CV)					
Anise	42.45 ± 0.26 b	−12.61 ± 0.43 a	17.54 ± 0.62 b	21.61 ± 0.75 b	125.76 ± 0.25
Cinnamon	42.84 ± 0.54 b	−13.03 ± 0.30 a	18.14 ± 0.78 b	22.35 ± 0.86 b	125.85 ± 0.62
Lemon	44.10 ± 0.36 a	−14.77 ± 0.68 b	20.46 ± 1.42 a	25.25 ± 1.55 a	126.03 ± 0.65
Stress (S)					
Control	43.83 ± 0.33	−14.37 ± 0.53	20.54 ± 0.88	25.08 ± 1.04	125.07 ± 0.28
Salt	42.43 ± 0.31	−12.57 ± 0.21	16.89 ± 0.20	21.06 ± 0.26	126.70 ± 0.35
CV × S					
Anise × Control	42.90 ± 0.29	−13.33 ± 0.61 ab	18.46 ± 0.92 b	22.77 ± 1.10 b	125.89 ± 0.31 ab
Anise × Salt	42.00 ± 0.23	−11.90 ± 0.21 a	16.63 ± 0.48 b	20.46 ± 0.50 b	125.64 ± 0.45 b
Cinnamon × Control	43.80 ± 0.57	−13.52 ± 0.47 b	19.57 ± 0.96 b	23.79 ± 1.27 b	124.66 ± 0.41 b
Cinnamon × Salt	41.88 ± 0.45	−12.55 ± 0.06 ab	16.71 ± 0.23 b	20.90 ± 0.15 b	127.04 ± 0.57 a
Lemon × Control	44.79 ± 0.09	−16.27 ± 0.17 c	23.59 ± 0.58 a	28.66 ± 0.58 a	124.65 ± 0.38 b
Lemon × Salt	43.41 ± 0.40	−13.27 ± 0.16 ab	17.34 ± 0.16 b	21.84 ± 0.21 b	127.41 ± 0.17 a
		Significance			
CV	**	***	***	***	ns
S	***	***	***	***	***
CV × S	ns	*	*	*	**

Data are mean values ± standard error, $n = 3$. Mean comparisons were performed by Tukey HSD test for CV and by *t*-Test for S. Different letters within each column indicate significant mean differences. ns, *, **, and *** denote non-significant or significant effects at $p \leq 0.05$, 0.01, and 0.001, respectively.

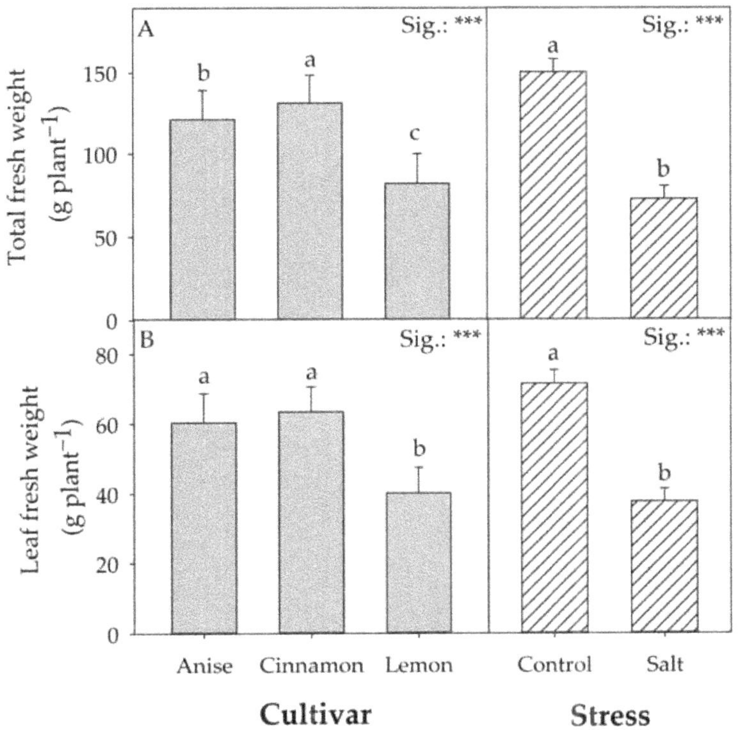

Figure 2. Effect of Cultivar and Stress on total fresh weight (**A**) and leaf fresh weight (**B**). Data are mean values ± standard error, $n = 3$. Mean comparisons were performed by Tukey HSD test for Cultivar and by *t*-Test for Stress. Different letters indicate significant mean differences. *** denotes significant effects at $p \leq 0.001$.

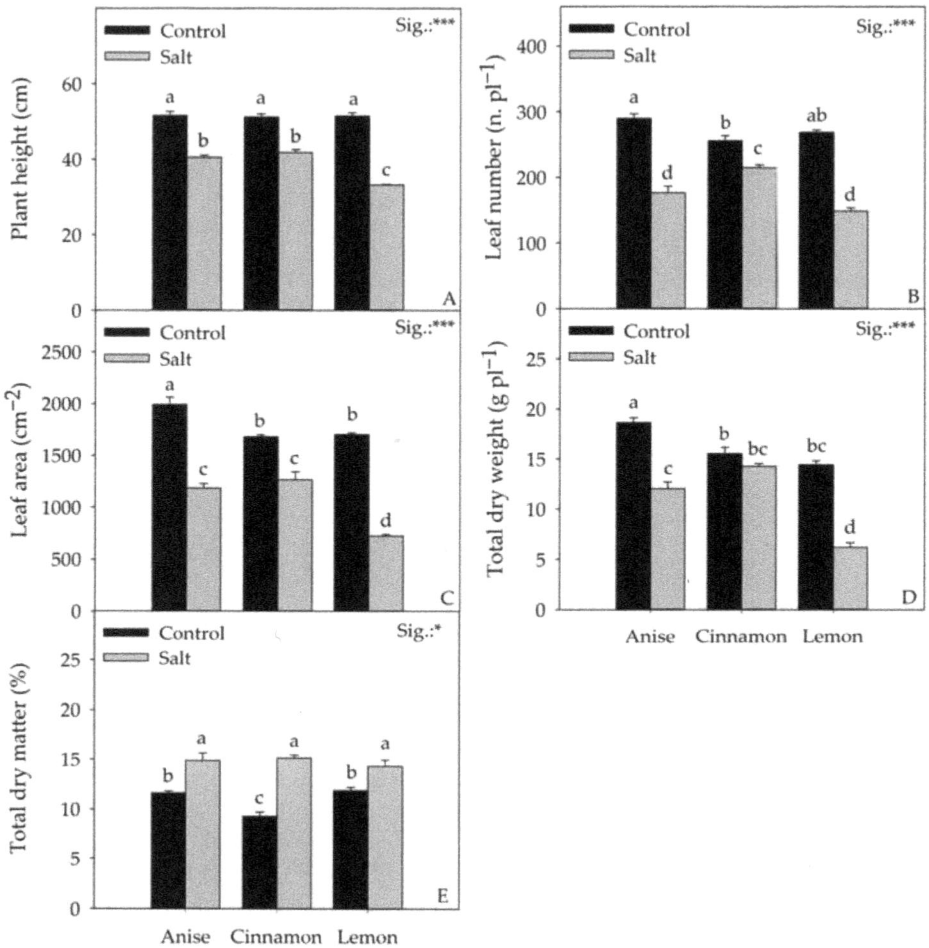

Figure 3. Effect of Cultivar × Stress on plant height (**A**), leaf number (**B**), leaf area (**C**), total dry weight (**D**), total dry matter (**E**). Data are mean values ± standard error, $n = 3$. Statistical significance of the CV × S interaction was determined by Tukey HSD test for Cultivar and by *t*-Test for Stress. Different letters indicate significant mean differences. * and *** denote significant effects at $p \leq 0.05$ and 0.001, respectively.

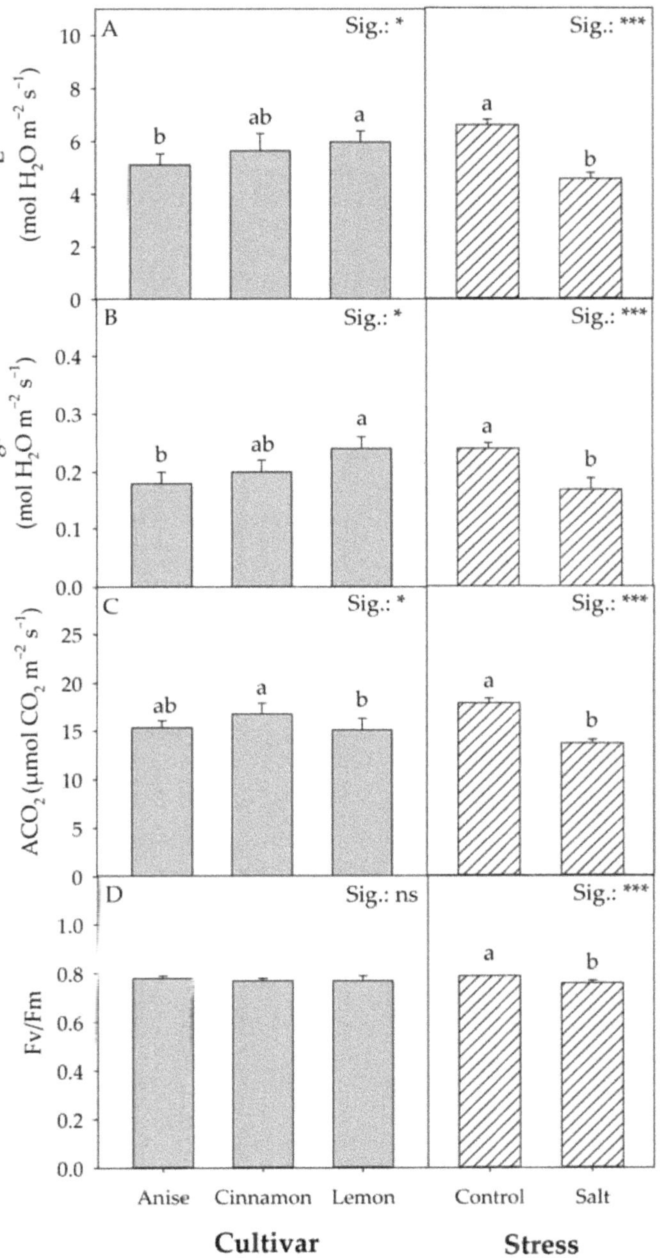

Figure 4. Effect of Cultivar and Stress on transpiration (**A**), stomatal conductance (**B**), net CO_2 assimilation rate (**C**), and chlorophyll fluorescence (**D**). Data are mean values ± standard error, $n = 3$. Mean comparisons were performed by Tukey HSD test for Cultivar and by t-Test for Stress. Different letters indicate significant mean differences. ns, *, and *** denote non-significant or significant effects at $p \leq 0.05$ and 0.001, respectively.

3.4. Leaf Pigments

Chlorophyll a, b, and total chlorophyll concentrations were not affected by the factors considered nor by their mutual interaction, unlike lutein and β-carotene (Table 2). The lutein concentration showed significant differences for the mean CV and S. Specifically, 'Lemon' had 25.75% lower lutein concentrations (on average) than the other cultivars. Regardless of the cultivar, salt stress increased the latter carotenoid by 22.85% compared to the control. Compared to the control, the interactions of the 'Anise' and 'Cinnamon' cultivars with salt treatment (Salt) increased β-carotene concentration by 39.17 and 47.67%, respectively. The same trend was also observed for antioxidant activity, where salinity increased by 16.18, 44.96, and 23.28% FRAP, DPPH, and ABTS essays, respectively (Figure 5). It should be noted that regardless of the antioxidant activity assay, the highest concentrations were recorded in 'Anise', followed by 'Lemon' and 'Cinnamon'.

Table 2. Effect of basil cultivars (CV) and stress (S) on Chlorophylls and carotenoids.

Treatment	Chlorophyll a	Chlorophyll b	Total Chlorophyll	Lutein	β-Carotene
		mg g^{-1} fw		mg kg^{-1} fw	
Cultivar (CV)					
Anise	1.21 ± 0.01	0.80 ± 0.03	2.01 ± 0.04	1305.20 ± 75.22 a	556.82 ± 40.88a
Cinnamon	1.22 ± 0.01	0.79 ± 0.01	2.01 ± 0.02	1243.47 ± 92.75 a	555.45 ± 47.88 a
Lemon	1.21 ± 0.01	0.89 ± 0.05	2.10 ± 0.05	953.05 ±6 3.54 b	511.73 ± 4.51 b
Stress (S)					
Control	1.22 ± 0.01	0.86 ± 0.04	2.08 ± 0.04	998.24 ± 50.62	478.50 ± 11.22
Salt	1.21 ± 0.01	0.79 ± 0.02	2.00 ± 0.02	1336.24 ± 61.77	604.17 ± 25.62
CV × S					
Anise × Control	1.23 ± 0.00	0.84 ± 0.05	2.07 ± 0.05	1139.44 ± 22.43	465.63 ± 5.39 d
Anise × Salt	1.19 ± 0.01	0.75 ± 0.03	1.95 ± 0.03	1470.97 ± 17.47	648.02 ± 2.88 a
Cinnamon × Control	1.20 ± 0.02	0.79 ± 0.01	1.99 ± 0.03	1042.12 ± 41.09	448.53 ± 5.11 e
Cinnamon × Salt	1.24 ± 0.01	0.80 ± 0.03	2.04 ± 0.03	1444.81 ± 28.04	662.38 ± 1.61 a
Lemon × Control	1.22 ± 0.01	0.96 ± 0.07	2.17 ± 0.08	813.16 ± 22.23	521.35 ± 2.27 b
Lemon × Salt	1.20 ± 0.01	0.82 ± 0.03	2.02 ± 0.03	1092.94 ± 11.14	502.11 ± 1.97 c
		Significance			
CV	ns	ns	ns	***	***
S	ns	ns	ns	***	***
CV × S	ns	ns	ns	ns	***

Data are mean values ± standard error, n = 3. Mean comparisons were performed by Tukey HSD test for CV and by t-Test for S. Different letters within each column indicate significant mean differences. ns and *** denote non-significant or significant effects at $p \leq 0.001$, respectively.

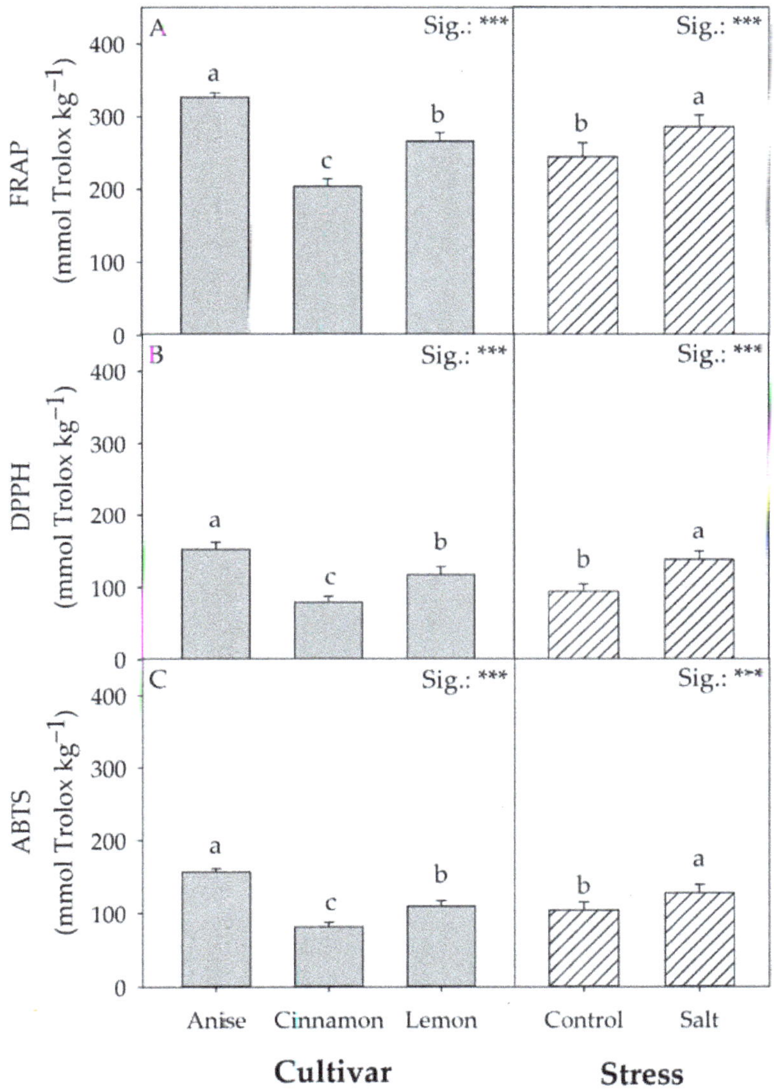

Figure 5. Effect of basil cultivars and stress on FRAP (**A**), DPPH (**B**), and ABTS (**C**) antioxidant activity. Data are mean values ± standard error, $n = 3$. Mean comparisons were performed by Tukey HSD test for CV and by t-Test for S. Different letters indicate significant mean differences. *** denotes significant effects at $p \leq 0.001$.

3.5. Phenolic Acids

Tables 3 and 4 show the phenolic acid and flavonoid derivatives of basil leaves, respectively, under salt stress.

Table 3. Effect of basil cultivars (CV) and stress (S) on phenolic acid derivatives.

Treatment	Salvianic Acid A	Salvianolic Acid k	Salvianolic Acid A	Salvianolic Acid L	Caftaric Acid	Rosmarinic Acid	Cichoric Acid	Caffeic Acid	Chlorogenic Acid	Feruloyl Tartaric Acid
	mg kg^{-1} dw									
Cultivar (CV)										
Anise	34.25 ± 2.34 b	100.50 ± 7.14	15.16 ± 0.69 a	75.18 ± 2.66 a	72.26 ± 4.48 a	1045.78 ± 51.34 a	2867.13 ± 143.05 a	128.18 ± 5.82 a	29.10 ± 1.13 a	234.67 ± 17.55 b
Cinnamon	39.53 ± 1.35 a	107.78 ± 6.93	2.73 ± 0.30 c	35.69 ± 2.26 c	21.28 ± 1.59 c	689.18 ± 35.78 c	2365.51 ± 125.86 b	45.78 ± 2.5 b	11.99 ± 0.82 c	334.54 ± 16.14 a
Lemon	6.00 ± 0.40 c	96.47 ± 6.87	7.02 ± 0.56 b	47.88 ± 2.87 b	28.30 ± 2.87 b	882.79 ± 33.32 b	2204.55 ± 62.98 b	48.03 ± 1.34 b	18.26 ± 1.08 b	114.86 ± 10.28 c
Stress (S)										
Control	23.77 ± 4.80	87.45 ± 2.48	7.19 ± 1.70	47.66 ± 5.89	34.38 ± 7.15	788.65 ± 49.42	2252.68 ± 90.84	67.13 ± 12.21	17.73 ± 2.47	197.23 ± 30.45
Salt	29.42 ± 5.68	115.71 ± 3.22	9.43 ± 1.95	58.18 ± 5.91	46.85 ± 8.90	956.51 ± 56.59	2705.44 ± 124.97	80.86 ± 14.96	21.84 ± 2.58	258.81 ± 33.66
CV × S										
Anise × Control	29.48 ± 1.76 c	86.79 ± 4.41	13.65 ± 0.15 b	69.97 ± 2.74	62.83 ± 0.77 b	944.97 ± 32.94	2589.89 ± 97.13	115.79 ± 2.16 b	26.94 ± 1.06	198.92 ± 15.52
Anise × Salt	39.02 ± 1.20 ab	114.21 ± 6.91	16.68 ± 0.31 a	80.40 ± 0.73	81.68 ± 3.34 a	1146.59 ± 43.95	3144.36 ± 126.56	140.57 ± 3.37 a	31.26 ± 0.77	270.41 ± 4.69
Cinnamon × Control	36.69 ± 0.36 b	93.65 ± 3.41	2.11 ± 0.19 f	31.09 ± 1.27	18.33 ± 1.25 d	611.74 ± 8.13	2089.00 ± 28.07	40.38 ± 0.92 d	10.34 ± 0.71	300.37 ± 7.99
Cinnamon × Salt	42.38 ± 0.94 a	121.91 ± 5.34	3.36 ± 0.18 e	40.29 ± 1.64	24.23 ± 1.57 d	766.62 ± 18.34	2642.02 ± 44.20	51.18 ± 1.09 c	13.64 ± 0.34	368.71 ± 8.49
Lemon × Control	5.13 ± 0.10 d	81.93 ± 2.87	5.80 ± 0.11 d	41.92 ± 2.31	21.96 ± 0.19 d	809.26 ± 8.96	2079.15 ± 59.19	45.23 ± 0.93 cd	15.90 ± 0.19	92.41 ± 2.54
Lemon × Salt	6.87 ± 0.22 d	111.01 ± 4.08	8.25 ± 0.21 c	53.84 ± 0.61	34.64 ± 0.93 c	956.32 ± 7.97	2329.94 ± 24.67	50.83 ± 0.54 c	20.62 ± 0.44	137.31 ± 4.20
					Significance					
CV	***	ns	***	***	***	***	***	***	***	***
S	***	***	***	***	***	***	***	***	***	***
CV × S	**	ns	**	ns	**	ns	ns	***	ns	ns

Data are mean values ± standard error, $n = 3$. Mean comparisons were performed by Tukey HSD test for CV and by t-Test for S. Different letters within each column indicate significant mean differences. ns, **, and *** denote non-significant or significant effects at $p \leq 0.01$ and 0.001, respectively.

Table 4. Effect of basil cultivars (CV) and stress (S) on flavonoids derivatives.

Treatment	Di-Hydroquercetin Glucoside	Rutin	Quercetin Glucoside
		mg kg^{-1} dw	
Cultivar (CV)			
Anise	3.66 ± 0.18 a	9.32 ± 0.78 b	9.91 ± 0.54 b
Cinnamon	0.87 ± 0.09 b	10.77 ± 0.79 a	3.78 ± 0.49 c
Lemon	0.52 ± 0.02 c	9.08 ± 0.41 b	30.59 ± 2.67 a
Stress (S)			
Control	1.48 ± 0.45	8.36 ± 0.26	12.33 ± 3.47
Salt	1.89 ± 0.54	11.09 ± 0.43	17.18 ± 4.75
CV × S			
Anise × Control	3.29 ± 0.06 b	7.69 ± 0.25	8.86 ± 0.54 cd
Anise × Salt	4.04 ± 0.10 a	10.95 ± 0.58	10.95 ± 0.25 c
Cinnamon × Control	0.68 ± 0.02 d	9.15 ± 0.40	2.74 ± 0.28 e
Cinnamon × Salt	1.06 ± 0.01 c	12.39 ± 0.56	4.82 ± 0.22 de
Lemon × Control	0.48 ± 0.03 d	8.23 ± 0.26	25.40 ± 2.62 b
Lemon × Salt	0.56 ± 0.01 d	9.92 ± 0.20	35.77 ± 1.40 a
		Significance	
CV	***	**	***
S	***	***	***
CV × S	***	ns	**

Data are mean values ± standard error, n = 3. Mean comparisons were performed by Tukey HSD test for CV and by t-Test for S. Different letters within each column indicate significant mean differences. ns, **, and *** denote non-significant or significant effects at $p \leq 0.01$ and 0.001, respectively.

As shown in Figure 6A, total phenolics were only affected by the average CV and S effects. Regardless of the stress condition, 'Anise' had the highest concentration of total phenolics (4625.09 mg kg^{-1} dw). Compared to the control, salt stress increased total phenolics by 21.63%.

Like total phenols, phenolic acid derivatives and total flavonoids showed a similar trend in response to salt stress (Figure 6B,C). Relative to the mean effect of the cultivar, 'Anise' showed the highest values of total phenolic acid derivatives (Figure 6B), in contrast to total flavonoid derivatives ('Lemon' > 'Anise' > 'Cinnamon'; Figure 6C).

Compared to total flavonoid derivatives, rutin was not significantly affected by the interaction of the factors considered (CV × S), unlike what was observed for di-hydroquercetin glucoside and quercitin glucoside. Regarding these glycoside compounds, salt stress increased their concentrations in 'Anise' and 'Cinnamon', while no significant differences were observed in 'Lemon'. In contrast, in 'Lemon', salt stress increased the concentration of quercetin glucoside (35.77 mg kg^{-1} dw; Table 4).

Regardless of the factors considered, cichoric acid was the predominant compound, followed by rosmarinic, feruloyl tartaric, salvianolic k, caffeic, salvianolic L, caftaric, salvianic A, chlorogenic, and salvianolic A acids. For chlorogenic, rosmarinic, and feruloyl tartaric acids, the CV × S interaction did not result in significant differences, unlike what was observed for the mean effects (CV and S). Specifically, salt stress, compared to the control, increased the concentrations of the above acids by 16.73, 21.28, and 31.22%, respectively. Relative to the mean effect of the cultivar, 'Anise' recorded the highest concentrations of cichoric and rosmarinic acids, while 'Cinnamon' had the highest content of feruloyl tartaric acid (Table 3).

Similarly to the findings for the above phenolic acids, the CV × S interaction was not significant for salvianolic K, salvianolic L, and chlorogenic acids. However, salt stress significantly increased the concentrations for the acids mentioned above compared to the control. Regardless of stress, 'Anise' was characterized by the highest concentration of salvianolic L and chlorogenic acids, contrary to what was observed for salvianolic K acid, which did not show a significant difference between cultivars (Table 3).

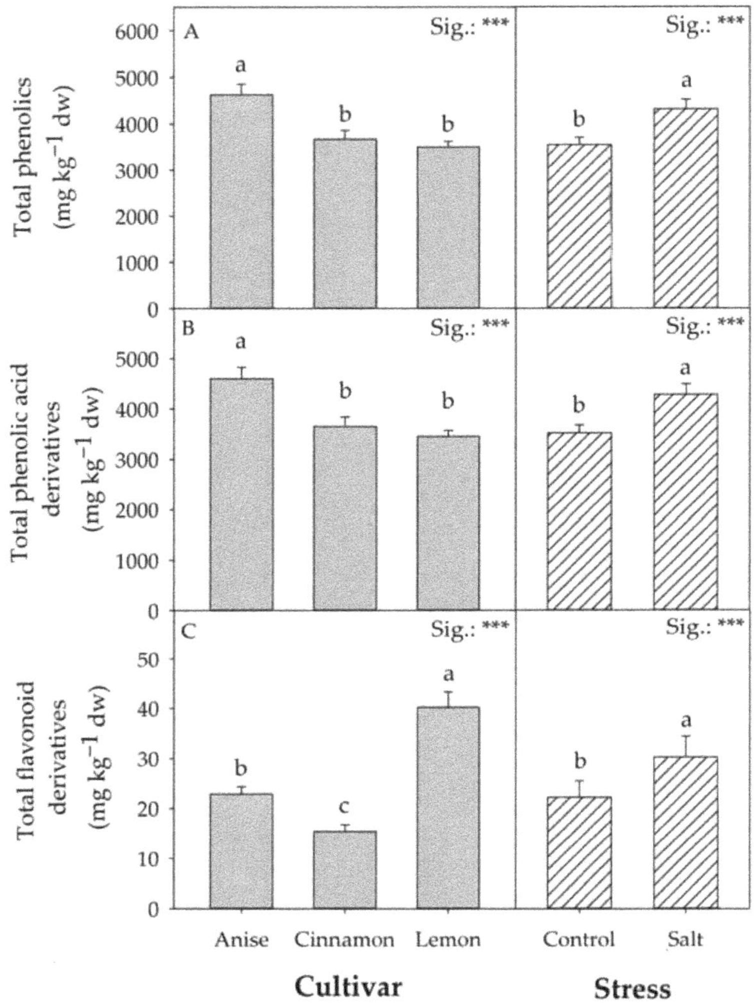

Figure 6. Effect of basil cultivars and stress on total phenolics (**A**), total phenolic acid derivatives (**B**), and total flavonoid derivatives (**C**). Data are mean values ± standard error, $n = 3$. Mean comparisons were performed by Tukey HSD test for CV and by *t*-Test for S. Different letters indicate significant mean differences. *** denotes significant effects at $p \leq 0.001$.

Unlike what was observed for the cichoric, rosmarinic, feruloyl tartaric, salvianolic K, caffeic, and salvianolic L acids, the concentrations of caftaric, salviananic acid A, chlorogenic, and salvianolic A acids were significantly affected by the CV × S interaction.

Salt stress increased salvianolinic acid A concentration in all cultivars, with the highest values obtained by Anise × Salt (16.68 mg kg^{-1} dw). The same interaction also resulted in the highest concentration of caftaric (81.68 mg kg^{-1} dw) and caffeic acids (140.57 mg kg^{-1} dw). In 'Anise' and 'Cinnamon', compared with the control, salt stress increased salvianic acid A concentrations by 32.36 and 15.50%, in contrast to 'Lemon', which did not show a significant difference.

3.6. Leaf Minerals

The effects of the factors considered (CV and S) and their interaction (CV × S) on the accumulation of minerals in basil leaves are shown in Table 5 and Figure 7. The concentrations of nitrate, Ca, Mg, Na, and Cl were influenced by the average CV and S effects and the CV × S interaction, except for K and P, where a significant difference was observed only for the average CV and S effects. Regardless of salt treatment, 'Lemon' had the highest P value (11.75 g kg^{-1} dw) while 'Anise' had an 8.7% higher K concentration than 'Cinnamon' (Table 5) Regardless of the cultivar, compared to the control, salt stress significantly reduced nitrate, K, and P concentrations (Table 5), while an opposite trend was observed for Cl (Figure 7). Compared to the CV × S interaction, the salt treatment cultivars 'Cinnamon' and 'Lemon' had significantly higher Na concentrations than the Control (Figure 7). The same treatment increased Ca and Mg concentrations in 'Anise' and 'Cinnamon' (Table 5).

Table 5. Effect of basil cultivars (CV) and stress (S) on Nitrate, K, P, Ca, and Mg leaf concentration.

Treatment	Nitrate	K	P	Ca	Mg
	mg g^{-1} fw	g kg^{-1} dw			
Cultivar (CV)					
Anise	2555.76 ± 785.29 a	38.25 ± 3.24 a	6.83 ± 0.43 b	9.19 ± 1.04 ab	5.97 ± 0.71 a
Cinnamon	1542.67 ± 361.53 b	31.72 ± 1.34 b	6.35 ± 0.40 b	9.32 ± 1.00 a	5.73 ± 0.84 a
Lemon	1403.80 ± 411.30 b	35.16 ± 2.59 ab	11.75 ± 0.45 a	7.53 ± 0.27 b	4.37 ± 0.14 b
Stress (S)					
Control	2987.44 ± 332.92	39.26 ± 1.95	8.99 ± 0.92	7.29 ± 0.26	4.26 ± 0.15
Salt	680.71 ± 66.37	30.82 ± 1.20	7.64 ± 0.87	10.07 ± 0.71	6.46 ± 0.56
CV × S					
Anise × Control	4303.80 ± 136.53 a	44.01 ± 3.76	7.24 ± 0.66	7.33 ± 0.67 b	4.51 ± 0.3 b
Anise × Salt	807.72 ± 95.40 c	32.49 ± 2.25	6.42 ± 0.59	11.05 ± 1.21 a	7.44 ± 0.51 a
Cinnamon × Control	2341.87 ± 39.22 b	33.24 ± 0.52	7.16 ± 0.39	7.18 ± 0.50 b	3.90 ± 0.24 b
Cinnamon × Salt	743.47 ± 115.19 c	30.19 ± 2.53	5.55 ± 0.12	11.47 ± 0.38 a	7.57 ± 0.32 a
Lemon × Control	2316.65 ± 100.39 b	40.54 ± 0.98	12.56 ± 0.34	7.36 ± 0.28 b	4.37 ± 0.11 b
Lemon × Salt	490.94 ± 49.49 c	29.78 ± 1.93	10.94 ± 0.51	7.70 ± 0.50 b	4.37 ± 0.29 b
			Significance		
CV	***	*	***	*	***
S	***	***	**	***	***
CV × S	***	ns	ns	*	***

Data are mean values ± standard error, n = 3. Mean comparisons were performed by Tukey HSD test for CV and by t-Test for S. Different letters within each column indicate significant mean differences. ns, *, **, and *** denote non-significant or significant effects at $p \leq 0.05$, 0.01 and 0.001, respectively.

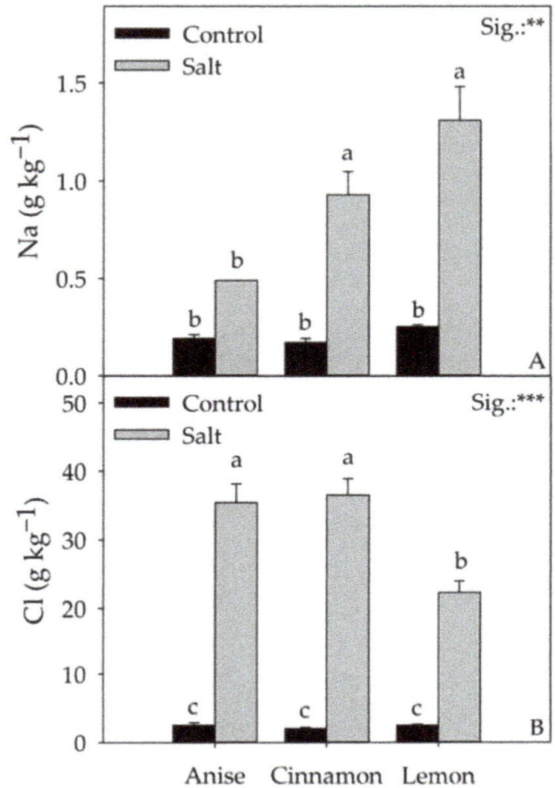

Figure 7. Effect of Cultivar × Stress on Na (**A**) and Cl (**B**) leaf concentration. Data are mean values ± standard error, n = 3. Statistical significance of the CV × S interaction was determined by Tukey HSD test for Cultivar and by t-Test for Stress. Different letters indicate significant mean differences. ** and *** denote non-significant or significant effects at $p \leq 0.01$ and 0.001, respectively.

4. Discussion

The significant difference recorded for total fresh weight in the basils evaluated in the present work underlines the strong impact of genotype. This variability was expected, assuming the phylogenetic gap between species and cultivars of the genus Ocimum. Regardless of salt treatment (Salt), 'Cinnamon' was the most productive cultivar, followed by 'Anise' and 'Lemon' (Figure 2). However, it should be noted that 'Anise' and 'Cinnamon' did not differ significantly in height, number of leaves, and leaf area, although the total fresh weight was different (Figure 3A–C). Relative to total fresh weight, the significant differences observed between cultivars did not result in a different response to salt stress, unlike what Ciriello et al. [27] observed in a recent work on basil. On average, salt treatment reduced the total fresh yield by 51.54% compared to the control; similar results were reported on *Salvia officinalis* L. [28] and *Mentha spicata* L. [29]. Regardless of cultivar, an adaptive response to salt stress was observed, culminating in reduced growth, an aspect that would allow plants to conserve energy by promoting the initiation of targeted defensive responses aimed at reducing permanent damage, as inferred from photosynthetic parameters shown in Table 2. Although Attia et al. [7] considered inappropriate to compare leaf production and photosynthetic activity (as it was related to a small number of leaves), in our study, salt stress reduced the net CO_2 assimilation rate and transpiration. Like Bekhradi et al. [30] and Attia et al. [7], salt-induced osmotic and nutritional stress would prompt plants to reduce stomatal conductance, affecting RuBisCo activity. Although salt stress can cause severe damage to photosystems due to poor management of excess excitation energy, the plants' morphophysiological and biochemical adaptations mitigated the deleterious effects induced by stress, as confirmed by F_v/F_m values that never reached critical values (Figure 4D). However, it is essential to note that, although F_v/F_m is the most widely used parameter for evaluating the performance of PSII under stress conditions [31], it may not be an acceptable parameter for assessing the physiological responses of plants under salt stress.

The analysis of morphological parameters of the salt-treated plants showed a lower sensitiveness of the Anise and Cinnamon cultivars than 'Lemon'. The latter reduced leaf area and height by 57.38 and 35.37%, compared to the control; in contrast, 'Anise' and 'Cinnamon' reduced, on average, the same parameters by 33.28 and 19.87%. The above is confirmed by the reduction in total dry weight in 'Lemon' (56.97%) in the salt treatment compared to the control. In contrast, no significant changes were observed in 'Cinnamon' for this key parameter. This result could be partially attributable to a constitutive higher tolerance to salt stress and partially to the increase in total dry matter, which increased from 9.27% (Cinnamon × Control) to 15.10% (Cinnamon × Salt) (Figure 3E). As Bernstein et al. [32] suggested, good hydration of plant tissues is a key characteristic of salt stress tolerance. However, in the present study, the unchanged values of total dry weight in 'Cinnamon' cannot be considered indicators of salt stress tolerance.

It is important to emphasize that the reduction in fresh weight in all cultivars cannot be attributed solely to the expansion of the leaf area, which plays a key role in the regulation of the transpiration process under osmotic stress. According to several authors [7,33], the reduction in fresh weight is also attributable to a significant and general reduction in leaf initiation (number of leaves). However, the smaller leaf number and area reduction in the Cinnamon × Salt interaction compared to the Anise × Salt and Lemon × Salt ones would elect the Cinnamon cultivar as the most salt tolerant (Figure 3B,C). In our experiment, controlled salt stress induced in plants for 26 days resulted in ionic and nutritional stress, attributed to the accumulation of Na^+ and Cl^- in transpiration fluxes, leading to a reduction in total fresh weight (Figures 2A and 7).

Although Attia et al. [34] showed that, in basil, Na^+ is partly transported by the xylem and accumulates in the roots, in our work, salt stress resulted in +355% Na^+ accumulation regardless of the cultivar (Figure 7). This result does not exclude the defensive response of the plants to salt since Na^+ values did not result in deleterious metabolic dysfunction. In general, the increase in dry matter percentage observed in all cultivars (+35.12%) would suggest that Na^+ accumulation was not totally internalized in the leaves, a condition that, as reported by Attia et al. [7], would explain the changes in leaf water content. Although it is frequently documented in the literature [35] that high Na^+ concentrations reduce K, Mg, and Ca uptake, only a reduction in K was observed in our experiment under salt stress; differentially, the concentration of Mg and Ca increased. As Scagel et al. [36] suggested, salt could have reduced Mg and Ca uptake while alternating the allocation of these elements, increasing their concentration. However, since Ca mediates different key processes of adaptation to stress conditions, including the Salt-Overly Sensitive (SOS) pathway, its increase could be related to a long-term response to salinity, as also argued by Mancarella et al. [5]. Similarly to what has been observed for Na^+, the use of 60 mM NaCl in the nutrient solution increased the Cl concentration in the leaves, reaching an average value of 36 g kg^{-1} dw in the 'Anise' and 'Cinnamon', which explaining the observed worsening production performance (Figures 2 and 3). In fact, the toxicity thresholds for salt-sensitive and salt-tolerant species are 7 and 50 g kg^{-1} dw, respectively [37]. As observed in green and red basil, the increase in Cl^- reduced the nitrate concentration (−77.21%), an antinutritional compound with a negative impact on human health. In salt treatment, the acknowledged antagonism between Cl^- and nitrate would have reduced the latter's uptake, slowing the plants' growth rate [38]. Although under salt stress, one of the most common physiological responses is chlorophyll degradation [30], regardless of the cultivar, we did not observe a reduction in chlorophyll a, b, or total chlorophyll concentrations (Table 2). Consistent with the observations of Bernstein et al. [32], this result suggests that this parameter should be considered an indicator of salt stress for basil. Partially in agreement with the above, the main colorimetric parameters in 'Anise' and 'Cinnamon' under salt stress did not change significantly from what was recorded in the Control (Table 1), considering that leaf color is often correlated with chlorophyll concentration [32]. However, for these two cultivars, salt stress did not significantly alter L, b, chroma, hue angle and especially a*, referred to in the literature as greenness; these results are probably attributable to the non-significant change in chlorophyll concentration under salt stress.

Unlike in 'Lemon', salt significantly altered colorimetric parameters (Table 1) compared to chlorophyll, and this result could be attributed to significant increases in lutein and β-carotene concentration (Table 2). In addition to their central role as supplementary pigments in photosynthetic processes, carotenoids are crucial for photoprotection due to their antioxidant properties [39]. As Bernstein et al. [32] hypothesized, an increase in carotenoids under salt stress would indicate a protective mechanism against stress, as it is universally recognized that environmental signals control the genetic regulation involved in their biosynthesis and bioaccumulation. Similarly to what was observed for carotenoids, salt stress, regardless of cultivar, increased the concentration of flavonoid derivatives (+35.17) and phenolic acids (+21.54%) and therefore total phenolic compounds (+21.63%; Figure 6). This result is not unexpected, as under NaCl salt stress, the energy stored during photosynthesis is used mainly for growth, with only a part used for synthesizing secondary metabolites. However, mainly by limiting growth, salt would prompt plants to allocate more energy to produce low-carbon secondary metabolites (such as polyphenols) that would help combat ROS [40]. Ghorbanpour and Varma [35] state that the increase in polyphenol concentration in response to salt is related to specific enzymes, including PAL (phenylalanine ammonia lyase). The drastic reduction in osmotic potential, a consequence of salt stress, increases L-proline levels, which is primarily responsible for the osmotic rebalancing. Although this amino acid is a primary metabolite, it may be related to the production of phenolic compounds due to its relationship with PAL [41]. Regardless of salt stress, it is worth noting again how the choice of genetic material significantly affects the bioaccumulation of phenolic compounds [17,42,43]. 'Anise' showed an average concentration of total phenolic compounds 29.12% higher than that recorded in 'Lemon' and 'Cinnamon'. Despite this, the most abundant phenolic acid was chicoric acid, which on average accounted for 63.51% of the phenolic acid derivatives. Although rosmarinic acid is listed in the literature as the most abundant phenolic acid in basil [44], a study of 15 basil cultivars confirmed the influence of genotype on both the quantity and quality of the phenolic profile [45].

'Lemon' was characterized by a 160.57% higher total concentration of flavonoid derivatives compared with 'Cinnamon'. This remarkable difference partially justifies the significant colorimetric differences observed between cultivars, as flavonoids also confer color to vegetables [46]; what has been observed may help to understand the results related to antioxidant activity (Figure 5). Regardless of the methodology used to determine antioxidant activity (DPPH, ABTS, and FRAP), salt increased antioxidant activity, probably due to increased production of compounds with antioxidant activity (carotenoids, phenolic acid derivatives, and flavonoids) [4]. It is well known that phenolic compounds contribute to the antioxidant activity of plant matrices. Therefore, according to Kwee and Niemeyer [45], the different phenolic profile of the three basil cultivars affected the antioxidant power and the activity of DPPH, ABTS, and FRAP (Figure 5). With a 29.50% (on average) concentration of phenolic acid derivatives higher than the other cultivars, 'Anise' showed the highest level of antioxidant activity. On the contrary, although 'Cinnamon' constitutively had a higher concentration of carotenoid (lutein and β-carotene) than 'Lemon', it had lower mean values of DPPH, ABTS, and FRAP. This result could be attributable to the higher constitutive concentration of flavonoids in 'Lemon', which by structural characteristics may have positively influenced the antioxidant activities of DPPH, ABTS, and FRAP.

5. Conclusions

Despite the phylogenetic distance and significant differences in yield and quality, the three basil types ('Anise', 'Cinnamon', and 'Lemon') tested had similar responses to NaCl salinization. Specifically, salinized plants enacted adaptive stress responses mainly based on reduced photosynthetic activity and growth (total fresh weight, fresh leaf weight, height, number of leaves, leaf area, and total dry weight) but at the same time increased secondary metabolites, increasing their antioxidant properties. Specifically, as shown in Figure 8, salt stress modulated the production of polyphenolic compounds in a cultivar-dependent manner. Although the moderate tolerance of basil to salinity is noted several times, we can state that the use of 60 mM NaCl in the nutrient solution is limiting for the production of the basil tested. In any case, considering the interest of the cosmetic, pharmaceutical and perfume industries in secondary metabolites with bioactive action, the feasibility of using salt stress as an enhancer of such compounds could be explored.

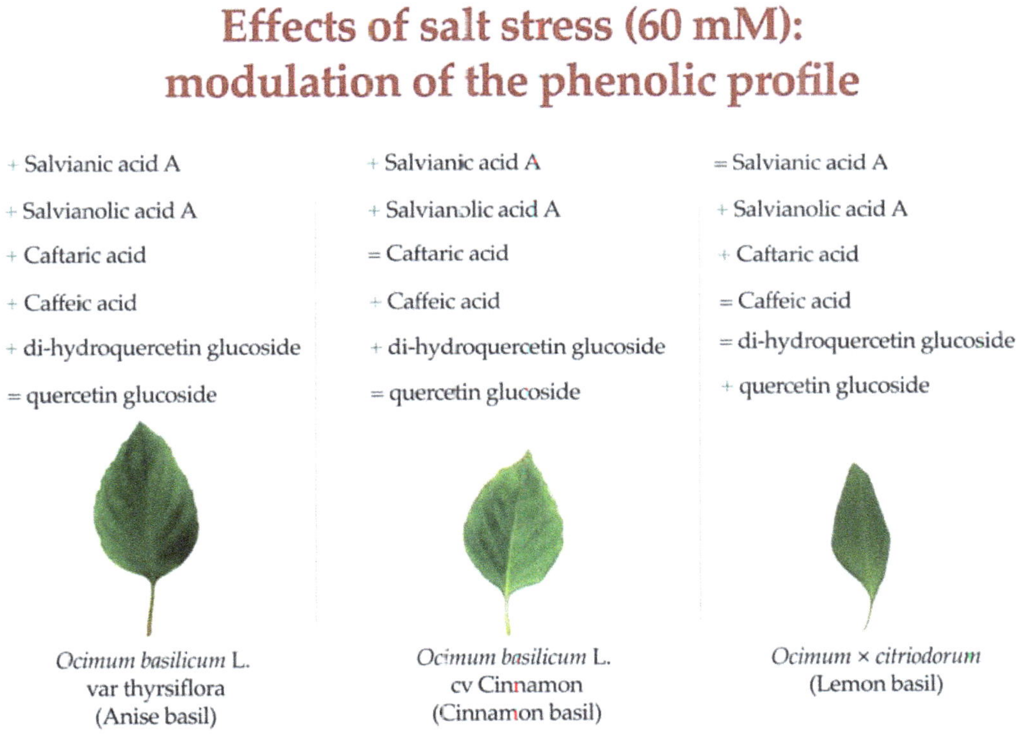

Figure 8. Effect of salt stress on representative phenolic compound of 'Anise', 'Cinnamon', and 'Lemon' basil.

Supplementary Materials: The following supporting information can be downloaded at: https://www.mdpi.com/article/10.3390/antiox11112207/s1, Table S1: Effect of basil cultivars (CV) and stress (S) on plant height, leaf number, leaf area, total dry weight, and total dry matter; Table S2: Effect of basil cultivars (CV) and stress (S) on leaf transpiration (E), leaf stomatal conductance (gs), CO_2 net assimilation rate, and Chlorophyll fluorescence (Fv/Fm).

Author Contributions: Conceptualization, M.C., S.D.P. and Y.R.; methodology, M.C. and L.F.; software, M.C. and L.F.; validation, M.C. and L.F.; formal analysis, M.C. and L.F.; investigation, M.C. and L.F.; resources, S.D.P. and Y.R.; data curation, M.C. and L.F.; writing—original draft preparation, M.C. and L.F.; writing—review and editing, M.C., L.F., M.C.K., P.C., L.S., S.D.P. and Y.R.; visualization, S.D.P. and Y.R.; supervision, S.D.P. and Y.R.; project administration, S.D.P. and Y.R.; funding acquisition, S.D.P. and Y.R. All authors have read and agreed to the published version of the manuscript.

Funding: This research received no external funding.

Institutional Review Board Statement: Not applicable.

Informed Consent Statement: Not applicable.

Data Availability Statement: The datasets generated for this study are available on request to the corresponding author.

Acknowledgments: We would like to acknowledge Alberto Ritieni and Giulia Graziani for providing the access to HPLC-DAD and UHPLC facilities and analysis. The authors are also grateful to Christophe El-Nakhel and Daniele Capurso for technical support in the field experiment.

Conflicts of Interest: The authors declare no conflict of interest.

Abbreviations

NS	nutrient solution
ACO_2	net assimilation rate of CO_2
E	transpiration
Gs	stomatal conductance
HPLC-DAD	high-performance liquid chromatography with diode array detection
DPPH	(1,1-diphenyl-2-picrylhydrazyl)
ABTS	(2,2′-azinobis-(3-ethylbenzothiazoline-6-sulfonate)
FRAP	(ferric ion reducing antioxidant power)
UHPLC	ultrahigh performance liquid chromatography
HRMS	high resolution mass spectrometry

References

1. Hasanuzzaman, M.; Hakim, K.; Nahar, K.; Alharby, H.F. *Plant Abiotic Stress Tolerance*; Springer: Berlin/Heidelberg, Germany, 2019.
2. Safriel, U.N. Status of desertification in the Mediterranean region. In *Water Scarcity, Land Degradation and Desertification in the Mediterranean Region*; Springer: Berlin/Heidelberg, Germany, 2009; pp. 33–73.
3. Siyal, A.; Siyal, A.; Abro, Z. Salt affected soils their identification and reclamation. *Pak. J. Appl. Sci.* **2022**, *2*, 537–540.
4. Tanaka, H.; Yamada, S.; Masunaga, T.; Yamamoto, S.; Tsuji, W.; Murillo-Amador, B. Comparison of nutrient uptake and antioxidative response among four *Labiatae herb* species under salt stress condition. *Soil Sci. Plant Nutr.* **2018**, *64*, 589–597. [CrossRef]
5. Mancarella, S.; Orsini, F.; Van Oosten, M.; Sanoubar, R.; Stanghellini, C.; Kondo, S.; Gianquinto, G.; Maggio, A. Leaf sodium accumulation facilitates salt stress adaptation and preserves photosystem functionality in salt stressed *Ocimum basilicum*. *Environ. Exp. Bot.* **2016**, *130*, 162–173. [CrossRef]
6. De Pascale, S.; Orsini, F.; Caputo, R.; Palermo, M.A.; Barbieri, G.; Maggio, A. Seasonal and multiannual effects of salinisation on tomato yield and fruit quality. *Funct. Plant Biol.* **2012**, *39*, 689–698. [CrossRef]
7. Attia, H.; Ouhibi, C.; Ellili, A.; Msilini, N.; Bouzaïen, G.; Karray, N.; Lachaâl, M. Analysis of salinity effects on basil leaf surface area, photosynthetic activity, and growth. *Acta Physiol. Plant.* **2011**, *33*, 823–833. [CrossRef]
8. Caliskan, O.; Kurt, D.; Temizel, K.E.; Odabas, M.S. Effect of salt stress and irrigation water on growth and development of sweet basil (*Ocimum basilicum* L.). *Open Agric.* **2017**, *2*, 589–594. [CrossRef]
9. Negrão, S.; Schmöckel, S.; Tester, M. Evaluating physiological responses of plants to salinity stress. *Ann. Bot.* **2017**, *119*, 1–11. [CrossRef]
10. Osakabe, Y.; Osakabe, K.; Shinozaki, K.; Tran, L.-S.P. Response of plants to water stress. *Front. Plant Sci.* **2014**, *5*, 86. [CrossRef]
11. Radácsi, P.; Inotai, K.; Sárosi, S.; Czövek, P.; Bernáth, J.; Németh, É. Effect of water supply on the physiological characteristic and production of basil (*Ocimum basilicum* L.). *Eur. J. Hortic. Sci.* **2010**, *75*, 193.
12. Valifard, M.; Mohsenzadeh, S.; Niazi, A.; Moghadam, A. Phenylalanine ammonia lyase isolation and functional analysis of phenylpropanoid pathway under salinity stress in '*Salvia*' species. *Aust. J. Crop Sci.* **2015**, *9*, 656–665.
13. Perin, E.C.; da Silva Messias, R.; Borowski, J.M.; Crizel, R.L.; Schott, I.B.; Carvalho, I.R.; Rombaldi, C.V.; Galli, V. ABA-dependent salt and drought stress improve strawberry fruit quality. *Food Chem.* **2019**, *271*, 516–526. [CrossRef] [PubMed]

14. Lee, C.-J.; Chen, L.-G.; Chang, T.-L.; Ke, W.-M.; Lo, Y.-F.; Wang, C.-C. The correlation between skin-care effects and phytochemical contents in *Lamiaceae* plants. *Food Chem.* **2011**, *124*, 833–841. [CrossRef]
15. Waller, S.B.; Cleff, M.B.; Serra, E.F.; Silva, A.L.; dos Reis Gomes, A.; de Mello, J.R.B.; de Faria, R.O.; Meireles, M.C.A. Plants from *Lamiaceae* family as source of antifungal molecules in humane and veterinary medicine. *Microb. Pathog.* **2017**, *104*, 232–237. [CrossRef] [PubMed]
16. Majdi, C.; Pereira, C.; Dias, M.I.; Calhelha, R.C.; Alves, M.J.; Rhourri-Frih, B.; Charrouf, Z.; Barros, L.; Amaral, J.S.; Ferreira, I.C. Phytochemical characterization and bioactive properties of cinnamon basil (*Ocimum basilicum* cv. 'Cinnamon') and lemon basil (*Ocimum* × *citriodorum*). *Antioxidants* **2020**, *9*, 369. [CrossRef]
17. Ciriello, M.; Formisano, L.; El-Nakhel, C.; Kyriacou, M.C.; Soteriou, G.A.; Pizzolongo, F.; Romano, R.; De Pascale, S.; Rouphael, Y. Genotype and successive harvests interaction affects phenolic acids and aroma profile of genovese basil for pesto sauce production. *Foods* **2021**, *10*, 278. [CrossRef]
18. Kheng, L.W. Color spaces and color-difference equations. *Color Res. Appl.* **2002**, *24*, 186–198.
19. Wellburn, A.R. The spectral determination of chlorophylls a and b, as well as total carotenoids, using various solvents with spectrophotometers of different resolution. *J. Plant Physiol.* **1994**, *144*, 307–313. [CrossRef]
20. Salomon, M.V.; Piccoli, P.; Fontana, A. Simultaneous determination of carotenoids with different polarities in tomato products using a C30 core-shell column based approach. *Microchem. J.* **2020**, *159*, 105390. [CrossRef]
21. Brand-Williams, W.; Cuvelier, M.-E.; Berset, C. Use of a free radical method to evaluate antioxidant activity. *LWT-Food Sci. Technol.* **1995**, *28*, 25–30. [CrossRef]
22. Re, R.; Pellegrini, N.; Proteggente, A.; Pannala, A.; Yang, M.; Rice-Evans, C. Antioxidant activity applying an improved ABTS radical cation decolorization assay. *Free Radic. Biol. Med.* **1999**, *26*, 1231–1237. [CrossRef]
23. Rajurkar, N.S.; Hande, S. Estimation of phytochemical content and antioxidant activity of some selected traditional Indian medicinal plants. *Indian J. Pharm. Sci.* **2011**, *73*, 146. [CrossRef] [PubMed]
24. Vallverdu-Queralt, A.; Medina-Remon, A.; Martínez-Huélamo, M.; Jauregui, O.; Andres-Lacueva, C.; Lamuela-Raventos, R.M. Phenolic profile and hydrophilic antioxidant capacity as chemotaxonomic markers of tomato varieties. *J. Agric. Food Chem.* **2011**, *59*, 3994–4001. [CrossRef] [PubMed]
25. El-Nakhel, C.; Pannico, A.; Graziani, G.; Kyriacou, M.C.; Gaspari, A.; Ritieni, A.; De Pascale, S.; Rouphael, Y. Nutrient supplementation configures the bioactive profile and production characteristics of three *Brassica* L. microgreens species grown in peat-based media. *Agronomy* **2021**, *11*, 346. [CrossRef]
26. Formisano, L.; Ciriello, M.; El-Nakhel, C.; De Pascale, S.; Rouphael, Y. Dataset on the effects of anti-insect nets of different porosity on mineral and organic acids profile of *Cucurbita pepo* L. fruits and leaves. *Data* **2021**, *6*, 50. [CrossRef]
27. Ciriello, M.; Formisano, L.; Soteriou, G.A.; Kyratzis, A.; De Pascale, S.; Kyriacou, M.C.; Rouphael, Y. Differential Response to NaCl Osmotic Stress in Sequentially Harvested Hydroponic Red and Green Basil and the Role of Calcium. *Front. Plant Sci.* **2022**, *13*, 799213. [CrossRef]
28. Hendawy, S.; Khalid, K.A. Response of sage (*Salvia officinalis* L.) plants to zinc application under different salinity levels. *J. Appl. Sci. Res.* **2005**, *1*, 147–155.
29. Chrysargyris, A.; Loupasaki, S.; Petropoulos, S.A.; Tzortzakis, N. Salinity and cation foliar application: Implications on essential oil yield and composition of hydroponically grown spearmint plants. *Sci. Hortic.* **2019**, *256*, 108581. [CrossRef]
30. Bekhradi, F.; Delshad, M.; Marín, A.; Luna, M.C.; Garrido, Y.; Kashi, A.; Babalar, M.; Gil, M.I. Effects of salt stress on physiological and postharvest quality characteristics of different Iranian genotypes of basil. *Hortic. Environ. Biotechnol.* **2015**, *56*, 777–785. [CrossRef]
31. Lazarević, B.; Šatović, Z.; Nimac, A.; Vidak, M.; Gunjača, J.; Politeo, O.; Carović-Stanko, K. Application of phenotyping methods in detection of drought and salinity stress in basil (*Ocimum basilicum* L.). *Front. Plant Sci.* **2021**, *12*, 629441. [CrossRef]
32. Bernstein, N.; Kravchik, M.; Dudai, N. Salinity-induced changes in essential oil, pigments and salts accumulation in sweet basil (*Ocimum basilicum*) in relation to alterations of morphological development. *Ann. Appl. Biol.* **2010**, *156*, 167–177. [CrossRef]
33. Ahmadi, M.; Souri, M.K. Growth and mineral content of coriander (*Coriandrum sativum* L.) plants under mild salinity with different salts. *Acta Physiol. Plant.* **2018**, *40*, 194. [CrossRef]
34. Attia, H.; Karray, N.; Ellili, A.; Msilini, N.; Lachaâl, M. Sodium transport in basil. *Acta Physiol. Plant.* **2009**, *31*, 1045–1051. [CrossRef]
35. Ghorbanpour, M.; Varma, A. *Medicinal Plants and Environmental Challenges*, 1st ed.; Springer: Cham, Switzerland, 2017; pp. 177–188.
36. Scagel, C.F.; Lee, J.; Mitchell, J.N. Salinity from NaCl changes the nutrient and polyphenolic composition of basil leaves. *Ind. Crops Prod.* **2019**, *127*, 119–128. [CrossRef]
37. Scagel, C.F.; Bryla, D.R.; Lee, J. Salt exclusion and mycorrhizal symbiosis increase tolerance to NaCl and $CaCl_2$ salinity in 'Siam Queen' basil. *HortScience* **2017**, *52*, 278–287. [CrossRef]
38. Borgognone, D.; Cardarelli, M.; Lucini, L.; Colla, G. Does $CaCl_2$ play a role in improving biomass yield and quality of cardoon grown in a floating system under saline conditions? *HortScience* **2014**, *49*, 1523–1528. [CrossRef]
39. Wurtzel, E.T. Changing form and function through carotenoids and synthetic biology. *Plant Physiol.* **2019**, *179*, 830–843. [CrossRef]
40. Vafadar, F.; Amooaghaie, R.; Ehsanzadeh, P.; Ghanadian, M. Salinity stress alters ion homeostasis, antioxidant activities and the production of rosmarinic acid, luteolin and apigenin in *Dracocephalum kotschyi* Boiss. *Biologia* **2020**, *75*, 2147–2158. [CrossRef]

41. Lattanzio, V.; Cardinali, A.; Ruta, C.; Fortunato, I.M.; Lattanzio, V.M.; Linsalata, V.; Cicco, N. Relationship of secondary metabolism to growth in oregano (*Origanum vulgare* L.) shoot cultures under nutritional stress. *Environ. Exp. Bot.* **2009**, *65*, 54–62. [CrossRef]
42. Ciriello, M.; Formisano, L.; El-Nakhel, C.; Corrado, G.; Pannico, A.; De Pascale, S.; Rouphael, Y. Morpho-Physiological Responses and Secondary Metabolites Modulation by Preharvest Factors of Three Hydroponically Grown Genovese Basil Cultivars. *Front. Plant Sci.* **2021**, *12*, 671026. [CrossRef]
43. Ciriello, M.; Pannico, A.; El-Nakhel, C.; Formisano, L.; Cristofano, F.; Duri, L.G.; Pizzolongo, F.; Romano, R.; De Pascale, S.; Colla, G. Sweet basil functional quality as shaped by genotype and macronutrient concentration reciprocal action. *Plants* **2020**, *9*, 1786. [CrossRef]
44. Ciriello, M.; Kyriacou, M.C.; De Pascale, S.; Rouphael, Y. An Appraisal of Critical Factors Configuring the Composition of Basil in Minerals, Bioactive Secondary Metabolites, Micronutrients and Volatile Aromatic Compounds. *J. Food Compos. Anal.* **2022**, *111*, 104582. [CrossRef]
45. Kwee, E.M.; Niemeyer, E.D. Variations in phenolic composition and antioxidant properties among 15 basil (*Ocimum basilicum* L.) cultivars. *Food Chem.* **2011**, *128*, 1044–1050. [CrossRef]
46. Scarano, A.; Chieppa, M.; Santino, A. Looking at flavonoid biodiversity in horticultural crops: A colored mine with nutritional benefits. *Plants* **2018**, *7*, 98. [CrossRef] [PubMed]

antioxidants

MDPI

Article

Transcriptome and Metabolome Analysis Revealed That Exogenous Spermidine-Modulated Flavone Enhances the Heat Tolerance of Lettuce

Wenjing Sun [1], Jinghong Hao [1], Shuangxi Fan [2], Chaojie Liu [1,*] and Yingyan Han [1,*]

1 Beijing Key Laboratory for Agricultural Application and New Technique, College of Plant Science and Technology, Beijing University of Agriculture, Beijing 102206, China
2 Beijing Vocational College of Agriculture, Beijing 102442, China
* Correspondence: cliu@bua.edu.cn (C.L.); yingyan.han@bua.edu.cn (Y.H.)

Abstract: Lettuce is sensitive to high temperature, and exogenous spermidine can improve heat tolerance in lettuce, but its intrinsic mechanism is still unclear. We analyzed the effects of exogenous spermidine on the leaf physiological metabolism, transcriptome and metabolome of lettuce seedlings under high-temperature stress using the heat-sensitive lettuce variety 'Beisansheng No. 3' as the material. The results showed that exogenous spermidine increased the total fresh weight, total dry weight, root length, chlorophyll content and total flavonoid content, increased the activities of antioxidant enzymes such as superoxide dismutase (SOD), peroxidase (POD) and catalase (CAT), and decreased malondialdehyde (MDA) content in lettuce under high temperature stress. Transcriptome and metabolome analyses revealed 818 differentially expressed genes (DEGs) and 393 metabolites between water spray and spermidine spray treatments under high temperature stress, and 75 genes from 13 transcription factors (TF) families were included in the DEGs. The Kyoto encyclopedia of genes and genomes (KEGG) pathway enrichment analysis of DEG contains pathways for plant–pathogen interactions, photosynthesis-antennal proteins, mitogen-activated protein kinase (MAPK) signaling pathway and flavonoid biosynthesis. A total of 19 genes related to flavonoid synthesis were detected. Most of these 19 DEGs were down-regulated under high temperature stress and up-regulated after spermidine application, which may be responsible for the increase in total flavonoid content. We provide a possible source and conjecture for exploring the mechanism of exogenous spermidine-mediated heat tolerance in lettuce.

Keywords: exogenous spermidine; heat stress; lettuce; metabolome; transcriptome

Citation: Sun, W.; Hao, J.; Fan, S.; Liu, C.; Han, Y. Transcriptome and Metabolome Analysis Revealed That Exogenous Spermidine-Modulated Flavone Enhances the Heat Tolerance of Lettuce. *Antioxidants* **2022**, *11*, 2332. https://doi.org/10.3390/antiox11122332

Academic Editor: Nafees A. Khan

Received: 25 October 2022
Accepted: 22 November 2022
Published: 25 November 2022

1. Introduction

Environmental temperatures are increasing and causing one of the most severe plant abiotic stresses. Global temperatures are projected to increase by 0.2 °C per decade, which will result in temperatures ranging from 1.8 to 4.0 °C above current levels by 2100 [1]. High-temperature stress is currently the main stress factor affecting plant growth and development. Due to the intensification of the greenhouse effect, the global temperature continues to rise, and agriculture is faced with serious challenges, which puts forward higher requirements for the heat resistance of plants [2]. High temperatures affect crop yield, shorten the life cycle of plants, accelerate senescence and cause a decrease in economic yield [2]. Heat stress leads to multiple changes in plant growth, development, physiological processes and yield, which are usually detrimental [3]. Heat stress affects the stability of various proteins, membranes, RNA species and cytoskeletal structures and alters enzymatic efficiency, impeding major physiological processes and causing metabolic imbalances in cellular responses [4–7].

In recent decades, exogenous applications of protectants, such as osmoprotectants, phytohormones, signaling molecules, and trace elements, have shown beneficial effects on plants grown at high temperatures because of their growth-promoting and antioxidant abilities [8]. Polyamines, mainly including putrescine (Put), spermidine (Spd), and spermidine (Spm), are low molecular weight natural compounds with aliphatic nitrogenous structures that are present in almost all organisms, from bacteria to plants and animals [9]. For plant growth and development, polyamines are widely involved in cell division and differentiation, root elongation, flower development, fruit ripening, leaf senescence, programmed cell death, DNA synthesis, gene transcription, protein translation, and chromatin organization [10–16]. Spermidine is an endogenous molecule that regulates plant growth and resists unfavorable environments [17]. Spd improves heat tolerance in tomato [18], lettuce [19], wheat [20], tall fescue [21], and rice [22]. Spd pretreatment inhibits membrane lipid peroxidation, increases antioxidant enzyme activities, superoxide dismutase (SOD), peroxidase (POD) catalase (CAT) and ascorbate peroxidase (APX), and reduces the accumulation of reactive oxygen species (ROS) and malondialdehyde (MDA) [23].

Lettuce (*Lactuca sativa* L.) is one of the most popular leafy vegetables in daily life. It originates from the Mediterranean coast and is highly intolerant to high temperatures, with 15~25 °C being the preferred growth temperature and growth being hindered above 30 °C. High temperatures are one of the main factors limiting the growth of lettuce. High temperatures always result in thinner leaves and longer internodes of lettuce, which leads to a decrease in its nutritional quality and commercial value [24].

In plants, flavonoids are important compounds that affect the color of leaves, flowers and fruits, and they exhibit antioxidant properties when plants are under stress. Flavonoids are considered to be an effective substance against various abiotic stresses because of their ability to reduce oxidative damage in organisms.

The development of transcriptomic and metabolomic technologies has increased the understanding of the network between genes, and RNA-seq has been widely used for various purposes in plant genetics research, especially transcriptome analysis, which has been used to explore a wide range of changes at the level of gene expression under various biotic and abiotic stresses and used to reveal differential gene expression in various biological processes. Additionally, metabolite analysis has become a new tool that can widely analyze components and is a powerful tool to study the changes in metabolites caused by various environmental changes [25]. Combining the joint analysis of different histologies to establish regulatory networks between genes and metabolites can provide a more comprehensive explanation of changes in species tolerance at multiple levels. The mechanism by which methyl jasmonate regulates monoterpene biosynthesis in grape berry skins was revealed by a combined transcriptome and metabolome analysis [26].

In our previous experiments, we demonstrated that exogenous spermidine could alleviate the damage to lettuce seedlings under high-temperature stress [27]; however, the molecular mechanism by which spermidine mediates the responses of lettuce to stress is not clear, so we attempted to explore the mechanistic model of its action through co-omics analysis. In this study, we further revealed the regulatory network between genes and metabolites by combining changes in the physiological properties of lettuce seedlings treated with spermidine under high-temperature stress with transcriptomic and metabolomic analyses and provided theoretical support for elucidating the mechanism of exogenous spermidine in enhancing the heat tolerance of lettuce.

2. Materials and Methods

2.1. Plant Material and Treatment

The heat-sensitive lettuce variety 'Beisansheng No. 3′ was selected from Beijing University of Agriculture [27]. The full-grained lettuce seeds were placed in filter paper-lined Petri dishes for 48 h and then sown in hydroponic seedling trays with built-in Hoagland nutrient solution and placed in an artificial climate chamber with a photoperiod of 14/10 h, diurnal temperature of 23 °C/17 °C, relative humidity of 70~75%, and light intensity of 350 ± 10 μmol·m^{-2}·s^{-1}, and seedlings with uniform growth were selected. Seedlings were transplanted into 10 L hydroponic tanks with Hoagland nutrient solution, and treatments were started when the seedlings reached four leaves. Lettuce leaves were sprayed with spermidine treatment at 9:00 a.m. each day to the extent that the leaves were moist but not dripping droplets. According to our previous experiments, we set the Spd spray concentration to 1 mmol·L^{-1} [27]. The standard for spraying is to spray the leaf surface and leaf back evenly, with the sprayed liquid adhering to the leaf surface so that it is all wet but not dripping.

Experimental material handling can be seen in Table 1. Three treatments were set up in this experiment: normal temperature control (CK): temperature of 22 °C/17 °C (day/night) and spraying distilled water; high temperature stress (H): temperature of 35 °C/30 °C (day/night) and spraying distilled water; high temperature stress spraying with spermidine (HS): temperature of 35 °C/30 °C (day/night) and 1 mmol·L^{-1} Spd.

Table 1. Experimental treatments.

Treatment	Temperature (Day/Night)	Type of Spraying
CK	22 °C/17 °C	distilled water
H	35 °C/30 °C	distilled water
HS	35 °C/30 °C	1 mmol·L^{-1} Spd

Each treatment was replicated three times. After 4 days of heat treatment of the lettuce seedlings, healthy plants were randomly selected, and the cotyledon portion of the leaves was collected. Samples were placed in liquid nitrogen at the time of collection and stored in a refrigerator at −80 °C for subsequent experiments

2.2. Measurement of the Physiological and Biochemical Parameters

The malondialdehyde (MDA) content was determined by the thiobarbituric acid method and calculated from the absorption values at 450, 532 and 600 nm. The chlorophyll content was determined by the method of Madhava Rao and Sresty [28]: 0.3 g of the sample was ground well with 2.5 mL of 95% ethanol, filtered and fixed in a 25 mL brown volumetric flask only.

The chloroplast pigment extract was poured into a cuvette, and 95% ethanol was used as a blank control. The optical density (OD) values were measured on a spectrophotometer (UV-5200, Shanghai, China) at 665 and 649 nm. Chlorophylls were calculated according to the following equations:

$$\text{Chl a content (mg/g)} = [13.95\text{OD}665 - 6.88\text{OD}649]\ \text{V}/1000\ \text{W} \quad (1)$$

$$\text{Chl b content (mg/g)} = [24.96\text{OD}649 - 7.32\text{OD}665]\ \text{V}/1000\ \text{W} \quad (2)$$

$$\text{Chl t content (mg/g)} = [18.08\text{OD}649 + 6.63\text{OD}665]\ \text{V}/1000\ \text{W} \quad (3)$$

The flavonoid content was determined by the sodium nitrite-aluminum nitrate chromogenic method. The flavonoid content was determined using a kit (Solarbio, Beijing, China). After the sample was dried to a constant weight, 0.1 g was dissolved in 1 mL of the extraction solution. The sample was extracted by ultrasonication at 60 °C for 30 min and then centrifuged at 12,000 rpm for 10 min, and the supernatant was taken for measurement. The absorbance value of the sample extract at 470 nm was measured.

2.3. Determination of Antioxidant Enzyme Activity

Place 0.5 g of lettuce sample in a cooled mortar and add in 5 mL of phosphate buffer (pH 7.0) for grinding. The homogenate was collected and centrifuged at 4 °C for ten minutes at a force of 11,000× *g*. The supernatant was collected, and the above steps were repeated twice.

Superoxide dismutase (SOD) activity was measured by the NBT (nitrogen blue tetrazolium) photochemical reduction method [29], with 50% inhibition of NBT photochemical reduction as one enzyme activity unit (U). Peroxidase (POD) activity was determined by the guaiacol method, with an increase of 1 per minute of OD 470 nm as one enzyme activity unit (U). Catalase (CAT) activity was determined by the hydrogen peroxide UV spectrophotometric method, with an increase of 0.1 per minute OD240 nm as one enzyme activity unit (U). Ascorbate peroxidase (APX) activity was determined by measuring the rate of ascorbate oxidation at 290 nm (ε = 2.8 mM^{-1} cm^{-1}) [30].

2.4. Metabolite Measurement and Quantification

The tissue sample (25 mg) was placed in a 1.5 mL tube with 800 µL of precooled precipitant (methanol: acetonitrile: distilled water = 2:2:1) and two small steel beads, then it was ground in a grinder (60 Hz, 4 min). After the removal of the beads, it was sonicated in an ice bath for 10 min (power 80 HZ), allowed to rest for 120 min at −20 °C, and then it was centrifuged for 15 min (25,000× *g*, 4 °C) and 600 µL of the supernatant was collected, and it was repeated once. The supernatant was placed in a freeze extractor to drain. Then, 600 µL of 10% methanol solution was added, it was placed in an ice bath and sonicated for 10 min (power 80 HZ) and then centrifuged for 15 min (25,000× *g*, 4 °C). The supernatant was collected, and 50 µL of each sample was injected into a QC using an ACQUITYUPLCHSST3 column (100 mm × 2.1 mm, 1.8 µm, Waters, Wilmslow, UK) for chromatographic separation, and the small molecules that eluted from the column were collected in positive and negative ion modes using high-resolution tandem mass spectrometry (Xevo G2-XS QTOF Waters, Wilmslow, UK). Peak extraction was mainly implemented by the commercial software Progenesis QI (version 2.2. Newcastle, UK), including peak alignment, peak extraction, normalization, deconvolution and compound identification steps. Three biological replicates were performed for each sample.

2.5. RNA Extraction and Quality Testing

The steps of total RNA extraction, RNA purity assessment, library construction, library quality control, and up sequencing were performed at UW (https://www.genomics.cn, accessed on 1 August 2019), following their standard procedures.

2.6. RNA-Seq Data Analysis

The high-throughput sequencing offline data (Raw Data) were filtered to obtain high-quality data (Clean Data), and the data were filtered using UW's self-developed filtering software SOAPnuke (v1.4.0. Shenzhen, China) for statistics and trimmomatic (v0.36. Dortmund, Germany) for filtering. The clean reads were compared to the reference genome sequence using HISAT (Hierarchical Indexing for Spliced Alignment of Transcripts, http://www.ccb.jhu.edu/software/hisat, accessed on 1 August 2019). The clean reads were aligned to the genomic sequences using Bowtie2, and then the gene expression levels were calculated for each sample using RSEM (http://deweylab.biostat.wisc.edu/rsem/rsem-calculate-expression.html, accessed on 1 August 2019). FPKM (Fragments Per Kilobases per Million reads) was used to express the gene expression levels. We defined genes with more than a two-fold difference and Q-value $\leq$ 0.001 to be screened as significantly differentially expressed genes (fold change $\geq$ 2 and adjusted p value $\leq$ 0.001). Based on the annotation results of GO (http://geneontology.org/, accessed on 1 August 2019) and KEGG (http://www.genome.jp/kegg, accessed on 1 August 2019) and the official classification, we functionally classified the differentially expressed genes and performed enrichment analysis using the phyper function in R software, with FDR correction for p values. Functions with Q values $\leq$ 0.05 were considered significantly enriched.

2.7. Quantitative Real-Time PCR

Total leaf RNA extraction was performed using a kit (Yueyang Hua, Yueyang, China, Cat:0416-50gk). The first cDNA strand was synthesized using a reverse transcription kit (Tiangen, Beijing, China. Cat:KR118-02). The primer sequences of the related genes were downloaded from the GenBank library of NCBI. The primer design is shown in Table S1. qRT-PCR analysis was performed using the reverse transcription product cDNA as the template and 18S as the internal reference gene. qRT-PCR was performed using the TB Green Premix Ex Taq II (2×) (Tli RNaseH Plus) kit (TaKaRa, Kyoto, Japan) in the CFX96 Real-Time PCR Detection System instrument (Bio-Rad Laboratories, Hercules, CA, USA). The reaction system was 20 μL, consisting of 9 μL TB Green Premix Ex Taq II (TliRNaseH Plus), 7 μL ddH_2O, 2 μL cDNA template and 1 μL forward and reverse primers. The cycling progression was as follows: 95 °C for 3 min; 95 °C for 10 s and 56 °C for 30 s, for a total of 39 cycles. Gene expression change ploidy analysis was calculated using $2^{-\Delta\Delta Ct}$, and relative mRNA expression levels were normalized using 18S. The qRT-PCR validation of the DEGs is shown in Figure S1.

2.8. Statistical Analysis

One-way ANOVA was performed on all data using SPSS 22.0 software (SPSS Inc., Chicago, IL, USA), and Duncan's test was used to test the significance of the differences between samples ($p < 0.05$).

3. Results

3.1. Effect of Exogenous Spermidine on the Growth and MDA and Chlorophyll Contents of Lettuce under High-Temperature Stress

As shown in Table 2, high-temperature stress significantly reduced the total fresh weight, shoot fresh weight, root length, total dry weight, and root-to-shoot ratio of lettuce, while the application of spermidine alleviated the damage to the total fresh weight, root length, and total dry weight under high-temperature stress, but had no significant effect on shoot fresh weight, root fresh weight, plant height and root-to-shoot ratio.

Table 2. Effects of exogenous spermidine on lettuce growth under high-temperature stress. Values above each vertical bar followed by different letters show significant differences ($p < 0.05$). The highest value was labeled as a, and those with significant differences were labeled as b, c in that order.

Treatments	Fresh Weight (g)	Shoot Fresh Weight (g)	Root Fresh Weight (g)	Root Length (cm)	Dry Weight (g)	Plant Height (cm)	Root-Shoot Ratio	Dry Weight (g)
CK	38.60 ± 1.46^{a}	22.78 ± 2.90^{a}	12.28 ± 0.68^{a}	25.83 ± 2.25^{a}	1.01 ± 0.23^{a}	15.33 ± 2.52^{a}	1.71 ± 0.25^{a}	1.01 ± 0.23^{a}
H	19.01 ± 0.93^{c}	7.66 ± 0.91^{b}	11.04 ± 1.26^{a}	9.61 ± 1.57^{c}	0.32 ± 0.02^{c}	16.57 ± 0.40^{a}	0.58 ± 0.10^{b}	0.32 ± 0.02^{c}
HS	24.32 ± 1.54^{b}	10.98 ± 0.85^{b}	11.22 ± 0.24^{a}	14.14 ± 1.36^{b}	0.61 ± 0.04^{b}	17.50 ± 1.00^{a}	0.81 ± 0.03^{b}	0.61 ± 0.04^{b}

CK: 22 °C/17 °C, distilled water; H: 35 still, distilled water; HS: 35 °C/30 °C, distilled 1 mM Spd.

From Figure 1A, it can be seen that the leaves of lettuce seedlings under high-temperature stress were elongated, appeared to be twitching, root growth was weak, fewer roots were produced, the main roots were short, biomass accumulation was reduced, which indicated that high temperature stress inhibited the growth condition and organic matter accumulation of lettuce seedlings. At the same time, the MDA content in the leaves increased under high temperature, and the increased MDA content reflected, to some extent, the increase in membrane lipid peroxidation. In the leaves of seedlings sprayed with spermidine, these effects were ameliorated to some extent. In addition, exogenous spermidine also affected the photosynthetic pigment content of the lettuce leaves. Although the total chlorophyll content of the leaves increased under high-temperature stress, the contents of chlorophyll a and chlorophyll b did not change significantly, while the application of spermidine significantly increased the contents of chlorophyll a and total chlorophyll (Figure 1C–E)

In conclusion, exogenous spermidine increased the total fresh weight, total dry weight, and root length, reduced the MDA content, and enhanced the chlorophyllscontent of the lettuce under high-temperature stress, alleviating the damage of high temperature on growth and physiological indexes of lettuce, providing a preliminary basis for us to further explore the possible mechanism of spermidine-mediated enhancement of heat tolerance of lettuce.

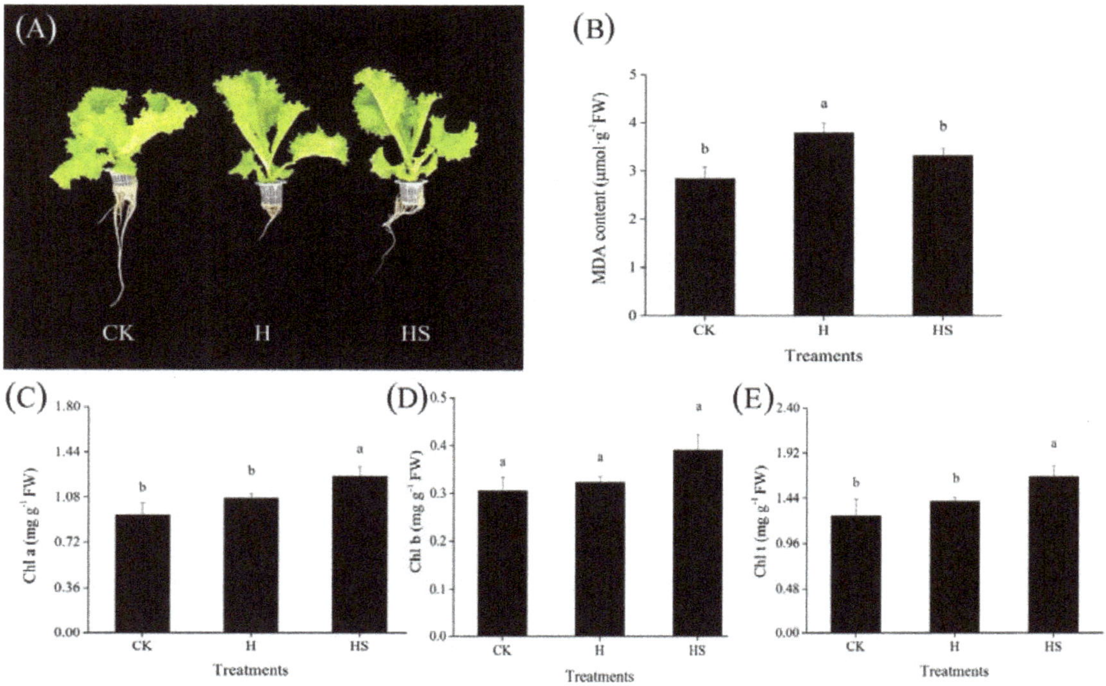

Figure 1. Effects of exogenous spermidine on the morphology, MDA content and chlorophyll content of lettuce under high temperature. (**A**) Phenotypic map. (**B**) MDA content. (**C**) Chlorophyll a content. (**D**) Chlorophyll b content. (**E**) Total chlorophyll content. Values above each vertical bar followed by different letters show significant differences ($p < 0.05$). The highest value was labeled as a, and those with significant differences were labeled as b in that order. CK: 22 °C/17 °C, distilled water; H: 35 °C/30 °C, distilled water; HS: 35 °C/30 °C, distilled 1 mM Spd.

3.2. *Effect of Spermidine on the Antioxidant Enzyme Activity of Lettuce under High-Temperature Stress*

Under high temperatures, SOD activity decreased in control plants sprayed with distilled water; however, SOD activity increased in plants treated with spermidine (Figure 2A). Similar to SOD activity, spermidine treatment increased CAT activity in lettuce under high temperature compared with deionized water spray (Figure 2C); compared with the control, high temperature stress had no significant effect on POD activity but spraying with spermidine under high temperature stress still significantly increased its activity compared with deionized water spray (Figure 2B). However, for APX, there was no significant effect of high temperature stress compared to the ambient control, and neither deionized water nor spermine spraying had any significant effect on its activity under high temperature stress (Figure 2D). This suggests that exogenous spermidine can withstand high temperature stress by regulating the activities of antioxidant enzymes such as SOD, CAT and POD, and may therefore attenuate oxidative damage in lettuce leaf cells.

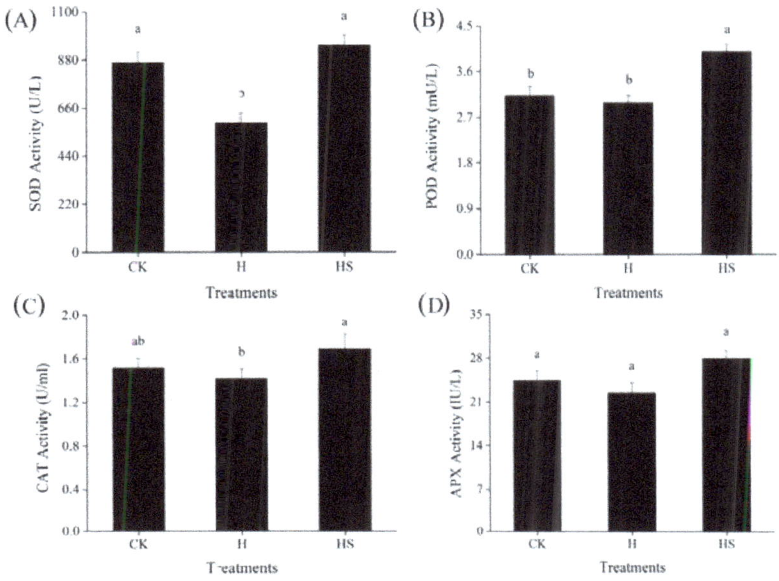

Figure 2. Effects of exogenous spermidine on the activities of antioxidant enzymes in lettuce under high-temperature stress. (**A**) SOD activity, (**B**) POD activity, (**C**) CAT activity, and (**D**) APX activity. Values above each vertical bar followed by different letters show significant differences ($p < 0.05$). The highest value was labeled as a, and those with significant differences were labeled as b in that order. CK: 22 °C/17 °C, distilled water; H: 35 °C/30 °C, distilled water; HS: 35 °C/30 °C, distilled 1 mM Spd.

3.3. Transcriptome Data Quality Analysis

We constructed nine sequencing libraries with three replicates of each treatment. The number of raw reads per sample ranged from 45.57 to 47.33 million, as shown in Table S2. Clean reads exceeded 42.33 million for all samples except for some low-quality reads, splices and fuzzy nucleotides. The number of clear reads obtained from the cDNA libraries of each experiment indicates that the gene abundance and transcript length are sufficient. In our libraries, the percentage of clean reads was not less than 90%, with Q20 and Q30 values exceeding 96.66% and 88.09% for all samples, respectively. This indicates that the data quality of the transcriptome is sufficient (Table S2).

3.4. Metabolomics Assay

The type and content of metabolites change under different stimuli leading to phenotypic changes. To investigate the effect of exogenous spermidine on changes in the lettuce metabolome under heat stress, we analyzed the major and minor metabolites in leaves. We quantified the relative levels of all filtered ions and performed PCA analysis (Figure 3B). As shown in the figure, the first two principal components explained 48% of the total variation, indicating a clear separation between water and spermine treatments under heat stress, suggesting that the identified metabolites play an important role in mitigating heat tolerance in lettuce under the influence of exogenous spermine(Figure 3B).Then, PLS-DA models (H vs. CK, HS vs. CK, HS vs. H) between each two groups were established using metaX software to screen out different metabolites, and when the values of model parameters (R2 and Q2) were high, it indicated that the current PLS-DA model was more reliable(Table 3). To further identify potential metabolites for water treatment and spermine treatment under high temperature stress, "candidate" metabolites with the following characteristics were included in the analysis with the screening conditions: (1) VIP $\geq$ 1; (2) fold change $\geq$ 1.2 or

≤ 0.8333; (3) p-value < 0.05, and the three were taken to intersect to obtain the common metabolite as the differential metabolite.

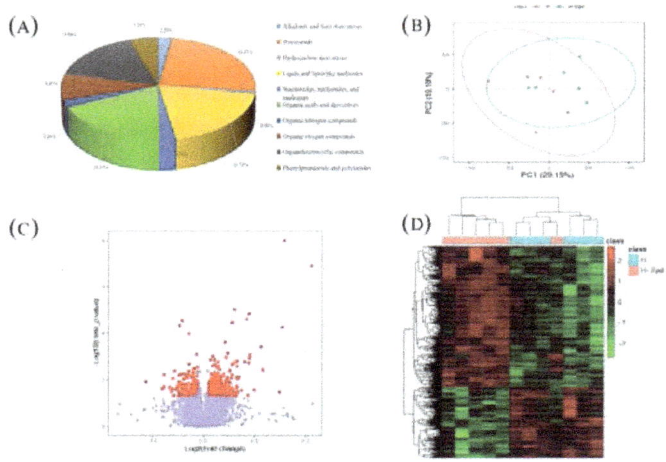

Figure 3. Differential metabolome analysis in HSvsH. (**A**) Differential metabolite composition. (**B**) PCA model for metabolic data. (**C**) Volcano plot of differential metabolites. (**D**) Heat map for cluster analysis of differential metabolites. CK: 22 °C/17 °C, distilled water; H: 35 °C/30 °C, distilled water; HS: 35 °C/30 °C, distilled 1 mM Spd.

Table 3. PLS-DA model parameters.

Group	R2	Q2
H vs. CK	0.992	0.735
HS vs. H	0.952	0.105
HS vs. CK	0.997	0.811

CK: 22 °C/17 °C, distilled water; H: 35 °C/30 °C, distilled water; HS: 35 °C/30 °C, distilled 1 mM Spd.

Compared with CK, 897 metabolites were identified under high temperature treatment and could be classified into 14 classes, while 1052 differential metabolites were identified in HSvsCK and classified into 15 classes; we obtained a total of 307 differential metabolites in HSvsH Supplementary Table S3). These metabolites included 7 alkaloids and derivatives, 75 benzenoids, 3 hydrocarbon derivatives, 60 lipids and lipid-like molecules, 8 nucleosides, nucleotides and analogues, 56 organic acids and derivatives, 7 organic nitrogen compounds, 26 organic oxygen compounds, 49 organic heterocyclic compounds and 16 phenylpropanoids and polyketides (Figure 3A).

In the comparison of the different treatments, "organic acids and their derivatives", "organic heterocyclic compounds", "lipids and lipid-like molecules" and "benzenes" were noted. Similarly, we found a number of "phenyl propane and polyketide compounds" in HS vs. H, and the phenyl propane pathway is the main pathway for the synthesis of flavonoids. It is noteworthy that we also found KEGG enrichment in the transcriptome for the flavonoid biosynthesis pathways, and the relative expression of most compounds was higher in the high-temperature applied spermidine treatment than in the high-temperature sprayed water treatment, as shown by clustering analysis and volcano plots (Figure 3C,D). This may further suggest that exogenous spermidine enhances the heat tolerance of lettuce by affecting flavonoid biosynthesis.

3.5. Analysis of DEGs

The number of DEG groups among the three treatments are shown, and overlapping parts indicate the intersection of different combinations (Figure 4B). There were 2101 genes upregulated and 960 genes downregulated in lettuce leaves under high-temperature stress compared to the control (Figure 4C). Comparing the treatment with spermidine spray to distilled water, 818 genes were upregulated, and 284 genes were downregulated. In addition, a total of 2489 genes were upregulated and 717 genes were downregulated in the high-temperature treatment with spermidine compared to the control. However, when comparing among the groups the overlapping relationships of the differentially expressed genes, there were 3061 genes, 1102 genes and 2306 DEGs expressed in the comparison of HvCK, HS vs. CK and HS vs. H, respectively, while 151 genes were expressed in common among the three treatment groups. These results suggest that spermidine affects the response of plant genes to high-temperature stress to some extent, leading to changes in the number and type of DEGs.

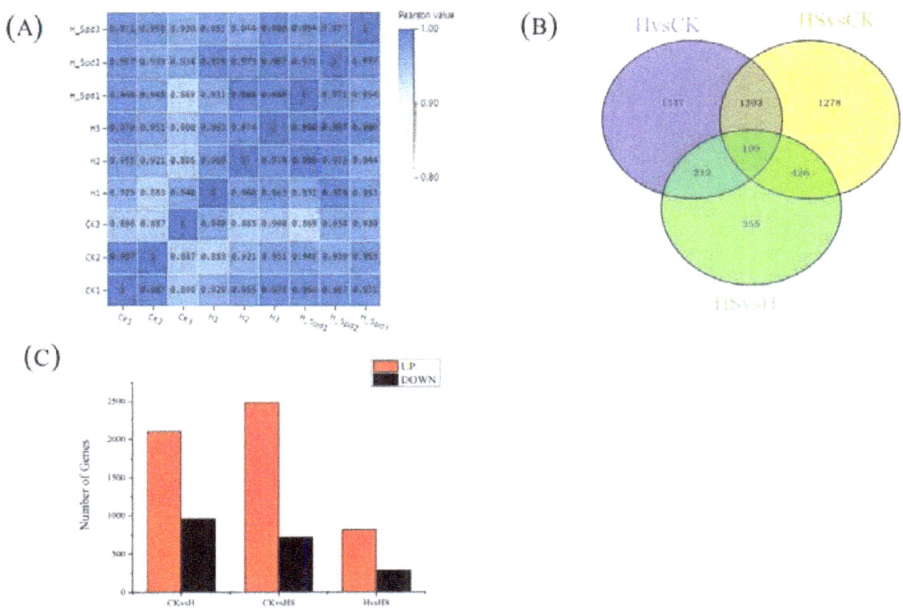

Figure 4. Analysis of DEGs in lettuce under high-temperature stress with exogenous spermidine. (**A**) Transcriptome sample correlation map. (**B**) Venn diagrams of DEGs. (**C**) Numbers of DEGs. CK: 22 °C/17 °C, distilled water; H: 35 °C/30 °C, distilled water; HS: 35 °C/30 °C, distilled 1 mM Spd.

3.6. GO Enrichment Analysis

To further understand the DEG functions of H and HS, GO functional annotation of DEGs was performed (Figure 5). A Q value (padj) < 0.05 was set as the threshold for significant enrichment. According to the GO database, genes can be classified according to biological process (BP), cytological component (CC) and molecular function (MF) categories. GO functional enrichment reveals the enrichment of distinct functional entries in differentially expressed genes, annotating differentially expressed genes to individual biological functions, so that we can understand the linkage between analyzing different differentially expressed genes and biological functions.

In this analysis, we significantly enriched the subcategories of differentially expressed genes in the GO database. In the DEGs of CK compared with H, 28.48%, 38.77% and 32.75% were categorized as biological processes, cellular components and molecular functions, respectively (Figure 5A). However, 30.32%, 37.29% and 32.39% of the DEGs in the HS versus H comparison were categorized as biological processes, cellular components and molecular functions, respectively (Figure 5B), these values were observed in 28.87%, 39.04% and 32.09% of the DEGs in the CK versus HS comparison (Figure 5C).

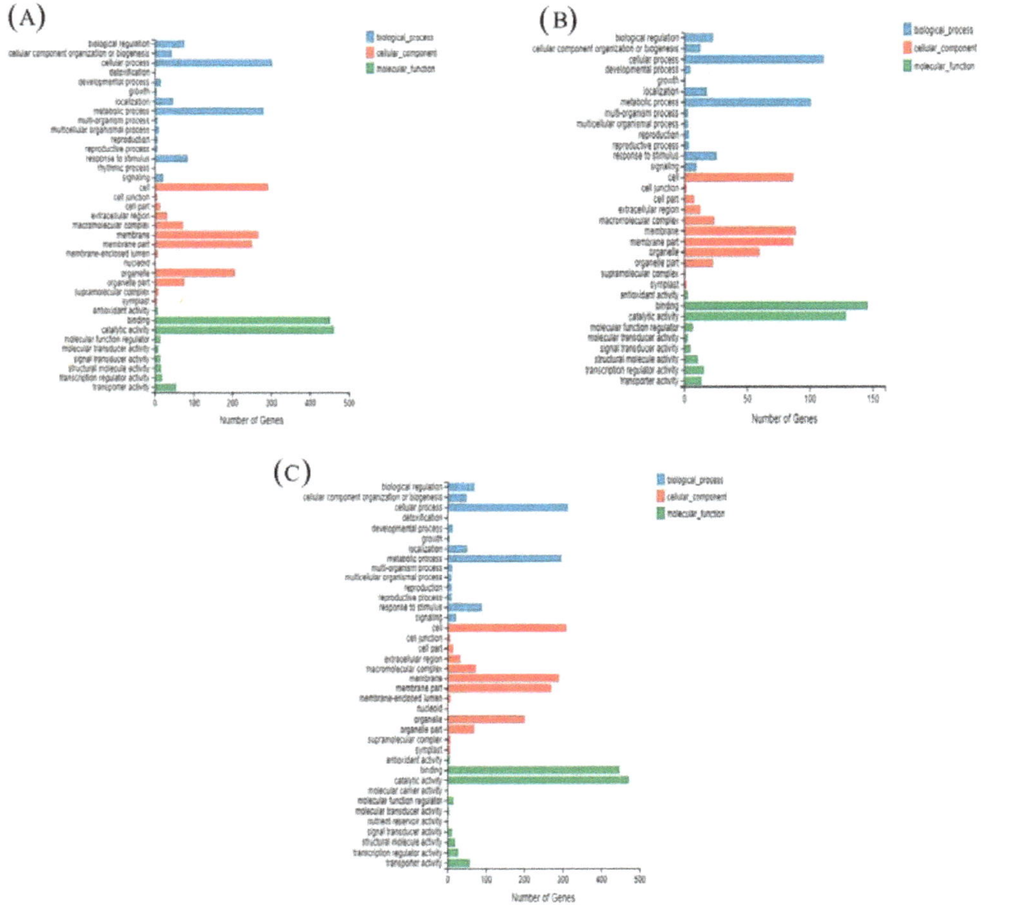

Figure 5. Gene Ontology (GO) classification and distribution of GO annotated genes. (**A**) H vs. CK; (**B**) HS vs. H; (**C**) HS vs. CK. CK: 22 °C/17 °C, distilled water; H: 35 °C/30 °C, distilled water; HS: 35 °C/30 °C, distilled 1 mM Spd.

3.7. KEGG Enrichment Analysis

To understand the metabolic or signaling pathways of spermidine involved in high-temperature stress, all differentially expressed genes were compared with the Kyoto encyclopedia of genes and genomes (KEGG) database to get the matched KEGG Orthology (KO). The KEGG enrichment analysis of different genes in each treatment group is shown in Figure 6. The vertical coordinates in the figure represent the KEGG pathway, and the horizontal coordinates are the enrichment factors. The larger the enrichment factor, the greater the degree of enrichment; the larger the point, the greater the number of differentially expressed genes enriched in the pathway; the bluer the point, the more significant the enrichment. The KEGG enrichment analysis of DEGs in the H and CK treatments is shown in Figure 6A. The DEGs between the H and CK treatments mainly focused on the biosynthesis of sesquiterpenes and triterpenes; biosynthesis of keratin, sulfites and waxes; fatty acid metabolism; biosynthesis of unsaturated fatty acids; starch and sucrose metabolism; degradation of other polysaccharides; phytohormone signaling; fatty acid degradation and biosynthesis; and biotin metabolism.

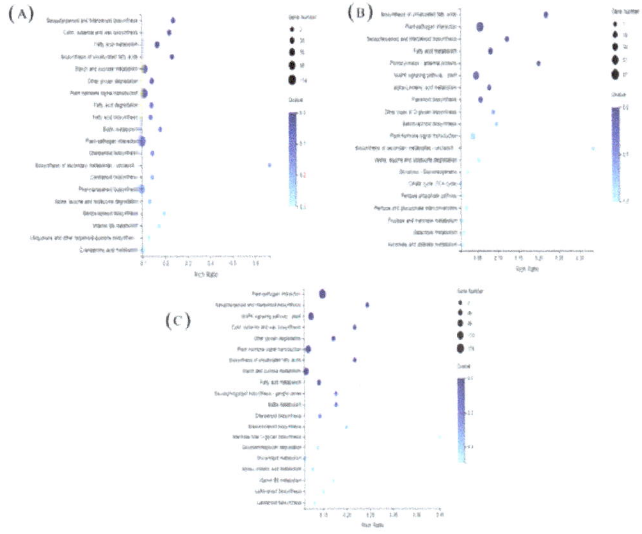

Figure 6. KEGG pathway enrichment analysis of DEGs in response to different stress treatments. (**A**) H vs. CK; (**B**) HS vs. H; (**C**) HS vs. CK. CK: 22 °C/17 °C, distilled water; H: 35 °C/30 °C, distilled water; HS: 35 °C/30 °C, distilled 1 mM Spd.

The KEGG analysis of DEGs between the HS and CK enrichment analyses is shown in Figure 6C. The DEGs between HS and CK were mainly involved in plant-pathogen interactions, sesquiterpene and triterpene biosynthesis, the MAPK signaling pathway, keratin, sulfite and wax biosynthesis, other polysaccharide degradation, phytohormone signaling, and unsaturated fatty acid biosynthesis.

The KEGG enrichment analysis of DEGs between HS and H is shown in Figure 6B. As shown in the figure, the DEGs between HS and H were mainly focused on unsaturated fatty acid biosynthesis, plant-pathogen interactions, sesquiterpene and triterpene biosynthesis, fatty acid metabolism, photosynthesis-antennal proteins, the MAPK signaling pathway, linolenic acid metabolism, and flavonoid biosynthesis. These results suggest that exogenous spermidine may protect against high temperature stress by improving photosynthesis, affecting and participating in signal transduction, and regulating flavonoid biosynthesis in lettuce.

3.8. Transcription Factors

The different transcription factors of differentially expressed genes (DEGs) were different in the different treatment comparisons (Figure 7). Seventeen transcription factor families, including 160 TF genes, were identified in the high-temperature treatment compared with control-treated lettuce seedlings, and the family with the highest number of transcription factors among the differentially expressed genes (DEGs) was AP2-EREBP, followed by the MYB, NAC, MADS, WRKY, SBP, and bHLH families (Figure 7A). In addition, a total of 172 TF genes from 28 transcription factor families, mainly including 37 AP2-EREBPs, 19 MYBs, 16 WRKYs, 14 NACs, and 12 bHLH, were found in the comparison of high-temperature spermidine treatment with the control treatment, while 75 TFs from 13 TF families were found in the comparison of the high-temperature spermidine treatment with the high-temperature control treatment. The family with the highest number of transcription factors among the differentially expressed genes (DEGs) was AP2-EREBP, followed by the WRKY, MYB, and NAM families (Figure 7B,C). Our experimental results revealed that most of the transcription factors were upregulated after spraying with spermidine under high-temperature stress (Figure 7D), suggesting that it may alleviate the damage caused by high-temperature stress mainly by upregulating the expression of AP2, MYB, WRKY and other transcription factors and thus further regulate the expression of the corresponding genes in lettuce.

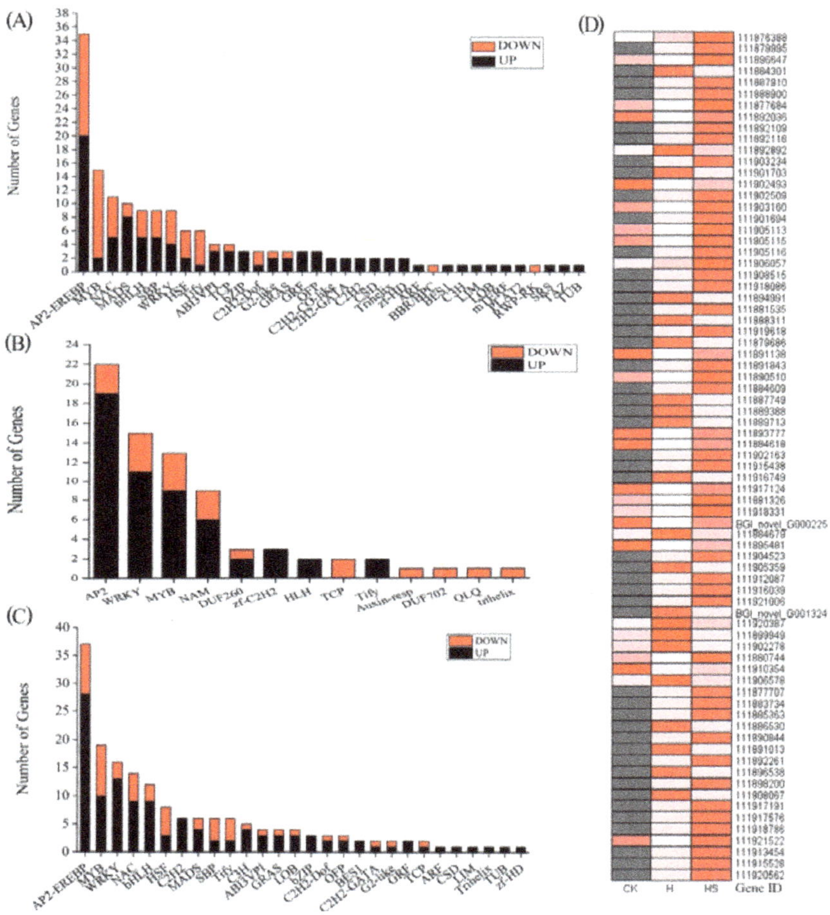

Figure 7. Regulation of DEGs of transcription factors. (**A**) H vs. CK; (**B**) HS vs. H; (**C**) HS vs. CK; (**D**) Expression trends of transcription factors in HS vs. H. CK: 22 °C/17 °C, distilled water; H: 35 °C/30 °C, distilled water; HS: 35 °C/30 °C, distilled 1 mM Spd.

3.9. Spermidine Regulates the Metabolism of Flavonoids under High-Temperature Stress

As shown in Figure 8A, spraying spermidine increased the total flavonoid content in leaves under high-temperature stress. A total of 19 DEGs rich in flavonoid-related metabolic pathways were identified in HS vs. H, encoding six enzymes related to flavonoid synthesis, including hydroxycinnamoyltransferase (HCT), flavonol synthase (FLS), caffeoyl coenzyme A methyltransferase (CCoAOMT) dihydroflavonol reductase (DFR), chalcone synthase (CHS), and colorless anthocyanin dioxygenase (LDOX), which are involved in the regulation of the synthesis of flavonoid substances such as dihydroflavonol, flavonol, colorless anthocyanin, and chalcone (Figure 8B). Among them, 12 genes significantly increased their expression under high-temperature stress after the application of spermidine, and half of these 12 genes enhanced their expression by more than 2-fold, which may be responsible for the increase in total flavonoid content. These results, in agreement with our previous results, suggest that exogenous spermidine regulates the synthesis of flavonoid substances in lettuce leaves, thereby affecting their tolerance to high-temperature stress.

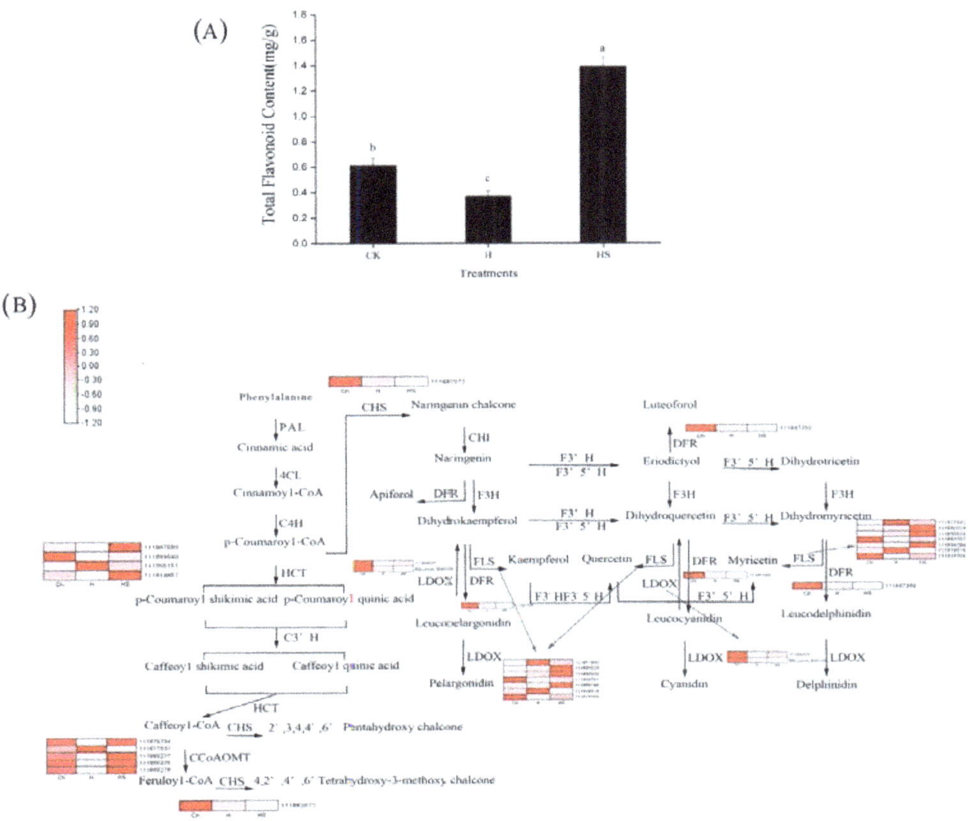

Figure 8. Flavonoid metabolic pathway association with differentially expressed genes (DEGs). (**A**) Total flavonoid content; (**B**) Expression of related DEGs in flavonoid metabolic pathways. Values above each vertical bar followed by different letters show significant differences ($p < 0.05$). The highest value was labeled as a, and those with significant differences were labeled as b, c in that order. CK: 22 °C/17 °C, distilled water; H: 35 °C/30 °C, distilled water; HS: 35 °C/30 °C, distilled 1 mM Spd.

4. Discussion

High temperature (HT) is a widespread environmental stress that affects most plants at all periods of growth and can limit plant growth and development and reduce productivity. Plant growth and development involve many temperature-sensitive biochemical responses [31]. One approach to dealing with the adverse effects of heat stress may involve exploring some molecules that have the potential to protect plants from the deleterious effects of HT. Polyamines (PAs) are low molecular weight aliphatic amines and organic polycations found in a variety of organisms from bacteria to plants and animals [32]. They also play an important role in plant responses to abiotic stresses. Our previous studies have shown that high-temperature stress limits normal growth and leads to the accumulation of unwanted ROS, altering the enzymatic and nonenzymatic antioxidant activity of lettuce seedlings [27], while exogenous spermidine restores growth and photosynthesis by improving carbon metabolism, thereby increasing stress tolerance [33].

In our experiments, heat stress resulted in wilting of lettuce, curling of leaves, a reduction in biomass, and an increase in malondialdehyde content. The application of spermidine improved the growth of lettuce leaves, reduced the malondialdehyde content under heat stress, and increased the chlorophyll content. Previous studies showed that Spd treatment significantly promoted FW and DW in white clover seedlings under PEG-simulated water stress conditions [34]. Zeng found a significant decrease in aboveground FW and DW of hybrid rice under low-temperature stress, while Spd-treated plants showed a significant increase in these indices by 30.48% and 18.19% under low temperature conditions [35]. Similar to these reports, our results confirm the role of spermidine in the protection of lettuce biomass.

Chlorophyll is an important indicator of the ability of leaves to maintain their green color, and the decrease in chlorophyll concentration is generally considered to be a response mechanism to reduced light uptake by chloroplasts under stress; furthermore, the decrease in chlorophyll content during stress may be related to impaired chlorophyll synthesis or pigment protein degradation and could be a result of ROS production [36]. The results in Figures 2 and 3 indicate that exogenous spermidine promoted the enhancement of antioxidant enzyme activity with increased chlorophyllin lettuce seedlings, which may be one of the reasons for their increased tolerance of heat stress. Our previous study also confirmed that spraying exogenous spermine at high temperatures increased chlorophyll a and b content in lettuce leaves and was able to avoid oxidative damage to chloroplasts [37]. These results suggest that Spd spraying can maintain chlorophyll and content to support light energy capture and transport in lettuce under high temperature stress. Higher photosynthesis rates promote flavonoid accumulation because flavonoids are synthesized in chloroplasts and flavonoids are hypothesized to be positively correlated with photosynthesis [38–40]. Therefore, we hypothesized that enhanced photosynthetic activity in lettuce leaves increased the production of primary and secondary metabolites, including flavonoids.

Crop tolerance to HT stress is associated with an increase in antioxidant capacity [41]. Previous studies have shown that Spm pretreatment increased antioxidant enzyme activities under HT and drought stress conditions, while the activity levels of CAT, POD and SOD were consistently higher in Spm pretreated seedlings than in the control [42]. In our experiments, high-temperature stress affected POD and SOD activities in lettuce leaves, while the application of spermidine increased CAT, POD and SOD activities without a significant effect on APX. This may be due to the fact that antioxidant enzymes differ in their perception of the degree and duration of temperature, and thus different antioxidant enzymes vary over different temperature ranges.

Upregulation of many genes has been reported to help plants resist stress conditions that lead to plant adaptation [43]. Under various biotic and abiotic stresses, plants are able to receive external signals and respond to the stress by associating various internal pathways to transmit information to downstream molecules through their respective methods. The main molecular mechanisms underlying the response to salt stress were previously revealed by transcriptome sequencing analysis [44]; physiological and transcriptomic approaches were used in grapes to explore the effect of exogenous monocrotaline lactones on drought stress at the transcriptional level, among others [45]. A combined transcriptomic and metabolomic analysis also provided an enhanced understanding of the regulatory network between certain specific genes and compounds in broccoli species under selenate treatment [46]. Therefore, in our study, we attempted to investigate the response of spermidine to differential gene expression patterns, expression of different transcription factors, and plant secondary metabolites and related genes in lettuce seedlings under high temperature stress by studying RNA-sequence analysis and metabolite components.

Signaling molecules are involved in the activation of many stress-responsive genes, and various signal transduction molecules associated with the activation of stress-responsive genes exist depending on the plant type and type of stress. Among them, widely used are Ca-dependent protein kinases (CDPKs), mitogen-activated protein kinases (MPKs), NO, sugar (as signaling molecules), and phytohormones [47]. These molecules, along with transcription factors, activate stress response genes. Once activated, stress response genes reactivate essential enzymes and structural proteins, which contribute to the detoxification of ROS (by activating detoxifying enzymes and free radical scavengers) to maintain cellular homeostasis [48]. The available data suggest that some signaling molecules may lead to an increase in cellular antioxidant capacity [49,50].

TFs stimulate and regulate multiple stress response pathways in plants subjected to high-temperature stress [48,49]. Flavonoids are plant polyphenol secondary metabolites with a wide range of physiological functions [51]. The flavonoid pathway is derived from the general phenylpropane pathway, and flavonoid biosynthetic genes are regulated by interactions between different TF families [44,46]. For example, genes involved in the anthocyanin and condensed tannin pathways are regulated by R2R3MYB, bHLH, and WD40 proteins (MYB-bHLH-WD40, MBW complex) [52]. Flavonoid synthesis genes, such as F3H, F3′H and FLS, are regulated by MYB genes [53]. In the present study, AP2-EREBP, WRKY, MYB, and NAM were the most abundantly expressed transcription factor families after spermidine treatment (Figure 6), and in agreement with previous studies, spermidine may have regulated the MYB family and thus the expression of genes related to the flavonoid metabolic pathway and, in this way, coordinated the total flavonoid content. However, the regulatory functions of these TFs in flavonoid biosynthesis need to be further investigated.

Flavonoids may protect plants from oxidative stress by blocking ROS production through their ability to chelate metal ions and scavenge ROS, thus achieving their antioxidant function [54–56]. Moreover, since the synthesis of flavonoids is performed in chloroplasts, the increased chlorophyll content and antioxidant enzyme activity in our experimental results suggest that spermidine may be able to maintain high light energy capture and transport by increasing chlorophyll content and avoiding oxidative loss of chloroplasts to ensure smooth synthesis of flavonoids, which in turn may protect organelles from oxidative damage [39,40]. It has been shown that excessive accumulation of flavonoids with higher free radical scavenging activity in transgenic Arabidopsis enhances tolerance to drought stress [57]. In our study, KEGG enrichment analysis of DEGs showed significant enrichment in the "flavonoid biosynthesis" pathway (Figure 6B) and six enzymes related to the flavonoid synthesis pathway (oxalate hydroxycinnamyltransferase (HCT), flavonol synthase (FLS), CCoAMOT (caffeoyl-coenzyme A methyltransferase), chalcocyanine, and chalcocyanine), chalcone synthase (CHS), and colorless anthocyanin dioxygenase (LDOX)) were differentially expressed. Similarly, most of the representative genes involved in these pathways were downregulated under high-temperature stress and upregulated under high-temperature spermidine treatment (Figure 8 and Supplementary Table S3), which may be responsible for the elevated total flavonoid concentration (Figure 8A). In other words, the elevated flavonoid concentration may have mitigated the effect of high-temperature stress on lettuce leaves to some extent.

5. Conclusions

Exogenous spermidine increased the content of chlorophylls and antioxidant enzymes and decreased the content of malondialdehyde under high-temperature stress. By integrating transcriptome and metabolite analysis, we found that spermidine may be involved in or induce the expression of TF families such as AP2-EREBP, WRKY, MYB, and NAM to transmit information, and combined with the metabolic pathways identified in KEGG enrichment, spermidine influenced the biosynthesis of flavonoid synthesis and ultimately alleviated the damage to lettuce from high-temperature stress. A hypothetical model of exogenous spermidine-mediated heat stress tolerance in lettuce is shown in Figure 9, but its specific regulatory mechanism needs further exploration and experimental corroboration. Exogenous spermidine may be a promising approach and may enhance the heat tolerance of lettuce by regulating various factors such as growth, physiology, molecular activities and metabolite accumulation under high temperature conditions.

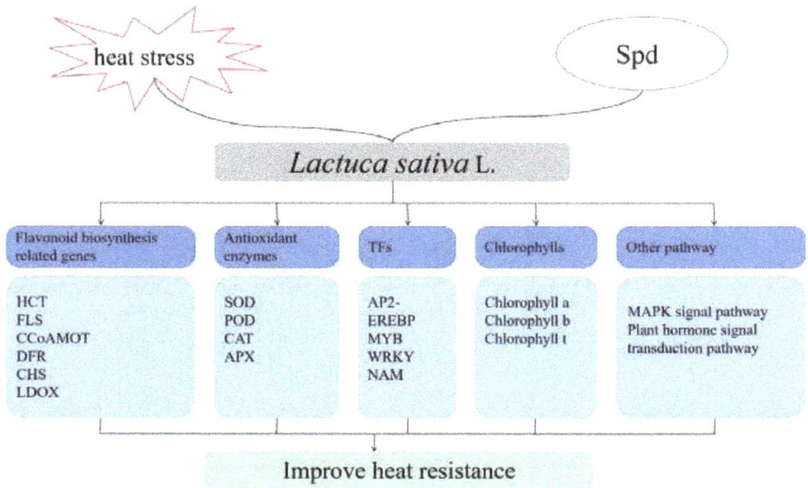

Figure 9. Hypothetical model of exogenous Spd to improve the heat tolerance of lettuce.

Supplementary Materials: The following supporting information can be downloaded at: https://www.mdpi.com/article/10.3390/antiox11122332/s1, Figure S1: Validation of the RNA-Seq results by qRT-PCR. Data are presented as the means of three replicates; Table S1: Sequence information of primers of genes of interest used in qRT-PCR assays; Table S2: Sequence reads of the transcriptome and their alignment with the reference genome; total of 307 differential metabolites in HSvsH.

Author Contributions: W.S.: investigation, writing—original draft; J.H.: writing—review and editing; S.F.: resources; C.L.: conceptualization, writing—review and editing, funding acquisition; Y.H.: conceptualization, funding acquisition. All authors have read and agreed to the published version of the manuscript.

Funding: This work was supported by the National Natural Science Foundation of China (Grant number: 32172607) and Beijing Joint Research Program for Germplasm Innovation and New Variety Breeding (Grant number: G20220628003).

Institutional Review Board Statement: Not applicable.

Informed Consent Statement: Not applicable.

Data Availability Statement: All of the data is contained within the article.

Conflicts of Interest: The authors declare no conflict of interest.

References

1. Org, W.M.; Strachan, P. Intergovernmental Panel on Climate Change (IPCC). *United Nations* **2007**, *114*, 48–56. [CrossRef]
2. Porter, J.R. Rising temperatures are likely to reduce crop yields. *Nature* **2005**, *436*, 174. [CrossRef]
3. Hasanuzzaman, M.; Hossain, M.A.; da Silva, J.A.T.; Fujita, M. Plant response and tolerance to abiotic oxidative stress: Antioxidant defense is a key factor. *Crop Stress Its Manag. Perspect. Strateg.* **2012**, *8*, 261–315. [CrossRef]
4. Ruelland, E.; Zachowski, A. How plants sense temperature. *Environ. Exp. Bot.* **2010**, *69*, 225–232. [CrossRef]
5. Suzuki, N.; Miller, G.; Morales, J.; Shulaev, V.; Torres, M.A.; Mittler, R. Respiratory burst oxidases: The engines of ROS signaling. *Curr. Opin. Plant Biol.* **2011**, *14*, 691–699. [CrossRef] [PubMed]
6. Suzuki, N.; Koussevitzky, S.; Mittler, R.; Miller, G. ROS and redox signalling in the response of plants to abiotic stress. *Plant Cell Environ.* **2012**, *35*, 259–270. [CrossRef]
7. Pagamas, P.; Nawata, E. Sensitive stages of fruit and seed development of chili pepper (*Capsicum annuum* L. var. Shishito) exposed to high-temperature stress. *Sci. Hortic.* **2008**, *117*, 21–25. [CrossRef]

8. Hasanuzzaman, M.; Hossain, M.A.; Fujita, M. Selenium-induced up-regulation of the antioxidant defense and methylglyoxal detoxification system reduces salinity-induced damage in rapeseed seedlings. *Biol. Trace Elem. Res.* **2011**, *143*, 1704–1721. [CrossRef]
9. Hussain, S.S.; Ali, M.; Ahmad, M.; Siddique, K.H. Polyamines: Natural and engineered abiotic and biotic stress tolerance in plants. *Biotechnol. Adv.* **2011**, *29*, 300–311. [CrossRef] [PubMed]
10. Su, G.X.; Zhang, W.H.; Liu, Y.L. Involvement of hydrogen peroxide generated by polyamine oxidative degradation in the development of lateral roots in Soybean. *J. Integr. Plant Biol.* **2006**, *48*, 426–432. [CrossRef]
11. Alcazar, R.; Bitrian, M.; Bartels, D.; Koncz, C.; Altabella, T.; Tiburcio, A.F. Polyamine metabolic canalization in response to drought stress in Arabidopsis and the resurrection plant *Craterostigma plantagineum*. *Plant Signal. Behav.* **2011**, *6*, 243–250. [CrossRef] [PubMed]
12. Feng, H.Y.; Wang, Z.M.; Kong, F.N.; Zhang, M.J.; Zhou, S.L. Roles of carbohydrate supply and ethylene, polyamines in maize kernel set. *J. Integr. Plant Biol.* **2011**, *53*, 388–398. [CrossRef] [PubMed]
13. Wimalasekera, R.; Tebartz, F.; Scherer, G.F. Polyamines, polyamine oxidases and nitric oxide in development, abiotic and biotic stresses. *Plant Sci.* **2011**, *181*, 593–603. [CrossRef] [PubMed]
14. Zhang, Y.; Wu, R.; Qin, G.; Chen, Z.; Gu, H.; Qu, L.J. Over-expression of WOX1 leads to defects in meristem development and polyamine homeostasis in Arabidopsis. *J. Integr. Plant Biol.* **2011**, *53*, 493–506. [CrossRef] [PubMed]
15. Alet, A.I.; Sanchez, D.H.; Cuevas, J.C.; Marina, M.; Carrasco, P.; Altabella, T.; Tiburcio, A.F.; Ruiz, O.A. New insights into the role of spermine in Arabidopsis thaliana under long-term salt stress. *Plant Sci.* **2012**, *182*, 94–100. [CrossRef] [PubMed]
16. Tavladoraki, P.; Cona, A.; Federico, R.; Tempera, G.; Viceconte, N.; Saccoccio, S.; Battaglia, V.; Toninello, A.; Agostinelli, E. Polyamine catabolism: Target for antiproliferative therapies in animals and stress tolerance strategies in plants. *Amino Acids* **2012**, *42*, 411–426. [CrossRef] [PubMed]
17. Pal, M.; Szalai, G.; Janda, T. Speculation: Polyamines are important in abiotic stress signaling. *Plant Sci.* **2015**, *237*, 16–23. [CrossRef]
18. Sang, Q.Q.; Shu, S.; Shan, X.; Guo, S.R.; Sun, J. Effects of exogenous spermidine on antioxidant system of tomato seedlings exposed to high temperature stress. *Russ. J. Plant Physiol.* **2016**, *63*, 645–655. [CrossRef]
19. Sun, J.; Lu, N.; Xu, H.; Maruo, T.; Guo, S. Root zone cooling and exogenous spermidine root-pretreatment promoting *Lactuca sativa* L. growth and photosynthesis in the high-temperature season. *Front. Plant Sci.* **2016**, *7*, 368. [CrossRef] [PubMed]
20. Goyal, M.; Asthir, B. Polyamine catabolism influences antioxidative defense mechanism in shoots and roots of five wheat genotypes under high temperature stress. *Plant Growth Regul.* **2009**, *60*, 13–25. [CrossRef]
21. Zhang, L.; Hu, T.; Amombo, E.; Wang, G.; Xie, Y.; Fu, J. The alleviation of heat damage to photosystem II and enzymatic antioxidants by exogenous spermidine in tall fescue. *Front. Plant Sci.* **2017**, *8*, 1747. [CrossRef]
22. Mostofa, M.G.; Yoshida, N.; Fujita, M. Spermidine pretreatment enhances heat tolerance in rice seedlings through modulating antioxidative and glyoxalase systems. *Plant Growth Regul.* **2013**, *73*, 31–44. [CrossRef]
23. Sagor, G.H.; Berberich, T.; Takahashi, Y.; Niitsu, M.; Kusano, T. The polyamine spermine protects Arabidopsis from heat stress-induced damage by increasing expression of heat shock-related genes. *Transgenic Res.* **2013**, *22*, 595–605. [CrossRef]
24. Hao, J.H.; Zhang, L.L.; Li, P.P.; Sun, Y.C.; Li, J.K.; Qin, X.X.; Wang, L.; Qi, Z.Y.; Xiao, S.; Han, Y.Y.; et al. Quantitative proteomics analysis of lettuce (*Lactuca sativa* L.) reveals molecular basis-associated auxin and photosynthesis with bolting induced by high temperature. *Int. J. Mol. Sci.* **2018**, *19*, 2967. [CrossRef]
25. Wu, X.; Cai, K.; Zhang, G.; Zeng, F. Metabolite profiling of barley grains subjected to water stress: To explain the genotypic difference in drought-induced impacts on malting quality. *Front. Plant Sci.* **2017**, *8*, 1547. [CrossRef]
26. Li, W.; Li, W.F.; Yang, S.J.; Ma, Z.H.; Zhou, Q.; Mao, J.; Han, S.Y.; Chen, B.H. Transcriptome and metabolite conjoint analysis reveals that exogenous methyl jasmonate regulates monoterpene synthesis in grape berry skin. *J. Agric. Food Chem.* **2020**, *68*, 5270–5281. [CrossRef] [PubMed]
27. Li, C.; Han, Y.; Hao, J.; Qin, X.; Liu, C.; Fan, S. Effects of exogenous spermidine on antioxidants and glyoxalase system of lettuce seedlings under high temperature. *Plant Signal. Behav.* **2002**, *15*, 1824697. [CrossRef] [PubMed]
28. Rao, K.M.; Sresty, T.V.S. Antioxidative parameters in the seedlings of pigeonpea (*Cajanus cajan* (L.) Millspaugh) in response to Zn and Ni stresses. *Plant Sci.* **2000**, *157*, 113–128. [CrossRef]
29. Giannopolitis, C.N.; Ries, S.K. Superoxide Dismutases: I. Occurrence in higher plants. *Plant Physiol.* **1977**, *59*, 309–314. [CrossRef]
30. Nakano, Y.; Asada, K. Hydrogen peroxide is scavenged by ascorbate-specific peroxidase in spinach chloroplasts. *Plant Cell Physiol.* **1981**, *22*, 867–880. [CrossRef]
31. Hasanuzzaman, M.; Nahar, K.; Alam, M.M.; Roychowdhury, R.; Fujita, M. Physiological, biochemical, and molecular mechanisms of heat stress tolerance in plants. *Int. J. Mol. Sci.* **2013**, *14*, 9643–9684. [CrossRef]
32. Alcázar, R.; Cuevas, J.C.; Patron, M.; Altabella, T.; Tiburcio, A.F. Abscisic acid modulates polyamine metabolism under water stress in Arabidopsis thaliana. *Physiol. Plant.* **2006**, *128*, 448–455. [CrossRef]
33. Yu, Q.; Sun, W.; Han, Y.; Hao, J.; Qin, X.; Liu, C.; Fan, S. Exogenous spermidine improves the sucrose metabolism of lettuce to resist high-temperature stress. *Plant Growth Regul.* **2022**, *96*, 497–509. [CrossRef]
34. Li, Z.; Peng, Y.; Zhang, X.Q.; Ma, X.; Huang, L.K.; Yan, Y.H. Exogenous spermidine improves seed germination of white clover under water stress via involvement in starch metabolism, antioxidant defenses and relevant gene expression. *Molecules* **2014**, *19*, 18003–18024. [CrossRef]

35. Zeng, Y.H.; Zahng, Y.P.; Xiang, J.; Wu, H.; Chen, H.Z.; Zhang, Y.K.; Zhu, D.F. Effects of chilling tolerance induced by spermidine pretreatment on antioxidative activity, endogenous hormones and ultrastructure of indica-japonica hybrid rice seedlings. *J. Integr. Agric.* **2016**, *15*, 295–308. [CrossRef]
36. Saravi, H.B.; Gholami, A.; Pirdashti, H.; Firouzabadi, M.B.; Asghari, H.; Yaghoubian, Y. Improvement of salt tolerance in Stevia rebaudiana by co-application of endophytic fungi and exogenous spermidine. *Ind. Crops Prod.* **2022**, *177*, 114443. [CrossRef]
37. Yang, X.; Han, Y.; Hao, J.; Qin, X.; Liu, C.; Fan, S. Exogenous spermidine enhances the photosynthesis and ultrastructure of lettuce seedlings under high-temperature stress. *Sci. Hortic.* **2022**, *291*, 110570. [CrossRef]
38. Xie, B.D.; Wang, H.T. Effects of light spectrum and photoperiod on contents of flavonoid and terpene in leaves of Ginkgo biloba. *L. J. Nanjing. Univ.* **2006**, *30*, 51–54.
39. Xu, M.; Dong, J.; Wang, H.; Huang, L. Complementary action of jasmonic acid on salicylic acid in mediating fungal elicitor-induced flavonol glycoside accumulation of Ginkgo biloba cells. *Plant Cell Environ.* **2009**, *32*, 960–967. [CrossRef]
40. Li, X.; Zhang, L.P.; Zhang, L.; Yan, P.; Ahammed, G.J.; Han, W.Y. Methyl salicylate enhances flavonoid biosynthesis in tea leaves by stimulating the phenylpropanoid pathway. *Molecules* **2019**, *24*, 362. [CrossRef]
41. Almeselmani, M.; Deshmukh, P.S.; Sairam, R.K.; Kushwaha, S.R.; Singh, T.P. Protective role of antioxidant enzymes under high temperature stress. *Plant Sci.* **2006**, *171*, 382–388. [CrossRef] [PubMed]
42. Fu, X.Z.; Xing, F.; Wang, N.Q.; Peng, L.Z.; Chun, C.P.; Cao, L.; Ling, L.L.; Jiang, C.L. Exogenous spermine pretreatment confers tolerance to combined high-temperature and drought stress in vitro in trifoliate orange seedlings via modulation of antioxidative capacity and expression of stress-related genes. *Biotechnol. Biotechnol. Equip.* **2014**, *28*, 192–198. [CrossRef]
43. Tuteja, N. Integrated calcium signaling in plants. In *Signaling in Plants*; Springer: Berlin/Heidelberg, Germany, 2009; pp. 29–49. [CrossRef]
44. Xu, Z.C.; Wang, M.; Ren, T.T.; Li, K.; Li, K.Y.; Marowa, P.; Zhang, C.S. Comparative transcriptome analysis reveals the molecular mechanism of salt tolerance in *Apocynum venetum*. *Plant Physiol. Biochem.* **2021**, *167*, 816–830. [CrossRef] [PubMed]
45. Wang, W.N.; Min, Z.; Wu, J.R.; Liu, B.C.; Xu, X.L.; Fang, Y.L.; Ju, Y.L. Physiological and transcriptomic analysis of Cabernet Sauvginon (*Vitis vinifera* L.) reveals the alleviating effect of exogenous strigolactones on the response of grapevine to drought stress. *Plant Physiol. Biochem.* **2021**, *167*, 400–409. [CrossRef]
46. Rao, S.; Gou, Y.Y.; Yu, T.; Cong, X.; Gui, J.Y.; Zhu, Z.Z.; Zhang, W.W.; Liao, Y.L.; Ye, J.B.; Cheng, S.Y.; et al. Effects of selenate on Se, flavonoid, and glucosinolate in broccoli florets by combined transcriptome and metabolome analyses. *Food Res. Int.* **2021**, *146*, 110463. [CrossRef]
47. Ahmad, P.; Bhardwaj, R.; Tuteja, N. Plant signaling under abiotic stress environment. In *Environmental Adaptations and Stress Tolerance of Plants in the Era of Climate Change*; Springer: New York, NY, USA, 2012; pp. 297–323.
48. Ciarmiello, L.F.; Woodrow, P.; Fuggi, A.; Pontecorvo, G.; Carillo, P. Plant genes for abiotic stress. *InTech* **2011**, *13*, 283–308. [CrossRef]
49. Gong, M.; Chen, S.N.; Song, Y.Q.; Li, Z.G. Effect of calcium and calmodulin on intrinsic heat tolerance in relation to antioxidant systems in maize seedlings. *Aust. J. Plant Physiol.* **1997**, *24*, 371–379. [CrossRef]
50. Dat, J.; Foyer, C.; Scott, I. Change in salicylic acid and antioxidants during induced thermo tolerance in mustard seedlings. *Plant Physiol.* **1998**, *118*, 1455–1461. [CrossRef]
51. Chinnusamy, V.; Schumaker, K.; Zhu, J.K. Molecular genetic perspectives on cross-talk and specificity in abiotic stress signalling in plants. *J. Exp. Bot.* **2004**, *55*, 225–236. [CrossRef]
52. Wang, H.; Wang, H.; Shao, H.; Tang, X. Recent advances in utilizing transcription factors to improve plant abiotic stress tolerance by transgenic technology. *Front. Plant Sci.* **2016**, *7*, 67. [CrossRef]
53. Koes, R.; Verweij, W.; Quattrocchio, F. Flavonoids: A colorful model for the regulation and evolution of biochemical pathways. *Trends Plant Sci.* **2005**, *10*, 236–242. [CrossRef] [PubMed]
54. Hichri, I.; Barrieu, F.; Bogs, J.; Kappel, C.; Delrot, S.; Lauvergeat, V. Recent advances in the transcriptional regulation of the flavonoid biosynthetic pathway. *J. Exp. Bot.* **2011**, *62*, 2465–2483. [CrossRef]
55. Falcone Ferreyra, M.L.; Rius, S.P.; Casati, P. Flavonoids: Biosynthesis, biological functions, and biotechnological applications. *Front. Plant Sci.* **2012**, *3*, 222. [CrossRef] [PubMed]
56. Williams, R.J.; Spencer, J.P.; Rice-Evans, C. Flavonoids: Antioxidants or signalling molecules? *Free Radic. Biol. Med.* **2004**, *36*, 838–849. [CrossRef] [PubMed]
57. Nakabayashi, R.; Yonekura-Sakakibara, K.; Urano, K.; Suzuki, M.; Yamada, Y.; Nishizawa, T.; Matsuda, F.; Kojima, M.; Sakakibara, H.; Shinozaki, K.; et al. Enhancement of oxidative and drought tolerance in Arabidopsis by overaccumulation of antioxidant flavonoids. *Plant J.* **2014**, *77*, 367–379. [CrossRef]

antioxidants

MDPI

Article

Salt Stress Induces Changes in Physiological Characteristics, Bioactive Constituents, and Antioxidants in Kenaf (*Hibiscus cannabinus* L.)

Ziggiju Mesenbet Birhanie [1,2,†], Dawei Yang [1,†], Mingbao Luan [1], Aiping Xiao [1], Liangliang Liu [1], Chao Zhang [1], Ashok Biswas [1], Susmita Dey [1], Yong Deng [1,*] and Defang Li [1,*]

1 Institute of Bast Fiber Crops, Chinese Academy of Agricultural Sciences, Changsha 410205, China
2 Department of Plant Science, College of Agriculture and Natural Resources, Debre Markos University, Debre Markos 251269, Ethiopia
* Correspondence: dengyong@caas.cn (Y.D.); lidefang@caas.cn (D.L.); Tel.: +86-(13)-808498640 (Y.D.); +86-(13)-873129468 (D.L.)
† These authors contributed equally to this work.

Abstract: Salinity stress is a major environmental threat in agricultural systems. Kenaf is a promising crop for the future for cultivation in salinity-affected soils because of its high phytoremediation potential. The current study aimed to investigate the effects of salt stress using six different sodium chloride (NaCl) concentrations (0, 50, 100, 150, 200, and 250 mM) on the plant growth, physiological characteristics, bioactive constituents, and antioxidant capacity of *H. cannabinus*. The results indicated that the NaCl stress induced significant reductions in plant height and in the dry and fresh weights of the leaf tissue. In addition, the K, Ca, Mg, and P concentrations in this tissue also decreased under NaCl stress treatment conditions. In contrast, the NaCl stress led to the accumulation of hydrogen peroxide (H_2O_2), superoxide anion ($O_2^{\bullet-}$), malondialdehyde (MDA), proline, total soluble sugar, and total soluble protein. Under NaCl stress, the levels of antioxidants, including phenolics and flavonoids, also increased. The gas chromatography–mass spectrometry (GC-MS) results showed that the volatile compounds, including heptacosane, 1-octadecanesulphonyl chloride, and tetratetracontane, were induced under the NaCl stress treatment. Furthermore, the salt stress significantly improved the antioxidant capacity of the leaf extracts. These findings may provide insight into how *H. cannabinus* plants respond to salt stress and may help improve its medicinal value under salt stress.

Keywords: *Hibiscus cannabinus*; salt stress; physiological changes; bioactive constituents

Citation: Birhanie, Z.M.; Yang, D.; Luan, M.; Xiao, A.; Liu, L.; Zhang, C.; Biswas, A.; Dey, S.; Deng, Y.; Li, D. Salt Stress Induces Changes in Physiological Characteristics, Bioactive Constituents, and Antioxidants in Kenaf (*Hibiscus cannabinus* L.). *Antioxidants* **2022**, *11*, 2005. https://doi.org/10.3390/antiox11102005

Academic Editor: Nafees A. Khan

Received: 20 September 2022
Accepted: 7 October 2022
Published: 10 October 2022

1. Introduction

Salinity is among the major abiotic stresses impacting plant growth and productivity [1,2]. It affects approximately 1125 million hectares of agricultural land globally [3]. In China, salinity affects about 36.7 million hectares of land. By 2050, it could damage more than 50% of the agricultural land [4]. The intensity of the salt stress affects the plants' morphological, physiological, and metabolic changes. Soil salinity can inhibit plant growth by causing ion toxicity, osmotic and oxidative stresses, pigment degradation, and photosynthesis inhibition [5–7]. Ion toxicity and osmotic stress cause nutritional imbalances and oxidative stress by restricting plants from extracting water from the soil and from inside the plants themselves [8,9]. Additionally, salt stress causes oxidative stress by increasing reactive oxygen species (ROS), such as hydrogen peroxide (H_2O_2), superoxide ($O_2^{\bullet-}$), and hydroxyl radicals ($OH^{\cdot}$) [10,11]. Salt stress can severely disrupt the equilibrium between producing and scavenging reactive oxygen species (ROS) [12]. Plants require a certain threshold level of reactive oxygen species (ROS) to function normally; any variation in the ROS concentration can have detrimental effects on a plant's physiology [13]. Specifically, excessive concentrations of radical species cause damage to plant cell components, resulting in cell death [14].

Plants have varied defense strategies against salt stresses, involving morphological, physiological, and molecular responses [15]. Plants can produce osmolytes, including soluble sugars, proteins, and proline, which protect plant cells from the adverse effects of salt stress [16,17]. Protecting cellular membranes via enzymatic antioxidants against salt-induced ROS over-production and membrane lipid peroxidation leads to salt tolerance [18–20]. Under salt stress, antioxidants such as phenolic and flavonoid compounds might also act as ROS scavengers [21]. Plants also defend themselves against biotic and abiotic stressors by emitting volatile organic molecules [22]. Multiple classes of terpenes, phenylpropanoids, and benzenoids, as well as volatile fatty acid and amino acid derivatives, are among the volatile chemicals produced in response to stress [23]. Moreover, salt-stressed plants regulate salt-stress-related genes and have signal transduction factors [24].

Kenaf (*Hibiscus cannabinus* L.) is a major annual fiber crop native to east-central Africa and widely grown in the Asia–Pacific region. *Hibiscus cannabinus* cultivation has increasingly shifted to saline land due to an increased demand for food crops and reduced available arable land [25]. Although kenaf is mostly used for fiber, the seeds, leaves, and flowers may be useful in the food industry [26]. It is also used as a cosmetic ingredient and in folk medicine. *Hibiscus cannabinus* contains bioactive components such as phenolics, flavonoids, terpenes, citric acid, and fatty acid derivatives, which have a variety of pharmacological activities. For example, phenolic compounds have antiaging [27], antiproliferative [28], antityrosinase [29], and antioxidant properties [30]. Flavonoid-rich products also have a variety of biological actions, such as antibacterial [30], anti-inflammatory [31], antioxidant [32], and antidiabetic activities [33]. Moreover, phytol, an acyclic diterpene alcohol, can be used as a precursor in producing synthetic vitamins E and K1 [34]. Hydroxycitric acid has been demonstrated to lower blood insulin levels [35]. Omega-3 polyunsaturated fatty acids are also responsible for lowering the risk of cardiovascular disease and the fracture risk [36]. The essential oil composition and phytotoxic and fungitoxic activity levels of kenaf leaves were investigated by Kobaisy et al. [22]. The oil was effective against *Colletotrichum gloeosporioides*, *Colletotrichum fragariae*, and *Colletotrichum accutatum*, while also being phytotoxic to bentgrass and lettuce. In addition, aqueous extracts of kenaf leaves have been shown to protect rats' livers from carbon tetrachloride and paracetamol-induced damage [37]. Diet-induced hyperlipidemia was mitigated by a hydroalcohol extract of *H. cannabinus* leaves [38]. Kenaf extract induces a cytoprotective molecule in activated macrophages, resulting in a significant immunomodulatory effect [39]. Secondary metabolites of *H. cannabinus*, namely phenolics, flavonoids, and phenolic acids, correlate strongly with the antioxidant capacity, and these compounds prevent oxidative damage to cells by lowering ROS levels under salt stress [40].

Most previous studies have concentrated on the phytochemical properties of *H. cannabinus* under normal conditions [32,41]. However, the acumination and synthesis of bioactive and nutritional compositions depend on abiotic stresses [16,42]. Many plants subjected to salinity stress exhibit changes in the composition of the phenolics [43], flavonoids [44], and saponins [45]. These alterations are dependent on the degree and duration of the stress. Thus, plants stressed by salinity may have the potential to be polyphenol sources. Furthermore, *Hibiscus cannabinus* can potentially be used in phytoremediation to remediate salt-affected soils due to its suitability for cultivation in salinity-affected soils [46,47]. Despite these promising features, the physiological and biochemical responses of *H. cannabinus* to salinity conditions have been scarcely studied. The current study was, therefore, carried out to investigate the effects of six different salt concentrations (0, 50, 100, 150, 200, and 250 mM of NaCl) on the plant growth, physiological traits, bioactive components, and antioxidant capacity of *H. cannabinus*.

2. Materials and Methods

2.1. Plant Material, Growth Conditions, and Salt Treatments

China kenaf 21, a typical kenaf variety from the Institute of Bast Fiber Crops, Chinese Academy of Agricultural Sciences, was chosen for this study. The kenaf seeds were placed on moist filter paper and were enabled to germinate for three days at a temperature of 25 °C in the dark after being soaked in sterile water for five hours. The germinated seeds were transferred to a $\frac{1}{4}$-strength Hoagland nutrient solution (pH 6.0) comprising 5.79 mmol L^{-1} Ca (NO_3), 2, 4.17 mmol L^{-1} MgSO4, 8.90 mol L^{-1} $MnSO_4$, 8.02 mmol L^{-1} KNO_3, 0.94 mol L^{-1} $ZnSO_4$, 1.35 mmol L^{-1} $NH_4H_2PO_4$, 0.20 mol $CuSO_4$, 0.015 μmol L^{-1} $(NH_4)_2MoO_4$, 48.3 μmol L^{-1} H_3BO_3, and 72.6 μmol L^{-1} Fe-EDTA for continued growth [48]. After 5 days, the seedlings were transferred to a $\frac{1}{4}$-strength Hoagland nutrient solution supplemented with 0 (control), 50, 100, 150, 200, and 250 mm NaCl solutions, with three replicates at each concentration level, replenished every two days. The seedlings were grown in a culture chamber with a 28/25 °C temperature regime, a photoperiod of 16 h/8 h (light/dark), relative humidity of around 60%, and a light intensity of 700 μmol $m^{-2}s^{-1}$. The plants were harvested 14 days after being subjected to salt stress because the plants treated with 200 and 250 mM of NaCl showed salt stress symptoms (leaf chlorosis and necrosis).

2.2. Chemicals and Reagents

Solar Bio-Science and Technology Co. (Beijing, China) supplied the 2,2-diphenyl-1-picrylhydrazyl (DPPH), 2,2′-azinobis (3-ethylbenzothiazoline-6-sulfonic acid (ABTS), rutin, and gallic acid. Coolaber Technology Co. (Beijing, China) provided the vanillin, 2,4,6-tripyridyl-s-triazine (TPTZ), and Folin–Ciocalteau reagent. The ferrozine, iron sulfate heptahydrate, and other chemicals were purchased from Shanghai Macklin Biochemical Technology Co. (Shanghai, China). All reagents used in the assay were of the highest analytical grade.

2.3. Parameters for Plant Growth

The plant height and fresh weight (FW) values were measured fourteen days after saline treatment. The plant material was oven-dried to a constant weight at 50 °C and the dry weight (DW) was recorded.

2.4. Determination of Mineral Contents

The plant leaves were powdered to a fine powder using a small crusher after drying for five days at 50 °C. In a 50 mL crucible, 0.1 g of each sample powder was digested with 5 mL of concentrated HNO_3 and 1 mL of $HClO_4$ (70%). The mixture was then heated to 150–200 °C on a hot plate until the digest became semi-dried. The cooled sample was dissolved in deionized water to a total volume of 20 mL before the analysis [49,50].

The content of K was measured using a flame photometer (Model: FP6431, Shanghai Yidian Analysis Instrument Co., Ltd., Shanghai, China). The Ca and Mg contents were determined using an atomic absorption spectrophotometer (Model: 3110, Thermo Scientific, Oxford, UK). The concentrations of P^{3+} were determined using UV–Vis spectrophotometry (UV-2007, Shimadzu Global Laboratory Consumables Co., Ltd., Shanghai, China). Inductively coupled plasma mass spectrometry (iCAP Q ICP-MS, Thermo Fisher Scientific, Germany) was also used to determine the content of Fe^{2+}. The ICP-MS operating conditions were as follows: the radio frequency (R.F.) power was 1550 W, the nebulizer gas flow was 1.01 L/min, the auxiliary gas flow was 0.8 L/min, and the cool gas drift was 14 L/min.

2.5. Quantification of Proline, Total Soluble Sugar, and Protein Contents

The Bates method, with minor modifications, was used for proline determinations [51]. The fresh leaves (0.5 g) were homogenized in 5 mL of sulfosalicylic acid (3%) and incubated at 100 °C for 10 min. The supernatant (2 mL) was mixed with 2 mL of ninhydrin reagent and 2 mL of glacial acetic acid. The mixture was then allowed to cool to room temperature and centrifuged for 10 min at 3000 rpm. The mixture was incubated at 100 °C for 1 h before being cooled in an ice bath for 15 min. As a reaction reagent, 4 mL of toluene was added to the previous mixture, and the absorbance at 520 nm was measured. The proline content was determined using the standard curve and expressed as g g^{-1} FW.

The total soluble sugars were determined using the phenol–sulfuric acid method [52]. The fresh samples (0.5 g) were homogenized in 10.0 mL of 80% ethanol and centrifuged for 20 min at 2000 rpm, then the supernatant was collected. The supernatant (0.1 mL) was mixed with 1.0 mL phenol (5%) and 5.0 mL of sulfuric acid (98%). The mixture was then allowed to stand in a 30 °C water bath for 20 min. Finally, the absorbance was measured at 490 nm. The amount of available soluble sugar was calculated using a glucose calibration curve (10–100 mg/mL) and expressed as mg/g FW.

The soluble protein was measured using the method used by Guy [53]. The fresh samples (0.5 g) were homogenized in a solution of 50 mM Tris-HCl (pH 7.5), 2 mM EDTA, and 0.04% (*v/v*) β-mercaptoethanol and centrifuged at 10,000× *g* for 15 min. After mixing 1 mL of supernatant with 1 mL of Coomassie Brilliant Blue, the absorbance was read at 595 nm.

2.6. ROS Determination—Hydrogen Peroxide (H_2O_2) and Superoxide Anion ($O_2^{\bullet -}$)

The H_2O_2 content was determined using the method developed by Okuda et al. [54]. The fresh leaf samples (200 mg) were milled in 2 mL of 200 mM perchloric acid in an ice bath and then centrifuged for 10 min at 12,000× *g*. After centrifugation, 4 M KOH was used to neutralize the perchloric acid in the supernatant. The insoluble potassium perchlorate was then removed by centrifugation at 500× *g* for 3 min. The supernatants (1 mL) were combined with 400 µL of 3-(dimethylamino) benzoic acid (12.5 mM) in phosphate buffer (0.375 M, pH 6.5), 80 µL of 3-methyl-2-benzothiazoline hydrazone, and 20 µL of peroxidase (0.25 unit). The reaction was started by adding peroxidase at 25 °C, then the absorbance was measured at 590 nm.

The superoxide radical ($O_2^{\bullet -}$) content was determined using the techniques used by Bu et al. [55] and Lang et al. [56] with minor modifications. The fresh leaf samples (0.2 g) were treated for 1 h with 1 mL of hydroxylamine hydrochloride. The mixture was then incubated at 25 °C for 20 min with 1 mL each of α-naphthylamine and 2-aminobenzenesulfonic acid. The absorbance of the solution at 530 nm was measured. The $O_2^{\bullet -}$ content was calculated using a NaNO2 calibration curve (10–100 mg/mL).

2.7. Measurement of the Lipid Peroxidation–MDA Content

The MDA content was determined using Heath and Packer's technique [57]. Roughly 0.5 g fresh leaf samples were pulverized in 5 mL of trichloroacetic acid (10%) and 2-thiobarbituric acid (0.65%). The mixture was then heated for 1 h at 100 °C, cooled to ambient temperature, and centrifuged at 10,000 rpm for 10 min, then the absorbances of the samples were measured in triplicate at 532, 600, and 450 nm. The MDA content of the reaction solution was determined using the following equation:

$$\text{MDA } (\mu\text{mol g} - 1 \text{ FW}) = 6.45\,(\text{A532} - \text{A600}) - (0.56\,\text{A450}) \quad (1)$$

The MDA concentration was calculated using an extinction coefficient of 155 mmol L^{-1} cm^{-1} and expressed as nmol of MDA g^{-1} FW.

2.8. Preparation of Ethanol Extracts

The collected leaves were vacuum-dried at 40 °C. The dried samples were ground into a powder using a small crusher (HX-200 A, Xian Hardware and Pharmacy Co., Ltd., Xian, China). For each treatment, 1 g of powdered leaves was extracted ultrasonically for 30 min with a 10 mL ethanol solution (80%). The mixtures were centrifuged for 3 min at 4 °C at 4000× *g* [58]. The supernatant was collected and stored at 4 °C for 48 h.

2.9. Determination of Total Phenolic and Flavonoid Contents

The phenolic content was determined using a Folin–Ciocalteau colorimetric assay [41]. Briefly, 1.5 mL of 20% (*v*/*v*) Folin–Ciocalteau reagent was mixed thoroughly into 0.2 mL of sample extract and allowed to stand for 5 min. The total volume was then built up to 10 mL with distilled water before being incubated in the dark for 90 min at ambient temperature.

The absorbance was measured at 760 nm against a prepared blank. The calibration curve of gallic acid, which ranged from 20 to 800 mg/L, was used to determine the phenolic content. The results are presented in milligrams of equivalent gallic acid per gram of sample dry weight (mg GAE/g DW).

The flavonoid content of each extract was determined using the aluminum chloride technique [32]. First, 0.2 mL of sample extract was mixed with 4 mL of distilled water, followed by 0.3 mL of $NaNO_2$ (5%), and reacted for 5 min. After this, 0.3 mL of $AlCl_3$ (10%) was added and left to react for 6 min. Then, after adding 2 mL of NaOH (4%) and filling it with up to 10 mL of distilled water, the absorbance of the solution was measured at 510 nm. The flavonoid content was determined using a rutin standard curve ranging from 50 to 400 mg/L. The results are given as milligrams of rutin equivalents (mg RE/g DW) for each gram of dry weight of the samples.

2.10. Determination of Total Saponin Content

The saponin content of the plant extract was determined using the vanillin–sulfuric acid colorimetric method [59]. In brief, 0.1 mL of sample extract was mixed with 0.5 mL of ethanol (50%), 0.5 mL of vanillin solution (8%), and 4.0 mL of sulfuric acid (77.5%). The solution was then cooled to room temperature after 15 min incubation in a water bath at 60 °C. Then, the absorbance was measured using a spectrophotometer at 545 nm. The results were expressed as milligrams of tea saponin equivalents per gram of dry weight of the sample (TSE/g DW) using a calibration curve ranging from 50 to 400 mg/L.

2.11. Gas Chromatography–Mass Spectrometry (GC–MS) Identification of Volatile Compounds

The plant extraction for the GC-MS analysis was performed as previously described [32]. Briefly, 2.5 g of the powdered leaf was extracted in 50 mL of hexane using ultrasonic-assisted extraction for 30 min for each treatment. The extracts were purified (Whatman no. 4), evaporated at reduced pressure and temperature using a rotary evaporator, weighed, and dissolved in hexane at 10 mg/mL. The samples were analyzed via GC-MS equipped with an HP-5ms capillary column (30 m × 0.25 mm × 0.25 μm). The carrier gas was pure helium with a purity greater than 99.99%, flowing at a rate of 1.2 mL/min. The sample was diluted with n-hexane at a 10% (v/v) concentration. The injection volume was 1 μL and the diversion ratio was 1:5. The temperatures for injection and detection were set to 250 and 230 °C, respectively. The chromatographic heating procedure was as follows: the temperature was initially set to 60 °C for 2 min, then raised to 280 °C at a rate of 5 °C/min for 9 min. The electron ionization mode of the mass spectrometry involved an electron energy of 70 eV, a scanning range of 40–400 (m/z), a scanning rate of 3.99 scans/s, and a solvent delay of 3 min. The retention time (RT) values and NIST05 mass spectral library were used to identify compounds. The relative peak area of each compound in the chromatogram was used to calculate the percentage of each compound.

2.12. Detection of Antioxidant Activity

The DPPH• scavenging assay was performed with minor modifications to the Brand–Williams method [60]. A DPPH solution in methanol (6×10^{-5} M) was prepared and mixed with 100 μL of each sample (3 mL). The sample absorbance (A1) was measured at 515 nm after the mixtures were incubated in the dark for 15 min at room temperature. The absorbance of a blank sample (A0) containing 100 μL of methanol was also measured. The scavenging ability of the triplicate experiments was estimated using the following equation:

$$\text{Inhibition (\%)} = [(A0 - A1)/A0] \times 100 \quad (2)$$

where A0 is the absorbance of the blank and A1 is the absorbance of the sample extract.

The ABTS assay was measured following the procedure described by Nisca et al. [61]. Furthermore, 100 μL of sample extract was mixed with 100 μL of ABTS reagent and left to react in the dark for 6 min. The absorbance of the sample was measured at 734 nm. The inhibition percentage was calculated using the above formula described in the DPPH method.

The FRAP was determined using a slightly modified Benzie and Strain [62] method. The fresh FRAP reagent working solution was prepared by mixing 20 mL of acetate buffer (300 mM, pH 3.6), 2 mL of TPTZ (10 mM) in 40 mM HCl, and 2 mL of $FeCl_3$, $6H_2O$ (20 mM). The mixture was then incubated in a water bath at 37 °C for 30 min. The samples (75 μL) were then vigorously mixed with 75 mL of FRAP reagent. The sample absorption was measured at 593 nm after 4 min. A ferrous sulfate solution (0.5–10 mg/mL) was used to create the standard curve. The results were expressed in millimoles of ferrous ion equivalent per gram dry weight of the sample (mmol Fe^{2+}/g DW).

The chelation ability of the ethanol extract was determined using a previously described ferrozine-based colorimetric assay [63]. The ethanol extract (50 μL) was mixed with 200 μL $FeSO_4$ (0.2 mM) and 200 μL ferrozine (0.5 mM). The mixture was shaken and left at room temperature for 10 min. Finally, the absorbance was measured at 562 nm. The inhibition percentage of the ferrozine–Fe^{2+} complex was calculated using the following equation:

$$\text{Inhibition (\%)} = [(\text{Ac} - \text{As})/\text{Ac}] \times 100 \quad (3)$$

where Ac is the absorbance of the control and As is the absorbance of the sample.

2.13. Statistical Analysis

The data for all parameters were subjected to a one-way ANOVA followed by Duncan's multiple comparisons ($p < 0.05$) test using SAS version 9.4 (SAS Inc., Cary, NC, USA). The results are presented as mean values ± standard deviations (SD). For graphical representations, OriginPro® version 9.8.0.200 software (Northampton, MA, USA) was used.

3. Results

3.1. Effect of Salt Stress on Growth Parameters

The effect of NaCl on the *H. cannabinus* plant growth was assessed by measuring the plant height as well as the fresh and dry weights of the leaf. The results revealed that the salinity significantly decreased the plant growth, as depicted in Figure 1. In detail, the treatments with 100, 150, 200, and 200 mM of NaCl significantly reduced the plant height by 14.80%, 22.81%, 29.78%, and 44.81%, respectively, compared to the control (Figure 2a). At 200 and 250 mM NaCl concentrations, the fresh leaf weight was reduced by 42.46 and 62.51%, respectively (Figure 2b). The leaf dry weight was also affected by the stressor, decreasing by 8.96 (150 mM), 30.45 (200 mM), and 50.11% (250 mM) (Figure 2c).

3.2. Impact of Salt Stress on Minerals in Leaves

Salinity stress significantly impacted the mineral concentration in the *H. cannabinus* leaves (Table 1). The salinity stress (100, 150, 200, and 250 mM) decreased the N (up to 8.49, 17.76, 22.63 and 24.85%, respectively) compared with the control. However, the N content reduction was not statistically significant at 50 mM of NaCl. The concentration of K significantly decreased at 50 mM, 200 mM, and 250 mM of NaCl by 6.52%, 5.74%, and 3.87%, respectively, as compared to the control, whereas the changes were slight at 100 and 150 mM of NaCl. Moreover, the concentrations of Ca and Mg decreased as the salinity intensified. However, the Mg content decreased slightly at 50 mM of NaCl. The concentration of *p* significantly decreased under salt stress. The low saline concentration (50 mM) significantly the increased Fe concentrations, whereas the medium and high saline concentrations significantly reduced the concentrations.

Figure 1. Changes in morphology of *H. cannabinus* seedlings grown under different salt stress conditions (0 mM of NaCl, 50 mM of NaCl, 100 mM of NaCl, 150 mM of NaCl, 200 mM of NaCl, and 250 mM of NaCl).

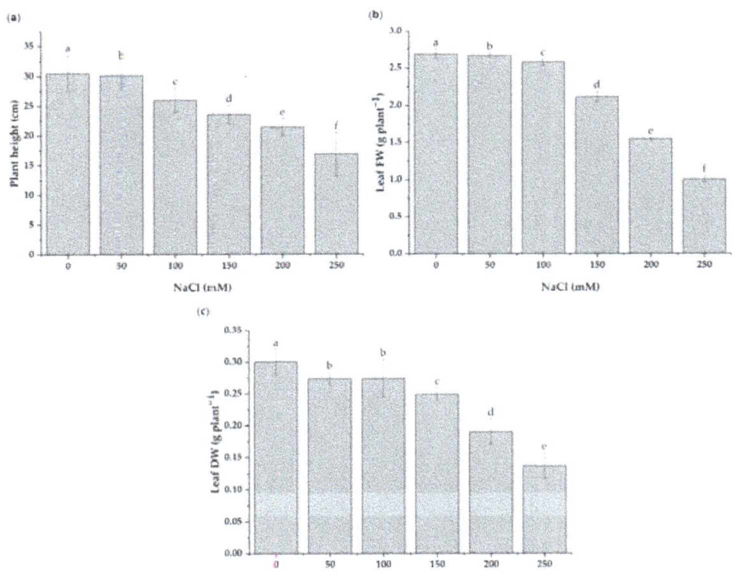

Figure 2. Effects of different levels of salinity stress (0, 50, 100, 150, 200, and 250 mM of NaCl) on the plant height (**a**) and fresh weight (FW) (**b**) and dry weight (DW) (**c**) of *H. cannabinus* leaves. The results are expressed in cm or g plant-1, as the means ± SD of different measurements (n = 15). Different letters (a–f) above the bars indicate a significant difference between treatments according to the Duncan test ($p < 0.05$).

Table 1. Compositions of nitrogen, potassium, calcium, magnesium, phosphorous, and iron in the leaves of *H. cannabinus* subjected to different levels of salt stress (0, 50, 100, 150, 200, and 250 mM of NaCl). The results are expressed in mg/g dry weight (DW), as means ± SD (n = 3). According to the Duncan test ($p < 0.05$), different letters within a column indicate significant differences.

Treatments (mM)	N (mg/g DW)	K (mg/g DW)	Ca (mg/g DW)	Mg (mg/g DW)	*p* (mg/g DW)	Fe (mg/g DW)
0	63.07 ± 0.04 ^{a}	29.65 ± 0.22 ^{a}	9.59 ± 0.32 ^{a}	1.14 ± 0.01 ^{a}	3.59 ± 0.05 ^{a}	0.13 ± 0.00 ^{b}
50	62.13 ± 0.04 ^{a}	27.72 ± 0.02 ^{d}	7.38 ± 0.14 ^{b}	1.08 ± 0.01 $^{a, b}$	3.31 ± 0.03 ^{b}	0.17 ± 0.00 ^{a}
100	57.43 ± 0.45 ^{b}	29.00 ± 0.19 $^{a, b}$	5.26 ± 0.06 ^{c}	1.01 ± 0.03 ^{b}	2.80 ± 0.09 ^{c}	0.10 ± 0.00 ^{c}
150	53.40 ± 1.33 ^{c}	28.87 ± 0.90 $^{a–c}$	4.15 ± 0.13 ^{d}	0.90 ± 0.07 ^{c}	2.10 ± 0.05 ^{d}	0.09 ± 0.00 ^{c}
200	48.80 ± 0.36 ^{d}	27.95 ± 0.10 $^{c, d}$	2.78 ± 0.04 ^{e}	0.86 ± 0.00 ^{c}	2.05 ± 0.05 ^{d}	0.08 ± 0.00 ^{c}
250	47.40 ± 0.46 ^{e}	28.50 ± 0.09 $^{b–d}$	5.10 ± 0.06 ^{c}	0.88 ± 0.02 ^{c}	2.16 ± 0.07 ^{d}	0.09 ± 0.00 ^{c}

3.3. Alterations of Proline, Total Soluble Sugar, and Soluble Protein Contents under Salt Stress

Proline is one of the most common osmotic adjustment substances. The results indicated that the plant proline content increased significantly with the salt concentration (Figure 3a). The proline levels increased by 432.76 and 527.38% under 150 and 200 mM NaCl concentrations, respectively, as compared with the control.

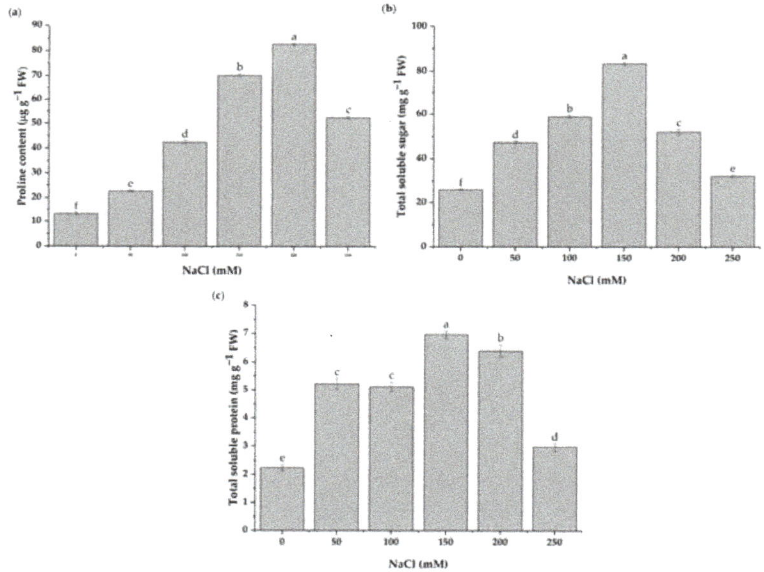

Figure 3. Effects of salt stress on the proline (**a**), total soluble sugar (**b**), and total soluble protein (**c**) contents in leaves of *H. cannabinus* seedlings subjected to 0, 50, 100, 150, 200, and 250 mM of NaCl. The values shown are means ± SD (n = 3). The different letters (a–f) above the bars indicate a significant difference according to the Duncan test ($p < 0.05$).

The results revealed that the NaCl concentrations increased the total soluble sugar content but to varying levels (Figure 3b). The total soluble sugar increased continuously from 50 to 150 mM of NaCl treatment. Treatments with 200 and 150 mM of NaCl increased the total soluble sugar by 101.63 and 222.59%, respectively, as compared to the control. Likewise, the application of 250 mM increased the total soluble sugar content by 24.38%, as compared to the control.

The content of soluble proteins was affected by salinity stress. The protein content increased significantly with the increasing salt concentration compared to the control (Figure 3c). The soluble protein levels were highest with 150 and 200 mM of NaCl. The soluble protein content was enhanced by 185.77 and 211.37% at 200 and 150 mM NaCl concentrations, respectively, as compared to the control treatment.

3.4. ROS Detection and Lipid Peroxidation

The concentration of $O_2^{\bullet-}$ in the leaves of *H. cannabinus* was altered by the salinity. In detail, the $O_2^{\bullet-}$ content significantly increased with the increasing salt concentrations, as shown in Figure 4a. The levels of $O_2^{\bullet-}$ were higher in seedlings treated with 200 and 250 mM of NaCl, with the values exceeding 136.07 and 291.80% of those found in the control plants, respectively. Likewise, in seedlings exposed to 50 mM of NaCl, the $O_2^{\bullet-}$ production increased by 19.67%. The H_2O_2 production rate increased progressively with the increase in salt concentration (Figure 4b). The H_2O_2 levels rose by 191.23 and 290.34% at 200 and 250 mM of NaCl, respectively, as compared to the control treatment. Moreover, 50 mM of NaCl increased the H_2O_2 concentration by 43.83% compared to the control.

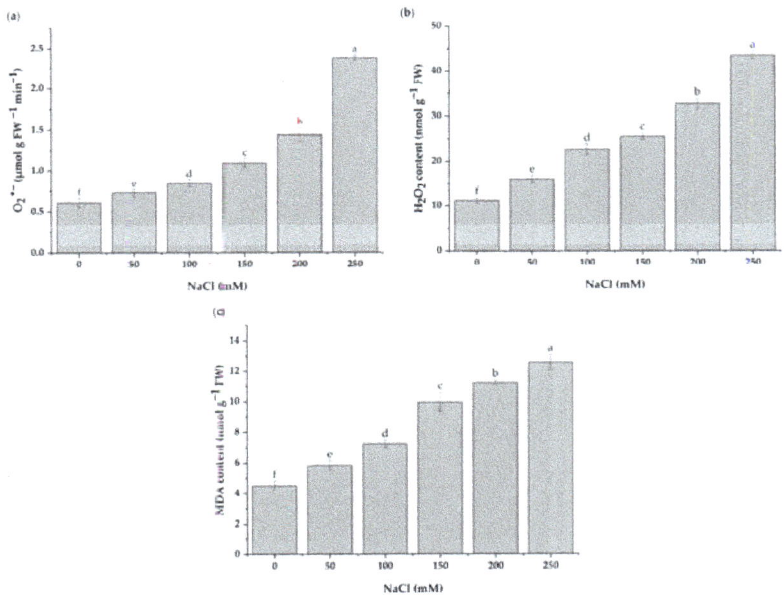

Figure 4. $O_2^{\bullet-}$ contents (**a**), H_2O_2 contents (**b**), and MDA contents (**c**) in leaves of *H. cannabinus* under various salt stress levels (0, 50, 100, 150, 200, and 250 mM of NaCl). The values presented are means ± SD ($n = 3$). Different letters (a–f) above the bars indicate a significant difference between treatments according to the Duncan test ($p < 0.05$).

As illustrated in Figure 4c, the MDA content is significantly influenced by the different salt concentrations. The results indicated that the MDA content increased as the salinity intensified. Compared to the control treatment, the MDA levels increased by 29.27 and 178.71% in response to 50 and 250 mM of NaCl, respectively.

3.5. Total Contents of Phenolics, Flavonoids, and Saponins

According to the findings, the salinity stress affected the total phenolic, flavonoid, and saponin contents. Under salt stress, the total phenolic content was found to be increased (Figure 5a). In detail, the total phenolic contents increased by 68.84 and 51.55% with 150 and 200 mM of NaCl, respectively, compared to the control. Similarly, the total flavonoid contents increased by 96.85 and 105.8% with 150 and 200 mM of NaCl, respectively, compared to the control (Figure 5b). There was no significant effect on the total flavonoid content with the 50 mM NaCl treatment. There was a considerable difference in saponin contents with increasing salt concentrations compared to the control (Figure 5c). The content of saponins increased steadily from the 50 to 200 mM NaCl treatments. Compared to the control, the saponin accumulation increased by 91.61% with 200 mM of NaCl but was significantly reduced by 47.81% with 250 mM of NaCl.

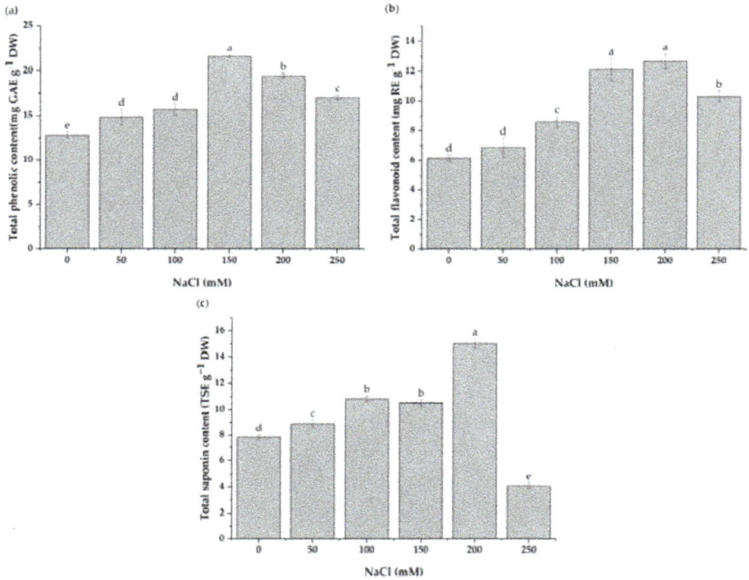

Figure 5. Changes in (**a**) the total phenolic contents, (**b**) total flavonoid contents, and (**c**) total saponin contents under different levels of salt stress (0, 50, 100, 150, 200, and 250 mM of NaCl). The values presented are means ± SD ($n = 3$). Different letters (a–e) above the bars indicate a significant difference between treatments according to the Duncan test ($p < 0.05$).

3.6. Alterations in the Compositions of Volatile Compounds under Salt Stress

The compositions of the leaf volatile compounds of *H. cannabinus* and their relative percentages (%) at different NaCl levels are shown in Table 2. Seventeen compounds were identified in the control. Phytol (18.64%) and 1-heptacosanol (18.23%) were the major components; other notable components were oleamide (9.36%), 12-methyl-E, E-2,13-octadecadien-1-ol (8.27%), alterungsschutzmittel BKF (7.30%), cis-vaccenic acid (6.80%), phthalic acid, hep-tyl undecyl ester (6.70%), and methyl linolelaidate (4.36%). When subjected to salt stress, the relative proportions of these components changed significantly. The application of NaCl reduced the phytol levels by 69.26%, 87.55%, and 88.79% at 150, 50, and 100 mM, respectively, compared with the control. This compound disappeared with 200 and 250 mM of NaCl. Compared to the control, the concentration of 1-heptacosanol increased significantly by 6.80% with 200 mM of NaCl; decreased substantially with 50, 100, and 250 mM of NaCl; and was not detected with 150 MM of NaCl. The biosynthesis of phthalic acid and heptyl undecyl ester was enhanced considerably with 100, 200, and 250 mM of NaCl, and it emerged as the first and most abundant compound. Also, the relative content of 3-(octadecyloxy) propyl ester increased at 50, 100, 150, and 200 mM NaCl concentrations and eventually became the second most abundant compound. The salt stress induced the biosynthesis of new compounds, including heptacosane, 1-octadecanesulphonyl chloride, and tetratetracontane. The biosynthesis of terpenes and alcohols was suppressed under salt stress. However, a high salt concentration stimulated the production of alkane. In addition, under all levels of salinity stress, the esters increased significantly and eventually became the most abundant chemical class in salt-treated plants.

Table 2. Changes in the compositions and relative percentages (%) of volatile compounds in the leaves of *H. cannabinus* under different NaCl concentrations (0, 50, 100, 150, 200, and 250 mM). The results, reported as percentages, are means ± SD (n = 3). According to the Duncan test (p < 0.05), different letters within a row differ significantly; nd, not detected.

No.	Name of the Compound	Content (%)					
		NaCl Concentration (mM)					
		0	50	100	150	200	250
1	Methyl linolelaidate	4.36 ± 0.49 ^{b}	3.43 ± 0.30 ^{c}	2.92 ± 0.36 ^{d}	9.15 ± 0.68 ^{a}	1.87 ± 0.19 ^{e}	nd
2	Phytol	18.64 ± 1.60 ^{a}	2.32 ± 0.45 ^{c}	2.09 ± 0.15 ^{d}	5.73 ± 0.10 ^{b}	nd	nd
3	Oleamide	9.36 ± 0.64 ^{c}	10.86 ± 0.10 ^{b}	5.02 ± 0.07 ^{f}	16.32 ± 1.18 ^{a}	7.92 ± 0.1 ^{d}	6.12 ± 0.05 ^{e}
4	Alterungsschutzmittel BKF	7.30 ± 0.74 ^{b}	6.25 ± 0.15 ^{c}	2.40 ± 0.17 ^{e}	10.30 ± 0.15 ^{a}	5.01 ± 0.20 ^{d}	2.06 ± 0.03 ^{d}
5	Diisooctyl phthalate	2.52 ± 0.70 ^{c}	5.25 ± 0.73 ^{a}	0.96 ± 0.14 ^{d}	Nd	4.10 ± 0.89 ^{b}	3.08 ± 0.23 ^{c}
6	Phthalic acid, heptyl undecyl ester	6.70 ± 1.38 ^{d}	6.55 ± 0.12 ^{e}	68.50 ± 2.50 ^{a}	11.43 ± 0.59 ^{b}	9.96 ± 1.54 ^{c}	4.76 ± 0.44 ^{f}
7	Heptacosane	nd	nd	nd	Nd	1.55 ± 0.91 ^{b}	2.78 ± 0.66 ^{a}
8	1-Octadecanesulphonyl chloride	nd	nd	nd	Nd	0.34 ± 0.14	nd
9	α-Glyceryl linolenate	2.72 ± 0.10 ^{c}	23.36 ± 0.94 ^{b}	0.68 ± 0.04 ^{d}	0.29 ± 0.06 ^{e}	nd	42.85 ± 1.10 ^{a}
10	β-Monoolein	2.06 ± 0.20 ^{c}	4.60 ± 0.14 ^{b}	0.57 ± 0.10 ^{d}	13.96 ±1.50 ^{a}	nd	0.55 ± 0.12 ^{e}
11	1-Heptacosanol	18.23 ±1.92 ^{b}	3.08 ± 0.32 ^{e}	4.61 ± 0.34 ^{d}	Nd	19.47 ± 1.44 ^{a}	13.97 ± 0.95 ^{c}
12	12-Methyl-E,E-2,13-octadecadien-1-ol	8.27 ± 0.15 ^{b}	20.73 ± 2.00 ^{a}	2.90 ± 0.52 ^{d}	Nd	4.35 ± 0.16 ^{c}	1.84 ± 0.15 ^{e}
13	Oleic acid, 3-(octadecyloxy)propyl ester	3.57 ± 0.03 ^{e}	6.16 ± 0.86 ^{c}	5.98 ± 0.12 ^{d}	31.67 ± 1.56 ^{b}	40.02 ± 0.10 ^{a}	2.87 ± 0.55 ^{f}
14	cis-Vaccenic acid	6.80 ± 0.25 ^{a}	1.64 ± 0.14 ^{b}	0.57 ± 0.08 ^{d}	1.16 ± 0.04 ^{c}	nd	nd
15	Tetratetracontane	nd	nd	nd	nd	0.71 ± 0.14 ^{b}	0.89 ± 0.59 ^{a}
16	Ethyl iso-allocholate	1.81 ± 0.03 ^{a}	0.50 ± 0.01 ^{b}	0.23 ± 0.09 ^{c}	nd	nd	0.15 ± 0.04 ^{d}
17	β-Sitosterol	0.92 ± 0.04 ^{c}	2.13 ± 0.13 ^{b}	0.22 ± 0.01 ^{e}	nd	0.34 ± 0.05 ^{d}	9.48 ± 0.50 ^{a}
18	β-Viscol	1.45 ± 0.05 ^{a}	1.12 ± 0.15 ^{b}	0.28 ± 0.02 ^{c}	nd	0.15 ± 0.02 ^{e}	0.16 ± 0.02 ^{d}
19	Unknown	3.90 ± 0.15 ^{b}	0.43 ± 0.04 ^{e}	1.93 ± 0.15 ^{d}	nd	3.32 ± 0.18 ^{c}	5.07 ± 0.25 ^{a}
20	Unknown	1.39 ± 0.14 ^{c}	1.60 ± 0.24 ^{b}	0.15 ± 0.05 ^{e}	nd	0.89 ± 0.14 ^{d}	3.62 ± 0.19 ^{a}
	Total	100	100	100	100	99.90	100
	Total identified classes						
	Alkanes	0.00 ^{c}	0.00 ^{c}	0.00 ^{c}	0.00 ^{c}	2.60 ± 0.21 ^{b}	3.67 ± 3.25 ^{a}
	Esters	19.41 ± 1.78 ^{f}	44.10 ± 2.62 ^{e}	78.65 ± 0.48 ^{a}	66.50 ± 2.24 ^{b}	51.85 ± 0.99 ^{c}	51.03 ± 0.79 ^{d}
	Phenols	7.30 ± 0.74 ^{b}	6.25 ± 0.15 ^{c}	2.40 ± 0.17 ^{e}	10.30 ± 0.15 ^{a}	5.01 ± 0.61 ^{d}	1.91 ± 0.03 ^{f}
	Amide	9.36 ± 0.64 ^{c}	10.86 ± 0.10 ^{b}	5.02 ± 0.07 ^{f}	16.32 ± 1.18 ^{a}	7.92 ± 0.10 ^{d}	6.12 ± 0.01 ^{e}
	Alcohols	27.42 ± 1.06 ^{a}	25.94 ± 1.13 ^{b}	7.73 ± 0.12 ^{e}	0.00 ^{f}	24.16 ± 1.61 ^{d}	25.29 ± 0.21 ^{c}
	Terpenes	20.09 ± 1.60 ^{a}	3.44 ± 0.45 ^{c}	2.37 ± 0.15 ^{d}	5.73 ± 0.42 ^{b}	0.15 ± 0.001 ^{e}	0.16 ± 0.002 ^{f}
	Steroid derivative	1.81 ± 0.03 ^{a}	0.50 ± 0.01 ^{b}	0.23 ± 0.09 ^{c}	0.00 ^{e}	0.00 ^{e}	0.15 ± 0.003 ^{d}
	Others	14.61 ± 0.20 ^{a}	8.92 ± 0.14 ^{c}	3.61 ± 0.04 ^{e}	1.16 ± 0.06 ^{f}	8.31 ± 0.02 ^{d}	11.77 ± 0.14 ^{b}

3.7. In Vitro Antioxidant Activities under Salt Stress

The antioxidant activity of *H. cannabinus* extracts was determined via DPPH, ABTS, FRAP, and ferrozine assays. Under salt stress, the antioxidant activity was significantly influenced (Figure 6). The plants treated with 150 mM of NaCl showed the highest antioxidant activity levels in DPPH, ABTS, and ferrozine assays, with values of 83.60%, 91.08%, and 63.68%, respectively. The FRAP results revealed the maximum antioxidant activity in the group treated with 100 mM of NaCl, with a value of 8.49 mmol Fe^{2+}/g.

The relative antioxidant capacity index (RACI) values were calculated by the merging antioxidant capacity values from different chemical methods to rank the samples' antioxidant capacities. The RACI values were calculated using the method described by Marić et al. [29] previously. The RACI is the mean value of transformed standard scores derived from initial data without unit or method restrictions. As shown in Figure 7, the group treated with 200 mM of NaCl exhibited the highest RACI value (0.36), followed by the group treated with 250 mM of NaCl (0.35). The lowest RACI value was observed at 150 mM of NaCl.

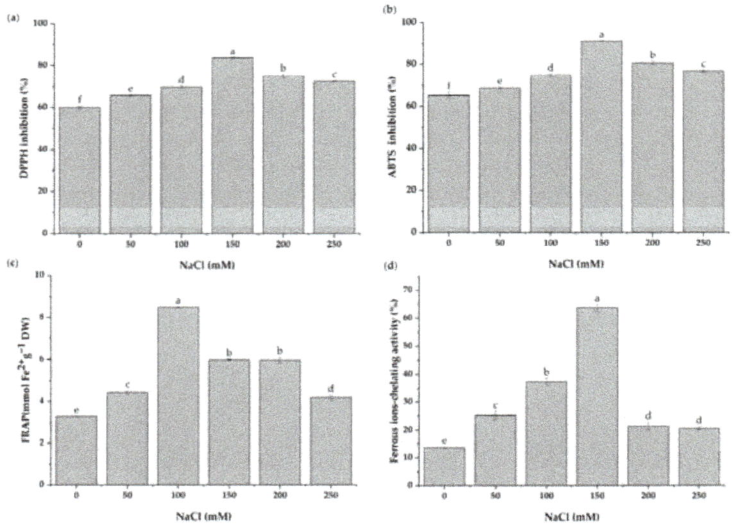

Figure 6. Changes in (**a**) DPPH inhibition, (**b**) ABTS inhibition, (**c**) ferrous-ion-chelating activity, and (**d**) FRAP antioxidant capacity levels in *H. cannabinus* leaves under different salt concentrations (0, 50, 100, 150, 200, and 250 mM of NaCl). The values presented are means ± SD (n = 3). Different letters (a–f) above the bars indicate a significant difference between treatments according to the Duncan test ($p < 0.05$).

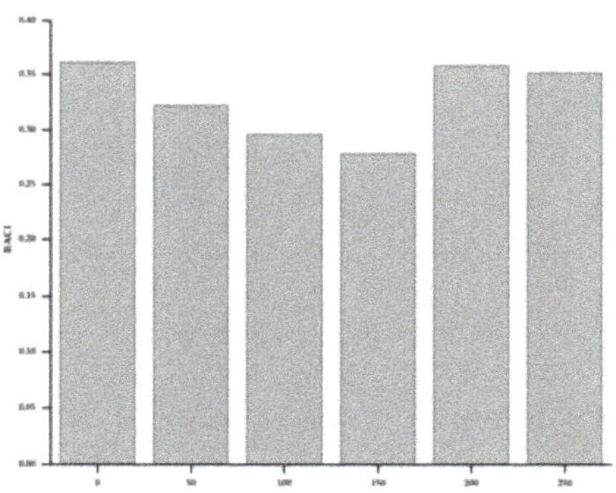

Figure 7. The relative antioxidant capacity index (RACI) was applied to combine the antioxidant capacity values from the various methods.

3.8. Correlation Analysis of Physiological and Biochemical Characteristics

The Pearson's correlations between physiological and biochemical character traits are illustrated in Figure 8. The DPPH and ABTS scavenging capacity levels correlated positively with the proline, total phenolic, and total flavonoid contents. In addition, soluble sugar was positively correlated with the ABTS scavenging capacity. MDA showed a significant positive correlation with the reactive oxygen species ($O_2^{\bullet -}$ and H_2O_2), proline, and total flavonoid contents. However, MDA was negatively correlated with the plant height and fresh and dry weight of the leaves.

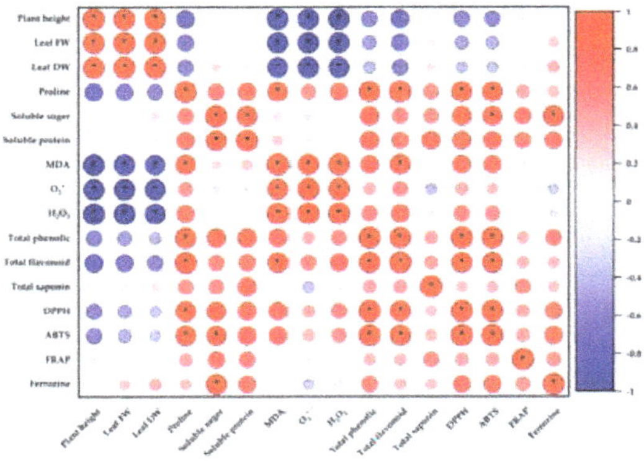

Figure 8. A correlation analysis of physiological, biochemical, and antioxidant activities of *H. cannabinus* seedlings under salt stress. * Indicates significance at the 5% level, while color depth denotes the correlation coefficient. The strength of the correlation is represented by the size of the circles.

4. Discussion

Soil salinization is now a major environmental threat to the long-term growth of global agriculture. It induces alterations in many physiological and metabolic processes, eventually reducing the crop yield, depending on the severity and duration of the stress [64,65]. Plants can tolerate or avoid saline conditions [65]. This study examined the growth parameters, physiology characteristics, bioactive constituents, and antioxidant capacity of *H. cannabinus* to assess its ability to deal with salinity.

The salt stress inhibited the *H. cannabinus* growth regarding the plant height and the fresh and dry weights of the leaves. Several authors have reported that various levels of salinity stress reduce plant growth in other medicinal plants [43,66,67]. In saline soils, the inhibition of plant growth is primarily caused by osmotic stress, which reduces the absorption of essential macro- and micronutrients [66]. In the current study, the salt stress decreased the concentrations of N, K, Ca, Mg, and P. However, there were no noticeable changes in the N and Mg contents with 50 mM of NaCl and only a slight change in the K content with 100 and 150 mM of NaCl. As previously demonstrated, the decreases in these minerals may be directly related to increased Na uptake by the roots [66]. Moreover, it has been found that NaCl treatment reduces Ca and Mg concentrations in plants [6,43]. However, adequate K, Ca, and Mg are needed to perform fundamental metabolic functions such as cellular K homeostasis, which is necessary for efficient photosynthetic system functioning and stomatal opening regulation [67]. Potassium plays a significant role in plant salinity resistance. Therefore, large quantities are required to reduce osmotic stress in a saline environment [68].

In response to salt stress, it is well established that osmolytes such as organic and inorganic solutes regulate the cellular osmotic potential of plants. The presence of more of these compounds aids in the selection of stress-tolerant cultivars [69,70]. The increased proline and soluble sugar contents protect cells from salt stress by maintaining the osmotic potential and ionic balance in the cytosol and outside of the cell, resulting in increased water and mineral absorption and cell membrane stability [71]. These increases are also commonly used to protect and stabilize enzyme structures against ROS [5,7]. Under salt stress, the proline content increased significantly in the current study. The present findings are consistent with previous research on *Brassica species* [44], *Phaseolus vulgaris* [72], and *Xanthoceras sorbifolia* [65], which revealed an increase in proline content under salt stress. The enhancement of the proline content might be due to increased activity of the pyrroline-5-carboxylate synthase (P5CS) of the proline biosynthetic pathway in *Hibiscus cannabinus* under salt stress. Higher enzymatic activities, which aid in regulating cellular structures and functions via interactions with macromolecules, could explain the increased total soluble sugars [73,74]. The current study found a significant increase in the total soluble sugar when exposed to salinity stress, consistent with [75,76]. Furthermore, soluble proteins act as osmotin, and their accumulation may play a role in the development of salt tolerance [6]. The current study revealed that under salt stress, the protein content increased significantly. The present findings are consistent with previous studies [6,7].

Salt stress may disrupt the proper balance between the induced ROS production and elimination, resulting in oxidative stress. Excessive levels of radical species, such as H_2O_2, $O_2^{\bullet-}$, and OH, damage plant cell components, resulting in cell death [77,78]. This study investigated the redox state of *Hibiscus cannabinus* seedlings by measuring the H_2O_2 and $O_2^{\bullet-}$ levels. The levels of hydrogen peroxide and superoxide anion increased significantly as the NaCl concentration increased. Abiotic-stress-induced increases in ROS generation often peroxidize cellular and organelle membrane lipids, resulting in membrane integrity losses [79]. MDA is commonly used to detect lipid peroxidation. It is a marker of oxidative damage induced by salinity stress, and the higher the level of MDA under stress, the greater the degree of membrane damage [69]. In this study, the MDA content under stress was significantly higher compared to the control. Similarly, a previous study found that increasing the NaCl concentration increased the MDA level in *H. cannabinus* [79]. Plants have a variety of defense systems against the harmful effects of oxygen radicals, including osmolytes and antioxidants [7,44].

Plants produce antioxidants such as phenolic and flavonoid compounds to scavenge or detoxify ROS [80]. In salt-exposed plants, the biosynthesis of such compounds is generally stimulated [81]. In this study, the phenolics and flavonoids increased significantly in salt-exposed seedlings compared to the control. Salinity alters the biosynthesis of primary and secondary metabolites in plants, as previously demonstrated in *Carthamus tinctorius* [82] and maize [83]. The increased phytochemicals with antioxidant properties in salt-stressed *H. cannabinus* improved the defense systems necessary to detoxify or prevent the detrimental effect of the increased production of ROS that occurs with stress conditions.

Alterations in the saponin content are reported in many plants subjected to salinity stress [84]. In this study, the saponin content was highest in plants treated with 200 mM of NaCl compared to the control, which then declined significantly with 250 mM of NaCl. Similarly, Mar and colleagues [85] found that the saponins level in *C. quinoa* treated with 200 mM of NaCl increased. Furthermore, the saponin content of cucumber increased with low and moderate salinity levels but decreased significantly with the highest concentration of salt [84]. The total saponin content showed changes under salt treatment, revealing the possible involvement of these compounds in the response of *Hibiscus cannabinus* to salt stress.

Previously, a GC-MS analysis of kenaf leaf hexane extract revealed 13 phytoconstituents [32]. The GC-MS analysis of the kenaf leaves revealed 19 compounds in our study. However, when exposed to salt stress, the relative percentages of the compounds changed significantly. The proportions of phytol decreased as the salt concentrations increased and then disappeared at higher levels of salinity stress. Phytol has antioxidant, antibacterial, anti-inflammatory, neuroprotective, analgesic, and anticancer properties [86]. The production of phthalic acid, heptyl undecyl ester, and β-monoolein disappeared under salt stress. The production of 1-heptacosanol was stimulated at low salinity but disappeared after the severe salt treatment. There have been several reports on the antibacterial and antioxidant activity of 1-heptacosanol [87,88]. Furthermore, a low level of salinity stress stimulated the production of oleic acid, 3-(octadecyloxy) propyl ester. However, a high level of salinity stress did not notably change the level of its content. Oleic acid, 3-(octadecyloxy) propyl ester has potent antifungal activity [89]. Some new compounds appeared under the salt stress, such as β-viscol, diisooctyl phthalate, and 3,7,11,15-tetramethyl-2E,6E,10E,14-hexadecatetraenyl acetate. Furthermore, alkanes such as heptacosane and tetratetracontane appeared at the highest salt concentration. These compounds may be responsible for antibacterial, anticancer, antiviral, and antifungal activities [90]. The biosynthesis of terpenes was inhibited and then disappeared with 150 and 200 mM of NaCl. These findings are consistent with a previous study in which salt inhibited terpene biosynthesis. Salt treatment can have an impact on the medicinal properties of *Hibiscus cannabinus*.

The present study revealed significant increases in the DPPH, ABTS, FRAP, and ferrozine antioxidant activity levels of *H. cannabinus* under NaCl stress. A previous study found that salt treatment increased the antioxidant capacity of sea lavender in terms of DPPH, ABTS, and FRAP [91]. Muscolo et al. [63] found that the DPPH, ABTS, FRAP, and ferrozine antioxidant activities in lentils increased significantly under NaCl stress. These antioxidant capacities could be associated with phenolic compound levels [92]. Similarly, the current study found that the antioxidant activity of *H. cannabinus* extracts was strongly related to the total phenolic content. The RACI and chemical assay results correlated, suggesting that the RACI could be used to measure food antioxidant power levels [93]. The salinity stress, thus, had a significant impact on the antioxidant capacity of the plant extracts. It could be indicated as a suitable strategy to increase the antioxidant activities of medicinal plants.

5. Conclusions

In *H. cannabinus*, the NaCl treatment increased the proline, soluble sugars, soluble protein, total phenolic content, and total flavonoid contents while lowering plant growth, K, Ca, Mg, and P levels. More osmolytes and antioxidants may improve its resistance to salinity stress. In addition, the NaCl treatment significantly enhanced the antioxidant capacity. The salt stress significantly affected the constituents of volatile compounds in *H. cannabinus*. The production of 1-heptacosanol was stimulated at low salinity but disappeared after the severe salt treatment. Moreover, the salt stress induced the biosynthesis of new compounds such as heptacosane, 1-octadecanesulphonyl chloride, and tetratetracontane. Therefore, the emergence of chemotypes at various salt concentrations may be an advantageous consequence of salinity stress in some plants, causing them to create substances with industrial and therapeutic significance.

Author Contributions: Conceptualization, Y.D., D.L. and Z.M.B.; data curation, D.Y., A.X., L.L., C.Z. and Z.M.B.; formal analysis, D.Y., A.X., L.L. and Z.M.B.; investigation, Z.M.B.; methodology, Y.D., D.L., A.X., L.L., A.B., S.D. and Z.M.B.; supervision, Y.D., M.L. and D.L.; validation, Y.D., M.L. and D.L. All authors have read and agreed to the published version of the manuscript.

Funding: This research was funded by the China Agriculture Research System of MOF and MARA (CARS-16-E5), and the Agricultural Science and Technology Innovation Program of the Chinese Academy of Agricultural Sciences (ASTIP-IBFC03).

Institutional Review Board Statement: Not applicable.

Informed Consent Statement: Not applicable.

Data Availability Statement: Not applicable.

Conflicts of Interest: The authors declare no conflict of interest.

References

1. Kashif, M.H.; Tang, D.; Li, Z.; Wei, F.; Liang, Z.; Chen, P. Comparative Cytological and Gene Expression Analysis Reveals Potential Metabolic Pathways and Target Genes Responsive to Salt Stress in Kenaf (*Hibiscus cannabinus* L.). *J. Plant Growth Regul.* **2020**, *39*, 1245–1260. [CrossRef]
2. Munns, R.; Day, D.A.; Fricke, W.; Watt, M.; Arsova, B.; Barkla, B.J.; Bose, J.; Byrt, C.S.; Chen, Z.; Foster, K.J.; et al. Energy Costs of Salt Tolerance in Crop Plants. *New Phytol.* **2020**, *225*, 1072–1090. [CrossRef] [PubMed]
3. Liu, M.; Pan, T.; Allakhverdiev, S.I.; Yu, M.; Shabala, S. Crop Halophytism: An Environmentally Sustainable Solution for Global Food Security. *Trends Plant Sci.* **2020**, *25*, 630–634. [CrossRef] [PubMed]
4. Li-ping, L.; Xiao-hua, L.; Hong-bo, S.; Zhao-Pu, L.; Ya, T.; Quan-suo, Z.; Jun-qin, Z. Ameliorants Improve Saline–Alkaline Soils on a Large Scale in Northern Jiangsu Province, China. *Ecol. Eng.* **2015**, *81*, 328–334. [CrossRef]
5. Rahneshan, Z.; Nasibi, F.; Moghadam, A.A. Effects of Salinity Stress on Some Growth, Physiological, Biochemical Parameters and Nutrients in Two Pistachio (*Pistacia vera* L.) Rootstocks. *J. Plant Interact.* **2018**, *13*, 73–82. [CrossRef]

6. Zhang, M.; Fang, Y.; Ji, Y.; Jiang, Z.; Wang, L. Effects of Salt Stress on Ion Content, Antioxidant Enzymes and Protein Profile in Different Tissues of *Broussonetia papyrifera*. *S. Afr. J. Bot.* **2013**, *85*, 1–9. [CrossRef]
7. Kumar, S.; Li, G.; Yang, J.; Huang, X.; Ji, Q.; Liu, Z.; Ke, W.; Hou, H. Effect of Salt Stress on Growth, Physiological Parameters, and Ionic Concentration of Water Dropwort (*Oenanthe javanica*) Cultivars. *Front. Plant Sci.* **2021**, *12*, 15. [CrossRef]
8. Singh, M. Plant Tolerance Mechanism Against Salt Stress: The Nutrient Management Approach. *Biochem. Pharm.* **2014**, *3*, e165. [CrossRef]
9. Rahman, A.; Hossain, M.S.; Mahmud, J.-A.; Nahar, K.; Hasanuzzaman, M.; Fujita, M. Manganese-Induced Salt Stress Tolerance in Rice Seedlings: Regulation of Ion Homeostasis, Antioxidant Defense and Glyoxalase Systems. *Physiol. Mol. Biol. Plants* **2016**, *22*, 291–306. [CrossRef]
10. Ali, Q.; Daud, M.K.; Haider, M.Z.; Ali, S.; Rizwan, M.; Aslam, N.; Noman, A.; Iqbal, N.; Shahzad, F.; Deeba, F.; et al. Seed Priming by Sodium Nitroprusside Improves Salt Tolerance in Wheat (*Triticum aestivum* L.) by Enhancing Physiological and Biochemical Parameters. *Plant Physiol. Biochem.* **2017**, *119*, 50–58. [CrossRef]
11. Sahin, U.; Ekinci, M.; Ors, S.; Turan, M.; Yildiz, S.; Yildirim, E. Effects of Individual and Combined Effects of Salinity and Drought on Physiological, Nutritional and Biochemical Properties of Cabbage (*Brassica oleracea* Var *Capitata*). *Sci. Hortic.* **2018**, *240*, 196–204. [CrossRef]
12. Nxele, X.; Klein, A.; Ndimba, B.K. Drought and Salinity Stress Alters ROS Accumulation, Water Retention, and Osmolyte Content in Sorghum Plants. *S. Afr. J. Bot.* **2017**, *108*, 261–266. [CrossRef]
13. Rahman, M.M.; Mostofa, M.G.; Das, A.K.; Anik, T.R.; Keya, S.S.; Ahsan, S.M.; Khan, M.A.R.; Ahmed, M.; Rahman, M.A.; Hossain, M.M.; et al. Ethanol Positively Modulates Photosynthetic Traits, Antioxidant Defense and Osmoprotectant Levels to Enhance Drought Acclimatization in Soybean. *Antioxidants* **2022**, *11*, 516. [CrossRef] [PubMed]
14. De Rossi, S.; Di Marco, G.; Bruno, L.; Gismondi, A.; Canini, A. Investigating the Drought and Salinity Effect on the Redox Components of *Sulla coronaria* (L.) Medik. *Antioxidants* **2021**, *10*, 1048. [CrossRef]
15. Shahid, M.A.; Sarkhosh, A.; Khan, N.; Balal, R.M.; Ali, S.; Rossi, L.; Gómez, C.; Mattson, N.; Nasim, W.; Garcia-Sanchez, F. Insights into the Physiological and Biochemical Impacts of Salt Stress on Plant Growth and Development. *Agronomy* **2020**, *10*, 938. [CrossRef]
16. Chen, C.; Wang, C.; Liu, Z.; Liu, X.; Zou, L.; Shi, J.; Chen, S.; Chen, J.; Tan, M. Variations in Physiology and Multiple Bioactive Constituents under Salt Stress Provide Insight into the Quality Evaluation of Apocyni Veneti Folium. *Int. J. Mol. Sci.* **2018**, *19*, 3042. [CrossRef]
17. Pang, J.; Cuin, T.; Shabala, L.; Zhou, M.; Mendham, N.; Shabala, S. Effect of Secondary Metabolites Associated with Anaerobic Soil Conditions on Ion Fluxes and Electrophysiology in Barley Roots. *Plant Physiol.* **2007**, *145*, 266–276. [CrossRef]
18. Hasanuzzaman, M.; Bhuyan, M.H.M.; Zulfiqar, F.; Raza, A.; Mohsin, S.; Mahmud, J.; Fujita, M.; Fotopoulos, V. Reactive Oxygen Species and Antioxidant Defense in Plants under Abiotic Stress: Revisiting the Crucial Role of a Universal Defense Regulator. *Antioxidants* **2020**, *9*, 681. [CrossRef]
19. Sachdev, S.; Ansari, S.A.; Ansari, M.I.; Fujita, M.; Hasanuzzaman, M. Abiotic Stress and Reactive Oxygen Species: Generation, Signaling, and Defense Mechanisms. *Antioxidants* **2021**, *10*, 277. [CrossRef]
20. Akram, N.A.; Shafiq, F.; Ashraf, M. Ascorbic Acid-A Potential Oxidant Scavenger and Its Role in Plant Development and Abiotic Stress Tolerance. *Front. Plant Sci.* **2017**, *8*, 613. [CrossRef]
21. Chen, W.; Feng, C.; Guo, W.; Shi, D.; Yang, C. Comparative Effects of Osmotic-, Salt- and Alkali Stress on Growth, Photosynthesis, and Osmotic Adjustment of Cotton Plants. *Photosynthetica* **2011**, *49*, 417–425. [CrossRef]
22. Kobaisy, M.; Tellez, M.R.; Webber, C.L.; Dayan, F.E.; Schrader, K.K.; Wedge, D.E. Phytotoxic and Fungitoxic Activities of the Essential Oil of Kenaf (*Hibiscus cannabinus* L.) Leaves and Its Composition. *J. Agric. Food Chem.* **2001**, *49*, 3768–3771. [CrossRef] [PubMed]
23. Zhou, Y.; Tang, N.; Huang, L.; Zhao, Y.; Tang, X.; Wang, K. Effects of Salt Stress on Plant Growth, Antioxidant Capacity, Glandular Trichome Density, and Volatile Exudates of Schizonepeta Tenuifolia Briq. *Int. J. Mol. Sci.* **2018**, *19*, 252. [CrossRef] [PubMed]
24. Rubio, M.C.; Bustos-Sanmamed, P.; Clemente, M.R.; Becana, M. Effects of Salt Stress on the Expression of Antioxidant Genes and Proteins in the Model Legume *Lotus Japonicus*. *New Phytol.* **2009**, *181*, 851–859. [CrossRef]
25. Li, H. RNA-Seq for Comparative Transcript Profiling of Kenaf under Salinity Stress. *J. Plant Res.* **2017**, *8*, 365–372. [CrossRef]
26. Kim, D.-G.; Ryu, J.; Lee, M.-K.; Kim, J.M.; Ahn, J.-W.; Kim, J.-B.; Kang, S.-Y.; Bae, C.-H.; Kwon, S.-J. Nutritional Properties of Various Tissues from New Kenaf Cultivars. *J. Crop Sci. Biotechnol.* **2018**, *21*, 229–239. [CrossRef]
27. Lin, D.; Xiao, M.; Zhao, J.; Li, Z.; Xing, B.; Li, X.; Kong, M.; Li, L.; Zhang, Q.; Liu, Y.; et al. An Overview of Plant Phenolic Compounds and Their Importance in Human Nutrition and Management of Type 2 Diabetes. *Molecules* **2016**, *21*, 1374. [CrossRef]
28. Maganha, E.G.; da Costa Halmenschlager, R.; Rosa, R.M.; Henriques, J.A.; de Paula Ramos, A.L.; Saffi, J. Pharmacological Evidences for the Extracts and Secondary Metabolites from Plants of the Genus *Hibiscus*. *Food Chem.* **2010**, *118*, 1–10. [CrossRef]
29. Xie, P.; Huang, L.; Zhang, C.; Ding, S.; Deng, Y.; Wang, X. Skin-Care Effects of Dandelion Leaf Extract and Stem Extract: Antioxidant Properties, Tyrosinase Inhibitory and Molecular Docking Simulations. *Ind. Crops Prod.* **2018**, *111*, 238–246. [CrossRef]
30. Sim, Y.Y.; Jess Ong, W.T.; Nyam, K.L. Effect of Various Solvents on the Pulsed Ultrasonic Assisted Extraction of Phenolic Compounds from *Hibiscus cannabinus* L. Leaves. *Ind. Crops Prod.* **2019**, *140*, 111708. [CrossRef]
31. Zhao, S.; Li, X.; Cho, D.; Arasu, M.; Al-Dhabi, N.; Park, S. Accumulation of Kaempferitrin and Expression of Phenyl-Propanoid Biosynthetic Genes in Kenaf (*Hibiscus cannabinus*). *Molecules* **2014**, *19*, 16987–16997. [CrossRef] [PubMed]

32. Ryu, J.; Kwon, S.-J.; Ahn, J.-W.; Jo, Y.D.; Kim, S.H.; Jeong, S.W.; Lee, M.K.; Kim, J.-B.; Kang, S.-Y. Phytochemicals and Antioxidant Activity in the Kenaf Plant (*Hibiscus cannabinus* L.). *J. Plant Biotechnol.* **2017**, *44*, 191–202. [CrossRef]
33. Survay, N. New Genera of Flavonols and Flavonol Derivatives As Therapeutic Molecules. *J. Korean Soc. Appl. Biol. Chem.* **2011**, *54*, 1–18. [CrossRef]
34. Byju, K.; Vasundhara, G.; Anuradha, V.; Nair, S.M.; Kumar, N.C. Presence of Phytol, a Precursor of Vitamin E in Chaetomorpha Antinnina. *Mapana J. Sci.* **2013**, *12*, 57–65. [CrossRef]
35. Da-Costa-Rocha, I.; Bonnlaender, B.; Sievers, H.; Pischel, I.; Heinrich, M. *Hibiscus sabdariffa* L.—A Phytochemical and Pharmacological Review. *Food Chem.* **2014**, *165*, 424–443.
36. Rajaram, S. Health Benefits of Plant-Derived α-Linolenic Acid. *Am. J. Clin. Nutr.* **2014**, *100*, 443S–448S. [CrossRef]
37. Agbor, G.; Oben, J.; Blaise, N.; Takala, J.; Jeanne, N. Hepatoprotective Activity of *Hibiscus cannabinus* (Linn.) Against Carbon Tetrachloride and Paracetamol Induced Liver Damage in Rats. *Pak. J. Biol. Sci.* **2005**, *8*, 1397–1401. [CrossRef]
38. Pradeep, K.; Gagandeep, K.; Nanjaian, M. Antihyperlipidemic Effect of Hydroalcoholic Extract of Kenaf (*Hibiscus cannabinus* L.) Leaves in High Fat Diet Fed Rats. *Ann. Biol. Res.* **2010**, *1*, 174–181.
39. Lee, Y.G.; Byeon, S.E.; Kim, J.Y.; Lee, J.Y.; Rhee, M.H.; Hong, S.; Wu, J.C.; Lee, H.S.; Kim, M.J.; Cho, D.H.; et al. Immunomodulatory Effect of Hibiscus Cannabinus Extract on Macrophage Functions. *J. Ethnopharmacol.* **2007**, *113*, 62–71. [CrossRef]
40. Brunetti, C.; Di Ferdinando, M.; Fini, A.; Pollastri, S.; Tattini, M. Flavonoids as Antioxidants and Developmental Regulators: Relative Significance in Plants and Humans. *Int. J. Mol. Sci.* **2013**, *14*, 3540–3555. [CrossRef]
41. Jin, C.H.; Ghimeray, A.; Wang, L.; Xu, M.L.; Piao, J.P.; Cho, D.H. Far Infrared Assisted Kenaf Leaf Tea Preparation and Its Effect on Phenolic Compounds, Antioxidant and ACE Inhibitory Activity. *J. Med. Plant Res.* **2013**, *7*, 1121–1128.
42. Munns, R.; Tester, M. Mechanisms of Salinity Tolerance. *Annu. Rev. Plant Biol.* **2008**, *59*, 651–681. [CrossRef] [PubMed]
43. Bistgani, Z.E.; Hashemi, M.; DaCosta, M.; Craker, L.; Maggi, F.; Morshedloo, M.R. Effect of Salinity Stress on the Physiological Characteristics, Phenolic Compounds and Antioxidant Activity of *Thymus vulgaris* L. and Thymus Daenensis Celak. *Ind. Crops Prod.* **2019**, *135*, 311–320. [CrossRef]
44. El-Badri, A.M.; Batool, M.; A. A. Mohamed, I.; Wang, Z.; Khatab, A.; Sherif, A.; Ahmad, H.; Khan, M.N.; Hassan, H.M.; Elrewainy, I.M.; et al. Antioxidative and Metabolic Contribution to Salinity Stress Responses in Two Rapeseed Cultivars during the Early Seedling Stage. *Antioxidants* **2021**, *10*, 1227. [CrossRef] [PubMed]
45. Wang, C.; Chen, L.; Cai, Z.; Chen, C.; Liu, Z.; Liu, X.; Zou, L.; Chen, J.; Tan, M.; Wei, L.; et al. Dynamic Variations in Multiple Bioactive Constituents under Salt Stress Provide Insight into Quality Formation of Licorice. *Molecules* **2019**, *24*, 3670. [CrossRef]
46. Deng, Y.; Li, D.; Huang, Y.; Huang, S. Physiological Response to Cadmium Stress in Kenaf (*Hibiscus cannabinus* L.) Seedlings. *Ind. Crops Prod.* **2017**, *107*, 453–457. [CrossRef]
47. Wei, F.; Tang, D.; Li, Z.; Kashif, M.H.; Khan, A.; Lu, H.; Jia, R.; Chen, P. Molecular Cloning and Subcellular Localization of Six HDACs and Their Roles in Response to Salt and Drought Stress in Kenaf (*Hibiscus cannabinus* L.). *Biol Res.* **2019**, *52*, 20. [CrossRef]
48. Hoagland, D.R.; Arnon, D.I. The Water Culture Method for Growing Plants without Soil. *Calif. Agric. Exp. Stn. Circ.* **1950**, *347*, 1–32.
49. Hseu, Z.-Y. Evaluating Heavy Metal Contents in Nine Composts Using Four Digestion Methods. *Bioresour. Technol.* **2004**, *95*, 53–59. [CrossRef] [PubMed]
50. Du Laing, G.; Tack, F.M.G.; Verloo, M.G. Performance of Selected Destruction Methods for the Determination of Heavy Metals in Reed Plants (*Phragmites australis*). *Anal. Chim. Acta* **2003**, *497*, 191–198. [CrossRef]
51. Bates, L.S.; Waldren, R.P.; Teare, I.D. Rapid Determination of Free Proline for Water-Stress Studies. *Plant Soil* **1973**, *39*, 205–207. [CrossRef]
52. Michel, D.; Gilles, K.A.; Hamilton, J.K.; Rebers, P.A.; Smith, F. Colorimetric Method for Determination of Sugars and Related Substances. *Anal. Chem.* **1956**, *28*, 350–356. [CrossRef]
53. Guy, C.; Haskell, D.; Neven, L.; Klein, P.; Smelser, C. Hydration-State-Responsive Protein Link Cold and Drought Stress in Spinach. *Planta* **1992**, *188*, 265–270. [CrossRef]
54. Okuda, T.; Matsuda, Y.; Yamanaka, A.; Sagisaka, S. Abrupt Increase in the Level of Hydrogen Peroxide in Leaves of Winter Wheat Is Caused by Cold Treatment. *Plant Physiol.* **1991**, *97*, 1265–1267. [CrossRef]
55. Bu, R.; Xie, J.; Yu, J.; Liao, W.; Xiao, X.; Lv, J.; Wang, C.; Ye, J.; Calderón-Urrea, A. Autotoxicity in Cucumber (*Cucumis sativus* L.) Seedlings Is Alleviated by Silicon through an Increase in the Activity of Antioxidant Enzymes and by Mitigating Lipid Peroxidation. *J. Plant Biol.* **2016**, *59*, 247–259. [CrossRef]
56. Lang, D.; Yu, X.; Jia, X.; Li, Z.; Zhang, X. Methyl Jasmonate Improves Metabolism and Growth of NaCl-Stressed Glycyrrhiza Uralensis Seedlings. *Sci. Hortic.* **2020**, *266*, 109287. [CrossRef]
57. Heath, R.L.; Packer, L. Photoperoxidation in Isolated Chloroplast I. Kinetics and Stoichiometry of Fatty Acid Peroxidation. *Arch. Biochem. Biophys.* **1968**, *125*, 189–198. [CrossRef]
58. Chrysargyris, A.; Panayiotou, C.; Tzortzakis, N. Nitrogen and Phosphorus Levels Affected Plant Growth, Essential Oil Composition and Antioxidant Status of Lavender Plant (*Lavandula angustifolia* Mill.). *Ind. Crops Prod.* **2016**, *83*, 577–586. [CrossRef]
59. He, J.; Wu, Z.; Zhang, S.; Zhou, Y.; Zhao, F.; Peng, Z.; Hu, Z. Optimization of Microwave-Assisted Extraction of Tea Saponin and Its Application on Cleaning of Historic Silks. *J. Surfact Deterg.* **2014**, *17*, 919–928. [CrossRef]
60. Brand-Williams, W.; Cuvelier, M.E.; Berset, C. Use of a Free Radical Method to Evaluate Antioxidant Activity. *LWT-Food Sci. Technol.* **1995**, *28*, 25–30. [CrossRef]

61. Nisca, A.; Ștefănescu, R.; Stegăruș, D.I.; Mare, A.D.; Farczadi, L.; Tanase, C. Comparative Study Regarding the Chemical Composition and Biological Activity of Pine (*Pinus nigra* and *P. sylvestris*) Bark Extracts. *Antioxidants* **2021**, *10*, 327. [CrossRef] [PubMed]
62. Benzie, I.F.F.; Strain, J.J. The Ferric Reducing Ability of Plasma (FRAP) as a Measure of "Antioxidant Power": The FRAP Assay. *Anal. Biochem.* **1996**, *239*, 70–76. [CrossRef]
63. Muscolo, A.; Calderaro, A.; Papalia, T.; Settineri, G.; Mallamaci, C.; Panuccio, M.R. Soil Salinity Improves Nutritional and Health Promoting Compounds in Three Varieties of Lentil (*Lens Culinaris* Med.). *Food Biosci.* **2020**, *35*, 100571. [CrossRef]
64. James, R.A.; Blake, C.; Byrt, C.S.; Munns, R. Major Genes for Na+ Exclusion, Nax1 and Nax2 (Wheat HKT1;4 and HKT1;5), Decrease Na+ Accumulation in Bread Wheat Leaves under Saline and Waterlogged Conditions. *J. Exp. Bot.* **2011**, *62*, 2939–2947. [CrossRef]
65. Zong, J.-W.; Zhang, Z.-L.; Huang, P.-L.; Chen, N.-Y.; Xue, K.-X.; Tian, Z.-Y.; Yang, Y.-H. Growth, Physiological, and Photosynthetic Responses of Xanthoceras Sorbifolium Bunge Seedlings Under Various Degrees of Salinity. *Front. Plant Sci.* **2021**, *12*, 730737. [CrossRef]
66. Oueslati, S.; Karray-Bouraoui, N.; Attia, H.; Rabhi, M.; Ksouri, R.; Lachaal, M. Physiological and Antioxidant Responses of Mentha Pulegium (Pennyroyal) to Salt Stress. *Acta Physiol Plant* **2010**, *32*, 289–296. [CrossRef]
67. Bettaieb Rebey, I.; Bourgou, S.; Rahali, F.Z.; Msaada, K.; Ksouri, R.; Marzouk, B. Relation between Salt Tolerance and Biochemical Changes in Cumin (*Cuminum cyminum* L.) Seeds. *J. Food Drug Anal.* **2017**, *25*, 391–402. [CrossRef]
68. Benito, B.; Haro, R.; Amtmann, A.; Cuin, T.; Dreyer, I. The Twins K+ and Na+ in Plants? *J. Plant Physiol.* **2014**, *171*, 723–731. [CrossRef]
69. Shabala, S.; Pottosin, I. Regulation of Potassium Transport in Plants under Hostile Conditions: Implications for Abiotic and Biotic Stress Tolerance. *Physiol Plant.* **2014**, *151*, 257–279. [CrossRef]
70. Mahouachi, J. Long-Term Salt Stress Influence on Vegetative Growth and Foliar Nutrient Changes in Mango (*Mangifera indica* L.) Seedlings. *Sci. Hortic.* **2018**, *234*, 95–100. [CrossRef]
71. Sarker, U.; Oba, S. The Response of Salinity Stress-Induced A. Tricolor to Growth, Anatomy, Physiology, Non-Enzymatic and Enzymatic Antioxidants. *Front. Plant Sci.* **2020**, *11*, 559876. [CrossRef] [PubMed]
72. Zulfiqar, F.; Akram, N.A.; Ashraf, M. Osmoprotection in Plants under Abiotic Stresses: New Insights into a Classical Phenomenon. *Planta* **2020**, *251*, 3. [CrossRef] [PubMed]
73. Gupta, B.; Huang, B. Mechanism of Salinity Tolerance in Plants: Physiological, Biochemical, and Molecular Characterization. *Int. J. Genom.* **2014**, *2014*, 701596. [CrossRef] [PubMed]
74. Al Hassan, M.; Morosan, M.; López-Gresa, M.; Prohens, J.; Vicente, O.; Boscaiu, M. Salinity-Induced Variation in Biochemical Markers Provides Insight into the Mechanisms of Salt Tolerance in Common (*Phaseolus vulgaris*) and Runner (*P. coccineus*) Beans. *Int. J. Mol. Sci.* **2016**, *17*, 1582. [CrossRef]
75. Sharif, P.; Seyedsalehi, M.; Paladino, O.; Van Damme, P.; Sillanpää, M.; Sharifi, A.A. Effect of Drought and Salinity Stresses on Morphological and Physiological Characteristics of Canola. *Int. J. Environ. Sci. Technol.* **2018**, *15*, 1859–1866. [CrossRef]
76. Ibrahimova, U.F. The Effect of Nacl on Some Physiological and Biochemical Parameters in *Triticum aestivum* L. Genotypes. *Plant Physiol.* **2019**, *24*, 370–375. [CrossRef]
77. El-Badri, A.M.A. Modulation of Salinity Impact on Early Seedling Stage via Nano-Priming Application of Zinc Oxide on Rapeseed (*Brassica napus* L.). *Plant Physiol. Biochem.* **2021**, *17*, 370–375. [CrossRef]
78. Alam, P.; Albalawi, T.H.; Altalayan, F.H.; Bakht, A.; Ahanger, M.A.; Raja, V.; Ashraf, M.; Ahmad, P. 24-Epibrassinolide (EBR) Confers Tolerance against NaCl Stress in Soybean Plants by Up-Regulating Antioxidant System, Ascorbate-Glutathione Cycle, and Glyoxalase System. *Biomolecules* **2019**, *21*, 640. [CrossRef]
79. Gill, S.S.; Tuteja, N. Reactive Oxygen Species and Antioxidant Machinery in Abiotic Stress Tolerance in Crop Plants. *Plant Physiol. Biochem.* **2010**, *48*, 909–930. [CrossRef]
80. Apel, K.; Hirt, H. Reactive oxygen species: Metabolism, Oxidative Stress, and Signal Transduction. *Annu. Rev. Plant Biol.* **2004**, *55*, 373–399. [CrossRef]
81. Li, Z.; Hu, Y.; Chang, M.; Kashif, M.H.; Tang, M.; Luo, D.; Cao, S.; Lu, H.; Zhang, W.; Huang, Z.; et al. 5-Azacytidine Pre-Treatment Alters DNA Methylation Levels and Induces Genes Responsive to Salt Stress in Kenaf (*Hibiscus cannabinus* L.). *Chemosphere* **2021**, *271*, 129562. [CrossRef] [PubMed]
82. Petridis, A.; Therios, I.; Samouris, G.; Tananaki, C. Salinity-Induced Changes in Phenolic Compounds in Leaves and Roots of Four Olive Cultivars (*Olea europaea* L.) and Their Relationship to Antioxidant Activity. *Environ. Exp. Bot.* **2012**, *79*, 37–43. [CrossRef]
83. Navarro, J.; Flores, P.; Garrido, C.; Martínez, V. Changes in the Contents of Antioxidant Compounds in Pepper Fruits at Different Ripening Stages, as Affected by Salinity. *Food Chem.* **2006**, *96*, 66–73. [CrossRef]
84. Salem, N.; Msaada, K.; Dhifi, W.; Limam, F.; Marzouk, B. Effect of Salinity on Plant Growth and Biological Activities of *Carthamus tinctorius* L. Extracts at Two Flowering Stages. *Acta Physiol. Plant* **2014**, *36*, 433–445. [CrossRef]
85. Panuccio, M.R.; Chaabani, S.; Roula, R.; Muscolo, A. Bio-Priming Mitigates Detrimental Effects of Salinity on Maize Improving Antioxidant Defense and Preserving Photosynthetic Efficiency. *Plant Physiol. Biochem.* **2018**, *132*, 465–474. [CrossRef] [PubMed]
86. Abdel-Farid, I.B.; Marghany, M.R.; Rowezek, M.M.; Sheded, M.G. Effect of Salinity Stress on Growth and Metabolomic Profiling of *Cucumis sativus* and *Solanum lycopersicum*. *Plants* **2020**, *9*, 1626. [CrossRef]

87. Caravaca, A.M.G.; Iafelice, G.; Lavini, A.; Pulvento, C.; Caboni, M.; Marconi, E. Phenolic Compounds and Saponins in Quinoa Samples (*Chenopodium Quinoa* Willd.) Grown under Different Saline and Nonsaline Irrigation Regimens. *J. Agric. Food Chem.* **2012**, *60*, 4620–4627. [CrossRef]
88. Rajesh, K.D.; Vasantha, S.; Panneerselvam, A.; Valsala Rajesh, N.; Jeyathilakan, N. Phytochemical analysis, in vitro antioxidant potential and gas chromatography-mass spectrometry studies of *Dicranopteris linearis*. *Asian J. Pharm. Clin. Res.* **2016**, *220*, 220–225. [CrossRef]
89. Faridha Begum, I.; Mohankumar, R.; Jeevan, M.; Ramani, K. GC–MS Analysis of Bio-Active Molecules Derived from Paracoccus Pantotrophus FMR19 and the Antimicrobial Activity Against Bacterial Pathogens and MDROs. *Indian J. Microbiol.* **2016**, *56*, 426–432. [CrossRef]
90. Al-Abd, N.M.; Mohamed Nor, Z.; Mansor, M.; Azhar, F.; Hasan, M.S.; Kassim, M. Antioxidant, Antibacterial Activity, and Phytochemical Characterization of *Melaleuca Cajuputi* Extract. *BMC Complement. Altern. Med.* **2015**, *15*, 385. [CrossRef]
91. Abubacker, M.N.; Kamala Devi, P. In Vitro Antifungal Potentials of Bioactive Compound Oleic Acid, 3-(Octadecyloxy) Propyl Ester Isolated from *Lepidagathis Cristata* Willd. (Acanthaceae) Inflorescence. *Asian Pac. J. Trop. Biomed.* **2014**, *4*, S661–S664 [CrossRef]
92. Thawabteh, A.; Juma, S.; Bader, M.; Karaman, D.; Scrano, L.; Bufo, S.; Karaman, R. The Biological Activity of Natural Alkaloids against Herbivores, Cancerous Cells and Pathogens. *Toxins* **2019**, *11*, 656. [CrossRef] [PubMed]
93. Rodrigues, M.J.; Soszynski, A.; Martins, A.; Rauter, A.; Neng, N.; Nogueira, J.; Varela, J.; Barreira, L.; Custódio, L. Unravelling the Antioxidant Potential and the Phenolic Composition of Different Anatomical Organs of the Marine Halophyte *Limonium Algarvense*. *Ind. Crops Prod.* **2015**, *77*, 315–322. [CrossRef]

antioxidants MDPI

Article

Ethanol Positively Modulates Photosynthetic Traits, Antioxidant Defense and Osmoprotectant Levels to Enhance Drought Acclimatization in Soybean

Md. Mezanur Rahman [1], Mohammad Golam Mostofa [1,2,*], Ashim Kumar Das [3], Touhidur Rahman Anik [4], Sanjida Sultana Keya [1], S. M. Ahsan [5], Md. Arifur Rahman Khan [6], Minhaz Ahmed [3], Md. Abiar Rahman [3], Md. Motaher Hossain [7] and Lam-Son Phan Tran [1,*]

1 Institute of Genomics for Crop Abiotic Stress Tolerance, Department of Plant and Soil Science, Texas Tech University, Lubbock, TX 79409, USA; mdmerahm@ttu.edu (M.M.R.); skeya@ttu.edu (S.S.K.)
2 Department of Biochemistry and Molecular Biology, Bangabandhu Sheikh Mujibur Rahman Agricultural University, Gazipur 1706, Bangladesh
3 Department of Agroforestry and Environment, Bangabandhu Sheikh Mujibur Rahman Agricultural University, Gazipur 1706, Bangladesh; ashimbsmrau@gmail.com (A.K.D.); minhaz@bsmrau.edu.bd (M.A.); abiar@bsmrau.edu.bd (M.A.R.)
4 Plant Pathology Division, Bangladesh Rice Research Institute, Gazipur 1701, Bangladesh; anikbge@gmail.com
5 Department of Agriculture, Bangabandhu Sheikh Mujibur Rahman Science and Technology University, Gopalganj 8100, Bangladesh; smvahsan@bsmrstu.edu.bd
6 Department of Agronomy, Bangabandhu Sheikh Mujibur Rahman Agricultural University, Gazipur 1706, Bangladesh; arif@bsmrau.edu.bd
7 Department of Plant Pathology, Bangabandhu Sheikh Mujibur Rahman Agricultural University, Gazipur 1706, Bangladesh; hossainmm@bsmrau.edu.bd
* Correspondence: mmostofa@ttu.edu (M.G.M.); son.tran@ttu.edu (L.-S.P.T.)

Citation: Rahman, M.M.; Mostofa, M.G.; Das, A.K.; Anik, T.R.; Keya, S.S.; Ahsan, S.M.; Khan, M.A.R.; Ahmed, M.; Rahman, M.A.; Hossain, M.M.; et al. Ethanol Positively Modulates Photosynthetic Traits, Antioxidant Defense and Osmoprotectant Levels to Enhance Drought Acclimatization in Soybean. *Antioxidants* **2022**, *11*, 516. https://doi.org/10.3390/antiox11030516

Academic Editor: Nafees A. Khan

Received: 5 February 2022
Accepted: 3 March 2022
Published: 8 March 2022

Abstract: Drought is a major environmental threat to agricultural productivity and food security across the world. Therefore, addressing the detrimental effects of drought on vital crops like soybean has a significant impact on sustainable food production. Priming plants with organic compounds is now being considered as a promising technique for alleviating the negative effects of drought on plants. In the current study, we evaluated the protective functions of ethanol in enhancing soybean drought tolerance by examining the phenotype, growth attributes, and several physiological and biochemical mechanisms. Our results showed that foliar application of ethanol (20 mM) to drought-stressed soybean plants increased biomass, leaf area per trifoliate, gas exchange features, water-use-efficiency, photosynthetic pigment contents, and leaf relative water content, all of which contributed to the improved growth performance of soybean under drought circumstances. Drought stress, on the other hand, caused significant accumulation of reactive oxygen species (ROS), such as superoxide and hydrogen peroxide, and malondialdehyde, as well as an increase of electrolyte leakage in the leaves, underpinning the evidence of oxidative stress and membrane damage in soybean plants. By comparison, exogenous ethanol reduced the ROS-induced oxidative burden by boosting the activities of antioxidant enzymes, including peroxidase, catalase, glutathione *S*-transferase, and ascorbate peroxidase, and the content of total flavonoids in soybean leaves exposed to drought stress. Additionally, ethanol supplementation increased the contents of total soluble sugars and free amino acids in the leaves of drought-exposed plants, implying that ethanol likely employed these compounds for osmotic adjustment in soybean under water-shortage conditions. Together, our findings shed light on the ethanol-mediated protective mechanisms by which soybean plants coordinated different morphophysiological and biochemical responses in order to increase their drought tolerance.

Keywords: antioxidant enzymes; gas exchange features; osmotic adjustment; oxidative damage; photosynthesis; reactive oxygen species; water deficiency; water-use-efficiency

1. Introduction

Water scarcity is undeniably the most critical environmental constraint limiting agricultural output worldwide. Moreover, growing water demand due to increased population pressure, ongoing climate change-mediated erratic precipitation patterns, and rising temperature will further intensify the drought situation in many regions of the world [1]. Drought stress can trigger a wide array of negative consequences in plants by altering their morphological, physiological, biochemical, cellular, and molecular responses, all of which impede plant growth and development [2,3]. A plethora of studies have reported that water scarcity reduces biomass production and stem elongation, disrupts cellular turgor pressure, restricts water uptake, and interrupts gas exchange performance and nutrient acquisition. Drought stress can also stimulate reactive oxygen species (ROS) accumulation and membrane lipid peroxidation, which ultimately leads to poor growth and even death of plants in severe cases [4–6].

Intriguingly, plants have evolved various adaptive strategies to fight against the detrimental consequences of drought [7]. These adaptive strategies include, but are not limited to, increased leaf succulence, enhanced root growth to acquire more water and essential nutrients from the deeper layer of soils, restricted transpirational water loss, retained optimum photosynthetic rate, and improved water-use-efficiency (WUE) [8,9]. In addition, plants can synthesize many compatible compounds, such as proline (Pro), amino acids, and sugars, to maintain osmotic adjustment under drought circumstances [10]. Moreover, plants have evolved complex oxidative stress protection mechanisms to avoid ROS-induced oxidative damage by increasing the synthesis of non-enzymatic antioxidants, such as carotenoids and flavonoids, and stimulating the activities of enzymatic antioxidants, such as glutathione peroxidase (GPX), peroxidase (POD), glutathione *S*-transferase (GST), ascorbate peroxidase (APX), superoxide dismutase (SOD), and catalase (CAT) [11].

Oilseeds have long been regarded as essential components of human diets and the vital raw materials of many industrial applications for the production of pharmaceutical products, oleochemicals, cosmetics, and biofuels [12]. Soybean (*Glycine max*), in particular, is the world's fourth most important grain crop, accounting for 59% of global oilseed production (www.soystats.com). Soybean acts as a source of 29% edible oil and 70% plant-derived proteins worldwide [6,13]. Importantly, being a legume crop, soybean plays a pivotal role in improving soil fertility through the process known as symbiotic nitrogen-fixation [14,15]. Drought is critical for soybean growth and development and is one of the leading reasons for the soybean yield penalty in arid and semi-arid areas of the world [6,13]. Many strategies, such as gene mining, genetic engineering, and molecular breeding, have been employed to develop soybean varieties with a heightened capacity to survive through water dearth conditions [16]. However, farmers in low-income countries prefer to practice an easy, cost-effective approach that provides immediate agronomic and economic benefits, because biotechnological and breeding research requires more investment and time for developing drought-resilience crops [6,17,18]. Considering these facts, treating plants with cost-effective signaling molecule(s) (SMs) has gained much attention for overcoming drought on numerous agricultural crops, including soybean [6]. Ethanol has emerged as an excellent representative of organic SMs that already showed promising effects in mitigating the adverse impacts of several abiotic stresses, such as chilling stress in rice (*Oryza sativa*) [19] and salt stress in soybean [20], rice, and *Arabidopsis* (*Arabidopsis thaliana*) [21]. These findings provide a strong rationale for testing the function of ethanol in alleviating the harmful impacts of drought on the economically valuable crop soybean.

In the current research, we intended to investigate whether ethanol could increase the resilience of soybean toward drought stress, as it did in the case of chilling and salt stress tolerance [19–21]. With this objective, we examined various morphophysiological and biochemical parameters, including (i) plant growth features and biomass production, (ii) leaf relative water content and succulence as an indicator of water status, (iii) gas exchange parameters, (iv) contents of different photosynthetic pigments, (v) drought-caused ROS generation and membrane lipid peroxidation, (vi) activities and/or levels of different enzymatic and non-enzymatic antioxidants, and (vii) accumulation of several osmoprotectants, to deduce ethanol-mediated drought tolerance mechanisms in soybean.

2. Materials and Methods

2.1. Plant Materials, Experimental Design, and Treatments

Seed germination and pot-culture of soybean (*Glycine max*) variety (BARI soybean #6) were carried out following the procedures described by Rahman et al. [6]. The average minimum and maximum temperatures during the experimental period were 17 and 34 °C, respectively, with a relative humidity of about 84%. Ten-day-old healthy soybean seedlings grown in pots (eight plants in each pot) were divided into four treatment groups, including (i) water-sprayed well-watered plants (WW), (ii) ethanol-sprayed well-watered plants (Eth), (iii) water-sprayed drought-exposed plants (D), and (iv) ethanol-sprayed drought-exposed plants (Eth + D). Following the methodology of Rahman et al. [6], drought stress was imposed by withholding water irrigation for 8 days, while the control plants were irrigated regularly during the whole experimental period. Plants from 'Eth' and 'Eth + D' treatment groups were sprayed (8-times in total) with 20-mM ethanol solution (20 mL to each pot), while plants from the 'WW' and 'D' treatment groups were sprayed (8-times in total) with an equal amount of water every day for a period of 8 days. The applied ethanol dose (20 mM) was selected based on the phenotypes obtained from a small-scale experiment (Supplementary Figure S1). After 8-days of drought exposure, the first trifoliate leaves of soybean plants (19-day-old plants) were harvested to determine numerous parameters associated with soybean morphology, physiology, and cellular biochemistry. The experiment was repeated thrice to ensure the accuracy of the experimental outcome.

2.2. Assessment of Growth Parameters

From each treatment, three randomly selected soybean plants were taken to evaluate the growth performance by measuring shoot height, shoot dry weight (DW), root DW, and total DW following the procedures described by Rahman et al. [22].

2.3. Estimation of Leaf Area, Succulence, Electrolyte Leakage, and Relative Water Content

Total leaf area per trifoliate was estimated according to the following formula reported by Carleton and Foote [23]:

$$\text{Leaf area (cm}^2) = \text{maximum length} \times \text{maximum width} \times 0.75 \text{ (correction factor).}$$

Leaf succulence was measured following the comprehensive procedure of Rahman et al. [17]. Leaf electrolyte leakage (EL) percentage was quantified following the protocol of Yang et al. [24] with slight modification. Briefly, 0.2 g of first trifoliate leaves were collected in a 50-mL Falcon tube containing 20 mL of tap water. Initial electrical conductivity (EC_1) was taken after incubating the samples at 32 °C for 2 h. The samples were heated at 100 °C for 30 min followed by cooling down at room temperature to record final EC (EC_2). EC of tap water was also measured and referred to as EC_0. Finally, the EL (%) was calculated using the following equation:

$$\text{EL (\%)} = (EC_1 - EC_0)/(EC_2 - EC_0) \times 100.$$

Leaf relative water content (RWC) was estimated following the procedure outlined by Das et al. [20].

2.4. Assessment of Gas Exchange Parameters

An infrared gas analyzer (LI-6400XT, LI-COR Inc., Lincoln, NE, USA) was utilized to estimate the net photosynthetic rate (Pn), the stomatal conductance to H_2O (g_s), the leaf temperature (LT), and the transpiration rate (E) as previously described by Rahman et al. [17]. Assessment of photosynthetic parameters was carried out under full sunlight between

11:00 AM and 12:30 PM. WUE parameters, including intrinsic WUE (WUEint) and instantaneous WUE (WUEins), were estimated using *Pn*, g_s, and *E* following the formulae reported in Rahman et al. [17].

2.5. Determination of Photosynthetic Pigment Contents

Freshly collected leaves were used to quantify the contents of different photosynthetic pigments, such as chlorophylls (Chls) (e.g., Chl *a*, Chl *b*, and total Chls) and carotenoids, following the protocol outlined by Arnon [25] and Lichtenthaler and Wellbura [26], respectively.

2.6. Quantification of the Content of Total Flavonoids

The method proposed by Das et al. [20] was followed to quantify the levels of total flavonoids in the leaf tissues of soybean plants.

2.7. Histochemical Analyses of ROS and the Estimation of Hydrogen Peroxide and Malondialdehyde Contents

Freshly harvested leaves were stained using the solutions of nitroblue tetrazolium (NBT) and 3, 3′-diaminobenzidine (DAB) to visualize the accumulations of superoxide ($O_2^{\bullet -}$) and hydrogen peroxide (H_2O_2), respectively, following previously described protocol [20]. The contents of H_2O_2 and malondialdehyde (MDA) in the leaf tissues were estimated using a spectrophotometer as outlined by Yu et al. [27] and Kim et al. [28], respectively.

2.8. Antioxidant Enzyme Extraction and Assessment of Enzyme Activities

Enzyme extracts were prepared from soybean leaf samples, and the activities of antioxidant enzymes, including CAT (EC: 1.11.1.6), GST (EC: 2.5.1.18), APX (EC: 1.11.1.11), and POD (EC: 1.11.1.7), were determined following the protocol described by Rahman et al. [17].

2.9. Measurements of the Levels of Water-Soluble Proteins, Proline, Soluble Sugars, Free Amino Acids, and Carbohydrates

The content of water-soluble proteins was quantified in the extracts used for enzyme activity determination following the method of Bradford [29] using bovine serum albumin as a protein standard. The level of proline (Pro) was determined spectrophotometrically by an acid ninhydrin-based technique, following the procedure reported by Bates et al. [30]. The method used by Somogyi [31] was followed for the quantification of total soluble sugars. The total free amino acid content of the leaf samples was determined with the aid of ninhydrin, in accordance with the protocol proposed by Lee and Takahashi [32]. Following the methodology of Dubios et al. [33], the phenol-sulfuric acid method was used for the determination of total carbohydrate contents in soybean leaves.

2.10. Statistical Analysis

Data obtained from four biological replicates per treatment were analyzed by one-way analysis of variance (ANOVA). The means were calculated from biological repeats and compared among the treatments using the least significant difference (LSD) test at $p < 0.05$ with the aid of Statistix 10 software. Different alphabetical letters symbolize significant variations among the control, drought, and ethanol treatments. Heatmap of the fold-change values of different phenotypical and biochemical parameters was created using R studio 1.4.1717.

3. Results

3.1. Application of 20-mM Ethanol Improved the Phenotypes and Growth Parameters of Soybean Plants Subjected to Drought Stress

To confirm whether ethanol could play a pivotal role in overcoming the drought-mediated pernicious impacts on growth attributes, we recorded plant phenotypes, the height of the shoots, shoot DW, root DW, total DW, trifoliate leaf area, and leaf succulence after 8-days of drought imposition (Figure 1A–J). Phenotypic observations indicated that drought stress caused substantial changes, such as yellowing and semi-drying of leaves in

'D' plants, compared with 'WW' plants (Figure 1A–C). On the other hand, 'Eth + D' plants displayed a noteworthy improvement in phenotypic appearance when compared with that of 'D' plants (Figure 1A–C). Shoot height was substantially reduced in 'D' plants relative to 'WW' plants, whereas external application of ethanol improved shoot height in 'Eth + D' plants when contrasted with 'D' plants (Figure 1D,J). Likewise, drought stress markedly reduced shoot DW, root DW, and total DW in 'D' plants relative to 'WW' plants, while ethanol application significantly improved all these DW parameters in 'Eth + D' plants in comparison with 'D' plants (Figure 1E–G,J). Similarly, relative to 'WW' plants, a conspicuous reduction in leaf area and leaf succulence was observed in 'D' plants (Figure 1H–J). On the other hand, ethanol application substantially improved leaf area and leaf succulence in 'Eth + D' plants when contrasted with 'D' plants (Figure 1H–J).

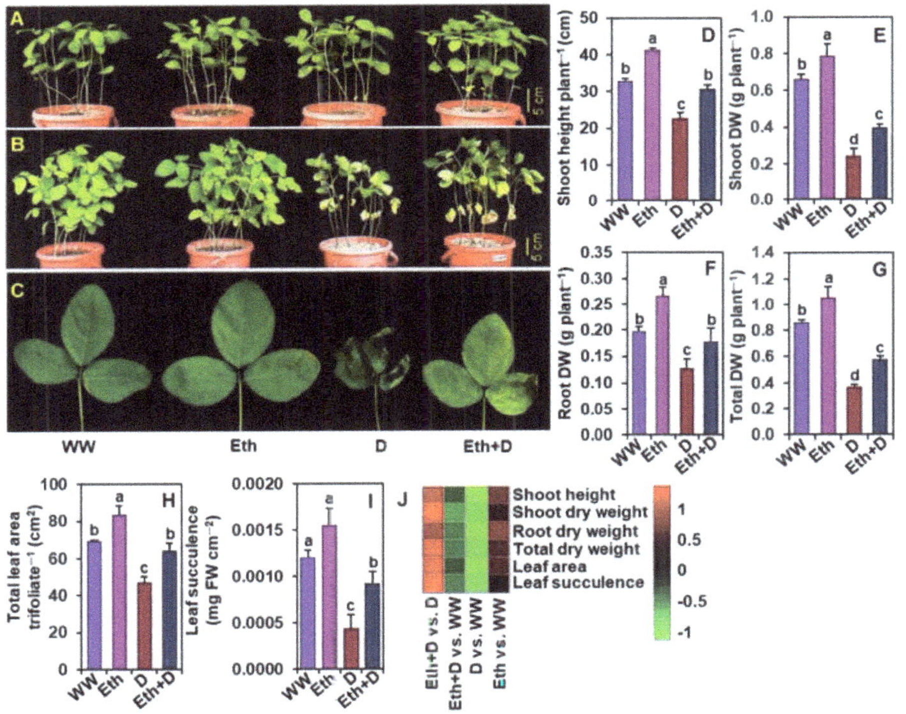

Figure 1. Effect of exogenously supplied 20-mM ethanol on soybean plants exposed to 8 days of water-withholding-induced drought stress. Photographs of soybean plants were taken before (**A**) and after (**B**) exposure to drought stress. (**C**) Close-up photographs of representative soybean leaves (first trifoliate from the bottom of the plant) from each treatment. (**D**) Shoot height, (**E**) shoot DW, (**F**) root DW, (**G**) total DW, (**H**) total leaf area per trifoliate, and (**I**) leaf succulence of soybean plants under different treatments. (**J**) Heatmap of the fold-change values of the aforementioned morphological parameters in soybean plants under different treatments. Bars represent means with standard errors ($n = 4$). Different letters shown above the bars are used to indicate significant differences among the treatments ($p < 0.05$). WW, Eth, D, and Eth + D indicate water-sprayed well-watered plants, ethanol-sprayed well-watered plants, water-sprayed drought-exposed plants, and ethanol-sprayed drought-exposed plants, respectively. DW, dry weight; FW, fresh weight.

3.2. Ethanol Improved Gas Exchange Parameters of Soybean Plants Subjected to Drought Stress

Pn, g_s, *E*, LT, WUEint, and WUEins were determined to assess whether ethanol has any role in modulating gas exchange parameters under drought stress (Figure 2A–G). We observed that drought stress caused a significant reduction in *Pn*, g_s, and *E*, and an increase in LT, WUEint, and WUEins in 'D' plants when compared with those of 'WW' plants (Figure 2A–G). By contrast, a noteworthy increase in *Pn*, *E*, WUEint, and WUEins, and a decrease in LT were observed in 'Eth + D' plants, relative to 'D' plants (Figure 2A,C–G). Nonetheless, 'Eth' plants displayed a substantial improvement in *Pn*, g_s, and *E*, and a decrease in LT compared with 'WW' plants (Figure 2A–D,G).

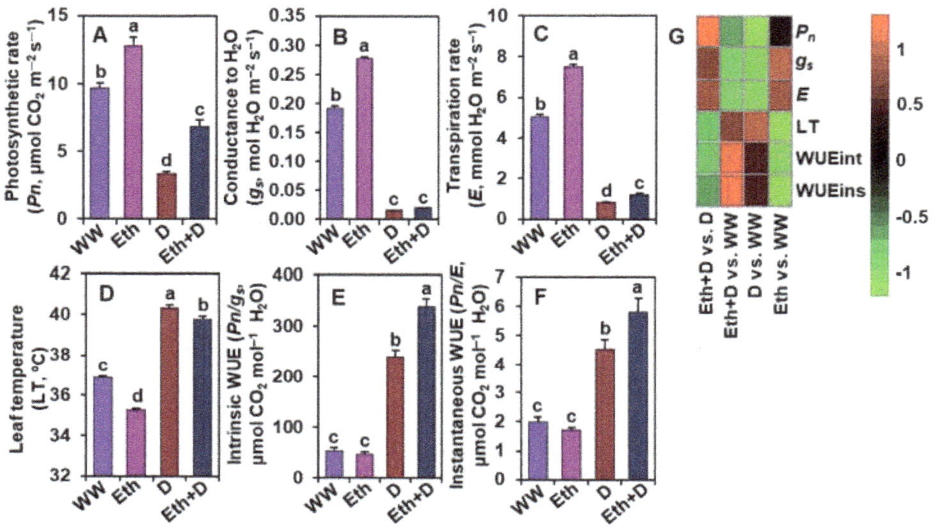

Figure 2. Effect of exogenously supplied 20-mM ethanol on (**A**) photosynthetic rate (*Pn*), (**B**) stomatal conductance to H_2O (g_s), (**C**) transpiration rate (*E*), (**D**) leaf temperature (LT), (**E**) intrinsic WUE (WUEint), and (**F**) instantaneous WUE (WUEins) of soybean leaves subjected to drought stress for a period of 8 days. (**G**) Heatmap of the fold-change values of the aforementioned parameters in soybean plants under different treatments. Bars represent means with standard errors ($n = 4$). Different letters shown above the bars are used to indicate significant differences among the treatments ($p < 0.05$). WW, Eth, D, and Eth + D indicate water-sprayed well-watered plants, ethanol-sprayed well-watered plants, water-sprayed drought-exposed plants, and ethanol-sprayed drought-exposed plants, respectively. WUE, water-use-efficiency.

3.3. Ethanol Improved Photosynthetic Pigment Contents in Soybean Plants Subjected to Drought Stress

The contents of different photosynthetic pigments (e.g., Chl *a*, *b*, total Chls, and carotenoids) were determined to examine whether ethanol improves these parameters under drought circumstances (Figure 3A–E). We found that drought stress caused a significant reduction of Chl *a*, *b*, total Chls, and carotenoid levels in 'D' plants in comparison with 'WW' plants (Figure 3A–E). Contrarily, a substantial improvement in the amounts of Chl *a*, *b*, total Chls, and carotenoids were observed in 'Eth + D' plants when equated to those in 'D' plants (Figure 3A–E). In comparison with 'WW' plants, exogenous ethanol also significantly augmented the levels of total Chls, Chl *a*, Chl *b*, and carotenoids in 'Eth' plants (Figure 3A–E).

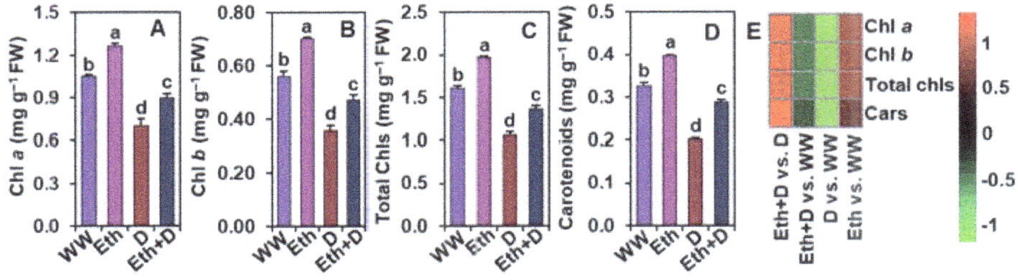

Figure 3. Effect of exogenously supplied 20-mM ethanol on (**A**) Chl *a*, (**B**) Chl *b*, (**C**) total Chls and (**D**) carotenoid contents in the trifoliate leaves of soybean plants subjected to drought stress for a period of 8 days. (**E**) Heatmap of the fold-change values of the aforementioned parameters in soybean plants under different treatments. Bars represent means with standard errors ($n = 4$). Different letters shown above the bars are used to indicate significant differences among the treatments ($p < 0.05$). WW, Eth, D, and Eth + D indicate water-sprayed well-watered plants, ethanol-sprayed well-watered plants, water-sprayed drought-exposed plants, and ethanol-sprayed drought-exposed plants, respectively. Chl, chlorophyll; Cars, carotenoids; FW, fresh weight.

3.4. Ethanol Protected Soybean Plants from Drought-Induced Oxidative Stress

To explore the role of ethanol in alleviating drought stress-mediated oxidative stress, we examined ROS generation in soybean leaves by executing histochemical staining of $O_2^{\bullet-}$ and H_2O_2, as well as quantifying the levels of H_2O_2, MDA, and EL (Figure 4A–F). In comparison with 'WW' plants, drought stress led to a significant accumulation of $O_2^{\bullet-}$ and H_2O_2 in 'D' plants, as evidenced by scattered but conspicuous dark-blue spots ($O_2^{\bullet-}$) and deep dark-brown polymerization patches (H_2O_2) (Figure 4A,B). In line with these results, 'D' plants also displayed significantly higher levels of H_2O_2, MDA, and EL percentage than 'WW' plants (Figure 4C–F). By comparison, 'Eth + D' plants exhibited substantially lower accumulation of ROS, as well as lower levels of H_2O_2, MDA, and EL percentage than 'D' plants (Figure 4A–F). 'Eth' plants also displayed reduced accumulation of ROS compared with 'WW' plants (Figure 4A–C,F). Nonetheless, comparable MDA levels and EL percentages were observed between 'Eth' and 'WW' plants (Figure 4D–F).

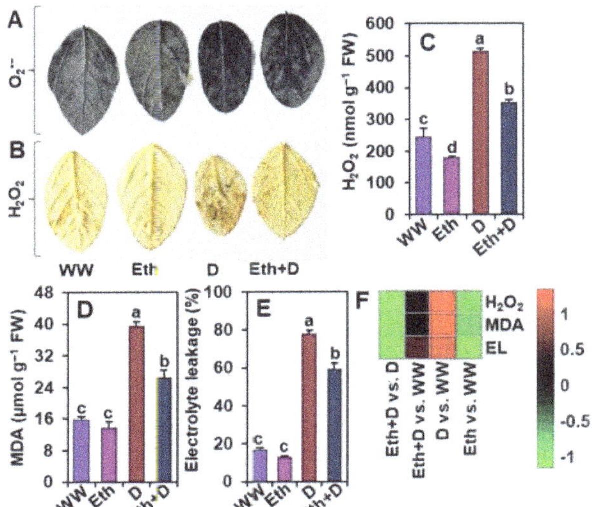

Figure 4. Effect of exogenously supplied 20-mM ethanol on ROS accumulation, and the contents

of hydrogen peroxide (H_2O_2), malondialdehyde (MDA), and electrolyte leakage in the leaves of soybean plants subjected to drought stress for a period of 8 days. (**A**) Nitroblue tetrazolium (NBT)-staining for detection of superoxide ($O_2^{\bullet-}$) and (**B**) diaminobenzidine (DAB)-staining for detection of H_2O_2. Estimated levels of (**C**) H_2O_2, (**D**) MDA, and (**E**) electrolyte leakage in the leaves of soybean plants. (**F**) Heatmap of the fold-change values of the aforementioned parameters in soybean plants under different treatments. Bars represent means with standard errors (n = 4). Different letters shown above the bars are used to indicate significant differences among the treatments ($p < 0.05$). WW, Eth, D, and Eth + D indicate water-sprayed well-watered plants, ethanol-sprayed well-watered plants, water-sprayed drought-exposed plants, and ethanol-sprayed drought-exposed plants, respectively. EL, electrolyte leakage; FW, fresh weight; ROS, reactive oxygen species.

3.5. Ethanol Improved Antioxidant Defense in Soybean Plants Subjected to Drought Stress

Next, we further examined the activities of some important antioxidant enzymes to evaluate the involvement of ethanol in improving antioxidant defense to alleviate oxidative stress (Figure 5A–F). We found that drought significantly reduced the CAT activity while enhancing the activities of GST and POD, and the number of total flavonoids in 'D' plants in comparison with 'WW' plants (Figure 5A,C–F). However, we did not observe any significant differences in the activity of APX between 'D' and 'WW' plants (Figure 5B,F). On the other hand, ethanol supplementation substantially increased the activities of CAT, APX, GST, and POD, and the contents of total flavonoids in 'Eth + D' plants in relation to 'D' plants (Figure 5A–F). Notably, the activities of CAT, APX, GST, and POD were remarkably improved in 'Eth' plants relative to 'WW' plants; however, the amounts of flavonoids did not significantly differ between 'Eth' and 'WW' plants (Figure 5A–F).

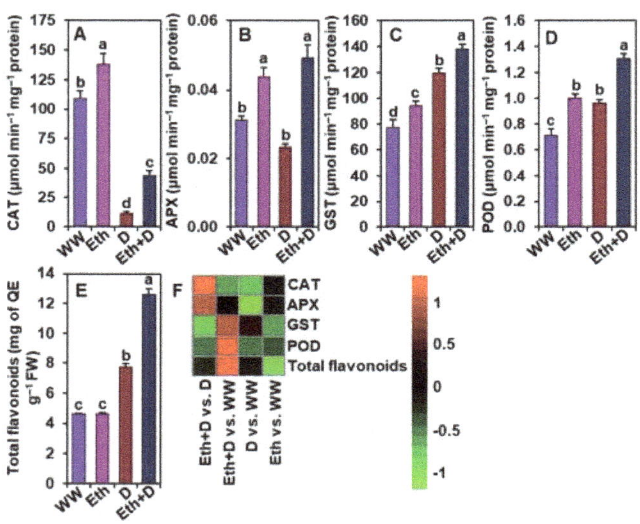

Figure 5. Effect of exogenously supplied 20-mM ethanol on antioxidant defense responses in the leaves of soybean plants subjected to drought stress for a period of 8 days. Activities of (**A**) CAT (catalase), (**B**) APX (ascorbate peroxidase), (**C**) GST (glutathione *S*-transferase), and (**D**) POD (peroxidase) and the content of (**E**) total flavonoids in soybean leaves under different treatment conditions. (**F**) Heatmap of the fold-change values of the aforementioned parameters in soybean plants under different treatments. Bars represent means with standard errors (n = 4). Different letters shown above the bars are used to indicate significant differences among the treatments ($p < 0.05$). WW, Eth, D, and Eth + D indicate water-sprayed well-watered plants, ethanol-sprayed well-watered plants, water-sprayed drought-exposed plants, and ethanol-sprayed drought-exposed plants, respectively. FW, fresh weight; QE, quercetin equivalent.

3.6. Ethanol Improved Water Status, Osmoprotectant Levels, and Water-Soluble Protein Contents in Soybean Plants Subjected to Drought Stress

To confirm whether ethanol assists in restraining water loss under drought stress, the levels of relative water content (RWC), osmoprotectants, water-soluble proteins, and total carbohydrates were determined in soybean plant leaves (Figure 6A–G). Upon drought exposure, 'D' plants had a significantly lower level of leaf RWC than 'WW' plants (Figure 6A,G). Interestingly, 'D' plants displayed higher levels of Pro than 'WW' plants (Figure 6B,G). By comparison, ethanol supplementation substantially improved the RWC without further enhancement of Pro in 'Eth + D' plants when compared with 'D' plants (Figure 6A,B,G). 'D' plants also had significantly higher levels of water-soluble proteins, total free amino acids, total soluble sugars, and total carbohydrates than 'WW' plants (Figure 6C–G). Notably, the levels of total free amino acids, total soluble sugars, and total carbohydrates were further escalated by exogenous application of ethanol in 'Eth + D' plants relative to 'D' plants (Figure 6C,E–G). 'Eth' plants displayed a significantly higher level of total soluble sugars and lower amount of water-soluble proteins than 'WW' plants; however, both 'Eth' and 'WW' plants showed comparable data for RWC, Pro, total free amino acids, and total carbohydrates (Figure 6A–G).

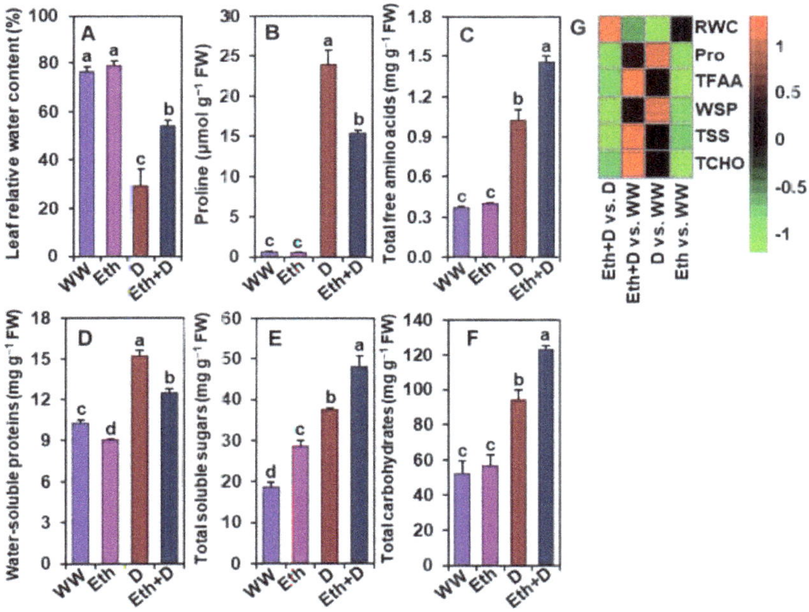

Figure 6. Effect of exogenously supplied 20-mM ethanol on the levels of (**A**) leaf relative water content, (**B**) proline, (**C**) total free amino acids, (**D**) water-soluble proteins, (**E**) total soluble sugars, and (**F**) total carbohydrates of soybean plants subjected to drought stress for a period of 8 days. (**G**) Heatmap of the fold-change values of the aforementioned parameters in soybean plants under different treatments. Bars represent means with standard errors ($n = 4$). Different letters shown above the bars are used to indicate significant differences among the treatments ($p < 0.05$). WW, Eth, D, and Eth + D indicate water-sprayed well-watered plants, ethanol-sprayed well-watered plants, water-sprayed drought-exposed plants, and ethanol-sprayed drought-exposed plants, respectively. FW, fresh weight; Pro, proline; RWC, relative water content; TFAA, total free amino acids; TSS, total soluble sugars; TCHO, total carbohydrates; WSP, water-soluble-proteins.

4. Discussion

Soybean growth and productivity are severely affected by drought episodes in many parts of the world [5]. Ethanol, an inexpensive chemical (e.g., $1.0 for 7.0 gallons of 20-mM ethanol; $290 for 4 L, Sigma-Aldrich), is known to protect soybean plants from the negative consequences of salinity and chilling stress [19–21]. In this study, we also provided evidence that ethanol supplementation enhanced drought tolerance in soybean plants by reducing drought-induced phenotypic aberrations, such as leaf yellowing and senescence, which corresponded with their better growth and biomass production (Figure 1A–G). Under a water-shortage scenario, plants need to forage water and nutrients from deeper layers of soil [34]. Thus, robust root growth can benefit plants by increasing their water absorption spheres under drought conditions. Our study demonstrated that drought stress attenuated total root biomass, whereas ethanol supplementation improved root biomass significantly (Figure 1F). The greater biomass of roots in ethanol-supplemented soybean plants might allow them to absorb more water from the soils [35], thereby contributing to improved soybean growth under water-deficit situations (Figure 7).

When roots perceive a reduction in soil water availability, they convey this environmental constraint as a stress signal toward the shoots [36]. Accordingly, shoots respond to the signal by producing stress hormones like abscisic acid (ABA) to induce stomatal closure for reducing drought-mediated transpirational water loss [37]. It is well known that the complete closure of stomata causes a sharp decline in the photosynthetic rate, which ultimately leads to growth and yield penalty in crops [38]. On the other hand, a partial stomatal closure might help maintain stomatal conductance and transpiration rates in favor of a properly reprogrammed photosynthesis under a water-shortage condition, allowing plants to maximize their WUE [37,39–41]. Our results revealed that exogenous ethanol treatment improved photosynthetic rate, stomatal conductance, and transpiration rate, resulting in enhanced WUE (carbon gain to water loss ratio), which might contribute to improving phenotypes and biomass production in drought-stressed soybean plants (Figure 1A–C,E–G and Figure 2A–C,E,F). Furthermore, an improvement in transpiration rate resulted in a decrease in LT, which helped keep the leaves cool, as evidenced by minimal leaf wilting symptoms (Figure 1C and 2D). Together, these results indicated that ethanol played a putative role in modulating gas exchange features to improve soybean drought acclimatization performance under water-limited conditions (Figure 7).

In support of these findings, ethanol-sprayed plants also displayed an improved level of photosynthetic pigments (e.g., Chl *a*, Chl *b*, total Chls, and carotenoids) under both well-watered and water-shortage conditions (Figure 3A–D). These findings suggest that ethanol might be involved in either promoting the synthesis or slowing down the degradation rate of photosynthetic pigments, or both, resulting in an improvement in the net photosynthetic rate and biomass production (Figure 1E–G, Figures 2A and 3A–D). In line with our findings, ethanol-mediated protection of photosynthetic pigments has also been reported in strawberry (*Fragaria ananasa*), soybean, and *Arabidopsis* plants [20,21,42]. It is also worth noting that the greater leaf area per trifoliate in ethanol-supplemented plants might play a positive role in increasing photosynthetic rate (Figures 1H and 2A). Leaf area directly influences plants' light interception capacity, and consequently, the overall photosynthetic rate and carbon assimilation process [43,44]. Our results highlighted that ethanol spraying significantly increased leaf area compared with water-sprayed plants, supporting the premise of a positive association between increased leaf area and increased photosynthetic rate (Figures 1H and 2A), which coincides with the previous findings of Rahman et al. [6] and Talbi et al. [45].

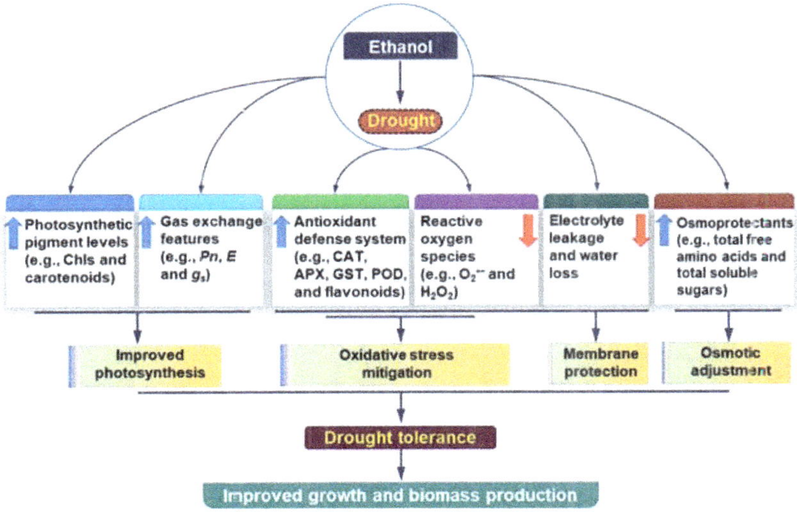

Figure 7. The regulatory roles of ethanol in alleviating drought-induced adverse effects in soybean plants. Foliar application of exogenous ethanol to drought-stressed soybean plants considerably rescued growth phenotypes, partly through the protection of photosynthetic pigments and improvement of gas exchange parameters, which ultimately increased the overall photosynthetic rate to improve growth performance. Exogenous ethanol also triggered the action of the antioxidant defense system by enhancing the activities of enzymatic antioxidants (e.g., CAT, APX, GST, and POD) and the levels of non-enzymatic antioxidants (e.g., total flavonoids), which together contributed to the protection of soybean plants from reactive oxygen species-induced oxidative stress and membrane damage. Additionally, the external application of ethanol enhanced the levels of osmoprotectants (e.g., free amino acids and soluble sugars) to maintain leaf water status for osmotic adjustment under drought circumstances. Upward (blue) and downward (red) arrows indicate increase and decrease, respectively. APX, ascorbate peroxidase; CAT, catalase; Chls, Chlorophylls; *E*, transpiration rate; GST, glutathione *S*-transferase; g_s, stomatal conductance to H_2O; H_2O_2, hydrogen peroxide; $O_2^{\bullet-}$, superoxide; POD, peroxidase; *Pn*, net photosynthetic rate.

A number of studies reported that drought-mediated biomass reduction was correlated with the induction of oxidative damage as a result of increased production of ROS, such as $O_2^{\bullet-}$ and H_2O_2 [46–48]. In the current study, 'D' plants displayed substantial levels of $O_2^{\bullet-}$, H_2O_2, MDA, and EL percentage in the leaves (Figure 4A–E), indicating that drought stress provoked serious oxidative stress and membrane damage in soybean plants. On the other hand, 'Eth + D' plants accumulated less $O_2^{\bullet-}$ and H_2O_2, as well as a reduced MDA level and EL percentage (Figure 4A–E), demonstrating that the application of external ethanol mitigated ROS-mediated cell membrane damage in drought-exposed leaves. In support of our results, previous reports also revealed that ethanol was involved in the reduction of salt stress-induced oxidative damage in the leaves of soybean, rice, and *Arabidoposis* [20,21]. In this study, we found a positive correlation of induction of the antioxidant defense with the reduced levels of ROS in 'Eth + D' soybean plants (Figure 5). We observed that 'Eth + D' soybean leaves maintained an increase in activities of CAT, APX, and POD, which likely contributed to the detoxification of drought-induced H_2O_2 [11] (Figure 5A,B,D). Additionally, the greater activity of GST in the leaves of 'Eth + D' plants further confirmed the activation of glutathione-dependent H_2O_2 removal [49] (Figure 5C). Non-enzymatic antioxidants, such as total flavonoids, were also found to be accumulated in 'Eth + D' plants (Figure 5E). Flavonoids are well-recognized for safeguarding cell membrane integrity from oxidative damage by quenching ROS during water-deficit conditions [6,34,50]. These results support that ethanol addition helped soybean plants to maintain a better status of flavonoids, conferring protection against drought-caused oxidative damage. Collectively, ethanol application boosted both enzymatic and non-enzymatic defense to trigger efficient ROS detoxification, thereby diminishing cellular damage for better soybean growth performance under drought stress (Figure 7).

Apart from a vibrant antioxidant defense mechanism, the biosynthesis of low-molecular-weight osmotic compounds, such as Pro, appears to be an important adaptive mechanism for conserving water status under water-shortage situations in plants [10,51]. Our results showed that 'D' plants accumulated more Pro but retained less RWC in their leaves than control plants (Figure 6A,B). These observations suggest that Pro accumulation in drought-stressed plants was not sufficient to retain water under severe water-deficient environments, which also corroborated with the findings of Dien et al. [52] and Rahman et al. [6]. Alternatively, 'Eth + D' plants replenished water loss in the leaves without a substantial increase of Pro contents, implying that Pro accumulation might be an indicator of soybean cellular dehydration (Figure 6A,B), as also observed in other plants under drought stress [53,54]. Interestingly, we also observed rising levels of total free amino acids, total soluble sugars, and total carbohydrates in 'Eth + D' plants, in contrast to 'D' plants, suggesting that ethanol might compensate water loss independently of Pro but dependently on free amino acids and total soluble sugars (Figure 6A–C,E,F). A number of previous studies have also reported that free amino acids and soluble sugars also played critical roles in maintaining the water status of plants in responses to abiotic stresses, including drought [6,17,22]. Moreover, increased amounts of amino acids and soluble sugars were also known to assure an adequate supply of nitrogen and carbon for sustaining the better metabolism of plants under stressful conditions [55–57]. Increased accumulations of total free amino acids and total soluble sugars, as well as their protective roles in counteracting drought-caused adverse effects, have also been reported by Du et al. [58] and Rahman et al. [6] in soybean and Zahoor et al. [59] in cotton (*Gossypium hirsutum*) plants.

5. Conclusions

Our study revealed that ethanol, in addition to its growth-promoting effects under normal conditions (Figure 1), improved soybean tolerance to water-deficit stress. We presented the first-ever evidence that supplementation of ethanol improved drought tolerance in soybean by improving root biomass, photosynthetic capacity, and WUE, protecting photosynthetic pigments, reducing ROS-triggered oxidative burst by strengthening antioxidant defense, and uplifting osmoprotectant levels (Figure 7). Nonetheless, it will be interesting to identify the major regulatory pathways that are targeted and modulated by ethanol for developing drought tolerance traits in crop plants. Furthermore, field trials and economic evaluations of ethanol application should also be taken into consideration to verify this cost-effective solution of minimizing drought-induced negative effects and reducing crop yield losses in water-limited adverse conditions.

Supplementary Materials: The following supporting information can be downloaded at: https://www.mdpi.com/article/10.3390/antiox11030516/s1, Figure S1. Effect of different concentrations of ethanol on soybean plants subjected to drought stress for a period of 8 days. D, drought; Eth, ethanol.

Author Contributions: Conceptualization, M.M.R., M.G.M. and L.-S.P.T.; methodology, M.M.R.; experimental work, M.M.R., A.K.D. and T.R.A.; software, M.M.R. and S.M.A.; formal analysis, M.M.R.; validation, M.M.R., M.G.M. and S.S.K.; investigation, M.M.R.; resources, M.A.R.K., M.A., M.A.R. and M.M.H.; data curation, M.M.R.; writing—original draft preparation, M.G.M., M.M.R., S.S.K. and S.M.A.; writing—review and editing, M.M.R., M.G.M. and L.-S.P.T.; supervision, M.G.M. and L.-S.P.T.; project administration, L.-S.P.T.; funding acquisition, L.-S.P.T. All authors have read and agreed to the published version of the manuscript.

Funding: The authors would like to extend their sincere appreciation to the United Soybean Board (USB project #2220-172-0148) for providing financial support to carry out the project works.

Institutional Review Board Statement: Not applicable.

Informed Consent Statement: Not applicable.

Data Availability Statement: Data is contained within the article and Supplementary Materials.

Acknowledgments: The authors would like to thank the Department of Crop Botany, Bangabandhu Sheikh Mujibur Rahman Agricultural University, Gazipur, Bangladesh for allowing us to use the spectrophotometer.

Conflicts of Interest: The authors declare no conflict of interest. The funders had no role in the design of the study; in the collection, analyses, or interpretation of data; in the writing of the manuscript, or in the decision to publish the results.

References

1. Bradford, J.B.; Schlaepfer, D.R.; Lauenroth, W.K.; Palmquist, K.A. Robust ecological drought projections for drylands in the 21st century. *Glob. Change Biol.* **2020**, *26*, 3906–3919. [CrossRef] [PubMed]
2. Kumar, R.; Berwal, M.K.; Saroj, P. Morphological, physiological, biochemical and molecular facet of drought stress in horticultural Crops. *Int. J. Bio-Resour. Stress Manag.* **2019**, *10*, 545–560. [CrossRef]
3. Seleiman, M.F.; Al-Suhaibani, N.; Ali, N.; Akmal, M.; Alotaibi, M.; Refay, Y.; Dindaroglu, T.; Abdul-Wajid, H.H.; Battaglia, M.L. Drought stress impacts on plants and different approaches to alleviate its adverse effects. *Plants* **2021**, *10*, 259. [CrossRef] [PubMed]
4. Tardieu, F.; Simonneau, T.; Muller, B. The physiological basis of drought tolerance in crop plants: A scenario-dependent probabilistic approach. *Annu. Rev. Plant Biol.* **2018**, *69*, 733–759. [CrossRef] [PubMed]
5. Nadeem, M.; Li, J.; Yahya, M.; Sher, A.; Ma, C.; Wang, X.; Qiu, L. Research progress and perspective on drought stress in legumes: A review. *Int. J. Mol. Sci.* **2019**, *20*, 2541. [CrossRef] [PubMed]
6. Rahman, M.; Mostofa, M.G.; Keya, S.S.; Rahman, A.; Das, A.K.; Islam, R.; Abdelrahman, M.; Bhuiyan, S.U.; Naznin, T.; Ansary, M.U. Acetic acid improves drought acclimation in soybean: An integrative response of photosynthesis, osmoregulation, mineral uptake and antioxidant defense. *Physiol. Plant.* **2021**, *172*, 334–350. [CrossRef] [PubMed]
7. Zia, R.; Nawaz, M.S.; Siddique, M.J.; Hakim, S.; Imran, A. Plant survival under drought stress: Implications, adaptive responses, and integrated rhizosphere management strategy for stress mitigation. *Microbiol. Res.* **2021**, *242*, 126626. [CrossRef]
8. Dunn, J.; Hunt, L.; Afsharinafar, M.; Meselmani, M.A.; Mitchell, A.; Howells, R.; Wallington, E.; Fleming, A.J.; Gray, J.E. Reduced stomatal density in bread wheat leads to increased water-use efficiency. *J. Exp. Bot.* **2019**, *70*, 4737–4748. [CrossRef]
9. Ilyas, M.; Nisar, M.; Khan, N.; Hazrat, A.; Khan, A.H.; Hayat, K.; Fahad, S.; Khan, A.; Ullah, A. Drought tolerance strategies in plants: A mechanistic approach. *J. Plant Growth Regul* **2021**, *40*, 926–944. [CrossRef]
10. Ozturk, M.; Turkyilmaz Unal, B.; García-Caparrós, P.; Khursheed, A.; Gul, A.; Hasanuzzaman, M. Osmoregulation and its actions during the drought stress in plants. *Physiol. Plant.* **2021**, *172*, 1321–1335. [CrossRef]
11. Laxa, M.; Liebthal, M.; Telman, W.; Chibani, K.; Dietz, K.-J. The role of the plant antioxidant system in drought tolerance. *Antioxidants* **2019**, *8*, 94. [CrossRef]
12. Song, J.-M.; Zhang, Y.; Zhou, Z.-W.; Lu, S.; Ma, W.; Lu, C.; Chen, L.-L.; Guo, L. Oil plant genomes: Current state of the science. *J. Exp. Bot.* **2021**, erab472. [CrossRef]
13. Cao, D.; Li, Y.; Liu, B.; Kong, F.; Tran, L.S.P. Adaptive mechanisms of soybean grown on salt-affected soils. *Land Degrad. Dev.* **2018**, *29*, 1054–1064. [CrossRef]
14. Li, R.; Chen, H.; Yang, Z.; Yuan, S. Research status of soybean symbiosis nitrogen fixation. *Oil Crop Sci.* **2020**, *5*, 6–10.
15. Sulieman, S.; Abdelrahman, M.; Tran, L.-S.P. Carbon metabolic adjustment in soybean nodules in response to phosphate limitation: A metabolite perspective. *Environ. Exp. Bot.* **2022**, *196*, 104810. [CrossRef]
16. Dubey, A.; Kumar, A.; Abd_Allah, E.F.; Hashem, A.; Khan, M.L. Growing more with less: Breeding and developing drought resilient soybean to improve food security. *Ecol. Indic.* **2019**, *105*, 425–437. [CrossRef]
17. Rahman, M.M.; Mostofa, M.G.; Rahman, M.A.; Islam, M.R.; Keya, S.S.; Das, A.K.; Miah, M.G.; Kawser, A.R.; Ahsan, S.; Hashem, A. Acetic acid: A cost-effective agent for mitigation of seawater-induced salt toxicity in mung bean. *Sci. Rep.* **2019**, *9*, 15186. [CrossRef]
18. Shahzad, A.; Ullah, S.; Dar, A.A.; Sardar, M.F.; Mehmood, T.; Tufail, M.A.; Shakoor, A.; Haris, M. Nexus on climate change: Agriculture and possible solution to cope future climate change stresses. *Environ. Sci. Pollut. Res.* **2021**, *28*, 14211–14232. [CrossRef]
19. Kato-Noguchi, H.; Kugimiya, T. Effects of ethanol on growth of rice seedlings. *Plant Growth Regul.* **2001**, *35*, 93–96. [CrossRef]
20. Das, A.K.; Anik, T.R.; Rahman, M.M.; Keya, S.S.; Islam, M.R.; Rahman, M.A.; Sultana, S.; Ghosh, P.K.; Khan, S.; Ahamed, T.; et al. Ethanol treatment enhances physiological and biochemical responses to mitigate saline toxicity in soybean. *Plants* **2022**, *11*, 272. [CrossRef]
21. Nguyen, H.M.; Sako, K.; Matsui, A.; Suzuki, Y.; Mostofa, M.G.; Ha, C.V.; Tanaka, M.; Tran, L.-S.P.; Habu, Y.; Seki, M. Ethanol enhances high-salinity stress tolerance by detoxifying reactive oxygen species in *Arabidopsis thaliana* and rice. *Front. Plant Sci.* **2017**, *8*, 1001. [CrossRef]

22. Rahman, M.M.; Rahman, M.A.; Miah, M.G.; Saha, S.R.; Karim, M.; Mostofa, M.G. Mechanistic insight into salt tolerance of *Acacia auriculiformis*: The importance of ion selectivity, osmoprotection, tissue tolerance, and Na^+ exclusion. *Front. Plant Sci.* **2017**, *8*, 155. [CrossRef]
23. Carleton, A.E.; Foote, W.H. A comparison of methods for estimating total leaf area of barley plants 1. *Crop Sci.* **1965**, *5*, 602–603. [CrossRef]
24. Yang, J.; Kloepper, J.W.; Ryu, C.-M. Rhizosphere bacteria help plants tolerate abiotic stress. *Trends Plant Sci.* **2009**, *14*, 1–4. [CrossRef]
25. Arnon, D.I. Copper enzymes in isolated chloroplasts. Polyphenoloxidase in Beta vulgaris. *Plant Physiol.* **1949**, *24*, 1. [CrossRef]
26. Lichtenthaler, H.K.; Wellburn, A.R. Determinations of total carotenoids and chlorophylls a and b of leaf extracts in different solvents. *Biochem. Soc. Trans.* **1983**, *11*, 591–592. [CrossRef]
27. Yu, C.-W.; Murphy, T.M.; Lin, C.-H. Hydrogen peroxide-induced chilling tolerance in mung beans mediated through ABA-independent glutathione accumulation. *Funct. Plant Biol.* **2003**, *30*, 955–963. [CrossRef]
28. Kim, T.Y.; Ku, H.; Lee, S.-Y. Crop enhancement of cucumber plants under heat stress by shungite carbon. *Int. J. Mol. Sci.* **2020**, *21*, 4858. [CrossRef] [PubMed]
29. Bradford, M.M. A rapid and sensitive method for the quantitation of microgram quantities of protein utilizing the principle of protein-dye binding. *Anal. Biochem.* **1976**, *72*, 248–254. [CrossRef]
30. Bates, L.S.; Waldren, R.P.; Teare, I.D. Rapid determination of free proline for water-stress studies. *Plant Soil* **1973**, *39*, 205–207. [CrossRef]
31. Somogyi, M. Notes on sugar determination. *J. Biol. Chem.* **1952**, *195*, 19–23. [CrossRef]
32. Lee, Y.-P.; Takahashi, T. An improved colorimetric determination of amino acids with the use of ninhydrin. *Anal. Biochem.* **1966**, *14*, 71–77. [CrossRef]
33. DuBois, M.; Gilles, K.A.; Hamilton, J.K.; Rebers, P.A.; Smith, F. Colorimetric method for determination of sugars and related substances. *Anal. Chem.* **1956**, *28*, 350–356. [CrossRef]
34. Xiong, L.; Wang, R.-G.; Mao, G.; Koczan, J.M. Identification of drought tolerance determinants by genetic analysis of root response to drought stress and abscisic acid. *Plant Physiol.* **2006**, *142*, 1065–1074. [CrossRef]
35. Zegada-Lizarazu, W.; Monti, A. Deep root growth, ABA adjustments and root water uptake response to soil water deficit in giant reed. *Ann. Bot.* **2019**, *124*, 605–615. [CrossRef]
36. Malcheska, F.; Ahmad, A.; Batool, S.; Müller, H.M.; Ludwig-Müller, J.; Kreuzwieser, J.; Randewig, D.; Hänsch, R.; Mendel, R.R.; Hell, R.; et al. Drought-enhanced xylem sap sulfate closes stomata by affecting ALMT12 and guard cell ABA synthesis. *Plant Physiol.* **2017**, *174*, 798–814. [CrossRef]
37. Bauer, H.; Ache, P.; Lautner, S.; Fromm, J.; Hartung, W.; Al-Rasheid, K.A.; Sonnewald, S.; Sonnewald, U.; Kneitz, S.; Lachmann, N.; et al. The stomatal response to reduced relative humidity requires guard cell-autonomous ABA synthesis. *Curr. Biol.* **2013**, *23*, 53–57. [CrossRef]
38. Cornic, G. Drought stress inhibits photosynthesis by decreasing stomatal aperture—Not by affecting ATP synthesis. *Trends Plant Sci.* **2000**, *5*, 187–188. [CrossRef]
39. Davies, W.J.; Wilkinson, S.; Loveys, B. Stomatal control by chemical signalling and the exploitation of this mechanism to increase water use efficiency in agriculture. *New Phytol.* **2002**, *153*, 449–460. [CrossRef]
40. Haworth, M.; Killi, D.; Materassi, A.; Raschi, A.; Centritto, M. Impaired stomatal control is associated with reduced photosynthetic physiology in crop species grown at elevated [CO_2]. *Front. Plant Sci.* **2016**, *7*, 1568. [CrossRef]
41. Lawson, T.; Vialet-Chabrand, S. Speedy stomata, photosynthesis and plant water use efficiency. *New Phytol.* **2019**, *221*, 93–98. [CrossRef]
42. Yavarpanah, Z.; Alizadeh, M.; Seifi, E. Effects of foliar and root applications of hydro-alcoholic solutions on physiological and biochemical attributes and fruit yield and weight of strawberry. *J. Plant Physiol. Breed.* **2015**, *5*, 47–54.
43. Weraduwage, S.M.; Chen, J.; Anozie, F.C.; Morales, A.; Weise, S.E.; Sharkey, T.D. The relationship between leaf area growth and biomass accumulation in *Arabidopsis thaliana*. *Front. Plant Sci.* **2015**, *6*, 167. [CrossRef]
44. Verma, K.K.; Song, X.-P.; Zeng, Y.; Li, D.-M.; Guo, D.-J.; Rajput, V.D.; Chen, G.-L.; Barakhov, A.; Minkina, T.M.; Li, Y.-R. Characteristics of leaf stomata and their relationship with photosynthesis in *Saccharum officinarum* under drought and silicon application. *ACS Omega* **2020**, *5*, 24145–24153. [CrossRef]
45. Talbi, S.; Rojas, J.A.; Sahrawy, M.; Rodríguez-Serrano, M.; Cárdenas, K.E.; Debouba, M.; Sandalio, L.M. Effect of drought on growth, photosynthesis and total antioxidant capacity of the saharan plant *Oudeneya africana*. *Environ. Exp. Bot.* **2020**, *176*, 104099. [CrossRef]
46. Halliwell, B. Reactive species and antioxidants. Redox biology is a fundamental theme of aerobic life. *Plant Physiol.* **2006**, *141*, 312–322. [CrossRef]
47. Noctor, G.; Mhamdi, A.; Foyer, C.H. The roles of reactive oxygen metabolism in drought: Not so cut and dried. *Plant Physiol.* **2014**, *164*, 1636–1648. [CrossRef]
48. De Rossi, S.; Di Marco, G.; Bruno, L.; Gismondi, A.; Canini, A. Investigating the drought and salinity effect on the redox components of *Sulla coronaria* (L.) Medik. *Antioxidants* **2021**, *10*, 1048. [CrossRef]
49. Ding, H.; Wang, B.; Han, Y.; Li, S. The pivotal function of dehydroascorbate reductase in glutathione homeostasis in plants. *J. Exp. Bot.* **2020**, *71*, 3405–3416. [CrossRef]

50. Gharibi, S.; Tabatabaei, B.E.S.; Saeidi, G.; Talebi, M.; Matkowski, A. The effect of drought stress on polyphenolic compounds and expression of flavonoid biosynthesis related genes in *Achillea pachycephala* Rech.f. *Phytochemistry* **2019**, *162*, 90–98. [CrossRef]
51. Zulfiqar, F.; Aisha, N.; Ashraf, M. Osmoprotection in plants under abiotic stresses: New insights into a classical phenomenon. *Planta* **2019**, *251*, 3. [CrossRef] [PubMed]
52. Dien, D.; Thu, T.T.P.; Moe, K.; Yamakawa, T. Proline and carbohydrate metabolism in rice varieties (*Oryza sativa* L.) under various drought and recovery conditions. *Plant Physiol. Rep.* **2019**, *24*, 376–387. [CrossRef]
53. Anjum, S.A.; Farooq, M.; Xie, X.-Y.; Liu, X.-J.; Ijaz, M.F. Antioxidant defense system and proline accumulation enables hot pepper to perform better under drought. *Sci. Hortic.* **2012**, *140*, 66–73. [CrossRef]
54. Bandurska, H.; Niedziela, J.; Pietrowska-Borek, M.; Nuc, K.; Chadzinikolau, T.; Radzikowska, D. Regulation of proline biosynthesis and resistance to drought stress in two barley (*Hordeum vulgare* L.) genotypes of different origin. *Plant Physiol. Biochem.* **2017**, *118*, 427–437. [CrossRef]
55. Gangola, M.P.; Ramadoss, B.R. Sugars Play a Critical Role in Abiotic Stress Tolerance in Plants. In *Biochemical, Physiological and Molecular Avenues for Combating Abiotic Stress Tolerance in Plants*; Academic Press: Cambridge, MA, USA, 2018; pp. 17–38.
56. Sharma, A.; Shahzad, B.; Kumar, V.; Kohli, S.K.; Sidhu, G.P.S.; Bali, A.S.; Handa, N.; Kapoor, D.; Bhardwaj, R.; Zheng, B. Phytohormones regulate accumulation of osmolytes under abiotic stress. *Biomolecules* **2019**, *9*, 285. [CrossRef]
57. Khan, N.; Ali, S.; Zandi, P.; Mehmood, A.; Ullah, S.; Ikram, M.; Shahid, M.; Babar, M. Role of sugars, amino acids and organic acids in improving plant abiotic stress tolerance. *Pak. J. Bot.* **2020**, *52*, 355–363. [CrossRef]
58. Du, Y.; Zhao, Q.; Chen, L.; Yao, X.; Zhang, W.; Zhang, B.; Xie, F. Effect of drought stress on sugar metabolism in leaves and roots of soybean seedlings. *Plant Physiol. Biochem.* **2020**, *146*, 1–12. [CrossRef]
59. Zahoor, R.; Zhao, W.; Abid, M.; Dong, H.; Zhou, Z. Potassium application regulates nitrogen metabolism and osmotic adjustment in cotton (*Gossypium hirsutum* L.) functional leaf under drought stress. *J. Plant Physiol.* **2017**, *215*, 30–38. [CrossRef]

antioxidants

MDPI

Article

Melatonin Improves Drought Stress Tolerance of Tomato by Modulating Plant Growth, Root Architecture, Photosynthesis, and Antioxidant Defense System

Muhammad Ahsan Altaf [1], Rabia Shahid [2], Ming-Xun Ren [3,4,*], Safina Naz [5], Muhammad Mohsin Altaf [4], Latif Ullah Khan [6], Rahul Kumar Tiwari [7,8], Milan Kumar Lal [7,8], Muhammad Adnan Shahid [9], Ravinder Kumar [7,8], Muhammad Azher Nawaz [10], Mohammad Shah Jahan [11], Basit Latief Jan [12] and Parvaiz Ahmad [13]

1 School of Horticulture, Hainan University, Haikou 570228, China; ahsanaltaf8812@gmail.com
2 School of Management, Hainan University, Haikou 570228, China; rabiashahidlhr@yahoo.com
3 Center for Terrestrial Biodiversity of the South China Sea, Hainan University, Haikou 570228, China
4 College of Ecology and Environment, Hainan University, Haikou 570228, China; mohsinaltaf641@gmail.com
5 Department of Horticulture, Bahauddin Zakariya University, Multan 60800, Pakistan; Safina_bzu@yahoo.com
6 College of Tropical Crop, Hainan University, Haikou 570228, China; latif.hainu3103@gmail.com
7 ICAR-Indian Agricultural Research Institute, New Delhi 110012, India; rahultiwari226@gmail.com (R.K.T.); milan2925@gmail.com (M.K.L.); chauhanravinder97@gmail.com (R.K.)
8 ICAR-Central Potato Research Institute, Shimla 171001, India
9 Department of Agriculture, University of New Hampshire, Durham, NC 03824, USA; muhammad.shahid@unh.edu
10 Department of Horticulture, College of Agriculture, University of Sargodha, Sargodha 171001, Pakistan; azher490@gmail.com
11 Department of Horticulture, Faculty of Agriculture, Sher-e-Bangla Agricultural University, Dhaka 1207, Bangladesh; shahjahansau@gmail.com
12 Department of Clinical Pharmacy, College of Pharmacy, King Saud University, Riyadh 11451, Saudi Arabia; Bjan@ksu.edu.sa
13 Botany and Microbiology Department, King Saud University, Riyadh 11451, Saudi Arabia; parvaizbot@yahoo.com
* Correspondence: renmx@hainanu.edu.cn

Citation: Altaf, M.A.; Shahid, R.; Ren, M.-X.; Naz, S.; Altaf, M.M.; Khan, L.U.; Tiwari, R.K.; Lal, M.K.; Shahid, M.A.; Kumar, R.; et al. Melatonin Improves Drought Stress Tolerance of Tomato by Modulation Plant Growth, Root Architecture, Photosynthesis, and Antioxidant Defense System. *Antioxidants* **2022**, *11*, 309. https://doi.org/10.3390/antiox11020309

Academic Editor: Nafees A. Khan

Received: 5 January 2022
Accepted: 26 January 2022
Published: 3 February 2022

Abstract: Tomato is an important vegetable that is highly sensitive to drought (DR) stress which impairs the development of tomato seedlings. Recently, melatonin (ME) has emerged as a nontoxic, regulatory biomolecule that regulates plant growth and enhances the DR tolerance mechanism in plants. The present study was conducted to examine the defensive role of ME in photosynthesis, root architecture, and the antioxidant enzymes' activities of tomato seedlings subjected to DR stress. Our results indicated that DR stress strongly suppressed growth and biomass production, inhibited photosynthesis, negatively affected root morphology, and reduced photosynthetic pigments in tomato seedlings. Per contra, soluble sugars, proline, and ROS (reactive oxygen species) were suggested to be improved in seedlings under DR stress. Conversely, ME (100 μM) pretreatment improved the detrimental-effect of DR by restoring chlorophyll content, root architecture, gas exchange parameters and plant growth attributes compared with DR-group only. Moreover, ME supplementation also mitigated the antioxidant enzymes [APX (ascorbate peroxidase), CAT (catalase), DHAR (dehydroascorbate reductase), GST (glutathione S-transferase), GR (glutathione reductase), MDHAR (monodehydroascorbate reductase), POD (peroxidase), and SOD (superoxide dismutase)], non-enzymatic antioxidant [AsA (ascorbate), DHA (dehydroascorbic acid), GSH (glutathione), and GSSG, (oxidized glutathione)] activities, reduced oxidative damage [EL (electrolyte leakage), H_2O_2 (hydrogen peroxide), MDA (malondialdehyde), and $O_2^{\bullet-}$ (superoxide ion)] and osmoregulation (soluble sugars and proline) of tomato seedlings, by regulating gene expression for *SOD*, *CAT*, *APX*, *GR*, *POD*, *GST*, *DHAR*, and *MDHAR*. These findings determine that ME pretreatment could efficiently improve the seedlings growth, root characteristics, leaf photosynthesis and antioxidant machinery under DR stress and thereby increasing the seedlings' adaptability to DR stress.

Keywords: tomato; photosynthesis; root growth; oxidative damage; melatonin; drought; gene expression

1. Introduction

In recent decades, water has been considered a vital environmental factor, and its deficiency is known as drought stress (DR) which leads to restricted plant growth, yield, and quality, mainly due to various morphological, physiological, anatomical, and biochemical responses. Compared with other environmental stresses, DR stress is particularly prominent in abiotic stress [1]. Water scarcity adversely influences plant growth by reducing relative water content (RWC), root length, seedling stem, leaf area, and leaf water potential [2]. Plants exhibit multiple and interconnected responses towards DR stress. DR stress can be fatal for underdeveloped root systems of seedlings in the spring season [3]. Earlier studies revealed that DR stress can cause reduced pigment content, imbalanced ion homeostasis, decreased transpiration, stomatal closure, cell enlargement reduction, reduce canopy size, and ultimately lead towards plant death [4]. The impact of DR stress varies with the intensity and the growth stage of plants [5].

The excessive accumulation of ROS in plants under water-deficient conditions can cause oxidative damage which might lead to significant damage to the cellular organelles [6]. The ROS molecules include H_2O_2, $O_2^{\bullet-}$, and other oxygen-containing molecules which interact with the membrane system of the and damage mainly the macromolecules present in the cell [7]. Excessive accumulation ROS cause oxidative damage to the electron transport chain, enhance lipid peroxidation in chloroplast and mitochondria, inactivates enzyme activity, proteins, and nucleic acids, and ultimately decreases photosynthesis and yield of crops [8]. The antioxidant enzymes (APX, CAT, DHAR, GR, GST, MDHAR, POD, and SOD) and non-enzymatic antioxidants [AsA, DHA, GSH, and GSSG] molecules effectively reduce the ROS accumulation, thereby balancing the ROS synthesis and signaling to provide enhance DR-stress tolerance in plants [9]. Additionally, the accumulation of osmolyte also helps effectively against DR stress to prevent oxidative meditative damage in the plant. Soluble sugars, proline, and proteins are the most common osmolytes which interact at the cellular level by decreasing membrane permeability under mild water scarcity, thus maintaining water balances of crops under DR stress [10].

Recent trends of climate change have significantly affected crop yield and among these various abiotic conditions, DR stress is one of the main culprits [11]. To circumvent this condition, various agronomic strategies are needed for improving drought tolerance in plants which include employing phytohormones or bio-stimulators in crop production. Such phytohormones pose a positive impact on growth regulation and resilience development in plants by boosting various physiological, biochemical and molecular processes [12]. Thus, for better tolerance towards DR stress, it is of great significance to explore the effect and mechanism of these phytohormones.

Melatonin is a natural, multifunctional, nontoxic, regulatory, and universal biomolecule, having low molecular weight with pleiotropic effects in the plant kingdom [13]. Hitherto, ME is evidenced to enhance the resistance to different abiotic stresses such as DR, heat, cold, salt, heavy metals, chemicals, and pathogens [14–16]. Furthermore, ME promoted root morphology seed germination, photosynthesis, seedlings growth, delay in leaf senescence, antioxidant efficiency, fruit maturation and rhizogenesis at cellular and tissue level [17–19]. Specifically, the increased ROS homeostasis and enhanced antioxidant capacity were related to higher ME content in DR-affected seedlings, such as *Actinidia deliciosa* [20], *Althaea rosea* [21] *Eriobotrya japonica* [22], and *Carthamus tinctorius* [23].

Previous studies revealed that under drought stress, ME enhanced plant growth attributes in tomato [24], improved the root architecture system in kiwifruit [25], protected mineral balance in *C. cathayensis* [26], increased photosynthetic efficiency in *E. japonica* [22], reduced lipid peroxidation and ROS accumulation in *Moringa oleifera* [27], promoted antioxidant enzymes system in fenugreek [7], increased glutathione, and ascorbic acid content in kiwifruit seedlings [20], enhanced osmoregulatory substances in tomato [28], protected grana lamella of chloroplast in wheat [29], preserved the chloroplast structure in *Brassica napus* [30], and significantly regulated the antioxidant enzymes (APX, CAT, DHAR, GST,

GR, MDHAR, POD, SOD) and non-enzymatic antioxidant (AsA, GSH) genes in *Solanum lycopersicum* and *Carya cathayensis* [19,26].

In the solanaceous family, after potato, tomato is the 2nd most important cultivated and extensively used vegetable around the world. Like other crops, water scarcity prominently causes an inhibitory effect on the physiological, morphological, biochemical, and anatomical structure of tomato [31]. Tomato is highly sensitive to water scarcity, particularly at the germination and seedling growth and seed germination stage [32]. However, information about how ME regulates the physiological, morphological, biochemical, and anatomical changes in tomato under water deficit conditions remains elusive. Therefore, in this study, we aimed to explore the impacts of ME in enhancing DR tolerance and analyze the mechanism of increased DR tolerance induced by ME. Our main focus was to investigate the photosynthetic performance, membrane damage of tomato seedlings, root architecture system, the chloroplast structure, and ROS homeostasis regulated by ME under DR stress. Precisely, we also examined enzymes related to the Haliwell Asada pathway and their genes expression. The present study has attempted to contribute to the elucidation of the biological function of ME in building resilience against DR stress.

2. Materials and Methods

2.1. Plant Material

In the present study, the tomato cultivar "Fenli" was collected from Minghao seed store, Hainan, China. The germinated seeds were sown in vermiculite-filled trays with a 50-cell plug in a controlled growth chamber. After the development of the second-true-leaf, equal size tomato seedlings were transferred into the vermiculite-filled black plastic pots (with height = 8.5 cm, top and bottom diameter = 10 and 7 cm, respectively), having one emerged seedling/pot. Subsequently, all the pots were placed in a greenhouse with specific conditions (25 °C ± 5 temperature; 16/8 h dark/light period; and 55–90% relative humidity). The pre-cultivation period was 10 days, with the aim of allowing seedlings get adapted to new conditions, with watering (80 mL per plant) at two days' interval using Hoagland's nutrient solution (pH 5.8 ± 0.1).

2.2. Experimental Design

The treatment plan was implemented after ten days, when the seedlings were attained at the four-true-leaf stage. The treatments are followed in the current study including: (1) CK (control) treatment included the seedlings with full water application during the complete span of the experiment, (2) DR (drought) treatment included the plants which were first given water fully for eight days, followed by up to two weeks' water-withholding, (3) ME + DR (melatonin + drought) was the treatment in which ME pretreatment was given to seedlings with 100 µM solution of ME (80 mL per plant) [33]. This pretreatment continued four times, with a 2-days' interval, followed by up to two weeks' water-withholding. Each treatment comprised of 3 replicates, with 8 plants per replicate. Further, 0 day was the day when irrigation-withholding was started. After two-weeks of DR treatment, collection of leaf samples was done to perform different analysis, including both biochemical and physiological tests.

2.3. Vegetative Growth

After two weeks of DR-stress, seedlings were washed with tap water. Roots and shoots were separated for measuring both fresh and dry weight. After measuring fresh weight, root and shoot were immediately placed in an individual paper bag and put in an oven for drying. The dry weight of root and shoot were recorded after drying of samples at 105 °C for 15 min, and 75 °C for three days [34]. Plant height was estimated using a measuring tape.

2.4. Leaf Gas Exchange and Pigments

Photosynthetic parameters including Pn (net photosynthetic rate), Ci (intercellular CO_2 concentration), Gs (stomatal conductance) and Tr (transpiration rate) were assayed (9 to 11 a.m.) by operating a portable photosynthesis system, namely CIRAS-3, bought from Hansatech Co., Amesbury, MA, USA, with controlled measuring chamber conditions (25 °C leaf temperature, 360 µM/mol CO_2 concentration, and 800 µM/m^2/s photosynthetic photon-flux density).

The method of Arnon [35] was followed to measure chlorophyll and carotenoid content. Concisely, slicing of leaf samples (0.1 g) was done, and sliced samples were stored in a dark place after placement in glass test tubes and addition of acetone (80%) per tube. The samples were stored in the dark until they were completely discolored, followed by the centrifugation of sample extract. Finally, Lambda 25 UV/VIS Spectrophotometer was employed to calculate chlorophyll a (Chl a), Chlorophyll b (Chl b), and carotenoid (Caro) via measuring absorbance at 645, 663, and 440 nm, respectively.

2.5. Root Morphology and Root Activity

After two-weeks of DR-stress, roots of three plants were taken and washed using tap water, to study root morphological traits. Scanning of roots was done using a root scanner, followed by evaluating the scanned images using a root image analysis software (Epson Expression 11000XL, Regent Instruments, Québec, QC, Canada), to obtain the root morphological parameters [36].

The TTC (triphenyl tetrazolium chloride) method, as explained by Comas et al. [37], was followed to estimate root activity (capacity of root deoxidization measured in mg g^{-1} h^{-1} FW). Briefly, root samples (500 mg) were dipped in 10 mL of a mixed solution of phosphate buffer and TTC (0.4%), which is mixed in a 1:1 ratio. After that, the sample was placed in the dark for one to three hours (37 °C), followed by the addition of 2 mL of H_2SO_4 (1 mol L^{-1}) to avoid the reaction of the solution. Finally, dipped roots were blended and transferred into 10 mL of ethyl acetate cleaning solution in a tube, and the spectrophotometer was used to record solution' absorption at 485 nm.

2.6. Leaf Relative Water Content and Soluble Sugar

According to the method of Barrs and Weatherley [38], the following formula was used to calculate leaves' *RWC* (relative water content) (Equation (1)):

$$RWC\ (\%) = \left[\frac{FW - DW}{SW - DW}\right] \times 100 \tag{1}$$

where FW = fresh weight; DW = dry weight; SW = saturated weight in water.

The anthrone method of Shi et al. [39] was employed to perform soluble sugar content. In brief, 80% (V/V) ethanol (2 mL) was added to 0.1 g sample at 80 °C, and kept for 30 min. Then, anthrone (2 mL) was mixed to extract (100 µL), followed by boiling for 10 min. Finally, after recording the absorption (at 630 nm), the calibration curve of the sucrose standard was used to calculate the soluble sugar content.

2.7. Oxidative Stress Markers, and Proline

For calculating H_2O_2, $O_2^{\bullet-}$, MDA, and proline contents, liquid nitrogen was used to crush frozen sample (0.1 g) into powder form, followed by isolation using 100 mM phosphate buffer (900 µL, pH 7.4) and guidelines labelled in kits (Nanjing Jiancheng Bioengineering Institute, located in Nanjing, China) of H_2O_2 (A064), $O_2^{\bullet-}$ (A052), MDA (A003-3), and proline (A107-1), at 405, 550, 530, and 520 nm, respectively. Leaves' EL was measured by following the method of Zhang et al. [40].

2.8. Antioxidant Enzymes Activity

To examine the antioxidant enzyme activities, powdered leaf sample (0.5 g) was homogenized using 100 mM phosphate buffer (900 μL, pH 7.4) by following kits' description. After that, centrifugation (at 4 °C temperature, 12,000× g revolutions for 15 min) of homogenized samples was done, and to analyze enzymatic activities, the supernatant was shifted to new falcon tube. The guidelines mentioned in kits (A001-1, A007-1, A123-1, AC62-1, A004, A084-3-1, BC0660, and BC0650) were followed to note the activities of SOD, CAT, APX, GR, GST, POD, DHAR, and MDHAR enzymes, at 550, 405, 290, 340, 412, 420, 412, and 340 nm wavelengths, respectively.

The method of Logan et al. [41] was followed to assess the levels of ascorbate (AsA) and oxidized ascorbate (DHA). The procedures of Griffith [42] were followed to measure reduced glutathione (GSH) and oxidized glutathione (GSSG) content.

2.9. Quantitative Real-Time PCR

Following the manufacturer's instructions, Trizole reagent was used to extract total RNA from individual treatment leaf samples. With the help of agarose gel electrophoresis, a NanoPhotometer® spectrophotometer (Implen, Westlake Village, CA, USA), the extracted RNA' purity and quality were examined. The manufacturer protocol of Vazyme HiScript II Q RT SuperMix for qPCR (+gDNA wiper) 1st strand cDNA synthesis kit (Vazyme, Nanjing, China) was followed to reverse-transcribe the extracted RNA, for complementary DNA (cDNA) synthesis. For qRT-PCR (quantitative real-time PCR) analysis, cDNA was used as templates. The Roche FastStart Essential DNA Green Master kit (Roche, Pleasanton, CA, USA) was employed in an Mx3000 P qPCR system (Agilent Technologies, Santa Clara, CA, USA), and 96-well plates were used to perform qRT-PCR. Supplementary Table S1 provides the detail of primers followed in this study Jahan et al. [43], and Actin was used as a reference gene. The formula of Livak and Schmittgen [44], i.e., $2^{-\Delta\Delta Ct}$ was used to calculate the changes in relative gene expression. Three biological replications were performed for each treatment, and three technical replicates were carried out for each biological replicate.

2.10. Statistical Analysis

The statistical package SPSS version 22.0 (IBM Corporation, Armonk, NY, USA) was used for the statistical analysis of data. One-way analysis of variance (ANOVA) was completed, and the treatment means were compared using the LSD (least significant difference) test (at $p \leq 0.05$).

3. Results

3.1. Effect of DR Stress on Tomato Seedlings Growth and Development

Drought treatment showed a significant decline in the growth of tomato seedlings compared with CK treatment (Figure 1). The tomato seedlings, when subjected to DR stress, significantly reduced in plant height (57.52%), fresh shoot weight (FSW; 61.11%), dry shoot weight (DSW; 63.58%), fresh root weight (FRW; 63.96%), and dry root weight (DRW; 64.74%), compared with CK (well-watered) plants (Figure 2) This decline in the growth of tomato seedlings was alleviated by the exogenous application of melatonin (Figures 1 and 2). After pretreatment with ME, growth limitations caused by DR stress were improved, and less reductions in plant height (26.43%), FSW (30.25%), DSW (37.06%), FRW (35.63%), and DRW (26.13%) was observed (Figure 2).

Figure 1. Tomato seedlings visual demonstration under ME and DR stress. Photographs of the tomato seedlings were taken.

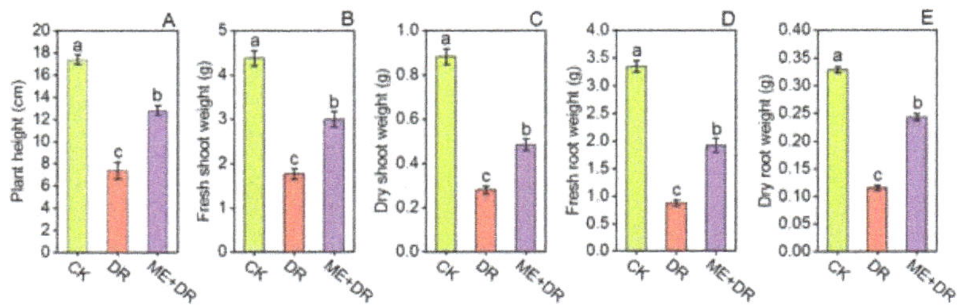

Figure 2. Exogenous supplementation of melatonin promoted the growth [plant height (**A**), fresh shoot weight (**B**), dry shoot weight (**C**), fresh root weight (**D**), and dry root weight (**E**)] of tomato seedlings under drought stress conditions. Means ± standard error, n = 3, significant differences are exhibited by lowercase letters ($p \leq 0.05$), according to LSD test.

3.2. Effect of DR Stress on Photosynthesis and Related Parameters

Under DR stress, Pn, Ci, Gs, and Tr were decreased by 71.58%, 43.42%, 66.89%, and 68.04%, respectively, compared with CK seedlings (Figure 3). Conversely, when seedlings were treated with ME, the reductions of these leaf gas exchange parameters were only 44.27%, 24.59%, 43.21%, and 46.93%, respectively, as compared to the CK plants.

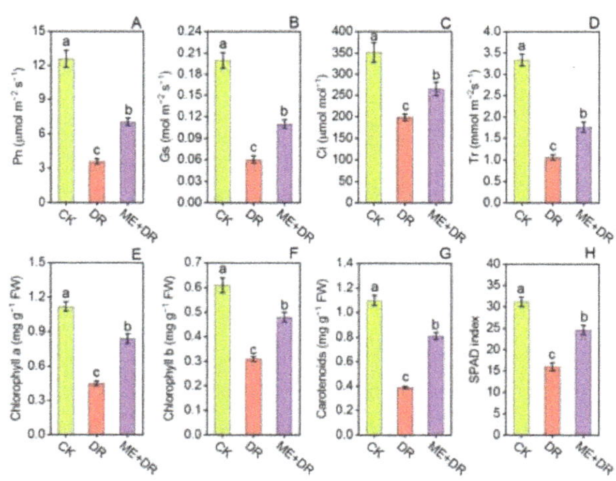

Figure 3. Exogenous supplementation of melatonin promoted leaf gas exchange [Pn (**A**), Gs (**B**), Ci (**C**) and Tr (**D**)] and level of pigment [chlorophyll a (**E**), chlorophyll b (**F**), carotenoids (**G**), and SPAD index (**H**)] of tomato leaves under drought stress conditions. Means ± standard error, n = 3, significant differences are exhibited by lowercase letters ($p \leq 0.05$), according to LSD test.

A similar trend of reduction in the pigment system was observed under the DR stress condition. Chl a, Chl b, and Carotenoid content in the leaves of tomatoes were reported to be sharply reduced by 59.94%, 48.63%, and 64.74%, respectively, under DR stress (Figure 3E–G). Per contra, ME-treated tomato seedlings when subjected to DR treatment, these pigments' content significantly increased—by 86.66%, 194.68%, and 110.34%, respectively—when compared with only DR-stressed seedlings (Figure 3E–G). In DR-stressed plants, the SFAD index showed a noticeable reduction. In contrast, the ME supplementation elevated SFAD index under DR-stress (Figure 3H).

3.3. Changes in Root Morphology under DR Stress

The present study showed that DR treatment remarkably diminished the root morphological parameters, including root length (68.46%), root volume (72.95%), root surface area (72.40%), root crossings (71.02%), root tips (70.47%), root forks (66.22%), average root diameter (64.61%), and projected area (62.31%) compared with CK tomato seedlings (Figure 4). Interestingly, compared with DR stressed plants, these root characteristics were improved by 70.40-, 75.30-, 82.98-, 92.89-, 65.60-, 73.70-, 86.02-, and 52.36%, respectively, in ME pretreated tomato plants subjected to DR stress (Figure 4).

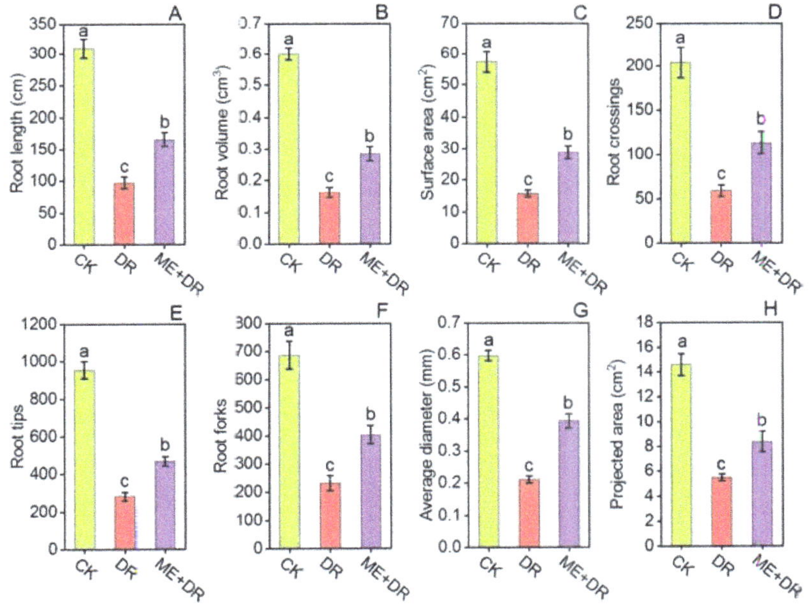

Figure 4. Exogenous supplementation of melatonin promoted root morphology [Root length (**A**), root volume (**B**), surface area (**C**), root crossings (**D**), root tips (**E**), root forks (**F**), average diameter (**G**), and projected area (**H**)] of tomato seedlings under drought stress conditions. Means ± standard error, $n = 3$, significant difference are exhibited by lowercase letters ($p \leq 0.05$), according to LSD test.

3.4. Relative Water Content, Proline, Soluble Sugars, and Root Activity Alternation Due to DR Stress

After two weeks of DR stress, the RWC of DR-stressed plants was noticeably reduced by 36.10%. Conversely, ME-treatment significantly improved the RWC by 34.12% compared with DR treatment (Figure 5A). As depicted in Figure 5D, the root activity was significantly reduced in DR-stressed plants by 43.22% in contrast to normal irrigated seedlings. Nonetheless, the supplementation of ME markedly enhanced the root activity by 29.33% compared with only DR-stressed seedlings (Figure 5B). proline and Soluble sugars content were remarkably increased (124.12%, 32.59%, respectively) under the DR group compared with CK plants. Per contra, ME treatment considerably reduced the proline content by 20.09% and 11. 24% compared with DR-stressed seedlings, respectively (Figure 5C,D).

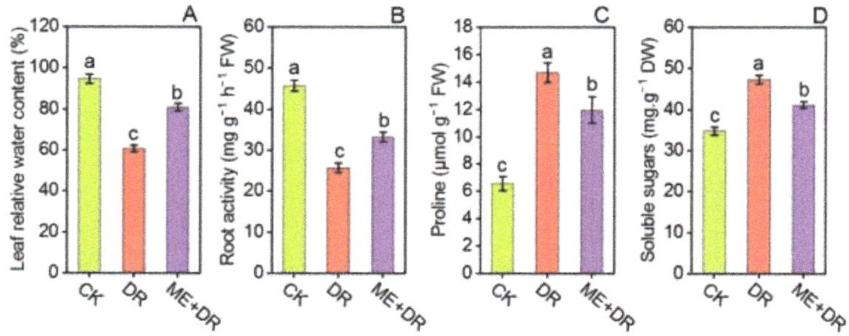

Figure 5. Exogenous supplementation of melatonin promoted level of relative water content (**A**), and root activity (**B**); reduced level of proline (**C**) and soluble sugar (**D**) content of tomato seedling under drought stress conditions. Means ± standard error, $n = 3$, significant difference are exhibited by lowercase letters ($p \leq 0.05$), according to LSD test.

3.5. *Oxidative Damage*

After two weeks of DR treatment, the H_2O_2, $O_2^{\bullet-}$, MDA, and EL levels were measured in the leaves of tomato seedlings (Figure 6). For instance, in the tomato plants under the DR environment, the H_2O_2, $O_2^{\bullet-}$, MDA, and EL levels significantly increased by 1.25-, 1.50-, 1.26-, and 0.99-fold, respectively, compared with well-water seedlings. Importantly, ME-pretreated plants subjected to DR-stress reduced this content only by 0.22-, 0.21-, 0.23-, and 0.22-fold, respectively, compared with the DR-stressed group (Figure 6).

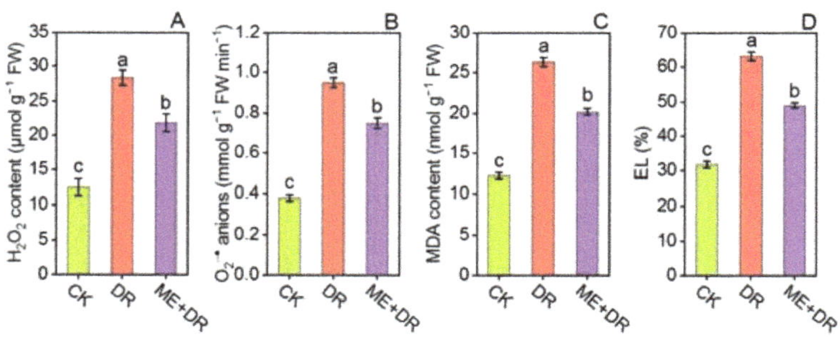

Figure 6. Exogenous supplementation of melatonin reduces oxidative damage biomarkers [H_2O_2 content (**A**), $O_2^{\bullet-}$ anions (**B**), MDA content (**C**), and EL (**D**)] of tomato seedling under drought stress conditions. Means ± standard error, $n = 3$, significant difference is exhibited by lowercase letters ($p \leq 0.05$), according to LSD test.

3.6. *Antioxidant Enzymes Activity and Gene Expression*

The antioxidant enzymes' (APX, CAT, GR, GST, POD, SOD, DHAR, and MDHAR) activities were measured in the leaves of tomato plants (Figures 7 and 8). By exposure to DR stress, the SOD, CAT, APX, GR, and POD, enzymes activity was enhanced by 47.15-, 85.39-, 14.03-, 51.54-, and 57.77%, respectively; activity of GST enzyme was decreased by 29.51% compared to CK seedlings. It is noteworthy that when ME-treated seedlings were subjected to DR-treatment, this further elevated these antioxidant enzymes by 27.24-, 17.63-, 35.24-, 38.64-, 40.01-, and 23.02% respectively, compared with the DR-stress group (Figure 7A–D and Figure 8A,B). Moreover, subject to the DR group, the DHAR and MDHAR activity noticeably increased by 42.74- and 29.18%, respectively, compared with well-watered seedlings (Figure 8C,D). Conversely, ME pretreatment along with DR-group, further increased the DHAR and MDHAR activities by 25.66- and 21.26%, respectively compared with DR-stressed seedlings.

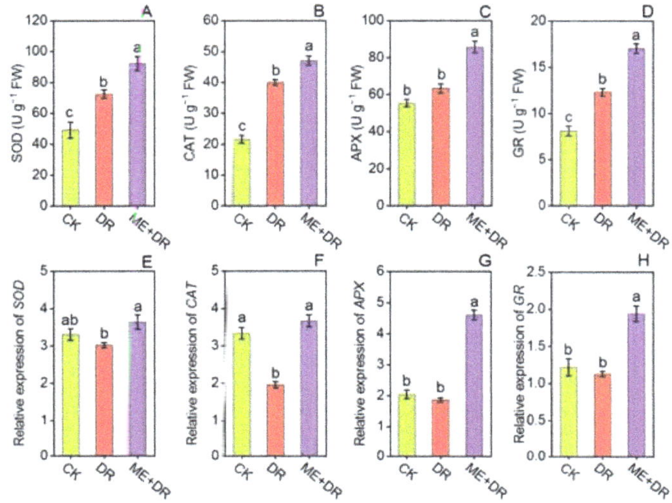

Figure 7. Exogenous supplementation of melatonin promoted antioxidant enzymes [SOD (**A**), CAT (**B**), APX (**C**), GR (**D**)] and their encoding genes [*SOD* (**E**), *CAT* (**F**), *APX* (**G**), *GR* (**H**)] of tomato seedlings under drought stress condition, as shown by leaf' analysis. Means ± standard error, $n = 3$, significant difference is exhibited by lowercase letters ($p \leq 0.05$), according to LSD test.

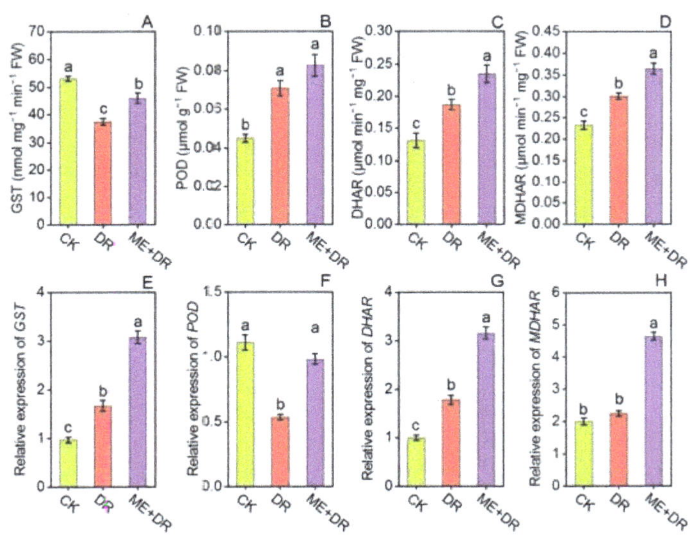

Figure 8. Exogenous supplementation of melatonin promoted antioxidant enzymes [GST (**A**), POD (**B**), DHAR (**C**), MDHAR (**D**)] and their encoding genes [*GST* (**E**), *POD* (**F**), *DHAR* (**G**), *MDHAR* (**H**)] of tomato seedlings under drought stress condition, as shown by leaf' analysis. Means ± standard error, $n = 3$, significant difference is exhibited by lowercase letters ($p \leq 0.05$), according to LSD test.

We further measured the transcriptional levels of genes related to antioxidant enzymes (APX, CAT, DHAR, GST, GR, MDHAR, POD, and SOD) (Figures 7 and 8E–H). The tomato plants exposed to DR-stress significantly amplified the relative gene expression level of these enzymes. Per contra, ME-treated plants exposed to DR-stress added more increments in the transcriptional levels of these enzyme-related genes, when compared to DR and CK seedlings (Figures 7 and 8E–H). These outcomes suggested that exogenous ME application alleviated DR-stress tolerance by enhancing the mRNA pattern of genes of these detoxifying enzymes.

We also performed a Pearson correlation analysis of eight enzymes and their related genes. Our results revealed that the expression of the antioxidant gene was positively correlated with the drought stress condition and its expression was higher when exogenous melatonin was applied to the tomato plant (Figure 9). Similarly, the activity of antioxidant enzymes of the Halliwell Asada pathway was reported to have a positive correlation with the drought condition (Figure 9). Moreover, the expression of genes *viz.*, *SOD*, *CAT*, *APX* and *GR* was reported to have a positive correlation (expressed in the blue color shade in Figure 9) with their respective enzymatic activity. However, there was also a negative but not strong correlation when SOD activity was compared with CAT, APX and GR activity. In most of the cases, the expression of eight antioxidant genes was reported to have a positive correlation with the enzymatic activities.

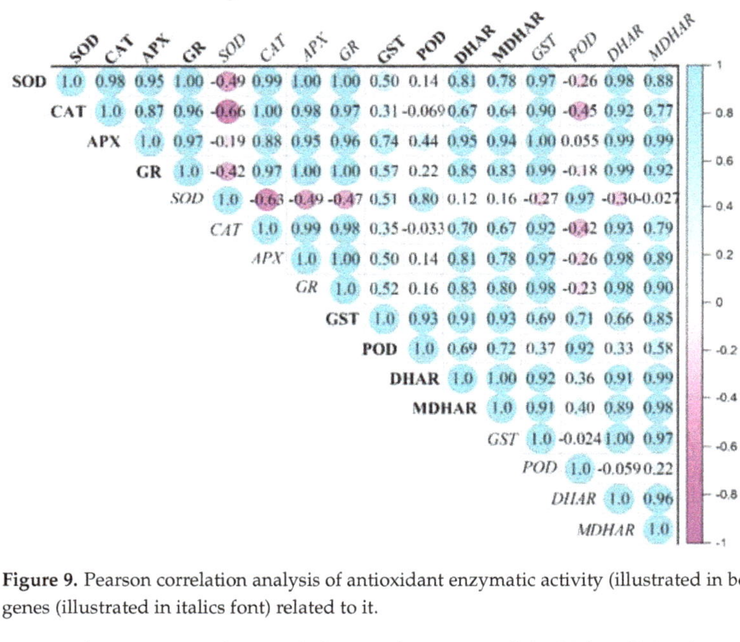

Figure 9. Pearson correlation analysis of antioxidant enzymatic activity (illustrated in bold font) and genes (illustrated in italics font) related to it.

Under DR stress, the metabolites and enzymes of the Haliwell Asada pathway were also reported to be enhanced significantly which affects the enzymatic and non-enzymatic antioxidant mechanisms in the tomato plant. Compared with CK seedlings, DR-treatment markedly increased the AsA, DHA, GSH, and GSSG content in leaves by 37.37%, 40.46, 43.51%, and 45.45%, respectively (Figure 10). Importantly, ME (100 μM) application subjected to the DR-group further increased the AsA, DHA, GSH, and GSSG content in leaves by 33.21%, 30.29%, 30.32%, and 29.16%, respectively, (Figure 10). These results suggested that antioxidant enzymes helped to reduce oxidative damage and increased the DR-stress tolerance.

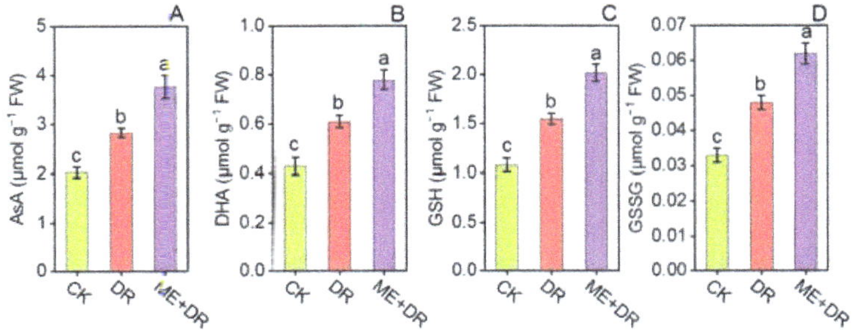

Figure 10. Exogenous supplementation of melatonin promoted non-enzymatic antioxidant [AsA (**A**), DHA (**B**), GSH (**C**), and GSSG (**D**)] system of tomato seedlings under drought stress conditions. Means ± standard error, $n = 3$, significant difference is exhibited by lowercase letters ($p \leq 0.05$), according to LSD test.

4. Discussion

The current study aimed to elucidate the functions applications of exogenous ME in tomato seedlings under DR stress. Plants could circumvent oxidative stress damages created by DR stress by improving their antioxidant capacity. However, if their antioxidant capacity is weak, and supplementation of some exogenous compounds having high antioxidant properties could help them to increase tolerance level to particular stress condition [18]. Amid such compounds, ME is the one posing antioxidant properties, thus it can increase stress resistance. As mentioned by Arnao and Hernández-Ruiz [45], under abiotic stresses, ME might act as a signal molecule, as it upregulates anti-stress genes and endogenous ME levels under stress conditions. Moreover, recently published reports suggested that ME has shown a protective role against DR stress in *E. japonica* [22], *A. rosea* [21], and *A. deliciosa* [20].

The tomato seedlings subjected to DR treatment significantly decreased growth characetristics. Conversely, ME supplementation notably enhanced plant growth attributes (Figure 2). Recent studies suggest that the growth and biomass production were markedly inhibited by heavy metal stress such as Ni stress in tomato seedlings. In contrast, the growth traits were markedly reinforced by ME application [17,46]. Foliar application of ME remarkably accelerated the vegetative growth of tomato seedlings under DR stress [24]. Moreover, exogenous ME ameliorated growth of *Cucumis sativus* [40] and *Trigonella foenum* [7], under DR stress condition.

During photosynthesis in the plant, carbohydrates are considered as the main supply, except for another substrate. As reported by Takahashi and Murata [47], under a stress environment, the rate of carbohydrates synthesis was reduction. Among the tomato seedlings exposed to DR stress, ME-pretreated seedlings showed an improvement in leaf photosynthesis, compared with those which were not pretreated with ME (Figure 3). Under drought stress, outcomes of the experiment carried out by Sharma et al. [26] on grafted Chinese hickory plants revealed improved photosynthetic efficiency, enhanced growth and successful recovery of chlorophyll content by the application of exogenous ME. The plethora of studies described that exogenous ME treatment recover leaf photosynthesis in *A. rosea*, *A. deliciosa* and *G. max* under DR stress [9,21,25], and *S. lycopersicum* under vanadium toxicity [43,48]. SPAD index and photosynthetic pigments including Chl a, Chl b, and Caro are necessary for the photosynthetic process which is suggested to be reduced significantly under DR stress. Interestingly, ME application robustly improved the content of the pigment of tomato seedlings (Figure 3). These results are supported by the fact that DR stress led to a reduction in leaf photosynthetic pigment, with ME foliar application alleviating these changes, and thus, ME-application might be a promising tool for mitigating DR stress in *A. rosea* [21], *Dracocephalum moldavica* [10], and *C. tinctorius* [23].

Plants obtain more energy due to increased photosynthetic capacity, which enables them to cope with environmental stresses.

Roots not only provide structural support to the aerial part of plants, but also supply nutrients and water. Thus, a plant's survival depends on its appropriate growth, development, and root functions. Drought stress significantly reduced the root growth of tomato [28]. In this work, DR treatment negatively affected the root morphological traits by decreasing root surface area, volume, length, root crossings, tips, forks, diameter, and projected area. Conversely, tomato roots pretreated with ME evidently enhanced root characteristics (Figure 4) contributing to better growth of tomato plants. Similarly, Altaf et al. [33] revealed that melatonin application dominantly enhanced the root architecture system of tomato seedling under NaCl stress. Moreover, the positive relationship between the ME application and root growth has been well-known in *S. lycopersicum* [17], *Citrullus lanatus* [49], and *Stevia rebaudiana* [50] under abiotic stresses. Interestingly, when the tomato seedlings were exposed to DR-stress, root activity declined remarkably, showing strict association to the nutrients uptake and water withholding. Per contra, ME pretreated plants repaired the roots from destruction, thus maintaining proper function of roots (Figure 5). Our study is concordant with the previously published reports where it was suggested that abiotic stress-mediated damage to the root was reported to be ameliorated by the application of exogenous ME [43].

Under drought stress, amid various accentuated responses, osmotic regulation is the most important [51]. The reduced leaf water content under DR stress leads to increment of two main osmoprotectants viz. proline and soluble sugars [52] in *S. lycopersicum* [24] and *C. cathayensis* [26]. Our results also affirm this finding (Figure 5). Nevertheless, the application of ME, particularly via root irrigation, declines levels of proline and soluble sugars. Thus, as indicated by our results, a positive turgor pressure and water balance may be maintained by ME.

Zhang et al. [40] elaborated that on the cell membrane, excess ROS caused peroxidation of pigments and lipids, ultimately upsurging the cell membranes' permeability and causing functional damages. The ROS (H_2O_2 and $O_2^{\bullet-}$) content, MDA, and EL have been used as oxidative damage biomarkers. In this study, the oxidative damage levels, as determined by H_2O_2, $O_2^{\bullet-}$, EL and MDA, were increased in tomato seedlings subjected to DR stress. Conversely, exogenous ME application effectively protected plant cells from oxidative damage (Figure 6). In a previous study, Sharma et al. [26] and Sadak et al. [27] exhibited significant lowering of MDA content and decreased oxidative stress by ME-pretreatment in *C. cathayensis* and *Moringa oleifera* plants, respectively, under DR stress. Similarly, Gao et al. [53] revealed that exogenous ME application strikingly declined the level of ROS and MDA content in peach fruit. The results of our study exhibited a reduction in oxidative damage by ME application under DR stress, which is also affirmed by extended literature on various plant species such as *Citrullus lanatus* [49], *Cucumis sativus* [54], and *Stevia rebaudiana* [50], under environmental stresses. Thus, it can be concluded that ME applications can lead to reduced oxidative damage and repairing of disrupted cellular membrane induced by salinity by balancing ROS.

Antioxidant enzymes play a principal role in the defense system of plants against biotic and abiotic stress conditions. Mittler [55] described that under different stresses, increased antioxidant enzymes activities lead to potential and specific ROS scavenging. Melatonin is a multi-regulatory molecule and is recognized as a universal antioxidant [56], because it strengthens plants' antioxidant defense system and enhances tolerance, mainly by detoxifying excess ROS, which is otherwise induced by environmental stresses [57]. Melatonin noticeably improved the activity of antioxidant enzymes (APX, CAT, DHAR, GR, GST, MDHAR, POD, and SOD) and their relative genes expression (Figures 7 and 8). Altaf et al. [58] revealed that in tomato seedlings, ME surprisingly enhanced the antioxidant machinery by reduction of over-accumulation of ROS, which is primarily due to enhanced resilience to nickel toxicity. Similarly, the upregulation in relative gene expression of *APX*, *CAT*, *DHAR*, *GR*, *GST*, *MDHAR*, *POD*, and *SOD* was observed in ME pretreated tomato seedlings under nickel toxicity [43]. Moreover, the literature exhibited significant enhancement of antioxidant enzymes' activities by ME under abiotic stresses in various plant species [26,47,59]. Furthermore, the production and accumulation balance of ROS is maintained by ME, because the performance of the antioxidative system gets boosted and the activity of antioxidative enzymes gets triggered by ME application [10,60].

On the other hand, in plant tissue, AsA and GSH is a well-known antioxidant, proving ME to be a dynamic antioxidant [61]. Wang et al. [60] reported that under environmental stresses, significant changes occur on AsA and GSH content. The present work revealed that ME supplementation predominantly elevated the AsA and GSH content. Furthermore, ME application significantly enhanced the DHA and GSSG content in tomato seedlings (Figure 10). Similarly, ME application sharply enhanced the AsA and GSH contents under Ni toxicity [58], under NaCl stress [62], and under $NaHCO_3$ stress [63] in *S. lycopersicum*. Summarizing the discussion, our results revealed that ME alleviated the negative impact of DR stress on tomato seedlings' growth by improving photosynthesis, root architecture, antioxidant defense system and by regulating the expression of antioxidants-related genes.

5. Conclusions

The present study explored the collective decline in drought stress resilience in plants due to impaired photosynthetic activities, severe growth retardation, excess ROS accumulation, and damaged root morphology under drought stress (Figure 11). Per contra, the stress induced by drought was effectively mitigated by exogenous application of melatonin, as oxidative damage is reduced (with refining in antioxidant defense system and inhibiting ROS production), root architecture is enhanced, photosynthesis efficiency is strengthened and thus growth attributes are recovered. In addition, the soluble sugars and proline content noticeably decline, and the formation of AsA-DHA and GSH-GSSG arere prominently improved in ME-pretreated tomato plants. This suggests that exogenous ME is an effective protectant that improves DR tolerance in tomato seedlings by enhancing antioxidant enzymes and reducing oxidative damages. Nevertheless, under drought stress and melatonin-mediated resilience in plants, and its mechanism should be investigated in-depth, along with the explicit study of molecular approaches.

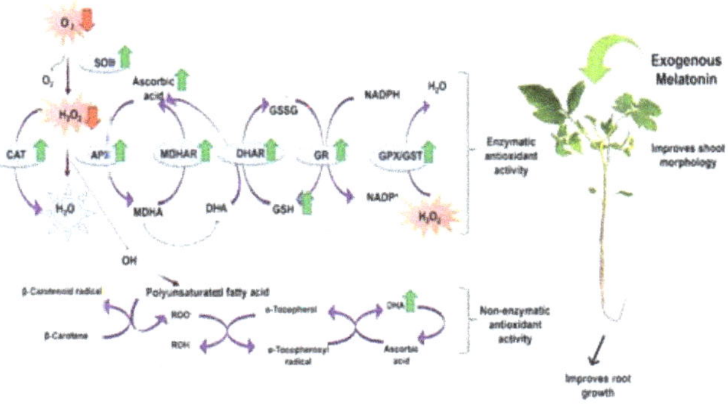

Figure 11. A proposed model showing how melatonin mitigates drought stress tolerance in tomato seedlings.

Supplementary Materials: The following are available online at https://www.mdpi.com/article/10.3390/antiox11020309/s1, Table S1: The detail of primers.

Author Contributions: Conceptualization, M.A.A. and M.-X.R.; methodology, M.A.A.; software, R.S. and B.L.J.; validation, S.N., M.-X.R. and M.A.A.; formal analysis, L.U.K.; investigation, M.A.A.; resources, M.-X.R.; data curation, M.M.A.; writing—original draft preparation, M.A.A.; review and editing, R.K.T., M.K.L., M.A.S., R.K., M.A.N., M.S.J. and P.A.; visualization, M.A.A.; supervision,

M.-X.R.; project administration, M.-X.R.; funding acquisition, M.-X.R., B.L.J. and P.A. All authors have read and agreed to the published version of the manuscript.

Funding: This program was financially supported by the Hainan Key Science and Technology Program (Grant No. ZDKJ202008-1-2), National Natural Science Foundation of China (41871041).

Institutional Review Board Statement: Not applicable.

Informed Consent Statement: Not applicable.

Data Availability Statement: Not applicable.

Acknowledgments: We would like to thank Wang Zhiwei from the School of Horticulture, Hainan University, Haikou for his kind guidance and laboratory equipment. The authors would also like to extend their sincere appreciation to the Researchers Supporting Project Number (RSP-2021/168), King Saud University, Riyadh, Saudi Arabia.

Conflicts of Interest: The authors declare no conflict of interest.

References

1. Huang, B.; Chen, Y.E.; Zhao, Y.Q.; Ding, C.B.; Liao, J.Q.; Hu, C. Exogenous melatonin alleviates oxidative damages and protects photosystem II in maize seedlings under drought stress. *Front. Plant Sci.* **2019**, *10*, 677. [CrossRef] [PubMed]
2. Zong, H.; Liu, S.; Xing, R.; Chen, X.; Li, P. Protective effect of chitosan on photosynthesis and antioxidative defense system in edible rape (*Brassica rapa* L.) in the presence of cadmium. *Ecotoxicol. Environ. Saf.* **2017**, *138*, 271–278. [CrossRef]
3. Kaczmarek, M.; Fedorowicz-Strońska, O.; Głowacka, K. CaCl2 treatment improves drought stress tolerance in barley (*Hordeum vulgare* L.). *Acta Physiol. Planta.* **2017**, *39*, 41. [CrossRef]
4. Kamran, M.; Wennan, S.; Ahmad, I.; Xiangping, M.; Wenwen, C.; Xudong, Z. Application of paclobutrazol affect maize grain yield by regulating root morphological and physiological characteristics under a semi-arid region. *Sci. Rep.* **2018**, *8*, 4818. [CrossRef] [PubMed]
5. Wu, W.; Ma, B.L.; Whalen, J.K. Enhancing rapeseed tolerance to heat and drought stresses in a changing climate: Perspectives for stress adaptation from root system architecture. *Adv. Agron.* **2018**, *151*, 87–157. [CrossRef]
6. Tiwari, R.K.; Lal, M.K.; Kumar, R.; Chourasia, K.N.; Naga, K.C.; Kumar, D. Mechanistic insights on melatonin-mediated drought stress mitigation in plants. *Physiol. Plant.* **2021**, *172*, 1212–1226. [CrossRef] [PubMed]
7. Zamani, Z.; Amiri, H.; Ismaili, A. Improving drought stress tolerance in fenugreek (*Trigonella foenum-graecum*) by exogenous melatonin. *Plant Biosyst. Int. J. Deal. Asp. Plant Biol.* **2020**, *154*, 643–655. [CrossRef]
8. Cui, G.; Zhao, X.; Liu, S.; Sun, F.; Zhang, C.; Xi, Y. Beneficial effects of melatonin in overcoming drought stress in wheat seedlings. *Plant Physiol. Biochem.* **2017**, *118*, 138–149. [CrossRef]
9. Cao, L.; Jin, X.J.; Zhang, Y.X. Melatonin confers drought stress tolerance in soybean (*Glycine max* L.) by modulating photosynthesis, osmolytes, and reactive oxygen metabolism. *Photosynthetica* **2019**, *57*, 812–819. [CrossRef]
10. Kabiri, R.; Hatami, A.; Oloumi, H.; Naghizadeh, M.; Nasibi, F.; Tahmasebi, Z. Foliar application of melatonin induces tolerance to drought stress in Moldavian balm plants (*Dracocephalum moldavica*) through regulating the antioxidant system. *Folia Hortic.* **2018**, *30*, 155–167. [CrossRef]
11. Swann, A.L.S. Plants and drought in a changing climate. *Curr. Clim. Change Rep.* **2018**, *4*, 192–201. [CrossRef]
12. Tiwari, R.K.; Lal, M.K.; Naga, K.C.; Kumar, R.; Chourasia, K.N.; Subhash, S. Emerging roles of melatonin in mitigating abiotic and biotic stresses of horticultural crops. *Sci. Hortic.* **2020**, *272*, 109592. [CrossRef]
13. Altaf, M.A.; Shahid, R.; Ren, M.X.; Mora-Poblete, F.; Arnao, M.B.; Naz, S. Phytomelatonin: An overview of the importance and mediating functions of melatonin against environmental stresses. *Physiol. Plant.* **2021**, *172*, 820–846. [CrossRef]
14. Kaya, C.; Higgs, D.; Ashraf, M.; Alyemeni, M.; Ahmad, P. Integrative roles of nitric oxide and hydrogen sulfide in melatonin-induced tolerance of pepper (*Capsicum annuum* L.) plants to iron deficiency and salt stress alone or in combination. *Physiol. Plant.* **2020**, *168*, 256–277. [CrossRef]
15. Devi, R.; Behera, B.; Raza, M.B.; Mangal, V.; Altaf, M.A.; Kumar, R.; Kumar, A.; Tiwari, R.K.; Lal, M.K.; Singh, B. An Insight into Microbes Mediated Heavy Metal Detoxification in Plants: A Review. *J. Soil Sci. Plant Nutr.* **2021**, 1–23. [CrossRef]
16. Tiwari, R.K.; Lal, M.K.; Kumar, R.; Mangal, V.; Altaf, M.A.; Sharma, S.; Singh, B.; Kumar, M. Insight into melatonin-mediated response and signaling in the regulation of plant defense under biotic stress. *Plant Mol. Biol.* **2021**, 1–15. [CrossRef]
17. Altaf, M.A.; Shahid, R.; Ren, M.X.; Altaf, M.M.; Khan, L.U.; Shahid, S. Melatonin alleviates salt damage in tomato seedling: A root architecture system, photosynthetic capacity, ion homeostasis, and antioxidant enzymes analysis. *Sci. Hortic.* **2021**, *285*, 110145. [CrossRef]
18. Nawaz, M.A.; Huang, Y.; Bie, Z.; Ahmed, W.; Reiter, R.J.; Niu, M. Melatonin: Current status and future perspectives in plant science. *Front. Plant Sci.* **2016**, *6*, 1230. [CrossRef]
19. Jahan, M.S.; Shu, S.; Wang, Y.; Hasan, M.; El-Yazied, A.A.; Alabdallah, N.M. Melatonin pretreatment confers heat tolerance and repression of heat-induced senescence in tomato through the modulation of ABA-and GA-mediated pathways. *Front. Plant Sci.* **2021**, *12*, 381. [CrossRef]

20. Xia, H.; Ni, Z.; Hu, R.; Lin, L.; Deng, H.; Wang, J. Melatonin alleviates drought stress by a non-enzymatic and enzymatic antioxidative system in kiwifruit seedlings. *Int. J. Mol. Sci.* **2020**, *21*, 852. [CrossRef]
21. Roy, R.; Sultana, S.; Begum, N.; Fornara, D.; Barmon, M.; Zhang, R. Exogenous melatonin reduces water deficit-induced oxidative stress and improves growth performance of *Althaea rosea* grown on coal mine spoils. *Environ. Sci. Poll. Res.* **2021**, 1–11. [CrossRef]
22. Wang, D.; Chen, Q.; Chen, W.; Guo, Q.; Xia, Y.; Wang, S. Physiological and transcription analyses reveal the regulatory mechanism of melatonin in inducing drought resistance in loquat (*Eriobotrya japonica* Lindl.) seedlings. *Environ. Exp. Bot.* **2021**, *181*, 104291. [CrossRef]
23. Heshmati, S.; Dehaghi, M.A.; Farooq, M.; Wojtyla, Ł.; Maleki, K.; Heshmati, S. Role of Melatonin seed priming on Antioxidant Enzymes and Biochemical Responses of *Carthamus tinctorius* L. under Drought Stress Conditions. *Plant Stress* **2021**, *2*, 100023. [CrossRef]
24. Ibrahim, M.F.; Elbar, O.H.A.; Farag, R.; Hikal, M.; El-Kelish, A.; El-Yazied, A.A. Melatonin counteracts drought induced oxidative damage and stimulates growth, productivity and fruit quality properties of tomato plants. *Plants* **2020**, *9*, 1276. [CrossRef] [PubMed]
25. Liang, D.; Ni, Z.; Xia, H.; Xie, Y.; Lv, X.; Wang, J. Exogenous melatonin promotes biomass accumulation and photosynthesis of kiwifruit seedlings under drought stress. *Sci. Hortic.* **2019**, *246*, 34–43. [CrossRef]
26. Sharma, A.; Wang, J.; Xu, D.; Tao, S.; Chong, S.; Yan, D. Melatonin regulates the functional components of photosynthesis, antioxidant system, gene expression, and metabolic pathways to induce drought resistance in grafted *Carya cathayensis* plants. *Sci. Total Environ.* **2020**, *713*, 136675. [CrossRef] [PubMed]
27. Sadak, M.S.; Abdalla, A.M.; Abd Elhamid, E.M.; Ezzo, M.I. Role of melatonin in improving growth, yield quantity and quality of *Moringa oleifera* L. plant under drought stress. *Bull. Nat. Res. Cent.* **2020**, *44*, 18. [CrossRef]
28. Liu, J.; Wang, W.; Wang, L.; Sun, Y. Exogenous melatonin improves seedling health index and drought tolerance in tomato. *Plant Growth Regul.* **2015**, *77*, 317–326. [CrossRef]
29. Kaya, C.; Okant, M.; Ugurlar, F.; Alyemeni, M.N.; Ashraf, M.; Ahmad, P. Melatonin-mediated nitric oxide improves tolerance to cadmium toxicity by reducing oxidative stress in wheat plants. *Chemosphere* **2019**, *225*, 627–638. [CrossRef]
30. Khan, M.N.; Zhang, J.; Luo, T.; Liu, J.; Rizwan, M.; Fahad, S. Seed priming with melatonin coping drought stress in rapeseed by regulating reactive oxygen species detoxification: Antioxidant defense system, osmotic adjustment, stomatal traits and chloroplast ultrastructure perseveration. *Ind. Crops Prod.* **2019**, *140*, 111597. [CrossRef]
31. Murshed, R.; Lopez-Lauri, F.; Sallanon, H. Effect of water stress on antioxidant systems and oxidative parameters in fruits of tomato (*Solanum lycopersicon* L, cv. Micro-tom). *Physiol. Mol. Biol. Plants* **2013**, *19*, 363–378. [CrossRef]
32. Foolad, M.R.; Zhang, L.; Subbiah, P. Genetics of drought tolerance during seed germination in tomato: Inheritance and QTL mapping. *Genome* **2003**, *46*, 536–545. [CrossRef]
33. Altaf, M.A.; Shahid, R.; Ren, M.X.; Naz, S.; Altaf, M.M.; Qadir, A. Exogenous melatonin enhances salt stress tolerance in tomato seedlings. *Biol. Plantar.* **2020**, *64*, 604–615. [CrossRef]
34. Altaf, M.A.; Shahid, R.; Altaf, M.A.; Ren, M.X.; Tan, K.; Xiang, W.Q.; Qadir, A.; Shakoor, A.; Altaf, M.M. Effect of NPK, organic manure and their combination on growth, yield and nutrient uptake of chilli (*Capsicum Annum* L.). *Hortic. Int. J.* **2019**, *3*, 217–222. [CrossRef]
35. Arnon, D.I. Copper enzymes in isolated chloroplasts. Polyphenoloxidase in *Beta vulgaris*. *Plant Physiol.* **1949**, *24*, 1. [CrossRef]
36. Altaf, M.M.; Diao, X.P.; ur Rehman, A.; Imtiaz, M.; Shakoor, A.; Altaf, M.A. Effect of vanadium on growth, photosynthesis, reactive oxygen species, antioxidant enzymes, and cell death of rice. *J. Soil Sci. Plant Nutr.* **2020**, *20*, 2643–2656. [CrossRef]
37. Comas, L.H.; Eissenstat, D.M.; Lakso, A.N. Assessing root death and root system dynamics in a study of grape canopy pruning. *New Phytol.* **2000**, *147*, 171–178. [CrossRef]
38. Barrs, H.; Weatherley, P. A re-examination of the relative turgidity technique for estimating water deficits in leaves. *Aust. J. Biol. Sci.* **1962**, *15*, 413–428. [CrossRef]
39. Shi, H.; Qian, Y.; Tan, D.X.; Reiter, R.J.; He, C. Melatonin induces the transcripts of CBF/DREB1s and their involvement in both abiotic and biotic stresses in *Arabidopsis*. *J. Pineal Res.* **2015**, *59*, 334–342. [CrossRef]
40. Zhang, N.; Zhao, B.; Zhang, H.J.; Weeda, S.; Yang, C.; Yang, Z.C. Melatonin promotes water stress tolerance, lateral root formation, and seed germination in cucumber (*Cucumis sativus* L.). *J. Pineal Res.* **2012**, *54*, 15–23. [CrossRef]
41. Logan, B.A.; Demmig-Adam, B.; Adams, W.W., III. Antioxidants and xanthophyll cycle-dependent energy dissipation in *Cucurbita pepo* L. and *Vinca major* L. upon a sudden increase in growth PPFD in the field. *J. Exp. Bot.* **1998**, *49*, 1881–1888. [CrossRef]
42. Griffith, O.W. Determination of glutathione and glutathione disulfide using glutathione reductase and 2-vinylpyridine. *Analyt. Biochem.* **1980**, *106*, 207–212. [CrossRef]
43. Jahan, M.S.; Guo, S.; Baloch, A.R.; Sun, J.; Shu, S.; Wang, Y. Melatonin alleviates nickel phytotoxicity by improving photosynthesis, secondary metabolism and oxidative stress tolerance in tomato seedlings. *Ecotoxicol. Environ. Saf.* **2020**, *197*, 110593. [CrossRef]
44. Livak, K.J.; Schmittgen, T.D. Analysis of relative gene expression data using real-time quantitative PCR and the $2^{-\Delta\Delta CT}$ method. *Methods* **2001**, *25*, 402–408. [CrossRef] [PubMed]
45. Arnao, M.B.; Hernández-Ruiz, J. Melatonin: Plant growth regulator and/or biostimulator during stress? *Trends Plant Sci.* **2014**, *19*, 789–797. [CrossRef]
46. Jahan, M.S.; Guo, S.; Sun, J.; Shu, S.; Wang, Y.; Abou El-Yazied, A. Melatonin-mediated photosynthetic performance of tomato seedlings under high-temperature stress. *Plant Physiol. Biochem.* **2021**, *167*, 309–320. [CrossRef]

47. Takahashi, S.; Murata, N. How do environmental stresses accelerate photo inhibition? *Trends Plant Sci.* **2008**, *13*, 178–182. [CrossRef] [PubMed]
48. Altaf, M.A.; Shahid, R.; Ren, M.X.; Khan, L.U.; Altaf, M.M.; Jahan, M.S.; Nawaz, M.A.; Naz, S.; Shahid, S.; Lal, M.K. Protective Mechanisms of Melatonin Against Vanadium Phytotoxicity in Tomato Seedlings: Insights into Nutritional Status, Photosynthesis, Root Architecture System, and Antioxidant Machinery. *J. Plant Growth Regul.* **2021**, 1–17. [CrossRef]
49. Nawaz, M.A.; Jiao, Y.; Chen, C.; Shireen, F.; Zheng, Z.; Imtiaz, M. Melatonin pretreatment improves vanadium stress tolerance of watermelon seedlings by reducing vanadium concentration in the leaves and regulating melatonin biosynthesis and antioxidant-related gene expression. *J. Plant Physiol.* **2018**, *220*, 115–127. [CrossRef] [PubMed]
50. Simlat, M.; Ptak, A.; Skrzypek, E.; Warchol, M.; Moranska, E.; Piorkowska, E. Melatonin significantly influences seed germination and seedling growth of *Stevia rebaudiana* Bertoni. *PeerJ* **2018**, *6*, e5009. [CrossRef]
51. Ali, B.; Tao, Q.; Zhou, Y.; Gill, R.A.; Ali, S.; Rafiq, M.T. 5-Aminolevolinic acid mitigates the cadmium-induced changes in *Brassica napus* as revealed by the biochemical and ultra-structural evaluation of roots. *Ecotoxicol. Environ. Saf.* **2013**, *92*, 271–280. [CrossRef]
52. Xiong, L.; Zhu, J.K. Molecular and genetic aspects of plant responses to osmotic stress. *Plant Cell Environ.* **2002**, *25*, 131–139. [CrossRef]
53. Gao, H.; Zhang, Z.K.; Chai, H.K.; Cheng, N.; Yang, Y. Melatonin treatment delays postharvest senescence and regulates reactive oxygen species metabolism in peach fruit. *Post. Biol. Technol.* **2016**, *118*, 103–110. [CrossRef]
54. Zhang, T.; Shi, Z.; Zhang, X.; Zheng, S.; Wang, J.; Mo, J. Alleviating effects of exogenous melatonin on salt stress in cucumber. *Sci. Hortic.* **2020**, *262*, 109070. [CrossRef]
55. Mittler, R. ROS are good. *Trends Plant Sci.* **2017**, *22*, 11–19. [CrossRef]
56. Ahammed, G.J.; Xu, W.; Liu, A.; Chen, S. Endogenous melatonin deficiency aggravates high temperature-induced oxidative stress in *Solanum lycopersicum* L. *Environ. Exp. Botany* **2019**, *161*, 303–311. [CrossRef]
57. Arnao, M.B.; Hernández-Ruiz, J. Functions of melatonin in plants: A review. *J. Pineal Res.* **2015**, *59*, 133–150. [CrossRef]
58. Altaf, M.A.; Shahid, R.; Ren, M.X.; Altaf, M.M.; Jahan, M.S.; Khan, L.U. Melatonin Mitigates Nickel Toxicity by Improving Nutrient Uptake Fluxes, Root Architecture System, Photosynthesis, and Antioxidant Potential in Tomato Seedling. *J. Soil Sci. Plant Nutr.* **2021**, *21*, 1842–1855. [CrossRef]
59. Zahedi, S.M.; Hosseini, M.S.; Abadía, J.; Marjani, M. Melatonin foliar sprays elicit salinity stress tolerance and enhance fruit yield and quality in strawberry (*Fragaria* × *ananassa* Duch.). *Plant Physiol. Biochem.* **2020**, *149*, 313–323. [CrossRef]
60. Wang, P.; Sun, X.; Li, C.; Wei, Z.; Liang, D.; Ma, F. Long-term exogenous application of melatonin delays drought-induced leaf senescence in apple. *J. Pineal Res.* **2013**, *54*, 292–302. [CrossRef]
61. Reiter, R.J.; Mayo, J.C.; Tan, D.X.; Sainz, R.M.; Alatorre-Jimenez, M.; Qin, L. Melatonin as an antioxidant: Under promises but over delivers. *J. Pineal Res.* **2016**, *61*, 253–278. [CrossRef]
62. Siddiqui, M.H.; Alamri, S.; Al-Khaishany, M.Y.; Khan, M.N.; Al-Amri, A.; Ali, H.M. Exogenous melatonin counteracts NaCl induced damage by regulating the antioxidant system, proline and carbohydrates metabolism in tomato seedlings. *Int. J. Mol. Sci.* **2019**, *20*, 353. [CrossRef]
63. Liu, N.; Jin, Z.; Wang, S.; Gong, B.; Wen, D.; Wang, X. Sodic alkaline stress mitigation with exogenous melatonin involves reactive oxygen metabolism and ion homeostasis in tomato. *Sci. Hortic.* **2015**, *181*, 18–25. [CrossRef]

antioxidants

MDPI

Article

Exogenous Betaine Enhances the Protrusion Vigor of Rice Seeds under Heat Stress by Regulating Plant Hormone Signal Transduction and Its Interaction Network

Xu Mo [1,†], Jingya Qian [1,†], Peng Liu [1], Hongli Zeng [1], Guanghui Chen [2,*] and Yue Wang [1,*]

1 Department of Agronomy, College of Agronomy, Hunan Agricultural University, Changsha 410128, China
2 The Key Laboratory of Crop Germplasm Innovation and Resource Utilization of Hunan Province, Hunan Agricultural University, Changsha 410128, China
* Correspondence: chenguanghui@hunau.edu.cn (G.C.); wangyue@hunau.edu.cn (Y.W.)
† These authors contributed equally to this work.

Abstract: Rice is an important food crop. Rice seedlings are mainly composed of root, coleoptile, mesocotyl and euphylla. The elongation of coleoptile and mesocotyl promotes the emergence of rice seedlings. Therefore, analyzing the mechanism of coleoptile and mesocotyl elongation is important for the cultivation of rice varieties. Due to global warming, heat stress is threatening rice yields. Betaine plays an important role in plant resistance to heat stress; however, we lack research on its regulation mechanism of rice seed germination under heat stress. Therefore, we explored the effects of soaking seeds with betaine at different concentrations on rice seed germination under heat stress. According to the results, soaking seeds with 10 mM of betaine could effectively improve the seeds' germination potential and rate under heat stress to promote the germination of rice seeds. To clarify the mitigation mechanism of betaine in heat stress, we measured the antioxidant enzyme activity, malondialdehyde content, soluble protein content and endogenous hormone content of seed protrusion under heat stress. We constructed the cDNA library for transcriptome sequencing. According to the results, 10 mM of betaine improved the activities of the superoxide dismutase, peroxidase and catalase of seed protrusion under heat stress to reduce the malondialdehyde content and increase the soluble protein content to alleviate the effect of heat stress on rice seed germination. The detection of the endogenous hormone content showed that soaking seeds with 10 mM of betaine increased the content of gibberellin and decreased the contents of auxin and abscisic acid of seed protrusion under heat stress. According to the transcriptome analysis, betaine can induce the expressions of key genes in the biosynthesis and metabolism of auxin, abscisic acid and gibberellins in the seed coleoptile and mesocotyl elongation stage, regulate the signal transduction of three hormones and promote the germination of rice seeds under heat stress. This study revealed, for the first time, the physiological and molecular regulation mechanism of betaine promotion of seed germination under heat stress.

Keywords: rice; heat stress; betaine; seed germination; physiology; biochemistry

Citation: Mo, X.; Qian, J.; Liu, P.; Zeng, H.; Chen, G.; Wang, Y. Exogenous Betaine Enhances the Protrusion Vigor of Rice Seeds under Heat Stress by Regulating Plant Hormone Signal Transduction and Its Interaction Network. *Antioxidants* **2022**, *11*, 1792. https://doi.org/10.3390/antiox11091792

Academic Editor: Nafees A. Khan

Received: 15 August 2022
Accepted: 7 September 2022
Published: 11 September 2022

1. Introduction

Rice (*Oryza sativa* L.) is one of the most important food crops, feeding more than half of the world's population [1]. Rice is a thermophilic crop, and temperature is one of the most important factors affecting its growth and development. However, greenhouse gas emissions are warming the planet's temperature because of a variety of human and natural forces. Researchers estimate that, by the end of the 21st century, the global average temperature will increase by 2–4 °C, and for every 1 °C increase in the minimum temperature in the growing season, the rice yield will be reduced by 10% [2,3]. According to estimates, the effects of global warming will cause a 41% reduction in the rice yield by the end of the 21st century [4].

The reduction in the rice yield caused by heat stress mainly affects the important period of rice grain formation. For example, heat stress at the booting stage of rice leads to the degeneration of spikelets, pollen abortion, a decline in pollen vitality and then to an increase in dry grains [5,6]. Heat stress during the tassel and filling stages can prevent pollination, increase the number of blighted grains in the spike and reduce the fruit set rate, which results in the reduced weight of a thousand seeds and, hence, a lower rice yield [7,8]. Therefore, in the future development of rice, high temperature will be one of the main factors limiting stable and high rice yields.

Temperature is one of the main factors affecting seed germination. The optimum temperature for the germination of rice seeds is 25 °C. When the temperature is 32 °C, the germination is fast and neat; however, when the germination period encounters heat stress of more than 35 °C, the germination of rice seeds is inhibited, which involves the accumulation of reactive oxygen species (ROS) and changes in plant hormone levels [9]. Heat stress decouples the enzymes and metabolic pathways of plants. Heat stress further induces and triggers an oxidative stress reaction by destroying the stability of the cell membrane through membrane lipid peroxidation and protein denaturation, which result in the accumulation of a large number of reactive oxygen species, such as singlet oxygen, superoxide radical, hydrogen peroxide (H_2O_2) and hydroxyl radical [10]. Efficient enzymatic and nonenzymatic antioxidant defense systems play an important role in scavenging and detoxifying ROS to reduce membrane lipid peroxidation and maintain ROS homeostasis and redox signaling [11]. Heat stress leads to a reduction in the activity of important members of the antioxidant enzyme system, such as superoxide dismutase (SOD), peroxidase (POD) and catalase (CAT), which can further lead to an imbalance in the ROS system, accelerating the accumulation of ROS and ultimately inhibiting seed germination [12,13].

In addition to the excessive accumulation of reactive oxygen species, the imbalance of endogenous hormone levels caused by heat stress is also an important reason for inhibiting seed germination. Abscisic acid (ABA) inhibits seed germination, and gibberellins (GAs) promote it. Abscisic acid and gibberellins act antagonistically with each other in the plant to regulate seed germination [14]. The inhibition of seed germination by heat stress is called the thermal inhibition of seed germination. Thermal inhibition inhibits the expression of GA biosynthesis genes and ABA inactivation genes; however, it promotes the expression of ABA synthesis genes [15]. The mitigation of heat stress by ABA is accomplished through synergistic effects with nitric oxide (NO) to increase the antioxidant enzyme activity, reduce the levels of H_2O_2 and thiobarbituric acid reactive substances (TBARS), and increase carbohydrates, adenosine triphosphate and heat-shock proteins in plants [16,17]. Auxin (IAA) is the second hormone after ABA that positively regulates seed dormancy [18]. In response to heat stress induction, the expression levels of the growth hormone synthesis genes *OsIAA13* and *OsIAA20* were increased, and the growth hormone response gene *SAUR* was also substantially induced to be expressed [19]. The elongation and growth of *Arabidopsis* hypocotyl under heat stress is related to the increase in the auxin content [20].

Betaine is a quaternary amine compound that is commonly found in bacteria, fungi, higher plants and animals [21]. It plays an extremely important role in plant resistance to heat stress. The heat tolerance of tomato lines transformed with the *codA* gene of the betaine synthesis gene was significantly higher than that of tomato lines without betaine [22]. During the germination of *Arabidopsis* seeds, transgenic plants with the *codA* gene increased the level of betaine, thereby enhancing the high-temperature tolerance of the seeds and promoting seed germination [23]. Endogenous betaine can enhance the high-temperature tolerance of plants. After spraying betaine on leaves, the photoinhibition caused by the heat stress of *Tagetes erecta* was reduced, and the CO_2 assimilation rate, stomatal conductance and transpiration rate improved. Betaine also reduced the levels of hydrogen peroxide, superoxide, lipid peroxidation and cell death [24]. Spraying betaine on tobacco leaves significantly increased the activities of SOD and POD, increased the contents of chlorophyll and proline, and decreased the contents of malondialdehyde (MDA), alleviating heat stress and promoting the accumulation of tobacco biomass under the combined stress of high temperature and drought [25]. Spraying leaves with 10mM betaine improved the average total number of florets obtained in summer maize Jingnongke 728 under the different treatments of early sowing date, normal sowing date and late sowing date, increasing by 5.09%, 4.70% and 2.27%, respectively. The yield of summer maize Jingnongke 728 after spraying betaine increased by 3.05–12.81% [26]. Rice is a typical non-betaine-accumulating species [27].

At present, researchers conduct investigations into betaine to alleviate the stress damage suffered by rice by introducing the betaine enzyme system into rice or by exogenous application through transgenic technology. Transforming the *BADH* gene encoding betaine aldehyde dehydrogenase from barley into rice promoted the accumulation of betaine in rice and significantly improved the tolerance of rice to salt, cold and heat stress [28]. The exogenous application of betaine can also effectively improve the adaptability of rice. Spraying 50 mM or 100 mM concentrations of betaine on the leaves effectively enhanced the osmotic adjustment ability of rice and improved its adaptability to osmotic stresses, such as salt, drought and heat stress [29–31]. These reports mostly investigate the effect of betaine on rice-plant morphogenesis; however, we lack research on the seed germination stage before seedling morphogenesis, and we do not yet understand the molecular mechanism of betaine to alleviate stress damage. In order to study the mechanism of exogenous betaine during rice seed germination, we took XZX45 as the test material, soaked seeds with different concentrations of betaine and carried out seed germination tests at different temperatures. We further measured the antioxidant enzyme activity, MDA content, soluble protein content and endogenous hormone content under heat stress. After heat stress at 38 °C for 24 h, we constructed the cDNA libraries for transcriptome sequencing. We identified differentially expressed genes (DEGs) related to plant endogenous hormone synthesis and metabolism, plant hormone signal transduction and reactive oxygen species accumulation, and we verified the credibility of these differentially expressed genes by qRT-PCR. Finally, we clarified the physiological characteristics of soaking seeds in betaine solutions to alleviate the seed coleoptile and mesocotyl elongation stage under heat stress, and we revealed the molecular mechanism of the betaine promotion of seed germination by regulating plant hormone synthesis and metabolism, as well as signal transduction.

2. Materials and Methods

2.1. Test Materials

The tested material was XZX45, which is a rice variety that is mainly planted in rice-growing areas in the middle and lower reaches of the Yangtze River in China. The high-quality variety Zhouyou 903 was used as the female parent to prepare an inter-variety hybrid combination with Zhefu 504, which had a short growth period and strong resistance. The offspring was named XZX45 after purification. The entire growth period of XZX45 is 106 days, which is characterized by strong lodging resistance and excellent rice quality. In China's provincial regional test of varieties, the average yield was 7573.5 kg ha^{-1}. It is also mainly used as the national grain reserve. All test seeds were harvested in 2021.

2.2. Experimental Design

We carried out seed germination in a germination box (length, width and height of 12 cm, 12 cm and 5 cm, respectively). According to the previous research results of this experiment, the betaine seed-soaking solution was set at 0 mM, 0.1 mM, 1 mM and 10 mM. We selected seeds with full grains, soaked them and disinfected them with 5% sodium hypochlorite for 30 min, and we then rinsed them 5 times with sterile distilled water. After soaking the seeds with betaine solutions of 0 mM, 0.1 mM, 1 mM and 10 mM concentrations for 24 h, we placed them in germination boxes with sterilized germination paper, placing 100 seeds in each germination box. We added 7 mL of sterile distilled water to each germination box for seed germination, and we repeated each treatment 20 times. We conducted the germination test in a light incubator. We set the temperature for the heat stress treatment at 26 °C, 32 °C, 35 °C and 38 °C in 16 h light/8 h dark, and the humidity was 70%. The germination of seeds is subject to the condition that the bud length is equal to half of the seed length, and the root length is equal to the seed length. We recorded the germination every 24 h after sowing, and we counted the final germination rate on the seventh day of germination. We referred to the method of Yu et al. (2022) to determine the germination potential and rate [32]. The germination potential and germination rate were investigated on days 3 and 7 after initiation, respectively. The germination potential was calculated as germination potential (%) = (number of seeds germinated on day 3/total number of experimental seeds) × 100. The germination rate was calculated as germination rate (%) = (number of seeds germinated on day 7/total number of experimental seeds) × 100. After 24 h of the high-temperature treatments of the seeds, we measured and sampled the antioxidant enzyme activity, MDA content, soluble protein content, endogenous hormone content, transcriptome determination and related gene expressions. We quickly put the samples into a 2 mL centrifuge tube with RNase inactivation. After quick freezing with liquid nitrogen, we placed them in a refrigerator at −80 °C for storage.

2.3. *Determination of Physiological Indexes*

We determined the activity of superoxide dismutase (SOD) according to the method of Dhindsa et al. [33]. We referred to the method of Kar et al. [34] to determine the activity of peroxidase (POD). We used the UV absorption method [35] to determine the activity of catalase (CAT). We determined the content of malondialdehyde (MDA) by using the method of Li et al. [36]. We determined the content of soluble protein by the Coomassie brilliant blue G–250 method [37]. We repeated each measurement 5 times.

2.4. *Determination of Endogenous Hormones*

We accurately weighed 50 mg of the sample that we had ground and crushed in liquid nitrogen. We conducted the extraction and quantification of the endogenous IAA, ABA and GAs according to the operation guidelines of Wuhan Metwell Biotechnology Co., Ltd., China. We performed data collection using ultra-performance liquid chromatography (UPLC) (the Shim–pack UFLCSHIMADZU CBM30A system, Kyoto, Japan) and tandem mass spectrometry (MS) (Applied Biosystems, Waltham, MA, USA). Wuhan Metwell Biotechnology Co., Ltd., Wuhan, China (http://www.metware.cn/) (accessed on 17 November 2021) analyzed the data. We performed three biological replicates for each treatment.

2.5. *RNA-Seq Analysis*

The samples used for RNA-seq were seed embryos soaked in 0 mM and 10 mM of betaine, heat-stressed at 38 °C for 24 h. We labeled the treatments BT and HT+BT, respectively, and we performed three biological replicates of each treatment. We extracted the total RNA from young ear tissue with Trizol (Invitrogen, Waltham, MA, USA) reagent. Wuhan Huada Gene Technology Co., Ltd. completed the RNA-seq-library construction and sequencing. The screening conditions of differential genes were a Q value ≤ 0.05 and a fold change ($|\log_2 \text{ratio}|$) >1.5. We used the FPKM values to analyze the differential gene expression levels, and we used gene ontology (GO) enrichment analysis (agriGO v2.0: a GO analysis toolkit for the agricultural community, 2017 update) to analyze the biological processes of the differential gene participation. Kyoto Encyclopedia of Genes and Genomes (KEGG) pathway functional enrichment analyses were performed via the KEGG Database (https://www.genome.jp/kegg (accessed on 14 August 2022)) [38]. We used the phyper function in the R software for enrichment analysis and to calculate the *p*-value, and then FDR corrected the *p*-value, considering a Q value ≤ 0.05 as significant enrichment.

2.6. *Quantitative Real-Time PCR*

A total of nine genes were differentially expressed and subjected to qRT-PCR verification. Total RNA was extracted using the RC411-01 RT-PCR reagent (Vazyme Biotech, Nanjing, China). The cDNA was synthesized according to the instructions of the M-MLV reverse transcriptase Kit (28025013) (Thermo Fisher Scientific, Waltham, MA, USA).

The qRT-PCR was FastStart Universal SYBR Green Master (Rox) superMIX reaction kit (4913850001) (Roche, Mannheim, Germany). The *Actin* gene was used as an internal reference. The primer pairs were designed using Primer Premier 5.0 software. The specific primers for genes involved in the hormone signaling pathway are listed in Table 1. Each biological sample was tested in triplicate, and the standard deviation (SD) values of the means were calculated using standard statistical methods. The expression of genes was analyzed using the DDCt data analysis method, and gene relative expression was calculated using the $2^{-\Delta\Delta Ct}$ method.

Table 1. Primers for qRT-PCR expression analysis.

Gene Name	Gene ID	Forward Primer	Reverse Primer
Os11g32520	AP014967	TCGTCCCCGTCATGAACAAG	GAAGTGGCGGCTCTTGTAGT
Os05g42150	AK101932	CCACCTACTTCAGCCCCAAG	GCCAAGCTATCACAGGTCGT
Os06g45950	XM_015786598	ATGAAGCGTCTCCTCAGGC	GAGGGAGGAGGAGGAGGAC
Os01g62450	XM_015763808	ATGGTCATGGATGCTGGGGT	CCGTCGCGAACCCATTGG
Os01g55240	XM_015793860	CGGGTTCTTCAAGGTCGTCA	ATGTCGCCATTGAACCCGAT
Os05g46040	XM_015784016	CATGTGGTGGTTGCCAACTG	TTCAATCCTTGCCCGCTCAT
Os03g44380	XM_015776052	GTCGTGCTCGACAAGGAGAA	CAGAGGTGGAAGCAGAAGCA
Os02g47470	AB277270	CCTCGCAACCAAGTACAGGT	ACTCCTGCTCGGTGTTCTTG
Os03g49630	XM_015774845	AGAAAGCGATGCCCTCCAAA	CCCTCGGCATGCACTATCAT

2.7. Data Processing and Analysis

We used Excel 2017 software for the data statistics, and the data drawing application Graphpad Prism 8.4.3. We used SPSS 21.0 (IBMCorp., Armonk, NY, USA) software to analyze the significant differences in the data based on the one-way analysis of variance (ANOVA) and Duncan methods, and the significant difference level was $p < 0.05$.

3. Results

3.1. Betaine Soaking Treatment Can Promote the Germination of Rice Seeds under Heat Stress

Germination potential and germination rate are two of the main indicators of seed quality. When germination rates are identical, seeds with high germination potential demonstrate greater vigor. Germination potential determines the regularity of seedling emergence [39]. Therefore, when the germination potential and germination rate of rice seeds are high, it indicates that the seeds emerge quickly and evenly. Meanwhile, the seedlings are robust. If rice seeds have low germination potential and a high germination rate, it indicates uneven seedling emergence and weaker seedlings. After soaking seeds in betaine at different concentrations, we conducted the germination test for rice seeds at different temperatures (Figure 1A,B). At 25 °C, soaking seeds with betaine at various concentrations had little effect on the germination rate and potential. At 32 °C, soaking seeds with betaine at all concentrations improved the germination rate and potential, which were 5.82–6.19% and 5.82–7.26% higher than the control (0 mM betaine), respectively. At 35 °C, soaking seeds with 0.1 mM, 1 mM and 10 mM of betaine had no significant effect on the germination potential and rate. At 38 °C, soaking seeds with 10 mM betaine significantly improved the germination potential and rate, which were 57.96 times and 43.77 times higher than the control (0 mM betaine), respectively.

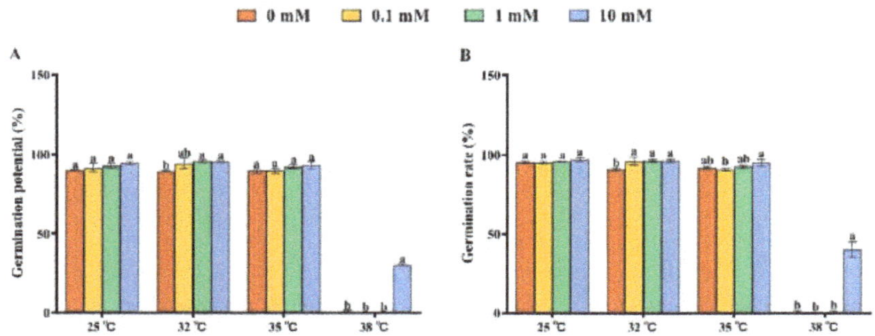

Figure 1. Effects of soaking seeds with different concentrations of betaine at 25 °C, 32 °C, 35 °C and 38 °C on germination characteristics of rice seeds. The difference in the average value of each letter indicates that there is a significant difference in the parameters ($p < 0.05$): (**A**) germination potential; (**B**) germination rate.

3.2. Betaine Increased the Antioxidant Enzyme Activity and Soluble Protein Content of Rice Seeds under Heat Stress

The seed germination activities of the 0 mM and 10 mM betaine soaking treatments were significantly different under heat stress for 24 h. By measuring the antioxidant enzyme activity, MDA content and soluble protein content, we found that the activities of SOD, POD and CAT in the HT + BT treatment were significantly higher than those in the HT treatment (Figure 2A–C) and were increased by 243.06%, 66.10% and 63.37%, respectively. The MDA content in the HT+BT treatment was significantly reduced by 70.69% compared with the HT treatment (Figure 2D), while its soluble protein content was significantly increased by 45.88%, compared with the HT treatment (Figure 2E).

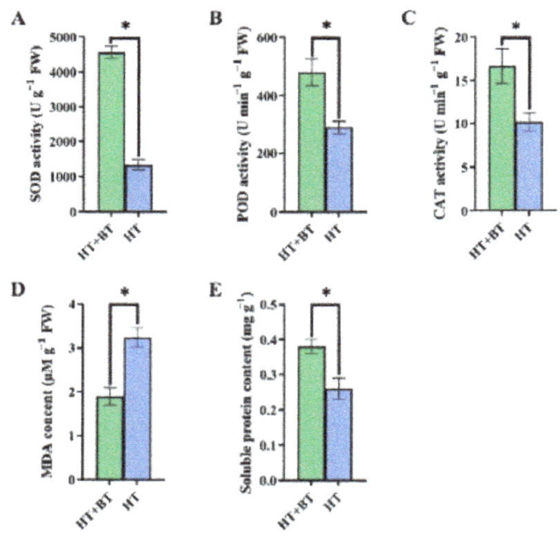

Figure 2. Effects of soaking seeds with 10 mM betaine on antioxidant enzyme system and soluble protein content of rice seeds under heat stress. Each parameter * indicates significant statistical difference, * ($p < 0.05$): (**A**) SOD activity; (**B**) POD activity; (**C**) CAT activity; (**D**) MDA content; (**E**) soluble protein content.

3.3. Hormone Content during Seed Germination

Seed germination is regulated by endogenous plant hormones, so we measured the ABA, IAA and GA contents of the seed protrusion of both the HT- and HT + BT-treated seed endosperm breakthroughs. Under heat stress, the contents of ABA and IAA in the seeds were significantly higher than those of the betaine treatment, of which the ABA was 1.7 times that of the betaine treatment (Figure 3A), and the levels of indole-3-carboxylic acid (ICA), indole-3-carboxaldehyde (ICAld), methyl indole-3-acetic acid ester (ME–IAA) and IAA were 2.49 times, 2.04 times, 3.28 times and 2.21 times that of the betaine treatment, respectively (Figure 3E,H). The contents of various GAs were significantly inhibited under heat stress. Except for gibberellic acid 15 (GA15), the contents of gibberellic acid 19 (GA19) and gibberellic acid 24 (GA24) in the betaine treatment were significantly higher than those in the heat-stress treatment and were 2.13 times and 1.47 times higher than those in the HT treatment, respectively (Figure 3B–D).

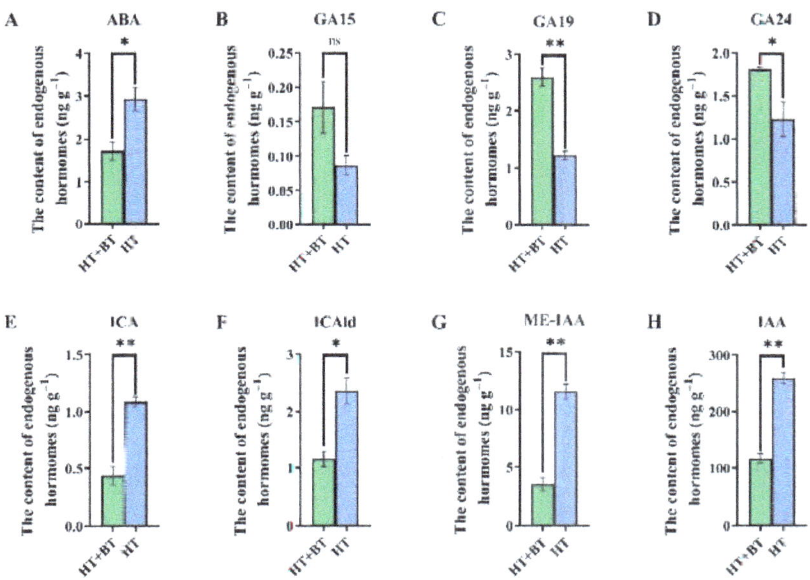

Figure 3. Comparison of hormone levels in HT- and HT + BT-treated seeds: (**A**) comparison of ABA content; (**B**) comparison of GA15 content; (**C**) comparison of GA19 content; (**D**) comparison of GA24 content; (**E**) comparison of ICA content; (**F**) comparison of ICAld content; (**G**) comparison of ME–IAA content; (**H**) comparison of IAA content. Each parameter * indicates significant statistical difference, * ($p < 0.05$), ** ($p < 0.01$).

3.4. *Transcriptome Analysis of the Effect of Betaine Treatment on Rice Seed Germination under Heat Stress*

During seed germination, we performed a transcriptomic analysis of the seed coleoptile and mesocotyl elongation stage after 24 h of immersion under heat stress at 38 °C. We performed three biological replicates of each treatment (Table 2). We present the sequencing data outputs and quality for the six samples in the table. Each sample produced an average of over 44 Mil. total raw reads. After quality control to remove low-quality reads, splice contamination and reads with high unknown base N content, we obtained an average of more than 6 Gb of total clean bases per sample, and all six samples had a Q30 of more than 85% with homogeneous base content.

Table 2. Output and quality of sample sequencing data.

Sample	Total Raw Reads (Mil.)	Total Clean Reads (Mil.)	Total Clean Bases (Gb)	Clean Reads Q20 (%)	Clean Reads Q30 (%)	Clean Reads Ratio (%)
HT1	49.08	45.37	6.81	96.97	89.08	92.45
HT2	49.08	45.2	6.78	97.14	89.56	92.10
HT3	49.08	45.05	6.76	97.08	89.46	91.80
HT + BT1	47.33	44.39	6.66	97.08	89.14	93.79
HT + BT2	44.19	41.11	6.17	97.11	89.40	93.03
HT + BT3	47.33	43.94	6.59	97.16	89.48	92.84

3.5. Comparative Analysis of Differential Genes

We performed principal component analysis (PCA) on the gene expressions of the six samples to identify outliers and to discriminate between clusters of samples with high similarity. We found significant differences between the HT and HT+BT groups, with the two groups forming separate clusters. Moreover, the spatial distribution of the three replicates within the groups was more concentrated, which indicates that the results are reproducible (Figure 4A). In the differential gene volcano plots, we can see that the 10 mM betaine treatment induced the differential expressions of a total of 528 genes under heat stress, of which 332 genes were downregulated, and 196 genes were upregulated. This implies that betaine under heat stress promotes rice seed germination through the differential expressions of these 528 genes (Figure 4B).

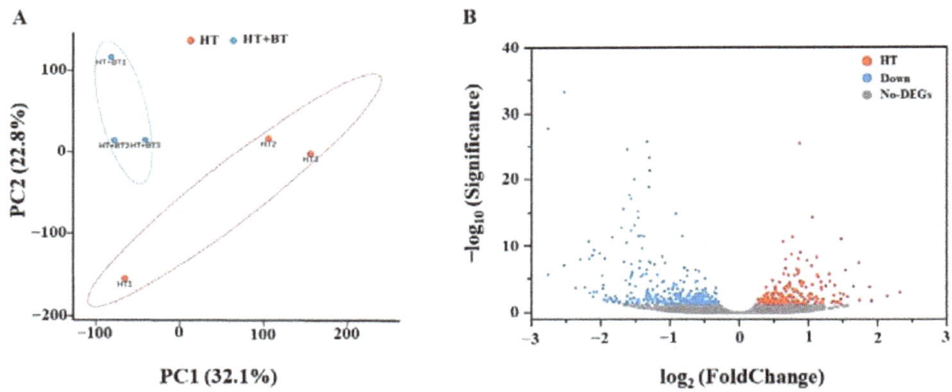

Figure 4. (**A**) HT and HT + BT treatments were the sources of gene expression variation under heat stress. We used principal component analysis to explore the impact on the variation. (**B**) Volcano plots of differentially expressed genes (DEGs) between HT and HT + BT; each point represents a gene. The upregulated genes are represented by red dots, the downregulated genes are represented by blue dots, and the genes with no significant differences are represented by gray dots. Under the corrected Q value, the DEG is considered significantly different.

3.6. GO Pathway Analysis

Functional annotation of transcriptome sequences is an important aspect of functional genomics. GO analysis is currently the main method of functional annotation. GO is a standard classification system of gene functions, which can comprehensively describe the properties of biological genes and gene products [40]. GO analysis includes the classification annotation of GO function and the significance enrichment analysis of GO function. The GO function classification annotation can obtain the gene list of the number statistics of genes with a certain GO function. The GO function significance enrichment analysis gives the GO items of genes that are significantly enriched compared with genomic background, which can obtain the biological functions of genes. The GO database is based on GO terms, which is a tree structure with redundancy. There are three root nodes in the GO database, which describe the molecular function, cellular component and biological process of genes. The GO term is a gene set, which is generally used to find functional changes caused by different genes [41,42]. For example, the molecular function of a gene may be catalytic activity. Its cellular components are localized to the cell membrane in the cell, and the biological process involved is protein trafficking. The three levels of GO are not related to each other, and there is no definition of the mutual relationship of genes. In order to visualize the data and obtain complete functional information, we performed a GO analysis to unify the gene attributes and classify differential genes into putative functional taxa. We divided the functional annotations into three broad categories: biological processes, cellular components and molecular functions. We annotated 528 significantly differentially expressed genes (DEGs) to 40 functional groups within the three broad functional categories (Figure 5A,B). A total of 16 were biological processes, 14 were cellular components and 10 were molecular functions. The majority of DEGs in biological processes were associated with cellular processes (29.06%) and metabolic processes (28.34%); the majority of DEGs in cellular components were associated with cells (34.29%), membranes (16.67%), membrane fractions (14.87%) and organelles (18.35%); the majority of DEGs in molecular functions were associated with binding (40.28%) and catalytic activity (38.89%).

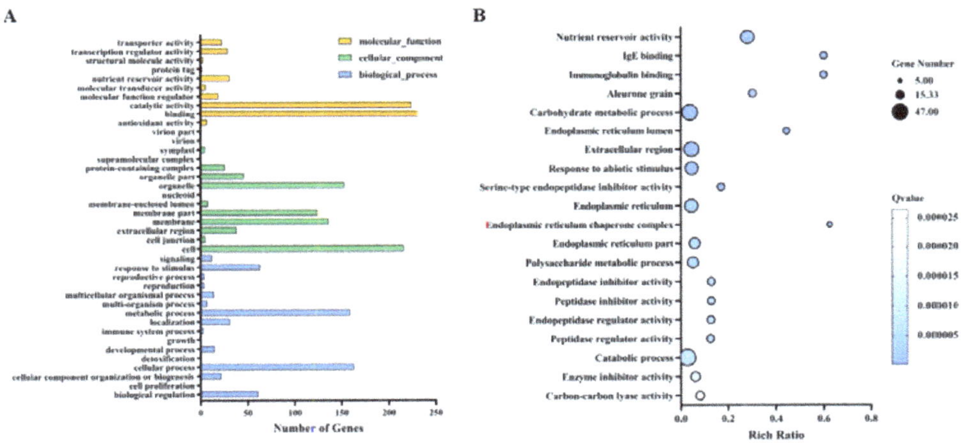

Figure 5. GO term comparison of differential genes with and without betaine soaking treatment under heat stress: (**A**) GO classification of differential genes; (**B**) bubble diagram of differential gene GO enrichment.

3.7. KEGG Pathway Analysis

KEGG is a signaling pathway database with an extremely rich mapping of signaling pathways and of the interactions between the genes contained in the pathways [38]. KEGG analysis allows us to visualize the expressions of genes and their regulatory pathways. According to the results, 290 DEGs were enriched in 90 pathways (Figure 6A,B). Seed germination is regulated by factors such as ambient temperature and endogenous hormones. In the signal transduction pathway of environmental information, we found that many different genes were enriched in plant hormone signal transduction, the plant mitogen-activated protein kinase (MAPK) signal pathway, carotenoid, diterpene biosynthesis and other plant hormone synthesis pathways.

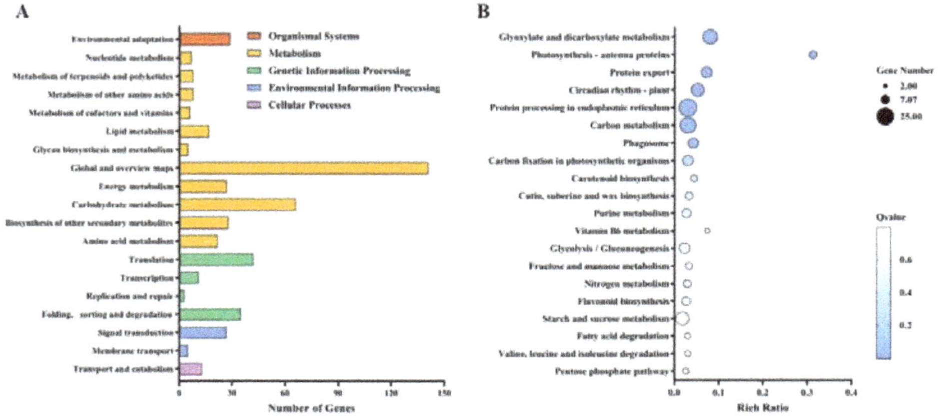

Figure 6. KEGG term comparison of differential genes with and without betaine soaking treatment under heat stress: (**A**) KEGG classification of differential genes; (**B**) bubble diagram of differential gene KEGG enrichment.

3.8. Differential Genes Related to Reactive Oxygen Species in MAPK Plant Signaling Pathway

As an important part of cell signal transduction, ROS are inevitably produced during the normal growth and development of plants and under stress. A certain concentration of ROS is necessary during plant growth, but if excessive ROS cannot be cleared in time, it will cause oxidative damage to cells [9]. H_2O_2, as a signal substance, can enter cells through aquaporins. In cells, H_2O_2 triggers reactive oxygen species (ROS) signals and responses, thus interfering with ROS stabilization [10]. MAPK signaling acts on receptor-like protein kinases (RLKs) and downstream of ROS signals to regulate ROS-related gene expression and programmed cell death (PCD) [43]. ROS signaling is closely related to the MAPK cascade pathway. H_2O_2 can induce and activate some MAPKs, and the MAPK cascade pathway can also regulate the dynamic balance of ROS [44]. In tobacco, the two signaling pathways, NPK1–MEK1–Ntf6 and Mek2–SIPK, can promote the accumulation of ROS by inducing the expression of the *NbRbohb* gene, thus enhancing its stress resistance [45]. In *Arabidopsis*, *AtMPK8* is also involved in regulating the dynamic balance of ROS, and the MKK1–MPK6 signaling pathway is involved in inducing *CAT1* expression and H_2O_2 production in the ABA signaling pathway [46,47].

Heat stress produces an oxidative stress response in seeds, which leads to the production of large amounts of reactive oxygen species, which ultimately inhibit seed germination. Therefore, reducing the accumulation of excess reactive oxygen species due to heat stress is an effective way to improve seed germination rates under stress. Our previous results show that 10 mM of betaine soaking significantly increased the activity of some antioxidant enzymes in seeds under heat stress and reduced the MDA content to promote rice seed germination. In addition to the enzymatic aspects, according to our transcriptomic results (Figure 7A), betaine infiltration significantly downregulated the expression level of *MEKK1* in the H_2O_2 signaling pathway in the plant MAPK signaling pathway. It further reduces the production of reactive oxygen species under heat stress and improves seed adaptation to high temperatures.

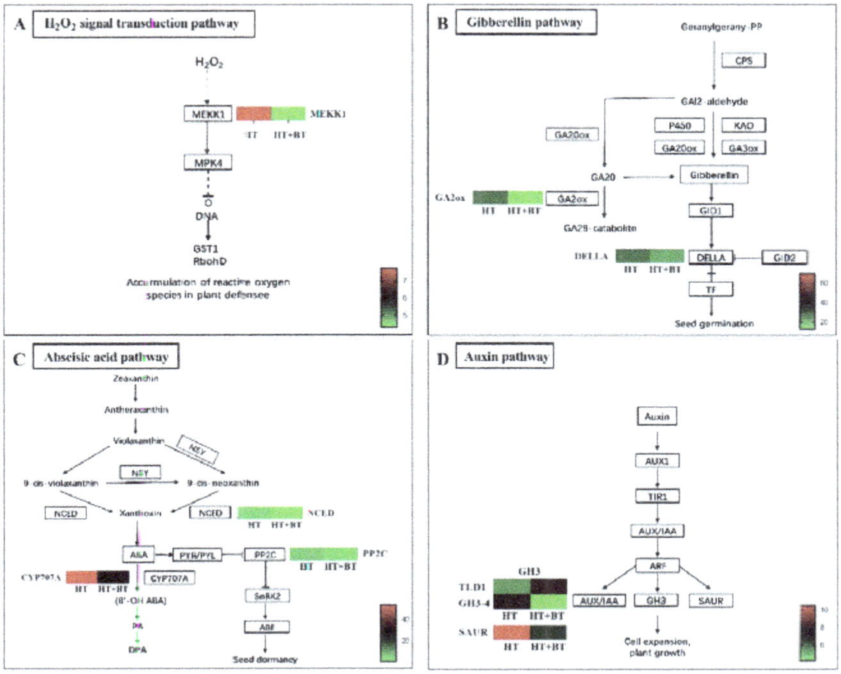

Figure 7. Expression profiles of DEGs in reactive oxygen species accumulation and three endogenous hormone-related pathways: (**A**) expression profiles of DEGs related to the H_2O_2 signal transduction pathway under HT and HT + BT treatments; (**B**) expression profiles of DEGs related to the GA pathway under HT and HT + BT treatments; (**C**) expression profiles of DEGs related to the ABA pathway under HT and HT + BT treatments; (**D**) expression profiles of DEGs related to the IAA pathway under HT and HT + BT treatments. The sample names are shown at the bottom of the figure. We indicate changes in expression levels by changes in color: green indicates a lower expression level, whereas red indicates a higher expression level. All data shown reflect the average mean of three biological replicates (n = 3). Means with different letters in each treatment represent a significant difference of $p \leq 0.05$.

3.9. *Differential Genes Related to Hormone Signaling Pathway*

To further explore the action pathway between betaine and plant hormones, we analyzed the differentially expressed genes related to plant hormones. Among the differential genes related to GAs, one is expressed in the GA synthesis pathway, and two are expressed in the GA signal transduction pathway (Figure 7B).

According to the gene expression heat map, betaine infiltration under heat stress significantly downregulated the expression level of gibberellin (*Gibberellin 2-beta-dioxygenase* (*OsGA2ox3*)) [48] and inhibited GA inactivation. We verified this by the significantly higher GA content in the betaine treatment compared with the heat-stress treatment. The betaine treatment decreased the expression level of *DELLA* [49] in the GA signal pathway, and it enhanced the output of the GA signal. In the ABA synthesis pathway, betaine induced the decrease in the expression level of *9-cis-epoxycarotenoid dioxygenase* (*NCED*) [50], the rate-limiting enzyme of ABA synthesis, and inhibited the synthesis of ABA (Figure 7C). However, at the same time, betaine also induced the downregulation of the *(+)—abscisic acid 8′–hydroxylase* (*CYP707A*) gene [51] in the ABA metabolic pathway, slowing down the metabolic rate of ABA

In addition, in the ABA signal transduction pathway, the expression levels of three genes of *protein phosphatase 2c* (*PP2C*) [52] were downregulated in the betaine treatment, and the ABA output signal was weakened. In the IAA signal transduction pathway, we observed the expressions of two IAA auxin early-response genes (*GH3*) [53], in which the expression level of *TLD1* [54] in the HT treatment was significantly lower than that in the HT+BT treatment, while *GH3–4* [53] was significantly higher than that in the HT+BT treatment (Figure 7D). Moreover, the expression level of *SAUR* [55] was significantly higher in the HT+BT treatment than in the HT treatment. The high expression of *TLD1* promotes IAA inactivation and reduces the level of IAA in seeds. These results are consistent with the assay results of the phytohormone levels.

3.10. Validation of DEGs by qRT-PCR

To verify the accuracy and reproducibility of the RNA-seq results, we randomly selected nine genes in the pathways related to seed germination for qRT-PCR validation (Figure 8). We calculated the expression levels of the selected genes using the $2^{-\Delta\Delta Ct}$ method. According to the qRT-PCR results, the expression levels of nine DEGs were consistent with the results of the RNA-seq, confirming the reproducibility of the data.

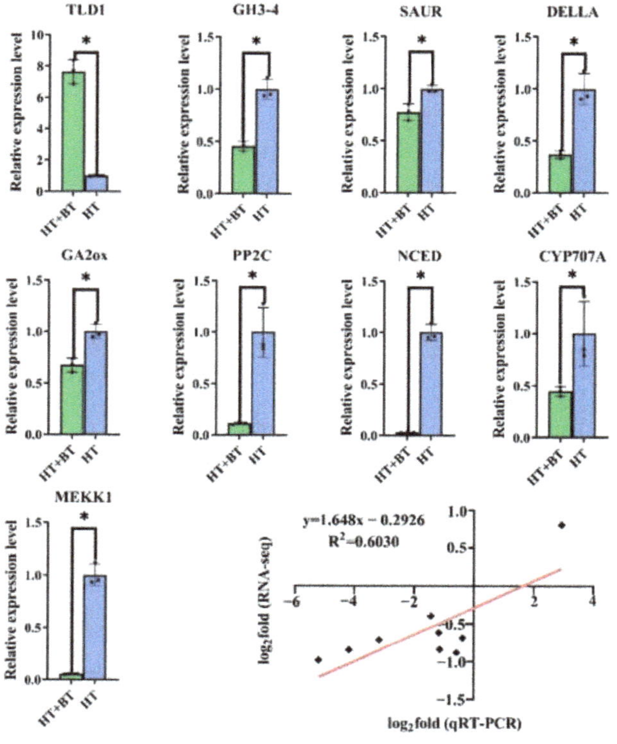

Figure 8. The expressions of 9 differentially expressed genes in the H_2O_2 pathway and 3 endogenous hormone pathways were randomly verified by qRT-PCR. The x-axis is the name of the treatment, in which HT is the 38 °C non-soaking treatment, and HT + BT is the 38 °C betaine seed-soaking treatment. The y-axis shows the relative expression of a specific gene relative to the reference gene actin. The last panel shows the results of the RNA-seq\qRT-PCR correlation analysis (lower right corner). The x-axis is the $\log_2$-fold-change value of the qRT-PCR, and the y-axis is the $\log_2$-fold-change value of the RNA-seq. According to the Pearson correlation (R^2 = 0.60; p < 0.05), there is a significant positive correlation between the multiple changes in expression. Each parameter * indicates significant statistical difference, * (p < 0.05).

4. Discussion

4.1. Seed Germination

Seed germination is an important stage in the life cycle of higher plants and is one of the periods with the greatest sensitivity to the external environment. Under the condition of sufficient water, seeds will break dormancy and start germination after sensing the appropriate ambient temperature. High temperatures above 35 °C will inhibit the germination of rice seeds. The germination performance of 304 lettuce materials cultured at 21 °C, 28 °C and 35 °C for 40 h were compared. The results showed that the seeds of all 304 materials germinated at 21 °C, with an average germination rate of 87.72%. The average germination rate decreased to 42.84% at 28 °C and 1.01% at 35 °C [56]. The germination percentage of *B. subalternans* weed seeds reached above 77% for a wide alternating temperature (15/20 °C to 30/35 °C night/day). The highest germination and uniformity occurred at 25/30 °C night/day. Only 11% of the seeds germinated at a temperature of 35/40 °C night/day [57].

According to our seed germination test, when the external ambient temperature was 35 °C or below, the germination of rice seeds was not greatly affected, and the germination rate of all the treated seeds could be maintained at more than 90%. In a previous study, a drying temperature of 35°C was safe for rice seeds with high initial moisture content, whereas higher drying temperatures (41 °C and 47 °C) remarkably reduced rice seed vigor. The metabolism of ROS, antioxidant enzymes, GA, ABA and α-amylase may be closely involved in the regulation of drying temperature on the seed vigor of rice seeds with high initial moisture content [58]. However, when the temperature reached 38 °C, the germination of most rice seeds was significantly inhibited, and there was almost no germination. Under heat stress, plants accumulate some soluble substances to adapt to the stress, and betaine is one of the quaternary amines that is effective at alleviating heat stress [23]. Because rice cannot accumulate betaine naturally, we investigated the effect of betaine on rice seed germination under heat stress by exogenous seed immersion. We found the best effect at a concentration of 10 mM, which demonstrated that the betaine immersion efficiently increased the rice seed germination under heat stress. We further investigated the regulatory mechanism of the betaine promotion of seed germination under heat stress by transcriptome analysis. Compared with the control treatment, the 10 mM betaine-soaked seeds significantly induced the differential expressions of 528 genes under heat stress, of which 332 genes were downregulated, and 196 genes were upregulated. The betaine regulated the seed germination mainly by downregulating related genes. According to the GO function analysis of the differential genes, the 10 mM betaine soaking treatment under heat stress mainly affected the cellular and metabolic processes in the biological processes; the cells, membranes, membrane parts and organelles in the cell components; binding in the molecular function; and the expressions of genes related to catalytic activity to promote seed germination.

4.2. Physiological Changes of Rice under Heat Stress

Heat stress will cause plants to produce a heat stress response, which disturbs the structural and metabolic integrity of cells to damage cell homeostasis and then causes irreversible damage to plants [59]. At the physiological level, heat stress leads to the excessive production of ROS in mitochondria and chloroplasts through disordered electron transmission, which induces lipid peroxidation to cause serious damage to the cell membrane and eventually causes cell death [60]. Therefore, enhancing the ability to eliminate excess ROS caused by heat stress is an effective defense measure to alleviate heat stress.

Soaking seeds in 10 mM of betaine considerably reduced the MDA content in rice seeds and the level of membrane lipid peroxidation under 38 °C heat stress. The activities of SOD, POD and CAT in the antioxidant enzyme system were significantly increased, which enhanced the ability to remove excess ROS. According to the results of this study's KEGG enrichment analysis, soaking rice seeds in betaine reduced the expression of *MEKK1* in the H_2O_2 signaling pathway and prevented the buildup of reactive oxygen species. On the one hand, a total of 10 mM of betaine can enhance the ability of scavenging reactive oxygen species by increasing the activity of antioxidant enzymes under heat stress. A previous study had confirmed that 10 mM exogenous betaine could effectively improve the germination potential and germination rate of *medicago sativa* seeds under drought stress, which were significantly increased by 46.8% and 30.5% compared with the control [61]. Some researchers also found that 10 mM exogenous betaine did not significantly promote the germination rate of *elymus nutans* seeds under 100 mM NaCl stress but could significantly increase the shoot length of seeds [62]. Previous research also confirmed that betaine, either applied exogenously or accumulated in vivo in *codA*-transgenic seeds, enhanced the expression of heat-shock genes and improved tolerance of high temperatures in tomato seeds during germination [22]. On the other hand, it inhibits the accumulation of reactive oxygen species by downregulating the expression of *MEKK1*. In addition, heat stress leads to a decline in the osmotic adjustment ability of crops.

Soluble protein is an important small molecular substance that participates in osmotic adjustment when crops are stressed and that participates in the relevant physiological repair after rehydration [63]. After soaking seeds with 10 mM of betaine, the content of soluble protein in the rice seeds substantially increased, and the osmotic adjustment ability was enhanced. Physiologically speaking, the 10 mM betaine treatment increased the activity of antioxidant enzymes and decreased the level of MDA, improving the removal of excess reactive oxygen species, thereby reducing the degree of membrane lipid peroxidation, and increased the content of soluble protein, enhancing osmotic regulation and promoting rice seed germination under heat stress.

4.3. Synthesis and Metabolism of IAA, GAs and ABA under Heat Stress

The germination of rice seeds is regulated by plant hormones. ABA is a typical plant defense hormone that can improve the stress resistance of plants by inducing resistance and that can also regulate seed dormancy and grain maturation [64]. GAs positively promote seed germination, and they regulate seed germination with ABA in an antagonistic manner. IAA was recently found to be the second hormone that positively regulates seed dormancy after ABA, and it also responds to the induction of heat stress. The hormone level in cells is regulated by the balance between biosynthesis and catabolism. According to the KEGG enrichment analysis, betaine soaking induced the differential expressions of some genes in carotenoid and diterpene biosynthesis and plant hormone signal transduction pathways under heat stress, which were related to ABA synthesis and metabolism, GA metabolism and IAA metabolism. *NCED* is a key rate-limiting enzyme in the ABA synthesis pathway. *NCED* catalyzes the cleavage of C40 9–cis–epoxy carotenoids, 90 CIS neoflavin and 9–cis–violet, which are considered to be the main regulatory steps of ABA biosynthesis [50]. Under heat stress, the ABA levels in seeds soaked with 10 mM betaine were significantly lower than those in the control. According to the transcriptome results, the betaine treatment significantly reduced the expression level of the *NCED* gene and inhibited the synthesis of ABA. ABA is metabolically inactivated by hydroxylation or coupling. The hydroxylation of 8′–methyl is the main step in the catabolic pathway of ABA, which is catalyzed by *CYP707A*, a member of cytochrome *P450* monooxygenase [51]. The betaine treatment not only downregulated the expression level of the *NCED* gene, but also inhibited the expression of *CYP707A*. The inhibition of *NCED* gene expression reduces the ABA level and then promotes seed germination. At the same time, a decrease in the *CYP707A* gene expression level means that the ABA metabolism has slowed down, which ensures that rice seeds still maintain a certain ABA level under heat stress to trigger the defense mechanism that requires ABA stimulation to reduce heat-stress damage. Betaine substantially increased the GA levels under heat stress. The *OsGA2ox3* gene belongs to a small polygenic family encoding GAs 2–oxidase, which is a major catabolic enzyme in plants and can convert active GAs into inactive GAs [65]. The betaine treatment substantially decreased the expression level of the *OsGA2ox3* gene and inhibited the inactivation of GAs. We did not find the expressions of differential genes in the synthesis pathway of GAs, which indicates that betaine mainly increases the level of GAs by inhibiting the inactivation of GAs, rather than by promoting their biosynthesis.

In response to the induction of heat stress, the IAA levels increased, while the betaine treatment substantially reduced them. In general, the IAA levels in plant cells are regulated in three different ways: dynamic balance, polar transport and auxin response [66]. In this experiment, there were no differential gene expressions in the biosynthesis and polar transport pathways of IAA, but we found that *GH3* family genes were differentially expressed in the IAA signal transduction pathway. *GH3–4* is very sensitive to changes in the IAA level. After the increase in the IAA level, the expression level of *GH3–4* is one of the most substantial [53]. *TLD1* encodes an IAA amination synthase. In the *TLD1–D* mutant, the expression of *TLD1* was significantly increased, which promoted the chelation of IAA with amino acids and reduced the content of free IAA [54]. The decrease in the IAA level in the betaine treatment under heat stress may be related to the high expression of *TLD1*, which suggests that betaine may regulate the level of IAA in plant cells through the auxin response. In terms of the regulation of seed germination by plant hormones, betaine reduces the level of ABA and increases the level of GAs by inhibiting the synthesis of ABA and the inactivation of GAs under heat stress, promotes the inactivation of IAA to reduce the level of IAA, and finally, promotes the germination of rice seeds.

4.4. Expression and Crosstalk of IAA, GA and ABA Signals

Plant hormones transmit endogenous and environmental signals through specific signal pathways to trigger output responses [67]. According to the KEGG enrichment analysis, the betaine induced the expressions of differential genes in the IAA, GA and ABA signal transduction pathways under heat stress. The expressions of these differential genes affect hormone signal transduction. In the IAA signal transduction pathway, the *GH3* expression was significantly upregulated in the betaine treatment, while the *SAUR* expression was significantly downregulated. The decrease in the IAA level is related to the upregulation of *TLD1* expression in the *GH3* gene family. *SAUR*, as an auxin early-response gene, participates in the regulation of cell elongation and growth. The overexpression of *SAUR* can promote the elongation of *Arabidopsis* cells [55]. *SAUR* promotes cell elongation and is related to *PP2C* in the ABA signal transduction pathway. *SAUR* inhibits the activity of *PP2C* and leads to the elevated activation of AHA2 phosphorylation, thereby derepressing PM H^+–ATPases (e.g., AHA2) to acidify the cell wall and ultimately promote elongation growth [52]. In this study, we found that the expression level of *SAUR* was downregulated in response to decreased growth hormone levels; however, at the same time, the expression level of *PP2C* was also downregulated by decreased ABA levels. The promotion of seed germination by betaine under heat stress indirectly indicates the promotion of elongated cell growth, which may be mainly caused by the downregulation of the *PP2C* expression levels.

Under heat stress, the expression level of *DELLA* in the GA signal transduction pathway was downregulated in the betaine treatment. DELLA protein is an inhibitor of GA signal transduction. In the presence of GAs, the GAs bind to the *GA–INSENSITIVE DWARF 1* (*GID1*) receptor and cause conformational changes that trigger *DELLA* degradation or that lead to its inactivation, ultimately allowing the output of the GA signaling pathway [49]. Notably, ABA can inhibit the expression of GA signaling by increasing the stability of *DELLA*, thereby inhibiting the growth of *Arabidopsis* roots [68]. In addition, the GA stimulation of *Arabidopsis* root elongation requires the involvement of growth hormone. GA-induced root elongation was inhibited when the stem tip, which is the main source of growth hormone, was removed; however, this inhibition was reversed when growth hormone was reapplied. The participation of auxin in GA-induced root elongation is completed by promoting the degradation and inactivation of DELLA protein, which is a prerequisite for GAs to stimulate root elongation [69]. This suggests that betaine can promote the degradation of DELLA protein under heat stress and enhance the output of the GA signal through three aspects: first, by increasing the level of GAs, promoting the combination of GAs and *GID1* to degrade or inactivate DELLA protein; second, by inhibiting the promoting effect of ABA on the stability of DELLA protein; third, through the DELLA protein degradation pathway participated in by IAA. According to our results, the crosstalk of plant hormones also occurs in the seed germination stage under stress, and it can coordinate the balance between signals through exogenous substances to adapt to changes in the external environment. This provides a feasible solution to alleviate the plant-growth crisis caused by global warming. The regulatory pathways of 10 mM exogenous betaine enhance the protrusion vigor of rice seeds under heat stress, as shown in Figure 9. In the future, we will explore the molecular regulatory mechanisms of betaine with respect to rice seed heat tolerance.

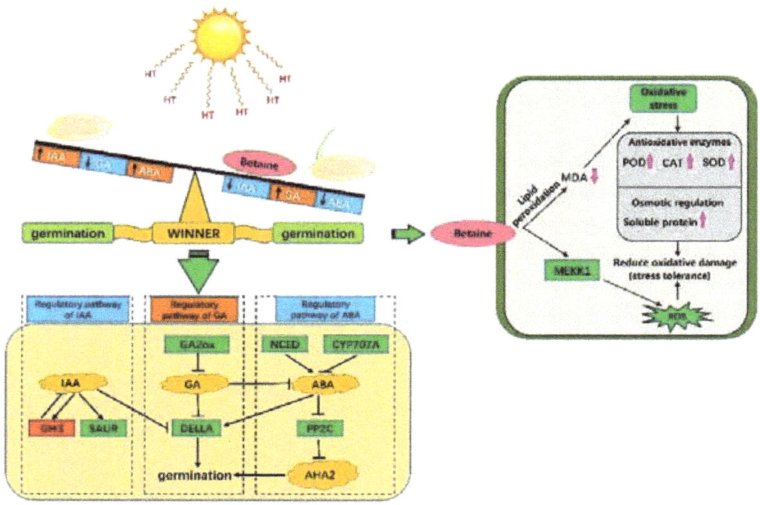

Figure 9. Regulatory pathways of 10 mM exogenous betaine enhance the protrusion vigor of rice seeds under heat stress.

5. Conclusions

In conclusion, according to our study, under 38 °C heat stress, the 10 mM betaine soaking treatment could alleviate the heat stress and promote the germination of rice seeds. In terms of physiology, the 10 mM betaine seed-soaking treatment promoted rice seed germination by increasing the SOD, POD and CAT antioxidant enzyme activities, decreasing the MDA content, increasing the soluble protein content to reduce ROS accumulation and mitigate membrane lipid peroxidation, and enhancing the osmoregulatory capacity to alleviate heat stress. By comparing the transcriptomes of the seed coleoptile and mesocotyl elongation stage, we obtained many DEGs involved in seed germination. According to the differential expression analysis, the betaine downregulated key genes in the H_2O_2 signaling pathway to reduce the accumulation of reactive oxygen species to mitigate the oxidative damage caused by heat stress. In addition, the betaine treatment affected the expressions of key genes related to the synthesis, metabolism and signal transduction of three endogenous hormones (GAs, IAA and ABA), increased the level of GAs, decreased the levels of IAA and ABA, and enhanced the output of the GA signal in rice seeds through the signaling crosstalk among the three hormones, which ultimately promoted the germination of rice seeds. The qRT-PCR results and the detection of the endogenous hormone levels further validated the expression patterns of these key genes. Our study reveals the molecular mechanism of betaine in regulating rice seed germination under heat stress, which provides an effective and feasible way to ensure the normal production of rice under the global warming trend.

Author Contributions: G.C. and Y.W. conceived and supervised the work; X.M. and J.Q. conducted the experiments, analyzed the data and prepared the figures; P.L. and H.Z. assisted in the data analysis; G.C. and Y.W. drafted the manuscript, together with X.M., J.Q., P.L. and H.Z. All authors have read and agreed to the published version of the manuscript.

Funding: This work was funded by the National Natural Science Foundation of China (31971923; 31301650); the National Key R&D Program of China (2017YFD0301501); the Hunan Provincial Natural Science Foundation of China (2020JJ4360); the Key Scientific Research Project of Hunan Provincial Education Department of China (19A220); and the Innovation and Entrepreneurship Training Program for college students of Hunan Agricultural University (XCX2021038).

Institutional Review Board Statement: Not applicable.

Informed Consent Statement: Not applicable.

Data Availability Statement: All of the data is contained within the article.

Acknowledgments: The authors thank Kai Wang and Zilu Liu of Hunan Agricultural University for all their help during the experiment.

Conflicts of Interest: The authors declare no conflict of interest.

References

1. Khush, G. What it will take to Feed 5.0 Billion Rice consumers in 2030. *Plant Mol. Biol.* **2005**, *59*, 1–6. [CrossRef] [PubMed]
2. IPCC (Intergovernmental Panel on Climate Change). Climate Change: The Physical Science Basis. In *Contribution of Working Group I to the Fourth Assessment Report of the Intergovernmental Panel on Climate Change*; Solomon, S., Qin, D., Manning, M., Chen, Z., Marquis, M., Averyt, K.B., Tignor, M., Miller, H.L., Eds.; Cambridge University Press: New York, NY, USA, 2007; p. 996.
3. Peng, S.B.; Huang, J.L.; Sheehy, J.E.; Laza, R.C.; Visperas, R.M.; Zhong, X.H.; Centeno, G.S.; Khush, G.S.; Cassman, K.G. Rice yields decline with higher night temperature from global warming. *Proc. Natl. Acad. Sci. USA* **2004**, *101*, 9971–9975. [CrossRef] [PubMed]
4. Ceccarelli, S.; Grando, S.; Maatougui, M.; Michael, M.; Slash, M.; Haghparast, R.; Rahmanian, M.; Taheri, A.; Al-Yassin, A.; Benbelkacem, A.; et al. Plant breeding and climate changes. *J. Agric. Sci.* **2010**, *148*, 627–637. [CrossRef]
5. Feng, B.; Zhang, C.; Chen, T.; Zhang, X.; Tao, L.; Fu, G. Salicylic acid reverses pollen abortion of rice caused by heat stress. *BMC Plant Biol.* **2018**, *18*, 245. [CrossRef] [PubMed]
6. Wang, Y.; Wang, L.; Zhou, J.; Hu, S.; Chen, H.; Xiang, J.; Zhang, Y.; Zeng, Y.; Shi, Q.; Zhu, D.; et al. Research Progress on Heat Stress of Rice at Flowering Stage. *Rice Sci.* **2019**, *26*, 1–10. [CrossRef]
7. Coast, O.; Murdoch, A.J.; Ellis, R.H.; Hay, F.R.; Jagadish, K.S. Resilience of rice (*Oryza* spp.) pollen germination and tube growth to temperature stress. *Plant Cell Environ.* **2016**, *39*, 26–37. [CrossRef]
8. Fu, G.; Feng, B.; Zhang, C.; Yang, Y.; Yang, X.; Chen, T.; Zhao, X.; Zhang, X.; Jin, Q.; Tao, L. Heat Stress Is More Damaging to Superior Spikelets than Inferiors of Rice (*Oryza sativa* L.) due to Their Different Organ Temperatures. *Front. Plant Sci.* **2016**, *7*, 1637. [CrossRef]
9. Duan, Y.; Yang, J.C. Research advances in the effect of high temperature on rice and its mechanism. *Chinese J. Rice Sci.* **2012**, *26*, 393–400. [CrossRef]
10. Asada, K. Production and scavenging of reactive oxygen species in chloroplasts and their functions. *Plant Physiol.* **2006**, *141*, 391–396. [CrossRef]
11. Gill, S.S.; Anjum, N.A.; Hasanuzzaman, M.; Gill, R.; Trivedi, D.K.; Ahmad, I.; Pereira, E.; Tuteja, N. Gluta-thione and glutathione reductase: A boon in disguise for plant abiotic stress defense operations. *Plant Physiol. Biochem.* **2013**, *70*, 204–212. [CrossRef]
12. Tian, X.J.; Lu, H.X. Effects of heat stress on seed vigor and antioxidant enzyme activity of rape seedlings. *Jiangsu Agri-Cult. Sci.* **2011**, *4*, 2. [CrossRef]
13. Sheng, W.; Wang, Y.F.; Yu, X.; Chen, Z.J.; Lu, J.W.; Cao, Z.F.; Wang, Q. Effects of priming treatment on the characteristics of germination and physiological and biochemical of lettuce seed under high temperature stress. *Seed* **2016**, *35*, 4.
14. Razem, F.A.; Baron, K.; Hill, R.D. Turning on gibberellin and abscisic acid signaling. *Curr. Opin. Plant Biol.* **2006**, *9*, 454–459. [CrossRef] [PubMed]
15. Zhou, F. Hormone signal interaction in seed germination. *North. Hortic.* **2017**, *6*, 4.
16. Iqbal, N.; Sehar, Z.; Fatma, M.; Umar, S.; Sofo, A.; Khan, N.A. Nitric Oxide and Abscisic Acid Mediate Heat Stress Tolerance through Regulation of Osmolytes and Antioxidants to Protect Photosynthesis and Growth in Wheat Plants. *Antioxidants* **2022**, *11*, 372. [CrossRef]
17. Li, G.; Zhang, C.; Zhang, G.; Fu, W.; Feng, B.; Chen, T.; Peng, S.; Tao, L.; Fu, G. Abscisic Acid Negatively Modulates Heat Tolerance in Rolled Leaf Rice by Increasing Leaf Temperature and Regulating Energy Homeostasis. *Rice* **2020**, *13*, 18. [CrossRef]
18. Shuai, H.W.; Meng, Y.J.; Luo, X.F.; Chen, F.; Qi, Y.; Yang, W.Y.; Shu, K. The roles of auxin in seed dormancy and germination. *Hereditas* **2016**, *38*, 314–322. [CrossRef]

19. Sharma, E.; Borah, P.; Kaur, A.; Bhatnagar, A.; Mohapatra, T.; Kapoor, S.; Khurana, J.P. A comprehensive tran-scriptome analysis of contrasting rice cultivars highlights the role of auxin and ABA responsive genes in heat stress re-sponse. *Genomics* **2021**, *113*, 1247–1261. [CrossRef]
20. Gray, W.M.; Östin, A.; Sandberg, G.; Romano, C.P.; Estelle, M. High temperature promotes auxin-mediated hypocotyl elongation in *Arabidopsis*. *Proc. Natl. Acad. Sci. USA* **1998**, *95*, 7197–7202. [CrossRef]
21. Li, Q.L.; Yang, H.; Gao, X.R.; Liu, D.W.; An, L.J. Molecular biology and genetic engineering of betaine synthase in plants. *China Biotechnol.* **2002**, *22*, 84–86. [CrossRef]
22. Li, S.; Li, F.; Wang, J.; Zhang, W.; Meng, Q.; Chen, T.H.; Murata, N.; Yang, X. Glycinebetaine enhances the tolerance of tomato plants to high temperature during germination of seeds and growth of seedlings. *Plant Cell Environ.* **2011**, *34*, 1931–1943. [CrossRef] [PubMed]
23. Hayashi, H.; Sakamoto, A.; Murata, N. Enhancement of the tolerance of *Arabidopsis* to high temperatures by genetic engineering of the synthesis of glycinebetaine. *Plant J.* **1998**, *16*, 155161. [CrossRef] [PubMed]
24. Sorwong, A.; Sakhonwasee, S. Foliar Application of Glycine Betaine Mitigates the Effect of Heat Stress in Three Marigold (*Tagetes erecta*) Cultivars. *J. Jpn. Soc. Hortic. Sci.* **2015**, *84*, 161–171. [CrossRef]
25. Lu, J.; Xing, X.J.; Zhu, L.Q.; Wang, Y.; Yan, H.; Yuan, J.J. Effects of exogenous glycine betaine and $CaCl_2$ on physiological response of tobacco plants under stresses of heat and drought. *J. Plant Nutr. Fertil.* **2011**, *17*, 1437–1443. [CrossRef]
26. Chen, J.; Ren, B.; Zhao, B.; Liu, P.; Zhang, J. Regulation of leaf-spraying glycine betaine on yield formation and antioxidation of summer maize sowed in different dates. *Acta Agron. Sin.* **2022**, *48*, 1502–1515. [CrossRef]
27. Ishitani, M.; Arakawa, K.; Mizuno, K.; Kishitani, S.; Takabe, T. Betaine Aldehyde Dehydrogenase in the Gramineae: Levels in Leaves Both Betaine-Accumulating and Nonaccumulating Cereal Plants. *Plant Cell Physiol.* **1992**, *34*, 493–495. [CrossRef]
28. Kishitani, S.; Takanami, T.; Suzuki, M.; Oikawa, M.; Yokoi, S.; Ishitani, M.; Alvarez-Nakase, A.M. Compatibility of gly-cinebetaine in rice plants: Evaluation using transgenic rice plants with a gene for peroxisomal betaine aldehyde dehy-drogenase from barley. *Plant Cell Environ.* **2000**, *23*, 107–114. [CrossRef]
29. Zulfiqar, F.; Ashraf, M.; Siddique, K.H.M. Role of Glycine Betaine in the Thermotolerance of Plants. *Agronomy* **2022**, *12*, 276. [CrossRef]
30. Hafez, E.M.; Gowayed, S.M.; Nehela, Y.; Sakran, R.M.; Rady, A.M.S.; Awadalla, A.; Omara, A.E.; Alowaiesh, B.F. Incorporated Biochar-Based Soil Amendment and Exogenous Glycine Betaine Foliar Application Ameliorate Rice (*Oryza sativa* L.) Tolerance and Resilience to Osmotic Stress. *Plants* **2021**, *10*, 1930. [CrossRef]
31. Tisarum, R.; Theerawitaya, C.; Samphumphung, T.; Takabe, T.; Cha-um, S. Exogenous Foliar Application of Glycine Betaine to Alleviate Water Deficit Tolerance in Two Indica Rice Genotypes under Greenhouse Conditions. *Agronomy.* **2019**, *9*, 138. [CrossRef]
32. Yu, Y.; Deng, L.; Zhou, L.; Chen, G.; Wang, Y. Exogenous Melatonin Activates Antioxidant Systems to Increase the Ability of Rice Seeds to Germinate under High Temperature Conditions. *Plants* **2022**, *11*, 886. [CrossRef] [PubMed]
33. Dhindsa, R.S.; Pamela, P.D.; Thorpe, T.A. Leaf Senescence: Correlated with Increased Levels of Membrane Permeability and Lipid Peroxidation, and Decreased Levels of Superoxide Dismutase and Catalase. *J. Exp. Bot.* **1981**, *1*, 93–101. [CrossRef]
34. Kar, R.K.; Choudhuri, M.A. Possible mechanisms of light-induced chlorophyll degradation in senescing leaves of Hydrilla verticillata. *Physiol. Plant* **1987**, *70*, 729–734. [CrossRef]
35. Aebi, H. Catalase in vitro. *Methods Enzymol.* **1984**, *105*, 121. [CrossRef]
36. Li, R.L.; Xie, Y.D.; Tang, Y. Effect of Application Accumulator Plant Straw on the Osmotic Adjustment Substances and Malondi-aldehyde Content of Lettuce. *IOP Conf. Ser. Earth Environ. Sci.* **2019**, *233*, 042025. [CrossRef]
37. Zou, Q. *Plant Physiology Experiment Guide*; China Agriculture Press: Beijing, China, 2000; pp. 125–130. (In Chinese)
38. Kanehisa, M.; Araki, M.; Goto, S.; Hattori, M.; Hirakawa, M.; Itoh, M.; Katayama, T.; Kawashima, S.; Okuda, S.; Tokimatsu, T.; et al. KEGG for linking genomes to life and the environment. *Nucleic Acids Res.* **2008**, *36*, 480–484. [CrossRef]
39. Zhang, H.; Wang, Z. *Seed Science*, 3rd ed.; Science Press: Beijing, China, 2021; pp. 172–173.
40. Wang, Z.; Fang, B.; Chen, J.; Zhang, X.; Luo, Z.; Huang, L.; Chen, X.; Li, Y. De novo assembly and characterization of root transcriptome using illumina paired-end sequencing and development of cSSR Markers in sweetpotato (*Ipomoea Batatas*). *BMC Genom.* **2010**, *11*, 726. [CrossRef]
41. Gene Ontology Consortium. The gene ontology (GO) database and informatics resource. *Nucleic Acids Res.* **2004**, *32*, 258–261. [CrossRef]
42. Liu, F.; Yang, W.; Sun, Q. Transcriptome sequencing data analysis and high-throughput GO annotation. *J. Anhui Agric. Sci.* **2018**, *46*, 88–91. [CrossRef]
43. Liu, Y.; He, C. A review of redox signaling and the control of MAP kinase pathway in plants. *Redox Biol.* **2017**, *11*, 192–204. [CrossRef]
44. Pitzschke, A.; Hirt, H. Disentangling the complexity of mitogen-activated protein kinases and reactive oxygen species signaling. *Plant Physiol.* **2009**, *149*, 606–615. [CrossRef] [PubMed]
45. Asai, S.; Ohta, K.; Yoshioka, H. MAPK signaling regulates nitric oxide and NADPH oxidase-dependent oxidative bursts in Nicotiana benthamiana. *Plant Cell* **2008**, *20*, 1390–1406. [CrossRef] [PubMed]
46. Takahashi, F.; Mizoguchi, T.; Yoshida, R.; Ichimura, K.; Shinozaki, K. Calmodulin-dependent activation of MAP kinase for ROS homeostasis in *Arabidopsis*. *Mol. Cell* **2011**, *4*, 649–660. [CrossRef]

47. Xing, Y.; Jia, W.; Zhang, J. *AtMKK1* mediates ABA-induced *CAT1* expression and H_2O_2 production via AtMPK6-coupled signaling in *Arabidopsis*. *Plant J.* **2008**, *54*, 440–451. [CrossRef] [PubMed]
48. Lo, S.; Yang, S.; Chen, K.; Hsing, Y.; Zeevaart, J.; Chen, L.; Yu, S. A novel class of gibberellin 2-oxidases control semi-dwarfism, tillering, and root development in rice. *Plant Cell* **2008**, *20*, 2603–2618. [CrossRef] [PubMed]
49. Ito, T.; Okada, K.; Fukazawa, J.; Takahashi, Y. DELLA-dependent and -independent gibberellin signaling. *Plant Signal Behav.* **2018**, *13*, e1445933. [CrossRef] [PubMed]
50. Chen, K.; Li, G.J.; Bressan, R.A.; Song, C.P.; Zhu, J.K.; Zhao, Y. Abscisic acid dynamics, signaling, and functions in plants. *J. Integr. Plant Biol.* **2020**, *62*, 25–54. [CrossRef] [PubMed]
51. Nambara, E.; Marion-Poll, A. Abscisic acid biosynthesis and catabolism. *Annu. Rev. Plant Biol.* **2005**, *56*, 165–185. [CrossRef]
52. Ren, H.; Gray, W.M. SAUR Proteins as Effectors of Hormonal and Environmental Signals in Plant Growth. *Mol. Plant* **2015**, *8*, 1153–1164. [CrossRef]
53. Jain, M.; Kaur, N.; Tyagi, A.K.; Khurana, J.P. The auxin-responsive *GH3* gene family in rice (*Oryza sativa*). *Funct Integr Genom.* **2006**, *6*, 36–46. [CrossRef]
54. Zhang, S.W.; Li, C.H.; Cao, J.; Zhang, Y.C.; Zhang, S.Q.; Xia, Y.F.; Sun, D.Y.; Sun, Y. Altered architecture and enhanced drought tolerance in rice via the down-regulation of indole-3-acetic acid by *TLD1/OsGH3.13* activation. *Plant Physiol.* **2009**, *151*, 1889–1901. [CrossRef] [PubMed]
55. Zhu, Y.B.; Kong, Y.Y.; Wang, J.H. Research advances in auxin-responsive *SAUR* genes . *Chin. Bull. Life Sci.* **2014**, *26*, 407–413. [CrossRef]
56. Wei, S.; Yang, X.; Hou, G.; Ge, G.; Liu, H.; Luo, L.; Hu, J.; Huang, D.; Long, P. Distinct metabolome changes during seed germination of lettuce (*Lactuca sativa* L.) in response to thermal stress as revealed by untargeted metabolomics analysis. *Int. J. Mol. Sci.* **2020**, *21*, 1481. [CrossRef] [PubMed]
57. Pamplona, J.; Souza, M.; Sousa, D.; Mesquita, H.; Freitas, C.; Lins, H.; Torres, S.; Silv, D. Seed germination of Bidens subalternans DC. exposed to different environmental factors. *PLoS ONE* **2020**, *15*, e0233228. [CrossRef]
58. Huang, Y.; Wu, W.; Zhao, T.; Lu, M.; Wu, H.; Cao, D. Drying temperature regulates seed vigor of high moisture rice seeds via involvement in phytohormone, ROS and relevant gene expression. *J. Sci. Food Agric.* **2020**, *101*, 2143–2155. [CrossRef]
59. Wahid, A.; Gelani, S.; Ashraf, M.; Foolad, M. Heat tolerance in plants: An overview. *Environ. Exp. Bot.* **2007**, *61*, 199–223. [CrossRef]
60. Suzuki, N.; Katano, K. Coordination Between ROS Regulatory Systems and Other Pathways Under Heat Stress and Pathogen Attack. *Front. Plant Sci.* **2018**, *9*, 490. [CrossRef]
61. Wang, J.; He, X.; Zhou, J.; Guan, Z.; Zhang, F. Effects of exogenous glycine betaine on seed germination and seedling growth of alfalfa under drought stress. *North. Hortic.* **2022**, *15*, 51–57. [CrossRef]
62. Tian, Y.; Gao, T.; Wang, X.; Luo, S.; Miao, Y. Effects of exogenous glycine betaine on seed germination of elymus nutans griseb under NaCl stress. *J. Plateau Agric.* **2022**, *6*, 1–10. [CrossRef]
63. Li, T.B.; Yang, S.Q.; Ren, G.X.; Feng, Y.Z.; Zhang, Q.; Li, P. Comparison of physiological changes and cold resistance of different gramineous grasses under low temperature treatment. *Acta Ecol. Sin.* **2009**, *29*, 1341–1347. [CrossRef]
64. Yang, Y.Y.; Chen, X.; Chen, Q.Z.; Lu, F.; Xu, C.; Yang, H.T.; Su, P.P.; Liu, X.L. Priming effects of abscisic acid on high temperature stress tolerance in rice at seed germination stage. *Acta Agric. Boreali-Sin.* **2021**, *36*, 185–194. [CrossRef]
65. Teshome, S.; Kebede, M. Analysis of regulatory elements in *GA2ox, GA3ox* and *GA20ox* gene families in *Arabidopsis thaliana*: An important trait. *Biotechnol. Biotec. Eq.* **2021**, *35*, 1603–1612. [CrossRef]
66. Qin, G.; Gu, H.; Zhao, Y.; Ma, Z.; Shi, G.; Yang, Y.; Pichersky, E.; Chen, H.; Liu, M.; Chen, Z.; et al. An indole-3-acetic acid carboxyl methyltransferase regulates *Arabidopsis* leaf development. *Plant Cell* **2005**, *17*, 2693–2704. [CrossRef] [PubMed]
67. Gazzarrini, S.; Tsai, A.Y. Hormone cross-talk during seed germination. *Essays Biochem.* **2015**, *58*, 151–164. [CrossRef]
68. Achard, P.; Cheng, H.; De Grauwe, L.; Decat, J.; Schoutteten, H.; Moritz, T.; Van Der Straeten, D.; Peng, J.; Harberd, N.P. Integration of plant responses to environmentally activated phytohormonal signals. *Science* **2006**, *311*, 91–94. [CrossRef]
69. Fu, X.D.; Harberd, N.P. Auxin promotes *Arabidopsis* root growth by modulating gibberellin response. *Nature* **2003**, *421*, 740–743. [CrossRef]

antioxidants

MDPI

Article

Genome-Wide Identification and Characterization of Chinese Cabbage S1fa Transcription Factors and Their Roles in Response to Salt Stress

Ali Anwar †, Shu Zhang †, Li-Xia Wang, Fengde Wang, Lilong He * and Jianwei Gao *

Shandong Branch of National Vegetable Improvement Center, Institute of Vegetables, Shandong Academy of Agricultural Sciences, Jinan 250100, China
* Correspondence: hllong1984@163.com (L.H.); scsgaojianwei@shandong.cn (J.G.)
† These authors contributed equally to this work.

Abstract: The S1fa transcription factor is part of a small family involved in plant growth and development and abiotic stress tolerance. However, the roles of the S1fa genes in abiotic stress tolerance in Chinese cabbage are still unclear. In this study, four S1fa genes in the Chinese cabbage genome were identified and characterized for abiotic stress tolerance. Tissue-specific expression analysis suggested that three of these four S1fa genes were expressed in all tissues of Chinese cabbage, while *Bra006994* was only expressed in the silique. Under Hg and Cd stresses, the S1fa genes were significantly expressed but were downregulated under NaCl stresses. The *Bra034084* and *Bra029784* overexpressing yeast cells exhibited high sensitivity to NaCl stresses, which led to slower growth compared with the wild type yeast cells (EV) under 1 M NaCl stress. In addition, the growth curve of the *Bra034084* and *Bra029784* overexpressing cells shows that the optical density was reduced significantly under salt stresses. The activities of the antioxidant enzymes, SOD, POD and CAT, were decreased, and the MDA, H_2O_2 and O_2^- contents were increased under salt stresses. The expression levels of cell wall biosynthesis genes *Ccw14p*, *Cha1p*, *Cwp2p*, *Sed1p*, *Rlm1p*, *Rom2p*, *Mkk1p*, *Hsp12p*, *Mkk2p*, *Sdp1p* and *YLR194c* were significantly enhanced, while *Bck1p*, and *Ptc1p* were downregulated under salt stresses. These results suggest that the *Bra034084* and *Bra029784* genes regulate cell wall biosynthesis and the defense regulatory system under salt stresses. These findings provide a fundamental basis for the further investigation of crop genetic modification to improve crop production and abiotic stress tolerance in Chinese cabbage.

Keywords: S1fa transcription factor; cell wall; yeast; antioxidant enzyme; ROS; salt stress

Citation: Anwar, A.; Zhang, S.; Wang, L.-X.; Wang, F.; He, L.; Gao, J. Genome-Wide Identification and Characterization of Chinese Cabbage S1fa Transcription Factors and Their Roles in Response to Salt Stress. *Antioxidants* 2022, *11*, 1782. https://doi.org/10.3390/antiox11091782

Academic Editor: Nafees A. Khan

Received: 5 July 2022
Accepted: 2 September 2022
Published: 9 September 2022

1. Introduction

Chinese cabbage (*Brassica rapa*) is a winter vegetable crop that originated in China and is mainly cultivated in north China [1,2]. It is the largest vegetable crop that is produced throughout the year, and because of its excellent nutritional value, it is highly consumed [3]. As a leafy vegetable, Chinese cabbage is more sensitive to environmental influences, including NaCl, heavy metals, low and high temperature, etc., which cause a series of physiological, molecular and biochemical changes that negatively affect plant growth and production [4,5]. To cope with different environmental influences, molecular, cellular and biochemical responses are regulated through many series of pathways, inducing antioxidant enzymes, hormones and transcription factors (TFs), to reduce the detrimental effects of stresses [6]. Under abiotic stresses, the plant produces ROS (reactive oxygen species), which is highly toxic and reactive, and hence causes oxidative damage and cell death [7]. Plant cells exhibit high efficiencies in scavenging ROS through the well-established coordination of antioxidant enzymes (SOD, POD, CAT, GS, APX and GR) and non-enzymatic antioxidants (ascorbic acid and glutathione) [8]. The overproduction of ROS initially causes damaged cells, hormonal imbalance and decreased metabolic and enzymatic activities, and thus interferes with signaling pathways that are involved in carbohydrate, protein, lipid, chlorophyll and photosynthetic machinery [6–8].

Plants have established a wide range of physiological and biochemical mechanisms to avoid the harmful effects of environmental influences [9,10]. The overproduction of ROS is diminished through the regulation of gene and protein expression levels in the mitochondria, nucleus, chloroplast and cell wall [11]. The cell wall is a complex structure, which is the first and central protective barrier for abiotic stress [11]. A number of signaling cascades have been reported to be involved in the regulation of stress responses in cell walls, including cell wall integrity (CWI), high-osmolality glycerol (HOG) and protein kinase A (PKA) [11]. *AtMyb41* regulates osmotic and salinity stresses through the activation of cell wall biosynthesis genes [12]. Rice *R2R3-type* MYB transcription factor, *OsMPS*, negatively regulates the expression of hormone signaling genes, EXPANSIN and biosynthesis genes in the cell wall to enhance NaCl stress tolerance [13]. Wall-associated kinases (WAKs), a potential sensor for the cell wall, regulate the pectic signaling network of the cell wall under abiotic stresses [14].

Transcription factors are crucial regulators of plant abiotic stresses, and are involved in the regulation of the defense system to stabilize ROS production [9,15,16]. Among different transcription factors, the S1fa transcription factor is highly conserved and belongs to the smallest family of the plant kingdom [17]. The members in this family have a small molecular weight (7 to 9 KD) and an average length of 70 to 80 aa, and are mainly localized in the nucleus [17]. Although usually no more than five S1fa proteins are found in most plants, such as maize, rice, tomato, soybean, *Arabidopsis* and Chinese cabbage, *Arachis duranensis* has 126 S1fa proteins [18]. Chinese cabbage has four members of the S1fa proteins, including *Bra003132*, *Bra034084*, *Bra006994* and *Bra029784*, which bind to the cis-element of the site 1 binding site, one of the three highly conserved binding sites (site 1, 2 and 3) located in the promoter region. Previous studies have reported that spinach S1fa has a nuclear localization signal peptide and a DNA recognition motif, which may function as a transcription factor [19]. The S1fa genes are mainly expressed in roots and etiolated seedlings rather than leaves, indicating their involvement in growth and development [17,18]. The S1fa gene plays a key role in abiotic stress tolerance. Under abiotic stresses, the S1fa genes are the most downregulated genes in cotton [17]. Moreover, in Chinese cabbage, the S1fa gene shows significant responses to salinity stress, suggesting that it may act as the upstream gene for salt responsive genes. Two genes of the S1fa family, *PtS1Fa1* and *PtS1Fa2*, have been characterized in *Populous trichocarpa* [17]. The results show that the *PtS1Fa2* overexpression lines of *P. trichocarpa* increase in fresh weight, root length and chlorophyll accumulation under drought stresses. However, the overexpression of *PtS1Fa1* has no obvious effect on the drought stress response. These findings suggest that *PtS1Fa2* plays a key role in the activation of antioxidant enzymes such as SOD and POD to reduce the MDA, H_2O_2 and O_2^- contents, and induces drought tolerance. The *OsS1fa* gene in rice confirms drought stress tolerance in *Arabidopsis* [18]. These results demonstrate that the *OsS1fa* gene is highly expressed in the leaf, culm and root. Drought tolerance-related genes, such as *LEA*, *GRF7*, *YODA*, *RD29A* and *CPK6*, are significantly expressed in the *OsS1fa* overexpression line under drought stresses, suggesting that *OsS1fa* plays a fundamental role in plant development and abiotic stress responses.

In this study, four members of the S1fa family were identified and characterized through the investigation of phylogeny, motif, gene structure, cis-element and miRNA in Chinese cabbage. Furthermore, the functions of the S1fa genes in response to abiotic stresses (Hg, Cd, Al, Co, Cu, mannitol (osmotic stress), salt and cold and heat stress) were investigated in yeast, which showed that two S1fa genes were highly sensitive to NaCl stress. The TPM value of the S1fa genes was measured in different plant tissues, such as the root, leaf, stem, flower, callus, silique and specific leaves of Chinese cabbage. The pRS416-GFP vector was used to test the subcellular localization of the S1fa genes under salinity stress in yeast. The significance of this study will be helpful for the understanding of the S1fa gene's function and boost the genetic modification of Chinese cabbage, which can improve crop production and adaptation to environmental cues. It will be more interesting to explore the role of S1fa genes in the hormone signaling pathway, interaction and cross-talk to identify novel genes in Chinese cabbage. Future research on S1fa will offer the possibility of genetic engineering of crop varieties with enhanced crop production.

2. Results

2.1. Identification and Characterization of the S1fa Family Genes in Chinese Cabbage

To identify and characterize the S1fa transcription factor genes in Chinese cabbage, we performed BLASTP searches against the Chinese cabbage genome database (http://brassicadb.cn (accessed on 22 March 2022)) using three *Arabidopsis* S1fa protein sequences (AT2G37120, AT3G53370 and AT3G09735) as query sequences, and confirmed four candidate genes of the S1fa family in Chinese cabbage, including *Bra003132*, *Bra034084*, *Bra006994* and *Bra029784*. The S1fa family genes are distributed on different chromosomes of Chinese cabbage; *Bra034084*, *Bra029784*, *Bra003132* and *Bra006994* are located on A01, A05, A07 and A09, respectively (Figure 1A). The protein 3D structure of S1fa gene showed similar structural homology in Chinese cabbage (Figure 1B). The length of the S1fa genes is between 70—88 aa, with a molecular weight ranging from 7.8 to 9.3 kDa (Table 1). The isoelectric point of Chinese cabbage S1fa proteins is relatively high (pI > 10.38), indicating that they are rich in alkaline amino acids (Table 1). Subcellular location analysis showed that all S1fa genes were localized in the nucleus.

Table 1. The characteristics of the S1fa genes in Chinese cabbage.

Gene ID	Protein Length (aa)	Molecular Weight (KD)	Chromosome Location	pI	Strand Direction	Subcellular Location
Bra003132	88	9.3	A07: 14791441..14792038	10.06	-	Nuclear
Bra034084	70	7.8	A01: 25080494..25080706	10.38	+	Nuclear
Bra006994	76	8.3	A09: 28232711..28233442	10.05	-	Nuclear
Bra029784	73	8.1	A05: 23013673..23014543	10.06	+	Nuclear

A

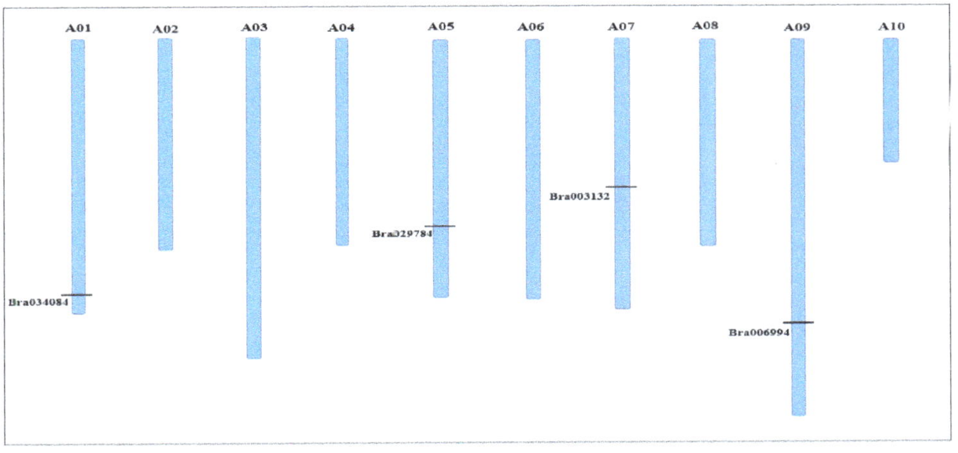

Figure 1. *Cont.*

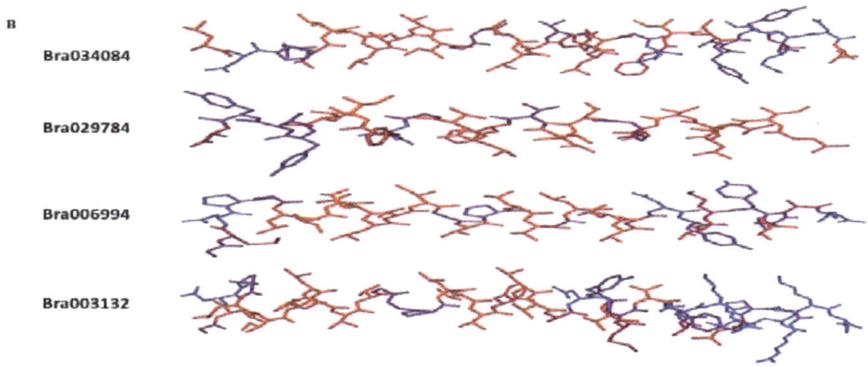

Figure 1. Chromosomal localization of the S1fa genes in Chinese cabbage and protein 3D structures. (**A**) Chinese cabbage has ten chromosomes, and the chromosome number (A01 to A10) is shown on the top of each chromosome. The position marked on the chromosome represents the location of the S1fa genes. (**B**) Protein 3D structure of Chinese cabbage S1fa genes.

2.2. Phylogenetic Analysis of the S1fa Genes in Chinese Cabbage

The phylogenetic analysis was used to investigate the evolution of the S1fa genes in Chinese cabbage. The S1fa proteins were compared with those in other species, including tomato, pepper, cotton, rice, *Arabidopsis*, cucumber, watermelon and rice, to investigate and explore the evolutionary relationships. A total of 27 S1fa proteins were clustered into three groups (I, II and III), which consisted of 6, 9 and 12 members, respectively. *Bra003132* was clustered into group I, while *Bra034084*, *Bra006994* and *Bra029784* were clustered into group II, indicating that Chinese cabbage S1fa genes have high homology with those in rice, cucumber and pepper (Figure 2A). Additionally, low bootstrap values in the phylogenetic tree are due to the divergence of the protein sequences that occur between Chinese cabbage and *Arabidopsis*, cotton and tomato during the evaluation. Multiple sequence alignments show that the amino acid sequences of the S1fa genes are highly conserved between Chinese cabbage and *Arabidopsis* (Figure 2B). The conserved domain of S1fa is highlighted in Figure 2B.

2.3. Cis-Element Analysis of S1fa

Cis-elements are the regions of non-coding DNA that regulate the transcription of the neighboring genes. The cis-elements of Chinese cabbage S1fa genes were identified in the promoter region as presented in Figure 3. The results show that the *Bra034084* gene is located on chromosome A07, which has 1 GATA-motif, 1 LTR, 1 TC-rich repeat, 1 TCA-element, 1 CGTCA-motif, 1 GT1-motif, 1 TGACG-motif, 2 AE-box, 3 AREs and 3 TCT-motifs. *Bra003132* has 1 LTR, 1 MBS, 1 AE-box, 1 CAT-box, 1 TCCC-motif, 1 TCT-motif, 1 LAMP-element, 2 CGTCA-motifs, 2 TGACG-motifs, 2 TGA-elements, 3 AREs, 3 MBSs, 4 ABREs and 6 G-boxes (Supplementary Table S1). *Bra006994* has 1 LTR, 1 TCA-element, 1 ARE, 1 Box II, 1 CAT-box, 1 TC-rich repeats, 2 TCT-motifs, 2 ABREs, 2 TCT-motifs and 3 O_2-sites. *Bra029784* has 1 TCA-element, 1 ARE, 1 AE-box, 1 G-box, 1 Box-II, 1 *chs*-Unit 1 m1, 2 TC-rich repeats, 2 MBSs, 2 ABREs and 2 TCT-motifs, which are involved in facilitating a plant's physiological and biochemical mechanisms under abiotic stresses.

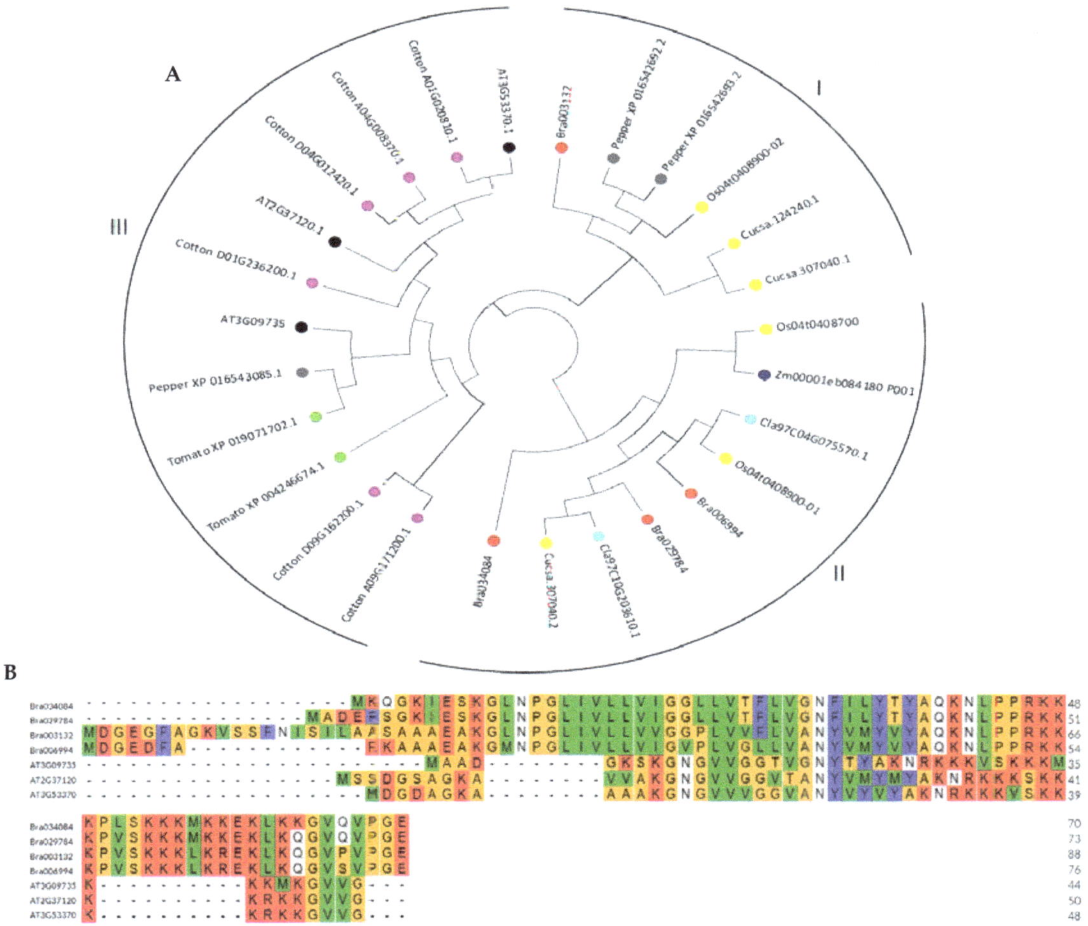

Figure 2. Phylogenetic analysis and multiple sequence alignment of the S1fa proteins. (**A**) Phylogenetic tree of the S1fa genes in Chinese cabbage, tomato, pepper, cotton, rice, *Arabidopsis*, cucumber, watermelon and rice. The neighbor-joining tree was generated using the MEGA7 software with 100 bootstrap replicates. The different colored dots represent different plant species. (**B**) Multiple sequence alignment of S1fa proteins from Chinese cabbage and *Arabidopsis*. Multiple sequence alignment was performed using the MEGA7 software. The highlighted amino acids are highly conserved.

2.4. Structure and Motif Analysis of the S1fa Genes

To explore the features of the S1fa genes, the conserved motifs of the genes in Chinese cabbage were analyzed. The results show that S1fa consists of three common motifs, namely, motif 1, 2 and 3, as presented in Figure 4. Motif 1 is the largest motif with a length of 55 aa, which is localized in the middle of the S1fa gene, followed by motif 2 and motif 3, respectively. Similarly, the exon–intron structures of the S1fa genes were analyzed (Figure 4). *Bra006994*, *Bra003132* and *Bra029784* have the same structure, while *Bra034084* has a different structure. The coding sequence of *Bra006994*, *Bra003132* and *Bra029784* is localized on the left and right borders of the UTR, and shares a similar gene structure, but *Bra034084* does not have a UTR.

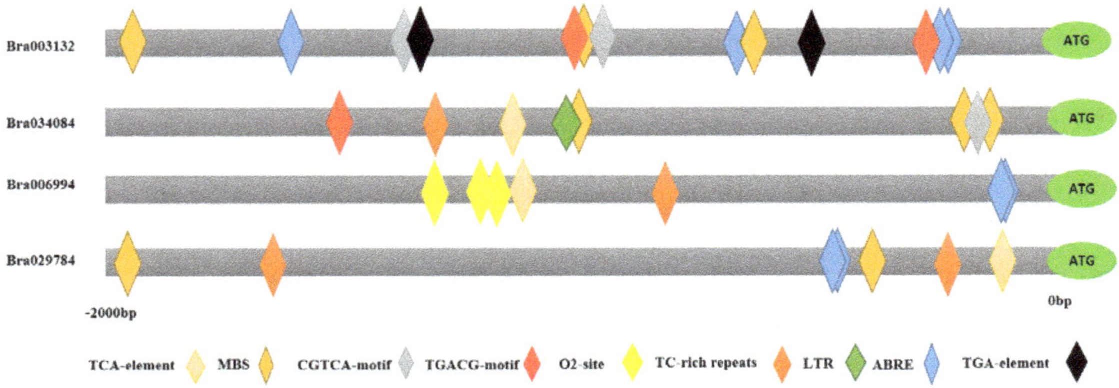

Figure 3. Cis-element analysis of the S1fa genes in Chinese cabbage. Different colors represent different cis-elements in the promoter region of the S1fa family genes in Chinese cabbage.

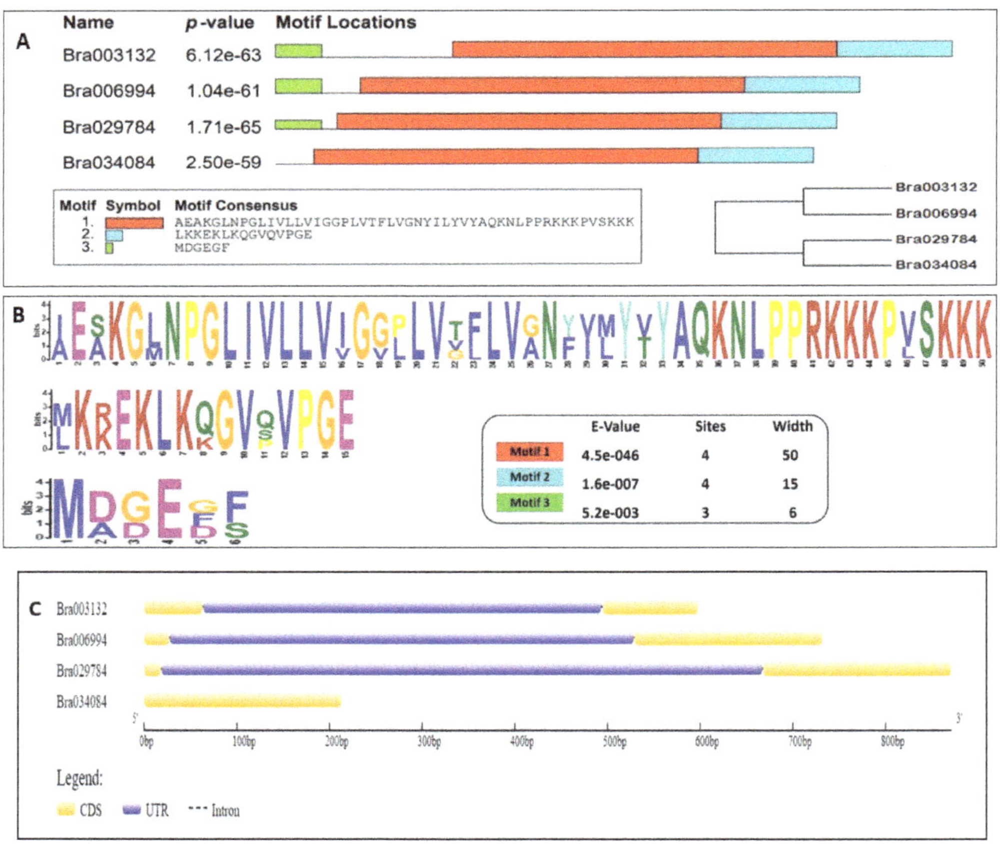

Figure 4. Analysis of motifs and gene structures of S1fa proteins in Chinese cabbage. (**A**) Protein motifs, location and phylogenetic trees in the S1fa family members. (**B**) The sequence of three identified motifs in Chinese cabbage. (**C**) Gene structure of the S1fa family member in Chinese cabbage.

2.5. Expression Profiles of the S1fa Genes in Different Tissues

To explore the potential functions of the S1fa genes in growth and development, the tissue-specific characteristics were obtained from the Chinese cabbage database (http://brassicadb.cn/#/ (accessed on 22 March 2022)). The results show that the TPM (Transcript per million) values of the S1fa genes varied in different plant tissues. The S1fa genes were highly expressed in the silique of Chinese cabbage (Figure 5), while the expression level was the lowest in the leaf tissue. Comparative analyses of the S1fa genes show that *Bra034084* had the highest expression, followed by *Bra003132* and *Bra006994*, respectively. *Bra029784* had the least expression in all tissues except the silique tissues compared with other members of the S1fa genes. Moreover, the expression of the S1fa genes was downregulated in the leaf and flower tissues, while *Bra006994* showed no expression in the leaf and flower tissues. Taken together, these findings indicate that the S1fa genes are actively expressed in Chinese cabbage, which could play vital roles in Chinese cabbage growth and developmental process. Thus, it is necessary to investigate the functions of the S1fa genes in abiotic stress tolerance.

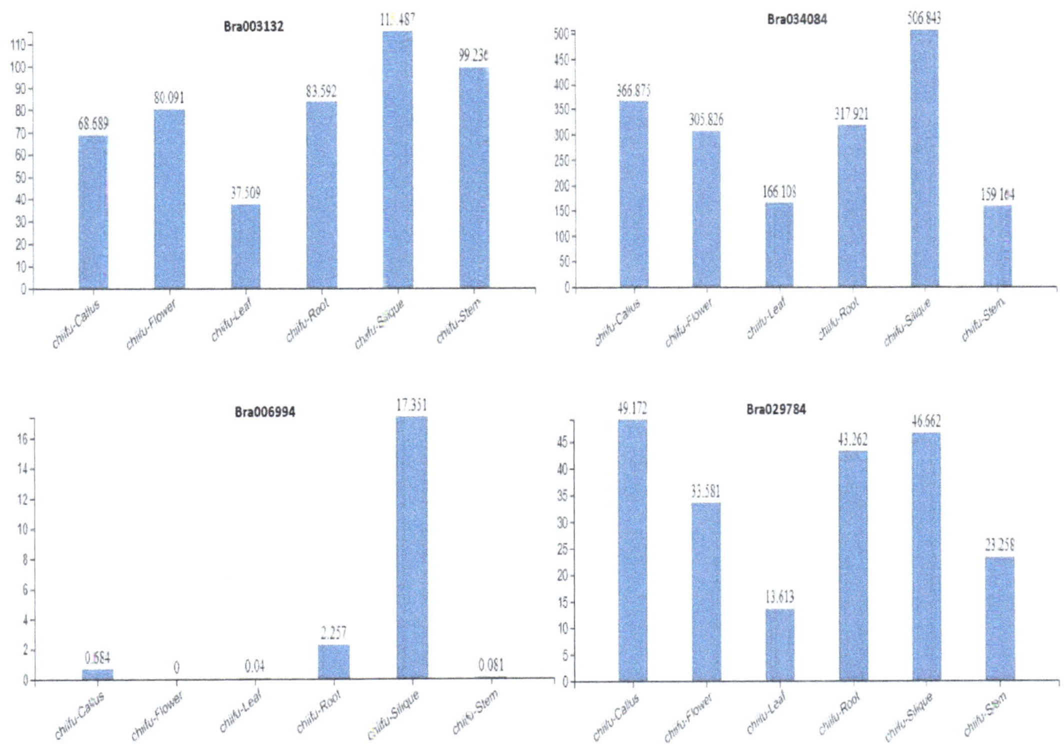

Figure 5. The transcription levels of the S1fa genes in different tissues, including root, leaves, flower, stem, callus and silique of Chinese cabbage. Data were obtained from Chinese cabbage database.

2.6. Expression Patterns of the S1fa Genes under Abiotic Stress

The S1fa transcription factor plays an important role in regulating plant growth and development, and abiotic stress tolerance. However, the involvement of the S1fa genes in response to abiotic stresses is not clear. To confirm the molecular mechanism of the S1fa genes in response to abiotic stresses, their transcript abundance was investigated under Hg, Cd and NaCl stresses (Figure 6). The expression levels of all four S1fa genes were investigated 24 h after the stress treatments. The S1fa genes were significantly expressed under abiotic stresses. Under Hg stress, the expressions of the *Bra034084* and *Bra029784* genes were significantly elevated compared with the control (CK) treatment, followed by *Bra003132*. Likewise, under Cd stress, the *Bra003132* and *Bra006994* genes showed high expression levels compared with CK, while NaCl stress significantly reduced the expression of the S1fa genes (Figure 6). These findings suggest that the S1fa genes are involved in and positively induced by various abiotic stresses in Chinese cabbage.

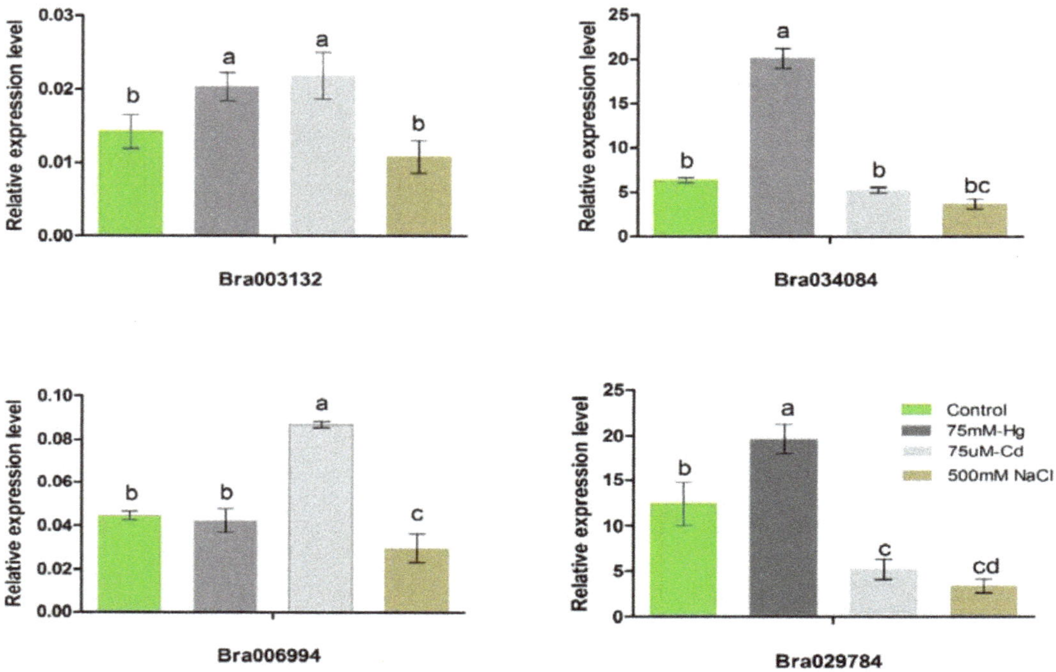

Figure 6. The expression levels of the S1fa genes under NaCl, Cd and Hg stresses. Different colors indicate different stresses and the letters above error bars represent significant differences at $p > 0.05$.

2.7. Prediction of miRNAs Targeting the S1fa Genes in Chinese Cabbage

miRNAs are a class of non-coding single strand RNA molecules of approximately 22 nucleotides, which are encoded by endogenous genes and may be involved in the activation of genes in response to abiotic stresses. A total of 39 miRNAs targeting the S1fa genes in Chinese cabbage are presented in Table 2, including 5 miRNAs targeting *Bra003132* (*ath-miR5661*, *ptc-miR397c*, *mtr-miR2641*, *hvu-miR6214* and *hme-miR-278*), 11 miRNAs targeting *Bra006994* (ptc-*miR397c*, *zma-miR399e-5p*, *aly-miR160c-3p*, *ath-miR5661*, *osa-miR160a-3p*, *osa-miR160b-3p*, *zma-miR160b-3p*, *zma-miR160g-3p*, *bdi-miR160b-3p*, *bdi-miR160c-3p* and *ata-miR160c-3p*), 11 miRNAs targeting *Bra029784* (*aly-miR838-3p*, *zma-miR399e-5p*, *ath-miR838*, *osa-miR3982-3p*, *bdi-miR398b*, *osa-miR2095-3p*, *aly-miR4248a*, *aly-miR4248b*, *aly-miR4248c*, *gma-miR4363* and *bdi-miR7757-3p.1*) and 12 miRNAs targeting *Bra034084* (*aly-miR838-3p*, *zma-miR399e-5p*, *ath-miR838*, *mtr-miR2673a*, *mtr-miR2673b*, *gma-miR4363*, *bdi-miR398b*, *osa-miR2055*, *osa-miR3982-3p*, *cca-miR6116-3p*, *stu-miR8050-3p* and *gma-miR9752*).

Table 2. Prediction of miRNAs targeting the S1fa genes of Chinese cabbage.

miRNA	Target S1fa	Target Site	miRNA Fragment	Alignment	Target Fragment	Inhibition
ath-miR5661	*Bra003132*	146–166	AGAGGUACAUCAUGUAGUCUG	::::.:::::::: : :	CCAACUACGUGAUGUACGUGU	Cleavage
ptc-miR397c	*Bra003132*	114–134	UCAAUGAGUGGAGCUUUGAUG	..::..:: :::.::..:::	UGUCGGAGGUCCGCUUGUUGU	Cleavage
mtr-miR2641	*Bra003132*	71–90	GUUUGAUCCUUUACGUUUAU	:.:: :::::::..:::	CUGAAGCCAAAGGAUUGAAC	Cleavage
hvu-miR6214	*Bra003132*	99–118	CGACGACGACGAGCACGACA	:::::: :.::.:::	AAUCGUGCUGCUUGUUGUCG	Translation
hme-miR-278	*Bra003132*	3–23	UCGGUGGGAUCUUCGUCCGUUU	..:.::: :::.:..:.:::.	GGAUGGA-GAAGGUUUCGCCGG	Cleavage
ptc-miR397c	*Bra006994*	78–98	UCAAUGAGUGGAGCUUUGAUG	.:::..::.:::::..:::.	UAUCGGAGUUCCACUUGUUGG	Cleavage
zma-miR399e-5p	*Bra006994*	148–168	GGGCUUCUCUUUCUUGGCAGG	::: : :.:::. :::::::	CCUCCAAGGAAGAAGAAGCCC	Cleavage
aly-miR160c-3p	*Bra006994*	91–110	GCGUACAAGGAGCCAAGCAUG	: ::.: :::::::: :::	CUUGUUGGGCUCCUUGU-CGC	Cleavage
ath-miR5661	*Bra006994*	110–130	AGAGGUACAUCAUGUAGUCUG	::::.:::::::: : :	CCAACUACGUGAUGUACGUGU	Cleavage
osa-miR160a-3p	*Bra006994*	91–110	GCGUGCAAGGAGCCAAGCAUG	: ::.: ::::::::. :::	CUUGUUGGGCUCCUUGU-CGC	Cleavage
osa-miR160b-3p	*Bra006994*	91–110	GCGUGCAAGGAGCCAAGCAUG	: ::.: ::::::::. :::	CUUGUUGGGCUCCUUGU-CGC	Cleavage
zma-miR160b-3p	*Bra006994*	91–110	GCGUGCAAGGAGCCAAGCAUG	: ::.: ::::::::. :::	CUUGUUGGGCUCCUUGU-CGC	Cleavage
zma-miR160g-3p	*Bra006994*	91–110	GCGUGCAAGGAGCCAAGCAUG	: ::.: ::::::::. :::	CUUGUUGGGCUCCUUGU-CGC	Cleavage
bdi-miR160b-3p	*Bra006994*	91–110	GCGUGCAAGGAGCCAAGCAUG	: ::.: ::::::::. :::	CUUGUUGGGCUCCUUGU-CGC	Cleavage
bdi-miR160c-3p	*Bra006994*	91–110	GCGUGCAAGGAGCCAAGCAUG	: ::.: ::::::::. :::	CUUGUUGGGCUCCUUGU-CGC	Cleavage
ata-miR160c-3p	*Bra006994*	91–110	GCGUGCAAGGAGCCAAGCAUG	: ::.: ::::::::. :::	CUUGUUGGGCUCCUUGU-CGC	Cleavage
aly-miR838-3p	*Bra029784*	161–181	UUUUCUUCUUCUUCUUGCACA	: : :::::::::: :::.:	UUUCCAAGAAGAAGAUGAAGA	Cleavage
zma-miR399e-5p	*Bra029784*	139–159	GGGCUUCUCUUUCUUGGCAGG	::: : :.:::. :::::::	CCUCCGAGGAAGAAGAAGCCC	Cleavage
ath-miR838	*Bra029784*	161–181	UUUUCUUCUACUUCUUGCACA	: : :::::: ::: :::.:	UUUCCAAGAAGAAGAUGAAGA	Translation
osa-miR3982-3p	*Bra029784*	86–106	AGUUGCCUACAUGGAGCGCCA	:: ::.::: ::::: :::	UGACGUUCCUUGUAGGAAACU	Cleavage
bdi-miR398b	*Bra029784*	75–96	CAGGAGUGUCACUGAGAACACA	: ::: .:::::..:::	AGGGUUGCUAGUGACGUUCCUU	Cleavage
osa-miR2095-3p	*Bra029784*	58–77	CUUCCAUUUAUGAUAAGUAU	.: ::: ::.:.: ::::.:	GUCCUUCUCGUGAUUGGAGG	Cleavage
aly-miR4248a	*Bra029784*	158–178	ACAUUUUAUUUUUGGCAAUCA	.:: :::.:: ::.:::	CCGUUUCCAAGAAGAAGAUGA	Cleavage
aly-miR4248b	*Bra029784*	158–178	ACAUUUUAUUUUUGGCAAUCA	.:: :::.:: ::.:::	CCGUUUCCAAGAAGAAGAUGA	Cleavage
aly-miR4248c	*Bra029784*	158–178	ACAUUUUAUUUUUGGCAAUCA	.:: :::.:: ::.:::	CCGUUUCCAAGAAGAAGAUGA	Cleavage
gma-miR4363	*Bra029784*	52–73	CGAUUACCAGAAGGCUUAUUAG	::.:: . ::::: ::.::.:	CUGAUCGUCCUUCUCGUGAUUG	Cleavage
bdi-miR7757-3p.1	*Bra029784*	192–212	GGUAGUUGAAUGUUUUGUUUA	::.:::..:.:::: : ::	GAAGCAAGGCGUUCAAGUUCC	Cleavage
aly-miR838-3p	*Bra034084*	152–172	UUUUCUUCUUCUUCUUGCACA	: : :::::::::: :::.:	UUUCCAAGAAGAAGAUGAAGA	Cleavage
zma-miR399e-5p	*Bra034084*	130–150	GGGCUUCUCUUUCUUGGCAGG	::: : :.:::. :::::::	CCUCCGAGGAAGAAGAAGCCC	Cleavage
ath-miR838	*Bra034084*	152–172	UUUUCUUCUACUUCUUGCACA	: : :::::: ::: :::.:	UUUCCAAGAAGAAGAUGAAGA	Translation
mtr-miR2673a	*Bra034084*	159–180	CCUCUUCCUCUUCCUCUUCCAC	::::: :::: :.::: :	GAAGAAGAUGAAGAAGGAGAAG	Cleavage
mtr-miR2673b	*Bra034084*	159–180	CCUCUUCCUCUUCCUCUUCCAC	::::: :::: :.::: :	GAAGAAGAUGAAGAAGGAGAAG	Cleavage
gma-miR4363	*Bra034084*	43–64	CGAUUACCAGAAGGCUUAUUAG	::.:: . ::::: ::.::::	CUGAUCGUCCUUCUUGUGAUCG	Cleavage
bdi-miR398b	*Bra034084*	66–87	CAGGAGUGUCACUGAGAACACA	: ::: .:::::..:::	AGGGUUGCUAGUGACGUUCCUU	Cleavage
osa-miR2055	*Bra034084*	144–163	UUUCCUUGGGAAGGUGGUUUC	:::.:: :::.:::: :.:	GAAGCC-CCUUUCCAAGAAGA	Cleavage
osa-miR3982-3p	*Bra034084*	77–97	AGUUGCCUACAUGGAGCGCCA	:: ::.::: :::.:: :::	UGACGUUCCUUGUGGGAAACU	Cleavage
cca-miR6116-3p	*Bra034084*	49–69	UCAUUUGAUCACAAGCAUGAG	:: : :::::::..: :.	GUCCUUCUUGUGAUCGGAGGG	Cleavage
stu-miR8050-3p	*Bra034084*	182–202	UGACUUGAGAUUCCUACUUGG	::: :::.: .::::::.	UGAAGAAGGGAGUUCAAGUUC	Translation
gma-miR9752	*Bra034084*	162–182	UGCUUCUUCUUUUCCCUGUUU	.:: : : :.:::.::::::	GAAGAUGAAGAAGGAGAAGCU	Cleavage

2.8. *S1fa Overexpression in Response to Abiotic Stresses in Yeast*

To elucidate the function of the S1fa genes in abiotic stress tolerance, we generated an overexpression model of yeast using the pRS416 vector. The yeast cells with S1fa overexpression were exposed to abiotic stresses (75 μM-Cd, 75 mM-Hg, 100 mM-Al, 50 mM-Cu, 100 mM-Co, 1M-NaCl, 2M-Mannitol, and cold and heat stresses (24 h stress followed by two days with the normal temperature)) (Figure 7(A,B)). The overexpression of *Bra034084*, *Bra003132*, *Bra029784* and *Bra006994* and EV (empty vector) showed no effects on Al, Mannitol, Co and cold stresses, However, the cells were sensitive to NaCl, Hg and Cd stresses, compared with EV (Figure 7(C,F,H)). Under NaCl stress, *Bra006994* and *Bra029784* were highly sensitive, when compared to EV (Figure 7(H)). Likewise, under heat stress (38 °C) and cold stress (4 °C), all cells with the overexpression of Chinese cabbage S1fa genes grew slightly slower than EV (Figure 7(A,B)). Cells overexpressed with *Bra006994* and *Bra029784* were more sensitive to these stresses (Figure 7(A,B)). These results indicate that the *Bra006994* and *Bra029784* genes play an important role in response to salt stresses.

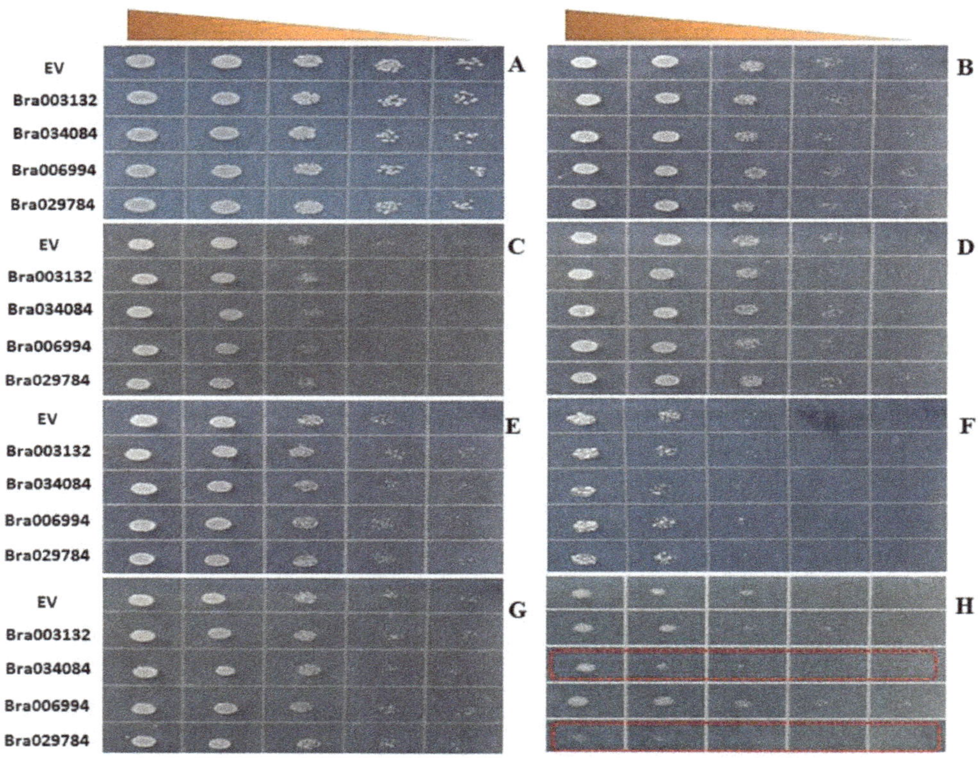

(a)

Figure 7. *Cont.*

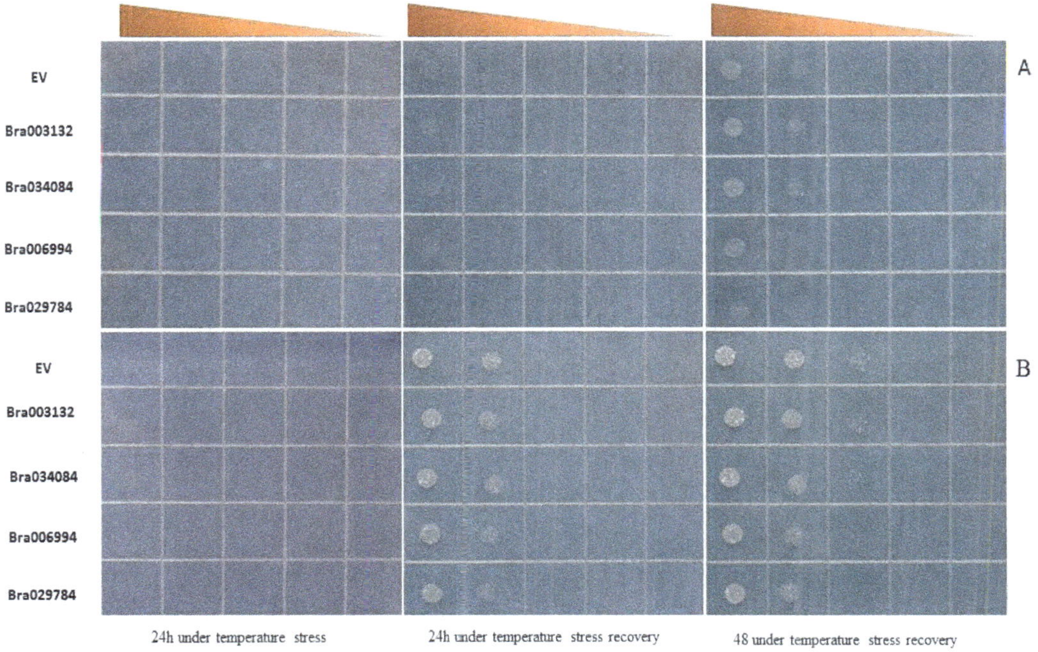

(b)

Figure 7. (**a**) Responses to abiotic stresses through yeast dilution bioassay with the wild type strain, S1fa transformed with pRS-416-GFP. Yeast wild type (EV) strain and S1fa overexpressing cells were grown in the URA liquid medium for 24 h at 28 °C. The cell solutions were diluted to an OD_{600} value of 0.3 and exposed to different types of abiotic stresses. (**A**) URA (control), (**B**) 50 mM Cu, (**C**) 75 mM Hg, (**E**) 100 mM Al, (**D**) 100 mM Co, (**F**) 75 µM Cd, (**G**) 2 M mannitol, (**H**) 1 M NaCl. Triangles represent the 10-fold serial dilutions (the starting OD_{600} is 0.3). (**b**) Responses to temperature stress tolerance through yeast dilution bioassay with the wild type strain, S1fa transformed with pRS-416-GFP. The yeast cells were exposed to (**A**) low temperature (4 °C) and (**B**) high temperature (38 °C) for 24 h, and then transferred to a place with the normal temperature (28 °C). Triangles represent the 10-fold serial dilutions (the starting OD_{600} is 0.3).

2.9. Growth Curve

To confirm the responses of the S1fa-overexpressing yeast cells to salinity, we conducted the growth curve of the yeast cells under salt stresses. The yeast cells were incubated under 28 °C until OD_{600} reached 0.3, and then were treated with 1 M NaCl in liquid URA medium. The OD_{600} values were observed after 12, 14, 16, 18, 20, 22 and 24 h, respectively. The results indicate that *Bra034084* and *Bra029784* overexpressing yeast cells were highly sensitive to NaCl stresses (Figure 8). Under normal conditions, the growth rate of the S1fa expressing cells was the same as that of EV, whereas under NaCl stresses, the optical density was significantly decreased compared with that of EV. The growth rate of the *Bra003132* and *Bra006994* overexpressing yeast cells showed no difference compared with that of EV, but was significantly higher than those of the *Bra034084* and *Bra029784* overexpressing yeast cells (Figure 8). These findings suggest that salinity stresses had detrimental effects on the growth of the *Bra034084* and *Bra029784* overexpressing yeast cells.

2.10. Subcellular Localization of the S1Fa Genes

Subcellular localization is considered a key parameter for transcription factor responses to abiotic stresses. However, no evidence was found to support the translocation of the S1fa genes into the nucleus under salt stresses.

To confirm the subcellular localization, the S1fa genes were transiently expressed in the fusion of GFP in yeast, and the fluorescence was observed using a confocal microscopy. Without salt stress, the S1fa genes were observed in the nucleus as dot-like structures as presented in Figure 9. When treated with salt, *Bra034084* and *Bra029784* were translocated into the cell wall.

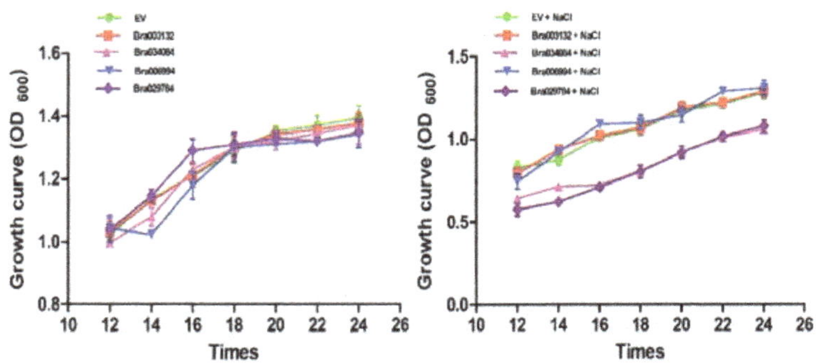

Figure 8. Growth curves of the S1fa gene overexpressing yeast cells under 1 M NaCl stress. EV (empty vector (yeast WT)) and cells with the overexpression *of Bra003132, Bra006994, Bra034084* and *Bra029784* were grown at 28 °C. Cell density was monitored after 12, 14, 16, 18, 20, 22 and 24 h after the treatment. The error bar represents the deviation of three independent replications.

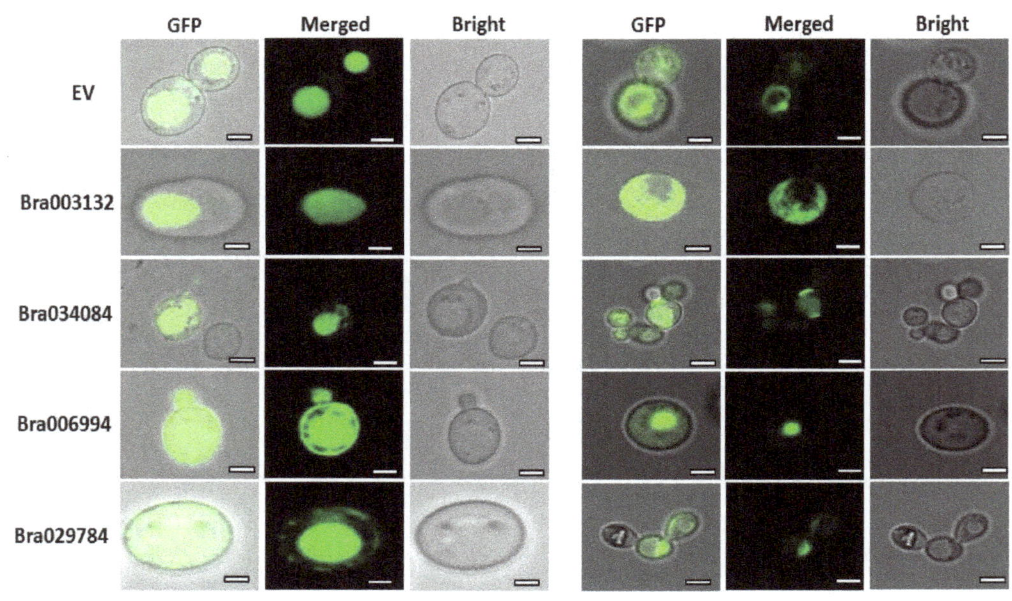

Figure 9. The subcellular localization of the S1fa genes and the empty vector (EV) tagged with GFP and transiently expressed in yeast cells treated with 1 M NaCl. The images were obtained from GFP, merged and bright channels. Scale bar: 10 μm.

2.11. Responses of Cell Wall Biosynthesis Genes to NaCl Stresses

In eukaryotic organisms, the biological integrity depends on the cell wall, which is considered an essential structure that not only maintains the morphology, but also participates in protecting the cell from environmental influences. Here, in this study, we investigated the expression levels of cell wall biosynthesis genes as presented in Figure 10. The results suggest that the S1fa genes *Bra034084* and *Bra029784* significantly enhanced the expression levels of cell wall biosynthesis genes compared with EV-Ck (no stress) and EV (with stress). The expression levels of cell wall biosynthesis genes, including *Ccw14p*, *Cha1p*, *Cwp2p*, *Sed1p*, *Rlm1p*, *Rom2p*, *Mkk1p*, *Hsp12p*, *Mkk2p*, *Sdp1p* and *YLR194c*, were significantly enhanced, while the expression levels of *Bck1p* and *Ptc1p* were downregulated under salt stresses in *Bra034084* and *Bra029784* overexpressing yeast cells. The expression of the *Pst1p* gene showed no significant difference. These findings suggest that *Bra034084* and *Bra029784* activate the signaling pathways regulating cell wall integrity by stimulating transcriptional and post-transcriptional genes in response to salt stresses.

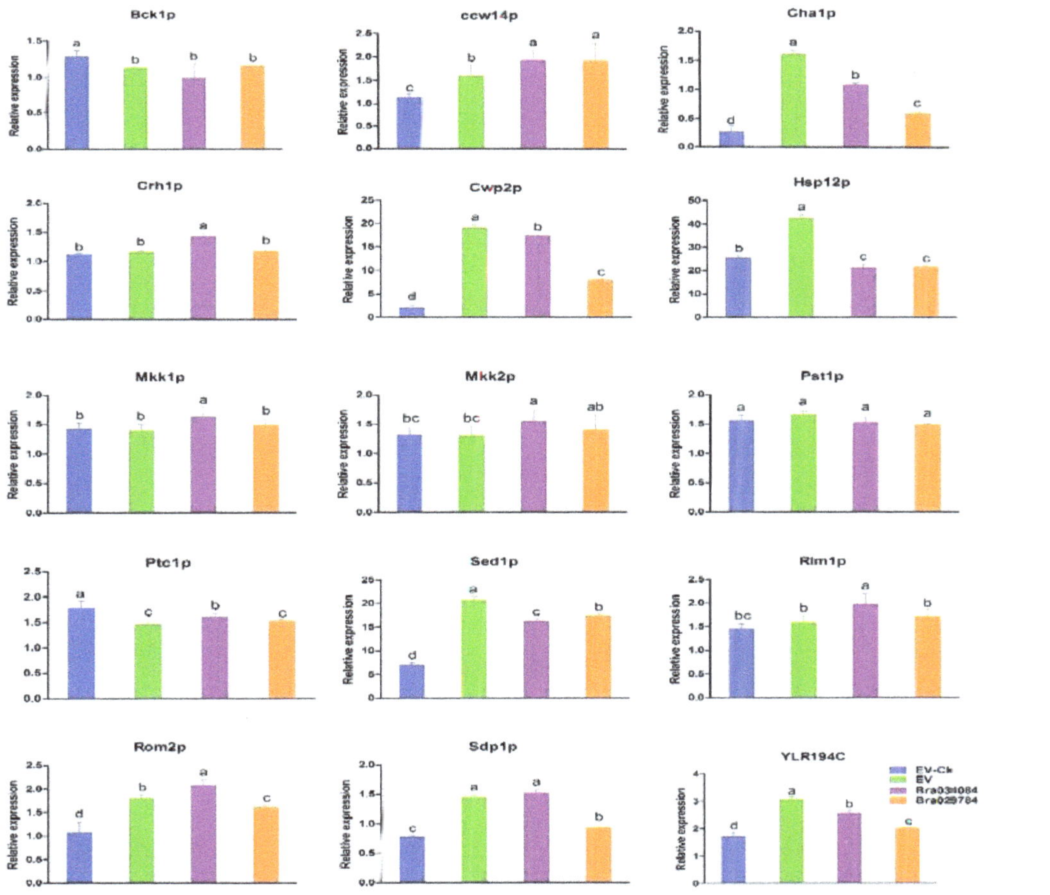

Figure 10. Quantitative reverse transcription polymerase chain reaction (qRT-PCR) analysis was used to assess the expression levels of cell wall biosynthesis genes under NaCl stresses. The error bar represents the deviation of three independent replications. The letters above the error bar represent significant differences at $p > 0.05$.

2.12. Antioxidant Enzyme Activities and ROS Accumulation under NaCl Stresses

The antioxidant enzyme activities, the MDA content and the ROS accumulations in the *Bra034084* and *Bra029784* overexpressing cells were analyzed after 14 h of exposure to NaCl stresses. The results suggest that the antioxidant enzyme activities were greatly reduced by NaCl stresses compared with EV, as presented in Figure 11. The activities of SOD and POD were reported to be significantly higher in the *Bra029784* overexpressing cells than those in the *Bra034084* overexpressing cells, whereas the activity of CAT was enhanced in the *Bra034084* overexpressing cells. Moreover, the activities of CAT, SOD and POD were significantly higher under NaCl stress than those in the control. The MDA content was higher in the *Bra034084* overexpressing cells, while the contents of H_2O_2 and O_2^- were not affected but were significantly higher than those of EV. The contents of ROS and MDA of NaCl stress were significantly higher than the control, as presented in Figure 11. These findings suggest that salinity stress negatively affected the activities of antioxidant enzymes but promoted the accumulation of ROS and MDA, which may activate the biosynthesis of the cell wall, and hence increase salt stress sensitivity.

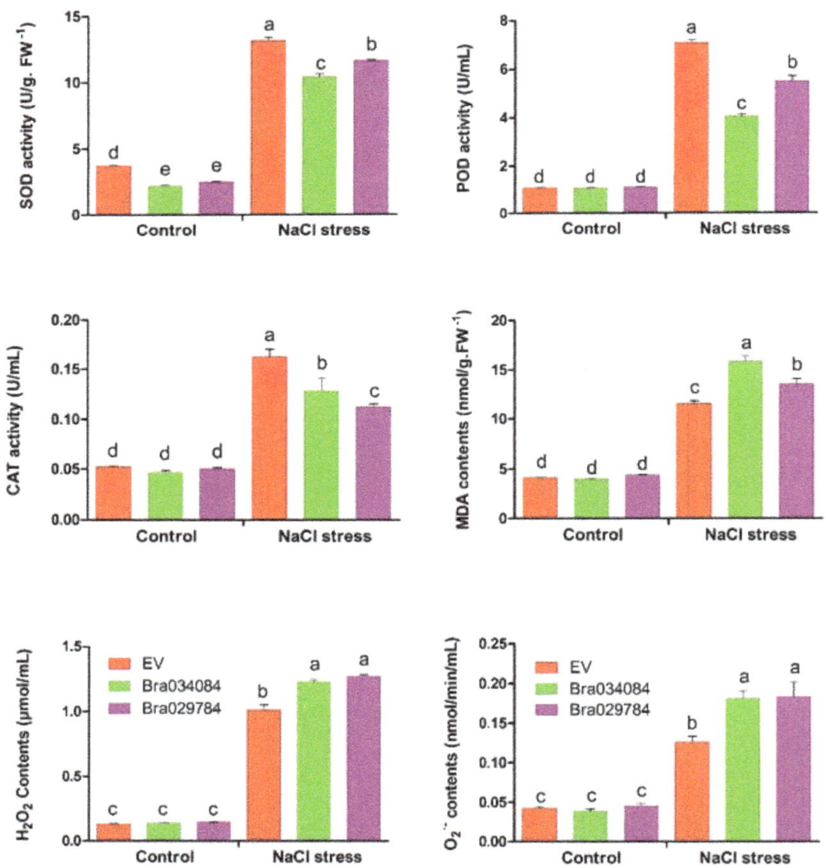

Figure 11. The effect of salt stress on antioxidant enzyme activities, MDA and ROS accumulations of Bra034084 and Bra029784 overexpressing yeast cells. *Bra034084* and *Bra029784* overexpressing cells were exposed to 1 M salt stress and control (without stress) for 14 h at 28 °C. Superoxide dismutase (SOD), peroxidase (POD), catalase (CAT), malondialdehyde (MDA), hydrogen peroxide (H_2O_2), and superoxide radicals (O_2^-). Different letters indicate significant differences at $p < 0.05$.

3. Discussion

Environmental influence impairs plant growth and development through oxidative stresses, which may change genome stability [10,20]. Plants have evolved a number of pathways to cope with ROS production and reduce the detrimental effects of abiotic stresses, including antioxidant enzymes, hormonal responses, the activation of transcription factors [8] and regulation of downstream genes [9,21].

Transcription factors (such as MYB, WRKY, ERF and NAC) play a crucial role in abiotic stress tolerance, and are involved in the transcriptional regulation of plant genes [22–25]. The S1fa transcription factor belongs to the smallest family of plants that are involved in plant growth and development [18]. However, no study has reported on their regulatory effects on abiotic stresses in Chinese cabbage. The present study was designed to investigate the physiological and molecular mechanism of the S1fa genes in abiotic stress tolerance. We identified and characterized four S1fa proteins in Chinese cabbage at the whole genome level, which were also compared with three *Arabidopsis* S1fa proteins. Systematic analyses including phylogenetic trees, gene and protein structures, motifs, physiochemical properties, miRNAs and cis-elements were conducted in the promoter region of the S1fa genes, and the effects of the S1fa genes on abiotic stress tolerance were investigated in yeast models. These findings provide novel insights into the functional characterization of the S1fa genes, which can be used in molecular breeding to enhance crop production and abiotic stress tolerance.

Based on phylogenetic analyses, the S1fa proteins were divided into three groups. Chinese cabbage S1fa proteins were classified in groups I and II, which share a high similarity with those in rice, pepper, cucumber and watermelon (Figure 2). The difference between these groups demonstrates that these proteins underwent great genetic variation after divergence, probably due to environmental influences, and that some gene fragments might have been lost during the evolutionary process [26]. The analysis of gene structure and motif suggested that the S1fa genes shared a similar exon–intron and motif structure, indicating there is a closer evolutionary relationship among the members in the same group but a functionally diversified relationship among the other group members [27–30]. The structure and motif analysis of Chinese cabbage S1fa genes showed a similar structure, which has three common motifs (Figure 4). Interestingly, based on motif analysis, the S1fa genes in Chinese cabbage can be divided into two subgroups, with *Bra003132* and *Bra006994* in one group, and *Bra029784* and *Bra034084* in the other (Figure 4). Protein analysis shows that motif 1 is highly conserved, which may be involved in or required for recognizing abiotic stress responsive cis-elements in response to stresses [31–33].

A gene's expression pattern can provide important indications for its biological functions. In the current study, the expression patterns of the S1fa genes were analyzed under abiotic stresses in Chinese cabbage (Figure 6). The S1fa genes showed different expression patterns in different tissues. The S1fa genes had the lowest expression in the leaves and the highest expression in the silique (Figure 5). *Bra006994* showed a minimum expression level in the leaf, flower and callus. The expression level of the S1fa genes was significantly different under different abiotic stresses (Figure 6). Under NaCl stress, *Bra034084* and *Bra029784* were significantly expressed compared with the other two S1fa genes. However, their expressions were significantly downregulated under Cd and Hg stresses (Figure 6). Taken together, these results suggest that the S1fa transcription factors are potentially involved in salinity stress tolerance, as well as plant growth and development [17,18].

The cis-element is a specific sequence on the promoter region of a given gene, which influences the expression of protein-coding and long non-coding RNA genes. Activation of the expression of the gene by binding with the cis-element is a common way of regulating developmental and physiological processes [33]. In plants, miRNAs act as a positive regulator in regulating related genes. Many studies have shown that miRNAs are involved in response to abiotic stresses [34]. *miR398b* negatively regulates the defense system in cotton and causes an adverse effect on plant growth. *miR1885* regulates plant growth and tolerance to viral infection through targeting *BraTNL1* and *BraCP24* genes in *Brassica* [35].

Bra034084 and *Bra029784* were targeted by miRNAs including miR398b and miR1885, which might be involved in salt stress responses (Figure 7). Under abiotic stresses, the cis-elements are involved in controlling the transcriptional regulation of the core gene network [31]. The S1fa genes of Chinese cabbage contain a number of cis-elements including light and abiotic stress responsive elements, ABA, GA, methyl jasmonate (MeJA), and low and high temperature responsive elements (Figure 3). Previous studies have reported that hormonal cis-elements, such as O_2-site, TGA-element, TGACG-motif, CGTCA-motif, TCA-element and ABRE motifs, are the key cis-regularity modules that stimulate the hormone signaling pathways under abiotic stresses [31,33]. The cis-elements activate specific transcription factors and their downstream genes, acting as a key cellular regulator in response to abiotic stresses [33]. These cis-elements may also be involved in salinity stress tolerance by regulating the specific hormonal signal transduction pathways [33].

The S1fa transcription factors have been reported to be involved in photomorphogenesis and abiotic stress tolerance [17]. However, their function has yet to be fully understood. Here, four Chinese cabbage S1fa genes were identified and cloned to investigate their function in abiotic stress tolerance using a yeast model (Figure 7). The results suggest that the *S1fa* genes did not respond to any abiotic stress. However, *Bra034084* and *Bra029784* were highly sensitive to NaCl stress (Figure 7(H)), suggesting that the S1fa transcription factors are involved in salinity stress tolerance. These findings are in line with a previous study, which shows that *OsS1fa* improves drought stress tolerance in *Arabidopsis* and increases the expression of the drought stress-related genes [18].

Plants exposed to abiotic stresses generate an excessive amount of ROS (H_2O_2 and O_2^-), which is highly toxic and detrimental to proteins, lipids, DNA and carbohydrates, eventually leading to cell death [6,36,37]. The plant possesses an antioxidant enzyme defense system to normalize the overproduction of ROS. In this study, the overexpression of *Bra034084* and *Bra029784* significantly decreased the activities of antioxidant enzymes including SOD, POD and CAT under salinity stresses in yeast (Figure 11). On the other hand, the overexpression of *Bra034084* and *Bra029784* enhanced the accumulation of H_2O_2, O_2^- and MDA in the *S1fa* overexpressing yeast cells compared with the wild type (EV) (Figure 11). These findings suggest that *Bra034084* and *Bra029784* inhibit the antioxidant enzyme activities, thereby leading to a higher hypersensitivity to salt stresses. *PsS1Fa2* overexpression in *Populus trichocarp* enhanced drought stress tolerance by increasing the activities of antioxidant enzymes (SOD and POD) and reducing the accumulation of MDA, H_2O_2 and O_2^- [17]. A similar report has been conducted, which shows *CaDHN4* can protect against cold and salt stresses by activating the antioxidant enzyme defense system [38].

In eukaryotic organisms, the cell wall plays a dominant role in the protection of the cell from environmental influences [11]. In yeast, transcriptional re-programing can alter the expression of key genes for cell wall biosynthesis, energy generation, signal transduction and stress [11,39]. Several signaling pathway cascades such as *MAPK*, *MAPKK1*, *MAPKK2* and *Slt2* have been reported to be involved in cell wall biosynthesis [40,41]. In this study, the overexpressed S1fa genes, *Bra034084* and *Bra029784*, were located in the cytoplasm, and were translocated under NaCl stresses (Figure 7), and they significantly enhanced the expression level of cell wall biosynthesis genes (Figure 11). The *PKC1* (protein kinase C) pathway plays an important role in cell wall biogenesis, maintenance and cell integrity [41], and is regulated by *Bck1p*, *Mkk1p*, *Slt2p* and *Rom2p* [11,41]. Our study showed that the expression of these factors was significantly increased in the S1fa overexpressing yeast cells under salinity stresses, as presented in Figure 10. Additionally, the overexpression of *Bra034084* and *Bra029784* enhanced the transcript level of *Bck1p*, *Ptc1p Ccw14p*, *Crh1p*, *Mkk1p*, *Mkk2p*, *Rlm2* and *Rom2*, which are involved in the signaling pathway regulating cell wall integrity [11,39,40]. Subcellular localization analysis suggests that the S1fa gene was translocated from the cytoplasm to the cell wall under salt stresses, as presented in Figure 9, which in turn might promote the expression of many cell wall biosynthesis genes. These findings are consistent with previous studies, which show that abiotic stress induces a significant reduction in the activities of antioxidant enzymes and increases the contents of

ROS and MDA [11,14], thereby regulating the expression of the genes controlling cell wall integrity [11]. In summary, it can be concluded that the S1fa transcription factors activate genes involved in cell wall integrity under salinity stresses. Future studies are required to investigate the linkage between cell wall integrity and S1fa genes under abiotic stresses, especially salinity stresses.

4. Materials and Methods

Chinese cabbage (Cv. Guangdongzao) seeds were soaked with 1% sodium hypochlorite for 3 min and then washed at least five times with ddH_2O to remove the excessive sodium hypochlorite. Then, the seeds were germinated in $\frac{1}{2}$MS media in a controlled growth chamber as described previously [42]. The uniform seedlings were transferred to a hydroponic culture and incubated for five more days before treated with 75 μM Cd, 75 mM Hg and 1 M NaCl, respectively. The samples were collected and ground in liquid nitrogen to extract the total RNA [43].

4.1. Identification of the S1fa Genes in Chinese Cabbage

In order to identify the S1fa genes in Chinese cabbage, the protein sequences of three S1fa genes from TAIR (www.Arabidopsis.org (accessed on 22 March 2022)) were downloaded and used as queries in Chinese cabbage genome database (http://brassicadb.cn (accessed on 22 March 2022)) with the BLASTP program [42]. All predicted Chinese cabbage S1fa proteins were confirmed through Pfam (http://pfam.xfam.org/ (accessed on 22 March 2022)) and SMART database. The physicochemical parameters including protein isoelectric point (pl), molecular weight (kDa) and length were calculated using the tools on the ExPASy server (http://web.expasy.org/compute_pi/ (accessed on 22 March 2022)). The chromosomal locations and strand directions were obtained from the BRAD database (http://brassicadb.cn (accessed on 22 March 2022)), and the subcellular location of each protein was investigated using CELLO 2.5 (http://cello.life.nctu.edu.tw/ (accessed on 22 March 2022)) [29].

4.2. Phylogenetic Trees and Sequence Alignment

The full length protein sequences from Chinese cabbage, tomato, pepper, cotton, *Arabidopsis*, cucumber, watermelon and rice were obtained from the genome database and aligned using MUSCLE (https://www.ebi.ac.uk/Tools/msa/muscle (accessed on 22 March 2022)), which were used to construct the evolutionary tree using the MAGA7 software by the neighbor-joining method, and protein sequence alignment was performed as described previously [29].

4.3. S1fa Structure and Conserved Motif

The S1fa gene structure was analyzed using the online MEME program (http://meme-suite.org/tools/meme (accessed on 22 March 2022)), and the maximum number of motifs was set at 10. The structures of the S1fa genes were designed by using the online program GSDS 2.0 (http://gsds.cbi.pku.edu.cn/ (accessed on 22 March 2022)) [29]. The online tool PlantCARE (http://bioinformatics.psb.ugent.be/webtools/plantcare/html/ (accessed on 22 March 2022)) was used to analyze the cis-elements as described previously [28,29,44].

4.4. S1fa Gene Promoter Analysis and miRNA Prediction

The promoter sequences of Chinese cabbage S1fa genes (2000 bp upstream) were obtained from Chinese cabbage genome database (http://brassicadb.cn (accessed on 22 March 2022)). The online tool PlantCARE (https://bioinformatics.psb.ugent.be/webtools/plantcare/html/ (accessed on 22 March 2022)) was used to analyze the cis-elements in the promoter sequence [17]. To predict the miRNAs, the coding sequence of the S1fa genes was submitted to the psRNATarget server (https://www.zhaolab.org/psRNATarget/ (accessed on 22 March 2022)) as the targeted candidate as described previously [32].

4.5. Total RNA Extraction and qRT-PCR Analysis

The total RNA was extracted for Chinese cabbage tissues using TRIzol, while yeast RNA was extracted using the M5 EASYspin yeast RNA rapid extraction kit, MF158-01 (Mei5 Biotechnology, Co., Ltd. Beijing China). For yeast RNA extraction, the cells were grown until the OD_{600} value reached 0.3 at 28 °C, and then treated with 1 M NaCl for 12 h before the total RNA was harvested [45]. The first-stand cDNA was synthesized using a PrimeScript and RT reagent kit with gDNA Eraser (TAKARA). The SYBR Premix Ex-Taq Kit (TAKARA) was used for quantitative real-time PCR. All experiments were performed with three independent biological replications. The transcript levels were calculated using the $2^{\Delta\Delta}$-CT method. The TMP values of Chinese cabbage tissues were obtained from the Chinese cabbage database (http://brassicadb.cn (accessed on 22 March 2022)) for each S1fa gene. The primers used for qRT-PCR are presented in Supplementary Table S1.

4.6. Yeast Constructs

To construct the yeast (*Saccharomyces cerevisiae*) overexpression vectors, the coding sequences of Chinese cabbage genes, *Bra034084*, *Bra003132*, *Bra029784* and *Bra006994*, were cloned separately into the pRS-416-GFP vector. The coding sequences of the S1fa genes were amplified from Chinese cabbage cDNA with specific primers (Supplementary Table S1) and then inserted into the SPE1 site on pRS-416-GFP using the infusion cloning kit (Catalog no. 011614; Clontech) [46]. The sequence insertions were confirmed through SANGER sequencing and then used for the investigation of abiotic stress tolerance in yeast. To determine the subcellular localization of the S1fa proteins, the S1fa genes were inserted into pRS-416-GFP. The subcellular localization of the fusion proteins was observed under a Zeiss Axiophot fluorescence microscope as described previously [45].

4.7. Tolerance Assay and Growth Curve

The final pRS-416-GFP vectors cultured in URA medium were diluted until the OD_{600} value was 0.1, and were incubated again until the OD_{600} reached 0.3. The cell culture was then five-fold diluted and treated with 75 µM Cd, 75 mM Hg, 100 mM Al, 50 mM Cu, 100 mM Co, 1 M NaCl and 2 M mannitol, respectively, and was incubated at 28 °C for five days [45]. No treatment was added for the control. The cold and heat stresses were applied at 4 °C and 38 °C, respectively, for 2 days before being transferred to a place at 28 °C to for 3 more days. The photos were taken after five days of incubation, and the experiment was repeated three times. The S1fa overexpressing yeast cells without and with the 1 M NaCl treatment were grown at 28 °C in liquid URA culture medium, and diluted until the OD_{600} value was about 0.1. The cells were incubated again, and when OD_{600} value reached 0.3, the OD_{600} was recorded every 2 h to prepare the growth curve of the cells [45].

4.8. Determination of Antioxidant Enzyme Activities and ROS Contents

To determine the antioxidant enzyme activities and ROS contents, the yeast cells were harvested 14 h after the NaCl treatment at 28 °C and stored at −80 °C. The antioxidant enzyme activities and ROS and MDA contents were analyzed using the service provided by Nanjing Ruiyuan Biotechnology company (https://bestofbest.top/).

5. Statistical Analysis

Three independent biological replications were used for each treatment, and the whole experiment was repeated three times. The data were statistically analyzed using an analysis of variance, and compared with the control using the LSD test ($p > 0.05$) by using the Statistix 8.1 software (https://www.statistix.com/). The Graphpad Prism 5 software was used for graphical presentation.

6. Conclusions

In this study, four S1fa family proteins were identified in Chinese cabbage, and the chromosomal location, structure, phylogenetic tree and physiochemical properties were analyzed, and molecular characterization performed in yeast, to understand their involvement in abiotic stress tolerance. The S1fa proteins have three highly conserved motifs. Moreover, the cis-elements and miRNAs targeting S1fa genes were predicted to understand the regulatory mechanism. The expression patterns of the S1fa gene in different tissues and their responses to abiotic stresses show that these genes may play a significant role in the growth and development of Chinese cabbage. In yeast, *Bra034084* and *Bra029784* was highly sensitive to salt stresses and might activate the expression of cell wall biosynthesis genes. For the first time, we elucidate the functions of the S1fa genes in Chinese cabbage and confirm their responses to salt stresses. The comprehensive understanding of the physiological and molecular mechanisms of *Bra034084* and *Bra029784* can serve as an important genetic resource for the improvement in salinity stress tolerance and the yield of Chinese cabbage.

Supplementary Materials: The following supporting information can be downloaded at: https://www.mdpi.com/article/10.3390/antiox11091782/s1, Table S1: Chinese cabbage S1fa family genes cis-elements.

Author Contributions: A.A., J.G. and S.Z. conceived and designed the study. A.A. and S.Z. performed the experiment and wrote the manuscript. L.H., F.W. and L.-X.W. helped to collect the materials and revise the manuscript. J.G. reviewed the manuscript. All authors have read and agreed to the published version of the manuscript.

Funding: This study was supported by the Natural Science Foundation of Shandong Province (ZR2020MC145); the Natural Science Foundation of China (32172591); the Key R & D Program of Shandong Province, China (2019GHZ014); Modern Agricultural Industrial Technology System Funding of Shandong Province, China (SDAIT-02-022-04); Agricultural Science and Technology Innovation Project of SAAS (CXGC2022E08); the Agricultural Science and Technology Innovation Project of SAAS, China (CXGC2022D01).

Institutional Review Board Statement: Not applicable.

Informed Consent Statement: Not applicable.

Data Availability Statement: Not applicable.

Conflicts of Interest: The authors declare no conflict of interest.

References

1. Mabuchi, R.; Tanaka, M.; Nakanishi, C.; Takatani, N.; Tanimoto, S. Analysis of Primary Metabolites in Cabbage (*Brassica oleracea* var. capitata) Varieties Correlated with Antioxidant Activity and Taste Attributes by Metabolic Profiling. *Molecules* **2019**, *24*, 4282. [CrossRef] [PubMed]
2. Gu, M.; Li, N.; Ty, S.; Xh, L.; Brestic, M.; Shao, H.; Li, J.; Rki, S. Accumulation capacity of ions in cabbage (*Brassica oleracea* L.) supplied with sea water. *Plant Soil Environ.* **2016**, *62*, 314–320.
3. Park, S.; Valan Arasu, M.; Lee, M.K.; Chun, J.H.; Seo, J.M.; Lee, S.W.; Al-Dhabi, N.A.; Kim, S.J. Quantification of glucosinolates, anthocyanins, free amino acids, and vitamin C in inbred lines of cabbage (*Brassica oleracea* L.). *Food Chem.* **2014**, *145*, 77–85. [CrossRef] [PubMed]
4. Yang, L.; Wu, Y.; Wang, X.; Lv, J.; Tang, Z.; Hu, L.; Luo, S.; Wang, R.; Ali, B.; Yu, J. Physiological Mechanism of Exogenous 5-Aminolevulinic Acid Improved the Tolerance of Chinese Cabbage (*Brassica pekinensis* L.) to Cadmium Stress. *Front. Plant Sci.* **2022**, *13*, 845396. [CrossRef] [PubMed]
5. Wang, A.; Hu, J.; Gao, C.; Chen, G.; Wang, B.; Lin, C.; Song, L.; Ding, Y.; Zhou, G. Genome-wide analysis of long non-coding RNAs unveils the regulatory roles in the heat tolerance of Chinese cabbage (*Brassica rapa* ssp. chinensis). *Sci. Rep.* **2019**, *9*, 5002. [CrossRef]
6. Anwar, A.; Kim, J.K. Transgenic Breeding Approaches for Improving Abiotic Stress Tolerance: Recent Progress and Future Perspectives. *Int. J. Mol. Sci.* **2020**, *21*, 2695. [CrossRef]
7. Anwar, A.; Liu, Y.; Dong, R.; Bai, L.; Yu, X.; Li, Y. The physiological and molecular mechanism of brassinosteroid in response to stress: A review. *Biol. Res.* **2018**, *51*, 46. [CrossRef]
8. Gill, S.S.; Tuteja, N. Reactive oxygen species and antioxidant machinery in abiotic stress tolerance in crop plants. *Plant Physiol. Biochem* **2010**, *48*, 909–930. [CrossRef]
9. Jiang, J.; Ma, S.; Ye, N.; Jiang, M.; Cao, J.; Zhang, J. WRKY transcription factors in plant responses to stresses. *J. Integr. Plant Biol.* **2017**, *59*, 86–101. [CrossRef]

10. Yang, Y.; Guo, Y. Unraveling salt stress signaling in plants. *J. Integr. Plant Biol.* **2018**, *60*, 796–804. [CrossRef]
11. Sanz, A.B.; García, R.; Rodríguez-Peña, J.M.; Arroyo, J. The CWI Pathway: Regulation of the Transcriptional Adaptive Response to Cell Wall Stress in Yeast. *J. Fungi* **2017**, *4*, 1. [CrossRef]
12. Lippold, F.; Sanchez, D.H.; Musialak, M.; Schlereth, A.; Scheible, W.R.; Hincha, D.K.; Udvardi, M.K. *AtMyb41* regulates transcriptional and metabolic responses to osmotic stress in Arabidopsis. *Plant Physiol.* **2009**, *149*, 1761–1772. [CrossRef] [PubMed]
13. Schmidt, R.; Schippers, J.H.; Mieulet, D.; Obata, T.; Fernie, A.R.; Guiderdoni, E.; Mueller-Roeber, B. MULTIPASS, a rice R2R3-type MYB transcription factor, regulates adaptive growth by integrating multiple hormonal pathways. *Plant J.* **2013**, *76*, 258–273. [CrossRef] [PubMed]
14. Zagorchev, L.; Kamenova, P.; Odjakova, M. The Role of Plant Cell Wall Proteins in Response to Salt Stress. *Sci. World J.* **2014**, *2014*, 764089. [CrossRef]
15. Ambawat, S.; Sharma, P.; Yadav, N.R.; Yadav, R.C. MYB transcription factor genes as regulators for plant responses: An overview. *Physiol. Mol. Biol. Plants* **2013**, *19*, 307–321. [CrossRef] [PubMed]
16. Nuruzzaman, M.; Sharoni, A.M.; Kikuchi, S. Roles of NAC transcription factors in the regulation of biotic and abiotic stress responses in plants. *Front. Microbiol.* **2013**, *4*, 248. [CrossRef] [PubMed]
17. Zhao, H.; Niu, Y.; Dong, H.; Jia, Y.; Wang, Y. Characterization of the Function of Two S1Fa-Like Family Genes From *Populus trichocarpa*. *Front. Plant Sci.* **2021**, *12*, 753099. [CrossRef]
18. Kim, S.-I.; Lee, K.H.; Kwak, J.S.; Kwon, D.H.; Song, J.T.; Seo, H.S. Overexpression of Rice *OsS1Fa1* Gene Confers Drought Tolerance in Arabidopsis. *Plants* **2021**, *10*, 2181. [CrossRef] [PubMed]
19. Zhou, D.X.; Li, Y.F.; Rocipon, M.; Mache, R. Sequence-specific interaction between S1F, a spinach nuclear factor, and a negative cis-element conserved in plastid-related genes. *J. Biol. Chem.* **1992**, *267*, 23515–23519. [CrossRef]
20. Van Zelm, E.; Zhang, Y.; Testerink, C. Salt Tolerance Mechanisms of Plants. *Annu. Rev. Plant Biol.* **2020**, *71*, 403–433. [CrossRef]
21. Choudhury, F.K.; Rivero, R.M.; Blumwald, E.; Mittler, R. Reactive oxygen species, abiotic stress and stress combination. *Plant J.* **2017**, *90*, 856–867. [CrossRef] [PubMed]
22. Wang, C.; Deng, P.; Chen, L.; Wang, X.; Ma, H.; Hu, W.; Yao, N.; Feng, Y.; Chai, R.; Yang, G.; et al. A wheat WRKY transcription factor *TaWRKY10* confers tolerance to multiple abiotic stresses in transgenic tobacco. *PLoS ONE* **2013**, *8*, e65120. [CrossRef] [PubMed]
23. Bo, W.; Zhaohui, Z.; Huanhuan, Z.; Xia, W.; Binglin, L.; Lijia, Y.; Xiangyan, H.; Deshui, Y.; Xuelian, Z.; Chunguo, W.; et al. Targeted Mutagenesis of NAC Transcription Factor Gene, *OsNAC041*, Leading to Salt Sensitivity in Rice. *Rice Sci.* **2019**, *26*, 98–108. [CrossRef]
24. Ohnishi, T.; Sugahara, S.; Yamada, T.; Kikuchi, K.; Yoshiba, Y.; Hirano, H.Y.; Tsutsumi, N. *OsNAC6*, a member of the NAC gene family, is induced by various stresses in rice. *Genes Genet. Syst.* **2005**, *80*, 135–139. [CrossRef]
25. Wei, Z.-Z.; Hu, K.-D.; Zhao, D.-L.; Tang, J.; Huang, Z.-Q.; Jin, P.; Li, Y.-H.; Han, Z.; Hu, L.-Y.; Yao, G.-F.; et al. *MYB44* competitively inhibits the formation of the *MYB340-bHLH2-NAC56* complex to regulate anthocyanin biosynthesis in purple-fleshed sweet potato. *BMC Plant Biol.* **2020**, *20*, 258. [CrossRef]
26. Quan, X.; Liu, J.; Zhang, N.; Xie, C.; Li, H.; Xia, X.; He, W.; Qin, Y. Genome-Wide Association Study Uncover the Genetic Architecture of Salt Tolerance-Related Traits in Common Wheat (*Triticum aestivum* L.). *Front. Genet.* **2021**, *12*, 663941. [CrossRef] [PubMed]
27. Wang, Y.; Zhang, Y.; Zhou, R.; Dossa, K.; Yu, J.; Li, D.; Liu, A.; Mmadi, M.A.; Zhang, X.; You, J. Identification and characterization of the bZIP transcription factor family and its expression in response to abiotic stresses in sesame. *PLoS ONE* **2018**, *13*, e0200850. [CrossRef]
28. Liu, H.; Wang, Y.X.; Li, H.; Teng, R.M.; Wang, Y.; Zhuang, J. Genome-Wide Identification and Expression Analysis of Calcineurin B-Like Protein and Calcineurin B-Like Protein-Interacting Protein Kinase Family Genes in Tea Plant. *DNA Cell Biol.* **2019**, *38*, 824–839. [CrossRef] [PubMed]
29. Li, S.; Miao, L.; Huang, B.; Gao, L.; He, C.; Yan, Y.; Wang, J.; Yu, X.; Li, Y. Genome-Wide Identification and Characterization of Cucumber BPC Transcription Factors and Their Responses to Abiotic Stresses and Exogenous Phytohormones. *Int J Mol Sci* **2019**, *20*, 5048. [CrossRef]
30. Liu, Q.-L.; Xu, K.-D.; Pan, Y.-Z.; Jiang, B.-B.; Liu, G.-L.; Jia, Y.; Zhang, H.-Q. Functional Analysis of a Novel Chrysanthemum WRKY Transcription Factor Gene Involved in Salt Tolerance. *Plant Mol. Biol. Rep.* **2014**, *32*, 282–289. [CrossRef]
31. Marand, A.P.; Schmitz, R.J. Single-cell analysis of cis-regulatory elements. *Curr. Opin. Plant Biol.* **2022**, *65*, 102094. [CrossRef] [PubMed]
32. Davoudi, M.; Chen, J.; Lou, Q. Genome-Wide Identification and Expression Analysis of *Heat Shock Protein 70* (*HSP70*) Gene Family in Pumpkin (*Cucurbita moschata*) Rootstock under Drought Stress Suggested the Potential Role of these Chaperones in Stress Tolerance. *Int. J. Mol. Sci.* **2022**, *23*, 1918. [CrossRef] [PubMed]
33. Schmitz, R.J.; Grotewold, E.; Stam, M. Cis-regulatory sequences in plants: Their importance, discovery, and future challenges. *Plant Cell* **2022**, *34*, 718–741. [CrossRef] [PubMed]
34. Miao, Y.; Chen, K.; Deng, J.; Zhang, L.; Wang, W.; Kong, J.; Klosterman, S.J.; Zhang, X.; Aierxi, A.; Zhu, L. *miR398b* negatively regulates cotton immune responses to *Verticillium dahliae* via multiple targets. *Crop J.* **2022**, *10*, 1026–1036. [CrossRef]
35. Cui, C.; Wang, J.J.; Zhao, J.H.; Fang, Y.Y.; He, X.F.; Guo, H.S.; Duan, C.G. A Brassica miRNA Regulates Plant Growth and Immunity through Distinct Modes of Action. *Mol. Plant* **2020**, *13*, 231–245. [CrossRef]

36. Collin, A.; Daszkowska-Golec, A.; Szarejko, I. Updates on the Role of *ABSCISIC ACID INSENSITIVE 5* (*ABI5*) and ABSCISIC ACID-RESPONSIVE ELEMENT BINDING FACTORs (ABFs) in ABA Signaling in Different Developmental Stages in Plants. *Cells* **2021**, *10*, 1996. [CrossRef]
37. Nath, M.; Bhatt, D.; Jain, A.; Saxena, S.C.; Saifi, S.K.; Yadav, S.; Negi, M.; Prasad, R.; Tuteja, N. Salt stress triggers augmented levels of Na^+, Ca_2^+ and ROS and alter stress-responsive gene expression in roots of *CBL9* and *CIPK23* knockout mutants of *Arabidopsis thaliana*. *Environ. Exp. Bot.* **2019**, *161*, 265–276. [CrossRef]
38. Zhang, H.-f.; Liu, S.-y.; Ma, J.-h.; Wang, X.-k.; Haq, S.U.; Meng, Y.-c.; Zhang, Y.-m.; Chen, R.-g. *CaDHN4*, a Salt and Cold Stress-Responsive Dehydrin Gene from Pepper Decreases Abscisic Acid Sensitivity in Arabidopsis. *Int. J. Mol. Sci.* **2020**, *21*, 26. [CrossRef]
39. Gerik, K.J.; Donlin, M.J.; Soto, C.E.; Banks, A.M.; Banks, I.R.; Maligie, M.A.; Selitrennikoff, C.P.; Lodge, J.K. Cell wall integrity is dependent on the PKC1 signal transduction pathway in *Cryptococcus neoformans*. *Mol. Microbiol.* **2005**, *58*, 393–408. [CrossRef]
40. He, Y.; Chen, Y.; Song, W.; Zhu, L.; Dong, Z.; Ow, D.W. A Pap1–Oxs1 signaling pathway for disulfide stress in Schizosaccharomyces pombe. *Nucleic Acids Res.* **2017**, *45*, 106–114. [CrossRef]
41. Techo, T.; Charoenpuntaweesin, S.; Auesukaree, C. Involvement of the Cell Wall Integrity Pathway of Saccharomyces cerevisiae in Protection against Cadmium and Arsenate Stresses. *Appl. Environ. Microbiol.* **2020**, *86*, e01339-20. [CrossRef]
42. Zhang, Y.; Li, J.; Zhou, D.; Song, J.; Gao, J. Nitrogen Uptake and Distribution in Different Chinese Cabbage Genotypes under Low Nitrogen Stress. *Int. J. Mol. Sci.* **2022**, *23*, 1573. [CrossRef] [PubMed]
43. Ying, S.; Zhang, D.F.; Fu, J.; Shi, Y.S.; Song, Y.C.; Wang, T.Y.; Li, Y. Cloning and characterization of a maize bZIP transcription factor, *ZmbZIP72*, confers drought and salt tolerance in transgenic Arabidopsis. *Planta* **2012**, *235*, 253–266. [CrossRef] [PubMed]
44. Ma, W.; Ren, Z.; Zhou, Y.; Zhao, J.; Zhang, F.; Feng, J.; Liu, W.; Ma, X. Genome-Wide Identification of the *Gossypium hirsutum NHX* Genes Reveals that the Endosomal-Type *GhNHX4A* is Critical for the Salt Tolerance of Cotton. *Int. J. Mol. Sci.* **2020**, *21*, 7712. [CrossRef] [PubMed]
45. Jing, Y.; Shi, L.; Li, X.; Zheng, H.; Gao, J.; Wang, M.; He, L.; Zhang, W. *OXS2* is Required for Salt Tolerance Mainly through Associating with Salt Inducible Genes, CA1 and Araport11, in Arabidopsis. *Sci. Rep.* **2019**, *9*, 20341. [CrossRef] [PubMed]
46. He, L.; Jing, Y.; Shen, J.; Li, X.; Liu, H.; Geng, Z.; Wang, M.; Li, Y.; Chen, D.; Gao, J.; et al. Mitochondrial Pyruvate Carriers Prevent Cadmium Toxicity by Sustaining the TCA Cycle and Glutathione Synthesis. *Plant Physiol.* **2019**, *180*, 198–211. [CrossRef]

antioxidants

MDPI

Article

Seed Priming with Spermine Mitigates Chromium Stress in Rice by Modifying the Ion Homeostasis, Cellular Ultrastructure and Phytohormones Balance

Farwa Basit [1,2], Javaid Akhter Bhat [3,*], Zaid Ulhassan [1], Muhammad Noman [1], Biying Zhao [3], Weijun Zhou [2], Prashant Kaushik [4], Ajaz Ahmad [5], Parvaiz Ahmad [6,*] and Yajing Guan [1,2,*]

1 Institute of Crop Science, Zhejiang Key Laboratory of Crop Germplasm, Zhejiang University, Hangzhou 310058, China
2 Hainan Research Institute, Zhejiang University, Sanya 572025, China
3 International Genome Center, Jiangsu University, Zhenjiang 212013, China
4 Instituto de Conservación y Mejora de la Agrodiversidad Valenciana, Universitat Politècnica de València, 46022 Valencia, Spain
5 Department of Clinical Pharmacy, College of Pharmacy, King Saud University, Riyadh 11451, Saudi Arabia
6 Department of Botany, GDC, Pulwama 192301, Jammu and Kashmir 192301, India
* Correspondence: javid.akhter69@gmail.com (J.A.B.); parvaizbot@yahoo.com (P.A.); ycguan@zju.edu.cn (Y.G.)

Abstract: Chromium (Cr) is an important environmental constraint effecting crop productivity. Spermine (SPM) is a polyamine compound regulating plant responses to abiotic stresses. However, SPM-mediated tolerance mechanisms against Cr stress are less commonly explored in plants. Thus, current research was conducted to explore the protective mechanisms of SPM (0.01 mM) against Cr (100 μM) toxicity in two rice cultivars, CY927 (sensitive) and YLY689 (tolerant) at the seedling stage. Our results revealed that, alone, Cr exposure significantly reduced seed germination, biomass and photosynthetic related parameters, caused nutrient and hormonal imbalance, desynchronized antioxidant enzymes, and triggered oxidative damage by over-accretion of reactive oxygen species (ROS), malondialdehyde (MDA) and electrolyte leakage in both rice varieties, with greater impairments in CY927 than YLY689. However, seed priming with SPM notably improved or reversed the above-mentioned parameters, especially in YLY689. Besides, SPM stimulated the stress-responsive genes of endogenous phytohormones, especially salicylic acid (SA), as confirmed by the pronounced transcript levels of SA-related genes (OsPR1, OsPR2 and OsNPR1). Our findings specified that SPM enhanced rice tolerance against Cr toxicity via decreasing accumulation of Cr and markers of oxidative damage (H_2O_2, $O_2^{\bullet-}$ and MDA), improving antioxidant defense enzymes, photosynthetic apparatus, nutrients and phytohormone balance.

Keywords: seed priming; spermine; chromium; reactive oxygen species; phytohormones

Citation: Basit, F.; Bhat, J.A.; Ulhassan, Z.; Noman, M.; Zhao, B.; Zhou, W.; Kaushik, P.; Ahmad, A.; Ahmad, P.; Guan, Y. Seed Priming with Spermine Mitigates Chromium Stress in Rice by Modifying the Ion Homeostasis, Cellular Ultrastructure and Phytohormones Balance. *Antioxidants* **2022**, *11*, 1704. https://doi.org/10.3390/antiox11091704

Academic Editor: Juan B. Barroso

Received: 8 July 2022
Accepted: 22 August 2022
Published: 30 August 2022

1. Introduction

Rice (*Oryza sativa*) is the second most abundant cereal crop and fulfils the nutritional needs of at least 50% of the global population [1]. Soil contamination with environmental pollutants (mainly of anthropogenic origin) such as heavy metals (HMs) has caused severe rice yield losses in Asian countries, including China [2]. Rice plants taking up HMs from contaminated soils and their depositing in plant parts, mainly grains, leads to reduction in crop yields [3]. Chromium (Cr) is the seventh most hazardous metal, which usually exists in Cr^{-3} and Cr^{+6} forms, while Cr^{+6} imposes more lethal symptoms on plants, relative to Cr^{+3} [4]. After taking up Cr ions from soils, plants generally accumulate an excessive amount of Cr in their tissues, severely hindering their growth and development [5–8]. At morpho-physiological, biochemical, metabolic and cellular levels, Cr inhibits the rate of seed germination, plant growth indices (length/height and biomass), photosynthesis process, and accumulation of mineral nutrients [1,7], induces oxidative stress, viz., lipid peroxidation, and reactive oxidative species (ROS), damages the inside of cellular membranes and ultra-structures [8,9], and provokes disorganization in antioxidant enzyme activities [10]. Plants boost their internal antioxidant defense system, which includes enzymatic and non-enzymatic antioxidants, by minimizing the overproduction of ROS and the induced oxidative cellular damage [7,11–13].

Polyamines (PAs) are aliphatic compounds and osmo-protectants, with low molecular weight, located within plant cells, and control diverse plant functioning such as regulation of seed germination, embryogenesis, flower expansion, fruit development and ripening, against both normal and stress conditions [6,14]. Moreover, PAs improve the plant's tolerance capacity by enhancing the endogenous accumulation of PAs under various environmental stresses, i.e., water-deficit [15], salt [16], chilling [17], heat [18] and heavy metals [19]. Previous findings have revealed that exogenous applications of PAs improved plants' growth indices by improving levels of chlorophyll pigments, photosystem II, the antioxidant enzymatic defense system and membrane protection, and reducing the HM-induced cellular oxidative damage, as noticed under cadmium [20], lead [21] and chromium [22] stress. Furthermore, PAs such as spermine (SPM) contribute to regulation of cellular functioning, cell division and phytohormonal signal transductions in cadmium and copper exposed wheat leaves [23]. The key mechanism is that the exogenous supply of PAs stimulates the endogenous PA content, which may help plants boost their intrinsic immunity in response to outside environmental stressors, improving the morpho-physiological and biochemical attributes of mung bean [20], wheat [24], rice [25], and maize [22]. However, the contribution of spermine when used as a priming agent in the detoxification of Cr in rice plants has been less commonly investigated. Therefore, the current study was performed to obtain insights into the protective roles of SPM in the alleviation of Cr stress in sensitive (CY927) and tolerant (YLY689) rice cultivars by targeting plant growth traits, photosynthetic apparatus (chlorophyll pigments and photosystem II), nutrients uptake, Cr-accumulation, oxidative damage, antioxidant enzyme activities, endogenous phytohormone production, and membrane or cellular ultra-structures. These biomarker studies may help plant scientists fully understand the mechanistic approaches utilized by SPM to detoxify Cr in rice, and possibly other cereals grown in HM-polluted soils.

2. Materials and Methods

2.1. Availability of Seeds

Herein, two various rice varieties, Chunyou 927 (CY927, sensitive) and Yliangyou 689 (YLY689, tolerant), obtained through the Zhejiang Nongke Seeds Co., Ltd. Hangzhou, Zhejiang Province, China were used in the present study.

2.2. Seed Priming and Germination Analysis

Initially, rice seeds were dipped into the solution for sterilization with 5% (w/v) of NaClO for 20 min, then quickly splashed with purified water (ddH_2O) to eradicate the remaining chloride. Secondly, these sterilized seeds were further primed by 0.01 mM SPM and water (H_2O) separately at 30 °C in the dark for 24 h. Then, to restore the seed's original moisture levels, they were dried at ambient temperature. The seeds primed with water (H_2O) were taken as controls. After priming, germination of seed assessment was conducted by using 50 seeds for each box (12 cm × 18 cm) with three repetitions for each treatment. Formerly, all germination boxes were retained inside a growth incubator with a fluctuating phase of 8 h light and 16 h of darkness at 25 °C for 14 days [26]. The chromium (0 and 100 μM) was applied to the incubated seeds. The Cr concentration was selected based on the preliminary experiments. The total seeds germinated, counted on the fifth day, were considered as the germination energy (GE) [26]. Furthermore, the total germinated seeds counted on day 14 were used for calculation of percentage of germination (GP), germination index (GI), vigor index (VI), and mean germination time (MGT), using the formulas used by Zheng [26].

2.3. Hydroponic Culture Treatments and Plant Growth Analysis

Subsequently, the primed seeds were placed into a 96 well black hydroponic box with one seed in each well. Each treatment was repeated six times and each repetition had 80 seeds. The hydroponic nutrient solution consisted of 0.5 μM potassium nitrate, 0.5 μM $Ca(NO_3)_2$, 0.5 μM magnesium sulfate, 2.5 μM monopotassium phosphate, 2.5 μM ammonium chloride, 100 μM ferric EDTA, 30 μM boric acid, 5 μM manganese sulfate, 1 μM copper sulfate, 1 μM zinc sulfate, and 1 μM ammonium hepta-molybdate ($(NH_4)_6Mo_7O_{24}$). The hydroponic boxes were repositioned daily within the growth chamber (30 °C with an alternation cycle of 12/12 h light/dark) by completely randomized design (CRD). After two weeks of hydroponics, three repeats of each treatment were exposed with 100 μM Cr for 7 days, as well as another three repeats without Cr, used as control. Sampling was carried out after the 7th day of Cr treatment.

The sampling of 21-day-old seedlings was conducted and they were washed with ddH_2O to exterminate the Cr residues. The height of seedlings was calculated by using a ruler and their fresh biomass was measured on a scale. To estimate the dry mass, leaves, and roots were dried out in an oven separately at 75 °C for 24 h.

2.4. Determination of Photosynthetic Pigments

The estimation of chlorophyll contents such as chlorophyll a (*Chl* a), chlorophyll b (*Chl* b), total chlorophyll as *Chl* (a + b) and carotenoids (Car) was carried out following the method of Lichtenthaler and Wellburn [27]. Concisely, fresh leaves tissues (0.2 g) were homogenized with pure water and soaked in 3 mL of 95 percent ethanol (v/v). The mixture was centrifuged for 10 min at 5000× g to separate the supernatant. After that, supernatant at the volume of 1 mL aliquot were mixed with 9 mL ethanol (95 percent, v/v). Thereafter, using an exhausting spectrophotometer, the mixture was measured using absorbance at 665, 649, and 470 nm wavelengths [28]. The following equation was used to determine the chlorophyll contents:

$$\text{Chlorophyll a } (Chl\text{ a}) = 13.95\text{ A665} - 6.88\text{ A649} \tag{1}$$

$$\text{Chlorophyll b } (Chl\text{ b}) = 24.96\text{ A649} - 7.32\text{ A665} \tag{2}$$

$$\text{Total chlorophyll content} = Chl\text{ a} + Chl\text{ b} \tag{3}$$

$$\text{Carotenoids } (C_{x+c}) = (1000\text{ A470} - 2.05\ C_a - 104\ C_b)/245 \tag{4}$$

The values of photosynthetic attributes were measured as milligrams (mg) per liter (L) of plant extract. Parameters related to gaseous exchange were investigated by the methodology of Zhou and Leul [29]. After 2 h of assimilation inside the growth cabinet at 18 °C, 1000 molm^{-2} s^{-1} light intensity, and 60% comparative moisture, the uppermost wholly extended leaves were used to calculate the rate of transpiration (Tr), net photosynthetic rate (Pn), stomatal conductance (gs), the intercellular level of CO_2 (Ci), and photochemical efficacy of PS II (Fv/Fm). ImagingWin software (IMAGING-PAM, Walz, Effeltrich, Germany) was used to examine the colored images for Fv/Fm and Fm levels.

2.5. Estimation of Cr Contents

The samples of dried root and shoot (0.2 g) per treatment were mixed with 5 mL concentrated HNO_3 and $HClO_4$ (5:1, v/v) on a stovetop at 70 °C for around 5 h. Dilution of the samples (digested) was subjected to dilution to a final amount of 10 mL with 2% HNO_3 before being tested, and three replicates per treatment were prepared. To examine the Cr along with microelements Na, Cu, K, P, Fe, Ca, Mn, and Zn, scum was measured by an atomic absorption spectrometer (iCAT-6000-6300, Thermo Scientific, Waltham, WA, USA).

2.6. Determination of Electrolyte Leakage and Total Soluble Sugar

The rice seedling is utilized to estimate electrolyte leakage (dSm^{-1}). Surface sterilization was carried out for 5 g seeds with $HgCl_2$ (1%) and speckled with ddH_2O, with four replications. Subsequently, seedlings were drenched in 25 mL ddH_2O, and were left at room temperature for one-day in an incubator. The sample was moved to a new void beaker, up to the volume of 25 mL by adding ddH_2O. Electrolyte leakage was reported in dSm^{-1} [30]. To estimate the total soluble sugar, samples of 0.5 g in the form of fresh shoots were crushed with pestle and mortar, in an extraction buffer that prepared from phosphate (50 mM, pH 7), ascorbate (1 mM), KCl (100 mM), glycerol (10%, v/v) and β-mercapto-ethanol (5 mM). Then, the supernatant was amassed into the micro-centrifuge tube through centrifugation at 12,000× *g* for 15 min. Later, the homogenate of three replicates was used to quantify the total soluble protein content [31] and total soluble sugar, by the procedure as for the phenol-sulfuric acid assay [32].

2.7. Estimation of Endogenous Abscisic Acid, Jasmonic Acid and Salicylic Acid Contents

The quantification of endogenous ABA contents was accomplished by using frozen samples following the protocol of Kim et al. [33] and Kamboj et al. [34]. The endogenous contents of JA were quantified following the detailed procedure of [35]. Extraction and quantification of free SA was carried out via the protocol of Yalpani et al. [36] and Fang et al. [37]. Three biological replicates were used to estimate the endogenous abscisic acid, jasmonic acid and salicylic acid contents.

2.8. Estimating MDA, H_2O_2 and $O_2^{\bullet-}$ Contents

The determination of MDA content was performed with 2-thiobarbituric acid (TBA). Homogenization of ~2 mL extract in 3 mL of TBA (5%) was performed, before being diluted inside 5% of trichloroacetic acid (TCA). Formerly, the grounded and mixed samples were preheated for 15 min at 95 °C, and earlier cooled to ice instantaneously and centrifuged at two different wavelengths, viz., 532 nm and 600 nm, via utilizing a UV–vis spectrophotometer (Hitachi U-2910) [38]. The value of MDA content was donated as nmol mg-1 protein. The hydrogen peroxide (H_2O_2) was determined by following the detailed protocol of Kwasniewski [39]. The H_2O_2 contents were estimated as μmol g^{-1} FW. Superoxide radical ($O_2^{\bullet-}$) contents were estimated following the method of Jiang and Zhang [40] with a few amendments. Three replications were used to estimate the MDA, H_2O_2 and $O_2^{\bullet-}$ contents.

2.9. Determination of Antioxidant Enzyme Activities

The supernatant attained from total soluble sugar was supplementarily used to determine the activities of antioxidants. Three replications were used to determine these antioxidant activities. To observe the SOD activity, the protocol utilized by Giannopolitis and Ries [41] was pursued with minor modifications. In addition, the SOD activity was estimated as U g^{-1} FW. Activity of catalase was estimated following the approach of Aebi [42]. The peroxidase (POD) activity was assessed by following Change and Maehly [43] detailed protocol. The enzyme activity (APX, POD, CAT) was deliberated as μ mol min^{-1} mg^{-1} protein at 25 ± 2 °C [44].

2.10. Analysis of Gene Expression

The transcript levels of SA-related genes were estimated by qRT-PCR. In a mortar and pestle, thawed shoot samples of both rice cultivars were crushed inside liquid nitrogen. Trizol method was used to extract the RNA, as already defined [45]. The NanoDrop 2000/2000 c (Thermo-Fisher Scientific, Waltham, MA, USA) was used to identify the purity of RNA. Subsequently, the synthesis of cDNA was accomplished by PrimeScript™ RT reagent kit. Three replications were used to analyze the gene expression level. The *PR1*, *PR2*, and *NPR1* gene primers were utilized to analyze the concerned gene expressions. The used primers are listed in Supplementary Table S1. The 2XSYBR Green Master Mix reagent (10 μL volume), (Applied Biosystems, Foster City, CA, USA), 200 nM gene-specific primers, and cDNA samples (6 μL volume) were utilized to prepare the 20 μL reaction mixture. The relative alteration inside the expression of genes was identified as documented [46]. The house-keeping gene (*OsActin*) was utilized as a control gene to standardize the other genes during internal calibration.

2.11. Transmission Electron Microscopic Analysis

The leaf ultrastructure changes were observed by the Sheteiwy et al. [47] procedure with minute modifications. The leaf samples without veins (7-10 per treatments) were randomly selected after treatment applications, then immersed in 2.5 percent (*v*/*v*) glutaraldehyde in 0.1 M PBS (sodium phosphate buffer, pH 7.4) and eroded thrice with the same PBS. Moreover, the 1% OsO_4 (osmium (VIII) oxide) was used to postfix the leaves for nearly 1 h. Further, it was washed three times in 0.1 M PBS with a 15 min gap between each wash. In addition, leaf samples were dehydrated using various classified categories of ethanol, such as 50%, 60%, 70%, 80%, 90%, 95%, and 100%, correspondingly, and washed by using absolute acetone for 20 min. Later, samples were immersed overnight in Spurr's resin. Thus, ultra-thin segments (80 nm) of samples were amended, then retained inside copper nets for visualization via transmission electron microscope (JEOLTEM-1230EX) at 60.0 kV.

2.12. Statistical Analysis

The experimental results were investigated by applying analysis of variance (One-way ANOVA) with the least significant differences (LSD) at $p < 0.05$ and 0.01 levels between mean values using Statistix (8.1) software. Three replications were used for each experiment and standard errors (S.E) were represented in figures.

3. Results

3.1. Priming Effect of SPM on Plant Phenotype, Seed Germination and Plant Growth Traits under Cr Exposure

In this study, seeds priming with water were utilized as a control. Under control conditions, plant height was visibly enhanced with SPM treatment as compared to seed primed control plants (Figure S1). The Cr treatments alone significantly decreased the plant height/length and yellowing/burning of leaves compared to their controls, while severe toxic symptoms were observed in rice cultivar CY927 than YLY689. Under Cr stress, seed priming with SPM reversed the Cr toxicity as noticed by the improved overall plant height/length and greenish texture of leaves (Figure S1). In control treatments, no significant difference in the GE, PG, GI and VI of rice seeds were observed. Nevertheless, the Cr-exposure alone caused a significant reduction in seed germination parameters such as GE, GP, GI, and VI relative to controls (Table 1). This decrease was more pronounced in CY927 as compared to YLY689 under Cr-alone treatments. The mean germination time (MGT) was noticeably increased in CY927 compared to YLY689 under Cr applications, whereas seed priming with SPM significantly reduced the MGT under Cr stress. In contrast, seeds primed with SPM noticeably enhanced the GE, GP, GI and VI of rice seedlings under Cr exposure (Table 1). It was observed that these germinations indices were more improved in YLY689 when compared to CY927.

Table 1. Effects of seed priming with SPM on germination energy (GE), germination percentage (GP), germination index (GI), mean germination time (MGT), Fresh weight (F/W), dry weight (D/W), vigor Index, shoot length (S/L), and root length (R/L) of rice seeds under Cr toxicity.

Genotypes	Treatments	GE (%)	GP (%)	GI	MGT (days)	F/W (g)	D/W (g)	V.I	S/L (cm)	R/L (cm)
CY927	CK	92.00 ± 5.29 [b]	98.67 ± 2.31 [a]	22.26 ± 0.61 [b]	2.70 ± 0.08 [d]	1.05 ± 0.04 [c]	0.10 ± 0.00 [a]	2.27 ± 0.09 [b]	16.47 ± 0.02 [b]	13.89 ± 0.02 [c]
	Spm	93.33 ± 1.15 [ab]	100.00 ± 0.00 [a]	23.36 ± 0.24 [ab]	2.72 ± 0.04 [d]	1.09 ± 0.02 [b]	0.10 ± 0.00 [a]	2.30 ± 0.01 [b]	16.79 ± 0.06 [b]	14.24 ± 0.06 [b]
	Cr	34.67 ± 2.31 [f]	35.33 ± 1.15 [e]	6.59 ± 0.51 [f]	3.59 ± 0.38 [a]	0.56 ± 0.01 [f]	0.06 ± 0.00 [d]	0.36 ± 0.02 [f]	8.25 ± 0.03 [f]	5.27 ± 0.03 [g]
	Spm + Cr	72.67 ± 1.15 [d]	87.33 ± 2.31 [c]	18.21 ± 0.96 [d]	3.29 ± 0.06 [b]	0.86 ± 0.01 [d]	0.08 ± 0.00 [c]	1.46 ± 0.04 [d]	12.04 ± 0.02 [e]	10.37 ± 0.02 [e]
YLY689	CK	96.67 ± 3.06 [a]	100.00 ± 0.00 [a]	25.69 ± 0.73 [a]	2.69 ± 0.10 [d]	1.18 ± 0.01 [a]	0.10 ± 0.00 [a]	2.66 ± 0.09 [a]	17.67 ± 0.02 [ab]	15.36 ± 0.02 [a]
	Spm	98.00 ± 2.00 [a]	100.00 ± 0.00 [a]	26.12 ± 0.46 [a]	2.71 ± 0.06 [d]	1.19 ± 0.02 [a]	0.10 ± 0.01 [a]	2.70 ± 0.10 [a]	18.57 ± 0.07 [a]	15.42 ± 0.07 [a]
	Cr	67.33 ± 1.15 [e]	69.33 ± 1.15 [d]	15.30 ± 0.05 [e]	3.47 ± 0.38 [ab]	0.69 ± 0.01 [e]	0.08 ± 0.00 [c]	1.17 ± 0.03 [e]	13.95 ± 0.02 [d]	9.76 ± 0.02 [f]
	Spm + Cr	79.33 ± 4.16 [c]	90.00 ± 2.00 [b]	20.43 ± 0.91 [c]	3.01 ± 0.06 [c]	0.88 ± 0.01 [d]	0.09 ± 0.00 [b]	1.80 ± 0.09 [c]	15.02 ± 0.12 [c]	12.21 ± 0.12 [d]

The same letters within a column designate that there was no significant difference at a 95% probability level at the $p < 0.05$ level according to LSD test, correspondingly.

Concerning the morphological changes, Cr-stress alone significantly reduced the plant height including shoot length (S/L), and root length (R/L), along with plant biomass including fresh weight (F/W) and dry weight (D/W). Compared with seeds primed with water (H_2O) in the nutrient only solution, seeds primed with SPM enhanced plant growth as well as biomass, especially in YLY689 under Cr stress (Table 1). These outcomes designated that seed priming with SPM mitigated the toxic effects of Cr on rice seedlings and improved the seed germination, plant growth and biomass efficiently under Cr-stress conditions, with more pronounced improvement noticed in YLY689 than CY927.

3.2. Priming Effect of SPM on Photosynthetic Pigments, Photosystem II and Gas Exchange Parameters under Cr Stress

Compared to control plants, seeds primed with SPM treatments marginally enhanced photosynthetic pigments such as chlorophyll-a (*Chl* a), chlorophyll-b (*Chl* b), chlorophyll-(a + b) (*Chl* (a + b)) and carotenoids in both rice cultivars, while treatments with Cr alone significantly reduced the *Chl* a (68.6% and 34.4%), *Chl* b (47.5% and 29.13%), *Chl* (a + b) (43.5% and 26.2%), and carotenoids (40.9% and 29.2%) in CY927 and YLY689, respectively (Figure 1). Seed priming with SPM enhanced the *Chl* a, *Chl* b, *Chl* (a + b), and carotenoids by 30.1%, 28.4%, 25.2%, 33.6% in CY927, and in 36.3%, 31.7%, 29.6%, 38.9% YLY689, respectively, under Cr stress (Figure 1a–d). Current investigations displayed that seed priming with SPM increased the *Chl* a, *Chl* b, *Chl* (a + b) and carotenoids in both rice cultivars as compared to respective controls in both treatments, with and without Cr toxicity. Likewise, the SPM application enhanced the *Fv*/*Fm* values as compared to relative controls, and the increment was 11.4% more in YLY689 than CY927 under Cr exposure (Table S2). In addition, this was further verified by taking visual images of *Fv*/*Fm*, and *Fm*. The rice leaves were exposed to 100 µM Cr stress more than untreated plants, showing as light blue/green, correspondingly, by decreasing the *Fv*/*Fm*, and *Fm* ratio (Figure 1e–h).

The exposure to Cr alone notably declined the values of gas exchange parameters including Pn, Tr, Gs, and Ci by 57.1%, 46.8%, 32.0%, 49.3% in CY927, and 38.7%, 33.1%, 23.9%, and 42.11% in YLY689 rice cultivars, correspondingly, but more decline was noticed in the values of the CY927 cultivar than YLY689 (Table S2). Nevertheless, seeds primed with SPM significantly improved the values of gas exchange indices under Cr toxicity in both CY927 and YLY689 rice cultivars. In control conditions, non-significant differences in the values of Pn, Tr, gs, Ci were noted between seeds primed with water and SPM of both cultivars. These findings revealed that seeds primed with SPM mitigated the toxic effects of Cr on photosynthetic pigments and gas exchange parameters in both cultivars, but this effect was more prominent in YLY689 relative to the CY927 cultivar.

3.3. Priming Effect of SPM on Electrolyte Leakage, Total Soluble Sugar and Total Soluble Protein under Cr Stress

In control conditions (seed priming with water and SPM without Cr toxicity), no noticeable changes in the values of electrolyte leakage (EL), total soluble sugar (TSS) and total soluble protein (TSP) were observed in the leaves of both rice cultivars (Figure 2a–c). However, Cr stress alone caused a prominent increase in the values of EL (78.4% in CY927, and 43.7% in YLY689) (Figure 2a) but decreased the TSS and TSP levels (49.2/36.5% in CY927, and 37.4/31.2% in YLY689, respectively) (Figure 2b,c), when compared to the control without Cr stress. The increasing or decreasing trends of EL, TSS and TSP values were more pronounced in CY927 (in the case of EL) than YLY689 (in the case of TSS and TSP), although results represented that seed priming with SPM reduced the EL (32.9% and 45.7%), and increased the TSS and TSP by 24.7/19.3% in CY927 and 32.2/21.6% in YLY689 cultivar, respectively, when compared to their relative controls under Cr stress (Figure 2a–c). These outcomes indicated that SPM notably minimized the EL but improved the TSS and TSP levels compared to seed priming with water, especially in YLY689 under Cr stress. These outcomes suggested that SPM notably reversed the Cr-mediated induction in EL and reduction in TSS, as well as TSP in rice seedlings.

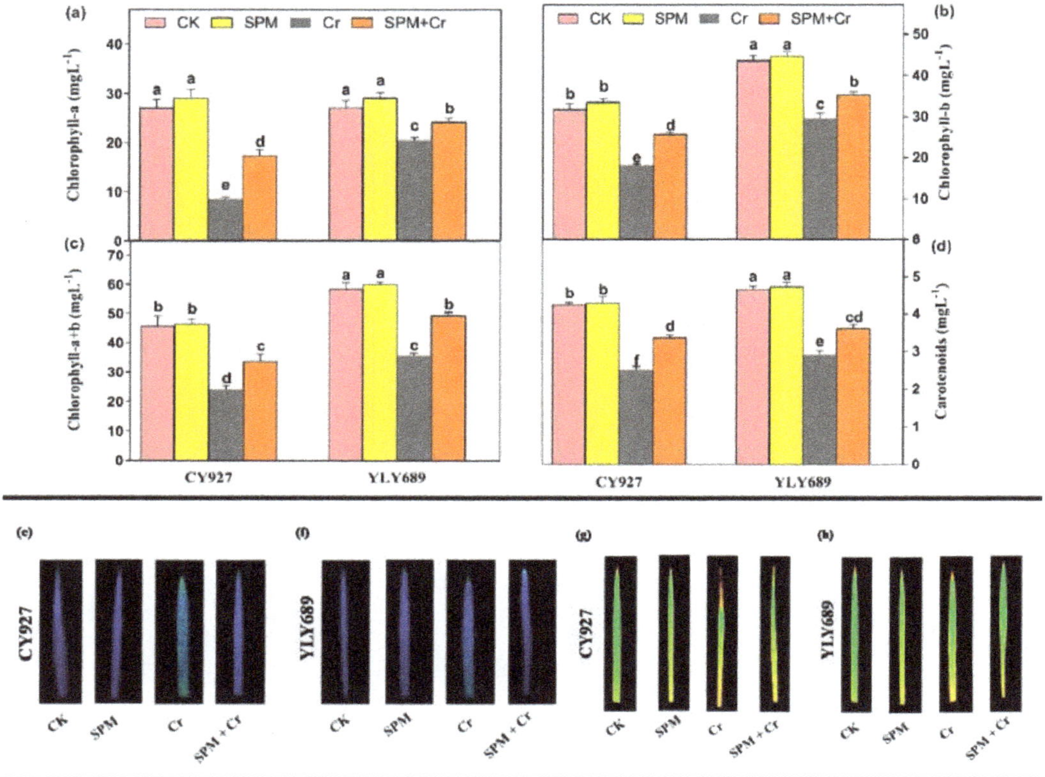

Figure 1. Effects of seed primed SPM on photosynthetic pigments (**a**) chlorophyll–a, (**b**) chlorophyll–b, (**c**) total chlorophyll–(a + b), and (**d**) carotenoids, (**e**,**f**) *Fv*/*Fm* levels in the leaves of two rice cultivars CY927 and YLY689, respectively and (**g**,**h**) *Fm* levels in the leaves of two rice cultivars CY927 and YLY689, correspondingly under chromium (Cr) stress. Values are mean ± SE of three independent replicates and different letters (a–f) above bars show a significant difference between treatments by LSD test at $p < 0.05$.

3.4. Priming Effect of SPM on Hydrogen Peroxide (H_2O_2), Superoxide ($O_2^{\bullet-}$) and Malondialdehyde (MDA) under Cr Stress

Under Cr alone treatments, the accumulation of H_2O_2 (91.2/72.5% and 67.1/55.8%), $O_2^{\bullet-}$ (76.9%/67.1% and 51.8/47.3%), and MDA (77.6/74.2% and 59.2/46.9%) were significantly induced in shoots/roots of CY927 and YLY689, correspondingly, as compared to control plants. Inclusively, a high accumulation of H_2O_2, $O_2^{\bullet-}$ and MDA was noticed in shoots rather than roots of both cultivars (Figure 2d–i). The accretion of H_2O_2, $O_2^{\bullet-}$ and MDA contents were more prominent in CY927 than in YLY689, though seed priming with SPM markedly decreased the accumulation of H_2O_2 (38.8/27.3% and 33.4/24.9%), $O_2^{\bullet-}$ (35.2/26.3% and 24.9/18.3%) and MDA (47.8/37.1% and 38.2/23.7%) in shoots/roots of both varieties (CY927 and YLY689), correspondingly, under Cr stress (Figure 2d–i). In control treatments, no significant difference was noticed among the H_2O_2, $O_2^{\bullet-}$ and MDA content values in the roots and shoots of both rice cultivars.

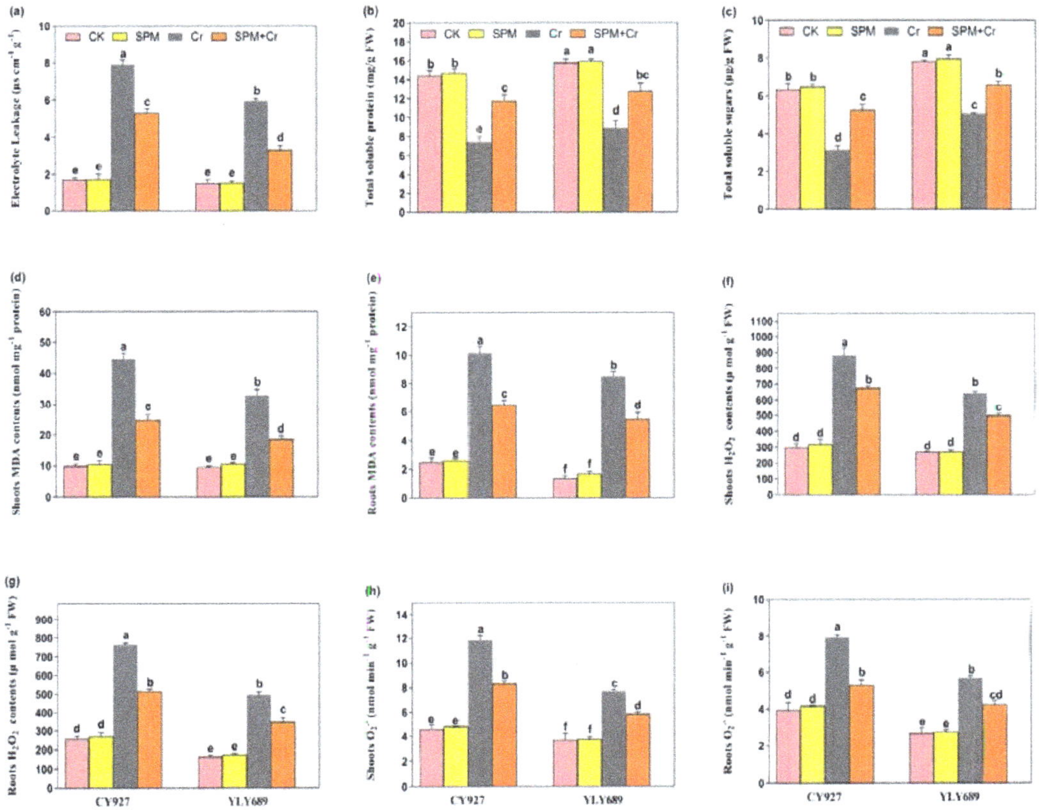

Figure 2. Effects of seed primed SPM on (**a**) electrolyte leakage (EL), (**b**) total soluble proteins, (**c**) total soluble sugar (TSP), (**d**) MDA contents in shoots, (**e**) MDA contents in roots, (**f**) H_2O_2, contents in shoots (**g**) H_2O_2, contents in roots (**h**) $O_2^{\bullet-}$ contents in the shoots and (**i**) $O_2^{\bullet-}$ contents in the roots of two rice varieties (CY927 and YLY689) against chromium (Cr) stress. Values are mean ± SE of three independent replicates and different letters (a–f) above bars show a significant difference between treatments by LSD test at $p < 0.05$.

To verify the accumulation of H_2O_2 and $O_2^{\bullet-}$ inside the shoots and roots of both cultivars of rice, leaves were stained with DAB and NBT in response to seed priming with water, SPM and Cr treatments (Figure 3a–h). Compared to untreated control, leaves treated with Cr only revealed dark brown as well as dark blue staining, correspondingly, for H_2O_2 and $O_2^{\bullet-}$. CY927 exhibited more dark staining colors than YLY689 indicating that CY927 accumulates more H_2O_2 or $O_2^{\bullet-}$ than YLY689. In contrast, control treatments (seed primed with water and SPM without Cr addition) displayed a slight staining intensity of DAB and NBT in both cultivars (Figure 3a–h). These differences revealed that seed priming with SPM sustained the plasma membrane integrity and reduced the oxidative damages against Cr-induced overproduction of ROS in both rice cultivars (particularly in YLY689) in comparison to seed priming with water against Cr toxicity conditions.

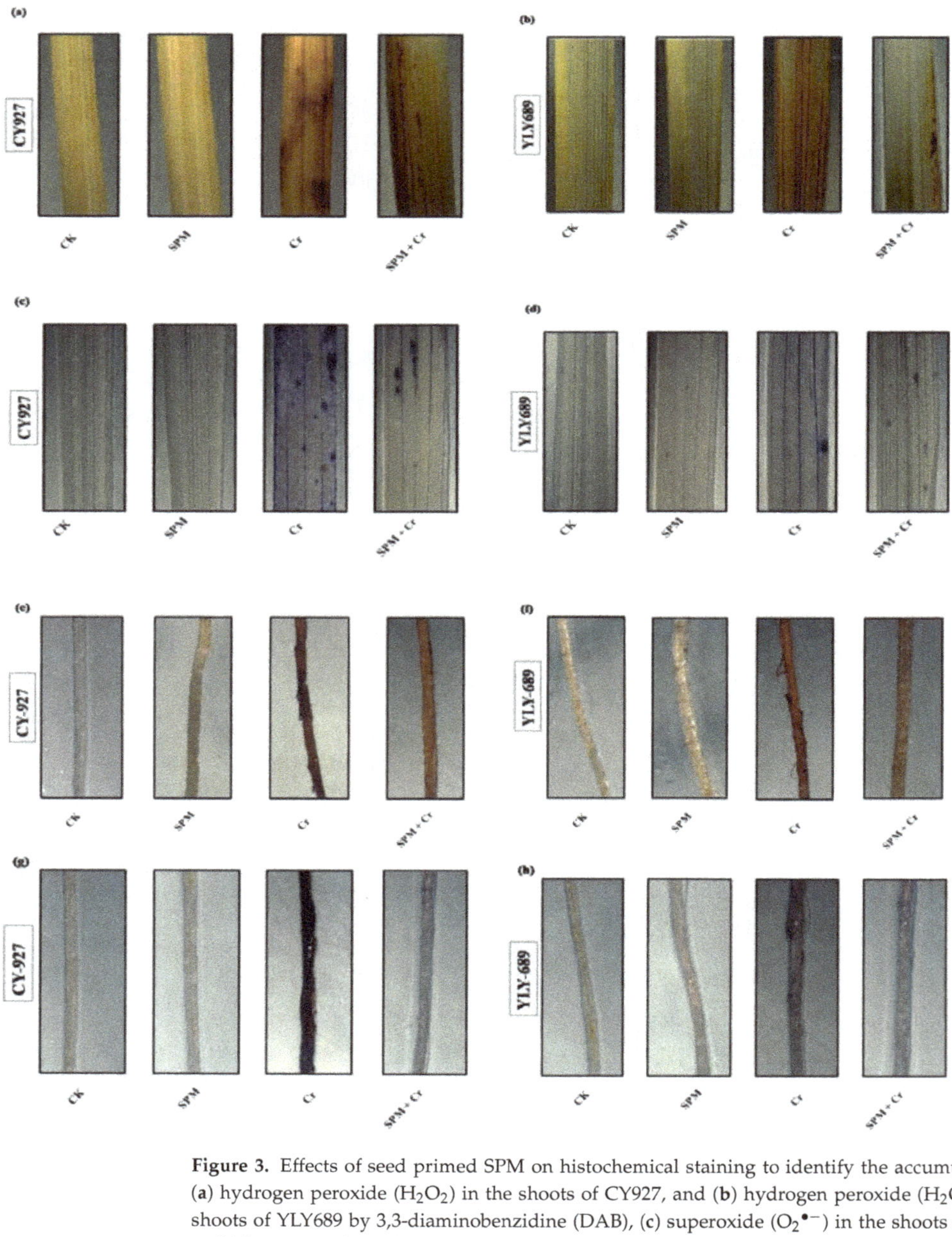

Figure 3. Effects of seed primed SPM on histochemical staining to identify the accumulation of (**a**) hydrogen peroxide (H_2O_2) in the shoots of CY927, and (**b**) hydrogen peroxide (H_2O_2) in the shoots of YLY689 by 3,3-diaminobenzidine (DAB), (**c**) superoxide ($O_2^{\bullet-}$) in the shoots of CY927, and (**d**) superoxide ($O_2^{\bullet-}$) in the shoots of YLY689 by nitro blue tetrazolium (NBT), (**e**) hydrogen peroxide (H_2O_2) in the roots of CY927, and (**f**) hydrogen peroxide (H_2O_2) in the roots of YLY689 by 3,3-diaminobenzidine (DAB), (**g**) superoxide ($O_2^{\bullet-}$) in the roots of CY927, (**h**) superoxide ($O_2^{\bullet-}$) in the roots of YLY689 by nitro blue tetrazolium (NBT).

3.5. Priming Effect of SPM on Cr Contents and Mineral Nutrients under Cr Stress

Under Cr alone treatments, the Cr accumulation was more pronounced in roots than shoots of both rice cultivars in seed priming with water under Cr stress. The Cr augmentation was reduced significantly in both shoots and roots in seed primed with SPM under Cr exposure. Micro- as well as macro-nutrients' (Na, Mg, K, P, Ca, Mn, Fe, Cu, and Zn) uptake and translocation imbalance were observed under Cr toxicity compared to their controls. The decrease was more noticeable in roots than shoots of both CY927 and YLY689 cultivars (Tables S3 and S4). The uptake of Cr was greater in YLY689 as compared to the CY927. Under Cr exposure, seeds primed with SPM remarkably enhanced the Mn, Zn, Cu, P, Fe, and K in both roots and shoots of both cultivars. Moreover, SPM treatment modulated the uptake of Na inside both roots and shoots of rice cultivars (Tables S3 and S4). These outcomes revealed that seed priming with SPM restricted the Cr uptake, accumulation and translocation inside rice plants' parts and further improved the macro- and micro-nutrient balance mandatory for plant development.

3.6. Priming Effect of SPM on Antioxidative Enzyme Activities under Cr Stress

Under Cr alone treatments, an upsurge in the activities of antioxidant (SOD, CAT, POD, and APX) activities were noticed inside the shoots and roots of both cultivars, but this increment was more conspicuous in shoots compared to roots. The outcomes demonstrated, compared to control plants, the Cr toxicity significantly surged the enzymatic activities i.e., SOD (58.1/47.9, 62.9/54.7%), APX (58.9/52.2, 65.9/63.1%), POD (70.4/58.5, 82.1/67.3%), and CAT (56.2/48.4, 64.9/53.2%) in shoots/roots of both cultivars CY927, and YLY689, respectively (Figure 4a–h). Interestingly, seed primed with SPM further enhanced the enzymatic activities such as SOD (29.1/16.2%, 38.7/19.5%), APX (35.7/26.6, 41.1/34.3%), and POD activity by 23.9/22.4, 28.6/25.9%, and CAT activity by 39.5/24.7%, 44.4/32.9% inside shoots/roots of both cultivars CY927, and YLY689, correspondingly (Figure 4a–h). Our outcomes revealed that seeds primed with SPM developed a higher possibility of scavenging the reactive oxygen species and increased plant tolerance capability with greater effect in YLY689 than CY927.

3.7. Priming Effect of SPM on the Production of Endogenous ABA, JA and Free SA Levels under Cr Stress

The ABA, JA and SA levels were examined in seeds primed with water or SPM, with and without Cr stress. The exposure of Cr alone led to elevated contents of ABA (2.83 and 2.49-fold), JA (3.32 and 3.29-fold) and SA (2.48 and 1.89-fold) in both varieties CY927 and YLY689, individually, when compared to controls (Figure 5a–c). Under Cr stress, seed priming with SPM applications significantly decreased the ABA level (Figure 5a), whereas no significant difference was observed in the levels of JA under Cr treatments (Figure 5b). Nevertheless, the level of SA was enhanced significantly (2.23 and 1.76-fold) in both rice cultivars (CY927 and YLY689), individually, by the applications of seed primed SPM under Cr stress (Figure 5c). A similar pattern was seen in the seedlings of both rice cultivars, with YLY689 displaying more obvious effects than CY927.

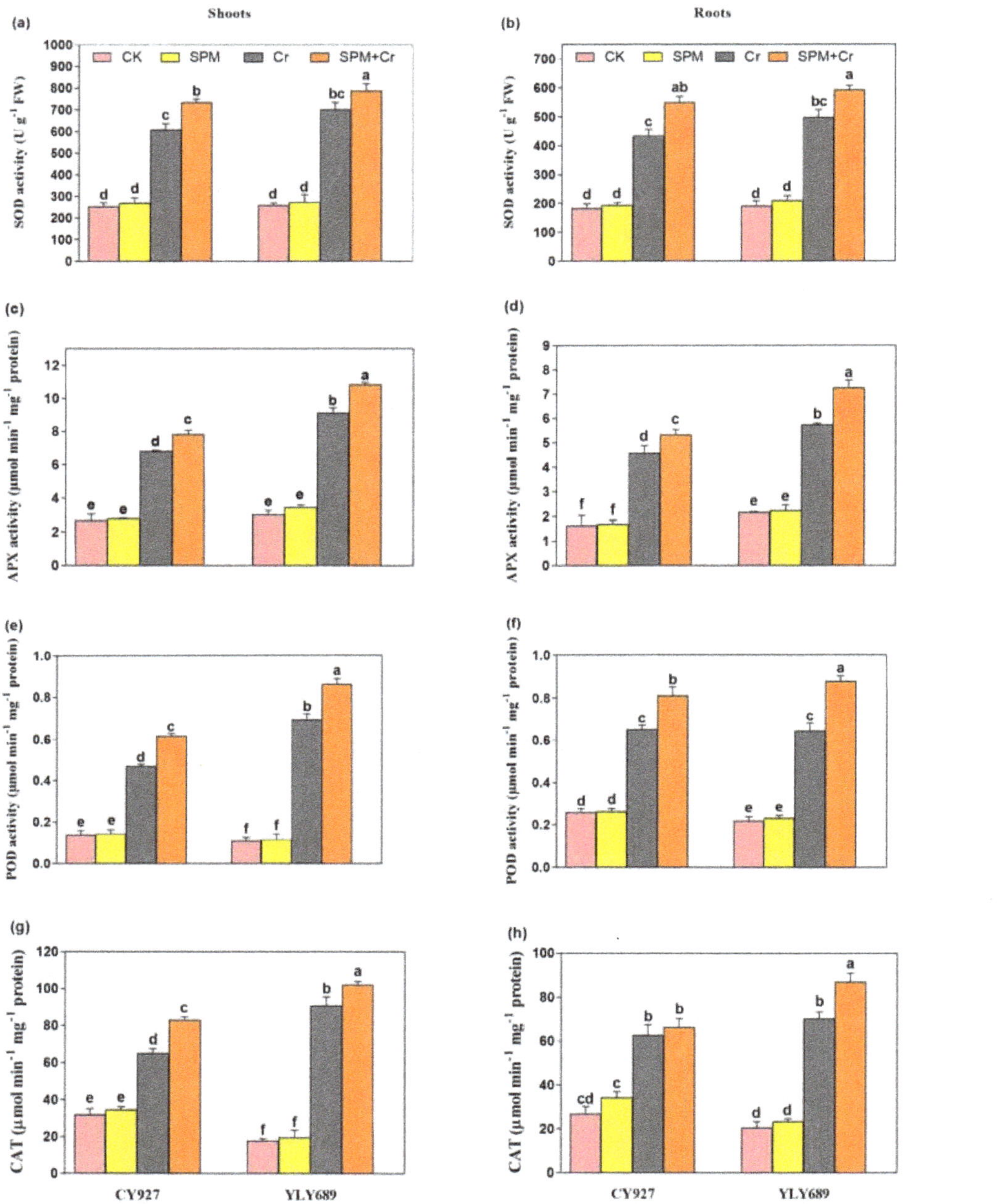

Figure 4. Effects of SPM on the activities of (**a**) superoxide dismutase (SOD) in shoots, (**b**) superoxide dismutase (SOD) in roots, (**c**) ascorbate peroxidase (APX) in shoots, (**d**) ascorbate peroxidase (APX) in roots, (**e**) peroxidase (POD) in shoots, (**f**) peroxidase (POD) in roots, (**g**) catalase (CAT) in shoots, and (**h**) catalase (CAT) in roots of both rice varieties (CY927 and YLY689) under chromium (Cr) stress. Values are mean ± SE of three independent replicates and different letters (a–f) above bars show a significant difference between treatments by LSD test at $p < 0.05$.

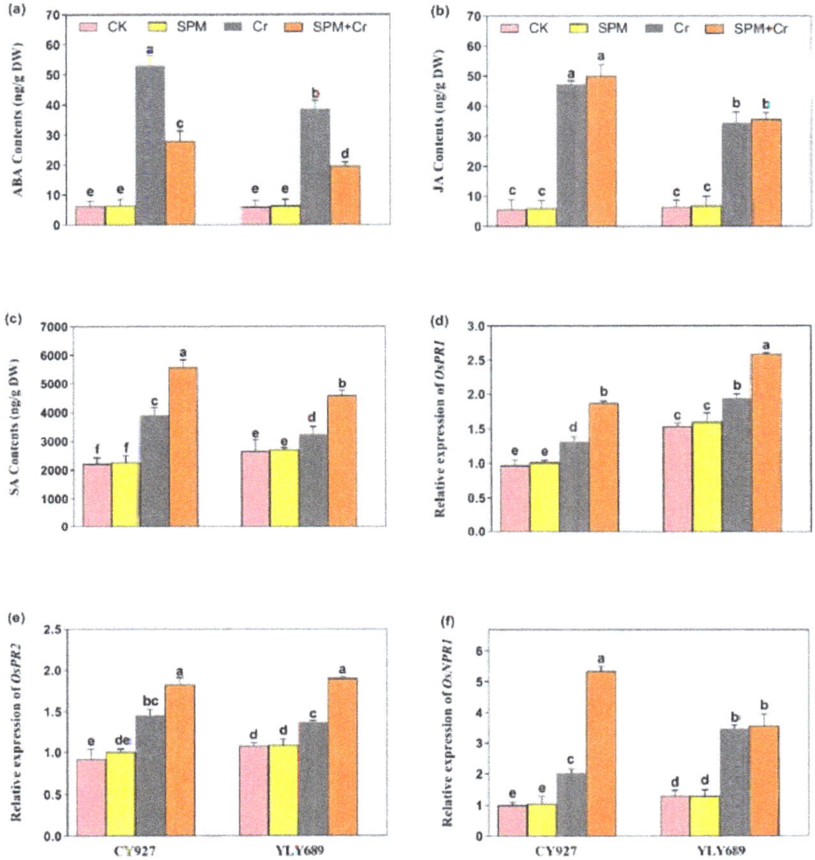

Figure 5. Effects of SPM on the endogenous production of phytohormones including (**a**) abscisic acid (ABA), (**b**) Jasmonic acid (JA), (**c**) salicylic acid (SA) and expressions of phytohormones-associated genes (**d**) *PR1*, (**e**) *PR2*, and (**f**) *NPR1* in two rice cultivars (CY927 and YLY689) under chromium (Cr) stress. Values are mean ± SE of three independent replicates and different letters (a–f) above bars show a significant difference between treatments by LSD test at $p < 0.05$.

3.8. Priming Effect of SPM on Hormone-Associated Gene Expression Analysis under Cr Stress

Under Cr stress, the transcript levels of *PR1*, and *PR2* genes were subsequently upregulated in both CY927 and YLY689 compared to relative controls. Furthermore, the transcription levels of both these genes were markedly stimulated within seeds primed by SPM against Cr-induced stress as compared to the treatments of Cr alone in both rice cultivars (Figure 5d,e). The expression levels were noticeably upregulated in CY927 as compared to YLY689 ($p < 0.01$). Likewise, the upregulation in *NPR1* levels was noticed under Cr stress in both rice cultivars. Nevertheless, their expression levels were further increased in the seeds treated with SPM priming under Cr exposure (Figure 5f). An identical drift was noted in both rice varieties. The upregulation was more pronounced in the expression of the *NPR1* gene as compared to *PR1*, and *PR2* genes. This indicated that *NPR1* has a dynamic role in stimulating the phytohormone synthesis and mitigation of Cr toxicity. In addition, gene expression analysis supported the notion that SPM mediated modifications in SA contents under Cr applications. Our findings clearly represented that SPM has extensive participation in rice tolerance against Cr stress and modulates the levels of transcription of certain stress-responsive genes associated with phytohormones.

3.9. Priming Effect of SPM on Cellular Ultrastructural Changes under Cr Stress

Within control group (seeds primed with water and SPM without Cr stress), the leaf ultrastructure of both cultivars CY927 and YLY689 displayed a well-shaped cell wall, vigorous chloroplast, mitochondria, vacuoles, peroxisomes and the usual organized granule thylakoids inside grana containing normal granum thylakoids, stroma thylakoids, starch grains and plastoglobuli (Figure 6A,B,E,F). However, the ultrastructural analysis of rice cultivar (CY92, seeds primed with water) compared with Cr toxicity indicated a ruptured nuclear membrane, expansion of double-layered nuclear membrane, swollen mitochondria and damaged chloroplast (Figure 6C). Relatively, seeds primed with SPM presented a developed nuclear membrane, normal structure of chloroplast, plastoglobuli, and less swollen mitochondria than seeds primed with water beneath Cr exposure in CY927 (Figure 6D). In YLY689, seeds priming water (with Cr disclosure) revealed structural abnormalities such as thylakoids disruption, slightly irregular structured starch grains, mitochondrial damages, and swollen chloroplast (Figure 6G) than their relative control groups (Figure 6E,F). Interestingly, the seeds primed with SPM under Cr treatments minimized the structural damages induced by Cr stress in the mesophyll cells of cultivar YLY689 as observed by the well-developed cell wall, peroxisomes, vacuoles, well-shaped stroma thylakoids, starch granules, chloroplast, granum thylakoids, plastoglobuli and matured mitochondria (Figure 6H).

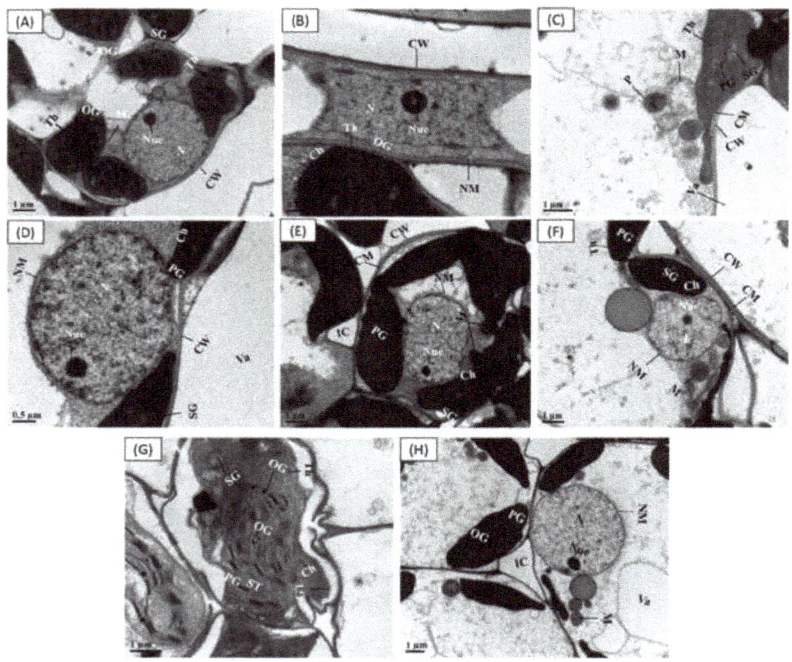

Figure 6. Effect of seeds primed with SPM on cellular ultra-structural changes in the leaves of two different rice cultivars (CY927 and YLY689) against chromium (Cr). (**A**) Leaf mesophyll cell of CY927 (primed with water) at control level, (**B**) Leaf mesophyll cell of CY927 (primed with Spm) at control level, (**C**) Leaf mesophyll cell of CY927 (primed with water) under Cr stress, (**D**) Leaf mesophyll cell of CY927 (primed with Spm) under Cr toxicity, (**E**) Leaf mesophyll cell of YLY689 (primed with water) at control level, (**F**) Leaf mesophyll cell of YLY689 (primed with Spm) at control level, (**G**) Leaf mesophyll cell of YLY689 (primed with water) under Cr stress, (**H**) Leaf mesophyll cell of YLY689 (primed with Spm) under Cr toxicity. Abbreviations: N (nucleus), SG (starch grain), NM (nuclear membrane), PG (plastogloboli), CW (cell wall), Va (vacuole), Ch (chloroplast), CM (cell membrane), (chloroplast), GT (granule thylakoids); M (mitochondria), ST (stomata thylakoids).

4. Discussion

Soil contamination with chromium (Cr) has become a serious concern for rice researchers due to its direct entry into food chains and severe health issues globally [48]. Thus, our prime focus was to investigate latent approaches to diminish Cr-accumulation and its associated phytotoxicity. For this purpose, we utilized seed priming with spermine (SPM) in minimizing the Cr-tempted toxic effects by targeting the seed germination indices, growth and biomass, photosynthetic apparatus, soluble protein levels, nutrients balance, oxidative damages, antioxidant enzymes' defense systems, phytohormone production and modifications in cellular ultrastructure. Our findings showed that the seed germination parameters, i.e., GE, GP, VI, GI and MGT, were significantly reduced by Cr exposure in both rice cultivars (Table 1) which ultimately minimized the plant development, biomass (fresh and dry weight, shoot or root lengths) (Table 1, Figure S1). Comparatively, seed primed SPM caused readjustments of seed germination indices, plant growth and biomass attributes under Cr toxicity (Tables 1 and S2; Figure S1). These alterations were more substantial in YLY689 (tolerant variety) than CY927 (sensitive variety). Recent studies revealed that Cr-stress inhibited the rates of seed germination indices [1]. This may be due to the alteration in seed coat structure and reactions with ions, electrons or intact radicals. Consequently, sugar supply to the seeds was compromised [49,50]; although SPM applications aided the conversion of starch hydrolysis into soluble sugar that improved the respiration of seeds and starch metabolism, and consequently enhanced plant growth and biomass production [51].

Furthermore, Cr also inhibited the biomass of rice seedling [1,11], maize [22], and *Brassica napus* [7] seedlings that might be associated with a disturbance of photosynthetic systems and nutrients imbalance. As predicted, Cr stress significantly reduced the chlorophyll pigments (Figure 1a–d), Fv/Fm and Fm values (Figure 1e–h), gas exchange parameters (Table S2) and nutrient accumulation (Table S3) in both rice varieties, especially CY927. The possible reason is that Cr caused inhibition inside the production of chlorophyll biosynthetic enzymes (δ-amino-levulinic acid dehydratase and protochlorophyllide reductase), which led to the impairment of chlorophyll biosynthetic pathways [52], while SPM priming mitigated Cr stress on both cultivars, and precisely in YLY689. Potentially, seed primed SPM boosted the plant growth traits (Table 1) by eliminating Cr accumulation via maintaining the nutrient uptake balance (Table S3), cell cycle and cell division, which led to the improvement in chlorophyll pigments (Figure 1a–d) and gas exchange indices (Table S2) under Cr exposure. Most probably, SPM application enhanced the levels of photosynthetic apparatus by regulating the nutrient uptake which might have modulated the expression of genes associated with metal homeostasis, lessening the Cr accretion under heavy metal stress [53,54]. It has been reported that photosynthetic efficiency maintains the level of total soluble sugar (TSS) and total soluble protein (TTP) that might be stimulating the metabolites' induction, and further helps to detoxify the Cr stress under Cr-induced toxicity [55,56]. Our upshots specified that the TSS and TSP are directly associated with the photosynthetic contents (Figure 2b,c) and the rates of gas exchange indices (Table S2). Seed priming with SPM might have improved the photosynthetic and biosynthetic enzyme activities (i.e., δ-aminolaevulinic acid and protochlorophyllide reductase) that regulated the photosynthetic biosynthesis pathways [57] and subsequently improved the TSS and TSP. Our results revealed that seed primed SPM minimized the Cr-indulged oxidative damages and minimized the electrolyte leakage.

In our current investigations, Cr exposure enhanced the extra accumulation of H_2O_2 and $O_2^{\bullet-}$ (Figure 2f–i), MDA contents (Figure 2d,e), electrolyte leakage (Figure 2a), particularly in CY927, and caused oxidative or membrane damages and lipid peroxidation as noticed in rice seedlings [57,58], Indian mustard [58] and tomato [59]. Seeds primed with SPM may have improved the ROS hemostasis by decreasing over-production of ROS (Figure 2d–i) that revealed the role of SPM in mitigating oxidative damages and subsequently improving photosynthetic performance. Probably, SPM improved the catabolism, maintained the cellular metabolism, and inhibited the overproduction of free radicals which controlled the level of MDA contents, electrolyte leakage, and ROS over-generation. Additionally, seed primed SPM treatment provided stability to cellular membranes by reducing the MDA contents (Figure 2d,e) and electrolyte leakage (Figure 2a) in rice seedlings under Cr-induced stress, thus enhancing the photosynthetic apparatus, plant growth and biomass production, specifically in YLY689. SPM capability in reducing ROS (H_2O_2 and $O_2^{\bullet-}$) over production was verified by histochemical straining, which matches recent observations under conditions of metal stress [21,60].

To cope with the Cr-triggered oxidative damage, plants stimulate their antioxidative defense mechanisms as indicated by the upregulation in enzymatic activities such as SOD, APX, POD and CAT in both rice cultivars (Figure 4a–h). Similar outcomes were documented in a recent investigation that plants increase their antioxidative defense mechanism to scavenge Cr-induced cellular oxidative damage [1]. Seed priming with SPM further stimulated the antioxidative enzyme activities and reduced the ROS or MDA accumulation, which revealed that SPM boosted the antioxidants' defense mechanisms and reduced the oxidative damage, thus maintaining ROS homeostasis in rice seedlings. Previous studies documented that PAs regulate antioxidant activities to eliminate the ROS extra generation, and hence maintain the redox balance, and thus improve the growth traits of wheat [21], *Brassica juncea* [61], and mung bean [20] under heavy metals stress. Expectedly, the sole treatment of Cr significantly restricted the accumulation of nutrients (Na, K, P, Ca, Mn, Zn, Cu and Fe) which eventually leads to lowered biomass production. The excessive Cr accumulation in plant parts may limit the accessibility of nutrients and disturb their hemostasis [1,7]. SPM applications displayed the capably to improve the balance of nutrients by lowering the Cr accretion via a vacuole sequestering approach and enhanced the tolerance capacity of YLY689 against Cr stress than CY927. Likewise, earlier investigations supported our results that the reduction in Cr-facilitation can improve the nutrients' bioavailability within the tissues of *Raphanus sativus* [62] and maize [22].

It has been reported that phytohormones, viz., SA and JA interactions with PAs including SPM and spermidine (SPD), modulated plant physiological attributes, and enhanced antioxidant enzyme activities in scavenging the extra accumulation of ROS under adverse environmental stresses [63–66]. Under heavy metal stress conditions, SA levels were enhanced inside plant parts that helped to reduce the extra accumulation of ROS [67,68]. Remarkably, seeds primed with SPM significantly improved the endogenous levels of SA rather than JA under Cr toxicity in both rice varieties, especially in YLY689 as compared to CY927 (Figure 5b,c). Hereafter, we targeted the gene expression of SA-related genes (*PR1*, *PR2*, and *NPR1*) to analyze the genetic responses under Cr stress (Figure 5d–f). We verified that Cr significantly stimulated the endogenous SA levels that clarified the potential roles of SA in alleviating Cr toxicity in rice seedlings. Nevertheless, Cr displayed no significant differences in the levels of endogenous JA (Figure 5b). This might be due to the fact that Cr stress, in addition to the consequent synthesis of JA, stimulated the lipases to deliver a substrate for lipoxygenase [33,69]. Still, the molecular basis of this phenomenon needed thorough exploration.

Notably, Cr-alone treatments enhanced the endogenous abscisic acid (ABA) levels, while seed priming with SPM significantly lowered them (Figure 5a). Conceivably, PAs may regulate the ABA levels to modulate the signal transduction pathways, and specific protein's activation and repression, which supplementarily controls the cell growth or cell death [70]. Therefore, PAs interact with phytohormones to maintain the plant's growth attributes, stomatal conductance, and chlorophyll contents to alleviate the nastiest effects of multiple toxicities [66].

The ultrastructural analysis revealed that Cr alone treatments severely damaged the cellular organelles (Figure 6A–H) and CY927 demonstrated more damaging effects than YLY689. The extra accumulation of Cr within cell organs ruptured the chloroplast structure by disturbing the granule thylakoid development, swollen mitochondria and damages to thylakoid membrane, as supported by earlier studies [71]. The Cr-induced chloroplast damage led to the reduction in chlorophyll contents which halted the photosynthetic rate and gas exchange parameters [71]. Conversely, seed primed SPM applications reestablished the structures of granule thylakoids, stroma, chloroplasts, thylakoid membrane, cell wall, vacuoles and mitochondria under Cr stress, specifically in YLY689 (Figure 6d,h). Possibly, SPM repaired the cellular ultrastructure (chloroplast and thylakoids) that helps to reduce the overproduction of RCS and improve the photosynthetic apparatus, thus leading to higher plant development and biomass against heavy metals' exposure [72].

5. Conclusions

A recent study reported that seeds primed with SPM exhibited a great capability to mitigate Cr-induced adverse effects in rice seedlings of both cultivars. Cr (100 µM) disclosure in nutrients media severely damaged the plant biomass/growth, impaired the photosynthetic pigment, imbalanced the micro- and macro-nutrients, augmented the cellular oxidative damage, altered the endogenous phytohormone level and cellular ultrastructure, and desynchronized the antioxidant defense system in rice seedlings, especially in CY927. Simultaneously, seeds primed with SPM ameliorated biomass/growth production, chlorophyll pigments, total soluble sugar, total soluble protein, maintained nutrient balance, and improved antioxidants' defense system by lowering the Cr uptake, accumulation, translocation, lipid peroxidation, oxidative damage, and electrolyte leakage. Moreover, SPM modulated the endogenous phytohormones content, as well as the transcription level of SA-related genes (*PR1*, *PR2*, and *NPR1*) under Cr stress. Henceforth, it was revealed that SPM can enhance plant tolerance and mitigate the Cr phytotoxicity in different rice varieties. Furthermore, Cr stress damaged the rice cultivar CY927 more than YLY689. YLY689 was found more tolerant to Cr stress and SPM further enhanced tolerance capacity, while CY927 displayed susceptibility to Cr stress whose negative effects were reversed by SPM.

Supplementary Materials: The following supporting information can be downloaded at: https://www.mdpi.com/article/10.3390/antiox11091704/s1, Figure S1. Effect of seed primed with SPM on phenotypic changes of two different rice cultivars under Cr toxicity. (a) Cr-induced phenotypical changes on the CY927 cultivar and mitigation effect of SPM under Cr stress. (b) Cr-induced phenotypical changes on the YLY689 cultivar and mitigation effect of SPM under Cr exposure, Table S1. Primers information.

Author Contributions: F.B. and Z.U. have equally contributed to this work as first author. Y.G., J.A.B. and P.A. were involved in conceptualization, Y.G., F.B. and Z.U. designed the experiment. F.B., Z.U., M.N. and A.A. performed the experiment and wrote the manuscript. F.B., Z.U., M.N., B.Z. and A.A. helped in writing, while W.Z. and Y.G. revised/edited the manuscript. B.Z. and P.K. performed the statistical analysis. All authors have read and agreed to the published version of the manuscript.

Funding: This research was supported by the Science and Technology Department of Zhejiang Province (2022C02034), Zhejiang Provincial Natural Science Foundation (No. LY21C130006), National Natural Science Foundation of China (No. 32072127), Dabeinong Funds for Discipline Development and Talent Training in Zhejiang University, and Collaborative Innovation Center for Modern Crop Production co-sponsored by Province and Ministry (CIC-MCP). National Natural Science Foundation of China (No. 31800386).

Institutional Review Board Statement: Not applicable.

Informed Consent Statement: Not applicable.

Data Availability Statement: The data are contained within the article and Supplementary Materials.

Acknowledgments: The authors would like to extend their sincere appreciation to the Researchers Supporting Project Number (RSP-2021/350), King Saud University, Riyadh, Saudi Arabia.

Conflicts of Interest: The authors declare no conflict of interest.

References

1. Basit, F.; Chen, M.; Ahmed, T.; Shahid, M.; Noman, M.; Liu, J.; An, J.; Hashem, A.; Fahad Al-Arjani, A.-B.; Alqarawi, A.A. Seed priming with brassinosteroids alleviates chromium stress in rice cultivars via improving ROS metabolism and antioxidant defense response at biochemical and molecular levels. *Antioxidants* **2021**, *10*, 1089. [CrossRef] [PubMed]
2. Miyan, M.A. Droughts in Asian least developed countries: Vulnerability and sustainability. *Weather Clim. Extrem.* **2015**, *7*, 8–23.
3. Ertani, A.; Mietto, A.; Borin, M.; Nardi, S. Chromium in agricultural soils and crops: A review. *Water Air Soil Pollut.* **2017**, *228*, 190.
4. Lopez-Luna, J.; Gonzalez-Chavez, M.; Esparza-Garcia, F.; Rodriguez-Vazquez, R. Toxicity assessment of soil amended with tannery sludge, trivalent chromium and hexavalent chromium, using wheat, oat and sorghum plants. *J. Hazard. Mater.* **2009**, *163*, 829–834. [CrossRef]
5. Ahmad, R.; Ali, S.; Hannan, F.; Rizwan, M.; Iqbal, M.; Hassan, Z.; Akram, N.A.; Maqbool, S.; Abbas, F. Promotive role of 5-aminolevulinic acid on chromium-induced morphological, photosynthetic, and oxidative changes in cauliflower (*Brassica oleracea botrytis* L.). *Environ. Sci. Pollut. Res.* **2017**, *24*, 8814–8824.
6. Basit, F.; Liu, J.; An, J.; Chen, M.; He, C.; Zhu, X.; Li, Z.; Hu, J.; Guan, Y. Brassinosteroids as a multidimensional regulator of plant physiological and molecular responses under various environmental stresses. *Environ. Sci. Pollut. Res.* **2021**, *28*, 44768–44779.
7. Ulhassan, Z.; Gill, R.A.; Huang, H.; Ali, S.; Mwamba, T.M.; Ali, B.; Huang, Q.; Hamid, Y.; Khan, A.R.; Wang, J. Selenium mitigates the chromium toxicity in Brassicca napus L. by ameliorating nutrients uptake, amino acids metabolism and antioxidant defense system. *Plant Physiol. Biochem.* **2019**, *145*, 142–152. [PubMed]
8. Yang, D.; Yang, Q.; Fu, N.; Li, S.; Han, B.; Liu, Y.; Tang, Y.; Guo, X.; Lv, Z.; Zhang, Z. Hexavalent chromium induced heart dysfunction via Sesn2-mediated impairment of mitochondrial function and energy supply. *Chemosphere* **2021**, *264*, 128547. [PubMed]
9. Singh, S.; Prasad, S.M. Regulation of chromium toxicity tolerance in tomato and brinjal by calcium and sulfur through nitric oxide: Involvement of enzymes of sulfur assimilation and the ascorbate-glutathione cycle. *Environ. Exp. Bot.* **2019**, *166*, 103789.
10. Handa, N.; Kohli, S.K.; Sharma, A.; Thukral, A.K.; Bhardwaj, R.; Abd_Allah, E.F.; Alqarawi, A.A.; Ahmad, P. Selenium modulates dynamics of antioxidative defence expression, photosynthetic attributes and secondary metabolites to mitigate chromium toxicity in *Brassica juncea* L. plants. *Environ. Exp. Bot.* **2019**, *161*, 180–192.
11. Yang, S.; Ulhassan, Z.; Shah, A.M.; Khan, A.R.; Azhar, W.; Hamid, Y.; Hussain, S.; Sheteiwy, M.S.; Salam, A.; Zhou, W. Salicylic acid underpins silicon in ameliorating chromium toxicity in rice by modulating antioxidant defense, ion homeostasis and cellular ultrastructure. *Plant Physiol. Biochem.* **2021**, *166*, 1001–1013. [PubMed]
12. Ulhassan, Z.; Gill, R.A.; Ali, S.; Mwamba, T.M.; Ali, B.; Wang, J.; Huang, Q.; Aziz, R.; Zhou, W. Dual behavior of selenium: Insights into physio-biochemical, anatomical and molecular analyses of four *Brassica napus* cultivars. *Chemosphere* **2019**, *225*, 329–341. [PubMed]
13. Kamran, M.; Wang, D.; Alhaithloul, H.A.S.; Alghanem, S.M.; Aftab, T.; Xie, K.; Lu, Y.; Shi, C.; Sun, J.; Gu, W. Jasmonic acid-mediated enhanced regulation of oxidative, glyoxalase defense system and reduced chromium uptake contributes to alleviation of chromium (VI) toxicity in choysum (*Brassica parachinensis* L.). *Ecotoxicol. Environ. Saf.* **2021**, *208*, 111758.
14. Takahashi, T. Plant Polyamines. 2020. Available online: https://www.mdpi.com/2223-7747/9/4/511 (accessed on 8 July 2022).
15. Li, L.; Gu, W.; Li, J.; Li, C.; Xie, T.; Qu, D.; Meng, Y.; Li, C.; Wei, S. Exogenously applied spermidine alleviates photosynthetic inhibition under drought stress in maize (*Zea mays* L.) seedlings associated with changes in endogenous polyamines and phytohormones. *Plant Physiol. Biochem.* **2018**, *129*, 35–55.
16. Rathinapriya, P.; Pandian, S.; Rakkammal, K.; Balasangeetha, M.; Alexpandi, R.; Satish, L.; Rameshkumar, R.; Ramesh, M. The protective effects of polyamines on salinity stress tolerance in foxtail millet (*Setaria italica* L.), an important C4 model crop. *Physiol. Mol. Biol. Plants* **2020**, *26*, 1815–1829. [PubMed]
17. Gao, C.; Sheteiwy, M.S.; Han, J.; Dong, Z.; Pan, R.; Guan, Y.; Alhaj Hamoud, Y.; Hu, J. Polyamine biosynthetic pathways and their relation with the cold tolerance of maize (*Zea mays* L.) seedlings. *Plant Signal. Behav.* **2020**, *15*, 1807722. [PubMed]
18. Jing, J.; Guo, S.; Li, Y.; Li, W. The alleviating effect of exogenous polyamines on heat stress susceptibility of different heat resistant wheat (*Triticum aestivum* L.) varieties. *Sci. Rep.* **2020**, *10*, 7467.
19. Shah, A.A.; Ahmed, S.; Ali, A.; Yasin, N.A. 2-Hydroxymelatonin mitigates cadmium stress in cucumis sativus seedlings: Modulation of antioxidant enzymes and polyamines. *Chemosphere* **2020**, *243*, 125308.
20. Nahar, K.; Rahman, M.; Hasanuzzaman, M.; Alam, M.M.; Rahman, A.; Suzuki, T.; Fujita, M. Physiological and biochemical mechanisms of spermine-induced cadmium stress tolerance in mung bean (*Vigna radiata* L.) seedlings. *Environ. Sci. Pollut. Res.* **2016**, *23*, 21206–21218.
21. Taie, H.A.; El-Yazal, M.A.S.; Ahmed, S.M.; Rady, M.M. Polyamines modulate growth, antioxidant activity, and genomic DNA in heavy metal–stressed wheat plant. *Environ. Sci. Pollut. Res.* **2019**, *26*, 22338–22350.
22. Naz, R.; Sarfraz, A.; Anwar, Z.; Yasmin, H.; Nosheen, A.; Keyani, R.; Roberts, T.H. Combined ability of salicylic acid and spermidine to mitigate the individual and interactive effects of drought and chromium stress in maize (*Zea mays* L.). *Plant Physiol. Biochem.* **2021**, *159*, 285–300. [PubMed]
23. Groppa, M.D.; Tomaro, M.L.; Benavides, M.P. Polyamines and heavy metal stress: The antioxidant behavior of spermine in cadmium-and copper-treated wheat leaves. *Biometals* **2007**, *20*, 185–195. [PubMed]
24. Paul, S.; Banerjee, A.; Roychoudhury, A. Role of polyamines in mediating antioxidant defense and epigenetic regulation in plants exposed to heavy metal toxicity. In *Plants under Metal and Metalloid Stress*; Springer: Singapore, 2018; pp. 229–247.

25. Pál, M.; Csávás, G.; Szalai, G.; Oláh, T.; Khalil, R.; Yordanova, R.; Gell, G.; Birinyi, Z.; Németh, E.; Janda, T. Polyamines may influence phytochelatin synthesis during Cd stress in rice. *J. Hazard. Mater.* **2017**, *340*, 272–280. [PubMed]
26. Zheng, Y.; Hu, J.; Zhang, S.; Gao, C.; Song, W. Identification of chilling-tolerance in maize inbred lines at germination and seedling growth stages. *J. Zhejiang Univ. (Agric. Life Sci.)* **2006**, *32*, 41–45.
27. Lichtenthaler, H.K.; Wellburn, A.R. Determinations of total carotenoids and chlorophylls a and b of leaf extracts in different solvents. *Analysis* **1983**, *11*, 591–592. [CrossRef]
28. Antonio Pasqualini, A.; Tadeu Paulino, V.; Maria Duarte, K.; Hansen Rosolen, F.; Menegante Neri, A. Chlorophyll evaluation methods in *Brachiaria brizantha brs Piatã* added with poultry litter compost. *B. Indústr. Anim.* **2013**, *70*.
29. Zhou, W.; Leul, M. Uniconazole-induced alleviation of freezing injury in relation to changes in hormonal balance, enzyme activities and lipid peroxidation in winter rape. *Plant Growth Regul.* **1998**, *26*, 41–47. [CrossRef]
30. Ista, L.K.; Callow, M.E.; Finlay, J.A.; Coleman, S.E.; Nolasco, A.C.; Simons, R.H.; Callow, J.A.; Lopez, G.P. Effect of substratum surface chemistry and surface energy on attachment of marine bacteria and algal spores. *Appl. Environ. Microbiol.* **2004**, *70*, 4151–4157. [CrossRef]
31. Bradford, M.M. A rapid and sensitive method for the quantitation of microgram quantities of protein utilizing the principle of protein-dye binding. *Anal. Biochem.* **1976**, *72*, 248–254. [CrossRef]
32. Dubois, M.; Gilles, K.; Hamilton, J.; Rebers, P.; Smith, F. A colorimetric method for the determination of sugars. *Nature* **1951**, *168*, 167. [CrossRef]
33. Kim, Y.-H.; Khan, A.L.; Kim, D.-H.; Lee, S.-Y.; Kim, K.-M.; Waqas, M.; Jung, H.-Y.; Shin, J.-H.; Kim, J.-G.; Lee, I.-J. Silicon mitigates heavy metal stress by regulating P-type heavy metal ATPases, *Oryza sativa* low silicon genes, and endogenous phytohormones. *BMC Plant Biol.* **2014**, *14*, 13. [CrossRef] [PubMed]
34. Kamboj, J.; Browning, G.; Blake, P.; Quinlan, J.; Baker, D. GC-MS-SIM analysis of abscisic acid and indole-3-acetic acid in shoot bark of apple rootstocks. *Plant Growth Regul.* **1999**, *28*, 21–27. [CrossRef]
35. McCloud, E.S.; Baldwin, I.T. Herbivory and caterpillar regurgitants amplify the wound-induced increases in jasmonic acid but not nicotine in *Nicotiana sylvestris*. *Planta* **1997**, *203*, 430–435. [CrossRef]
36. Yalpani, N.; Enyedi, A.J.; León, J.; Raskin, I. Ultraviolet light and ozone stimulate accumulation of salicylic acid, pathogenesis-related proteins and virus resistance in tobacco. *Planta* **1994**, *193*, 372–376. [CrossRef]
37. Fang, Y.; Umasankar, Y.; Ramasamy, R.P. Plant volatile sensor: Enzymatic transducer for selective and sensitive determination of methyl salicyalte. In Proceedings of the ECS Meeting Abstracts, Vancouver, BC, Canada, May 29–2 June 2022; p. 414. [CrossRef]
38. Heath, R.L.; Packer, L. Photoperoxidation in isolated chloroplasts: I. Kinetics and stoichiometry of fatty acid peroxidation. *Arch. Biochem. Biophys.* **1968**, *125*, 189–198. [CrossRef]
39. Kwasniewski, M.; Chwialkowska, K.; Kwasniewska, J.; Kusak, J.; Siwinski, K.; Szarejko, I. Accumulation of peroxidase-related reactive oxygen species in trichoblasts correlates with root hair initiation in barley. *J. Plant Physiol.* **2013**, *170*, 185–195.
40. Jiang, M.; Zhang, J. Effect of abscisic acid on active oxygen species, antioxidative defence system and oxidative damage in leaves of maize seedlings. *Plant Cell Physiol.* **2001**, *42*, 1265–1273. [CrossRef]
41. Giannopolitis, C.N.; Ries, S.K. Superoxide dismutases: I. Occurrence in higher plants. *Plant Physiol.* **1977**, *59*, 309–314. [CrossRef]
42. Aebi, H. Catalase in vitro. *Methods Enzymol.* **1984**, *105*, 121–126.
43. Change, B.; Maehly, A. Assay of catalases and peroxidase. *Methods Enzym.* **1955**, *2*, 764–775.
44. Nakano, Y.; Asada, K. Hydrogen peroxide is scavenged by ascorbate-specific peroxidase in spinach chloroplasts. *Plant Cell Physiol.* **1981**, *22*, 867–880.
45. Sah, S.K.; Kaur, G.; Kaur, A. Rapid and reliable method of high-quality RNA extraction from diverse plants. *Am. J. Plant Sci.* **2014**, *5*, 3129. [CrossRef]
46. Livak, K.J.; Schmittgen, T.D. Analysis of relative gene expression data using real-time quantitative PCR and the $2^{-\Delta\Delta CT}$ method. *Methods* **2001**, *25*, 402–408. [CrossRef] [PubMed]
47. Sheteiwy, M.; Shen, H.; Xu, J.; Guan, Y.; Song, W.; Hu, J. Seed polyamines metabolism induced by seed priming with spermidine and 5-aminolevulinic acid for chilling tolerance improvement in rice (*Oryza sativa* L.) seedlings. *Environ. Exp. Bot.* **2017**, *137*, 58–72. [CrossRef]
48. Hussain, A.; Ali, S.; Rizwan, M.; ur Rehman, M.Z.; Hameed, A.; Hafeez, F.; Alamri, S.A.; Alyemeni, M.N.; Wijaya, L. Role of zinc–lysine on growth and chromium uptake in rice plants under Cr stress. *J. Plant Growth Regul.* **2018**, *37*, 1413–1422. [CrossRef]
49. Bewley, J.D.; Black, M. *Physiology and Biochemistry of Seeds in Relation to Germination: Volume 2: Viability, Dormancy, and Environmental Control*; Springer Science & Business Media: Berlin, Germany, 2012.
50. Devi, P.; Kumar, P. Enhancement effect of biofertilizers on germination percentage and plant height in maize grown under chromium toxic soil. *J. Pharmacogn. Phytochem.* **2020**, *9*, 702–707.
51. Hussain, S.; Khan, F.; Hussain, H.A.; Nie, L. Physiological and biochemical mechanisms of seed priming-induced chilling tolerance in rice cultivars. *Front. Plant Sci.* **2016**, *7*, 116. [CrossRef]
52. Ganesh, K.S.; Baskaran, L.; Rajasekaran, S.; Sumathi, K.; Chidambaram, A.; Sundaramoorthy, P. Chromium stress induced alterations in biochemical and enzyme metabolism in aquatic and terrestrial plants. *Colloids Surf. B BioInterfaces* **2008**, *63*, 159–163. [CrossRef]
53. Mustafa, G.; Komatsu, S. Toxicity of heavy metals and metal-containing nanoparticles on plants. *Biochim. Biophys. Acta (BBA)-Proteins Proteom.* **2016**, *1864*, 932–944. [CrossRef]

54. Faizan, M.; Bhat, J.A.; Hessini, K.; Yu, F.; Ahmad, P. Zinc oxide nanoparticles alleviates the adverse effects of cadmium stress on *Oryza sativa* via modulation of the photosynthesis and antioxidant defense system. *Ecotoxicol. Environ. Saf.* **2021**, *220*, 112401. [CrossRef]
55. Mostofa, M.G.; Rahman, M.; Ansary, M.; Uddin, M.; Fujita, M.; Tran, L.-S.P. Interactive effects of salicylic acid and nitric oxide in enhancing rice tolerance to cadmium stress. *Int. J. Mol. Sci.* **2019**, *20*, 5798. [CrossRef] [PubMed]
56. Todorenko, D.; Timofeev, N.; Kovalenko, I.; Kukarskikh, G.; Matorin, D.; Antal, T. Chromium effects on photosynthetic electron transport in pea (*Pisum sativum* L.). *Planta* **2020**, *251*, 11. [CrossRef] [PubMed]
57. Marco, F.; Busó, E.; Lafuente, T.; Carrasco, P. Spermine confers stress resilience by modulating abscisic acid biosynthesis and stress responses in Arabidopsis plants. *Front. Plant Sci.* **2019**, *10*, 972. [CrossRef] [PubMed]
58. Kohli, S.K.; Handa, N.; Sharma, A.; Gautam, V.; Arora, S.; Bhardwaj, R.; Wijaya, L.; Alyemeni, M.N.; Ahmad, P. Interaction of 24-epibrassinolide and salicylic acid regulates pigment contents, antioxidative defense responses, and gene expression in *Brassica juncea* L. seedlings under Pb stress. *Environ. Sci. Pollut. Res.* **2018**, *25*, 15159–15173. [CrossRef] [PubMed]
59. Alam, P.; Balawi, T.H.; Altalayan, F.H.; Hatamleh, A.A.; Ashraf, M.; Ahmad, P. Silicon attenuates the negative effects of chromium stress in tomato plants by modifying antioxidant enzyme activities, ascorbate–glutathione cycle and glyoxalase system. *Acta Physiol. Plant.* **2021**, *43*, 1–17. [CrossRef]
60. Nahar, K.; Hasanuzzaman, M.; Alam, M.M.; Rahman, A.; Suzuki, T.; Fujita, M. Polyamine and nitric oxide crosstalk: Antagonistic effects on cadmium toxicity in mung bean plants through upregulating the metal detoxification, antioxidant defense and methylglyoxal detoxification systems. *Ecotoxicol. Environ. Saf.* **2016**, *126*, 245–255. [CrossRef]
61. Hussain, A.; Nazir, F.; Fariduddin, Q. 24-epibrassinolide and spermidine alleviate Mn stress via the modulation of root morphology, stomatal behavior, photosynthetic attributes and antioxidant defense in *Brassica juncea*. *Physiol. Mol. Biol. Plants* **2019**, *25*, 905–919. [CrossRef]
62. Choudhary, S.P.; Kanwar, M.; Bhardwaj, R.; Yu, J.-Q.; Tran, L.-S.P. Chromium stress mitigation by polyamine-brassinosteroid application involves phytohormonal and physiological strategies in *Raphanus sativus* L. *PLoS ONE* **2012**, *7*, e33210. [CrossRef]
63. He, Y.; Liu, Y.; Cao, W.; Huai, M.; Xu, B.; Huang, B. Effects of salicylic acid on heat tolerance associated with antioxidant metabolism in Kentucky bluegrass. *Crop Sci.* **2005**, *45*, 988–995. [CrossRef]
64. Shi, Q.; Zhu, Z. Effects of exogenous salicylic acid on manganese toxicity, element contents and antioxidative system in cucumber. *Environ. Exp. Bot.* **2008**, *63*, 317–326. [CrossRef]
65. Khan, N.; Bano, A.; Ali, S.; Babar, M.A. Crosstalk amongst phytohormones from planta and PGPR under biotic and abiotic stresses. *Plant Growth Regul.* **2020**, *90*, 189–203. [CrossRef]
66. Podlešáková, K.; Ugena, L.; Spíchal, L.; Doležal, K.; De Diego, N. Phytohormones and polyamines regulate plant stress responses by altering GABA pathway. *New Biotechnol.* **2019**, *48*, 53–65. [CrossRef] [PubMed]
67. Wani, A.B.; Chadar, H.; Wani, A.H.; Singh, S.; Upadhyay, N. Salicylic acid to decrease plant stress. *Environ. Chem. Lett.* **2017**, *15*, 101–123. [CrossRef]
68. Sharma, A.; Sidhu, G.P.S.; Araniti, F.; Bali, A.S.; Shahzad, B.; Tripathi, D.K.; Brestic, M.; Skalicky, M.; Landi, M. The role of salicylic acid in plants exposed to heavy metals. *Molecules* **2020**, *25*, 540. [CrossRef]
69. Upchurch, R.G. Fatty acid unsaturation, mobilization, and regulation in the response of plants to stress. *Biotechnol. Lett.* **2008**, *30*, 967–977. [CrossRef]
70. Pál, M.; Tajti, J.; Szalai, G.; Peeva, V.; Végh, B.; Janda, T. Interaction of polyamines, abscisic acid and proline under osmotic stress in the leaves of wheat plants. *Sci. Rep.* **2018**, *8*, 12839. [CrossRef]
71. Eder, C.S.; Roberta, d.P.M.; Alexandra, L.; Marcelo, M.; Paulo, A.H.; Zenilda, L.B. Effects of cadmium on growth, photosynthetic pigments, photosynthetic performance, biochemical parameters and structure of chloroplasts in the agarophyte Gracilaria domingensis (Rhodophyta, Gracilariales). *Am. J. Plant Sci.* **2012**, *3*, 1077–1084.
72. Spormann, S.; Soares, C.; Teixeira, J.; Fidalgo, F. Polyamines as key regulatory players in plants under metal stress—A way for an enhanced tolerance. *Ann. Appl. Biol.* **2021**, *178*, 209–226. [CrossRef]

antioxidants

Article

Role of Plasma Membrane NADPH Oxidase in Response to Salt Stress in Cucumber Seedlings

Katarzyna Kabała *, Małgorzata Reda, Anna Wdowikowska and Małgorzata Janicka

Department of Plant Molecular Physiology, Faculty of Biological Sciences, University of Wrocław, Kanonia 6/8, 50-328 Wrocław, Poland
* Correspondence: katarzyna.kabala@uwr.edu.pl

Abstract: Plasma membrane NADPH oxidases (RBOHs, EC 1.6.3.1) are known as the main ROS generators involved in plant adaptation to stress conditions. In the present work, regulation of NADPH oxidase was analyzed in cucumber (*Cucumis sativus* L. var. Krak) seedlings exposed to salinity. RBOH activity and gene expression, as well as H_2O_2 content, were determined in the roots of plants treated with 50 or 100 mM NaCl for 1 h, and 50 mM NaCl for 1 or 6 days. It was found that enzyme activity increased in parallel with an enhancement in the H_2O_2 level in roots exposed to 100 mM NaCl for 1 h, and to 50 mM NaCl for 1 day. The expression of some *CsRboh* genes was induced by salt. Moreover, an increase in the activity of G6PDH, providing the substrate for the NADPH oxidase, was observed. In seedlings subjected to salinity for a longer time, antioxidant enzymes—including superoxide dismutase, catalase, and ascorbate peroxidase—were activated, participating in maintaining a steady-state H_2O_2 content in the root cells. In conclusion, NADPH oxidase and endogenous H_2O_2 up-regulation seem to be early events in cucumber response to salinity.

Keywords: ascorbate peroxidase; catalase; hydrogen peroxide; NADP dehydrogenases; NADPH oxidase; salt stress; superoxide dismutase

Citation: Kabała, K.; Reda, M.; Wdowikowska, A.; Janicka, M. Role of Plasma Membrane NADPH Oxidase in Response to Salt Stress in Cucumber Seedlings. *Antioxidants* **2022**, *11*, 1534. https://doi.org/10.3390/antiox11081534

Academic Editor: Nafees A. Khan

Received: 14 July 2022
Accepted: 4 August 2022
Published: 6 August 2022

1. Introduction

In plants, reactive oxygen species (ROS), including the superoxide radical (O_2^-) and hydrogen peroxide (H_2O_2), are known to act as harmful compounds responsible for the oxidation and damage of numerous cellular components. On the other hand, when present in low concentrations, they are essential signaling molecules involved in the regulation of multiple physiological processes that affect plant growth and development [1]. ROS production may occur through both enzymatic and non-enzymatic pathways, among which plasma membrane NADPH oxidases (respiratory burst oxidase homologs, RBOHs) are the most studied of enzymes considered to be the main ROS generators [2]. Both ROS and RBOHs have been shown to participate in cell growth via regulation of cell wall loosening and stiffening. RBOH-dependent spatial and temporal control of ROS production is necessary for proper cell elongation. It has been demonstrated that *Arabidopsis* and barley *Rboh* mutants display impaired growth [3].

RBOHs are integral membrane proteins containing several functional domains, including haem cofactors, FAD, and calcium-binding (EF-hand) motifs. They catalyze an electron transfer from cytosolic donor NADPH to extracellular acceptor O_2 via FAD and two haems, forming O_2^-, followed by its dismutation to H_2O_2 [4]. Studies using *Arabidopsis* have revealed that NADPH oxidase activity may be regulated at post-translational level synergistically by phosphorylation and Ca^{2+} binding [5]. More recently, it has been indicated that NADPH oxidases remain under the control of differential cellular effectors, such as small GTPases, calcium sensors CBL-CIPKs (calcineurin B-like protein-CBL-interacting protein kinases), and phosphatidic acid, the lipid product of phospholipase D [2]. Plant NADPH oxidases are encoded by small multigenic families. Ten *Rboh* genes have been identified in

Arabidopsis, whereas nine isogenes are present in the genomes of rice and cucumber [3,6]. Differential *Rboh* expression, both in plant organs/tissues and at distinct developmental stages, suggests specific or housekeeping functions for individual RBOH isoforms [3].

In addition to the plasma membrane NADPH oxidases, apoplastic H_2O_2 generation can also be mediated by oxidative degradation of polyamines (PAs) [7]. The cell wall PA-catabolizing amine oxidases are classified into copper-containing diamine oxidases (DAOs, also known as CuAOs) and flavin-dependent polyamine oxidases (PAOs). DAOs oxidize mainly putrescine, whereas the apoplastic PAOs terminally degrade spermidine and spermine, both yielding aminoaldehydes and H_2O_2 [7]. Consequently, an increase in the extracellular H_2O_2 pool appears. It has been suggested that H_2O_2 generated as a product of PA oxidation may be directly involved in cell signaling processes, as well as in plant adaptation to abiotic stresses [8]. It is worth noting that PAO is highly expressed mainly in monocots, whereas DAO is active at high levels in dicotyledons [8].

In their natural environment, plants are exposed to unfavorable factors, including drought, salt, and extreme temperatures, which are perceived by cells. Tolerance of plants to salinity is governed by numerous physiological and molecular adaptive mechanisms [9]. Some of them involve ROS signaling [3,9]. Several studies have suggested that the plasma membrane NADPH oxidase may play an important role in plant acclimation to salinity, participating in both transcriptional and post-translational regulation. It has been shown, that under salinity, NADPH oxidase activity and endogenous ROS production are involved in the elevation of Ca^{2+} levels [10], the stabilization of the plasma membrane Na^+/H^+ antiporter, AtSOS1, mRNA [11], the regulation of Na^+/K^+ homeostasis via activation of plasma membrane Ca^{2+}-permeable channels as well as inwardly rectifying K^+ channels [12], and NaCl-induced activation of antioxidant enzymes [13].

Other studies have revealed that salinity induces the PA catabolism pathway into the apoplast, generating H_2O_2, and thereby activating tolerance mechanisms [7]. Tanou, et al. [14], observed that PAO and DAO encoding genes were remarkably induced by salinity treatments. They showed that PA degradation by DAO and PAO represents an active source for H_2O_2 generation. H_2O_2 produced by PA catabolism may activate antioxidative defense responses [14]. More recently, the interplay between amine oxidases and NADPH oxidases was examined in tobacco subjected to salt stress [15]. RBOH and apoplastic PAO were found to cooperate in controlling the accumulation of H_2O_2 and superoxides O_2^-. Both enzymes are a part of the same ROS regulatory network, in which NADPH seems to act upstream of apoplastic PAO [15].

The aim of the present study was to understand the role of the plasma membrane NADPH oxidase in the regulation of the H_2O_2 level, and its interrelation with antioxidant enzymes (superoxide dismutase, catalase, and ascorbate peroxidase) in the roots of cucumber seedlings exposed to salinity. Roots are the first organ exposed to unfavorable factors that get into plants from the soil. It was expected that under conditions of different time-exposure to salt, the plasma membrane NADPH oxidase activity would be modulated in a manner dependent on its gene expression and substrate (NADPH) delivery, and that, consequently, the content of the H_2O_2 would be modified. Moreover, it was assumed that changes in NADPH oxidase activity are related to the action of antioxidant enzymes.

2. Materials and Methods

Cucumber (*Cucumis sativus* L. var. Krak) seeds (originating from W. Legutko, Jutrosin, Poland) were germinated in darkness at 25 °C for 2 days. Seedlings were grown in containers (one container for each treatment) filled with nutrient solution (pH 6.5) consisting of 1.7 mM KNO_3, 1.7 mM $Ca(NO_3)_2 \cdot 4H_2O$, 0.33 mM KH_2PO_4, 0.33 mM $MgSO_4 \cdot 7H_2O$, and micronutrients: 75 μM ferric citrate, 10 μM $MnSO_4 \cdot 5H_2O$, 5 μM H_3BO_4, 1 μM $CuSO_4 \cdot 5H_2O$, 0.01 μM $ZnSO_4 \cdot 7H_2O$, 0.05 μM $Na_2MoO_4 \cdot 2H_2O$ without (control) or with the addition of salt. The plants were treated with 50 or 100 mM NaCl for 1 h immediately before analysis (50/1H and 100/1H, respectively), 50 mM NaCl for 1 day (50/1D; the seedlings were grown for 5 days without NaCl, and then transferred to an NaCl-containing medium for the next 24 h) or 50 mM NaCl for 6 days (50/6D; the seedlings were grown for 6 days in an NaCl-containing medium). Concentration of 5 mM H_2O_2 was added to the control nutrient solution for 24 h. The plants were grown hydroponically under a 16-h photoperiod at temperatures of 25 °C/22 °C (day/night), photon flux density 180 mol $m^{-2}s^{-1}$, and about 70% relative humidity. Whole roots were collected from all the cucumber plants in the container (200 plants/1 dm^3 or 20 plants/100 cm^3 of nutrient solution, depending on the biochemical measurement), and used for analysis.

H_2O_2 was quantified according to the method of Velikova, et al. [16], with some modifications as described by Kabała, et al. [17], based on the oxidation of potassium iodide (KI). After incubation in darkness at room temperature for 60 min, the absorbance of the reaction product, triiodide (I_3^-), was measured at 390 nm.

The plasma membrane fractions were isolated from the cucumber roots in accordance with the method of Larsson [18], as modified by Kłobus [19], using a 6.2% two-phase system containing PEG (polyethylene glycol) 3350 and dextran T500. The upper phase was enriched in highly purified plasma membrane vesicles.

The catalytic activity of the NADPH oxidase (EC 1.6.3.1) was determined in the plasma membrane fractions, according to the modified method of Sagi and Fluhr [20], as described by Jakubowska, et al. [6], using 0.5 mM XTT (2,3-bis(2-methoxy-4-nitro-5-sulfophenyl)-2H-tetrazolium-5-carboxanilide inner salt) and 0.1 mM NADPH as an electron donor. XTT reduction by an NADPH oxidase-generated O_2^- radical was determined at 470 nm in the presence and absence of 50 units of CuZn-SOD. NADPH oxidase in-gel assays were performed according to the method described by Sagi and Fluhr [20]. The plasma membrane fractions were subjected to Native PAGE, using 7.5% (*w*/*v*) polyacrylamide gel. Electrophoresis was conducted at 110 V and 20 mA for 60 min. The NADPH-dependent O_2^- production was assayed in gel by a modified NBT (nitroblue tetrazolium) reduction method. After the NADPH addition, the appearance of blue formazan bands was monitored.

Diamine oxidase (DAO, EC 1.4.3.6) activity was assayed in seedling roots spectrophotometrically, using putrescine as a substrate, in accordance with the method of Holmsted, et al. [21], as modified by Federico and Angelini [22]. During the reaction, Δ-pyrroline, formed by the enzymatic oxidation of putrescine, reacted with 2-aminobenzaldehyde to produce a yellowish-colored dihydroquinazolinium derivative. The absorbance was measured at 430 nm ($\varepsilon = 1.86\ 10^3\ mol^{-1}\ cm^{-1}$).

The activities of enzymes generating NADPH, 6PGDH (6-phosphogluconate dehydrogenase, EC 1.1.1.44), G6PDH (glucose-6-phosphate dehydrogenase, EC 1.1.1.49), NADP-ICDH (NADP-isocitrate dehydrogenase, EC 1.1.1.42), and NADP-ME (NADP-malic enzyme, EC. 1.1.1.40), were determined in the cucumber roots in accordance with the method of Li, et al. [23], as described by Jakubowska, et al. [6]. Generation of the NADPH was monitored spectrophotometrically by measuring the reduction of the NADP at 340 nm.

Catalase (CAT, EC 1.11.1.6) activity was determined according to the method of Aebi [24], as described by Janicka-Russak, et al. [25]. H_2O_2 decomposition was monitored spectrophotometrically by measuring the reduction in absorbance at 240 nm. One unit of CAT is defined as the amount of enzyme that breaks down 1 µmol of H_2O_2 per min.

Ascorbate peroxidase (APX, EC 1.11.1.11) activity was determined in accordance with the method of Chen and Asada [26], as described by Janicka-Russak et al. [25]. The H_2O_2-dependent oxidation of ascorbic acid was followed by monitoring the decrease of absorbance at 290 nm, assuming an absorption coefficient value of 2.8 $mM^{-1}\ cm^{-1}$. One unit of APX is defined as the amount of enzyme that oxidizes 1 µmol of ascorbic acid per min.

Superoxide dismutase (SOD, EC 1.15.1.1) activity was measured in accordance with the method (xanthine/xanthine oxidase test) of Beauchamp and Fridovich [27]. The photoreduction of NBT and the formation of blue formazan was measured at 560 nm; an inhibition curve was made against different volumes of extract (5–20 µL). One unit of SOD is defined as the enzyme activity that inhibits the reduction of NBT to blue formazan by 50%.

Protein content was determined according to the method of Bradford [28], using bovine serum albumin (BSA) as the standard.

The total RNA was isolated from the cucumber roots using Tri Reagent (Sigma, St. Louis, MO, USA), and cDNA was synthetized with a High-Capacity cDNA Reverse Transcription Kit (Applied Biosystems, Waltham, Massachusetts, USA), in accordance with the manufacturer's instructions, as described by Jakubowska, et al. [6].

The expression profile of the NADPH oxidase and SOD encoding genes in the cucumber roots was analyzed with the LightCycler 480 system (Roche, Basel, Switzerland). The cDNA was used as a template in a real-time PCR reaction with the Real-Time 2× PCR Master Mix SYBR kit (A&A Biotechnology, Gdańsk, Poland), performed in accordance with the manufacturer's instructions, as described by Jakubowska, et al. [6]. The amplification conditions were as follow: 30 s at 95 °C; 40–45 cycles of 10 s at 95 °C; 10 s at 58 °C (for SOD genes) and 57 or 59 °C (for RBOH genes); 12 s at 72 °C; and final melting for 15 s at 65 °C. Genes encoding the cucumber tonoplast intrinsic protein, *CsTIP41*, and the clathrin adaptor complex subunit, *CsCACS*, were used as the internal standards [29]. Primer sequences specific to amplified genes, designed by LightCycler ProbeDesign software (Roche, Basel, Switzerland), are listed in Table 1. A melting curve analysis was performed to confirm the specificity of amplification and the absence of nonspecific by-products.

Table 1. The list of primers used in qRT-PCR reactions.

Gene	Forward Primer	Reverse Primer
CsTIP41	CAACAGGTGATATTGGATTATGATTATAC	GCCAGCTCATCCTCATATAAG
CsRbohB	AGAGTCGGCTTCAGCGG	TCTCCTCCTCTGATGTTCTGAACGG
CsRbohD	TCTTCTTCTTCTTCCTCCCTCAAAGCC	GAAAGTTCAGGGTCTTCAAGAGAGTTG
CsRbohE	TCACTTTGGAAGTTGAGGATGATTCTGTT	GTCGAGAGAGGAGGCAGTGG
CsRbohF1	GCTTCGATTTCGAGATCGCCGAC	AAATCCATTTCCACCACCAGAAGAATGT
CsRbohF2	TCGCATCGGTTTCGGCAG	AGTTCACGGTTGAAGATGACGC
CsRbohF3	AAGAGAAGTCCGAGAATGAGGTTTAT	CTCCTGAGATTCTGACATTCCAATGC
CsRbohH1	AGATTCCGATGTGATAGATGTCATGGT	TTGGACTGTCTTCTTCCAAGCCTC
CsRbohH2	AAATCCAGGAGAATGCCCAC	TGAGAAGTCTCAGCGACGG
CsRbohJ	CTGACGATGGGATTACTCTGCAACA	GAAGCAGTCCACTTCAGTATGGTCT
CsCACS	TGGGAAGATTCTTATGAAGTGC	CTCGTCAAATTTACACATTGGT
CsCSD1	AAAGATGGTGAAGGCTGTGG	CATGTTGTTTTCCAGCAG
CsCSD2	CCCATTCTCCAATTCTTCAT	TTGGGTGAGCGTGACAA
CsCSD3	CCAAAAGAGGGGGAATTTTT	GCCTCTGACGTTGGAATCTC
CsMSD	CGCTTCGAATTCTAGGCAGG	CCTTTGTTGATGGCCTCGTG
CsFSD2	TAACACGGTTGACCATCC	TTCTGTAGCCTCAAGTCT
CsFSD3	TCTGATGGAACCACAAGAG	TGTGGCCATATAGAATATCATTT

All data presented are expressed as the means of at least three biological replicates (each performed with two or three analytical replications) ± standard deviation (SD). One-way ANOVA and Duncan's post hoc analysis (significant at $p < 0.05$) were used for statistical analysis, performed with Statistica 13.3 (TIBCO Software Inc., Palo Alto, CA, USA; 2017).

3. Results

Environmental stress factors are known to elevate endogenous levels of small signaling molecules, including hydrogen peroxide, in plant tissues. Accordingly, H_2O_2 content was measured in the roots of cucumber seedlings exposed to 50 or 100 mM NaCl for 1 hour, and to 50 mM NaCl for 1 or 6 days. It was observed that 1-h-treatment of plants with 100 mM salt resulted in a visible increase in the H_2O_2 level, whereas 50 mM salt had no significant effect. The H_2O_2 achieved 130% and 110%, respectively, of the control (Figure 1A). As a result of longer treatment of seedlings with 50 mM NaCl concentration, the H_2O_2 content increased, reaching about 137% of the control value after 24 h, and then decreased in 6-day exposed plants, to the level observed in 1-h-treated seedlings, i.e., 110% of the control (Figure 1B).

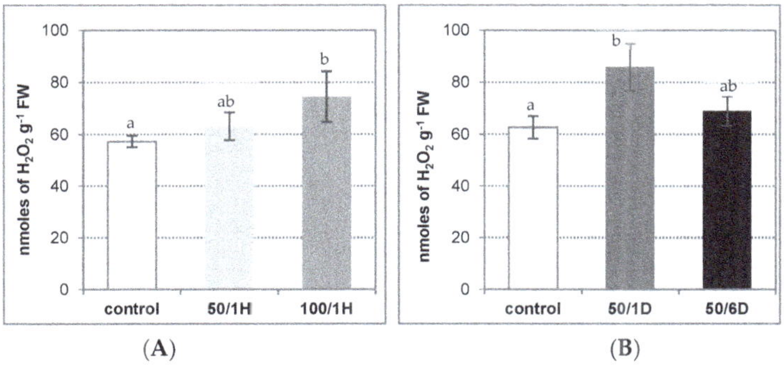

Figure 1. H_2O_2 level in roots of cucumber seedlings treated with: (**A**) 50 mM or 100 mM NaCl for 1 h (50/1H and 100/1H, respectively); (**B**) 50 mM NaCl for 1 or 6 days (50/1D and 50/6D, respectively). The H_2O_2 level was determined in the roots as described in Materials and Methods. The presented results are the means ± SD of three independent experiments, each run in triplicate. Different letters show different homogeneous groups of means calculated by Duncan's multiple range test (significant at $p < 0.05$).

The plant NADPH oxidases are the main enzymes that participate in H_2O_2 production. Using two methods, we demonstrated that the treatment of seedlings with NaCl stimulated NADPH oxidase activity in the roots. The activity achieved about 126% and 145% of the control level in plants exposed to 50 mM and 100 mM NaCl, respectively, for 1 h (Figure 2A,C). When plants were treated with 50 mM NaCl for a longer time, the enzyme activity also increased, reaching about 149% of the control after 24 h, and consequently decreased to nearly the control value (115% of the control) in roots exposed to salt for 6 days (Figure 2B,C). Thus, the observed increase in the H_2O_2 level occurred in parallel with the enhanced NADPH oxidase activity. To verify whether the observed modulations of enzyme activity were a result of changes at transcriptional level, the *Rboh* gene expression was measured in the roots of seedlings exposed to salt stress. It was demonstrated that two of nine cucumber genes, *CsRbohD* and *CsRbohF1*, were significantly up-regulated by treatment with 100 mM NaCl for 1 h, reaching a 4,5-fold and 2,2-fold increase, respectively, in comparison to the control (Figure 2D). When the plants were exposed to salt for a longer time, the most pronounced effect of 6-day treatment with 50 mM NaCl on *CsRbohH1* expression (a 2,4-fold increase) was observed (Figure 2E).

Apoplastic diamine oxidase (DAO) and polyamine oxidase (PAO) appeared to be the other enzymes responsible for H_2O_2 generation. Our results indicated that DAO activity remained at a similar level in both the control and the salt-stressed plants (Figure 3A,B). On the other hand, in the cucumber roots, the PAO activity was maintained at a very low level, several times lower than the DAO, regardless of the growing conditions (personal communications: Janicka, M., 2021, available in the repository) and, for this reason, it has not been studied. This suggests that the elevation in the H_2O_2 amount could be related to the plasma membrane NADPH activity.

The activity of the plasma membrane NADPH oxidase is related to the action of the superoxide dismutase (SOD). It was found in cucumber that SOD activity was enhanced by salt, similarly to NADPH oxidase. However, a significant increase, by about 60%, was observed in the roots of the plants treated with salt for 6 days (Figure 4A,B). Analysis of six genes encoding SOD in the cucumber genome—including *CsCSD1*, 2, and 3 belonging to Cu/Zn-SODs, *CsMSD* representing Mn-SODs as well as *CsFSD2* and 3 identified as Fe-SOD—indicated that the expression of one of them, *CsMSD*, was upregulated about twice in seedlings exposed to 50 mM NaCl for both 1 h and 6 days. On the other hand, the transcript level of *CsCSD2* significantly increased (a 2,5-fold increase) only after plant exposure to salt for 6 days (Figure 4C,D).

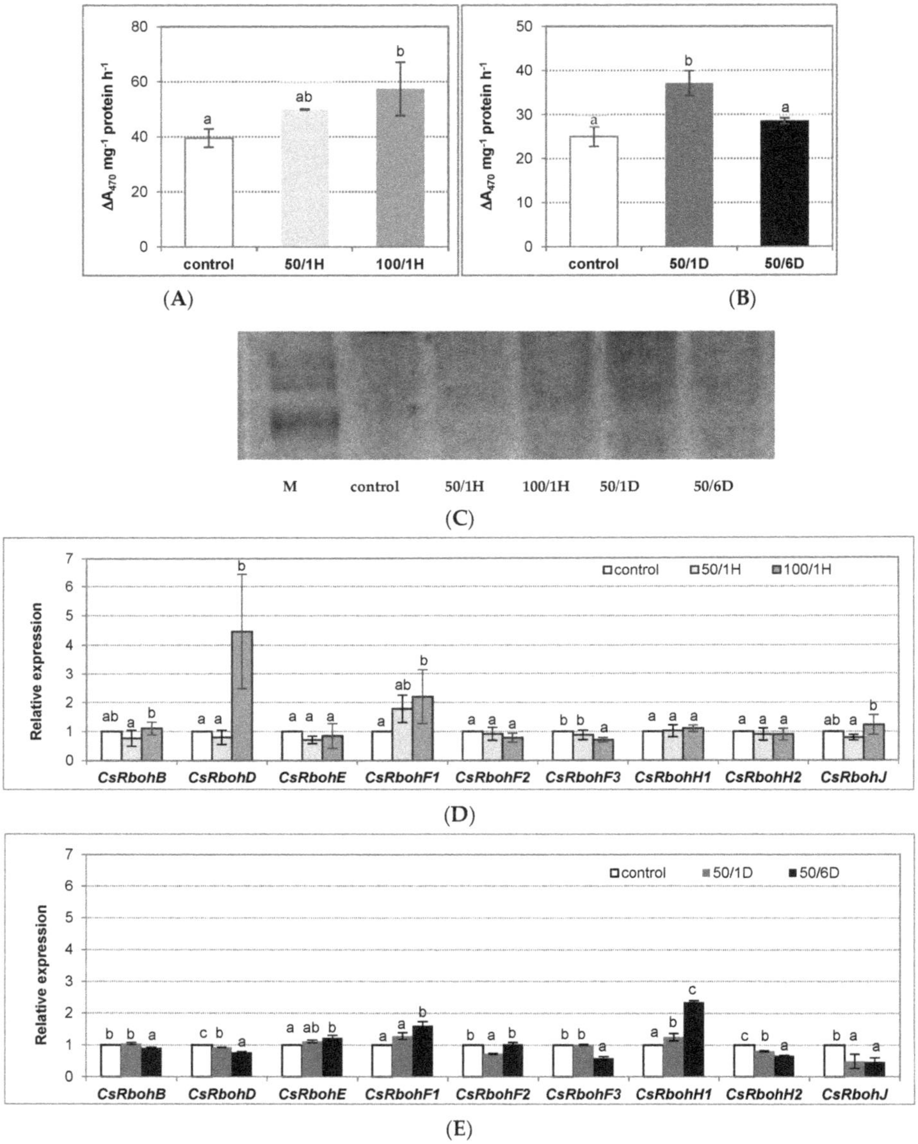

Figure 2. NADPH oxidase catalytic activity (**A**,**B**), in-gel activity(**C**), and relative expression of *CsRboh* genes (**D**,**E**) in the tissues of cucumber roots exposed to salinity. The cucumber seedlings were treated with: (**A**,**C**,**D**) 50 mM or 100 mM NaCl for 1 h (50/1H and 100/1H, respectively) and (**B**,**C**,**E**) 50 mM NaCl for 1 or 6 days (50/1D and 50/6D, respectively). The plasma membrane vesicles were isolated from the cucumber roots to determine both catalytic (**A**,**B**) and in-gel (**C**) NADPH oxidase activity, as described in Materials and Methods. (**A**,**B**) The presented results are the means ± SD of three independent experiments, with each experiment performed in triplicate, and are expressed as the ΔA_{470} of formazan XTT mg^{-1} protein h^{-1}. (**C**) Gel image with blue formazan bands detected using

the NBT method. The plasma membrane proteins were subjected to native PAGE, and examined for their ability to generate O_2^- by reduction of nitroblue tetrazolium. The bands detected in the gel were due to NADPH-dependent NBT reduction by O_2^- radicals. The gel image shown is representative of three independent replications of the experiment; M—molecular weight marker. (**D**,**E**) The total RNA was isolated from the cucumber roots, and real-time PCR analysis was performed with *CsTIP41* as a reference gene, as described in Materials and Methods. The results are shown as means ± SD of three replications. Different letters show different homogeneous groups of means calculated by Duncan's multiple range test (significant at $p < 0.05$).

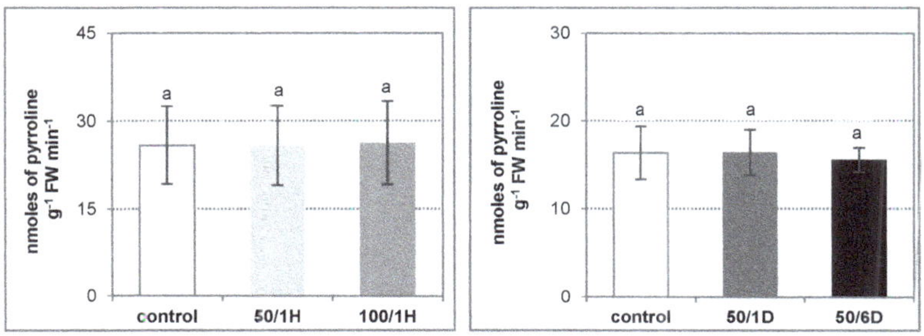

(**A**) (**B**)

Figure 3. Diamine oxidase activity in the tissues of cucumber roots treated with: (**A**) 50 mM or 100 mM NaCl for 1 h (50/1H and 100/1H, respectively); (**B**) 50 mM NaCl for 1 or 6 days (50/1D and 50/6D, respectively). DAO activity was assayed in the roots, as described in Materials and Methods. The presented results are the means ± SD of three independent experiments, with each experiment performed in triplicate, and are expressed as nmoles of Δ^1-pyrroline g^{-1} FW min^{-1}. Means with the same letter are not significantly different ($p < 0.05$).

The activity of NADPH oxidase can be dependent on NADPH as its metabolic substrate. For this reason, the role of four NADPH-generating enzymes, i.e., glucose-6-phosphate dehydrogenase (G6PDH), 6-phosphogluconate dehydrogenase (6PGDH), NADP-isocitrate dehydrogenase (NADP-ICDH), and NADP-malic enzyme (NADP-ME), was examined in cucumber plants exposed to salinity (Figure 5). Among the enzymes tested, G6PDH was found to be significantly stimulated by salt applied for a longer period. In roots stressed with 50 mM NaCl for 1 and 6 days, the enzyme activity reached about 160–170% of the control value (Figure 5A,B). On the other hand, the activities of both 6PGDH and NADP-ICDH were relatively unaffected by NaCl, achieving similar levels in the roots of control as well as salt-treated plants (Figure 5C–F). In contrast to the above results, NADP-ME was visibly inhibited under longer salinity stress. The observed decrease in its activity was closely related to the time of NaCl treatment, and reached about 50% of the control in roots stressed with salt for 6 days (Figure 5G,H). Thus, it seems very likely that G6PDH is responsible for providing the substrate for NADPH oxidase under salt stress condition.

To minimize hydrogen peroxide accumulation, plants activate the H_2O_2-metabolizing enzymes, including catalase (CAT) and ascorbate peroxidase (APX). In the cucumber root tissues, CAT activity was reduced by about 30% after one day of plant treatment with 50 mM NaCl, as compared to the control plants. Longer plant exposure to this salt concentration (6 days) caused a significant increase in enzyme activity, which achieved above 150% of the control value. CAT activity did not change significantly in the roots treated with salt for 1 h (Figure 6A,B). On the other hand, APX activity was stimulated in the cucumber seedlings stressed with NaCl for a longer time period, reaching its highest level (140% of the control) in roots exposed to salt for 6 days (Figure 6C,D).

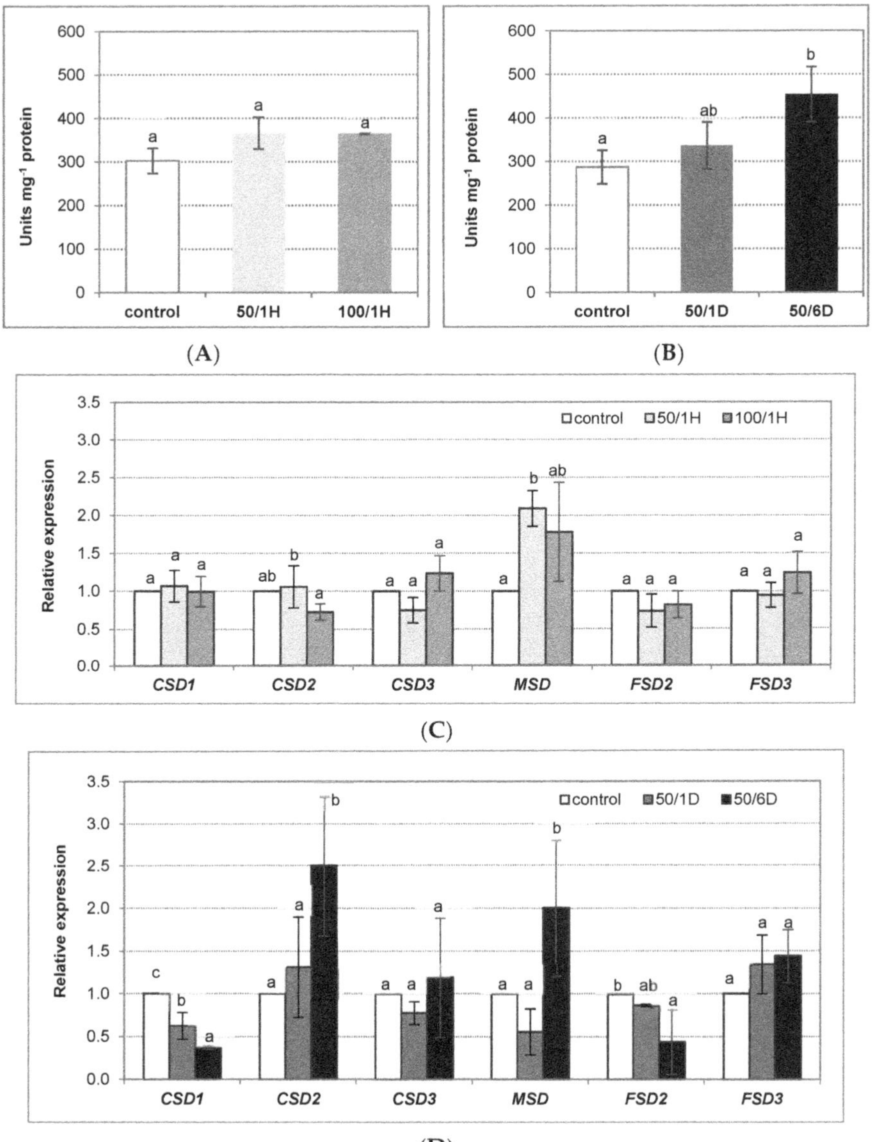

Figure 4. Superoxide dismutase activity (**A**,**B**) and relative expression of SOD encoding genes (**C**,**D**) in the tissues of cucumber roots exposed to salinity. The cucumber seedlings were treated with: (**A**,**C**) 50 mM or 100 mM NaCl for 1 h (50/1H and 100/1H, respectively); (**B**,**D**) 50 mM NaCl for 1 or 6 days (50/1D and 50/6D, respectively). (**A**,**B**) SOD activity was determined in the roots, as described in Materials and Methods. The presented results are the means ± SD of three independent experiments, each run in triplicate, and are expressed as units g^{-1} FW. (**C**,**D**) The total RNA was isolated from the cucumber roots, and real-time PCR analysis was performed, with *CsCACS* as a reference gene, as described in Materials and Methods. The results are shown as the means ± SD of three replications. Different letters show different homogeneous groups of means calculated by Duncan's multiple range test (significant at $p < 0.05$).

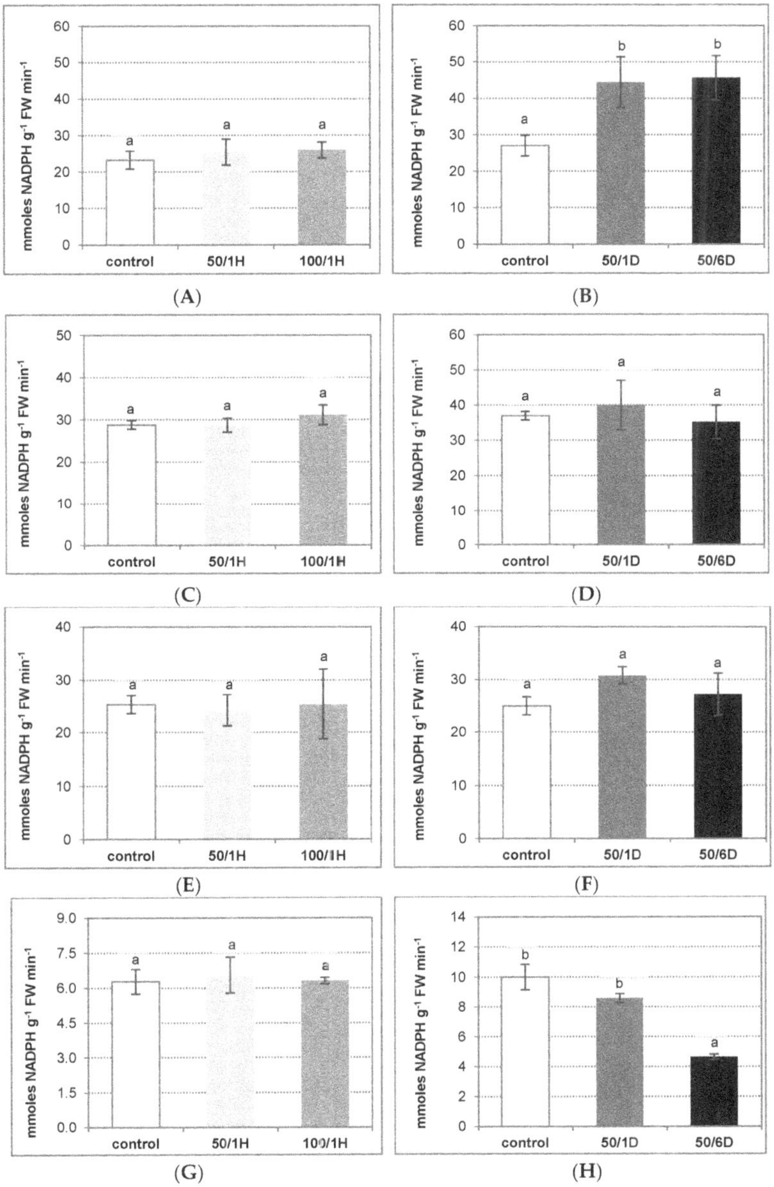

Figure 5. Activity of NADPH-generating enzymes: G6PDH (**A**,**B**), 6PGDH (**C**,**D**), NADP-ICDH (**E**,**F**), and NADP-ME (**G**,**H**) in the roots of cucumber seedlings exposed to salinity. The cucumber seedlings were treated with: (**A**,**C**,**E**,**G**) 50 mM or 100 mM NaCl for 1 h (50/1H and 100/1H, respectively); (**B**,**D**,**F**,**H**) 50 mM NaCl for 1 or 6 days (50/1D and 50/6D, respectively). The activities of NADP dehydrogenases were determined in the roots, as described in Materials and Methods. The presented results are the means ± SD of three independent experiments, each run in triplicate, and are expressed as nmoles of NADPH g^{-1} FW min^{-1}. Different letters show different homogeneous groups of means calculated by Duncan's multiple range test (significant at $p < 0.05$).

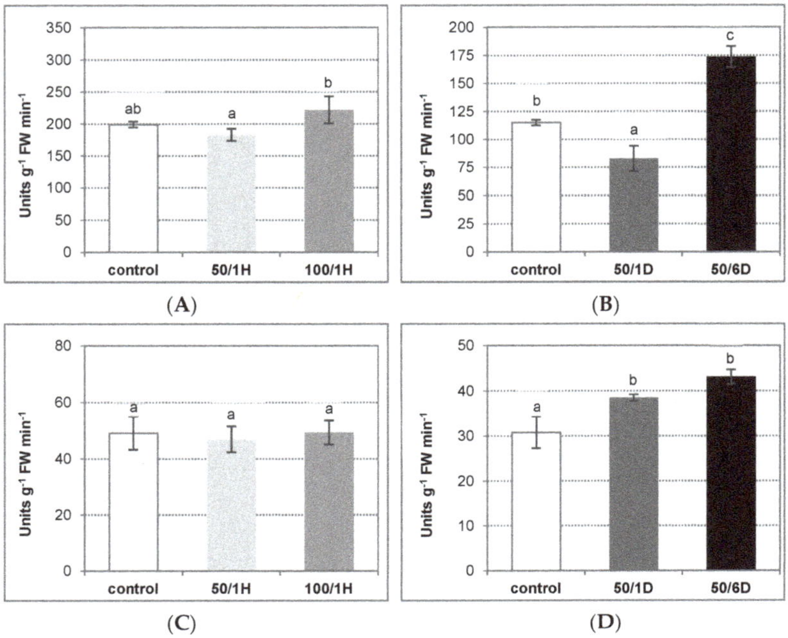

Figure 6. Catalase (**A**,**B**) and ascorbate peroxidase (**C**,**D**) activity in the tissues of cucumber roots exposed to salinity. The cucumber seedlings were treated with: (**A**,**C**) 50 mM or 100 mM NaCl for 1 h (50/1H and 100/1H, respectively); (**B**,**D**) 50 mM NaCl for 1 or 6 days (50/1D and 50/6D, respectively). The activities of antioxidant enzymes were determined in the roots, as described in Materials and Methods. The presented results are the means ± SD of three independent experiments, each run in triplicate, and are expressed as units g^{-1} FW min^{-1}. Different letters show different homogeneous groups of means calculated by Duncan's multiple range test (significant at $p < 0.05$).

The possibility that NADPH oxidase can be subjected to regulation by H_2O_2 was also studied. The study demonstrated that exogenously applied H_2O_2 significantly stimulated the enzyme. The activity of the NADPH oxidase increased 1.9-fold in the roots of seedlings treated with this signaling molecule for 24 h (Figure 7A). Furthermore, it was indicated that the H_2O_2-induced increase in the enzyme activity was not related to the activation of *CsRboh* gene expression in roots (Figure 7B).

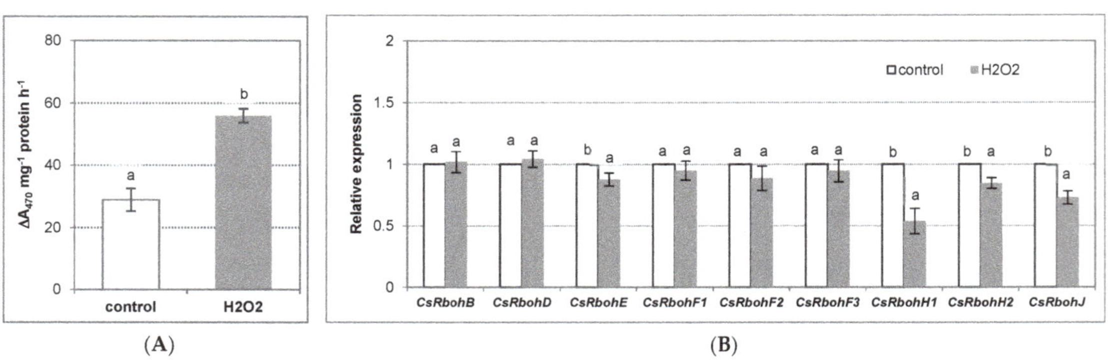

Figure 7. NADPH oxidase catalytic activity (**A**) and relative expression of *CsRboh* genes (**B**) in cucumber roots treated with H_2O_2. The cucumber seedlings were treated with 5 mM H_2O_2 for 24 h,

as described in Materials and Methods. (**A**) The plasma membrane vesicles were isolated from the cucumber roots, as described in Materials and Methods. The presented results are the mean ± SD of three independent experiments, with each experiment performed in triplicate, and are expressed as the ΔA_{470} of formazan XTT mg^{-1} protein h^{-1}. (**B**) The total RNA was isolated from the cucumber roots, and real-time PCR analysis was performed, with *CsTIP41* as a reference gene, as described in Materials and Methods. The results are shown as the means ± SD of three replications. Different letters show different homogeneous groups of means calculated by Duncan's multiple range test (significant at $p < 0.05$).

4. Discussion

To survive in soils contaminated with high salt concentration, plants have developed different adaptive mechanisms, in which signaling molecules seem to play an essential role. Many studies have indicated that, in response to salinity, enhanced levels of ROS, including hydrogen peroxide, are generated in plant tissues [30]. Consequently, oxidative damage of cellular components occurs, affecting plant cell functioning [30]. On the other hand, an increase in ROS content detected after exposure to salt can act as an early component of salinity sensing (acclimation signal), activating downstream H_2O_2-dependent signaling pathways, and leading to stress responses [9]. Among others, ROS were found to activate the MAPK cascade and potassium channels, and to modulate ion fluxes in the cells [9]. For this reason, it has been suggested that the coordinated ROS generating and scavenging system plays a crucial role in plant adaptation to salt stress [31].

Although knowledge about salinity sensing by plant cells is still limited, several sensory mechanisms have been proposed [32]. Moreover, it is supposed that different systems can operate in the same cell and at the same time, including the plasma membrane NADPH oxidase [33]. Accordingly, it has been shown in several plant species that salt stress rapidly induces a transient increase in NADPH oxidase activity, and up-regulates the expression of *Rboh* genes. This increase in enzyme activity is accompanied by H_2O_2 accumulation [34,35]. *Arabidopsis* mutants with defect in functional RBOH isoforms (AtrbohD and AtrbohF) exhibit increased Na^+ hypersensitivity, and lack of ROS accumulation in the xylem vessels [12,36]. Thus, it has been proposed that RBOH could be involved in the control of sodium loading to the xylem, protecting leaves from the adverse effects of salinity [37].

In contrast to the above results, NaCl can also negatively affect the NADPH oxidase. Srivastava, et al. [38], have demonstrated that enzyme activity decreases in a glycophyte *Brassica juncea* (low salt accumulator), and that it remains unaffected in a halophyte *Sesuvium portulacastrum* (high salt accumulator) under salt stress conditions. Decrease in the ROS level related to inhibited RBOH activity has been shown in maize seedlings treated with NaCl [39]. Similarly, in sugar beet, H_2O_2 content was significantly lower in salt-treated tissues compared to the controls. This correlated with decreased NADPH oxidase activity and down-regulation of *RBOH* gene transcription [40].

Our results indicate that an enhanced H_2O_2 level occurred in cucumber roots stressed with 100 mM NaCl for 1 h and 50 mM NaCl for 1 day. At the same time, NADPH oxidase activity was significantly stimulated. The observed increase in enzyme activity was related to the enhanced expression of two *Rboh* genes, *CsRbohD* and *CsRbohF1*. Similarly, Niu, et al. [33], using 75 mM NaCl concentration, indicated that the H_2O_2 content is rapidly elevated in the roots of salt-treated pumpkin-grafted cucumber plants, reaching a peak at 3 h and decreasing during the 3–12 h period. This increase is related to the NaCl-induced NADPH oxidase activity, as well as enhanced expression of *RbohD* and *RbohF* transcription level.

The generation of ROS in the apoplastic space could be related not only to the activation of the cell membrane-bound RBOH but also the apoplastic polyamine or diamine oxidases [41]. In salt-stressed maize leaves, RBOH activity is inhibited. Under such condition, the apoplastic polyamine oxidase, which activity increases, is responsible for the elevation in H_2O_2 content required for cell wall loosening and leaf growth [42]. In our study, DAO activity remained unchanged in the cucumber seedlings, regardless of the salt concentration and treatment time. Furthermore, PAO activity maintained a very low level in the cucumber roots (personal communications: Janicka M, 2021, available in the repository). This may suggest that RBOH participates in H_2O_2 generation in the apoplast in cucumber roots under salt stress. As we demonstrated earlier, the H_2O_2 content only slightly increased in 1-day stressed roots, whereas it was very high in 6-day treated roots of *Cucumis sativus* var. Wisconsin under salinity condition [43]. It is interesting that such visible differences occurred in two cultivars of the same plant species, suggesting some distinct stress responses. Duan, et al. [44], using two cucumber cultivars with different salt tolerance—Changchun mici (tolerant) and Jinchun No. 2 (more sensitive)—exposed to salinity for 7 days, demonstrated that 50 mM NaCl caused a significant increase in DAO activity in the roots of the former, but only a slight increase (only observed after 3 days of stress) in the roots of the latter. Similar to our results, salt treatment had no effect on DAO activity in the roots of rice seedlings [45]. The authors concluded that DAO does not contribute to H_2O_2 production in the cell wall of stressed plants. In contrast, salt stress was found to both strongly promote DAO activity and to stimulate polyamine degradation in the root tissues of maize seedlings [46] and lupine seedlings [47]. Thus, it seems obvious that different mechanisms function in plant cells to enhance H_2O_2 production, depending on the plant species/cultivar and its sensitivity to salt as well as treatment conditions.

The activity of plasma membrane NADPH oxidase can be regulated in several ways [2]. Research using *Arabidopsis* has revealed some regulatory mechanisms of RBOH proteins, which depend on different signaling effectors. Thus, it has been suggested that differences in regulatory mechanisms may be responsible for diverse RBOH functions in plant cells [48]. A few studies have shown that, under salinity conditions, cytosolic Na^+ accumulation followed by Ca^{2+} elevation can lead to the activation of RBOH and, consequently, to the accumulation of apoplastic H_2O_2 [37]. In our research, treatment of plants with H_2O_2 clearly activated NADPH oxidase, indicating that this enzyme remains under feedback control. However, the observed effect did not include the gene expression level, suggesting some post-translational regulatory mechanisms or additional signaling events, such as nitrosylation [49], H_2S-dependent persulfidation [50], and reversible phosphorylation [51].

Substrate availability appears to be another factor involved in controlling NADPH oxidase activity. Several enzymes are responsible for NADPH delivery function in plant cells [52]. It has been proposed that various NADPH-generating dehydrogenases could provide this essential cofactor for different NADPH-dependent processes. Moreover, there is evidence that one or more NADP dehydrogenases are regulated under specific stress conditions [52]. In *Arabidopsis* seedlings, the NADP-ICDH, which activity increased, appeared to act as a barrier in the defense mechanisms activated against salt stress [53]. In contrast, a decrease in the activities of the main NADPH-generating enzymes, especially NADP-ICDH, has been demonstrated in tomato roots under salinity conditions [54]. Among the NADP dehydrogenases (G6PDH, 6PGDH, NADP-ME, and NADP-ICDH) analyzed in the present work, the NADP-ICDH activity remained relatively unaffected, whereas the G6PDH activity was importantly enhanced in the roots of 1-week-old cucumber seedlings treated with NaCl for 1 and 6 days, suggesting that this protein may serve as a source of NADPH for RBOH under stress conditions. On the other hand, treatment of 5-week-old cucumber plants with 100 mM NaCl for 1–3 days significantly reduced G6PDH activity in roots, but stimulated it in leaves [55]. As it was reported in *A. thaliana* exposed to high salinity, the cytosolic G6PDH isozyme (G6PD6) was subjected to phosphorylation, and this modification was responsible for the observed enzyme activation [56]. Studies using *Arabidopsis* mutants revealed that both the G6PD enzymatic activity and the expression of *G6PD5* and *G6PD6* decreased in *atrbohD1*, *atrbohF1*, and *atrbohD1/F1* mutants in comparison to the WT plants, confirming the involvement of cytosolic G6PDH in RBOH-dependent ROS generation [57]. Scharte, et al. [58], have proposed that cytosolic G6PDH functions as an essential element of plant defense reactions, and that its modulation is important to enhance stress tolerance.

Plasma membrane-localized NADPH oxidases are a major source of cellular O_2^{-} radicals, which are transformed by superoxide dismutases to produce less reactive H_2O_2. Related to the metal cofactor at the active center, three distinct isoenzymes function in plant cells: Mn-SODs are present in the mitochondria and peroxisomes; Cu/Zn-SODs mainly in the cytosol, mitochondria, and plastids; while Fe-SODs are localized in the cytosol, mitochondria, chloroplasts, and peroxisomes [59]. Many studies have confirmed that salt stress triggers an increase in total SOD activity, especially in the leaves of plants treated with high doses of NaCl for several days [59]. In cucumber root cells, total SOD activity as well as *CsMSD* and *CsCSD2* expression increased significantly in the roots of seedlings exposed to 50 mM NaCl for 6 days, when NADPH oxidase activity and H_2O_2 level were lowered to nearly control values. This suggests that hydrogen peroxide generated by SOD may be effectively decomposed by H_2O_2-metabolizing systems. Hossain et al [40], have shown that in a highly salt-tolerant sugar beet stressed with 300 mM NaCl for up to 14 days, H_2O_2 content is always significantly lower in the salt-treated leaves in comparison to the control. This was correlated with the suppressed NADPH oxidase activity and gene expression (with the exception of *RBOHB*), as well as with the up-regulation of SOD activity and increased transcription of genes encoding this enzyme (with the exception of *FeSOD1*). Such a system, together with enhanced expression of genes for alternative oxidase and plastid terminal oxidase, is responsible for maintaining the low ROS level under severe salinity [40].

It is assumed that early production of H_2O_2 acts as a signal to induce antioxidant systems and prevent subsequent ROS generation and ROS-dependent cell damage [37]. In *Arabidopsis* under short-term salt treatments, RbohD and RbohF were suggested to be required in this process [13]. The main enzymes able to break down H_2O_2 are catalase, with the highest turnover rates, and ascorbate peroxidase, exhibiting higher affinity for H_2O_2 than CAT [60,61]. In accordance with the above assumption, in this study, both the up-regulation of APX and CAT activity were shown in the roots of cucumber seedlings exposed to salinity for 6 days, confirming the involvement of both enzymes in maintaining the low H_2O_2 level under prolonged salt exposure. In contrast, catalase activity was inhibited in roots treated with salt for 1 day. Stimulation of both APX and CAT in response to NaCl has been reported in many plant species. It has been shown to be associated with decreased oxidative damage and improved tolerance to salinity. Similar to SOD, a greater increase in CAT activity was observed in salt-tolerant plants/genotypes than in more sensitive ones [60]. Furthermore, in some glycophytes, including rice, the activity of this enzyme decreased after NaCl treatment [62].

5. Conclusions

At the early stages of salt stress (1 h and 1-day-treatment), the plasma membrane NADPH oxidase was activated in the cucumber (*Cucumis sativus* var. Krak) seedlings. This activation correlated with an observed enhancement in H_2O_2 content, suggesting the involvement of NADPH oxidase in the generation of H_2O_2, which may function as a signaling molecule participating in stress response. The observed stimulation of NADPH oxidase may have been at least partially related to the induction of some *CsRboh* genes, as well as the increased activity of G6PDH, providing NADPH. Another factor which may contribute to positive RBOH regulation at the post-translational level, could be H_2O_2. This proposal, however, requires additional studies concerning direct or indirect H_2O_2 action.

After 6-day exposure to salt stress, the plasma membrane NADPH oxidase activity, and consequently the H_2O_2 level, were diminished. This protected the cells against the toxic effect of high H_2O_2 concentrations. Additionally, antioxidant enzymes, including SOD, APX, and CAT, were significantly activated to maintain low H_2O_2 content in cells, and to avoid ROS-dependent damages. However, to explain the precise mechanism of

NADH oxidase regulation, and its interrelation with the antioxidant system in cucumber seedlings subjected to salinity, further analysis is needed.

Author Contributions: Conceptualization, M.J.; methodology, K.K., M.R., A.W., and M.J.; formal analysis, K.K., M.R., A.W., and M.J; writing—original draft preparation, K.K.; writing—review and editing, M.R., A.W., and M.J; supervision, M.J. All authors have read and agreed to the published version of the manuscript.

Funding: This research received no external funding.

Institutional Review Board Statement: Not applicable.

Informed Consent Statement: Not applicable.

Data Availability Statement: The data presented are available in this manuscript and in the repository at https://www.repozytorium.uni.wroc.pl/dlibra/publication/139358/edition/128721# (accessed on 10 September 2021).

Acknowledgments: We thank Beata Kuligowska for her excellent technical assistance.

Conflicts of Interest: The authors declare no conflict of interest.

References

1. Apel, K.; Hirt, H. Reactive oxygen species: Metabolism, oxidative stress, and signal transduction. *Annu. Rev. Plant. Biol.* **2004**, *55*, 373–399. [CrossRef] [PubMed]
2. Qu, Y.; Yan, M.; Zhang, Q. Functional regulation of plant NADPH oxidase and its role in signaling. *Plant Signal. Behav.* **2017**, *12*, e1356970. [CrossRef] [PubMed]
3. Marino, D.; Dunand, C.; Puppo, A.; Pauly, N. A burst of plant NADPH oxidases. *Trends Plant Sci.* **2012**, *17*, 9–15. [CrossRef]
4. Sagi, M.; Fluhr, R. Production of reactive oxygen species by plant NADPH oxidases. *Plant Physiol.* **2006**, *141*, 336–340. [CrossRef] [PubMed]
5. Ogasawara, Y.; Kaya, H.; Hiraoka, G.; Yumoto, F.; Kimura, S.; Kadota, Y.; Hishinuma, H.; Senzaki, E.; Yamagoe, S.; Nagata, K.; et al. Synergistic activation of the *Arabidopsis* NADPH oxidase AtrbohD by Ca^{2+} and phosphorylation. *J. Biol. Chem.* **2008**, *283*, 8885–8892. [CrossRef] [PubMed]
6. Jakubowska, D.; Janicka-Russak, M.; Kabała, K.; Migocka, M.; Reda, M. Modification of plasma membrane NADPH oxidase activity in cucumber seedling roots in response to cadmium stress. *Plant Sci.* **2015**, *234*, 50–59. [CrossRef]
7. Angelini, R.; Cona, A.; Federico, R.; Fincato, P.; Tavladoraki, P.; Tisi, A. Plant amine oxidases "on the move": An update. *Plant Physiol. Biochem.* **2010**, *48*, 560–564. [CrossRef]
8. Moschou, P.N.; Paschalidis, A.K.; Delis, I.D.; Andriopoulou, A.H.; Lagiotis, G.D.; Yakoumakis, D.I.; Roubelakis-Angelakis, K.A. Spermidine exodus and oxidation in the apoplast induced by abiotic stress is responsible for H_2O_2 signatures that direct tolerance responses in tobacco. *Plant Cell* **2008**, *20*, 1708–1724. [CrossRef]
9. Isayenkov, S.V.; Maathuis, F.J.M. Plant salinity stress: Many unanswered questions remain. *Front. Plant Sci.* **2019**, *10*, e80. [CrossRef]
10. Yang, Y.L.; Xu, S.J.; An, L.Z.; Chen, N.L. NADPH oxidase dependent hydrogen peroxide production, induced by salinity stress, may be involved in the regulation of total calcium in roots of wheat. *J. Plant Physiol.* **2007**, *164*, 1429–1435. [CrossRef]
11. Chung, J.S.; Zhu, J.K.; Bressan, R.A.; Hasegawa, P.M.; Shi, H.Z. Reactive oxygen species mediate Na^+-induced SOS1 mRNA stability in Arabidopsis. *Plant J.* **2008**, *53*, 554–565. [CrossRef] [PubMed]
12. Ma, L.; Zhang, H.; Sun, L.; Jiao, Y.; Zhang, G.; Miao, C.; Hao, F. NADPH oxidase AtrbohD and AtrbohF function in ROS-dependent regulation of Na^+/K^+ homeostasis in *Arabidopsis* under salt stress. *J. Exp. Bot.* **2012**, *63*, 305–317. [CrossRef] [PubMed]
13. Ben Rejeb, K.; Benzarti, M.; Debez, A.; Bailly, C.; Savouré, A.; Abdelly, C. NADPH oxidase-dependent H_2O_2 production is required for salt-induced antioxidant defense in *Arabidopsis thaliana*. *J. Plant Physiol.* **2015**, *174*, 5–15. [CrossRef]
14. Tanou, G.; Ziogas, V.; Belghazi, M.; Christou, A.; Filippou, P.; Job, D.; Fotopoulos, V.; Molassiotis, A. Polyamines reprogram oxidative and nitrosative status and the proteome of citrus plants exposed to salinity stress. *Plant Cell Environ.* **2014**, *37*, 864–885. [CrossRef]
15. Gémes, K.; Kim, Y.J.; Park, K.Y.; Moschou, P.N.; Andronis, E.; Valassaki, C.; Roussis, A.; Roubelakis-Angelakis, K.A. An NADPH-Oxidase/Polyamine Oxidase feedback loop controls oxidative burst under salinity. *Plant Physiol.* **2016**, *172*, 1418–1431. [CrossRef] [PubMed]
16. Velikova, V.; Yordanov, I.; Edreva, A. Oxidative stress and some antioxidant systems in acid rain-treated bean plants. Protective role of exogenous polyamines. *Plant Sci.* **2000**, *151*, 59–66. [CrossRef]
17. Kabała, K.; Zboińska, M.; Głowiak, D.; Reda, M.; Jakubowska, D.; Janicka, M. Interaction between the signaling molecules hydrogen sulfide and hydrogen peroxide and their role in vacuolar H^+-ATPase regulation in cadmium-stressed cucumber roots. *Physiol. Plant.* **2019**, *166*, 688–704. [CrossRef] [PubMed]

18. Larsson, C. Plasma membranes. In *Cell Components*; Linskens, H.F., Jackson, J.F., Eds.; Springer: Berlin/Heidelberg, Germany, 1985; pp. 85–104.
19. Kłobus, G. The role of plasma membrane-bound activities in nitrate transport into sealed plasma membrane vesicles from *Cucumis sativus* L. roots. In *Developments in Plant and Soil Science*; Baluška, F., Ciamporova, M., Gasparicova, O., Barlow, P.W., Eds.; Kluwer Academic Publishers: Dordrecht, The Netherlands, 1995; Volume 58, pp. 133–140.
20. Sagi, M.; Fluhr, R. Superoxide production by plant homologues of the gp91(phox)NADPH oxidase. Modulation of activity by calcium and by tobacco mosaic virus infection. *Plant Physiol.* **2001**, *126*, 1281–1290. [CrossRef]
21. Holmsted, B.; Larsson, L.; Tham, R. Further studies on spectrophotometric method for the determination of amine oxidase activity. *Biochim. Biophys. Acta* **1961**, *48*, 182–186. [CrossRef]
22. Federico, R.; Angelini, R. Distribution of polyamines and their related catabolic enzyme in etiolated and light-grown leguminosae seedlings. *Planta* **1988**, *173*, 317–321. [CrossRef]
23. Li, J.; Chen, G.; Wang, X.; Zhang, Y.; Jia, H.; Bi, Y. Glucose-6-phosphate dehydrogenase-dependent hydrogen peroxide production is involved in the regulation of plasma membrane H^+-ATPase and Na^+/H^+ antiporter protein in salt-stressed callus from *Carex moorcroftii*. *Physiol. Plant.* **2011**, *141*, 239–250. [CrossRef] [PubMed]
24. Aebi, H. Catalase in vitro. *Methods Enzymol.* **1984**, *105*, 121–126. [PubMed]
25. Janicka-Russak, M.; Kabała, K.; Burzyński, M. Different effect of cadmium and copper on H^+-ATPase activity in plasma membrane vesicles from *Cucumis sativus* roots. *J. Exp. Bot.* **2012**, *63*, 4133–4142. [CrossRef]
26. Chen, G.X.; Asada, K. Ascorbate peroxidase in tea leaves: Occurrence of two isozymes and the differences in their enzymatic and molecular properties. *Plant Cell Physiol.* **1989**, *30*, 987–998.
27. Beauchamp, C.; Fridovich, I. Superoxide dismutase: Improved assays and an assay applicable to acrylamide gels. *Anal. Biochem.* **1971**, *44*, 276–287. [CrossRef]
28. Bradford, M.M. A rapid and sensitive method for quantitation of microgram quantities of protein utilizing the principles of protein dye binding. *Anal. Biochem.* **1976**, *72*, 248–254. [CrossRef]
29. Migocka, M.; Papierniak, A. Identification of suitable reference genes for studying gene expression in cucumber plants subjected to abiotic stress and growth regulators. *Mol. Breed.* **2010**, *28*, 343–357. [CrossRef]
30. Gupta, B.; Huang, B. Mechanism of salinity tolerance in plants: Physiological, biochemical, and molecular characterization. *Int. J. Genom.* **2014**, e701596. [CrossRef]
31. Kaur, N.; Dhawan, M.; Sharma, I.; Pati, P.K. Interdependency of Reactive Oxygen Species generating and scavenging system in salt sensitive and salt tolerant cultivars of rice. *BMC Plant Biol.* **2016**, *16*, e131. [CrossRef]
32. Shabala, S.; Wu, H.; Bose, J. Salt stress sensing and early signalling events in plant roots: Current knowledge and hypothesis. *Plant Sci.* **2015**, *241*, 109–119. [CrossRef]
33. Niu, M.; Huang, Y.; Sun, S.; Sun, J.; Cao, H.; Shabala, S.; Bie, Z. Root respiratory burst oxidase homologue-dependent H_2O_2 production confers salt tolerance on a grafted cucumber by controlling Na^+ exclusion and stomatal closure. *J. Exp. Bot.* **2017**, *69*, 3465–3476. [CrossRef] [PubMed]
34. Xie, Y.J.; Xu, S.; Han, B.; Wu, M.Z.; Yuan, X.X.; Han, Y.; Gu, Q.; Xu, D.K.; Yang, Q.; Shen, W.B. Evidence of *Arabidopsis* salt acclimation induced by up-regulation of HY1 and the regulatory role of RbohD-derived reactive oxygen species synthesis. *Plant J.* **2011**, *66*, 280–292. [CrossRef] [PubMed]
35. Liu, L.; Huang, L.; Lin, X.; Sun, C. Hydrogen peroxide alleviates salinity-induced damage through enhancing proline accumulation in wheat seedlings. *Plant Cell Rep.* **2020**, *39*, 567–575. [CrossRef] [PubMed]
36. Jiang, C.; Belfield, E.J.; Mithani, A.; Visscher, A.; Ragoussis, J.; Mot, R.; Smith, J.A.; Harberd, N.P. ROS-mediated vascular homeostatic control of root-to-shoot soil Na delivery in *Arabidopsis*. *EMBO J.* **2012**, *31*, 4359–4370. [CrossRef]
37. Kurusu, T.; Kuchitsu, K.; Tada, Y. Plant signaling networks involving Ca^{2+} and Rboh/Nox-mediated ROS production under salinity stress. *Front. Plant Sci.* **2015**, *6*, e427. [CrossRef]
38. Srivastava, A.K.; Srivastava, S.; Lokhande, V.H.; D'Souza, S.F.; Suprasanna, P. Salt stress reveals differential antioxidant and energetics responses in glycophyte (*Brassica juncea* L.) and halophyte (*Sesuvium portulacastrum* L.). *Front. Environ. Sci.* **2015**, *3*, e19. [CrossRef]
39. Rodríguez, A.A.; Ramiro Lascano, H.; Bustos, D.; Taleisnik, E. Salinity-induced decrease in NADPH oxidase activity in the maize leaf blade elongation zone. *J. Plant Physiol.* **2007**, *64*, 223–230. [CrossRef]
40. Hossain, M.S.; ElSayed, A.I.; Moore, M.; Dietz, K.J. Redox and reactive oxygen species network in acclimation for salinity tolerance in sugar beet. *J. Exp. Bot.* **2017**, *68*, 1283–1298. [CrossRef]
41. Hossain, M.S.; Dietz, K.J. Tuning of redox regulatory mechanisms, reactive oxygen species and redox homeostasis under salinity stress. *Front. Plant Sci.* **2016**, *7*, e548. [CrossRef]
42. Rodríguez, A.A.; Maiale, S.J.; Menéndez, A.B.; Ruiz, O.A. Polyamine oxidase activity contributes to sustain maize leaf elongation under saline stress. *J. Exp. Bot.* **2009**, *60*, 4249–4262. [CrossRef]
43. Janicka, M.; Reda, M.; Czyżewska, K.; Kabała, K. Involvement of signalling molecules NO, H_2O_2 and H_2S in modification of plasma membrane proton pump in cucumber roots subjected to salt or low temperature stress. *Funct. Plant Biol.* **2018**, *45*, 428–439. [CrossRef] [PubMed]
44. Duan, J.; Li, J.; Guo, S.; Kang, Y. Exogenous spermidine affects polyamine metabolism in salinity-stressed *Cucumis sativus* roots and enhances short-term salinity tolerance. *J. Plant Physiol.* **2008**, *165*, 1620–1635. [CrossRef] [PubMed]

45. Lin, C.C.; Kao, C.H. Cell wall peroxidase activity, hydrogen peroxide level and NaCl-inhibited root growth of rice seedlings. *Plant Soil* **2001**, *230*, 135–143. [CrossRef]
46. Kongkiattikajorn, J. Effect of salinity stress on degradation of polyamines and amine oxidase activity in maize seedlings. *Kasetsart J. (Nat. Sci.)* **2009**, *43*, 28–33.
47. Legocka, J.; Sobieszczuk-Nowicka, E.; Ludwicki, D.; Lehmann, T. Putrescine catabolism via DAO contributes to proline and GABA accumulation in roots of lupine seedlings growing under salt stress. *Acta Soc. Bot. Pol.* **2017**, *86*, e3549. [CrossRef]
48. Baxter, A.; Mittler, R.; Suzuki, N. ROS as key players in plant stress signaling. *J. Exp. Bot.* **2014**, *65*, 1229–1240. [CrossRef] [PubMed]
49. Yun, B.W.; Feechan, A.; Yin, M.; Saidi, N.B.; Le Bihan, T.; Yu, M.; Moore, J.W.; Kang, J.G.; Kwon, E.; Spoel, S.H.; et al. S-nitrosylation of NADPH oxidase regulates cell death in plant immunity. *Nature* **2011**, *478*, 264–268. [CrossRef]
50. Shen, J.; Zhang, J.; Zhou, M.; Zhou, H.; Cui, B.; Gotor, C.; Romero, L.C.; Fu, L.; Yang, J.; Foyer, C.H.; et al. Persulfidation-based modification of cysteine desulfhydrase and the NADPH oxidase RBOHD controls guard cell abscisic acid signaling. *Plant Cell* **2020**, *32*, 1000–1017. [CrossRef]
51. Wu, F.; Chi, Y.; Jiang, Z.; Xu, Y.; Xie, L.; Huang, F.; Wan, D.; Ni, J.; Yuan, F.; Wu, X.; et al. Hydrogen peroxide sensor HPCA1 is an LRR receptor kinase in *Arabidopsis*. *Nature* **2020**, *578*, 577–581. [CrossRef]
52. Corpas, F.J.; Barroso, J.B. NADPH-generating dehydrogenases: Their role in the mechanism of protection against nitro-oxidative stress induced by adverse environmental conditions. *Front. Environ. Sci.* **2014**, *2*, e55. [CrossRef]
53. Leterrier, M.; Barroso, J.B.; Valderrama, R.; Palma, J.M.; Corpas, F.J. NADP-dependent isocitrate dehydrogenase from *Arabidopsis* roots contributes in the mechanism of defence against the nitro-oxidative stress induced by salinity. *Sci. World J.* **2012**, *6*, e694740.
54. Manai, J.; Gouia, H.; Corpas, F.J. Redox and nitric oxide homeostasis are affected in tomato (*Solanum lycopersicum*) roots under salinity-induced oxidative stress. *J. Plant Physiol.* **2014**, *171*, 1028–1035. [CrossRef] [PubMed]
55. Hýsková, V.; Plisková, V.; Červený, V.; Ryšlavá, H. NADP-dependent enzymes are involved in response to salt and hypoosmotic stress in cucumber plants. *Gen. Physiol. Biophys.* **2017**, *36*, 247–258. [CrossRef]
56. Dal Santo, S.; Stampfl, H.; Krasensky, J.; Kempa, S.; Gibon, Y.; Petutschnig, E.; Rozhon, W.; Heuck, A.; Clausen, T.; Jonak, C. Stress-induced GSK3 regulates the redox stress response by phosphorylating glucose-6-phosphate dehydrogenase in *Arabidopsis*. *Plant Cell* **2012**, *24*, 3380–3392. [CrossRef]
57. Yang, L.; Wang, X.; Chang, N.; Nan, W.; Wang, S.; Ruan, M.; Sun, L.; Li, S.; Bi, Y. Cytosolic glucose-6-phosphate dehydrogenase is involved in seed germination and root growth under salinity in *Arabidopsis*. *Front. Plant Sci.* **2019**, *10*, e182. [CrossRef] [PubMed]
58. Scharte, J.; Schön, H.; Tjaden, Z.; Weis, E.; von Schaewen, A. Isoenzyme replacement of glucose-6-phosphate dehydrogenase in the cytosol improves stress tolerance in plants. *Proc. Natl. Acad. Sci. USA* **2009**, *106*, 8061–8066. [CrossRef]
59. Szőllősi, R. Superoxide dismutase (SOD) and abiotic stress tolerance in plants: An overview. In *Oxidative Damage to Plants. Antioxidant Networks and Signaling*; Ahmad, P., Ed.; Elsevier Academic Publishers: Amsterdam, The Netherlands, 2014; pp. 89–129.
60. Sofo, A.; Scopa, A.; Nuzzaci, M.; Vitti, A. Ascorbate peroxidase and catalase activities and their genetic regulation in plants subjected to drought and salinity stresses. *Int. J. Mol. Sci.* **2015**, *16*, 13561–13578. [CrossRef] [PubMed]
61. Filiz, E.; Ozyigit, I.I.; Saracoglu, I.A.; Uras, M.E.; Sen, U.; Yalcin, B. Abiotic stress-induced regulation of antioxidant genes in different Arabidopsis ecotypes: Microarray data evaluation. *Biotech. Biotechnol. Equip.* **2019**, *33*, 128–143. [CrossRef]
62. Kibria, M.G.; Hossain, M.; Murata, Y.; Hoque, M.A. Antioxidant defense mechanisms of salinity tolerance in rice genotypes. *Rice Sci.* **2017**, *24*, 155–162. [CrossRef]

antioxidants

Article

Hydrogen Sulfide, Ethylene, and Nitric Oxide Regulate Redox Homeostasis and Protect Photosynthetic Metabolism under High Temperature Stress in Rice Plants

Harsha Gautam, Mehar Fatma, Zebus Sehar, Iqbal R. Mir and Nafees A. Khan *

Plant Physiology and Biochemistry Laboratory, Department of Botany, Aligarh Muslim University, Aligarh 202002, India; harshagautam99@gmail.com (H.G.); meharfatma30@gmail.com (M.F.); seharzebus5779@gmail.com (Z.S.); m3riqbal@gmail.com (I.R.M.)

* Correspondence: naf9.amu@gmail.com

Abstract: Rising temperatures worldwide due to global climate change are a major scientific issue at present. The present study reports the effects of gaseous signaling molecules, ethylene (200 μL L^{-1}; 2-chloroethylphosphonic acid; ethephon, Eth), nitric oxide (NO; 100 μM sodium nitroprusside; SNP), and hydrogen sulfide (H_2S, 200 μM sodium hydrosulfide, NaHS) in high temperature stress (HS) tolerance, and whether or not H_2S contributes to ethylene or NO-induced thermo-tolerance and photosynthetic protection in rice (*Oryza sativa* L.) cultivars, i.e., Taipei-309, and Rasi. Plants exposed to an HS of 40 °C for six h per day for 15 days caused a reduction in rice biomass, associated with decreased photosynthesis and leaf water status. High temperature stress increased oxidative stress by increasing the content of hydrogen peroxide (H_2O_2) and thiobarbituric acid reactive substance (TBARS) in rice leaves. These signaling molecules increased biomass, leaf water status, osmolytes, antioxidants, and photosynthesis of plants under non-stress and high temperature stress. However, the effect was more conspicuous with ethylene than NO and H_2S. The application of H_2S scavenger hypotaurine (HT) reversed the effect of ethylene or NO on photosynthesis under HS. This supports the findings that the ameliorating effects of Eth or SNP involved H_2S. Thus, the presence of H_2S with ethylene or NO can enhance thermo-tolerance while also protecting plant photosynthesis.

Keywords: ethylene; hydrogen sulfide; nitric oxide; photosynthesis; rice

Citation: Gautam, H.; Fatma, M.; Sehar, Z.; Mir, I.R.; Khan, N.A. Hydrogen Sulfide, Ethylene, and Nitric Oxide Regulate Redox Homeostasis and Protect Photosynthetic Metabolism under High Temperature Stress in Rice Plants. *Antioxidants* **2022**, *11*, 1478. https://doi.org/10.3390/antiox11081478

Academic Editor: Stanley Omaye

Received: 13 June 2022
Accepted: 25 July 2022
Published: 28 July 2022

1. Introduction

High temperature stress (HS) is significant environmental stress that restricts growth, metabolism, and crop production worldwide. Numerous biochemical processes susceptible to heat stress are involved in the growth and development of plants. Heat stress leads to the excess production of reactive oxygen species (ROS) that alter the cellular membrane protein structure and functions. Heat stress alters the expression of genes involved in direct protection from heat stress at the molecular level [1–3]. Crop production is currently a major concern due to HS, and methods for maintaining high crop yields under heat stress are prime agricultural objectives. Rice, a cereal crop belonging to the Poaceae family, is consumed by the majority of the world's population. It can provide high productivity and a prominent position in the international rice trade of food grains [4]. Rice crops are frequently subjected to HS, which impacts their quality and use around the world. In defense, plants respond to heat stress in several ways, including accumulating solutes that can arrange proteins and cellular structures, maintaining cell turgor by osmotic adjustment, modifying the antioxidant system to re-establish cellular redox balance and homeostasis, and involving complex regulatory signaling molecules for protection from oxidative stress [3,5–7]. Phytohormones, as signaling molecules, are the crucial molecules for coordinating a wide range of plant growth and development processes. They are important as endogenous signaling molecules that play a role in mediating various physiological reactions under heat stress by activating stress-responsive regulatory genes involved in plant growth [3,8,9].

Ethylene, nitric oxide (NO), and hydrogen sulfide (H_2S) have been identified as essential gaseous signaling molecules in plants and have recently attracted attention due to their participation in a number of physiological, biochemical, and cellular processes [4,7,10–12]. Ethylene is a plant hormone that regulates abiotic stress responses [13]. The treatment of ethephon (Eth; ethylene source) activates various stress-related proteins to preserve plant cell functional integrity and stability under heat stress [14]. Ethylene accumulation at varying concentrations is linked to plant responses to heat stress challenges that affect growth and development [15,16]. Ethylene signaling also enhances thermo-tolerance in plants by maintaining chlorophyll content and mitigating heat stress-induced adversity by reducing oxidative stress [17]. Recently, it was suggested that ethylene reduced glucose sensitivity and induced glutathione production, resulting in the increased expression of *psbA* and *psbB* genes to protect the pigment system (PS) II and photosynthesis under salt stress [9].

Nitric oxide has also emerged as a signaling molecule that masks the adverse effects of abiotic stresses such as heat, drought, salinity, ultraviolet (UV) radiation, heavy metals, etc., and received attention from the plant science community [7,18–21]. According to evidence, it appears to be a major signaling molecule in modulating various plant responses under heat stress, including photosynthesis, oxidative defense, gene expression, and protein changes. Nitric oxide modulates the heat stress transcription factors and DNA binding activity and acts upstream of AtCaM3 in heat stress signaling [22]. Nitric oxide plays a protective effect in PS II recovery in *Festuca arundinacea* under heat stress [18]. It scavenges ROS in plants [4,23] and increases the gene expression of *psbA* in maize [24], while CP43 and CP47 decrease under heat stress in rice [25]. Nitric oxide also upregulated the activities and expression of SOD, CAT, and APX genes in chickpea plants and mitigated the adverse effect of high salinity [26]. The consequences of NO's regulation and the genetic and molecular evidence for its function in improving heat and cold stress tolerance and adaptation have led to the discovery of potential new techniques to deal with future environmental difficulties [27]. These studies emphasized the protective roles of NO against heat stress-induced direct damage to the crops.

Few recent studies emphasize the role of H_2S with a diverse range of functions similar to NO in plants involved in various growths and development processes [7,21,28]. Recent research has linked H_2S, an endogenously-produced signaling molecule, to the regulation of autophagy in both plants and mammals by persulfidating particular targets [29]. The close proximity of two modifications—persulfidation and phosphorylation—could influence one another and serve as integration points for the H_2S- and ABA-signaling pathways [30]. Depending on the concentrations, both signaling molecules, NO and H_2S, work synergistically or antagonistically in plants. The gap between NO and H_2S is rapidly closing, and H_2S is emerging as a critical signal mediator involved in various biological processes, including the modulation of multiple stress responses [31]. However, the function of NO and H_2S in photosynthetic recovery processes during heat stress is still ambiguous. Hydrogen sulfide protects the crops and is involved in various physiological processes such as seed germination, root growth, stomatal movement, leaf wilting, fruit ripening, etc., under adverse environmental stress [11,28,32]. Additionally, H_2S protects plants from heavy metals, salinity, drought, and extreme temperature stresses [33,34]. The study of Li et al. [35] suggested that H_2S alleviated alkaline salt stress by regulating the expression of micro-RNAs through changes in the root architecture of *Malus hupehensis*. Hydrogen sulfide was influential in the thermo-tolerance in plants, and sodium hydrosulfide (NaHS)-pretreated seedlings (a H_2S donor) decreased oxidative stress by increasing the action and gene expression of antioxidant enzymes, as well as soluble sugar levels in wheat [36]. Melatonin and H_2S work together to protect against heat stress-induced photosynthetic inhibition by regulating carbohydrate metabolism, according to a study [11]. A few studies have also shown interactions between ethylene and H_2S. A report observed that endogenous H_2S is required for ethylene-mediated hexavalent chromium stress reduction in two pulse crops [37]. Similarly crucial for ethylene-induced stomatal closure in response to osmotic stress is ethylene-induced H_2S, which is a downstream component of osmotic stress signaling [38]. In *Solanum lycopersicum*, ethylene and H_2S co-treatment increased the expression of antioxidant-encoding genes *SlAPX2*, *SlCAT1*, *SlPOD12*, and *SlCuZnSOD* compared to ethylene treatment alone [39].

Thus, it was hypothesized that under HS, ethylene, NO, and H_2S may play a critical role in plant defense and cause considerable improvements in thermo-tolerance in plants by influencing multiple pathways. However, until now, there is no study available that correlates the study of ethylene, NO, and H_2S under HS. Thus, the present study highlights the impact of HS on ethylene, NO, and H_2S-mediated mechanisms and traits associated with thermo-tolerance and the involvement of H_2S in ethylene or NO-induced management strategies for oxidative stress-signaling and defense systems in rice plants.

2. Materials and Methods

2.1. Plant Material, Growth Conditions, and Experimental Design

Rice (*Oryza sativa* L.) cultivars, Taipei-309 (HS-tolerant) and Rasi (HS-non tolerant), obtained from the Indian Agricultural Research Institute, New Delhi were selected for the study. Their tolerance to HS was determined after their screening for changes in photosynthesis, growth, and yield parameters relative to controls, according to our earlier work [4]. After sterilizing the seeds with $HgCl_2$ (0.01%) for 2 min and rinsing repetitively with double distilled water, they were soaked in distilled water for 12–24 h and then incubated at 30 °C. Following incubation, the seeds were sown in 23 cm diameter pre-sterilized earthen pots containing 4 kg of acid-washed sand. Ten seeds of each cultivar were initially sown per pot, and later three seedlings were maintained after thinning. The pots were placed in an environmental growth chamber (Khera Instruments, New Delhi, India) with a day/night regime of 16/8 h, photosynthetically active photon flux density (PPFD) of 200 μmol m^{-2} s^{-1} at plant level, the temperature of 28 °C in the light and 22 °C in the dark, and relative humidity of 65 ± 5%.

In the experimentation, the plants were subjected to HS by exposing them to 40 °C temperature for six h daily, and the heat treatment was administered ten days after sowing (DAS). The heat treatment was maintained for 15 days for the same duration. After that, the plants were allowed to grow for five days at 28 °C (optimum temperature). The experimentation continued for 30 days. Control plants were kept at 28 °C for the duration of the experiment (30 days). A concentration of 200 μL L^{-1} 2-chloroethyl phosphonic acid (Eth; as an ethylene donor), 100 μM sodium nitroprusside (SNP; as a NO donor), and 200 μM NaHS was applied to the foliage of both HS-treated and non-treated plants with a hand sprayer at 15 DAS. Moreover, 100 μM hypotaurine (HT; H_2S scavenger), 100 μM norbornadiene (NBD; ethylene action inhibitor) and 100 μM 2-4-carboxyphenyl-4,4,5,5 -tetramethylimidazoline-1-oxyl-3-oxide (cPTIO; NO scavenger) were also applied to heat-stressed plants. The concentration of 100 μM cPTIO [7], 100 μM NBD [40], and 100 μM HT [11] used was based on our earlier studies. A surfactant teepol (0.5%) was added with the control and other treatment solutions. Our experimental design consisted of twelve treatments as follows: (i) control, (ii) HS, (iii) Eth, (iv) SNP, (v) NaHS, (vi) HS + Eth, (vii) HS + SNP, (viii) HS + NaHS, (ix) HS + Eth + HT, (x) HS + SNP + HT, (xi) HS + NaHS + NBD, and (xii) HS + NaHS + cPTIO. The hydrolysis of ethephon releases ethylene and phosphate, and the yield of phosphate is equivalent to ethylene [41,42]. Thus, the phosphate effect was nullified by adjusting the phosphate available from 200 μL L^{-1} Eth as single super phosphate. The arrangement of the treatments was a complete randomized block design with four replicates for each treatment (n = 4). The sampling of the plants was performed at 30 DAS to record various parameters of interest.

2.2. Measurement of Photosynthetic and Growth Characteristics

The Infrared Gas Analyzer (CID-340, Photosynthesis System, Bio-science, Washington, WI, USA) was used to measure gas exchange photosynthetic parameters (net photosynthetic rate, stomatal conductance, and intercellular CO_2 concentration) in the fully expanded upper leaves. At the time of measurements (between 11.00 and 12.00 h), the atmospheric CO_2 concentration was 380 ± 5 µmol mol^{-1}, the relative humidity was 70%, the photosynthetic active radiation was 780 µmol m^{-2} s^{-1}, and the air temperature was 28 °C. The chlorophyll content was measured in intact upper leaves of the plants with a SPAD chlorophyll meter (SPAD 502 DL PLUS, Spectrum Technologies, Plainfield, IL, USA) in the early morning hours. The dry weight of shoots and roots was recorded after separating the plants into roots and shoots, washed with water, and blotted with a soft paper towel to remove excess moisture. The separated shoots and roots were dried in a hot air oven (80 °C) for 72 h until a constant weight was achieved.

2.3. Chlorophyll Fluorescence Measurement

Fully expanded leaves were allowed to adapt under dark for 30 min before chlorophyll fluorescence measurements using Junior-PAM chlorophyll fluorometer (Heinz Walz GmbH, Eichenring, Effeltrich, Germany) were taken. The actual efficiency of PS II (ΦPSII), maximal efficiency of PS II (Fv/Fm), intrinsic efficiency of PS II (Φesc), photochemical quenching (qP), non-photochemical quenching (NPQ), and electron transport rate (ETR) were calculated. The details of the procedure are given in Supplementary File S1.

2.4. Leaf Relative Water Content (RWC) Determination

Leaf RWC was measured following the method of Barrs and Weatherley [43]. Fresh leaves were collected, weighed instantly by a standardized weighing balance, and then dipped into water in separate Petri-dishes for 12 h. The turgid weight was calculated by weighing the wet leaves. Afterwards, the leaf samples were oven-dried at 80 °C for 48 h, and the dry weight was recorded. The RWC was calculated using the following formula:

$$\text{RWC (\%)} = [(\text{Fresh weight} - \text{Dry weight})/(\text{Turgid weight} - \text{Dry weight})] \times 100.$$

2.5. Estimation of the Contents of Proline, Glycine Betaine (GB), Trehalose and Soluble Sugars

Proline content was determined by adopting the ninhydrin method [44]. Briefly, fresh leaf tissues (300 mg) were homogenized in 3 mL of 3% sulphosalicylic acid, and the homogenate filtrate was reacted with 1 mL each of acid ninhydrin and glacial acetic acid for 1 h in a test tube placed in a water bath at 100 °C. The mixture was extracted with toluene, and the absorbance was measured on a spectrophotometer at 520 nm using L-proline as a standard.

Glycine betaine was determined by estimating the betaine-periodite complex [45] in a sample from 500 mg dried leaf powder mechanically shaken for 24 h at 25 °C with 20 mL of deionized water. After filtering the samples, the filtrates were diluted (1:1) with 2 N H_2SO_4. A portion (0.5 mL) was taken and centrifuged before cooling in ice water for one hour. After adding 0.2 mL of cold KI-I_2 reagent, the reactants were gently stirred. The tubes were kept at 4 °C for 16 h and were centrifuged at 10,000× *g* for 15 min at 0 °C. After carefully aspirating the supernatant, the absorbance at 365 nm was measured after two hours. In 2 N H_2SO_4, reference standards for GB (50–200 µg mL^{-1}) were created. The trehalose content was determined following the protocol given by Trevelyan and Harrison [46]. Dried leaf powder (500 mg) was extracted in 80% ethanol, followed by centrifugation at 5000× *g* for 15 min at 4 °C. A 100 µL sample of the supernatant was combined with 4 mL of anthrone reagent and 2.0 mL of trichloroacetic acid (TCA). The absorbance was read at 620 nm. A standard curve was plotted using glucose.

The method developed by Xu et al. [47] was used to measure the amount of soluble sugars. A total of 100 mg of the dried sample powder was extracted with 10 mL of 80% ethanol and incubated at 80–85 °C for 30 min. Three additional extractions were performed after centrifuging the extract and transferring the supernatant to a 100 mL volumetric flask.

At 80–85 °C, alcohol extract was evaporated over a water bath. Following the addition of 100 mL distilled water, all three supernatants were poured into the flask. An aliquot of the extract was used to measure the amount of soluble sugars using the anthrone reagent, and the reaction mixture's absorbance was observed at 630 nm using a spectrophotometer.

2.6. Measurement of Hydrogen Peroxide (H_2O_2) and Lipid Peroxidation

The Okuda et al. [48] method was used for the H_2O_2 assay and is explained earlier [49]. The details of the method are given in Supplementary File S1. Fresh leaf tissues (500 mg) were macerated in ice-cold 200 mM perchloric acid before being spun at 1200× *g* for 10 min. Later, 4 M KOH was used to neutralize the supernatant. In order to remove the insoluble potassium perchlorate from the homogenate, it was further centrifuged at 500× *g* for three minutes. The reaction mixture (1.5 mL) included 20 µL of peroxidase (0.25 unit), 400 µL of 12.5 mM 3-(dimethylamino) benzoic acid in 0.375 M phosphate buffer (pH 6.5), 80 µL of 3-methyl-2-benzothiazoline hydrazone, and 1 mL of the eluate. Peroxidase was added and the reaction was started at 25 °C. On a spectrophotometer, the increase in absorbance was calculated at 590 nm. The content of thiobarbituric acid reactive substances (TBARS) was estimated to determine lipid peroxidation as described by Dhindsa et al. [50] and explained earlier [49].

2.7. Assay of Antioxidant Enzymes Activities

The activity of superoxide dismutase (SOD), ascorbate peroxidase (APX), and glutathione reductase (GR) was measured using the methods of Beyer and Fridovich [51], Giannopolitis and Ries [52], Nakano and Asada [53], and Foyer and Halliwell [54], respectively. The details of the methods are given in Supplementary File S1.

2.8. Determination of Nitric Oxide, Hydrogen Sulfide, and Ethylene

The method of Zhou et al. [55] was used for determining NO content with a slight modification by estimating nitrite content. Using a pre-chilled mortar and pestle, 500 mg of healthy leaves were ground in 3.0 mL of 50 mM ice-cold acetic acid buffer (pH 3.6) containing 4% zinc acetate. The mixture was then centrifuged at 11,500× *g* for 15 min at 4 °C. The supernatant was collected, and the pellets were rinsed in extraction buffer (1.0 mL) before being centrifuged again. After adding 100 mg of charcoal, the supernatants from the two spins were neutralized. The filtrate was collected after a brief vortex. Greiss reagent (0.1% N-1-naphthyl ethylenediamine dihydrochloride and 1% sulphanilamide in 5% H_2PO_4 solution) was added to each 1.0 mL filtrate and mixed in a 1:1 ratio before incubation at room temperature for 30 min. The NO content was determined using a calibration curve with sodium nitrite as a standard, and the absorbance was measured at 540 nm. Methylene blue formation from dimethyl-p-phenylenediamine in HCl was used to estimate the content of leaf H_2S as described by Xie et al. [56] with minor modifications. The fresh leaf samples (700 mg) were homogenized in 2.5 mL of Tris-HCl buffer (20 mM L^{-1}, pH 6.8) containing 10 mM L^{-1} ethylene diamine tetraacetic acid (EDTA). The homogenate was centrifuged for 15 min at 4 °C and 12,000× *g*. For H_2S trapping, 0.2 mL of 1% (*w*/*v*) zinc acetate was added to the supernatant (0.75 mL). After 30 min of development, 0.1 mL of 30 mM L^{-1} ferric chloride in 1.2 mol L^{-1} of HCl and 0.1 mL of 20 mM L^{-1} dimethyl-p-phenylenediamine dissolved in 7.2 mol L^{-1} of HCl were added. At 670 nm, spectrophotometric analysis was used to determine the methylene blue formation. As a standard curve, different concentrations of NaHS were used, expressed as nmol g^{-1} fresh weight (FW).

The ethylene evolution in leaves was measured using a gas chromatograph following the procedure as described earlier by Fatma et al. [12]. The details are given in Supplementary File S1.

2.9. RNA Isolation and cDNA Synthesis

Following the manufacturer's instructions, total RNA was extracted from rice leaves using the TRIzol reagent (Ambion, Life Technologies, Austin, TX, USA). With the help of a Nanodrop spectrophotometer (Thermo Scientific, Waltham, MA, USA), the extracted RNA was quantified. The details of the procedure are given in Gautam et al. [49] and presented in Supplementary File S1.

2.10. Quantitative Real-Time PCR Analysis

Real-time PCR (RT-PCR) was performed in a 96-well reaction plate (Roche, Mannheim, Germany) containing 20 μL reaction mixture of ×10 reaction buffer, 2 mM dNTPs, 1 mM $MgCl_2$, 0.35 μM each of forward and reverse primers, 1 μL Sybr green (×10), 10 μg cDNA template, and 5 U Taq polymerase on a thermal cycler (Light cycler 480 II, Roche, Germany). The details of the procedure are given in Gautam et al. [49] and presented in Supplementary File S1. Primer pairs used for the quantitative RT-PCR are listed in Supplementary Table S1. The data was interpreted as the differential expression of the target gene in the treated sample versus the untreated control in relation to the internal control.

2.11. Statistical Analysis

A two-way analysis of variance (ANOVA) was performed with SPSS software version 17.0 for Windows to analyze the data and presented as a treatment mean ± SE (n = 4). The least significant difference (LSD) was calculated for the significant data at $p < 0.05$. The data bars with the same letter were not significantly different by the LSD test at $p < 0.05$. The principal component analysis (PCA) and Pearson correlation analyses ($p < 0.05$, $p < 0.01$ and $p < 0.001$) were carried out using OriginPro software. To create biplots, the first two components (PC1 and PC2) showing the maximum variance in the datasets were considered.

3. Results

3.1. Growth Parameters

The rice plants' growth was analyzed in terms of shoot and root dry weight. The plants subjected to HS showed a reduction in shoot dry weight by (25.4%) in Taipei-309 and (28.0%) in Rasi, and root dry weight by (26.7%) in Taipei-309 and (30.3%) in Rasi, in comparison to control. However, Eth, SNP or NaHS application resulted in the reversal of the adverse effects of HS on rice growth; as the shoot dry weight recovered by Eth (29.0% and 28.0%), SNP (25.4% and 22.0%), and NaHS (21.8% and 20.0%) in Taipei-309 and Rasi, respectively, compared to control plants (Figure 1).

In Taipei-309 and Rasi, foliar applications of Eth, SNP, or NaHS lowered the effect of HS on root dry weight by (25.3% and 19.6%), (21.1% and 16.6%), or (18.3% and 13.6%), respectively, compared to controls.

This suggests that Eth, SNP, or NaHS can reduce the negative effects of HS on the biomass of both rice cultivars. Moreover, the application of HT exacerbated the deleterious effects of HS in Taipei and Rasi cultivars, resulting in reduced shoot dry weight by (12.7% and 14.0%) and by (16.3% and 18.0%) in the presence of Eth and SNP, respectively. Similarly, the root dry weight was reduced by (14.0% and 16.6%) and by (16.9% and 19.6%) in Eth and SNP-treated Taipei-309 and Rasi, respectively, relative to control plants. Therefore, combining HT with Eth or SNP completely reversed the mitigating effects of Eth or SNP. The application of NBD and cPTIO reduced shoot and root dry weights in the presence of NaHS in heat-treated plants in both the cultivars, reflecting that the application of NBD or cPTIO with NaHS did not entirely reverse the mitigating effects of NaHS on growth characteristics.

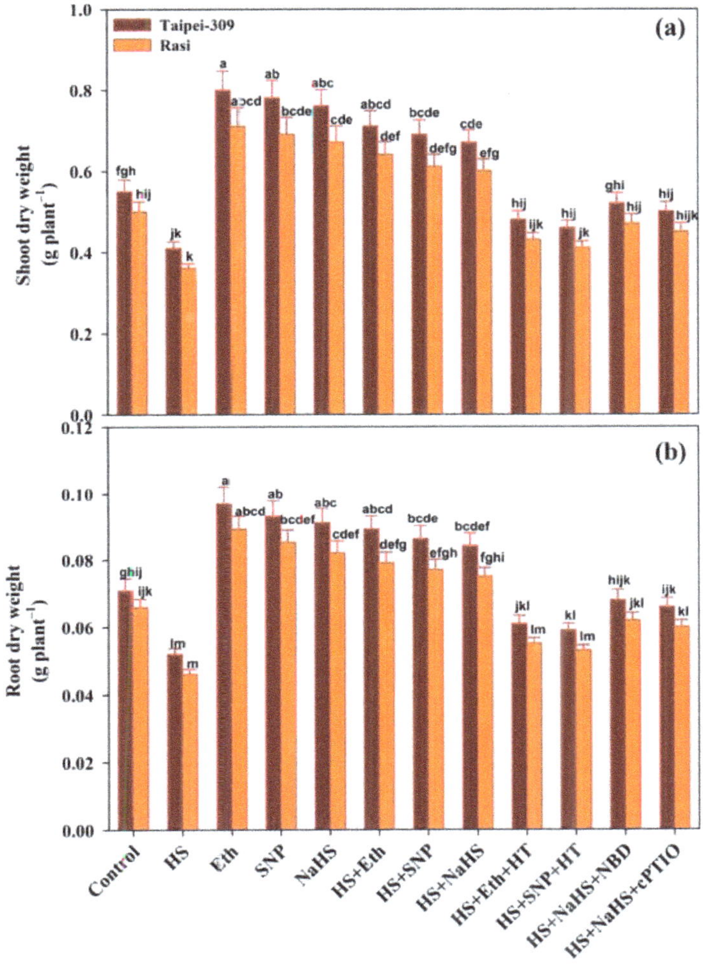

Figure 1. (**a**) Shoot dry weight and (**b**) root dry weight of rice (*Oryza sativa* L.) cultivars Taipei-309 and Rasi under control and high temperature stress (HS) supplied with 200 µL L^{-1} ethephon (Eth), 100 µM sodium nitroprusside (SNP), 200 µM sodium hydrosulfide (NaHS) or 100 µM hypotaurine (HT), 100 µM 2-4-carboxyphenyl-4,4,5,5 -tetramethylimidazoline-1-oxyl-3-oxide (cPTIO) or 100 µM norbornadiene (NBD) scavengers of hydrogen sulfide (H_2S), nitric oxide (NO), and ethylene action inhibitors, respectively. Data are presented as treatments mean ± SE (n = 4). The values followed by the same letters did not differ significantly by LSD test at $p < 0.05$.

3.2. Gas-Exchange Parameters and Chlorophyll Content

Net photosynthesis rate (Pn), stomatal conductance (Gs), intercellular CO_2 concentration (Ci), and SPAD value decreased under HS by 35.0%, 26.9%, 23.2%, and 28.4% in Taipei-309 and 36.1%, 28.2%, 24.2%, and 29.2% in Rasi, respectively, compared to control plants (Table 1). In comparison to control and heat-stressed plants, the individual treatment of Eth, SNP, and NaHS increased these parameters significantly. The individual applications of Eth, SNP, or NaHS under HS showed a substantial increase in *Pn* (29.1% and 98.9%, 27.1% and 95.9%, 24.5% and 91.8%), *Gs* (25.6% and 71.9%, 25.0% and 71.8%, 24.4% and 70.3%), *Ci* (21.3% and 57.9%, 20.2% and 56.6%, 19.9% and 56.1%), and SPAD value (37%

and 91.7% and 35.9% and 90%, 32% and 84.6%) in Taipei-309 and *Pn* (25.6% and 96.7%, 24.3% and 94.5%, 20.8% and 89%), *Gs* (23.1% and 71.6%, 21.9% and 70%, 21.4% and 69.3%), *Ci* (18.7% and 56.8%, 17.8% and 55.6%, 16.2% and 53.4%), and SPAD value (35.8% and 91.9%, 31.4% and 85.6%, 27.6% and 80.2%) in Rasi, respectively. Exogenously-applied Eth considerably alleviated more compared to SNP or NaHS, the decrement in the levels of *Pn*, *Gs*, *Ci*, and SPAD caused by HS. However, the potential effects of Eth or SNP on these parameters were significantly minimized by the H_2S scavenger, HT. The treatment of NBD or cPTIO moderately reversed the mitigating effects of NaHS.

Table 1. Net photosynthetic rate (μmol CO_2 m^{-2} s^{-1}), stomatal conductance (mmol m^{-2} s^{-1}), intercellular CO_2 concentration (μmol mol^{-1}), and chlorophyll content (SPAD value) of rice (*Oryza sativa* L.) cultivars Taipei-309 and Rasi after foliar treatment of plants with 200 μL L^{-1} ethephon (Eth), 100 μM sodium nitroprusside (SNP) or 200 μM sodium hydrosulfide (NaHS) grown with or without high temperature stress (HS; 40 °C) or 100 μM hypotaurine (HT), 100 μM 2-4-carboxyphenyl-4,4,5,5-tetramethylimidazoline-1-oxyl-3-oxide (cPTIO) or 100 μM norbornadiene (NBD) scavengers of hydrogen sulfide (H_2S), nitric oxide (NO), and ethylene action inhibitors, respectively, with HS at 15 days after sowing. Data are presented as treatments mean ± SE (*n* = 4). The values followed by the same letters did not differ significantly by LSD test at $p < 0.05$.

	Taipei-309				Rasi			
Treatments	**Net Photosynthetic Rate**	**Stomatal Conductance**	**Intercellular CO_2 Concentration**	**SPAD**	**Net Photosynthetic Rate**	**Stomatal Conductance**	**Intercellular CO_2 Concentration**	**SPAD**
Control	15.1 ± 1.08^{defg}	367.2 ± 15.0^{de}	260.1 ± 13.8^{de}	33.7 ± 1.30^{g}	14.4 ± 1.04^{fgh}	355.3 ± 14.0^{ef}	251.2 ± 12.4^{ef}	31.5 ± 1.25^{gh}
HS	9.8 ± 0.66^{ij}	268.3 ± 12.2^{gh}	199.8 ± 11.6^{gh}	24.1 ± 1.10^{kl}	9.2 ± 0.55^{j}	254.8 ± 11.0^{h}	190.3 ± 10.2^{h}	22.3 ± 1.0^{l}
Eth	21.7 ± 1.45^{a}	475.2 ± 16.5^{a}	346.3 ± 14.9^{a}	50.6 ± 1.76^{a}	20.1 ± 1.28^{abc}	451.1 ± 15.7^{abc}	328.2 ± 14.5^{abc}	46.2 ± 1.65^{bcd}
SNP	21.3 ± 1.40^{ab}	471.2 ± 16.4^{ab}	340.3 ± 14.8^{ab}	49.3 ± 1.69^{ab}	19.7 ± 1.25^{abc}	445.4 ± 15.6^{abc}	320.4 ± 14.2^{abc}	45.1 ± 1.51^{bcde}
NaHS	20.9 ± 1.34^{abc}	469.3 ± 16.0^{ab}	339.6 ± 14.6^{ab}	48.7 ± 1.60^{abc}	19.1 ± 1.19^{abc}	442.4 ± 15.1^{abc}	319.7 ± 13.8^{abc}	44.3 ± 1.40^{cdef}
HS + Eth	19.5 ± 1.24^{abc}	461.3 ± 15.9^{abc}	315.6 ± 14.4^{abc}	46.2 ± 1.55^{bcd}	18.1 ± 1.13^{bcd}	437.4 ± 14.9^{bcd}	298.3 ± 13.7^{bcd}	42.8 ± 1.37^{def}
HS + SNP	19.2 ± 1.13^{abc}	459.3 ± 15.9^{abc}	312.9 ± 14.1^{abc}	45.8 ± 1.45^{bcd}	17.9 ± 1.10^{bcde}	433.2 ± 14.8^{bcd}	296.2 ± 13.6^{bcd}	41.4 ± 1.35^{ef}
HS + NaHS	18.8 ± 1.11^{abc}	457.1 ± 15.4^{abc}	311.9 ± 14.0^{abc}	44.5 ± 1.40^{cde}	17.4 ± 1.06^{cdef}	431.6 ± 14.2^{cd}	292.0 ± 12.4^{cd}	40.2 ± 1.30^{f}
HS + Eth + HT	13.1 ± 0.90^{ghi}	311.6 ± 13.8^{efgh}	230.8 ± 12.9^{efgh}	29.4 ± 1.19^{hij}	12.2 ± 0.85^{ghij}	294.4 ± 12.9^{efgh}	220.8 ± 11.3^{efgh}	26.9 ± 1.13^{ijk}
HS + SNP + HT	12.2 ± 0.83^{ghij}	302.6 ± 13.7^{efgh}	226.3 ± 12.5^{efgh}	28.1 ± 1.15^{hijk}	11.4 ± 0.77^{hij}	289.1 ± 12.5^{fgh}	214.3 ± 11.2^{fgh}	25.4 ± 1.11^{jkl}
HS + NaHS + NBD	14.6 ± 0.96^{efgh}	361.7 ± 14.9^{ef}	247.7 ± 13.7^{ef}	31.9 ± 1.25^{gh}	13.7 ± 0.91^{gh}	341.2 ± 13.7^{efgh}	233.8 ± 12.1^{efgh}	29.1 ± 1.22^{hij}
HT + NaHS + cPTIO	14.3 ± 0.92^{fgh}	340.5 ± 14.7^{efg}	242.7 ± 13.4^{efg}	30.6 ± 1.20^{ghi}	13.3 ± 0.88^{gh}	323.1 ± 13.6^{efgh}	227.8 ± 11.4^{efgh}	28.2 ± 1.19^{hijk}

3.3. Chlorophyll Fluorescence Parameters

The measurement of chlorophyll fluorescence parameters was taken under both stress and without stress in leaves of rice cultivars (Tables 2 and 3). Heat stress exposure reduced the studied fluorescence parameters compared to control, but treatment with Eth, SNP, or NaHS increased these parameters under stress and no stress conditions. In contrast, NPQ increased under HS but decreased significantly with Eth, SNP, or NaHS application under no stress compared to control. Eth, SNP or NaHS treatments proved effective in improving (ΦPS II, Fv/Fm, Φesc, qP, and ETR) in heat-treated plants compared to controls. The data revealed that Eth, SNP, or NaHS treatments were essential to mitigate the negative effects of HS on the parameters mentioned above. Still, Eth was more effective than SNP or NaHS. The application of NBD or cPTIO reversed the positive effects of NaHS on the chlorophyll fluorescence parameters under HS; however, the addition of HT (100 μM) along with Eth or SNP entirely reversed the positive effects of Eth or SNP on these parameters in heat-stressed conditions.

Table 2. Actual efficiency of PSII, maximal efficiency of PSII, intrinsic efficiency of PSII, photochemical quenching, non-photochemical quenching, and electron transport rate of rice (*Oryza sativa* L.) cultivar Taipei-309 after foliar treatment of plants with 200 µL L^{-1} ethephon (Eth), 100 µM sodium nitroprusside (SNP), or 200 µM sodium hydrosulfide (NaHS) grown with or without high temperature stress (HS; 40 °C) or 100 µM hypotaurine (HT), 100 µM 2-4-carboxyphenyl-4,4,5,5-tetramethylimidazoline-1-oxyl-3-oxide (cPTIO) or 100 µM norbornadiene (NBD) scavengers of hydrogen sulfide (H_2S), nitric oxide (NO), and ethylene action inhibitors, respectively, with HS at 15 days after sowing. Data are presented as treatments mean ± SE (n = 4). The values followed by the same letters did not differ significantly by LSD test at $p < 0.05$.

Treatments	Actual Efficiency of PSII	Maximum Efficiency of PSII	Intrinsic Efficiency of PSII	Photochemical Quenching	Non-Photochemical Quenching	Electron Transport Rate
Control	0.621 ± 0.036 bcdef	0.808 ± 0.051 bcdef	0.731 ± 0.045 abcde	0.684 ± 0.045 abcdefghi	0.573 ± 0.032 cdefgh	164.6 ± 9.4 e
HS	0.497 ± 0.021 gh	0.622 ± 0.029 gh	0.620 ± 0.025 de	0.569 ± 0.029 i	0.778 ± 0.052 ab	132.8 ± 6.8 fg
Eth	0.751 ± 0.060 a	0.876 ± 0.071 a	0.849 ± 0.068 a	0.831 ± 0.067 a	0.455 ± 0.021 h	241.3 ± 11.2 a
SNP	0.744 ± 0.059 a	0.869 ± 0.069 a	0.841 ± 0.066 a	0.824 ± 0.063 ab	0.463 ± 0.025 gh	237.6 ± 10.2 ab
NaHS	0.732 ± 0.052 ab	0.861 ± 0.065 ab	0.837 ± 0.061 ab	0.816 ± 0.060 abc	0.469 ± 0.029 gh	231.8 ± 10.1 abc
HS + Eth	0.711 ± 0.048 abc	0.849 ± 0.062 abc	0.796 ± 0.058 abc	0.796 ± 0.058 abcde	0.582 ± 0.034 cdefgh	220.7 ± 9.7 abcd
HS + SNP	0.707 ± 0.045 abc	0.836 ± 0.059 abc	0.789 ± 0.051 abc	0.787 ± 0.055 abcdef	0.591 ± 0.036 cdefg	216.5 ± 9.5 abcd
HS + NaHS	0.701 ± 0.040 abc	0.830 ± 0.053 abc	0.781 ± 0.048 abcd	0.773 ± 0.051 abcdefg	0.598 ± 0.038 cdefg	212.4 ± 9.5 abcd
HS + Eth + HT	0.563 ± 0.028 efgh	0.777 ± 0.039 efgh	0.695 ± 0.035 abcde	0.642 ± 0.036 efghi	0.651 ± 0.046 bc	147.6 ± 8.4 efg
HS + SNP + HT	0.547 ± 0.024 fgh	0.769 ± 0.034 fgh	0.686 ± 0.033 abcde	0.631 ± 0.033 fghi	0.660 ± 0.048 bc	141.9 ± 7.7 efg
HS + NaHS + NBD	0.598 ± 0.033 cdefg	0.795 ± 0.047 cdefg	0.727 ± 0.042 abcde	0.671 ± 0.043 bcdefghi	0.623 ± 0.040 cd	159.3 ± 8.8 ef
HS + NaHS + cPTIO	0.582 ± 0.030 defgh	0.786 ± 0.042 defgh	0.721 ± 0.039 abcde	0.664 ± 0.040 cdefghi	0.639 ± 0.043 c	152.4 ± 8.6 ef

Table 3. Actual efficiency of PSII, maximal efficiency of PSII, intrinsic efficiency of PSII, photochemical quenching, non-photochemical quenching, and electron transport rate of rice (*Oryza sativa* L.) cultivar Rasi after foliar treatment of plants with 200 µL L^{-1} ethephon (Eth), 100 µM sodium nitroprusside (SNP), or 200 µM sodium hydrosulfide (NaHS) grown with or without high temperature stress (HS; 40 °C) or 100 µM hypotaurine (HT), 100 µM 2-4-carboxyphenyl-4,4,5,5 -tetramethylimidazoline-1-oxyl-3-oxide (cPTIO) or 100 µM norbornadiene (NBD) scavengers of hydrogen sulfide (H_2S), nitric oxide (NO), and ethylene action inhibitors, respectively, with HS at 15 days after sowing. Data are presented as treatments mean ± SE (n = 4). The values followed by the same letters did not differ significantly by LSD test at $p < 0.05$.

Treatments	Actual Efficiency of PSII	Maximum Efficiency of PSII	Intrinsic Efficiency of PSII	Photochemical Quenching	Non-Photochemical Quenching	Electron Transport Rate
Control	0.611 ± 0.025 bcdefg	0.797 ± 0.042 bcdefg	0.721 ± 0.038 abcde	0.671 ± 0.037 bcdefghi	0.582 ± 0.035 cdefgh	153.7 ± 8.6 ef
HS	0.472 ± 0.015 h	0.601 ± 0.021 h	0.598 ± 0.018 e	0.543 ± 0.021 i	0.799 ± 0.055 a	121.4 ± 6.4 g
Eth	0.725 ± 0.040 ab	0.854 ± 0.067 ab	0.826 ± 0.066 ab	0.811 ± 0.056 abcd	0.471 ± 0.027 fgh	220.1 ± 10.7 abcd
SNP	0.719 ± 0.036 ab	0.847 ± 0.062 ab	0.820 ± 0.063 ab	0.801 ± 0.054 abcd	0.483 ± 0.031 efgh	215.8 ± 10.3 abcd
NaHS	0.709 ± 0.034 abc	0.839 ± 0.060 abc	0.814 ± 0.059 abc	0.798 ± 0.051 abcde	0.492 ± 0.033 defgh	210.6 ± 9.5 bcd
HS + Eth	0.687 ± 0.032 abcd	0.825 ± 0.058 abcd	0.769 ± 0.051 abcd	0.775 ± 0.048 abcdefg	0.597 ± 0.037 cdefg	202.5 ± 9.2 cd
HS + SNP	0.679 ± 0.029 abcde	0.819 ± 0.052 abcde	0.758 ± 0.047 abcde	0.769 ± 0.045 abcdefg	0.606 ± 0.039 cdef	198.3 ± 9.1 d
HS + NaHS	0.674 ± 0.026 abcde	0.811 ± 0.047 abcde	0.749 ± 0.042 abcde	0.754 ± 0.040 abcdefgh	0.611 ± 0.041 cde	191.4 ± 8.9 d
HS + Eth + HT	0.538 ± 0.021 fgh	0.760 ± 0.028 fgh	0.676 ± 0.026 bcde	0.624 ± 0.030 ghi	0.667 ± 0.050 bc	136.5 ± 8.1 efg
HS + SNP + HT	0.529 ± 0.018 fgh	0.751 ± 0.023 fgh	0.654 ± 0.021 cde	0.611 ± 0.027 hi	0.679 ± 0.052 abc	130.8 ± 7.4 fg
HS + NaHS + NBD	0.576 ± 0.023 defgh	0.780 ± 0.038 defgh	0.705 ± 0.033 abcde	0.656 ± 0.034 defghi	0.635 ± 0.044 c	146.6 ± 8.4 efg
HS + NaHS + cPTIO	0.563 ± 0.022 efgh	0.771 ± 0.030 efgh	0.694 ± 0.030 abcde	0.642 ± 0.032 efghi	0.642 ± 0.046 c	141.2 ± 8.2 efg

3.4. *Leaf RWC*

The HS decreased the leaf RWC of both the cultivars by (18.5%) in Taipei-309 and (20.1%) in Rasi compared to control plants (Figure 2a). The results revealed that Eth-, SNP-, or NaHS-spraying treatments improved the RWC in no stress and HS conditions. In heat-stressed plants, Eth recovered RWC by (17.5% and 44.2%), SNP by (15.7% and 42%) or NaHS by (13.1% and 38.8%) in Taipei-309; and by (15.7% and 44.8%), (12.4% and 40.7%) or by (11.1% and 39.1%) in Rasi compared to control and heat-stressed plants, respectively. Overall, maximum RWC was recorded in plants treated with Eth, followed by SNP or NaHS spraying treatments under control and heat-stressed conditions. Furthermore, in heat-exposed plants, the combined application of (NBD and NaHS) and (cPTIO and NaHS) reduced RWC, relative to control plants. The addition of HT along with Eth or SNP under HS more drastically declined RWC, reversing the beneficial effects of Eth or SNP treatments on RWC.

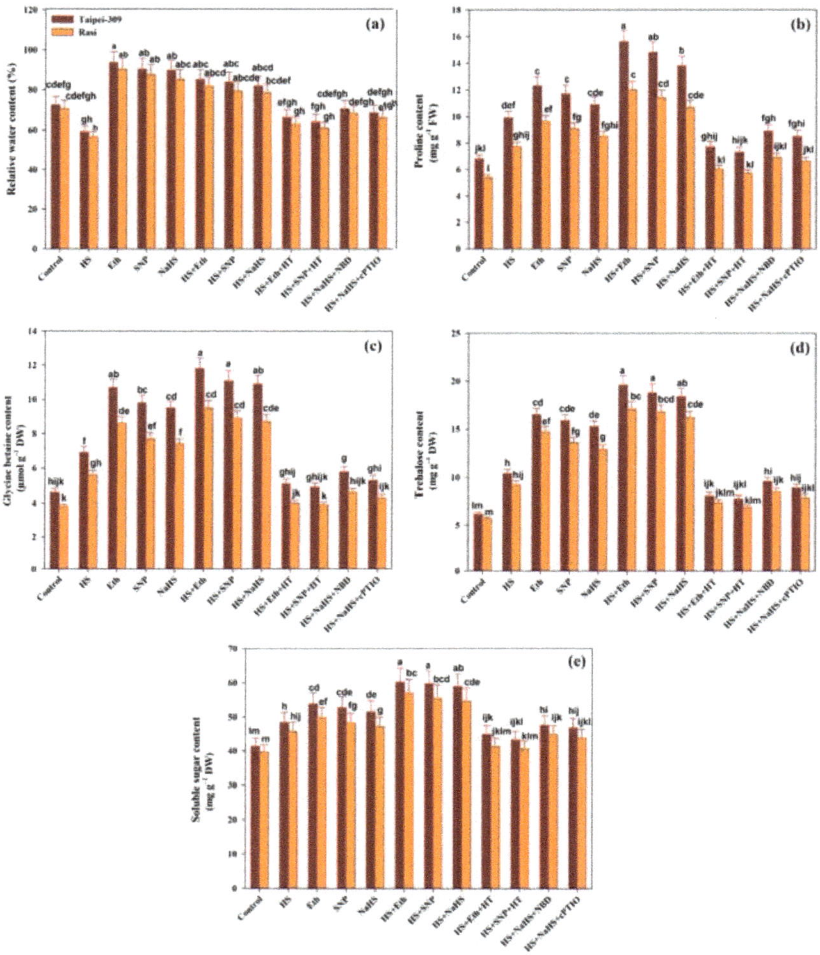

Figure 2. (**a**) Relative water content, (**b**) proline content, (**c**) glycine betaine content, (**d**) trehalose content, and (**e**) soluble sugars content of rice (*Oryza sativa* L.) cultivars Taipei-309 and Rasi under control and high temperature stress (HS) supplied with 200 µL L^{-1} ethephon (Eth), 100 µM sodium nitroprusside (SNP), 200 µM sodium hydrosulfide (NaHS), or 100 µM hypotaurine (HT), 100 µM 2-4-carboxyphenyl-4,4,5,5 -tetramethylimidazoline-1-oxyl-3-oxide (cPTIO) or 100 µM norbornadiene (NBD) scavengers of hydrogen sulfide (H_2S), nitric oxide (NO), and ethylene action inhibitors, respectively. Data are presented as treatments mean ± SE (n = 4). The values followed by the same letters did not differ significantly by LSD test at $p < 0.05$.

3.5. Accumulation of Osmolytes

Plants accumulate osmolytes or compatible solutes to protect the cellular machinery from various environmental stresses. Glycine betaine, sugars (trehalose), and proline are the most well-known osmolytes. The treatment of HS increased the proline and GB content by 45.5% and 50% in Taipei-309 and 42.5% and 47.3% in Rasi, respectively, compared to control plants (Figure 2b,c). With Eth, SNP, or NaHS application under no stress, the levels of proline and GB increased appreciably. The application of Eth improved proline content (129.4% and 57.5%) in Taipei-309 and (122.2% and 55.8%) in Rasi, and GB content (156.5% and 71.0%) in Taipei-309 and (150% and 69.6%) in Rasi, respectively, under stressful condition compared to control and heat-stressed plants.

Improvements in proline and GB contents were also observed by SNP application under HS conditions (117.6% and 111.1%) in Taipei-309 and (141.3% and 134.2%) in Rasi, respectively, compared to control plants. Foliar-applied NaHS enhanced proline content (102.9% and 98.1%) and GB content (136.9% and 128.9%) in Taipei-309 and Rasi, respectively, compared to controls. Under HS stress, the treatment of HT completely suppressed the beneficial effects of Eth or SNP on proline and GB accumulation. At the same time, NBD or cPTIO could not considerably inhibit the effects induced by NaHS on proline and GB content in the leaves of heat-stressed plants. A considerable increase in trehalose and soluble sugar accumulation was recorded in heat-stressed plants of both rice cultivars, with Taipei-309 (67.7%) and (17.1%) with a higher accumulation than Rasi (61.4%) and (15.4%), respectively, compared to control plants (Figure 2d,e). Under no stress conditions, foliar treatment with Eth, SNP, or NaHS alone further increased the trehalose and soluble sugars content, although the increase was cultivar specific. The trehalose and soluble sugars content also increased considerably in both rice cultivars due to exogenously-applied Eth, SNP, or NaHS alone under HS.

Eth application increased trehalose and soluble sugars content (216.1% and 200.0%) and (45.7% and 43.6%) while SNP increased (203.2% and 194.7%) and (44.5% and 39.8%), and NaHS (196.7% and 184.2%) and (42.6% and 37.8%) in Taipei-309 and Rasi, respectively, under heat-stressed conditions, compared to control plants. However, even NBD and cPTIO did not completely inhibit NaHS from accumulating trehalose and soluble sugars. Furthermore, trehalose and soluble sugars accumulation by Eth and SNP was not sustained on the addition of HT to Eth and SNP-supplemented heat-stressed plants, which supported the role of H_2S in ethylene and NO-induced osmolytes accumulation under HS.

3.6. Oxidative Stress

The oxidative stress was measured as H_2O_2 and TBARS. The plants subjected to HS showed a rise in the content of H_2O_2 by 3.0 fold in Taipei-309 and by 3.2 fold in Rasi, as well as TBARS content by 2.0 fold in Taipei-309 and by 2.1 fold in Rasi, compared to control plants (Tables 4 and 5). However, the spraying of Eth, SNP, or NaHS reduced the contents of H_2O_2 and TBARS in both heat-stressed and non-stressed conditions. The Eth application reduced heat-induced oxidative stress, as evidenced by the observed reductions in the levels of H_2O_2 (66.0% and 67.5%) and TBARS (50.6% and 50.9%) in Taipei-309 and Rasi, respectively, compared to heat-stressed plants. Under HS, the application of SNP or NaHS decreased the contents of H_2O_2 by 65.1% and 64.3% in Taipei-309 and by 66.3% and 66.0% in Rasi, as well as TBARS content by 49.3% and 47.9% in Taipei-309 and by 49.0% and 47.1% in Rasi, respectively, relative to the values of heat-treated plants. These findings suggest that the individual treatment of Eth, SNP, or NaHS mitigated heat-induced oxidative stress by lowering the accumulation of H_2O_2 and TBARS. However, the addition of HT further stimulated H_2O_2 and TBARS in both rice cultivars under HS. Furthermore, Eth and SNP did not rescue the negative effects of HT on H_2O_2 and TBARS accumulation. Therefore, the application of HT completely reversed the alleviating effects of both Eth and SNP. The application of NBD or cPTIO reversed the reduced oxidative stress induced by NaHS.

Table 4. Hydrogen peroxide (H_2O_2), thiobarbituric acid reactive substances (TBARS) content, and activities of superoxide dismutase (SOD), ascorbate peroxidase (APX), and glutathione reductase (GR) in the leaves of rice (*Oryza sativa* L.) cultivar Taipei-309 after foliar treatment of plants with 200 μL L^{-1} ethephon (Eth), 100 μM sodium nitroprusside (SNP) or 200 μM sodium hydrosulfide (NaHS) grown with or without high temperature stress (HS; 40 °C) or 100 μM hypotaurine (HT), 100 μM 2-4-carboxyphenyl-4,4,5,5-tetramethylimidazoline-1-oxyl-3-oxide (cPTIO) or 100 μM norbornadiene (NBD) scavengers of hydrogen sulfide (H_2S), nitric oxide (NO), and ethylene action inhibitors, respectively, with HS at 15 days after sowing. Data are presented as treatments mean ± SE (n = 4). The values followed by the same letters did not differ significantly by LSD test at $p < 0.05$. FW, fresh weight.

Treatments	H_2O_2 Content	TBARS Content	SOD Activity	APX Activity	GR Activity
	(nmol g^{-1} FW)		(U mg^{-1} Protein min^{-1})		
Control	28.3 ± 1.26 ^{ghi}	10.6 ± 1.22 ^{efghijk}	7.41 ± 0.43 ^{mn}	1.35 ± 0.10 ^{mn}	0.19 ± 0.01 ^{jkl}
HS	86.7 ± 4.14 ^{b}	22.1 ± 2.07 ^{a}	11.0 ± 0.55 ^{fghij}	1.94 ± 0.14 ^{jklm}	0.27 ± 0.017 ^{ghi}
Eth	21.1 ± 1.17 ^{i}	7.1 ± 0.79 ^{k}	13.9 ± 0.76 ^{bcd}	2.97 ± 0.23 ^{bcdef}	0.38 ± 0.028 ^{bcd}
SNP	23.5 ± 1.20 ^{hi}	7.9 ± 0.81 ^{jk}	13.1 ± 0.62 ^{cde}	2.73 ± 0.22 ^{cdefgh}	0.36 ± 0.026 ^{bcd}
NaHS	24.7 ± 1.22 ^{ghi}	8.2 ± 0.97 ^{ijk}	12.8 ± 0.50 ^{cdef}	2.59 ± 0.20 ^{efghi}	0.34 ± 0.025 ^{cdef}
HS + Eth	29.4 ± 1.72 ^{ghi}	10.9 ± 1.20 ^{efghijk}	15.8 ± 0.98 ^{a}	3.74 ± 0.29 ^{a}	0.46 ± 0.034 ^{a}
HS + SNP	30.2 ± 2.11 ^{gh}	11.2 ± 1.25 ^{efghijk}	14.9 ± 0.81 ^{ab}	3.53 ± 0.26 ^{ab}	0.42 ± 0.032 ^{ab}
HS + NaHS	30.9 ± 2.51 ^{gh}	11.5 ± 1.34 ^{defghijk}	14.1 ± 0.79 ^{bc}	3.32 ± 0.24 ^{abc}	0.40 ± 0.030 ^{abc}
HS + Eth + HT	61.3 ± 3.16 ^{cde}	14.2 ± 1.55 ^{bcdefg}	10.8 ± 0.49 ^{ghij}	1.91 ± 0.16 ^{jklm}	0.25 ± 0.015 ^{hij}
HS + SNP + HT	64.1 ± 3.37 ^{cd}	14.7 ± 1.65 ^{bcdef}	9.7 ± 0.40 ^{jkl}	1.87 ± 0.13 ^{jklm}	0.24 ± 0.011 ^{hijk}
HS + NaHS + NBD	51.8 ± 2.70 ^{f}	13.3 ± 1.36 ^{bcdefgh}	12.3 ± 0.65 ^{cdefgh}	2.31 ± 0.19 ^{ghijk}	0.29 ± 0.023 ^{efgh}
HS + NaHS + cPTIO	53.2 ± 2.80 ^{ef}	13.9 ± 1.49 ^{bcdefg}	12.1 ± 0.58 ^{defgh}	2.19 ± 0.18 ^{hijkl}	0.28 ± 0.020 ^{fgh}

Table 5. Hydrogen peroxide (H_2O_2), thiobarbituric acid reactive substances (TBARS) content, and activities of superoxide dismutase (SOD), ascorbate peroxidase (APX), and glutathione reductase (GR) in the leaves of rice (*Oryza sativa* L.) cultivar Rasi after foliar treatment of plants with 200 μL L^{-1} ethephon (Eth), 100 μM sodium nitroprusside (SNP) or 200 μM sodium hydrosulfide (NaHS) grown with or without high temperature stress (HS; 40 °C) or 100 μM hypotaurine (HT), 100 μM 2-4-carboxyphenyl-4,4,5,5-tetramethylimidazoline-1-oxyl-3-oxide (cPTIO), or 100 μM norbornadiene (NBD) scavengers of hydrogen sulfide (H_2S), nitric oxide (NO), and ethylene action inhibitors, respectively, with HS at 15 days after sowing. Data are presented as treatments mean ± SE (n = 4). The values followed by the same letters did not differ significantly by LSD test at $p < 0.05$. FW, fresh weight.

Treatments	H_2O_2 Content	TBARS Content	SOD Activity	APX Activity	GR Activity
	(nmol g^{-1} FW)		(U mg^{-1} Protein min^{-1})		
Control	30.5 ± 1.85 ^{gh}	12.2 ± 1.13 ^{cdefghij}	6.38 ± 0.21 ^{n}	1.24 ± 0.07 ^{n}	0.16 ± 0.005 ^{l}
HS	97.9 ± 5.46 ^{a}	26.1 ± 2.11 ^{a}	9.31 ± 0.34 ^{jkl}	1.76 ± 0.12 ^{klmn}	0.21 ± 0.011 ^{ijkl}
Eth	24.6 ± 1.19 ^{ghi}	8.5 ± 0.73 ^{hijk}	11.7 ± 0.55 ^{efghi}	2.67 ± 0.19 ^{defghi}	0.29 ± 0.017 ^{efgh}
SNP	26.9 ± 1.24 ^{ghi}	9.7 ± 0.87 ^{ghijk}	10.9 ± 0.51 ^{ghij}	2.44 ± 0.17 ^{fghij}	0.28 ± 0.015 ^{fgh}
NaHS	28.0 ± 1.32 ^{ghi}	10.0 ± 1.02 ^{fghijk}	10.5 ± 0.45 ^{hijk}	2.32 ± 0.16 ^{ghijk}	0.26 ± 0.011 ^{hi}
HS + Eth	31.8 ± 2.46 ^{gh}	12.8 ± 1.22 ^{bcdefghi}	13.2 ± 0.66 ^{bcde}	3.27 ± 0.24 ^{abcd}	0.38± 0.026 ^{bcd}
HS + SNP	32.9 ± 2.60 ^{g}	13.3 ± 1.30 ^{bcdefgh}	12.6 ± 0.61 ^{cdefg}	3.10 ± 0.21 ^{bcde}	0.35 ± 0.023 ^{cde}
HS + NaHS	33.2 ± 2.88 ^{g}	13.8 ± 1.53 ^{bcdefg}	11.9 ± 0.57 ^{efghi}	2.90 ± 0.20 ^{cdefg}	0.33 ± 0.020 ^{defg}
HS + Eth + HT	65.8 ± 3.56 ^{cd}	16.9 ± 1.92 ^{bc}	8.8 ± 0.30 ^{klm}	1.71 ± 0.13 ^{klmn}	0.19 ± 0.01 ^{jkl}
HS + SNP + HT	69.7 ± 4.05 ^{c}	17.3 ± 2.04 ^{b}	8.0 ± 0.25 ^{lmn}	1.67 ± 0.11 ^{lmn}	0.18 ± 0.01 ^{kl}
HS + NaHS + NBD	57.4 ± 2.90 ^{def}	15.4 ± 1.72 ^{bcde}	10.1 ± 0.40 ^{ijk}	2.05 ± 0.18 ^{ijkl}	0.24 ± 0.015 ^{hijk}
HS + NaHS + cPTIO	59.3 ± 3.17 ^{def}	16.2 ± 1.81 ^{bcd}	9.7 ± 0.35 ^{jkl}	1.96 ± 0.16 ^{jklm}	0.23 ± 0.011 ^{hijk}

3.7. Antioxidants Enzyme Activity

The activity of antioxidants, SOD, APX, and GR was studied to investigate the regulatory role of ethylene, NO, or H_2S in the alleviation of HS-induced oxidative stress (Tables 4 and 5). The increases in the activity of SOD (48.4% and 45.9%), APX (43.7% and 41.9%), and GR (42.1% and 31.2%) in Taipei-309 and Rasi, respectively, were noted under HS compared to control plants. Applying Eth to heat-treated plants showed a higher increase in SOD, APX, and GR activity than SNP compared to control or heat-stressed plants in both cultivars. Similarly, NaHS stimulated SOD, APX, and GR activity in both cultivars compared to control or heat-stressed plants.

The combined application of (NBD and NaHS) and (cPTIO and NaHS) in heat-treated plants did not entirely counteract the beneficial effect of NaHS on antioxidative enzyme activity. In addition, HT application in heat-stressed plants completely reversed the positive effects of Eth and SNP on the antioxidant defense system. Thus, HT supplementation resulted in reduced antioxidative enzyme activity in Eth and SNP-treated heat-stressed plants.

3.8. Hydrogen Sulfide and NO Content and Ethylene Production

The endogenous content of H_2S and NO in rice cultivar leaves was examined to evaluate the effects of HS on H_2S and NO regulation. Figure 3a,b depicts increased NO and H_2S contents in the leaves of rice cultivar plants subjected to HS. The treatment of HS increased NO levels by 49.2% and 46.7% and H_2S levels by 36.3% and 34.0% in Taipei-309 and Rasi, respectively, compared to control plants. Exogenously-applied Eth, SNP, or NaHS enhanced the level of both NO and H_2S in rice cultivar leaves compared to control and heat-stressed plants. The treatment of HT and Eth or SNP under HS reduced endogenous NO and H_2S levels relative to heat-treated plants. Similarly, applying cPTIO and NaHS under HS reduced NO levels compared to stressed and controlled plants.

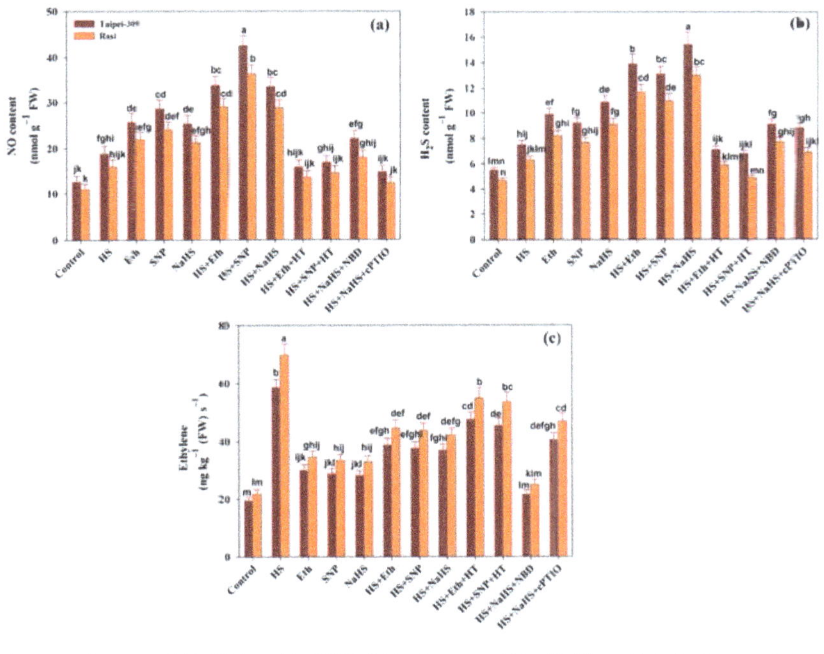

Figure 3. Content of (**a**) Nitric oxide (NO), (**b**) hydrogen sulfide (H_2S), and (**c**) ethylene evolution in rice (*Oryza sativa* L.) cultivars Taipei-309 and Rasi under control and high temperature stress (HS) supplied with 200 µL L^{-1} ethephon (Eth), 100 µM sodium nitroprusside (SNP), 200 µM sodium hydrosulfide (NaHS) or 100 µM hypotaurine (HT), 100 µM 2-4-carboxyphenyl-4,4,5,5-tetramethylimidazoline-1-oxyl-3-oxide (cPTIO), or 100 µM norbornadiene (NBD) scavengers of hydrogen sulfide (H_2S), NO, and ethylene action inhibitors, respectively. Data are presented as treatments mean ± SE (n = 4). The values followed by the same letters did not differ significantly by LSD test at $p < 0.05$.

The individual application of NBD and cPTIO and NaHS under HS did not affect H_2S levels compared to control and heat-stressed plants.

Ethylene production in leaves of rice cultivars exposed to HS is depicted in Figure 3c. Compared to control plants, HS elevated ethylene levels by 200.5% and 222.1% in Taipei-309 and Rasi, respectively. Although SNP, NaHS, or Eth individually increased ethylene emission, the increase was less than that of plants that were subjected to high temperatures. Compared to heat-treated plants, plants that received Eth, SNP, or NaHS under stressful conditions showed a reduction in ethylene emission. The inhibition of H_2S and NO using inhibitors HT and cPTIO, respectively, increased ethylene levels relative to control plants. Ethylene action inhibitor NBD application along with NaHS under HS decreased ethylene level compared to heat-stressed plants.

3.9. Expression of Photosynthesis-Related Genes and Genes Encoding Antioxidant Enzymes

The expression of two genes relevant to the photosynthetic system was investigated in rice cultivars under HS (Figure 4a,b). The treatment of HS downregulated the expression of *psbA* and *psbB* in rice cultivar leaves, whereas heat-treated plants supplemented with Eth, SNP, or NaHS had higher levels of *psbA* and *psbB* transcription than control plants. In addition, under HS, exogenous HT with Eth or SNP significantly down-regulated the expression of *psbA* and *psbB* compared to the heat-treated Eth or SNP-supplemented plants. The expression of *psbA* and *psbB* was reduced in NBD or cPTIO with NaHS-treated heat-stressed plants, but the transcription was lowered more sharply in HT-treated plants.

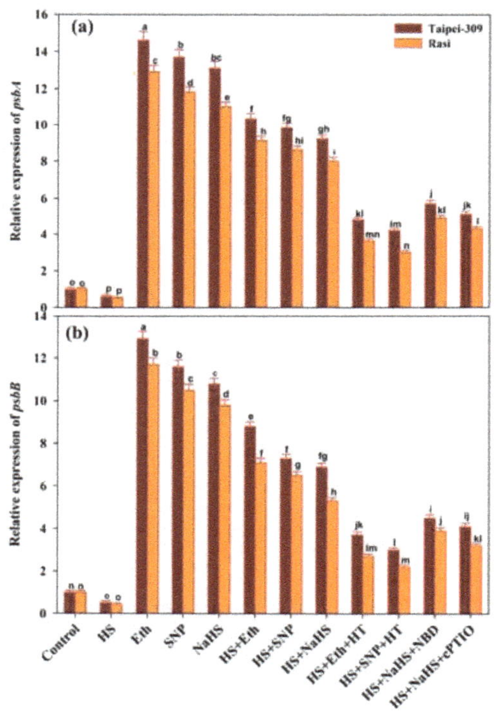

Figure 4. Relative expression of (**a**) *psbA* and (**b**) *psbB* of rice (*Oryza sativa* L.) cultivars Taipei-309 and Rasi under control and high temperature stress (HS) supplied with 200 μL L^{-1} ethephon (Eth), 100 μM sodium nitroprusside (SNP), 200 μM sodium hydrosulfide (NaHS) or 100 μM hypotaurine (HT), 100 μM 2-4-carboxyphenyl-4,4,5,5-tetramethylimidazoline-1-oxyl-3-oxide (cPTIO), or 100 μM norbornadiene (NBD) scavengers of hydrogen sulfide (H_2S), nitric oxide (NO), and ethylene action inhibitors, respectively. Data are presented as treatments mean ± SE (n = 4). The values followed by the same letters did not differ significantly by LSD test at $p < 0.05$.

The relative expression analysis based on a qRT-PCR method for three SOD isoforms containing *Mn-SOD*, *Cu-SOD*, and *Fe-SOD* and *APX* was carried out on two rice cultivars (Figure 5). The results revealed that the expression of SOD isoforms and *APX* was upregulated in HS-treated rice cultivars compared to control plants. The individual application of Eth, SNP, or NaHS further enhanced the expression of SOD isoforms and APX in rice cultivars exposed to HS compared to heat-treated plants alone.

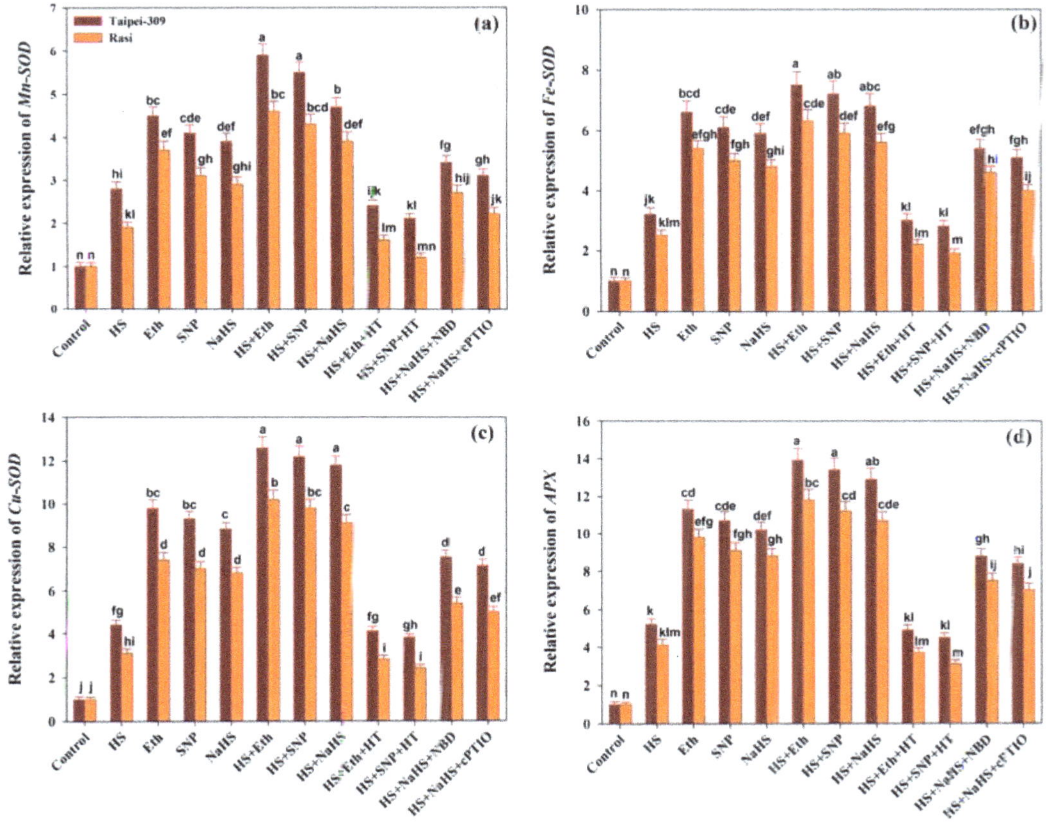

Figure 5. Relative expression of (**a**) *Mn-SOD*, (**b**) *Fe-SOD*, (**c**) *Cu-SOD*, and (**d**) *APX* of rice (*Oryza sativa* L.) cultivars Taipei-309 and Rasi under control and high temperature stress (HS) supplied with 200 μL L^{-1} ethephon (Eth), 100 μM sodium nitroprusside (SNP), 200 μM sodium hydrosulfide (NaHS) or 100 μM hypotaurine (HT), 100 μM 2-4-carboxyphenyl-4,4,5,5-tetramethylimidazoline-1-oxyl-3-oxide (cPTIO), or 100 μM norbornadiene (NBD) scavengers of hydrogen sulfide (H_2S), nitric oxide (NO), and ethylene action inhibitors, respectively. Data are presented as treatments mean ± SE (n = 4). The values followed by the same letters did not differ significantly by LSD test at $p < 0.05$.

Meanwhile, in HT treatment with Eth or SNP under HS, no significant difference was observed in the expression levels of SOD isoforms and *APX* compared to control plants. Unlike the above, Eth, SNP, or NaHS treatments significantly increased the transcription level of these genes in heat-exposed plants. In contrast, HT application appears to affect antioxidant defense-related gene expression.

3.10. Determination of Interaction among Morpho-Physiological, Biochemical and Molecular Parameters through PCA

The PCA was carried out to determine the degree of data variation and the relationship between the different treatments and parameters in both rice cultivars (Figure 6). The two components (PC1 and PC2) described 94.70% of data variability in Taipei-309 under the influence of different treatments (Figure 6). PC1, the first component, contributed 79.14% of the total variation, and the second component, PC2, accounted for 15.56% of the total variation. On the other hand, in the Rasi cultivar, the PCA showed 94.67% (PC1 = 78.83% and PC2 = 15.84%) data variance under the influence of different treatments (Figure 6. Shoot and root dry weight, SPAD value, parameters of chlorophyll fluorescence (ETR, qP, ΦPSII, Φesc, and Fv/Fm), net photosynthesis, intercellular CO_2 concentration, stomatal conductance, photosynthesis-related genes (*psbA* and *psbB*), and RWC were grouped together and exhibited a positive correlation. Antioxidant enzymes (SOD, APX, and GR) and genes (*Mn-SOD*, *Cu-SOD*, *Fe-SOD*, and *APX*), contents of proline, GB, soluble sugars and trehalose, NO, and H_2S content were grouped together and were also found to bepositively correlated with plant dry mass, chlorophyll fluorescence parameters, and photosynthesis parameters. However, H_2O_2 and TBARS, NPQ, and ethylene contents were grouped together and showed a negative correlation with other groups of parameters.

3.11. Pearson Correlation

The Pearson correlation heatmap of Taipei-309 and Rasi showed a strong linear correlation of NO, ethylene, and H_2S with growth, leaf water status, osmolytes, antioxidants, and the photosynthesis of plants (Figure 7). The plant growth and photosynthesis variables showed a negative correlation with the high temperature-induced oxidative stress biomarkers like H_2O_2 and TBARS content. The content of NO, H_2S, GB, Tre, SG, and Pro showed a significant ($p \leq 0.05$, $p \leq 0.01$, and $p \leq 0.001$) positive correlation with plant growth (SDW and RDW), photosynthetic parameters (SPAD, PN, gs, Ci, Actual PSII, Maximum PSII, qP, and ETR), the expression of photosynthesis-related genes (PSB A and PSB B), and the level of antioxidant enzymes. On the other hand, endogenous Eth showed a strong correlation with oxidative stress, suggesting its potential role and generation during stress conditions. The photosynthesis-related genes showed a strong dependency upon the relative expression of *Mn-SOD*, *Cu-SOD*, and *Fe-SOD* and *APX*. Therefore, these connections portray a nearby association between NO, Eth, and H_2S and plant response to thermo-tolerance in Taipei-309 and Rasi cultivars.

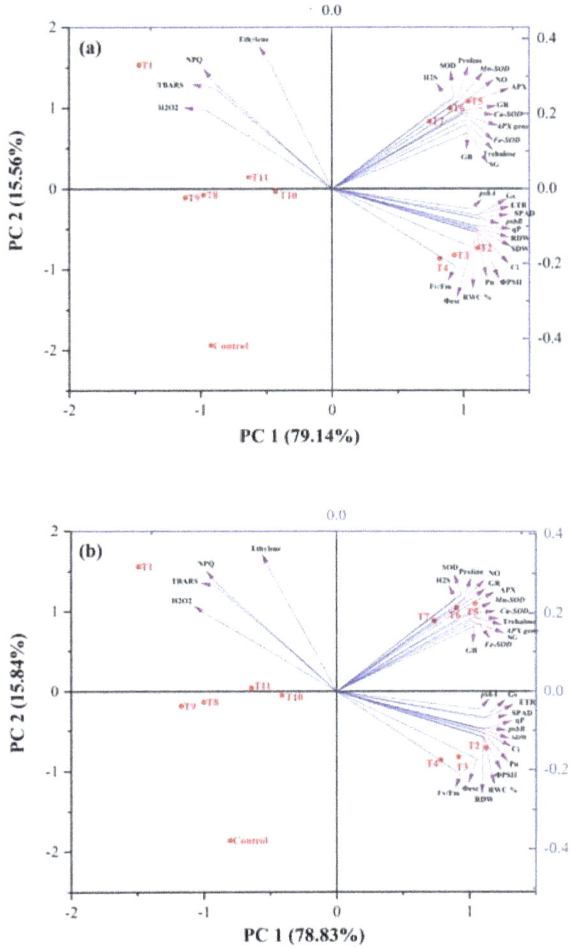

Figure 6. Biplots of principal component analysis (PCA) represents the relationship among different treatments and variables of two rice cultivars, Taipei-309 (**a**) and Rasi (**b**) grown under different conditions such as control, high temperature stress; HS (T1), Eth; Eth (T2), sodium nitroprusside; SNP (T3), sodium hydrosulfide; NaHS (T4), HS + Eth (T5), HS + SNP (T6), HS + NaHS (T7), HS + Eth + hypotaurine; HT (T8), HS + SNP + HT (T9), HS + NaHS + norbornadiene; NBD (T10), and HS + NaHS + 2-4-carboxyphenyl-4,4,5,5 -tetramethylimidazoline-1-oxyl-3-oxide; cPTIO (T11). The variables included H_2O_2 (hydrogen peroxide), TBARS (thiobarbituric acid reactive substances), non-photochemical quenching (NPQ), ethylene evolution, SOD (superoxide dismutase), APX (ascorbate peroxidase), GR (glutathione reductase) activity, gene expression of (*Mn-SOD*, *Cu-SOD*, *Fe-SOD*, and *APX*), contents of H_2S (hydrogen sulfide), NO (nitric oxide), proline, trehalose, SG (soluble sugar) and GB (glycine betaine), gene expression of *psbA* and *psbB*, *Pn* (net photosynthesis), *Gs* (stomatal conductance), *Ci* (intercellular CO_2 concentration), SPAD value, Fv/Fm (maximum efficiency of PSII), ΦPSII (actual efficiency of PSII), Φesc (intrinsic efficiency of PSII), qP (photochemical quenching), NPQ (non-photochemical quenching), ETR (electron transport rate), RDW (root dry weight), SDW (shoot dry weight), and RWC% (relative water content).

(a)

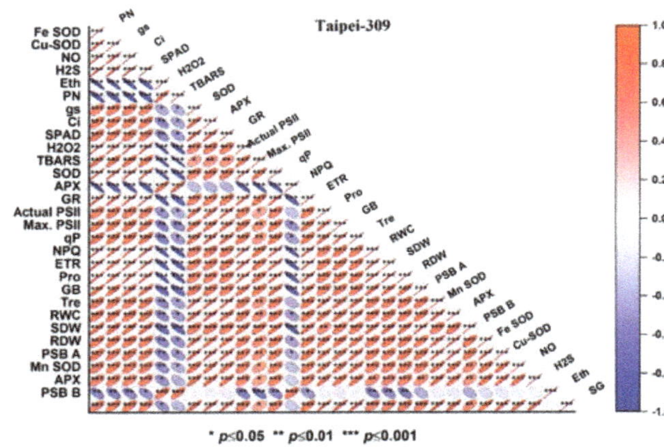

(b)

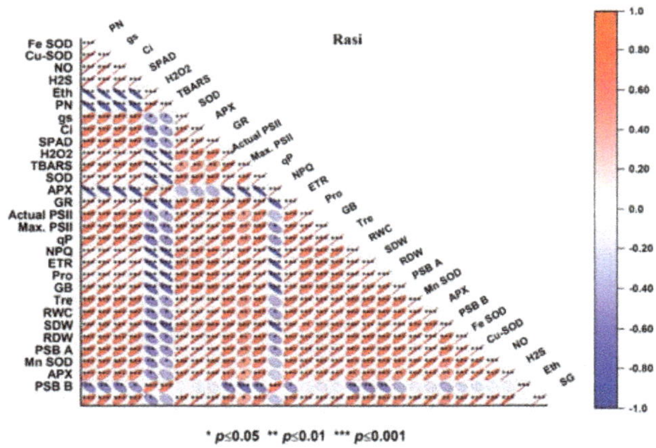

Figure 7. Pearson correlation represents the relationship among different variables of two rice cultivars, Taipei-309 (**a**) and Rasi (**b**) grown under different conditions such as control, high temperature stress; HS (T1), Eth; Eth (T2), sodium nitroprusside; SNP (T3), sodium hydrosulfide; NaHS (T4), HS + Eth (T5), HS + SNP (T6), HS + NaHS (T7), HS + Eth + hypotaurine; HT (T8), HS + SNP + HT (T9), HS + NaHS + norbornadiene; NBD (T10) and HS + NaHS + 2-4-carboxyphenyl-4,4,5,5-tetramethylimidazoline-1-oxyl-3-oxide; cPTIO (T11). The variables included H_2O_2 (hydrogen peroxide), TBARS (thiobarbituric acid reactive substances), non-photochemical quenching (NPQ), ethylene evolution, SOD (superoxide dismutase), APX (ascorbate peroxidase), GR (glutathione reductase) activity, gene expression of (*Mn-SOD*, *Cu-SOD*, *Fe-SOD*, and *APX*), contents of H_2S (hydrogen sulfide), NO (nitric oxide), proline, Tre (trehalose), SG (soluble sugar) and GB (glycine betaine), gene expression of *psbA* and *psbB*, *Pn* (net photosynthesis), *Gs* (stomatal conductance), *Ci* (intercellular CO_2 concentration), SPAD value, maximum efficiency of PSII, actual efficiency of PSII, qP (photochemical quenching), NPQ (non-photochemical quenching), ETR (electron transport rate), RDW (root dry weight), SDW (shoot dry weight), and RWC (relative water content).

4. Discussion

High temperature stress, one of the most common types of abiotic stresses that plants face in nature, has an independent mode of action on the physiology and metabolism of plant cells. Previous studies have documented the effects of the application of several signaling molecules and growth regulators on various plant species; however, there aren't many reports on the comparative actions of ethylene, NO, and H_2S on rice cultivars under HS. In the current study, we evaluated the efficacy of ethylene, NO, and H_2S in modulating photosynthesis, growth, osmolytes, antioxidant metabolism, and the potential to ameliorate oxidative stress-induced impairments in rice cultivars subjected to HS. Among the various treatments used, 200 μL L^{-1} ethylene treatments were the most effective, followed by 100 μM SNP and 200 μM NaHS. Meanwhile, we also explored the influence of H_2S in ethylene or NO-mediated tolerance of HS in rice cultivars.

The findings of the present study revealed that HS negatively impacts growth parameters and photosynthesis, which could be linked to an elevated level of oxidative stress indicators in rice plants, as evident from PCA and Pearson correlation. High temperature stress reduced the SPAD value, stomatal conductance, intercellular CO_2 concentration, and net photosynthetic rate. Heat stress has been observed to cause changes in plant growth, pigment concentrations, and photosynthesis in various plants [3,4,11,49]. Furthermore, previous research has found that severe heat stress in plants can result in cellular injury, cell death, and a reduction in the total dry weight of plants [57,58]. To examine whether or not Eth, SNP, or NaHS can mitigate the detrimental effect of HS on plant growth attributes and photosynthetic activity, they were sprayed onto the foliage of rice plants. The result showed that Eth, SNP, or NaHS relatively relieved reduced plant growth and photosynthesis. Overall, the maximum improvement in plant growth and photosynthesis was recorded from the plants treated with Eth followed by SNP and NaHS treatments. In comparison to SNP and NaHS, Eth might be an active growth regulator involved in the heat tolerance of rice cultivars. To understand more about whether H_2S plays a role in ethylene or NO-induced heat tolerance of rice cultivars, HT, an H_2S scavenger, was given to the plants treated with Eth or SNP under HS stress. The results of this study showed that when HT was applied along with Eth or SNP, these treatments were ineffective in enhancing plant growth and photosynthesis when exposed to HS stress. HT reversed the availability of H_2S, and Eth or NO were unable to efficaciously sustain heat tolerance in rice cultivars. Thus, the findings imply that ethylene or NO causes H_2S to be produced in heat-stressed plants and that H_2S increases heat tolerance in rice cultivars. Eth, NO, and H_2S are important gaseous signaling molecules that regulate each other's behavior, and H_2S might work as a downstream signaling agent of NO and Eth on photosynthetic and growth under heat stress. In the present study, the supplementation of Eth, NO, and H_2S effectively alleviated heat stress, which was reversed by the supplementation of HT, a H_2S scavenger, suggesting that H_2S works as a downstream signaling agent in NO and Eth-mediated heat stress tolerance. However, in the signaling cascade, H_2S may act either upstream or downstream of NO, and there are complex relationships between NO, Eth, and H_2S that are engaged in a variety of physiological processes and pathways.

Exogenously-applied H_2S has been shown to promote growth and reduce lead (Pb) accumulation in *Zea mays* plants under Pb stress [59], improve photosynthesis, protect chloroplast structure, and promote growth in *Oryza sativa* under Ni stress [60], and improve the content of photosynthetic pigments and seedling biomass in *Cucurbita pepo* under nickel (Ni) stress [61]. H_2S and NO are known to increase plant resistance to a variety of stresses, and they might serve as secondary signals to activate signal pathways downstream. [62,63]. The H_2S has previously been shown to act as a downstream signal in the NO-induced enhanced adaptability of heat in maize plants [64]. The interaction of ethylene and H_2S in heat stress tolerance has been shown [65].

Photosystem II is a pigment-protein complex with many components that are important for water splitting, oxygen evolution, and plastoquinone reduction. The photosystem PS II is more sensitive to environmental stress than PS I in chloroplasts [66,67]. Chlorophyll fluorescence parameters have been proven to be an effective measure of stress intensity [68]. The result of the present study demonstrated that chlorophyll fluorescence parameters were reduced in heat-exposed rice plants, which contributed to the decrease in net photosynthesis. Under HS stress, there was a decrease in ΦPS II, Fv/Fm, Φesc, ETR, and qP, as well as an increase in NPQ. This indicates that heat stress-induced ROS production causes a decrease in PS II reaction center activity and renders the reaction center unable to use light energy efficiently. Havaux [69] observed an irreversible decline in the photochemical efficiency after 90 min of exposure of *Solanum tuberosum* plants to 39.5 °C, whereas Camejo et al. [70] showed that heat-sensitive tomato (*Lycopersicon esculentum* Mill. cv. Campbell-28) plants exhibited a decline in Rubisco activity and PS II performance after exposure of plants to 45 °C for 2 h. Heat stress not only enhances thylakoid membrane fluidity but also causes protein complexes and photosystems to reorganize and even dissociate [71,72]. Extreme heat stress causes structural changes in protein complexes, photosystem degradation, and a loss of oxygen-evolving activity, all of which impair the photosystem's ability to transfer electrons [73]. The application of Eth, SNP, or NaHS reversed photo-inhibition and the impairment of photosynthetic characteristics caused by HS. Furthermore, the application of HT along with Eth or SNP enhanced the detrimental effects of HS and reversed the mitigation effects of Eth or SNP on chlorophyll fluorescence attributes, implying that H_2S plays an important role in regulating the impact of HS on photosynthetic attributes. As a result, H_2S is implicated in the augmentation of PS II reaction center activity via ethylene or NO in rice cultivars subjected to heat stress, and it participated with ethylene or NO to improve light energy utilization efficiency. It has previously been demonstrated that ethylene application contributes to waterlogging stress reduction by strengthening photosynthetic pigment or improving electron transport [74]. According to Shi et al. [75], the Fv/Fm and ETR were higher in the presence of SNP, indicating that NO partially alleviated photodamage in UV-B-stressed bean leaves. In a previous study on the mung bean cultivars, it was reported that ethylene and H_2S can protect photosynthesis against hexavalent chromium stress [37].

In this study, we assessed the RWC of rice cultivars under various treatments. Under HS stress, RWC declined significantly in both rice cultivars, with Rasi experiencing a greater decrease. Heat stress may reduce the water status of the leaves by reducing the hydraulic conductance, resulting in a decrease in water absorption, or by lowering stomatal conductance [76]. In this study, the heat-sensitive cultivar Rasi was shown to lose more water than Taipei-309, the heat-tolerant cultivar. Individual applications of Eth, SNP, or NaHS significantly altered the RWC of rice cultivars in which Eth showed better results than SNP and NaHS treatments. Furthermore, the applications of Eth, SNP, or NaHS enhanced the RWC of leaves under HS, thereby reducing heat-related plant damage. Intriguingly, the treatment of HT with Eth or SNP reversed this effect, suggesting that H_2S was involved in ethylene or NO-induced changes in RWC in rice cultivars under HS stress. NO treatment, according to Khan et al. [77], benefitted mustard plants in retaining more water when subjected to salt stress. Similarly, Li et al. [78] observed that H_2S maintained leaf RWC in cadmium (Cd)-stressed seedlings of *Brassica rapa*. Higher leaf RWC may have enhanced stomatal conductivity and, as a result, photosynthetic activity and biomass production [79]. Tomato plants treated with NaHS or Eth showed no decrease in RWC in response to low osmotic stress but did show a slight decrease in response to severe osmotic stress [38].

Many organisms counteract the environmental challenges by accumulating low molecular weight water-soluble substances known as osmolytes. Under heat stress, the accumulation of osmolytes aids in osmotic adjustment increases the concentration of cell protoplasm to maintain proper membrane function and quenches ROS in plants. [80,81]. The reason for higher proline levels is related to the synthesis and accumulation of free amino acids under stressed conditions [82]. Furthermore, GB accumulated in transgenic tobacco plants improved PS II thermo-tolerance from heat stress [83]. Osmolytes serve as stress markers and hence play an important role in stress reduction. *Zea mays* L. plants subjected to copper (Cu) and Pb exhibited higher levels of proline, which protected them from an oxidative burst and helped to maintain cell structures [84]. Heat tolerance, a crucial physiological trait for heat resistance, necessitates the accumulation of sugars in plants and the availability of carbohydrates [85]. A non-reducing disaccharide known as trehalose accumulated in Arabidopsis plants exposed to heat stress and served as a ROS scavenger in heat-exposed wheat plants [86,87]. Trehalose is essential for maintaining growth under adverse conditions because it controls how efficiently most plants use water and stomatal movement [88]. According to Li et al. [89] trehalose serves as an osmoprotectant during water deficit, which aids in stabilizing dehydrated enzymes, proteins, and membrane lipids and guards against damage to biological structures.

The results of the present study showed that heat stress treatment increased the levels of osmolytes such as proline, GB, trehalose, and soluble sugars; however, these increased levels of osmolytes were unable to counteract heat stress and settled the stressed rice cultivars with decreased water status. However, the application of Eth, SNP, and NaHS to stressed rice cultivars augmented proline, GB, and trehalose levels, which reduced heat stress and enabled plant cells with increased osmotic pressure to take in more water as evidenced by enhanced RWC. On the other hand, the application of HT with Eth or SNP under heat stress lowered the amount of these osmolytes, confirming the role of H_2S in the ethylene- or NO-mediated osmotic adjustment of plants. Previously, it was shown that NO-induced H_2S generated an increase in proline and GB, which protected wheat plants from osmotic stress-induced oxidative stress [90]. The observed augmentation of proline, GB, total soluble sugars, and total soluble proteins in response to NO application potentially improved salt tolerance through osmotic regulation [77]. Under heat stress and after applying ABA and NO, trehalose accumulation increased even more [91]. In *P. eryngiivar. tuoliensis* under heat stress, trehalose accumulation increased with NO [92]. Heat-stressed rice cultivar leaves accumulated higher levels of H_2O_2 and TBARS contents. Increased levels of ROS may be attributed to the altered photosynthetic process in rice cultivars under heat stress. The enhanced ROS levels were accompanied by increased lipid peroxidation. However, Eth, SNP, or NaHS application resulted in a reduction in the levels of H_2O_2 and TBARS in both heat-stressed cultivars, more effectively in Taipei-309. Therefore, the application of Eth, SNP, or NaHS could be a useful strategy to prevent plants from oxidative damage brought on by HS. These results show that Eth treatment alleviated heat-induced oxidative damage more effectively than SNP and NaHS in rice cultivars by lowering H_2O_2 and TBARS contents. An excess in ROS could cause severe damage to lipids and proteins, which is a major cause of plant growth reduction [93]. However, using HT with Eth or SNP dramatically reversed this effect and resulted in considerable cell membrane damage exhibited as a significant increase in TBARS and H_2O_2 levels. The elimination of H_2S via scavenging resulted in the production of ROS again. It reveals that ethylene or NO-induced H_2S were involved in reducing stress intensity in plants by scavenging ROS and decreasing lipid peroxidation, hence minimizing oxidative damage. Moreover, investigations have shown that individual applications of ethylene or H_2S have the capacity to reduce the levels of ROS in plants under abiotic stress [94,95]. The treatment of NO has an important role in enhancing endogenous H_2S production, which helps plants resist abiotic stress-induced oxidative stress by reducing ion leakage, H_2O_2, O_2^-, and TBARS levels [96–98]. NO has been found to reduce lipid peroxidation and ROS production in plants grown in Cd and Cu-enriched environments [23,99].

The result showed that HS resulted in oxidative stress as observed by the excessive production of ROS. Overproduction of ROS occurs in stressed cells when the cellular antioxidant defense mechanism is slower than the ROS synthesis that causes oxidative stress. In the present study, under HS, the activities of antioxidant enzymes, APX, GR, and SOD were enhanced; simultaneously, the H_2O_2 and TBARS content also increased in both the cultivars. As a result, increased levels of antioxidant enzymes in heat-stressed plants were insufficient to detoxify ROS, resulting in an excess of H_2O_2 and TBARS accumulation. Meanwhile, the application of Eth, SNP, or NaHS to heat-stressed plants increased antioxidant enzyme activity to the point where it was capable of detoxifying ROS by significantly lowering the levels of H_2O_2 and TBARS. Furthermore, in heat-challenged plants, the treatment of HT with Eth or SNP caused an alteration in the antioxidant enzyme activities induced by Eth or SNP. According to the present study, the exogenous application of SNP or Eth induced H_2S production and improved heat stress tolerance, which could be altered by treatment with an HT, suggesting that NO or ethylene-activated H_2S might be required for heat stress response in rice plants. Under stress, the levels of ROS are tightly regulated by enzymatic and non-enzymatic antioxidants, determining the stressed plant's fate. The enzyme SOD is well-known for dismutating superoxide (O_2^- radicals to hydrogen peroxide (H_2O_2), whereas APX and GR transform H_2O_2 to water and oxygen. An increase in H_2O_2 levels due to an inhibition in APX activity could damage lipids and proteins [98]. Exogenous NO may stimulate the synthesis of endogenous NO, that can function as a signaling molecule or ROS scavenger under intense stress circumstances by controlling and improving the activities of antioxidant enzymes [100,101]. Through explorations into the effects of exogenous H_2S in wheat during flooding-induced hypoxic stress, it has been reported that NaHS application positively enhanced the activity of certain enzymes, including APX and GR [102]. In arsenate-only treated seedlings, the addition of NaHS raised NO levels, implying that both (H_2S and NO) cause the upregulation of the ascorbate-glutathione (AsA and GSH) cycle to counterbalance ROS-mediated damage, resulting in enhanced pea seedling growth according to Singh et al. [98]. Additionally, NO plays a crucial role in encouraging endogenous H_2S production, which increases the activity of antioxidant enzymes and aids wheat plants in tolerating oxidative stress brought on by osmotic stress [92]. Furthermore, NO and H_2S promote protein post-translational modifications via S-nitrosylation and tyrosine nitration. The altered protein function and activity caused by such Pb modifications may have given plants greater tolerance to abiotic stress [103,104]. Consistent with the accumulation of antioxidant enzymes in rice cultivars subjected to heat stress, either alone or in combination with Eth, SNP, or NaHS, the expression of SOD isoforms (*Mn-SOD*, *Fe-SOD*, *Cu-SOD*) and *APX* genes were also upregulated in treated rice cultivars. This suggested that up-regulation of *SOD* isoforms and *APX* genes could improve the activities of the SOD and APX enzymes, hence protecting cells from oxidative damage caused by HS stress. The activity of SOD isoenzymes is increased by ethylene in Arabidopsis plants under Cd stress, which affects root morphology [105]. In EIN2-1 mutant plants, higher transcription levels of Cu/Zn SOD2 and CAT3 resulted in higher SOD and CAT enzyme activity when compared to control plants [106,107]. In contrast, other investigations have revealed that inducing ethylene under abiotic stress could be harmful to plants, decreasing the activity of antioxidative enzymes and increasing the accumulation of ROS [108,109]. The hormone, NO's antioxidant property may be due to its direct interaction with ROS, which is then neutralized by several cellular processes, or NO could boost the antioxidant potential of cells by enhancing antioxidant enzyme activities [110]. At the post-translational level, NO modulates APX through S-nitrosation of cysteine residues, which enhances its activity, and metal nitrosation and tyrosine nitrosation, which both decrease its activity [111,112]. In a prior study, NaHS root pretreatment boosted the gene expression of antioxidant enzymes (cAPX, CAT, Mn-SOD, and GR), heat shock proteins (HSP70, HSP80, and HSP90), and aquaporins (PIP) [113]. Furthermore, we found that HT treatment with Eth or SNP under HS stress reversed Eth- or SNP-induced upregulation of *SOD* isoforms and *APX* genes. In *Solanum lycopersicum*, ethylene and H_2S fumigation sustained higher levels of *SlAPX1*, *SlAPX2*, and *SlCAT3* expression [39].

A previous study indicated that HS stress may enhance NO synthesis in tobacco [114] and higher plants [115]. Similarly, in this study, HS stress increased NO levels in rice cultivar leaves. Furthermore, plants under HS stress had higher levels of H_2S in their leaves. Similar to these results, increased H_2S generation was seen in wheat [36] and maize [116] exposed to heat stress, as well as bermudagrass exposed to cold, salt, and osmotic stresses [87]. In our experimental conditions, a rise in both NO and H_2S levels under HS stress was detected, which is in good agreement with these reports. In the present study, donors and inhibitors of H_2S and NO were applied to rice cultivars subjected to heat stress in order to better understand the interaction between H_2S and NO. In this study, NaHS,

SNP, and Eth treatment raised NO content in rice cultivar leaves in both heat-stressed and no-stress conditions; such an increase has previously been documented in barley and wheat plants [117,118]. Under HS stress, the decrease in NO content was greatest when HT was combined with Eth or SNP, which affected stress alleviation. Research indicates an interaction between NO and H_2S and it was recently reviewed [119]. Similarly, SNP, Eth, or NaHS treatment improved the H_2S content in leaves of rice cultivars in control and stressed plants, but significantly with NaHS treatment. Furthermore, the suppression of ethylene and NO in the presence of NaHS under HS stress using their inhibitors NBD and cPTIO, respectively, had no significant influence on H_2S levels.

Ethylene is produced in response to a variety of environmental stresses, implying that it acts as a connection between environmental change and developmental adaptability [120]. Ethylene increases photosynthesis and dry matter accumulation in plants under optimal and stressful environments. However, ethylene homeostasis is important for plant response and stress tolerance since excess ethylene formation under stress condition negatively impacts plant physiological and metabolic functions and plant growth. The involvement of ethylene in heat stress tolerance has been investigated earlier [17,49]. Previous studies have shown that heat stress, particularly in the 30–38 °C range, causes an increase in ethylene production in plants such as *Phaseolus vulgaris* [121] and *Triticum aestivum* [122]. Salt tolerance depends on ethylene production, and ethylene signaling is crucial for plants to self-correct quickly in response to salinity stress and to adapt better to the stress condition [123]. Under heat stress conditions, plants release stress ethylene by the same process that produces ethylene during normal development. In the present study, the ethylene level in heat-stressed plants was higher than control plants because of the burst of ethylene that occurred under stress. The application of ethephon following heat treatment resulted in ethylene release that showed protective functions at this stage and induced mechanisms of the activation of the antioxidant system to scavenge ROS and relieve plants from the stress. As the plants were relieved from the stress, the burst of stress ethylene was minimized resulting in a lower level of ethylene compared to heat-stressed plants. It has been shown that when plants are exposed to conditions that threaten their ability to survive, the same mechanism that produces ethylene for normal development instead functions to produce what is known as stress ethylene [124]. The paradoxical effects of stress ethylene on plants were shown emphasizing the fact that in stressed plant tissues, there is an initial small peak of ethylene close in time to the onset of stress and the second much larger peak some time later. The first small peak shows the protective response of plant. The second peak is so large that processes that are inhibitory to plant survival are initiated [124]. Thus, the modulation of ethylene production could reduce the stress-related injuries. According to studies, the signaling molecule NO modulates endogenous ethylene levels at different levels by altering a variety of pathways, leading to post-climacteric biochemical changes related to fruit quality [125]. In the current study, the application of SNP or NaHS resulted in lower ethylene levels more notably when there was no stress, compared to HS plants. It was recently proposed that H_2S counteracts the effect of ethylene action in banana fruit ripening and senescence [126]. As a result, H_2S might be able to resist ethylene function. Furthermore, H_2S, which is similar to NO, inhibits 1-amino-cyclopropane carboxylic acid oxidase (ACO) activity in tomato leaves [127]. In this study, we used HT to investigate the mechanism of H_2S and its effect on ethylene levels. Under HS stress, the application of HT with Eth or SNP increased the level of ethylene.

Furthermore, HS conditions have been shown to affect the light-harvesting complex, water-oxidizing complex, and PS II reaction center [128]. Chloroplast gene expression and responses to environmental stress may be related [31]. The chloroplast genes *psbA* and *psbB* encode the D1 protein of PS II and the PS II chlorophyll-binding protein (CP47), respectively [129,130]. In the present study, the qRT-PCR analysis revealed that heat stress downregulated the expression of the *psbA* and *psbB* genes, which were linked to PS II inactivation. Reduced photosynthetic pigments and organic solutes, such as soluble sugars, sucrose, and proline, were associated with the deleterious consequences of HS stress [131].

Salt stress was reported to cause the degradation of D1 protein (encoded by the *psbA* gene) in *Avena sativa* plants, as well as the downregulation of *psbA*, *psbB*, *psbC*, and *psbD* [132]. Meanwhile, Eth, SNP, or NaHS treatments upregulated the *psbA* and *psbB* gene expression of heat-stressed rice cultivar leaves, which may be responsible for PS II stability under heat stress. Notably, the expression of the investigated genes was significantly higher in Eth-treated leaves compared to SNP and NaHS leaves. Photosystem II tolerance to high light would be improved by enhanced *psbA* transcription and translation [31]. On the contrary, the application of HT along with Eth or SNP under HS resulted in the downregulation of *psbA* and *psbB* gene expression. It was investigated that H_2S was involved in the ethylene- or NO-mediated protection of photosynthetic machinery during HS stress, hence enhancing photosynthetic efficiency and mitigating the negative consequences of HS stress.

In the present study, the data were also examined using PCA in order to identify and classify the enormous data set in terms of growth, physio-biochemical and molecular characteristics into a manageable set of dynamically interrelated variables [133–135]. The PCA explained 94.70% and 94.67% of the data variability in Taipei-309 and Rasi, respectively, which accords with Sneath and Sokal [136], who considered that data must account for at least 70% of the total variability. The lines originating from the central point of the biplot represent correlations between various variables, with the closeness of the lines indicating the strength of the correlation with a specific treatment. Treatments such as (HS + Eth + HT), (HS + SNP + HT), and control were distributed in left quadrant, i.e., in the negative direction. However, treatments such as Eth, SNP, and NaHS with no stress and HS stress were present away from the origin in a positive direction. Growth and physiological attributes clustered opposite to oxidative stress attributes and NPQ. Biplot categorized the traits into three groups. The first group provided three treatments, Eth, SNP, and NaHS, under control conditions. These treatments were correlated to the variables including the dry weight of root and shoot, chlorophyll fluorescence, photosystem II genes, photosynthetic traits, and leaf RWC. This suggests that under control conditions, these treatments significantly increased these traits. Afterward, Eth, SNP, and NaHS under HS were grouped together. These treatments have a significant association with antioxidant enzyme activity (SOD, APX and GR), antioxidant genes (*Mn-SOD*, *Cu-SOD*, *Fe-SOD*, and *APX*), and osmolytes (proline, GB, soluble sugar, and trehalose), NO, and H_2S content. In addition, HS treatment shows a significant association with oxidative stress markers (H_2O_2 and TBARS), NPQ, and ethylene levels. However, treatments with inhibitors and scavengers such as HT (HS + Eth + HT) and (HS + SNP + HT), NBD (HS + NaHS + NBD) and cPTIO (HS + Eth + cPTIO) seemed to have no significant association with any parameter. This analysis confirmed our results that HS stress reduced the majority of morpho-physiological and biochemical traits while increasing H_2O_2 and TBARS content and NPQ in the two cultivars tested. On the other hand, Eth, SNP, and NaHS increased most of the traits under control and heat conditions. In contrast, the application of HT did not demonstrate any significant association with any parameters A proposed model to show the significance of ethylene, NO and H_2S in high temperature stress tolerance is presented in Figure 8.

To further confirm the role of ethylene and NO in H_2S-mediated thermotolerance, we run our dataset for Pearson correlation. The correlation between all paired attributes of antioxidant enzymes, their relative expression and osmolytes were positively significant with growth and photosynthesis parameters and negatively correlated with oxidative stress markers.

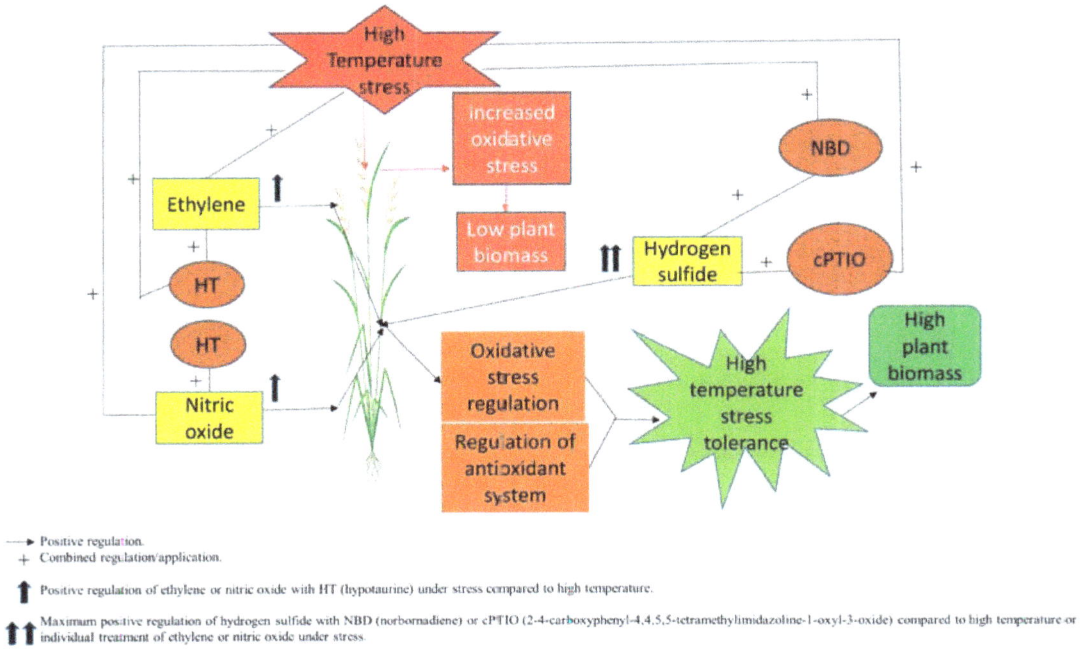

Figure 8. Proposed model for the role of ethylene, NO and H_2S in the alleviation of high temperature stress in rice (*Oryza sativa* L.) NBD; norbornadiene, cPTIO; 2-4-carboxyphenyl-4,4,5,5-tetramethylimidazoline-1-oxyl-3-oxide, HT; hypotaurine.

5. Conclusions

In summary, signaling molecules such as ethylene, H_2S, and NO have been found as promising for enhancing rice plants' thermo-tolerance, as well as growth and photosynthesis, particularly more in Taipei than Rasi. However, in both rice cultivars, Eth was more effective than SNP and NaHS in alleviating HS. The activation of antioxidant enzymes, the detoxification of ROS, and the higher accumulation of osmolytes caused by the use of signaling molecules equipped the plants in combating the negative effects of heat stress. In addition, the scavenging of H_2S by HT subsequently damaged the rice plants in the presence of Eth or SNP, confirming that the beneficial action of Eth and SNP is, at least to some extent, reliant on H_2S.

Supplementary Materials: The following are available online at https://www.mdpi.com/article/10.3390/antiox11081478/s1, File S1: Material and methods details, Table S1: Primer pairs used for quantitative RT-PCR.

Author Contributions: Conceptualization: N.A.K.; Investigation and data curation: H.G.; M.F.; Biochemical analysis: H.G., Z.S., I.R.M., and M.F.; Original draft preparation: H.G., M.F.; Editing and content improvement: N.A.K. All authors have read and agreed to the published version of the manuscript.

Funding: This research received no external funding.

Institutional Review Board Statement: Not applicable.

Informed Consent Statement: Not applicable.

Data Availability Statement: The data presented in this study and the Supplementary Materials are available in the graphs provided in the manuscript.

Acknowledgments: The research work on the mechanisms of heat stress tolerance in the laboratory of N.A.K. is supported by the Council of Scientific and Industrial Research, New Delhi (38(1473)/19/EMR-II).

Conflicts of Interest: The authors declare no conflict of interest.

References

1. Shinozaki, K.; Yamaguchi-Shinozaki, K. Gene networks involved in drought stress response and tolerance. *J. Exp. Bot.* **2007**, *58*, 221–227. [CrossRef] [PubMed]
2. Zhang, S.S.; Yang, H.; Ding, L.; Song, Z.T.; Ma, H.; Chang, F.; Liu, J.X. Tissue-specific transcriptomics reveals an important role of the unfolded protein response in maintaining fertility upon heat stress in Arabidopsis. *Plant Cell* **2017**, *29*, 1007–1023. [CrossRef] [PubMed]
3. Fatma, M.; Iqbal, N.; Sehar, Z.; Alyemeni, M.N.; Kaushik, P.; Khan, N.A.; Ahmad, P. Methyl jasmonate protects the PS II system by maintaining the stability of chloroplast d1 protein and accelerating enzymatic antioxidants in heat-stressed wheat plants. *Antioxidants* **2021**, *10*, 1216. [CrossRef] [PubMed]
4. Gautam, H.; Sehar, Z.; Rehman, M.T.; Hussain, A.; AlAjmi, M.F.; Khan, N.A. Nitric oxide enhances photosynthetic nitrogen and sulfur-use efficiency and activity of ascorbate-glutathione cycle to reduce high temperature stress stress-induced oxidative stress in rice (*Oryza sativa* L.) plants. *Biomolecules* **2021**, *11*, 305. [CrossRef] [PubMed]
5. Janská, A.; Maršík, P.; Zelenková, S.; Ovesná, J. Cold stress and acclimation—What is important for metabolic adjustment? *Plant Biol.* **2010**, *12*, 395–405. [CrossRef]
6. Hasanuzzaman, M.; Nahar, K.; Alam, M.; Roychowdhury, R.; Fujita, M. Physiological, biochemical, and molecular mechanisms of heat stress tolerance in plants. *Int. J. Mol. Sci.* **2013**, *14*, 9643–9684. [CrossRef]
7. Iqbal, N.; Umar, S.; Khan, N.A.; Corpas, F.J. Nitric oxide and hydrogen sulfide coordinately reduce glucose sensitivity and decrease oxidative stress via ascorbate-glutathione cycle in heat-stressed wheat (*Triticum aestivum* L.) plants. *Antioxidants* **2021**, *10*, 108. [CrossRef]
8. Jha, U.C.; Nayyar, H.; Palakurthi, R.; Jha, R.; Valluri, V.; Bajaj, P.; Thudi, M. Major QTLs and potential candidate genes for heat stress tolerance identified in chickpea (*Cicer arietinum* L.). *Front. Plant Sci.* **2021**, *12*, 655103. [CrossRef]
9. Sehar, Z.; Iqbal, N.; Khan, M.I.R.; Masood, A.; Rehman, M.T.; Hussain, A.; Khan, N.A. Ethylene reduces glucose sensitivity and reverses photosynthetic repression through optimization of glutathione production in salt-stressed wheat (*Triticum aestivum* L.). *Sci. Rep.* **2021**, *11*, 12650. [CrossRef]
10. Shivaraj, S.M.; Vats, S.; Bhat, J.A.; Dhakte, P.; Goyal, V.; Khatri, P.; Deshmukh, R. Nitric oxide and hydrogen sulfide crosstalk during heavy metal stress in plants. *Physiol. Plant.* **2020**, *168*, 437–455. [CrossRef]
11. Iqbal, N.; Fatma, M.; Gautam, H.; Umar, S.; Sofo, A.; D'Ippolito, I.; Khan, N.A. The crosstalk of melatonin and hydrogen sulfide determines photosynthetic performance by regulation of carbohydrate metabolism in wheat under heat stress. *Plants* **2021**, *10*, 1778. [CrossRef]
12. Fatma, M.; Iqbal, N.; Gautam, H.; Sehar, Z.; Sofo, A.; D'Ippolito, I.; Khan, N.A. Ethylene and sulfur coordinately modulate the antioxidant system and ABA accumulation in mustard plants under salt stress. *Plants* **2021**, *10*, 180. [CrossRef]
13. Gautam, H.; Sehar, Z.; Khan, N.A. Ethylene: A small signaling molecule with diverse roles. In *The Plant Hormone Ethylene. Stress Acclimation and Agricultural Applications*; Khan, N.A., Ferrante, A., Munne-Bosch, S., Eds.; Elsevier: Amsterdam, The Netherlands, 2022; pp. 1–10.
14. Jegadeesan, S.; Chaturvedi, P.; Ghatak, A.; Pressman, E.; Meir, S.; Faigenboim, A.; Firon, N. Proteomics of heat-stress and ethylene-mediated thermotolerance mechanisms in tomato pollen grains. *Front. Plant Sci.* **2018**, *9*, 1558. [CrossRef]
15. Poór, P.; Nawaz, K.; Gupta, R.; Ashfaque, F.; Khan, M.I.R. Ethylene involvement in the regulation of heat stress tolerance in plants. *Plant Cell Rep.* **2021**, *41*, 675–698. [CrossRef]
16. Singh, G.; Sarkar, N.K.; Grover, A. Tango between ethylene and HSFA2 settles heat tolerance. *Trends Plant Sci.* **2021**, *26*, 429–432. [CrossRef]
17. Wu, Y.S.; Yang, C.Y. Ethylene-mediated signaling confers thermotolerance and regulates transcript levels of heat shock factors in rice seedlings under heat stress. *Bot. Stud.* **2019**, *60*, 23. [CrossRef]
18. Chen, K.; Chen, L.; Fan, J.; Fu, J. Alleviation of heat damage to photosystem II by nitric oxide in tall fescue. *Photosynth. Res.* **2013**, *116*, 21–31. [CrossRef]
19. Fatma, M.; Masood, A.; Per, T.S.; Khan, N.A. Nitric oxide alleviates salt stress inhibited photosynthetic performance by interacting with sulfur assimilation in mustard. *Front. Plant Sci.* **2016**, *7*, 521. [CrossRef]
20. Parankusam, S.; Adimulam, S.S.; Bhatnagar-Mathur, P.; Sharma, K.K. Nitric oxide (NO) in plant heat stress tolerance: Current knowledge and perspectives. *Front. Plant Sci.* **2017**, *8*, 1582. [CrossRef]
21. Rather, B.A.; Masood, A.; Sehar, Z.; Majid, A.; Anjum, N.A.; Khan, N.A. Mechanisms and role of nitric oxide in phytotoxicity-mitigation of copper. *Front. Plant Sci.* **2020**, *11*, 675. [CrossRef]
22. Xuan, Y.; Zhou, S.; Wang, L.; Cheng, Y.; Zhao, L. Nitric oxide functions as a signal and acts upstream of AtCaM3 in thermotolerance in Arabidopsis seedlings. *Plant Physiol.* **2010**, *153*, 1895–1906. [CrossRef]

23. Rather, B.A.; Mir, I.R.; Masood, A.; Anjum, N.A.; Khan, N.A. Nitric oxide pre-treatment advances seed germination and alleviates copper-induced photosynthetic inhibition in Indian mustard. *Plants* **2020**, *9*, 776. [CrossRef]
24. Graziano, M.; Beligni, M.V.; Lamattina, L. Nitric oxide improves internal iron availability in plants. *Plant Physiol.* **2002**, *130*, 1852–1859. [CrossRef]
25. Vani, B.; Saradhi, P.P.; Mohanty, P. Alteration in chloroplast structure and thylakoid membrane composition due to in vivo heat treatment of rice seedlings: Correlation with the functional changes. *J. Plant Physiol.* **2001**, *158*, 583–592. [CrossRef]
26. Ahmad, P.; Abdel Latef, A.A.; Hashem, A.; Abd Allah, E.F.; Gucel, S.; Tran, L.S.P. Nitric oxide mitigates salt stress by regulating levels of osmolytes and antioxidant enzymes in chickpea. *Front. Plant Sci.* **2016**, *7*, 347. [CrossRef] [PubMed]
27. Sánchez-Vicente, I.; Lorenzo, O. Nitric oxide regulation of temperature acclimation: A molecular genetic perspective. *J. Exp. Bot.* **2021**, *72*, 5789–5794. [CrossRef] [PubMed]
28. Corpas, F.J.; González-Gordo, S.; Cañas, A.; Palma, J.M. Nitric oxide and hydrogen sulfide in plants: Which comes first? *J. Exp. Bot.* **2019**, *70*, 4391–4404. [CrossRef] [PubMed]
29. Aroca, A.; Yruela, I.; Gotor, C.; Bassham, D.C. Persulfidation of ATG18a regulates autophagy under ER stress in Arabidopsis. *Proc. Natl. Acad. Sci. USA* **2021**, *118*, e2023604118. [CrossRef]
30. Chen, S.; Wang, X.; Jia, H.; Li, F.; Ma, Y.; Liesche, J.; Li, J. Persulfidation-induced structural change in SnRK2. 6 establishes intramolecular interaction between phosphorylation and persulfidation. *Mol. Plant* **2021**, *14*, 1814–1830. [CrossRef]
31. Zhang, J.; Zhou, M.; Zhou, H.; Zhao, D.; Gotor, C.; Romero, L.C.; Xie, Y. Hydrogen sulfide, a signaling molecule in plant stress responses. *J. Integr. Plant Biol.* **2021**, *63*, 146–160. [CrossRef]
32. Papanatsiou, M.; Scuffi, D.; Blatt, M.R.; García-Mata, C. Hydrogen sulfide regulates inward-rectifying K+ channels in conjunction with stomatal closure. *Plant Physiol.* **2015**, *168*, 29–35. [CrossRef]
33. Pandey, A.K.; Gautam, A. Stress responsive gene regulation in relation to hydrogen sulfide in plants under abiotic stress. *Physiol. Plant.* **2020**, *168*, 511–525. [CrossRef]
34. Huang, D.; Huo, J.; Liao, W. Hydrogen sulfide: Roles in plant abiotic stress response and crosstalk with other signals. *Plant Sci.* **2021**, *302*, 110733. [CrossRef]
35. Li, H.; Yu, T.T.; Ning, Y.S.; Li, H.; Zhang, W.W.; Yang, H.Q. Hydrogen Sulfide Alleviates Alkaline Salt Stress by Regulating the Expression of MicroRNAs in *Malus hupehensis* Rehd. Roots. *Front. Plant Sci.* **2021**, *12*, 663519. [CrossRef]
36. Min, Y.A.N.G.; Qin, B.P.; Ping, W.A.N.G.; Li, M.L.; Chen, L.L.; Chen, L.T.; Yin, Y.P. Foliar application of sodium hydrosulfide (NaHS), a hydrogen sulfide (H2S) donor, can protect seedlings against heat stress in wheat (*Triticum aestivum* L.). *J. Integr. Agric.* **2016**, *15*, 2745–2758.
37. Husain, T.; Suhel, M.; Prasad, S.M.; Singh, V.P. Ethylene needs endogenous hydrogen sulfide for alleviating hexavalent chromium stress in *Vigna mungo* L. and *Vigna radiata* L. *Environ. Pollut.* **2021**, *290*, 117968. [CrossRef]
38. Jia, H.; Chen, S.; Liu, D.; Liesche, J.; Shi, C.; Wang, J.; Li, J. Ethylene-induced hydrogen sulfide negatively regulates ethylene biosynthesis by persulfidation of ACO in tomato under osmotic stress. *Front. Plant Sci.* **2018**, *9*, 1517. [CrossRef]
39. Yao, G.F.; Wei, Z.Z.; Li, T.T.; Tang, J.; Huang, Z.Q.; Yang, F.; Zhang, H. Modulation of enhanced antioxidant activity by hydrogen sulfide antagonization of ethylene in tomato fruit ripening. *J. Agric. Food Chem.* **2018**, *66*, 10380–10387. [CrossRef]
40. Khan, M.I.R.; Khan, N.A. Ethylene reverses photosynthetic inhibition by nickel and zinc in mustard through changes in PS II activity, photosynthetic nitrogen use efficiency, and antioxidant metabolism. *Protoplasma* **2014**, *251*, 1007–1019. [CrossRef]
41. Yang, S. Ethylene evolution from 2-chloroethylphosphonic acid. *Plant Physiol.* **1969**, *44*, 1203. [CrossRef]
42. Biddle, E.; Kerfoot, D.G.; Kho, Y.H.; Russell, K.E. Kinetic studies of the thermal decomposition of 2-chloroethylphosphonic acid in aqueous solution. *Plant Physiol.* **1976**, *58*, 700–702. [CrossRef]
43. Barrs, H.D.; Weatherley, P.E. A re-examination of the relative turgidity technique for estimating water deficits in leaves. *Aust. J. Biol. Sci.* **1962**, *15*, 413–428. [CrossRef]
44. Bates, L.S.; Waldren, R.P.; Teare, I.D. Rapid determination of free proline for water-stress studies. *Plant Soil* **1973**, *39*, 205–207. [CrossRef]
45. Grieve, C.M.; Grattan, S.R. Rapid assay for determination of water soluble quaternary ammonium compounds. *Plant Soil* **1983**, *70*, 303–307. [CrossRef]
46. Trevelyan, W.E.; Harrison, J. Studies on yeast metabolism. 5. The trehalose content of baker's yeast during anaerobic fermentation. *Biochem. J.* **1956**, *62*, 177. [CrossRef]
47. Xu, W.; Cui, K.; Xu, A.; Nie, L.; Huang, J.; Peng, S. Drought stress condition increases root to shoot ratio via alteration of carbohydrate partitioning and enzymatic activity in rice seedlings. *Acta Physiol. Plant* **2015**, *37*, 9. [CrossRef]
48. Okuda, T.; Matsuda, Y.; Yamanaka, A.; Sagisaka, S. Abrupt increase in the level of hydrogen peroxide in leaves of winter wheat is caused by cold treatment. *Plant Physiol.* **1991**, *97*, 1265–1267. [CrossRef]
49. Gautam, H.; Fatma, M.; Sehar, Z.; Iqbal, N.; Albaqami, M.; Khan, N.A. Exogenously-Sourced Ethylene Positively Modulates Photosynthesis, Carbohydrate Metabolism, and Antioxidant Defense to Enhance Heat Tolerance in Rice. *Int. J. Mol. Sci.* **2022**, *23*, 1031. [CrossRef]
50. Dhindsa, R.S.; Plumb-Dhindsa, P.; Thorpe, T.A. Leaf senescence: Correlated with increased levels of membrane permeability and lipid peroxidation, and decreased levels of superoxide dismutase and catalase. *J. Exp. Bot.* **1981**, *32*, 93–101. [CrossRef]
51. Beyer Jr, W.F.; Fridovich, I. Assaying for superoxide dismutase activity: Some large consequences of minor changes in conditions. *Anal. Biochem.* **1987**, *161*, 559–566. [CrossRef]

52. Giannopolitis, C.N.; Ries, S.K. Superoxide dismutases: I. Occurrence in higher plants. *Plant Physiol.* **1977**, *59*, 309–314. [CrossRef] [PubMed]
53. Nakano, Y.; Asada, K. Hydrogen peroxide is scavenged by ascorbate-specific peroxidase in spinach chloroplasts. *Plant Cell Physiol.* **1981**, *22*, 867–880.
54. Foyer, C.H.; Halliwell, B. The presence of glutathione and glutathione reductase in chloroplasts: A proposed role in ascorbic acid metabolism. *Planta* **1976**, *133*, 21–25. [CrossRef] [PubMed]
55. Zhou, B.; Guo, Z.; Xing, J.; Huang, B. Nitric oxide is involved in abscisic acid-induced antioxidant activities in *Stylosanthes guianensis*. *J. Exp. Bot.* **2005**, *56*, 3223–3228. [CrossRef]
56. Xie, Y.; Zhang, C.; Lai, D.; Sun, Y.; Samma, M.K.; Zhang, J.; Shen, W. Hydrogen sulfide delays GA-triggered programmed cell death in wheat aleurone layers by the modulation of glutathione homeostasis and heme oxygenase-1 expression. *J. Plant Physiol.* **2014**, *171*, 53–62. [CrossRef]
57. Nievola, C.C.; Carvalho, C.P.; Carvalho, V.; Rodrigues, E. Rapid responses of plants to temperature changes. *Temperature* **2017**, *4*, 371–405. [CrossRef]
58. Wahid, A.; Gelani, S.; Ashraf, M.; Foolad, M.R. Heat tolerance in plants: An overview. *Environ. Exp. Bot.* **2007**, *61*, 199–223. [CrossRef]
59. Zanganeh, R.; Jamei, R.; Rahmani, F. Role of salicylic acid and hydrogen sulfide in promoting lead stress tolerance and regulating free amino acid composition in *Zea mays* L. *Acta Physiol. Plant.* **2019**, *41*, 94. [CrossRef]
60. Rizwan, M.; Mostofa, M.G.; Ahmad, M.Z.; Zhou, Y.; Adeel, M.; Mehmood, S.; Liu, Y. Hydrogen sulfide enhances rice tolerance to nickel through the prevention of chloroplast damage and the improvement of nitrogen metabolism under excessive nickel. *Plant Physiol. Biochem.* **2019**, *138*, 100–111. [CrossRef]
61. Valivand, M.; Amooaghaie, R.; Ahadi, A. Seed priming with H_2S and Ca^{2+} trigger signal memory that induces cross-adaptation against nickel stress in zucchini seedlings. *Plant Physiol. Biochem.* **2019**, *143*, 286–298. [CrossRef]
62. Rodríguez-Serrano, M.; Romero-Puertas, M.C.; Pazmino, D.M.; Testillano, P.S.; Risueño, M.C.; Del Río, L.A.; Sandalio, L.M. Cellular response of pea plants to cadmium toxicity: Cross talk between reactive oxygen species, nitric oxide, and calcium. *Plant Physiol.* **2009**, *150*, 229–243. [CrossRef]
63. Louis, X.L.; Murphy, R.; Thandapilly, S.J.; Yu, L.; Netticadan, T. Garlic extracts prevent oxidative stress, hypertrophy and apoptosis in cardiomyocytes: A role for nitric oxide and hydrogen sulfide. *BMC Complement. Altern. Med.* **2012**, *12*, 140. [CrossRef]
64. Li, Z.G.; Yang, S.Z.; Long, W.B.; Yang, G.X.; Shen, Z.Z. Hydrogen sulphide may be a novel downstream signal molecule in nitric oxide-induced heat tolerance of maize (*Zea mays* L.) seedlings. *Plant Cell Environ.* **2013**, *36*, 1564–1572. [CrossRef]
65. Sehar, Z.; Gautam, H.; Iqbal, N.; Alvi, A.F.; Jahan, B.; Fatma, M.; Albaqami, M.; Khan, N.A. The functional interplay between ethylene, hydrogen sulfide, and sulfur in plant heat stress tolerance. *Biomolecules* **2022**, *12*, 678. [CrossRef]
66. Zhuang, J.; Wang, Y.; Chi, Y.; Zhou, L.; Chen, J.; Zhou, W.; Ding, J. Drought stress strengthens the link between chlorophyll fluorescence parameters and photosynthetic traits. *PeerJ* **2020**, *8*, e10046. [CrossRef]
67. Lu, Y. Identification and roles of photosystem II assembly, stability, and repair factors in Arabidopsis. *Front. Plant Sci.* **2016**, *7*, 168. [CrossRef]
68. Baker, N.R.; Rosenqvist, E. Applications of chlorophyll fluorescence can improve crop production strategies: An examination of future possibilities. *J. Exp. Bot.* **2004**, *55*, 1607–1621. [CrossRef]
69. Havaux, M. Rapid photosynthetic adaptation to heat stress triggered in potato leaves by moderately elevated temperatures. *Plant Cell Environ.* **1993**, *16*, 461–467. [CrossRef]
70. Camejo, D.; Rodríguez, P.; Morales, M.A.; Dell'Amico, J.M.; Torrecillas, A.; Alarcón, J.J. High temperature stress effects on photosynthetic activity of two tomato cultivars with different heat susceptibility. *J. Plant Physiol.* **2005**, *162*, 281–289. [CrossRef]
71. Mathur, S.; Agrawal, D.; Jajoo, A. Photosynthesis: Response to high temperature stress. *J. Photochem. Photobiol. B* **2014**, *137*, 116–126. [CrossRef]
72. Yamamoto, Y. Quality control of photosystem II: The mechanisms for avoidance and tolerance of light and heat stresses are closely linked to membrane fluidity of the thylakoids. *Front. Plant Sci.* **2016**, *7*, 1136. [CrossRef] [PubMed]
73. Marutani, Y.; Yamauchi, Y.; Kimura, Y.; Mizutani, M.; Sugimoto, Y. Damage to photosystem II due to heat stress without light-driven electron flow: Involvement of enhanced introduction of reducing power into thylakoid membranes. *Planta* **2012**, *236*, 753–761. [CrossRef] [PubMed]
74. Kim, Y.; Seo, C.W.; Khan, A.L.; Mun, B.G.; Shahzad, R.; Ko, J.W.; Lee, I.J. Exo-ethylene application mitigates waterlogging stress in soybean (*Glycine max* L.). *BMC Plant Biol.* **2018**, *18*, 254. [CrossRef] [PubMed]
75. Shi, S.; Wang, G.; Wang, Y.; Zhang, L.; Zhang, L. Protective effect of nitric oxide against oxidative stress under ultraviolet-B radiation. *Nitric Oxide* **2005**, *13*, 1–9. [CrossRef]
76. Morales, D.; Rodríguez, P.; Dell'Amico, J.; Nicolas, E.; Torrecillas, A.; Sánchez-Blanco, M.J. high temperature stress preconditioning and thermal shock imposition affects water relations, gas exchange and root hydraulic conductivity in tomato. *Biol. Plant.* **2003**, *47*, 203–208. [CrossRef]
77. Khan, M.N.; Siddiqui, M.H.; Mohammad, F.; Naeem, M. Interactive role of nitric oxide and calcium chloride in enhancing tolerance to salt stress. *Nitric Oxide* **2012**, *27*, 210–218. [CrossRef]

78. Li, G.; Shah, A.A.; Khan, W.U.; Yasin, N.A.; Ahmad, A.; Abbas, M.; Safdar, N. Hydrogen sulfide mitigates cadmium induced toxicity in *Brassica rapa* by modulating physiochemical attributes, osmolyte metabolism and antioxidative machinery. *Chemosphere* **2021**, *263*, 127999. [CrossRef]
79. Duan, B.; Ma, Y.; Jiang, M.; Yang, F.; Ni, L.; Lu, W. Improvement of photosynthesis in rice (*Oryza sativa* L.) as a result of an increase in stomatal aperture and density by exogenous hydrogen sulfide treatment. *Plant Growth Regul.* **2015**, *75*, 33–44. [CrossRef]
80. Szabados, L.; Savouré, A. Proline: A multifunctional amino acid. *Trends Plant Sci.* **2010**, *15*, 89–97. [CrossRef]
81. Kaur, G.; Asthir, B.J.B.P. Proline: A key player in plant abiotic stress tolerance. *Biol. Plant.* **2015**, *59*, 609–619. [CrossRef]
82. Parmar, P.; Kumari, N.; Sharma, V. Structural and functional alterations in photosynthetic apparatus of plants under cadmium stress. *Bot. Stud.* **2013**, *54*, 45. [CrossRef]
83. Yang, X.; Wen, X.; Gong, H.; Lu, Q.; Yang, Z.; Tang, Y.; Lu, C. Genetic engineering of the biosynthesis of glycinebetaine enhances thermotolerance of photosystem II in tobacco plants. *Planta* **2007**, *225*, 719–733. [CrossRef]
84. Rizvi, A.; Khan, M.S. Heavy metal induced oxidative damage and root morphology alterations of maize (*Zea mays* L.) plants and stress mitigation by metal tolerant nitrogen fixing *Azotobacter chroococcum*. *Ecotoxicol. Environ. Saf.* **2018**, *157*, 9–20. [CrossRef]
85. Liu, X.; Huang, B. Carbohydrate accumulation in relation to heat stress tolerance in two creeping bentgrass cultivars. *J. Am. Soc. Hortic. Sci.* **2000**, *125*, 442–447. [CrossRef]
86. Kaplan, F.; Kopka, J.; Haskell, D.W.; Zhao, W.; Schiller, K.C.; Gatzke, N.; Guy, C.L. Exploring the temperature-stress metabolome of Arabidopsis. *Plant Physiol.* **2004**, *136*, 4159–4168. [CrossRef]
87. Luo, Y.; Li, F.; Wang, G.P.; Yang, X.H.; Wang, W. Exogenously-supplied trehalose protects thylakoid membranes of winter wheat from heat-induced damage. *Biol. Plant.* **2010**, *54*, 495–501. [CrossRef]
88. Kosar, F.; Akram, N.A.; Sadiq, M.; Al-Qurainy, F.; Ashraf, M. Trehalose: A key organic osmolyte effectively involved in plant abiotic stress tolerance. *J. Plant Growth Regul.* **2019**, *38*, 606–618. [CrossRef]
89. Li, Z.G.; Luo, L.J.; Zhu, L.P. Involvement of trehalose in hydrogen sulfide donor sodium hydrosulfide-induced the acquisition of heat tolerance in maize (*Zea mays* L.) seedlings. *Bot. Stud.* **2014**, *55*, 20. [CrossRef]
90. Khan, M.N.; Mobin, M.; Abbas, Z.K.; Siddiqui, M.H. Nitric oxide-induced synthesis of hydrogen sulfide alleviates osmotic stress in wheat seedlings through sustaining antioxidant enzymes, osmolyte accumulation and cysteine homeostasis. *Nitric Oxide* **2017**, *68*, 91–102. [CrossRef]
91. Iqbal, N.; Sehar, Z.; Fatma, M.; Umar, S.; Sofo, A.; Khan, N.A. Nitric Oxide and Abscisic Acid Mediate Heat Stress Tolerance through Regulation of Osmolytes and Antioxidants to Protect Photosynthesis and Growth in Wheat Plants. *Antioxidants* **2022**, *11*, 372. [CrossRef]
92. Kong, W.W.; Huang, C.Y.; Chen, Q.; Zou, Y.J.; Zhao, M.R.; Zhang, J.X. Nitric oxide is involved in the regulation of trehalose accumulation under heat stress in *Pleurotus eryngii* var. *tuoliensis*. *Biotechnol. Lett.* **2012**, *34*, 1915–1919. [CrossRef]
93. Awasthi, R.; Bhandari, K.; Nayyar, H. Temperature stress and redox homeostasis in agricultural crops. *Front. Environ. Sci.* **2015**, *3*, 11. [CrossRef]
94. Shi, H.; Ye, T.; Chan, Z. Exogenous application of hydrogen sulfide donor sodium hydrosulfide enhanced multiple abiotic stress tolerance in bermudagrass (*Cynodon dactylon* (L). Pers.). *Plant Physiol. Biochem.* **2013**, *71*, 226–234. [CrossRef]
95. Khan, M.I.R.; Jahan, B.; AlAjmi, M.F.; Rehman, M.T.; Khan, N.A. Ethephon mitigates nickel stress by modulating antioxidant system, glyoxalase system and proline metabolism in Indian mustard. *Physiol. Mol. Biol. Plants* **2020**, *26*, 1201–1213. [CrossRef]
96. Shi, H.; Ye, T.; Chan, Z. Nitric oxide-activated hydrogen sulfide is essential for cadmium stress response in bermudagrass (*Cynodon dactylon* (L). Pers.). *Plant Physiol. Biochem.* **2014**, *74*, 99–107. [CrossRef]
97. Peng, R.; Bian, Z.; Zhou, L.; Cheng, W.; Hai, N.; Yang, C.; Wang, C. Hydrogen sulfide enhances nitric oxide-induced tolerance of hypoxia in maize (*Zea mays* L.). *Plant Cell Rep.* **2016**, *35*, 2325–2340. [CrossRef]
98. Singh, V.P.; Singh, S.; Kumar, J.; Prasad, S.M. Hydrogen sulfide alleviates toxic effects of arsenate in pea seedlings through up-regulation of the ascorbate–glutathione cycle: Possible involvement of nitric oxide. *J. Plant Physiol.* **2015**, *181*, 20–29. [CrossRef]
99. Per, T.S.; Masood, A.; Khan, N.A. Nitric oxide improves S-assimilation and GSH production to prevent inhibitory effects of cadmium stress on photosynthesis in mustard (*Brassica juncea* L.). *Nitric Oxide* **2017**, *68*, 111–124. [CrossRef] [PubMed]
100. Hao, G.P.; Xing, Y.; Zhang, J.H. Role of nitric oxide dependence on nitric oxide synthase-like activity in the water stress signaling of maize seedling. *J. Integr. Plant Biol.* **2008**, *50*, 435–442. [CrossRef]
101. Zhao, M.G.; Chen, L.; Zhang, L.L.; Zhang, W.H. Nitric reductase-dependent nitric oxide production is involved in cold acclimation and freezing tolerance in Arabidopsis. *Plant Physiol.* **2009**, *151*, 755–767. [CrossRef] [PubMed]
102. Shan, C.J.; Zhang, S.L.; Li, D.F.; Zhao, Y.Z.; Tian, X.L.; Zhao, X.L.; Liu, R.Q. Effects of exogenous hydrogen sulfide on the ascorbate and glutathione metabolism in wheat seedlings leaves under water stress. *Acta Physiol. Plant.* **2011**, *33*, 2533. [CrossRef]
103. Ziogas, V.; Tanou, G.; Filippou, P.; Diamantidis, G.; Vasilakakis, M.; Fotopoulos, V.; Molassiotis, A. Nitrosative responses in citrus plants exposed to six abiotic stress conditions. *Plant Physiol. Biochem.* **2013**, *68*, 118–126. [CrossRef] [PubMed]
104. Ziogas, V.; Tanou, G.; Belghazi, M.; Filippou, P.; Fotopoulos, V.; Grigorios, D.; Molassiotis, A. Roles of sodium hydrosulfide and sodium nitroprusside as priming molecules during drought acclimation in citrus plants. *Plant Mol. Biol.* **2015**, *89*, 433–450. [CrossRef] [PubMed]
105. Abozeid, A.; Ying, Z.; Lin, Y.; Liu, J.; Zhang, Z.; Tang, Z. Ethylene improves root system development under cadmium stress by modulating superoxide anion concentration in *Arabidopsis thaliana*. *Front. Plant Sci.* **2017**, *8*, 253. [CrossRef]

106. Cao, S.; Jiang, S.; Zhang, R. Evidence for a role of Ethylene-Insensitive 2 gene in the regulation of the oxidative stress response in Arabidopsis. *Acta Physiol. Plant.* **2006**, *28*, 417–425. [CrossRef]
107. Cao, Y.; Song, F.; Goodman, R.M.; Zheng, Z. Molecular characterization of four rice genes encoding ethylene-responsive transcriptional factors and their expressions in response to biotic and abiotic stress. *J. Plant Physiol.* **2006**, *163*, 1167–1178. [CrossRef]
108. Li, C.H.; Wang, G.; Zhao, J.L.; Zhang, L.Q.; Ai, L.F.; Han, Y.F.; Sun, Y. The receptor-like kinase SIT1 mediates salt sensitivity by activating MAPK3/6 and regulating ethylene homeostasis in rice. *Plant Cell* **2014**, *26*, 2538–2553. [CrossRef]
109. Sharma, A.; Kumar, V.; Sidhu, G.P.S.; Kumar, R.; Kohli, S.K.; Yadav, P.; Bhardwaj, R. Abiotic stress management in plants: Role of ethylene. In *Molecular Plant Abiotic Stress: Biology and Biotechnology*; Wiley: London, UK, 2019; pp. 185–208.
110. Neill, S.; Barros, R.; Bright, J.; Desikan, R.; Hancock, J.; Harrison, J.; Wilson, I. Nitric oxide, stomatal closure, and abiotic stress. *J. Exp. Bot.* **2008**, *59*, 165–176. [CrossRef]
111. Begara-Morales, J.C.; Sánchez-Calvo, B.; Chaki, M.; Valderrama, R.; Mata-Pérez, C.; López-Jaramillo, J.; Barroso, J.B. Dual regulation of cytosolic ascorbate peroxidase (APX) by tyrosine nitration and S-nitrosylation. *J. Exp. Bot.* **2014**, *65*, 527–538. [CrossRef]
112. Clark, D.; Durner, J.; Navarre, D.A.; Klessig, D.F. Nitric oxide inhibition of tobacco catalase and ascorbate peroxidase. *Mol. Plant Microbe Interact.* **2000**, *13*, 1380–1384. [CrossRef]
113. Christou, A.; Filippou, P.; Manganaris, G.A.; Fotopoulos, V. Sodium hydrosulfide induces systemic thermotolerance to strawberry plants through transcriptional regulation of heat shock proteins and aquaporin. *BMC Plant Biol.* **2014**, *14*, 42. [CrossRef]
114. Locato, V.; Gadaleta, C.; De Gara, L.; De Pinto, M.C. Production of reactive species and modulation of antioxidant network in response to heat shock: A critical balance for cell fate. *Plant Cell Environ.* **2008**, *31*, 1606–1619. [CrossRef]
115. Ya'acov, Y.L.; Wills, R.B.; Ku, V.V.V. Evidence for the function of the free radical gas–nitric oxide (NO•)—As an endogenous maturation and senescence regulating factor in higher plants. *Plant Physiol. Biochem.* **1998**, *36*, 825–833.
116. Li, Z.G.; Yi, X.Y.; Li, Y.T. Effect of pretreatment with hydrogen sulfide donor sodium hydrosulfide on heat tolerance in relation to antioxidant system in maize (*Zea mays*) seedlings. *Biol. Plant.* **2014**, *69*, 1001–1009. [CrossRef]
117. Chen, J.; Wang, W.H.; Wu, F.H.; He, E.M.; Liu, X.; Shangguan, Z.P.; Zheng, H.L. Hydrogen sulfide enhances salt tolerance through nitric oxide-mediated maintenance of ion homeostasis in barley seedling roots. *Sci. Rep.* **2015**, *5*, 12516. [CrossRef]
118. Kaya, C.; Ashraf, M.; Alyemeni, M.N.; Ahmad, P. Responses of nitric oxide and hydrogen sulfide in regulating oxidative defence system in wheat plants grown under cadmium stress. *Physiol. Plant.* **2020**, *168*, 345–360. [CrossRef]
119. He, H.; Li, Y.; He, L.F. Role of nitric oxide and hydrogen sulfide in plant aluminum tolerance. *Biometals* **2019**, *32*, 1–9. [CrossRef]
120. Dubois, M.; Van den Broeck, L.; Inzé, D. The pivotal role of ethylene in plant growth. *Trends Plant Sci.* **2018**, *23*, 311–323. [CrossRef]
121. Sauter, K.J.; Davis, D.W.; Li, P.H.; Wallerstein, I.S. Leaf ethylene evolution level following high temperature stress in common bean. *Hort. Sci.* **1990**, *25*, 1282–1284. [CrossRef]
122. Khan, M.I.R.; Iqbal, N.; Masood, A.; Per, T.S.; Khan, N.A. Salicylic acid alleviates adverse effects of heat stress on photosynthesis through changes in proline production and ethylene formation. *Plant Signal. Behav.* **2013**, *8*, e26374. [CrossRef]
123. Tao, J.J.; Chen, H.W.; Ma, B.; Zhang, W.K.; Chen, S.Y.; Zhang, J.S. The role of ethylene in plants under salinity stress. *Front. Plant Sci.* **2015**, *6*, 1059. [CrossRef]
124. Steams, J.C.; Glick, B.R. Transgenic plants with altered ethylene biosynthesis or perception. *Biotechnol. Adv.* **2003**, *21*, 193–210.
125. Manjunatha, G.; Gupta, K.J.; Lokesh, V.; Mur, L.A.; Neelwarne, B. Nitric oxide counters ethylene effects on ripening fruits. *Plant Signal. Behav.* **2012**, *7*, 476–483. [CrossRef]
126. Ge, Y.; Hu, K.D.; Wang, S.S.; Hu, L.Y.; Chen, X.Y.; Li, Y.H.; Zhang, H. Hydrogen sulfide alleviates postharvest ripening and senescence of banana by antagonizing the effect of ethylene. *PLoS ONE* **2017**, *12*, e0180113. [CrossRef]
127. Liu, H.; Wang, J.; Liu, J.; Liu, T.; Xue, S. Hydrogen sulfide (H_2S) signaling in plant development and stress responses. *Abiotech* **2021**, *2*, 32–63. [CrossRef]
128. Lípová, L.; Krchňák, P.; Komenda, J.; Ilík, P. Heat-induced disassembly and degradation of chlorophyll-containing protein complexes in vivo. *Biochim. Biophys. Acta (BBA)-Bioenerg.* **2010**, *1797*, 63–70. [CrossRef]
129. Mulo, P.; Sakurai, I.; Aro, E.M. Strategies for psbA gene expression in cyanobacteria, green algae and higher plants: From transcription to PSII repair. *Biochim. Biophys. Acta (BBA)-Bioenerg.* **2012**, *1817*, 247–257. [CrossRef]
130. Barber, J.; Nield, J.; Morris, E.P.; Zheleva, D.; Hankamer, B. The structure, function and dynamics of photosystem two. *Physiol. Plant.* **1997**, *100*, 817–827. [CrossRef]
131. Hassan, H.; Alatawi, A.; Abdulmajeed, A.; Emam, M.; Khattab, H. Roles of Si and SiNPs in Improving Thermotolerance of Wheat Photosynthetic Machinery via Upregulation of PsbH, PsbB and PsbD Genes Encoding PSII Core Proteins. *Sci. Hortic.* **2021**, *7*, 16. [CrossRef]
132. Varghese, N.; Alyammahi, O.; Nasreddine, S.; Alhassani, A.; Gururani, M.A. Melatonin positively influences the photosynthetic machinery and antioxidant system of *Avena sativa* during salinity stress. *Plants* **2019**, *8*, 610. [CrossRef]
133. Aziz, A.; Mahmood, T.; Mahmood, Z.; Shazadi, K.; Mujeeb-Kazi, A.; Rasheed, A. Genotypic variation and genotype × environment interaction for yield-related traits in synthetic hexaploid wheats under a range of optimal and heat-stressed environments. *Crop Sci.* **2018**, *58*, 295–303. [CrossRef]

134. Ali, Q.; Perveen, R.; El-Esawi, M.A.; Ali, S.; Hussain, S.M.; Amber, M.; Ahmad, P. Low doses of *Cuscuta reflexa* extract act as natural biostimulants to improve the germination vigor, growth, and grain yield of wheat grown under water stress: Photosynthetic pigments, antioxidative defense mechanisms, and nutrient acquisition. *Biomolecules* **2020**, *10*, 1212. [CrossRef] [PubMed]
135. Basit, F.; Chen, M.; Ahmed, T.; Shahid, M.; Noman, M.; Liu, J.; Guan, Y. Seed priming with brassinosteroids alleviates chromium stress in rice cultivars via improving ROS metabolism and antioxidant defense response at biochemical and molecular levels. *Antioxidants* **2021**, *10*, 1089. [CrossRef] [PubMed]
136. Sneath, P.H.; Sokal, R.R. *Numerical Taxonomy: The Principles and Practice of Numerical Classification*; W. H. Freeman and Company: San Francisco, CA, USA, 1973; p. 573.

antioxidants

MDPI

Article

BAG9 Confers Thermotolerance by Regulating Cellular Redox Homeostasis and the Stability of Heat Shock Proteins in *Solanum lycopersicum*

Huamin Huang [1,†], Chenxu Liu [1,†], Chen Yang [1], Mukesh Kumar Kanwar [1], Shujun Shao [1], Zhenyu Qi [2,3] and Jie Zhou [1,2,4,5,*]

1 Department of Horticulture, Zhejiang Provincial Key Laboratory of Horticultural Plant Integrative Biology, Zhejiang University, Yuhangtang Road 866, Hangzhou 310058, China; 22016059@zju.edu.cn (H.H.); chenxuliu@zju.edu.cn (C.L.); 20210107@hznu.edu.cn (C.Y.); kanwar@zju.edu.cn (M.K.K.); ssjun@zju.edu.cn (S.S.)
2 Hainan Institute, Zhejiang University, Sanya 572025, China; qizhenyu@zju.edu.cn
3 Agricultural Experiment Station, Zhejiang University, Hangzhou 310058, China
4 Key Laboratory of Horticultural Plants Growth, Development and Quality Improvement, Ministry of Agriculture and Rural Affairs of China, Yuhangtang Road 866, Hangzhou 310058, China
5 Shandong (Linyi) Institute of Modern Agriculture, Zhejiang University, Linyi 276000, China
* Correspondence: jie@zju.edu.cn
† These authors contributed equally to this work.

Citation: Huang, H.; Liu, C.; Yang, C.; Kanwar, M.K.; Shao, S.; Qi, Z.; Zhou, J. BAG9 Confers Thermotolerance by Regulating Cellular Redox Homeostasis and the Stability of Heat Shock Proteins in *Solanum lycopersicum*. *Antioxidants* **2022**, *11*, 1467. https://doi.org/10.3390/antiox11081467

Academic Editor: Nafees A. Khan

Received: 17 June 2022
Accepted: 25 July 2022
Published: 27 July 2022

Abstract: The Bcl-2-associated athanogene (BAG) family, a group of co-chaperones that share conservative domains in flora and fauna, is involved in plant growth, development, and stress tolerance. However, the function of tomato *BAG* genes on thermotolerance remains largely unknown. Herein, we found that the expression of *BAG9* was induced during heat stress in tomato plants. Knockout of the *BAG9* gene by CRISPR/Cas9 reduced, while its overexpression increased thermotolerance in tomato plants as reflected by the phenotype, photosynthesis rate, and membrane peroxidation. Heat-induced reactive oxygen species and oxidative/oxidized proteins were further increased in *bag9* mutants and were normalized in *BAG9* overexpressing plants. Furthermore, the activities of antioxidant enzymes, ascorbic acid (AsA)/dehydroascorbic acid (DHA), and reduced glutathione (GSH)/oxidized glutathione (GSSG) were reduced in *bag9* mutants and were increased in *BAG9* overexpressing plants under heat stress. Additionally, BAG9 interacted with Hsp20 proteins in vitro and in vivo. Accumulation of Hsp proteins induced by heat showed a reduction in *bag9* mutants; meanwhile, it was increased in *BAG9* overexpressing plants. Thus, BAG9 played a crucial role in response to heat stress by regulating cellular redox homeostasis and the stability of heat shock proteins.

Keywords: antioxidants; BAG9; Hsps; *Solanum lycopersicum*; thermotolerance

1. Introduction

Global warming exacerbates the occurrence of extreme weather, among which high temperature is a major environmental threat to crop yields [1]. Under heat stress, the ultrastructure and function of chloroplasts and mitochondria suffer damage, resulting in a burst of reactive oxygen species (ROS), such as singlet oxygen, superoxide anion, hydrogen peroxide, and hydroxyl [2]. The accumulation of ROS leads to the damage of nucleotides, membrane lipid peroxidation, and protein denaturation [3,4]. Furthermore, protein denaturation induced by high temperature results in oxidation, misfolding, and aggregation of proteins. The gathering of these proteins leads to cell death in the absence of chaperones, proteasomes, and autophagy systems [5].

Molecular chaperones help in maintaining protein homeostasis under heat by restoring the native conformation of proteins and preventing the aggregation of non-native proteins for later folding or assembling [6]. Five groups of molecular chaperones heat shock proteins

(Hsps) have been identified, including small heat shock proteins (sHsps)/Hsp20, Hsp60, Hsp70, Hsp90, and Hsp100 [7,8]. They not only protect proteins, but also increase the stability of lipid membranes, membrane proteins such as the photosystems, and soluble proteins [9]. Small Hsps are distinguished from other Hsps since they work in an ATP-independent manner to form a complex with non-native proteins preventing the harmful aggregation of proteins under stress [10]. Overexpressing *OsHsp18.2* in *Arabidopsis* highly enhanced the activity of seeds and the percentage of germination under heat stress [11]. Hsp60 especially improved the thermotolerance of plastid proteins such as Rubisco and retarded cell death [12,13]. Hsp90 interacted with the FK506 binding proteins (FKBPs) regulating thermotolerance [14]. In Hsp100 class, Hsp101 exhibited significant heat resistance and functioned well in recovery from heat shock [15,16]. Co-operation between Hsp100 and heat stress-associated 32-KD protein (HSA32) promoted the effects of heat acclimation in rice seedlings [17].

Among Hsps, Hsp70 regulating mechanism has been widely researched [18]. The work of hsp70 is assisted by a large chaperone system [19–21]. Under cell stress, ATP hydrolysis is indispensable for the binding of Hsp70 to polypeptide chains in non-native protein structures [22]. J-proteins are significant components in the Hsp70 chaperone system, which involve in heat stress response by regulating ATP activity, thus enhancing the binding affinity of Hsp70 with unfolded peptides or other substrates [23]. Nucleotide exchange factors (NEFs) are also necessary co-chaperones in the Hsp70 system [24]. Bcl-2-associated athanogene (BAG) has been identified as a NEF chaperone family, which contains a BAG domain interacting with Hsp70 on its ATPase domain, influencing nucleotide exchange by assisting ATP to bind with Hsp70 and releasing ADP, enhancing protein quality control. The BAG family may establish an association between the Hsp chaperone system and its substrates [25].

As chaperones, the BAG family in plants plays various roles in response to multiple stresses such as heat, freezing, salinity, drought, and ultraviolet (UV) [26,27]. For temperature resistance, *Atbag2* or *Atbag6* mutants survived worse under heat [28]. Upon sensing heat, the processed AtBAG7 entered the nucleus from the endoplasmic reticulum (ER) to interact with WRKY29, initiating unfolded protein response (UPR) pathway to enhance thermotolerance [29,30]. For pathogen resistance, BAG6 activated autophagy by being cleaved by aspartyl protease (APCB1) upon recognizing an intrusive pathogen in *Arabidopsis thaliana* [31]. Similarly in rice, enhanced blight and blast resistance 1 (EBR1) targeted OsBAG4, ubiquitinating and degrading it for immunity regulation and extensive defense against disease [32]. For inhibiting senescence, the signal complex calmodulin-like motif (CaM)/AtBAG5/heat shock cognate 70 (Hsc70) upregulated a high level of Ca^{2+} in mitochondria to inhibit senescence [33]. Likewise in tomato, BAG2 and BAG5b improved the resistance to dark-induced leaf senescence [34]. Various abiotic stresses induced AtBAG4 and regulated ion channels and stomatal motion by interacting with and adjusting KAT1 [35,36].

Tomato is one of the main economic crops in protected cultivation. Heat stress deranges metabolic imbalance in tomato, highly decreasing the quality and production [37]. However, the mechanism of BAGs affecting the thermotolerance of tomato is unclear. To further explore the role of the BAG chaperone family under heat stress and its relationship with Hsps, we generated *BAG9* overexpressing lines and *bag9* mutants and treated them with high temperature. We observed the phenotypes and measured a range of resistance indicators. Results showed that *bag9* was more sensitive to heat stress compared to the wild type (WT), while *BAG9* overexpressing plants showed the opposite tendency. It indicated a positive regulatory effect of BAG9 in temperature tolerance.

2. Materials and Methods

2.1. Phylogenetic Analysis and Structural Domain Prediction of BAG Family

The amino-acid sequences of BAG family proteins in *Solanum lycopersicum*, *Arabidopsis thaliana*, *Oryza sativa*, and *Nicotiana tabacum* were obtained from the Ensembl Plants database (http://plants.ensembl.org (accessed on 4 May 2020)). The set of protein sequences was imported into the Molecular Evolutionary Genetics Analysis tool (MEGA 11) and multiplexed using the ClustalW method and exported in MEGA format. The phylogenetic tree was constructed using the maximum likelihood tree (ML) method and the bootstrap analysis was applied with 1000 replicates/iterations. Finally, the constructed phylogenetic tree was polished with Evolview (http://evolgenius.info (accessed on 5 May 2020)). Structural domains of the BAG family in tomato were analyzed using the native InterProScan program (http://www.ebi.ac.uk/interpro/ (accessed on 5 May 2020)). The structural domain sequences were obtained from the Pfam database and the structural schematics were manufactured using Domain Graph (DOG) software (http://dog.biocuckoo.org/ (accessed on 5 May 2020)).

2.2. Plant Material, Growth Condition, and Heat Treatment

Ailsa Craig (AC) of tomato from Tomato Genetics Resource Center (TGRC) was used as a wild type (WT). Peat and vermiculite were mixed in a suitable ratio (7:3, v:v) for seedling growth. Hoagland nutrient solutions were used twice a week to supplement the tomato with nutrients. The growing conditions of plants were ensured according to the following criteria: photoperiod was performed by 14 h/10 h (day/night), the ambient air temperature was kept by 25 °C/20 °C (day/night), and photosynthetic photon flux density was arranged to 400 μmol m^{-2} s^{-1}. Plants at the five-week seedling stage were used for the following experiments. Two groups of AC, OE-*BAG9*, and *bag9* plants were separated. The control group and the heat stress group were treated for 10 h at 25 °C and 45 °C, respectively, in growth chambers (Qiushi, Hangzhou, China). Except for the temperature in the growth chambers, other environmental parameters remained the same as previously described. Leaf samples were collected at different times from heated or unheated tomato plants, then frozen rapidly in liquid nitrogen and stored at −80 °C before analysis for gene expression, malondialdehyde (MDA), antioxidant, enzyme activity, and immunoblotting. While after being treated for 7 h, leaf samples were collected from the control group and the heat stress group and then immediately analyzed for a maximum quantum yield of PSII (Fv/Fm) and 3,3′-diaminobenzidine (DAB) and nitroblue tetrazolium (NBT) staining.

2.3. Construction of Plant Expression Vector and Tomato Transformation

To generate the *BAG9*-overexpressing lines, *BAG9* full-length coding sequence (CDS) was amplified with the forward primer (5′-gggcgcgccgatatcgtcgacATGGAGAATCTCTTCAATTGG TCC-3′) and reverse primer (5′-aacatcgtatgggtaggtaccGCTGCCGGAAACAATGGAG-3′) using tomato complementary DNA (cDNA) as the template. To insert the PCR product into the pFGC1008-HA vector behind the cauliflower mosaic virus (CaMV) 35S promoter, the product was digested with *Asc*I and *Kpn*I. As described previously, CRISPR/Cas9 vectors were constructed and used to generate *bag9* mutants [38]. Using the CRISPR-P web tool (http://crispr.hzau.edu.cn/ (accessed on 11 September 2020)), the target sequences (5′-GCTCGCCGTCGCTATTCCTC-3′) were achieved and subsequently introduced into the *Bbs*I site of the AtU6-sgRNA-AtUBQ-Cas9 vectors following annealing into the double strands. The fragments of the AtU6-sgRNA-AtUBQ-Cas9 were fused to the *Kpn*I and *Hind*III sites of the pCAMBIA1301 binary vectors. The final vectors were introduced into tomato AC via *A. tumefaciens*-mediated transformation. A homozygous T2 *BAG9* overexpressing line was used for experiments and identified by Western blot using an anti-HA (26183, Thermo Fisher Scientific, Waltham, MA, USA) monoclonal antibody (Figure S1A). *bag9* mutant contained mutations near the protospacer adjacent motif (PAM), which induced mismatched amino-acid sequence and terminated translation (Figure S1B).

2.4. Total RNA Extraction and Gene-Expression Analysis

RNA extraction kits were used for obtaining total RNA (DP419, Tiangen, Beijing, China). The HiScript Q RT SuperMix for the quantitative real-time PCR (+gDNA wiper) Kit (R223, Vazyme, Nanjing, China) was used to produce first-strand cDNA from 500 ng of total RNA. ChamQ Universal SYBR qPCR Master Mix (Q711, Vazyme, Nanjing, China) and Light Cycler® 480 II Real-Time PCR detection system (Roche, Basel, Switzerland) were used in the RT-qPCR. In this program, predenaturation at 95 °C for 3 min, followed by 40 cycles of denaturation at 95 °C for 30 s, annealing at 58 °C for 15 s and 72 °C for 30 s, and a final extension at 72 °C for 30 s. Table S2 listed primers used for RT-qPCR, as well as tomato *Actin* as an internal control. To calculate relative gene expression, the $2^{-\Delta\Delta CT}$ method was used, and a heat-map analysis was conducted using MEV version 4.9 (http://www.mev.tm4.org/ (accessed on 10 June 2020)). At the bottom, the intensity of the color bar showed the intensity of expression.

2.5. Gas Exchange and Chlorophyll Fluorescence Measurements

The infrared gas analyzer-based portable photosynthesis system (LI-6400T, Li-Cor Inc., Lincoln, NE, USA) was applied for measuring the net photosynthetic rate (*Pn*) in plants under heat or controlled environment. The measurements were carried out at 1000 μmol m^{-2} s^{-1} photosynthetic photon flux density (PPFD), 400 μmol mol^{-1} atmospheric carbon dioxide (CO_2) concentrations, and 25 °C leaf temperature, respectively. Fluorescence measurements for chlorophyll were conducted using a MAXI Version of the Imaging-PAM M-Series fluorescence system (Heinz-Walz, Effeltrich, Germany). For 30 min prior to measurement, plants were kept in the dark. According to previous descriptions, the maximum quantum yield of PSII (*Fv/Fm*) was measured and calculated [39].

2.6. Analysis of H_2O_2, $O_2^{\bullet -}$ and Malondialdehyde (MDA)

In order to observe the accumulation of hydrogen peroxide (H_2O_2) and superoxide anion ($O_2^{\bullet -}$) on leaves, the DAB and NBT staining were performed as previously described with minor modifications [40].

For $O_2^{\bullet -}$ staining, leaf samples were stained with 0.5 mg mL^{-1} NBT in 25 mM N-2-hydroxyethylpiperazine-N-ethane-sulphonic acid (HEPES) (pH 7.8) and incubated in the dark under 25 °C for 6 h. For H_2O_2 staining, leaf samples were stained with 1 mg mL^{-1} DAB in 50 mM Tris-HCl (pH 3.8) and incubated at 25 °C for 12 h in the dark. In both cases, leaf samples were washed in 95% (v:v) ethanol for 10 min at 95 °C, kept in lactic acid/phenol/water (1:1:1; v:v:v), and photographed.

The H_2O_2 concentration in the leaves was quantified based on the method described previously with minor modifications [41]. In brief, a 0.3 g leaf sample was taken for analysis. After being ground with 3 mL 0.2 M $HClO_4$ in liquid nitrogen, the material was centrifuged at 6000 g for 10 min at 4 °C. A total of 4 M KOH was used to neutralize the pH to about 6–7. 0.05 g activated carbon was added and the solution was centrifuged at 12,000× *g* for 5 min at 4 °C. The 0.22 μm filter membrane was used to filter the supernatant into a new centrifuge tube to obtain extracting solution. A total of 100 mM potassium acetate buffer (pH 4.4, containing 1 mM ABTS) was used as the reaction buffer. For the nonenzymatic tube reaction system, 1 mL H_2O_2 sample and 1 mL reaction buffer were mixed and the absorption peak at 412 nm was determined. For the enzyme tube reaction system, 1 mL H_2O_2 sample, 996 μL reaction buffer, and 4 μL horseradish peroxidase (POD) were mixed. Finally, the absorption peak at 412 nm was determined to measure the content of H_2O_2. The content of MDA in the leaves was measured according to a previous protocol [39]. Extracted leaves were heated at 95 °C for 25 min with trichloroacetic acid containing 0.65% 2-thiobarbituric acid (TBA). By subtracting the absorbance at 532 nm of a TBA-free solution containing the plant extract, non-MDA compounds were corrected.

2.7. Antioxidant and Enzyme Activity Assays

For nonenzymatic antioxidant assays, approximately 100 mg of leaf sample was powdered in liquid nitrogen and extracted into 1 mL 0.2 M HCl. The solution was centrifugated by 12,000 g for 10 min under 4 °C and then 0.2 M NaOH was used to neutralize the mixed solution to pH 4–5 containing 500 μL supernatant of the last step and 100 μL 0.2 M phosphate buffer (pH 5.6). Finally, spectrophotometric assays were used to measure the extracting solution for ascorbic acid (AsA)/dehydroascorbic acid (DHA), and reduced glutathione (GSH)/oxidized glutathione (GSSG) according to previous methods [42].

To measure antioxidant enzyme activity, 300 mg leaf sample was milled with 3 mL of ice-cold enzyme buffer containing 25 mM HEPES, 0.2 mM ethylene diamine tetraacetic acid (EDTA), 2 mM AsA, and 2% polyvinylpolypyrrolidone (w:v) (pH 7.8). The extracting solution was centrifugated at 12,000× *g* for 10 min under 4 °C and then the supernatants were kept for measurement. Subsequently, SHIMADZU UV-2410PC spectrophotometer (Shimadzu, Kyoto, Japan) was employed to detect enzyme activity. The activities of antioxidant enzymes catalase (CAT), ascorbate peroxidase (APX), glutathione reductase (GR), and dehydroascorbate reductase (DHAR) were analyzed according to the previous protocol with minor modifications [43]. For analyzing CAT activity, 100 μL of enzyme solution, 1700 μL of 25 mM phosphate buffer Solution (PBS) (PH 7.0, containing 0.1 mM EDTA), and 200 μL of 100 mM hydrogen peroxide were mixed. The kinetic changes of OD240 were determined according to the kinetic program, and the enzymatic reaction rate was calculated by taking the kinetic changes of 10 s. For analyzing APX activity, 100 μL of enzyme solution, 1700 μL of 25 mM PBS (pH 7.0, containing 0.1 mM EDTA), 100 μL of 20 mM H_2O_2, and 100 μL of 5 mM AsA were mixed together at 25 °C. The kinetic changes of OD290 were determined according to the kinetic program, and the enzymatic reaction rate was calculated by taking the kinetic changes of 10 s. The reaction rate without H_2O_2 was used as blank control. For analyzing GR activity, 100 μL of enzyme solution, 1700 μL of 25 mM PBS buffer (PH7.8, containing 0.2 mM EDTA), 100 μL of 10 mM GSSG, and 100 μL of 2.4 mM NADPH were mixed together at 25 °C. The kinetic changes of OD340 were measured, and the enzymatic reaction rate was calculated by taking the kinetic changes of 10 s. For analyzing DHAR activity, 100 μL of enzyme solution, 1700 μL of 25 mM PBS (pH 7.0, containing 0.1 mM EDTA), 100 μL of 70 mM GSH, and 100 μL of 8 mM DHA were mixed together. The kinetic changes of OD265 were measured, and the kinetic change of 10 s was taken to calculate the enzymatic reaction rate. The enzyme activities of superoxide dismutase (SOD) and peroxidase (POD) were detected according to the previous protocol with minor modifications [44]. For analyzing SOD activity, 50 μL of enzyme solution and 3 mL reaction solution (containing 50 mM PBS (pH 7.8), 15 mM methionine, 65 mM NBT, 2 μM riboflavin, 0.1 mM EDTA) were mixed. After 15 min illumination at 25 °C, 4000 lx, the absorbance was measured at 560 nm. For analyzing POD activity, 100 μL of enzyme solution, 1700 μL of 25 mM PBS (pH 7.0, containing 0.1 mM EDTA), 100 μL of 10 mM H_2O_2, and 100 μL of 1% guaiacol were mixed together at 25 °C. The kinetic changes of OD470 were determined according to the kinetic program, and the kinetic changes of 10 s were taken to calculate the enzymatic reaction rate.

2.8. Immunoblotting Assay

Following the manufacturer's instructions, the oxidized protein fractions extracted from the soluble protein were tested with an OxyBlot Protein Oxidation Detection Kit (Chemicon International, Temecula, CA, USA).

For immunoblotting assay, the protein extraction and Western blotting assay were modified by protocol described previously [45]. A 0.1 g leaf sample was grinded in liquid nitrogen and added with the extraction buffer (100 mM Tris-HCl, pH 8.0, 10 mM NaCl, 1 mM EDTA, 1% Triton X-100, 1 mM phenylmethylsulphonyl fluoride, and 0.2% β-mercaptoethanol). The Bio-Rad protein assay kit was used to measure the protein concentration and the total protein concentration of all samples were adjusted to 6 μg/μL. After denaturation by 95 °C for 10 min, the protein samples were detected by sodium dodecyl sulfate-polyacrylamide gel electrophoresis (SDS-PAGE) and were subsequently transferred to nitrocellulose membrane (GE Healthcare Biosciences, Piscataway, NJ, USA). Antibodies of cytosolic Hsp90 (AS08 346, Agrisera, Vännäs, Sweden), Hsp70 (PHY0034S, Phytoab, San Jose, CA, USA), Hsp101 (AS07 253, Agrisera, Vännäs, Sweden) and Hsp17.6 (PHY0149S, Phytoab, San Jose, CA, USA) were used to detect proteins. Afterwards, the goat anti-rabbit horseradish peroxidase-linked antibody (7074, Cell Signaling Technology, Boston, MA, USA) was used as the secondary antibody for these analyses.

2.9. Yeast Two-Hybrid (Y2H) Screen and Assays, and Bimolecular Fluorescence Complementation (BiFC) Assay

In order to find out BAG9-interacting proteins in tomato, the coding sequences of *BAG9* were cloned into the pGBKT7 vector using gene-specific promoters (Table S3) and subsequently transferred into the AH109 yeast strain. The cDNA library building and Y2H screening were implemented as the manufacturer's protocol described (Takara, Shiga, Japan). SD-Trp-Leu-Ade-His plates were used for Y2H screening. Hsp20s in tomato were identified as BAG9-interacting proteins from Y2H screens. The coding sequences of Hsp20s were amplified by PCR using specific primers (Table S4) and cloned into a pGADT7 vector. Cotransformed bait-and-prey constructs were plated onto a selection medium lacking Trp, Leu, Ade, and His to analyze interactions. Before this study, pFGC-N-YFP and pFGC-C-YFP had been described for the BiFC vectors [46]. Gene-specific primers were used to amplify the full-length sequences of BAG9 and Hsp20s in PCRs and clone them into pFGC-N-YFP or pFGC-C-YFP vectors (Figure S5). To infiltrate *N. benthamiana*, plasmids were infectively introduced into *A. tumefaciens* GV3101 strains, according to previously described procedures [46]. During 48 h after infiltration, fluorescent signals from infected tissues were analyzed by a Zeiss LSM 780 confocal microscope (Zeiss LSM 780, Oberkochen, Germany) using appropriate filter sets (excitation wavelengths 488 nm and emission between 500 nm and 530 nm).

2.10. Statistical Analysis

Each determination was repeated at least three times independently. Based on the results of independent biological replicates, the data were presented as means ± standard deviations. Analyzing the bioassays was accomplished using SPSS 25 statistics 25 (SPSS Inc., Chicago, IL, USA). In the analysis of treatment differences, Tukey's test was used at 0.05 for significance.

3. Results

3.1. Identification of BAG Homologs in Plants

Previous studies have demonstrated that the BAG protein family was evolutionarily conserved and highly similar in structure and function in eukaryotes [26]. Phylogenetic analysis of the *BAG* gene family across species was significant for understanding the differences in function or predicting similarities between tomato and other species. We identified 10 *BAG* genes in the tomato genome using the SGN database (https://solgenomics.net/ (accessed on 19 April 2020)) and named *BAG1-10* based on homology and evolutionary analysis with the *Arabidopsis* protein sequences (Figure 1A, Table S1). In light of the function of BAGs, we performed a phylogenetic analysis of BAG proteins from three dicot plants, *Arabidopsis*, tomato, and tobacco (*Nicotiana tabacum*), and a monocot plant, rice. Based on the resultant phylogenetic tree, the BAG proteins of the four species were divided into three subfamilies (Figure 1A). BAG5, BAG6, BAG8, and BAG9 belonged to the first group, BAG7 belonged to the second group, and BAG1, BAG2, BAG3, BAG4, and BAG10 belonged to the third group.

Then, we further analyzed the structural domains of the BAG proteins (Figure 1B). Results showed that all BAG proteins contained a conservative BAG domain. Furthermore, BAG1-4 and BAG10 contained extra ubiquitin-like (UBL) structural domains at the N-terminus, while BAG6, BAG8, and BAG9 each comprised an extra CaM-binding motif. In addition, BAG7 protein was distinguished since it had no other kinds of motifs but triple BAG domains. In terms of the length of BAG proteins, BAG5 was the shortest, while BAG6 had the longest sequence length.

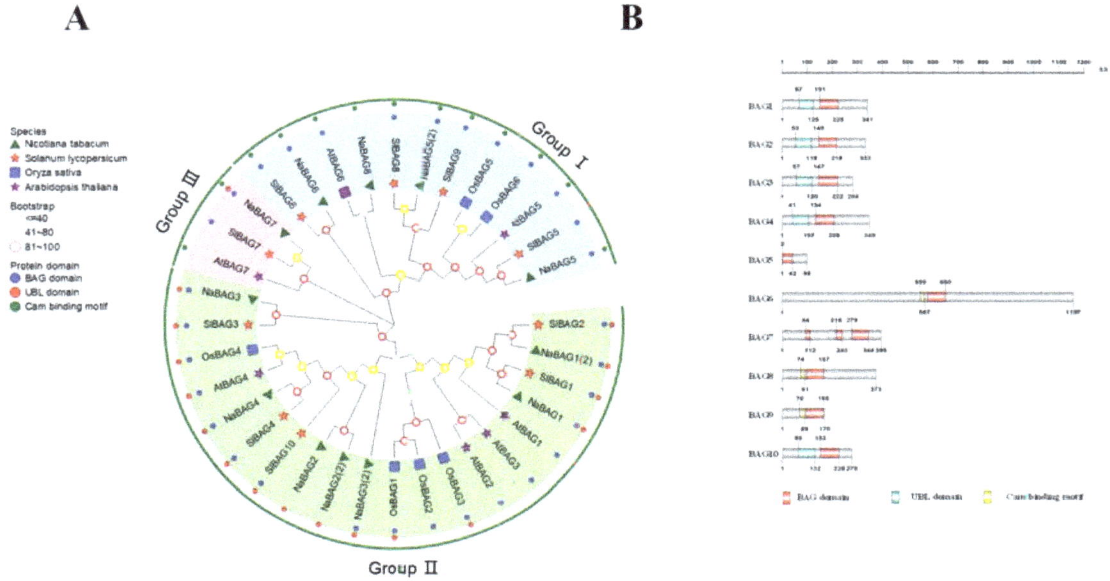

Figure 1. Phylogenetic tree construction of BAGs from different plants and protein structures of BAGs. (**A**) Phylogenetic tree of BAG proteins. The different colored circles on the outside of the protein names represented the types of structural domains possessed by amino-acid sequences (blue for BAG domains, red for UBL domains, and green for calmodulin-binding (CaM) motifs). The symbols on the inside of the protein names represented different species (purple stars for *Arabidopsis thaliana*, green triangles for *Nicotiana tabacum*, blue squares for *Oryza sativa*, and red stars for *Solanum lycopersicum*). (**B**) Schematic diagram of the domains of BAG proteins in tomato. The protein lengths were shown in grey.

3.2. Involvement of BAG9 in Tomato Thermotolerance

Transcript analysis of 10 *SlBAGs* under heat stress was conducted to determine whether heat stress induced BAG gene expression.

Figure 2 displayed that exposure to heat within 1 h can quickly induced *BAG6*, *BAG8*, and *BAG9* and whose transcript levels subsequently reached a maximum after 3 h. Nevertheless, the expression levels of other *BAGs* were not changed or decreased after heat stress (Figure 2). These results suggested that *BAG6*, *BAG8*, and *BAG9* may be important in regulating tomato response to heat stress.

Then, we analyzed the cis-elements in promoters of *BAG* genes and found that only the *BAG9* promoter contained the heat shock element (HSE), which was transcriptionally regulated by heat shock factors under heat stress (Figure S2). To investigate whether BAG9 was involved in the regulation of plant thermotolerance, we generated the *bag9* mutants and *BAG9* overexpressing plants as described in the "Materials and Methods" section (Figure S1). As shown in Figure 3A, the phenotypes of *bag9* mutants and *BAG9* overexpressing plants were similar to WT plants, when they were grown under normal conditions (Figure 3A).

To examine how BAG9 functions in tomato under heat, *bag9* mutants, WT plants, and *BAG9* overexpressing plants grown for about 5 weeks were kept in a 45 °C growth chamber for 10 h. The exposure of tomato plants to heat stress resulted in plant withering and decreased *Fv/Fm* value, more significantly in *bag9* mutants compared with WT plants (Figure 3). In contrast, thermotolerance was significantly increased in *BAG9* overexpressing plants with higher *Fv/Fm* value (Figure 3). Moreover, heat stress inhibited photosynthesis in tomato plants. Net photosynthetic rate (*Pn*) was decreased by 42.4% in *bag9* mutants but was increased by 100.1% in *BAG9* overexpressing plants compared with WT plants (Figure 3D). Additionally, MDA accumulation was aggravated in *bag9* mutants, while alleviated in *BAG9* overexpressing plants compared with WT plants (Figure 3E). Thus, these results suggested that BAG9 played a positive role in tomato response to heat stress.

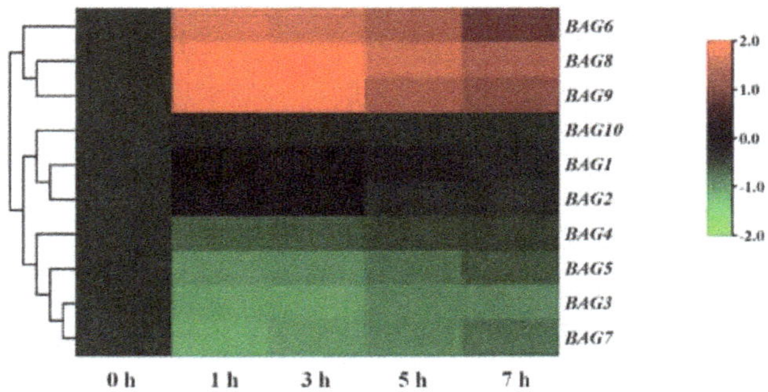

Figure 2. Transcripts of *BAG* genes in response to heat stress. Cluster analysis of expression patterns of *BAGs* under heat stress at 0 h, 1 h, 3 h, 5 h, and 7 h. The heat map was manufactured using $\log_2$ logarithmic-transformed expression values. The color transition from red to green on behalf of high to low expression levels. According to the expression, the *BAGs* were clustered in the figure. The data represented the means ± SD of three biological replicates.

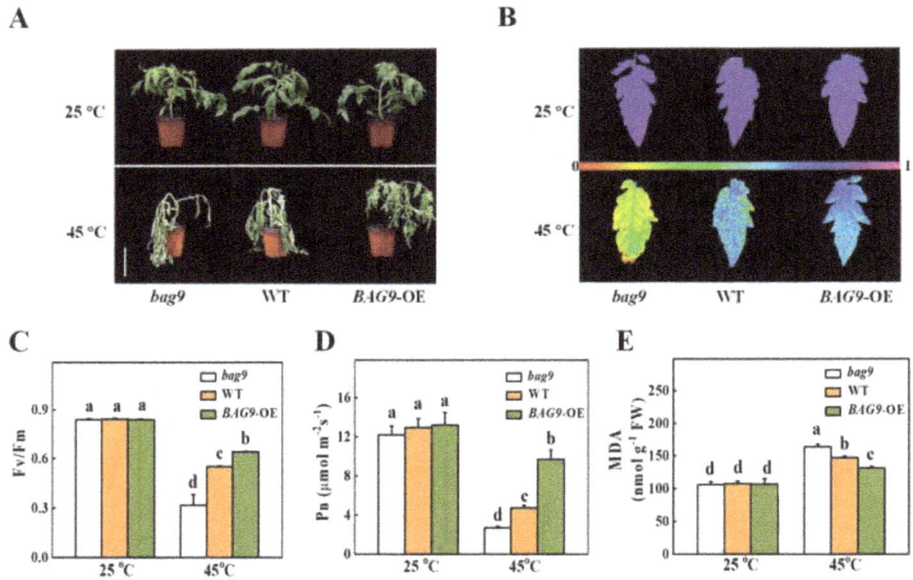

Figure 3. Influence of BAG9 on tomato thermotolerance. (**A**) Representative images of *bag9* mutants, wild type (WT), and *BAG9* overexpressing (*BAG9*-OE) plants without or with heat stress. Bar = 10 cm. The plants were subjected to normal temperature (25 °C) or high temperature (45 °C) treatment for 10 h, photographs of plants were then taken. (**B**,**C**) After undergoing different temperature treatments for 7 h, images of representative leaves showed the maximum photochemical efficiency of photosystem II (*Fv/Fm*). At the bottom, a color gradient showed the strength of the fluorescence signal depicted by each color. (**D**) Net photosynthetic (*Pn*) efficiency at 7 h under heat. (**E**) MDA content at 7 h under heat or without heat stress. Data were the means ± SD of three biological replicates. Different letters represented significant differences ($p < 0.05$) according to Tukey's test.

3.3. BAG9 Alleviates Heat-Stress-Induced ROS Accumulation

ROS production and scavenging keep homeostasis balanced in plants under normal conditions [42]. However, this homeostasis will be disturbed after heat-stress exposure [2]. To verify the effect of BAG9 on heat-induced oxidative stress, we first detected H_2O_2 and $O_2^{\bullet-}$ accumulation. Tomato leaves were stained with DAB dye for histochemical detection of H_2O_2 and with NBT dye for $O_2^{\bullet-}$ detection. As shown in Figure 4, heat stress induced H_2O_2 and $O_2^{\bullet-}$ production in the leaves of WT plants. Interestingly, H_2O_2 and $O_2^{\bullet-}$ production was significantly induced in *bag9* mutants, whereas it was reduced in *BAG9* overexpressing plants (Figure 4A). Similarly, the H_2O_2 content was quantitatively analyzed in support of the observation that H_2O_2 was more accumulated in *bag9* mutants, but significantly reduced in *BAG9* overexpressing plants compared with WT plants (Figure 4B).

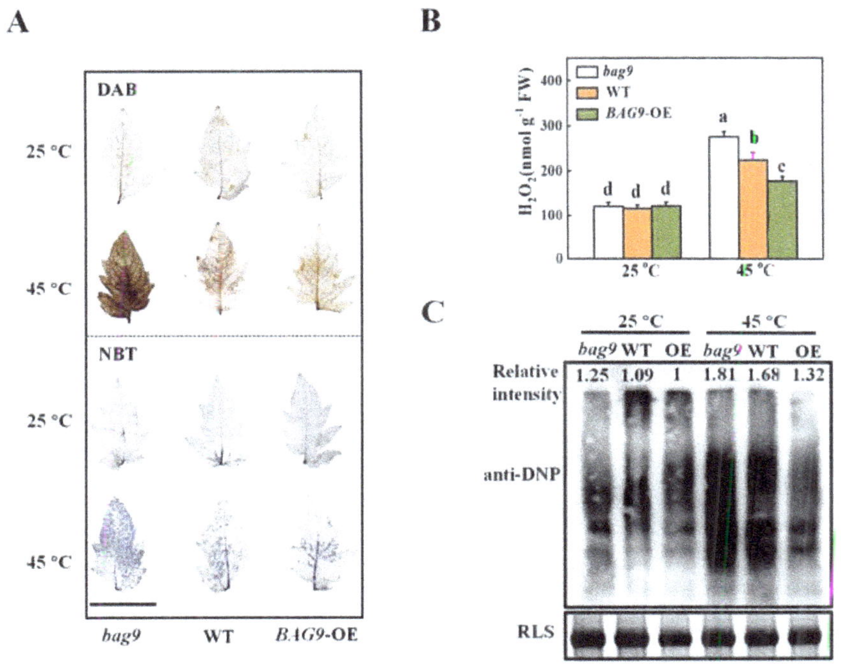

Figure 4. The accumulation of reactive oxygen species (ROS) and oxidative proteins in tomato plants under heat stress. (**A**) Representative images of H_2O_2 and $O_2^{\bullet-}$ accumulation were detected by DAB and NBT staining, respectively. Bar = 5 cm. (**B**) Quantification of H_2O_2 at 7 h under heat. (**C**) Oxidative proteins. An anti-DNP antibody was used to detect total proteins on SDS-PAGE. Coomassie Blue staining (CBB) was applied to indicate the protein input, and on the top of the image was the relative intensity of oxidative proteins. Three independent experiments were performed with similar results. Data were the means ± SD of three biological replicates. Different letters represented significant differences ($p < 0.05$) according to Tukey's test. WT, wild type; *BAG9*-OE, *BAG9* overexpressing plants; RLS, Rubisco large subunit.

To further investigate whether heat-induced oxidative stress caused the oxidation of functional proteins, SDS-PAGE was used to analyze protein oxidation among proteins isolated from total proteins. Figure 4C illustrated that the accumulation of oxidative proteins was similar in *bag9* mutants, WT, and *BAG9* overexpressing plants under normal conditions. Mutants *bag9* and plants overexpressing *BAG9*, however, had increased and decreased levels of oxidative proteins, respectively, compared to wild-type plants. Thus, these results suggested that BAG9 reduced the accumulation of ROS and the oxidation of protein caused by heat.

3.4. BAG9 Enhances Antioxidant Capacity under Heat Stress

Antioxidant defense mechanisms contain antioxidant enzymes such as SOD, APX, GR, CAT, DHAR, POD, and antioxidants such as ASA and GSH to trap and scavenge free radicals and ROS, thereby protecting plant cells and organelles from destruction and increasing stress resistance [47]. As shown in Figure 5, heat stress increased all six antioxidant enzyme activities in WT and *BAG9* overexpressing plants. However, in *bag9* mutants, POD, APX, GR, DHAR, and CAT activities between control and heat treatment showed no significant difference (Figure 5). The enzyme activities in *BAG9* overexpressing tomato were higher than those in WT. According to these results, BAG9 promoted the activities of antioxidant enzymes under heat stress.

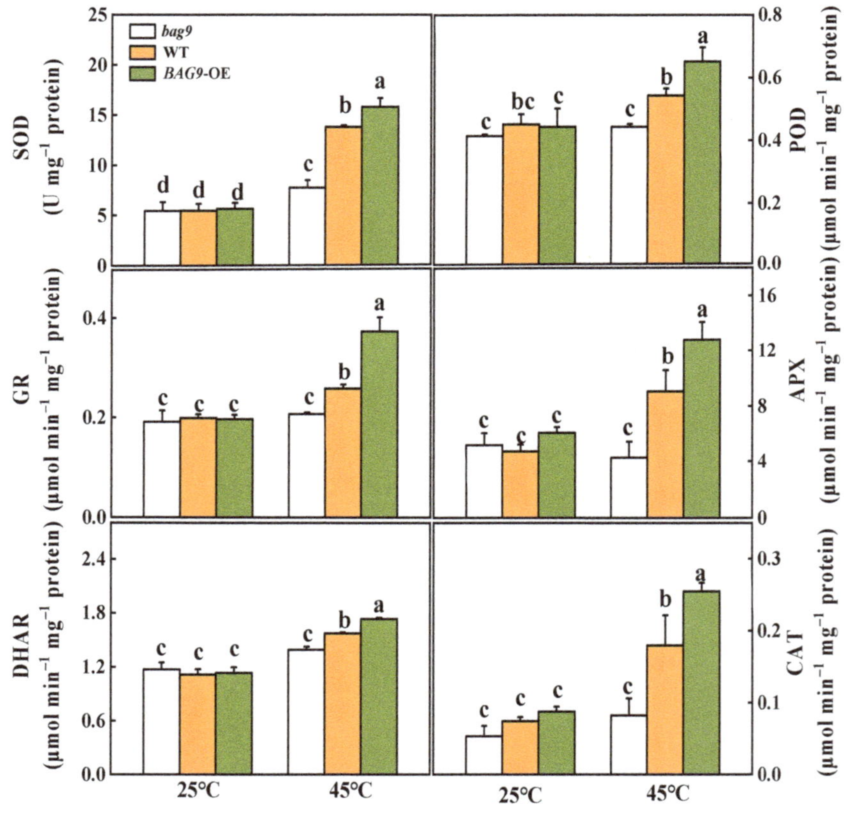

Figure 5. Activities of SOD, POD, APX, GR, CAT, and DHAR with or without heat stress in tomato leaves. Data were the means ± SD of three biological replicates. Different letters represented significant differences ($p < 0.05$) according to Tukey's test. WT, wild type; *BAG9*-OE, *BAG9* overexpressing plants.

To determine whether BAG9-induced thermotolerance was related to the state of cellular redox, the variation of contents and ratios of AsA/DHA and GSH/GSSG were examined (Figure 6). Heat stress had little effect on the AsA and GSH levels but significantly increased the DHA and GSSG contents, leading to significant declines in the AsA/DHA and GSH/GSSG ratios in all plants compared with control. Under heat stress, the DHA and GSSG contents were considerably increased in *bag9* mutants but reduced in *BAG9* overexpressing plants compared with WT plants. Meanwhile, ratios of AsA/DHA and GSH/GSSG were lower in *bag9* mutants but higher in *BAG9* overexpressing plants compared with WT plants (Figure 6).

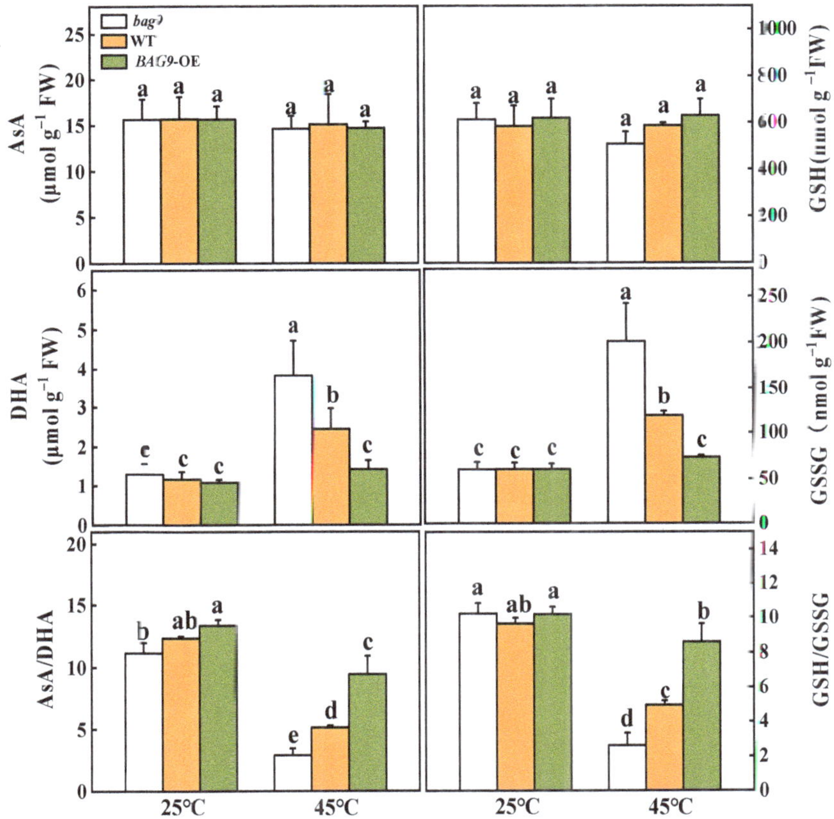

Figure 6. Effects of heat stress on AsA and GSH pools in tomato leaves. Data were the means ± SD of three biological replicates. Different letters represented significant differences ($p < 0.05$) according to Tukey's test. WT, wild type; *BAG9*-OE, *BAG9* overexpressing plants.

3.5. BAG9 Interacts with Hsp20s and Maintains Hsps Stability under Heat Stress

We next identified BAG9-interacting proteins by applying yeast two-hybrid screens. Choosing the fused BAG9 protein as baits, we screened 6×10^6 independent transformants of a tomato cDNA prey library and identified more than twenty clones. The proteins encoded by these positive clones included four Hsp20s (Hsp17.7A, Solyc06g076520; Hsp17.7B, Solyc09g015020; Hsp17.6B, Solyc06g076560; Hsp17.6C, Solyc06g076570). Then, we performed yeast two-hybrid assays to explore whether BAG9 interacted with Hsp20s. By co-transforming the bait and prey vectors, we found that BAG9 interacted with four Hsp20 proteins in yeast (Figure 7A).

To determine whether BAG9 and Hsp20s interact in vivo, we performed a BiFC assay in *A. tumefaciens-infiltrated* tobacco. BAG9 was fused to the C-YFP vector (BAG9-C-YFP) and Hsp20s were fused to the N-YFP vectors (Hsp 17.7A, Solyc06g076520; Hsp17.7B, Solyc09g015020; Hsp17.6B, Solyc06g076560; Hsp17.6C, Solyc06g076570). When the BAG9-C-YFP was co-expressed with four Hsp-N-YFP in tobacco leaves, YFP signals were observed in tobacco cells that had been transformed (Figure 7B). All these experiments revealed that BAG9 interacted with four Hsp proteins.

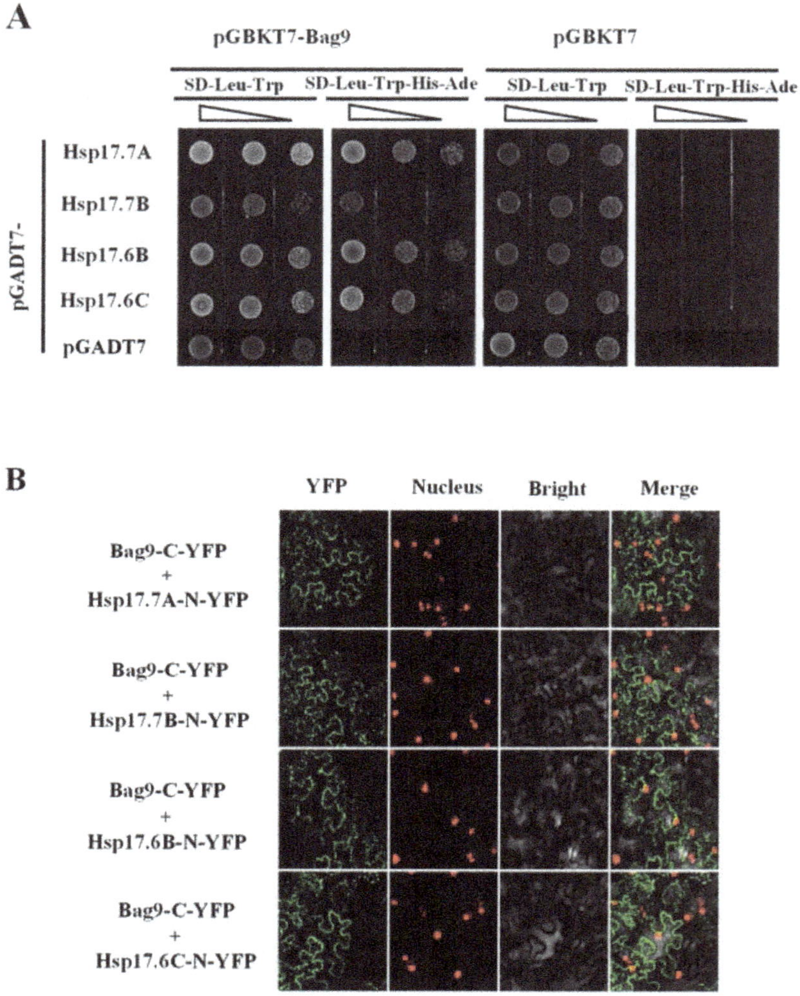

Figure 7. BAG9 interacted with Hsp20s. (**A**) Yeast two-hybrid assay showed interactions between BAG9 and Hsp17.7A, Hsp17.7B, Hsp17.6B, and Hsp17.6C. By growing yeast cells at different concentrations lacking Trp (T), Leu (L), Ade (A), and His (H), the interaction of proteins has been evaluated. (**B**) BiFC analysis showed that the interaction between BAG9 and Hsp20s took place in the cytoplasm. Spliced YFP fusion constructs were transiently coexpressed in *N. benthamiana* leaves for 2 d. The YFP fluorescence signals were obtained by confocal microscopy.

BAG9 and Hsps are both chaperones. To investigate whether BAG9 affects the stability of Hsps under heat stress, we examined the accumulation of Hsps by Western blotting. As shown in Figure 8, there was almost no difference in the accumulation under normal conditions. While heat stress induced a great accumulation of Hsp20, Hsp70, Hsp90, and Hsp101 in all genotypes. However, compared with WT, the accumulation of these four Hsps was still lower in *bag9* mutants, while higher in *BAG9* overexpressing plants (Figure 8). Thus, BAG9 promoted the stability of Hsps under heat stress.

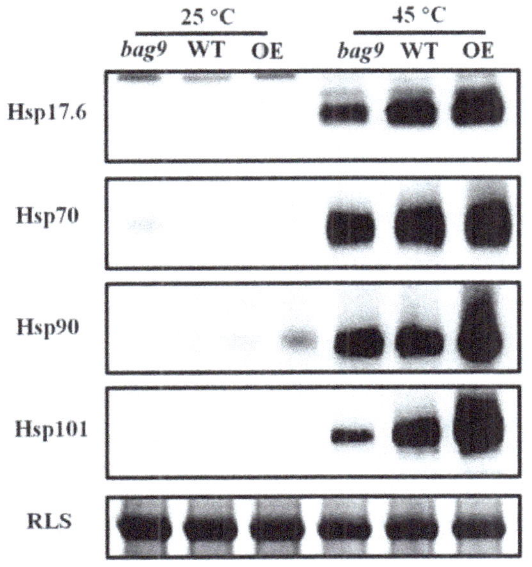

Figure 8. The accumulation of Hsps with or without heat stress in tomato leaves. Hsp17.6, Hsp70, Hsp90, and Hsp101 were detected by immunoblot analysis. After exposing to heat for 7 h, the leaf samples were obtained for experiments. The protein input was indicated by Coomassie Blue staining (CBB). Three independent experiments were performed with similar results. WT, wild type; *BAG9*-OE, *BAG9* overexpressing plants; RLS, Rubisco large subunit.

4. Discussion

In this study, we found that the expression of *BAG9* was highly induced under heat stress in tomato. *Bag9* mutants reduced thermotolerance while overexpressing *BAG9* increased thermotolerance as reflected by antioxidant assays. We also found that BAG9 interacted with Hsp20 proteins in vitro and in vivo. Overexpressing *BAG9* enhanced the accumulation of Hsp proteins induced by heat, while the mutants had the opposite tendency. Thus, BAG9 played a crucial role in response to heat stress by regulating cellular redox homeostasis and the stability of heat shock proteins.

Similar to our study, the transcript levels of *OsBAGs* and *BAG* family members in grapes were significantly increased under heat exposure [48,49]. Considering that *BAG9* contained the HSE in the promoter region, it was selected to conduct further research for its potential significance in thermotolerance. BAG9 contained a conserved BAG domain and a CaM binding motif. The BAG domain combined with Hsc70 for decomposing incorrectly folded or translocated chloroplast proteins in *Arabidopsis* [50]. The phylogenetic analysis revealed that BAG9 was most close to OsBAG5, OsBAG6, AtBAG5, and AtBAG6. According to previous research and evolutionary relationships, we speculated that BAG9 may function in temperature protection, especially heat stress by binding with Hsps and maintaining cellular stability or involving in the Ca^{2+} sensing [28,48].

Various kinds of BAG proteins functioning in plant thermotolerance have been identified [28,29]. Heat shock-induced gene 1 (HSG1), a grape Bcl-2-associated athanogene, enhanced heat tolerance and activated *CONSTANS* (CO) expression in transgenic *Arabidopsis* plants [51]. In *Arabidopsis*, heat shock transcription factor (HsfA2) directly bound to HSE motif of *AtBAG6*, which dramatically increased its relative expression under heat stress [52]. AtBAG2 enhanced survival under heat by clearing ROS in plants [28]. AtBAG7 played a key role in mediating the heat-induced UPR pathway [29]. Studies in the BAG family showed that BAG9 stimulated burning symptoms under heat and reduced the thermotolerance of tomato, which did not occur in our experiment [53]. By overexpressing *BAG9* in *Arabidopsis*, the sensitivity to water scarcity, salinity, as well as ABA during the germination of seeds and the growth of seedlings were increased [54]. *BAG5b* (Solyc10g084170, namely *BAG9*) in leaves was activated by various adversity stimuli (extreme temperatures, salinity, and UV light) as well as treatment with phytohormones. Specifically, it improved the resistance to dark-induced leaf senescence by eliminating ROS and downgrading genes associated with leaf senescence [34].

In this study, *Pn*, a typical indicator of photosystem I (PSI), was decreased in *bag9* mutants but was highly increased in *BAG9* overexpressing plants compared with WT plants. Similarly, *BAG9* overexpressing plants showed higher *Fv/Fm* values, and the mutants showed compromised *Fv/Fm* values than WT plants. Our results indicated that BAG9 promoted the stability of photosynthesis under heat exposure. Photosynthesis is a thermosensitive physiological process since the photochemical reactions and the carbon metabolism are susceptible to damage under heat exposure [55]. The disruption of the thylakoid membranes inhibits the rate of photosynthesis and PSII activity is also greatly reduced or even stopped under heat stress [56]. Chaperones protect and enhance photosynthesis under stressful environments [56]. The thermal resistance of photosystem II is upregulated by constitutive overexpression of a small Hsp, which suggests that sHsps prevent the damaging of photosynthetic apparatus from high temperature [57]. Hsp90 in the chloroplast was also an irreplaceable chaperone for protein translocating from the membrane into the organelles and served a significant role in heat resistance in photosynthetic organisms [58,59]. Similar to previous research, BAG9 served as a chaperone protein that may protect photosynthesis as shown in this study.

In previous studies, ROS is used as an indicator of plant resistance [60]. Overexpressing *AtBAG4* into the rice and exposing it to osmotic stress revealed that ROS accumulation was significantly reduced in its overexpressing plants [61]. The mutants *Atbag2* and *Atbag6* also showed higher ROS levels and less survival after heat treatment than WT [28]. Similar to the previous study, our results showed that *BAG9* overexpressing plants accumulated less ROS (H_2O_2, $O^{2\bullet-}$) and less protein carbonylation (which is a hallmark of protein oxidation), indicating a better resistance to high temperature. MDA is one of the products of ROS-induced membrane damage, whose amount represents the degree of cell membrane lipid peroxidation [60]. The continuous accumulation of MDA is positively correlated to high temperatures [62]. This study discovered that *BAG9* overexpressing plants showed less accumulation of MDA than WT, which indicated that BAG9 may protect biomembrane from being damaged under heat stress.

To mitigate elevated ROS-induced damage, plants have established a well-organized antioxidant-defense mechanism [62]. Antioxidants in plants have been classified into two main types: enzymatic and nonenzymatic antioxidants. The significant antioxidant enzymes in plant cells contain SOD, CAT, POD, and so on [63]. GSH and AsA are vital nonenzymatic antioxidants in plants. Meanwhile, APX, DHAR, and GR serve as significant enzymes in the AsA-GSH cycle [64]. Antioxidants are involved in multiple plant abiotic stresses, including heat stress [63]. Treating seedlings of *Broussonetia papyrifera* at high temperature, the activities of SOD, POD, and CAT were significantly increased [65]. The antioxidant enzyme activities in *Cruciferae* were closely related to high temperature, since its SOD, CAT, and GR activities under high-temperature (32 °C) stress were all higher than those of the control plants (20 °C) [66]. In *Brassica napus*, the developed activities of MDAR, DHAR, and GR under sub-high-temperature treatment (30 °C) elevated the levels of AsA and GSH, resulting in enhanced thermotolerance [67]. Similarly, our results illustrated that *BAG9* overexpressing plants upregulated the activities of antioxidant enzymes (SOD, CAT, POD, APX, DHAR, GR) and ratios of AsA/DHA and GSH/GSSG. All results indicated that the higher thermotolerance in *BAG9* overexpressing plants was probably achieved by enhanced activities of various antioxidants.

Hsps exist widely in plants to prevent stress from inducing damage to cells [68]. Previous studies showed that Hsp70 functioned in a chaperone cycle by Hsp70 chaperone systems [20]. BAG family is a kind of NEF that establishes direct interactions with the

ATPase domain of Hsp70 [25]. In tomato, results showed that BAG1 and BAG2 interacted with Hsp70 protein [69].

However, there have been no other Hsp–BAG interactions reported. In this study, we discovered that BAG9 interacted with Hsp20s (Hsp17.7A, Hsp17.7B, Hsp17.6B, Hsp17.6C) in the cytoplasm. Hsp20 is the predominant and most abundant class of proteins in many species induced by heat stress [70]. High temperature significantly induced the upregulation of *TaHsp17.4*, *TaHsp17.7A*, *TaHsp19.1*, and *TaHsp23.7* in wheat [71]. *OsHsp20* overexpressing plants had longer root length and higher germination rates than the control under heat and showed better resistance to high temperature [70]. Nonetheless, how BAG9 works under heat stress by interacting with Hsp20s requires further study.

Our results also witnessed the increase in the accumulation of Hsps (Hsp20, Hsp70, Hsp90, Hsp101) in *BAG9* overexpressing plants, indicating that BAG9 stimulated Hsps for enhancing thermotolerance. Hsp90 bound with Hsp70, establishing multiple complexes of chaperones and functioning well in sense signaling [72]. Hsp101, the most functional member in Hsp100, not only increased heat tolerance but also helped in recovery from heat shock [15]. However, the co-operations of BAG9, Hsp90, and Hsp101 need to be studied further.

5. Conclusions

In conclusion, we identified that BAG9 was involved in tomato thermotolerance. *BAG9* was highly induced under high temperature. *bag9* mutants were sensitive, while *BAG9* overexpressing plants were resistant under heat stress compared with WT. By analyzing the antioxidant and photosynthetic systems, we found that overexpressing *BAG9* may help in the removal of ROS and protect photosynthesis under heat stress. BAG9 interacted with Hsp20 proteins and protected Hsps accumulation under heat stress. In a word, BAG9 was probably significant for thermotolerance by regulating cellular redox homeostasis and the stability of heat shock proteins. Our findings further illustrated the functions of *BAGs* in adversity modulation, especially temperature stress.

Supplementary Materials: The following supporting information can be downloaded at: https://www.mdpi.com/article/10.3390/antiox11081467/s1, Figure S1: Identification of *BAG9*-OE plants and *bag9* mutants; Figure S2: Analysis of *cis*-acting elements in the promoter region of *BAGs*; Table S1: Gene information of BAG family in tomato; Table S2: Primer sequences designed for RT-qPCR assays; Table S3: Primer sequences designed for pGBKT7 vectors construction; Table S4: Primer sequences designed for pGADT7 vectors construction; Table S5: Primer sequences designed for p2YC and p2YN vectors construction.

Author Contributions: J.Z. designed the research; H.H., C.L., C.Y., S.S. and Z.Q. performed the experiments; H.H. and J.Z. analyzed the data; J.Z., H.H., M.K.K. and C.L. wrote the manuscript. All authors have read and agreed to the published version of the manuscript.

Funding: This work was supported by the National Key Research and Development Program of China (2019YFD1000300), the National Natural Science Foundation of China (31922078 and 31872089), the Public Projects of Zhejiang Province (LGN20C150011), the Fundamental Research Funds for the Central Universities (226-2022-00122), and the Starry Night Science Fund of Zhejiang University Shanghai Institute for Advanced Study (SN-ZJU-SIAS-0011).

Institutional Review Board Statement: Not applicable.

Informed Consent Statement: Not applicable.

Data Availability Statement: Data is contained within the article and Supplementary Materials.

Conflicts of Interest: The authors report no conflict of interest.

References

1. Mittler, R.; Finka, A.; Goloubinoff, P. How Do Plants Feel the Heat? *Trends Biochem. Sci.* **2012**, *37*, 118–125. [CrossRef] [PubMed]
2. Shafi, A.; Hassan, F.; Khanday, F.A. Reactive Oxygen and Nitrogen Species: Oxidative Damage and Antioxidative Defense Mechanism in Plants under Abiotic Stress. *Plant Abiotic Stress Physiol.* **2022**, *1*, 71.

3. Narayanan, S.; Tamura, P.J.; Roth, M.; Prasad, P.V.V.; Welti, R. Wheat Leaf Lipids during Heat Stress: I. High Day and Night Temperatures Result in Major Lipid Alterations. *Plant Cell Environ.* **2016**, *39*, 787–803. [CrossRef] [PubMed]
4. Sharkey, T.D.; Zhang, R. High Temperature Effects on Electron and Proton Circuits of Photosynthesis. *J. Integr. Plant Biol.* **2010**, *52*, 712–722. [CrossRef] [PubMed]
5. Zhou, J.; Wang, J.; Yu, J.; Chen, Z. Role and Regulation of Autophagy in Heat Stress Responses of Tomato Plants. *Front. Plant Sci.* **2014**, *5*, 174. [CrossRef]
6. Nakamoto, H.; Akter, T. Molecular Chaperones and Acquisition of Thermotolerance in Plants. *Handb. Plant Crop Stress* **2020**, *18*, 343–359.
7. Gupta, S.C.; Sharma, A.; Mishra, M.; Mishra, R.K.; Chowdhuri, D.K. Heat Shock Proteins in Toxicology: How Close and How Far? *Life Sci.* **2010**, *86*, 377–384. [CrossRef]
8. Hartl, F.U.; Bracher, A.; Hayer-Hartl, M. Molecular Chaperones in Protein Folding and Proteostasis. *Nature* **2011**, *475*, 324–332. [CrossRef]
9. Wu, J.; Gao, T.; Hu, J.; Zhao, L.; Yu, C.; Ma, F. Research Advances in Function and Regulation Mechanisms of Plant Small Heat Shock Proteins (SHSPs) under Environmental Stresses. *Sci. Total Environ.* **2022**, *825*, 154054. [CrossRef]
10. Haslbeck, M.; Vierling, E. A First Line of Stress Defense: Small Heat Shock Proteins and Their Function in Protein Homeostasis. *J. Mol. Biol.* **2015**, *427*, 1537–1548. [CrossRef]
11. Kaur, H.; Petla, B.P.; Kamble, N.U.; Singh, A.; Rao, V.; Salvi, P.; Ghosh, S.; Majee, M. Differentially Expressed Seed Aging Responsive Heat Shock Protein OsHSP18.2 Implicates in Seed Vigor, Longevity and Improves Germination and Seedling Establishment under Abiotic Stress. *Front. Plant Sci.* **2015**, *6*, 713. [CrossRef]
12. Wang, W.; Vinocur, B.; Shoseyov, O.; Altman, A. Role of Plant Heat-Shock Proteins and Molecular Chaperones in the Abiotic Stress Response. *Trends Plant Sci.* **2004**, *9*, 244–252. [CrossRef]
13. Ishikawa, A.; Tanaka, H.; Nakai, M.; Asahi, T. Deletion of a Chaperonin 60β Gene Leads to Cell Death in the *Arabidopsis Lesion Initiation 1* Mutant. *Plant Cell Physiol.* **2003**, *44*, 255–261. [CrossRef] [PubMed]
14. Breiman, A. Plant Hsp90 and Its Co-Chaperones. *Curr. Protein Pept. Sci.* **2014**, *15*, 232–244. [CrossRef]
15. McLoughlin, F.; Basha, E.; Fowler, M.E.; Kim, M.; Bordowitz, J.; Katiyar-Agarwal, S.; Vierling, E. Class I and II Small Heat Shock Proteins Together with HSP101 Protect Protein Translation Factors during Heat Stress. *Plant Physiol.* **2016**, *172*, 536. [CrossRef]
16. Merret, R.; Carpentier, M.-C.; Favory, J.-J.; Picart, C.; Descombin, J.; Bousquet-Antonelli, C.; Tillard, P.; Lejay, L.; Deragon, J.-M.; Charng, Y. Heat Shock Protein HSP101 Affects the Release of Ribosomal Protein mRNAs for Recovery after Heat Shock. *Plant Physiol.* **2017**, *174*, 1216–1225. [CrossRef] [PubMed]
17. Lin, M.; Chai, K.; Ko, S.; Kuang, L.; Lur, H.; Charng, Y. A Positive Feedback Loop between HSP101 and HSA32 Modulates Long-Term Acquired Thermotolerance Illustrating Diverse Heat Stress Responses in Rice Varieties. *Plant Physiol.* **2014**, *164*, 2045–2053. [CrossRef] [PubMed]
18. Colvin, T.A.; Gabai, V.L.; Gong, J.; Calderwood, S.K.; Li, H.; Gummuluru, S.; Matchuk, O.N.; Smirnova, S.G.; Orlova, N.V.; Zamulaeva, I.A.; et al. Hsp70-Bag3 Interactions Regulate Cancer-Related Signaling Networks. *Cancer Res.* **2014**, *74*, 4731–4740. [CrossRef]
19. Genevaux, P.; Georgopoulos, C.; Kelley, W.L. The Hsp70 Chaperone Machines of *Escherichia Coli*: A Paradigm for the Repartition of Chaperone Functions. *Mol. Microbiol.* **2007**, *66*, 840–857. [CrossRef] [PubMed]
20. Mayer, M.P. Gymnastics of Molecular Chaperones. *Mol. Cell* **2010**, *39*, 321–331. [CrossRef]
21. Mayer, M.P.; Bukau, B. Hsp70 Chaperones: Cellular Functions and Molecular Mechanism. *Cell. Mol. Life Sci.* **2005**, *62*, 670–684. [CrossRef] [PubMed]
22. Song, J.; Takeda, M.; Morimoto, R.I. Bag1–Hsp70 Mediates a Physiological Stress Signalling Pathway That Regulates Raf-1/ERK and Cell Growth. *Nat. Cell Biol.* **2001**, *3*, 276–282. [CrossRef] [PubMed]
23. Wang, G.; Kong, F.; Zhang, S.; Meng, X.; Wang, Y.; Meng, Q. A Tomato Chloroplast-Targeted DnaJ Protein Protects Rubisco Activity under Heat Stress. *J. Exp. Bot.* **2015**, *66*, 3027–3040. [CrossRef]
24. Bracher, A.; Verghese, J. The Nucleotide Exchange Factors of Hsp70 Molecular Chaperones. *Front. Mol. Biosci.* **2015**, *2*, 10. [CrossRef] [PubMed]
25. Brive, L.; Takayama, S.; Briknarová, K.; Homma, S.; Ishida, S.K.; Reed, J.C.; Ely, K.R. The Carboxyl-Terminal Lobe of Hsc70 ATPase Domain Is Sufficient for Binding to BAG1. *Biochem. Biophys. Res. Commun.* **2001**, *289*, 1099–1105. [CrossRef] [PubMed]
26. Kabbage, M.; Dickman, M.B. The BAG Proteins: A Ubiquitous Family of Chaperone Regulators. *Cell. Mol. Life Sci.* **2008**, *65*, 1390–1402. [CrossRef] [PubMed]
27. Takayama, S.; Reed, J.C. Molecular Chaperone Targeting and Regulation by BAG Family Proteins. *Nat. Cell Biol.* **2001**, *3*, E237–E241. [CrossRef] [PubMed]
28. Arif, M.; Li, Z.; Luo, Q.; Li, L.; Shen, Y.; Men, S. The *BAG2* and *BAG6* Genes Are Involved in Multiple Abiotic Stress Tolerances in *Arabidopsis thaliana*. *Int. J. Mol. Sci.* **2021**, *22*, 5856. [CrossRef]
29. Li, Y.; Williams, B.; Dickman, M. Arabidopsis B-Cell Lymphoma2 (Bcl-2)-Associated Athanogene 7 (BAG7)-Mediated Heat Tolerance Requires Translocation, Sumoylation and Binding to WRKY29. *New Phytol.* **2017**, *214*, 695–705. [CrossRef] [PubMed]
30. Williams, B.; Kabbage, M.; Britt, R.; Dickman, M.B. *AtBAG7*, an *Arabidopsis* Bcl-2-Associated Athanogene, Resides in the Endoplasmic Reticulum and Is Involved in the Unfolded Protein Response. *Proc. Natl. Acad. Sci. USA* **2010**, *107*, 6088–6093. [CrossRef]

31. Li, Y.; Kabbage, M.; Liu, W.; Dickman, M. Aspartyl Protease-Mediated Cleavage of BAG6 Is Necessary for Autophagy and Fungal Resistance in Plants. *Plant Cell* **2016**, *28*, 233–247. [CrossRef] [PubMed]
32. You, Q.; Zhai, K.; Yang, D.; Yang, W.; Wu, J.; Liu, J.; Pan, W.; Wang, J.; Zhu, X.; Jian, Y.; et al. An E3 Ubiquitin Ligase-BAG Protein Module Controls Plant Innate Immunity and Broad-Spectrum Disease Resistance. *Cell Host Microbe* **2016**, *20*, 758–769. [CrossRef] [PubMed]
33. Fu, S.; Li, L.; Kang, H.; Yang, X.; Men, S.; Shen, Y. Chronic Mitochondrial Calcium Elevation Suppresses Leaf Senescence. *Biochem. Biophys. Res. Commun.* **2017**, *487*, 672–677. [CrossRef]
34. He, M.; Wang, Y.; Jahan, M.S.; Liu, W.; Raziq, A.; Sun, J.; Shu, S.; Guo, S. Characterization of *SlBAG* Genes from *Solanum Lycopersicum* and Its Function in Response to Dark-Induced Leaf Senescence. *Plants* **2021**, *10*, 947. [CrossRef]
35. Doukhanina, E.V.; Chen, S.; van der Zalm, E.; Godzik, A.; Reed, J.; Dickman, M.B. Identification and Functional Characterization of the BAG Protein Family in *Arabidopsis thaliana*. *J. Biol. Chem.* **2006**, *281*, 18793–18801. [CrossRef] [PubMed]
36. Locascio, A.; Marqués, M.C.; García-Martínez, G.; Corratgé-Faillie, C.; Andrés-Colás, N.; Rubio, L.; Fernández, J.A.; Véry, A.-A.; Mulet, J.M.; Yenush, L. BCL2-Associated Athanogene4 Regulates the KAT1 Potassium Channel and Controls Stomatal Movement. *Plant Physiol.* **2019**, *181*, 1277–1294. [CrossRef] [PubMed]
37. McClung, C.R.; Davis, S.J. Ambient Thermometers in Plants: From Physiological Outputs towards Mechanisms of Thermal Sensing. *Curr. Biol.* **2010**, *20*, R1086–R1092. [CrossRef]
38. Chi, C.; Xu, X.; Wang, M.; Zhang, H.; Fang, P.; Zhou, J.; Xia, X.; Shi, K.; Zhou, Y.; Yu, J. Strigolactones Positively Regulate Abscisic Acid-Dependent Heat and Cold Tolerance in Tomato. *Hortic. Res.* **2021**, *8*, 237. [CrossRef]
39. Zhou, J.; Wang, J.; Shi, K.; Xia, X.; Zhou, Y.H.; Yu, J. Hydrogen Peroxide Is Involved in the Cold Acclimation-Induced Chilling Tolerance of Tomato Plants. *Plant Physiol. Biochem.* **2012**, *60*, 141–149. [CrossRef]
40. Hu, Z.; Ma, Q.; Foyer, C.H.; Lei, C.; Choi, H.W.; Zheng, C.; Li, J.; Zuo, J.; Mao, Z.; Mei, Y.; et al. High CO_2- and Pathogen-driven Expression of the Carbonic Anhydrase βCA3 Confers Basal Immunity in Tomato. *New Phytol.* **2021**, *229*, 2827–2843. [CrossRef]
41. Zhou, J.; Wang, J.; Li, X.; Xia, X.; Zhou, Y.; Shi, K.; Chen, Z.; Yu, J. H_2O_2 Mediates the Crosstalk of Brassinosteroid and Abscisic Acid in Tomato Responses to Heat and Oxidative Stresses. *J. Exp. Bot.* **2014**, *65*, 4371–4383. [CrossRef] [PubMed]
42. Noctor, G.; Mhamdi, A.; Foyer, C.H. Oxidative Stress and Antioxidative Systems: Recipes for Successful Data Collection and Interpretation: Methods in Oxidative Stress Research. *Plant Cell Environ.* **2016**, *39*, 1140–1160. [CrossRef] [PubMed]
43. Xia, X.-J.; Wang, Y.-J.; Zhou, Y.-H.; Tao, Y.; Mao, W.-H.; Shi, K.; Asami, T.; Chen, Z.; Yu, J.-Q. Reactive Oxygen Species Are Involved in Brassinosteroid- Induced Stress Tolerance in Cucumber. *Plant Physiol.* **2009**, *150*, 801–814. [CrossRef] [PubMed]
44. Cheng, F.; Yin, L.-L.; Zhou, J.; Xia, X.-J.; Shi, K.; Yu, J.-Q.; Zhou, Y.-H.; Foyer, C.H. Interactions between 2-Cys Peroxiredoxins and Ascorbate in Autophagosome Formation during the Heat Stress Response in *Solanum lycopersicum*. *J. Exp. Bot.* **2016**, *67*, 1919–1933. [CrossRef]
45. Liu, C.-C.; Chi, C.; Jin, L.-J.; Zhu, J.; Yu, J.-Q.; Zhou, Y.-H. The bZip Transcription Factor HY5 Mediates *CRY1a*-Induced Anthocyanin Biosynthesis in Tomato. *Plant Cell Environ.* **2018**, *41*, 1762–1775. [CrossRef] [PubMed]
46. Zhou, J.; Wang, Z.; Wang, X.; Li, X.; Zhang, Z.; Fan, B.; Zhu, C.; Chen, Z. Dicot-Specific ATG8-Interacting ATI3 Proteins Interact with Conserved UBAC2 Proteins and Play Critical Roles in Plant Stress Responses. *Autophagy* **2018**, *14*, 487–504. [CrossRef] [PubMed]
47. Devireddy, A.R.; Tschaplinski, T.J.; Tuskan, G.A.; Muchero, W.; Chen, J.-G. Role of Reactive Oxygen Species and Hormones in Plant Responses to Temperature Changes. *Int. J. Mol. Sci.* **2021**, *22*, 8843. [CrossRef]
48. Rana, R.M.; Dong, S.; Ali, Z.; Khan, A.I.; Zhang, H.S. Identification and Characterization of the Bcl-2- Associated Athanogene (BAG) Protein Family in Rice. *Afr. J. Biotechnol.* **2012**, *11*, 88–98. [CrossRef]
49. Hu, L.; Chen, J.; Guo, J.; Zhao, Y.; Zhu, L.; Huang, Y. Functional Divergence and Evolutionary Dynamics of *BAG* Gene Family in Maize (*Zea Mays*). *Int. J. Agric. Biol.* **2013**, *15*, 200–206.
50. Lee, D.; Kim, S.; Oh, Y.; Choi, B.; Lee, J.; Hwang, I. *Arabidopsis* BAG1 Functions as a Cofactor in Hsc70-Mediated Proteasomal Degradation of Unimported Plastid Proteins. *Mol. Plant* **2016**, *9*, 1428–1431. [CrossRef]
51. Kobayashi, M.; Takato, H.; Fujita, K.; Suzuki, S. HSG1, a Grape Bcl-2-Associated Athanogene, Promotes Floral Transition by Activating *CONSTANS* Expression in Transgenic *Arabidopsis* Plant. *Mol. Biol. Rep.* **2012**, *39*, 4367–4374. [CrossRef] [PubMed]
52. Nishizawa-Yokoi, A.; Yoshida, E.; Yabuta, Y.; Shigeoka, S. Analysis of the Regulation of Target Genes by an *Arabidopsis* Heat Shock Transcription Factor, HsfA2. *Biosci. Biotechnol. Biochem.* **2009**, *73*, 890–895. [CrossRef] [PubMed]
53. Ding, H.; Qian, L.; Jiang, H.; Ji, Y.; Fang, Y.; Sheng, J.; Xu, X.; Ge, C. Overexpression of a Bcl-2-Associated Athanogene *SlBAG9* Negatively Regulates High-Temperature Response in Tomato. *Int. J. Biol. Macromol.* **2022**, *194*, 695–705. [CrossRef] [PubMed]
54. Jiang, H.; Ji, Y.; Sheng, J.; Wang, Y.; Liu, X.; Xiao, P.; Ding, H. Genome-Wide Identification of the Bcl-2 Associated Athanogene (BAG) Gene Family in *Solanum Lycopersicum* and the Functional Role of *SlBAG9* in Response to Osmotic Stress. *Antioxidants* **2022**, *11*, 598. [CrossRef] [PubMed]
55. Ashraf, M.; Harris, P.J.C. Photosynthesis under Stressful Environments: An Overview. *Photosynthetica* **2013**, *51*, 163–190. [CrossRef]
56. Hasanuzzaman, M.; Nahar, K.; Alam, M.; Roychowdhury, R.; Fujita, M. Physiological, Biochemical, and Molecular Mechanisms of Heat Stress Tolerance in Plants. *Int. J. Mol. Sci.* **2013**, *14*, 9643–9684. [CrossRef]
57. Nakamoto, H.; Suzuki, N.; Roy, S.K. Constitutive Expression of a Small Heat-Shock Protein Confers Cellular Thermotolerance and Thermal Protection to the Photosynthetic Apparatus in Cyanobacteria. *FEBS Lett.* **2000**, *483*, 169–174. [CrossRef]

58. Inoue, H.; Li, M.; Schnell, D.J. An Essential Role for Chloroplast Heat Shock Protein 90 (Hsp90C) in Protein Import into Chloroplasts. *Proc. Natl. Acad. Sci. USA* **2013**, *110*, 3173–3178. [CrossRef]
59. Tanaka, N.; Nakamoto, H. HtpG Is Essential for the Thermal Stress Management in Cyanobacteria. *FEBS Lett.* **1999**, *458*, 117–123. [CrossRef]
60. Hu, Z.; Li, J.; Ding, S.; Cheng, F.; Li, X.; Jiang, Y.; Yu, J.; Foyer, C.H.; Shi, K. The Protein Kinase CPK28 Phosphorylates Ascorbate Peroxidase and Enhances Thermotolerance in Tomato. *Plant Physiol.* **2021**, *186*, 1302–1317. [CrossRef]
61. Hoang, T.M.L.; Moghaddam, L.; Williams, B.; Khanna, H.; Dale, J.; Mundree, S.G. Development of Salinity Tolerance in Rice by Constitutive-Overexpression of Genes Involved in the Regulation of Programmed Cell Death. *Front. Plant Sci.* **2015**, *6*, 175. [CrossRef] [PubMed]
62. Suzuki, N.; Koussevitzky, S.; Mittler, R.; Miller, G. ROS and Redox Signalling in the Response of Plants to Abiotic Stress: ROS and Redox Signalling in Plants. *Plant Cell Environ.* **2012**, *35*, 259–270. [CrossRef] [PubMed]
63. Wang, P.; Luo, Q.; Yang, W.; Ahammed, G.J.; Ding, S.; Chen, X.; Wang, J.; Xia, X.; Shi, K. A Novel Role of Pipecolic Acid Biosynthetic Pathway in Drought Tolerance through the Antioxidant System in Tomato. *Antioxidants* **2021**, *10*, 1923. [CrossRef] [PubMed]
64. Saed-Moucheshi, A.; Shekoofa, A.; Pessarakli, M. Reactive Oxygen Species (ROS) Generation and Detoxifying in Plants. *J. Plant Nutr.* **2014**, *37*, 1573–1585. [CrossRef]
65. Wu, Y.; Ye, B. Effects of Combined Elevated Temperature and Drought Stress on Anti-Oxidative Enzyme Activities and Reactive Oxygen Species Metabolism of *Broussonetia papyrifera* Seedlings. *Acta Ecol. Sin.* **2016**, *36*, 403–410.
66. Soengas, P.; Rodríguez, V.M.; Velasco, P.; Cartea, M.E. Effect of Temperature Stress on Antioxidant Defenses in *Brassica oleracea*. *ACS Omega* **2018**, *3*, 5237–5243. [CrossRef]
67. Zou, M.; Yuan, L.; Zhu, S.; Liu, S.; Ge, J.; Wang, C. Response of Osmotic Adjustment and Ascorbate-Glutathione Cycle to Heat Stress in a Heat-Sensitive and a Heat-Tolerant Genotype of Wucai (*Brassica campestris* L.). *Sci. Hortic.* **2016**, *211*, 87–94. [CrossRef]
68. Cabrera, M.; Boronat, S.; Marte, L.; Vega, M.; Perez, P.; Ayte, J.; Hidalgo, E. Chaperone-Facilitated Aggregation of Thermo-Sensitive Proteins Shields Them from Degradation during Heat Stress. *Cell Rep.* **2020**, *30*, 2430–2443.e4. [CrossRef]
69. Irfan, M.; Kumar, P.; Ahmad, I.; Datta, A. Unraveling the Role of Tomato Bcl-2-Associated Athanogene (BAG) Proteins during Abiotic Stress Response and Fruit Ripening. *Sci. Rep.* **2021**, *11*, 21734. [CrossRef] [PubMed]
70. Guo, L.; Li, J.; He, J.; Liu, H.; Zhang, H. A Class I Cytosolic HSP20 of Rice Enhances Heat and Salt Tolerance in Different Organisms. *Sci. Rep.* **2020**, *10*, 1383. [CrossRef]
71. Muthusamy, S.K.; Dalal, M.; Chinnusamy, V.; Bansal, K.C. Genome-Wide Identification and Analysis of Biotic and Abiotic Stress Regulation of Small Heat Shock Protein (*HSP20*) Family Genes in Bread Wheat. *J. Plant Physiol.* **2017**, *211*, 100–113. [CrossRef] [PubMed]
72. Pratt, W.B.; Toft, D.O. Regulation of Signaling Protein Function and Trafficking by the Hsp90/Hsp70-Based Chaperone Machinery. *Exp. Biol. Med.* **2003**, *228*, 111–133. [CrossRef] [PubMed]

antioxidants

Article

Comparative Physiological and Transcriptomic Analyses Reveal Mechanisms of Exogenous Spermidine-Induced Tolerance to Low-Iron Stress in *Solanum lycopersicum* L.

Yu Shi [1,†], Yihong Zhao [1,†], Qi Yao [1], Feng Liu [2], Xiumin Li [2], Xiu Jin [1], Yi Zhang [1,*] and Golam Jalal Ahammed [3,*]

[1] College of Horticulture, Shanxi Agricultural University, Jinzhong 030801, China; shiyu@sxau.edu.cn (Y.S.); s20212238@stu.sxau.edu.cn (Y.Z.); z20203455@stu.sxau.edu.cn (Q.Y.); s20192167@stu.sxau.edu.cn (X.J.)
[2] Research Institute of Vegetables, Hunan Academy of Agricultural Sciences, Changsha 410125, China; liufengrich@126.com (F.L.); M929989171@163.com (X.L.)
[3] College of Horticulture and Plant Protection, Henan University of Science and Technology, Luoyang 471023, China
* Correspondence: yyzy@sxau.edu.cn (Y.Z.); ahammed@haust.edu.cn (G.J.A.)
† These authors contributed equally to this work.

Abstract: Iron (Fe) deficiency in plants is a major problem in agriculture. Therefore, we investigated both the physiological features and molecular mechanisms of plants' response to low-Fe (LF) stress along with the mitigation of LF with exogenous spermidine (Spd) in tomato plants. The results showed that exogenous Spd foliar application relieved the suppressing effect of LF stress on tomato plants by regulating the photosynthetic efficiency, chlorophyll metabolism, antioxidant levels, organic acid secretion, polyamine metabolism and osmoregulatory systems. Analysis of transcriptomic sequencing results revealed that the differentially expressed genes of iron-deficiency stress were mainly enriched in the pathways of phytohormone signaling, starch and sucrose metabolism and phenyl propane biosynthesis in both leaves and roots. Moreover, Spd-induced promotion of growth under LF stress was associated with upregulation in the expression of some transcription factors that are related to growth hormone response in leaves (*GH3*, *SAUR*, *ARF*) and ethylene-related signaling factors in roots (*ERF1*, *ERF2*). We propose that traits associated with changes in low-iron-tolerance genes can potentially be used to improve tomato production. The study provides a theoretical basis for dealing with the iron deficiency issue to develop efficient nutrient management strategies in protected tomato cultivation.

Keywords: polyamine; tomato; iron-deficiency; oxidative stress; transcriptomics

Citation: Shi, Y.; Zhao, Y.; Yao, Q.; Liu, F.; Li, X.; Jin, X.; Zhang, Y.; Ahammed, G.J. Comparative Physiological and Transcriptomic Analyses Reveal Mechanisms of Exogenous Spermidine-Induced Tolerance to Low-Iron Stress in *Solanum lycopersicum* L. *Antioxidants* **2022**, *11*, 1260. https://doi.org/10.3390/antiox11071260

Academic Editor: Nafees A. Khan

Received: 7 June 2022
Accepted: 23 June 2022
Published: 27 June 2022

1. Introduction

Iron (Fe) is a trace mineral element necessary for the normal life activities of almost all living organisms including plants. It is the fourth most abundant element in the earth's crust. Despite the high total iron content in soils, the soluble iron (Fe^{2+}) fraction is easily fixed to the insoluble form (Fe^{3+}) in an alkaline environment, which seriously affects the normal uptake of iron by plants. Iron deficiency impairs photosynthetic efficiency, plant growth and biomass yield [1,2]. As a redox-active metal, Fe is engaged in photosynthesis, mitochondrial respiration, nitrogen anabolism, hormone (ethylene, gibberellic acid, jasmonic acid) synthesis and pathogen defense [3]. Iron also acts as the cofactor of many antioxidant enzymes, and thus iron deficiency has a regulatory effect on the antioxidant mechanisms, including the activities of superoxide dismutase (SOD), peroxidase (POD) and catalase (CAT) in plants [4], which are in charge of protecting the biological system against the harmful effects of reactive oxygen species (ROS) [5]. ROS are produced in all forms of aerobic life under stress or normal conditions. The excessive production of ROS causes oxidative damage that has a negative impact on the function of important macromolecules [6]. Thus, a better understanding of the mechanisms of plant response to iron stress can be useful to improve crop stress resilience and enhance crop yield and quality [7].

When plants are exposed to a low-Fe environment, insufficient iron uptake causes retarded growth, interveinal chlorosis and reduced plant fitness. In severe iron deficiency, chloroplasts are dissociated or vesiculated, thus affecting chlorophyll formation [8]. To ensure the normal growth of plants, phytohormone auxin accumulates in large amounts in the roots, promoting the development of lateral roots and positively regulating the transcriptional expression of the *FIT1* and *AHA2* genes. Thus, growth factors are involved in the plant Fe-deficiency response network through different pathways [9].

Under low-iron stress, plants enhance Fe uptake by the root system through two strategies: One strategy based on the reduction that occurs in all dicotyledons and non-grass monocotyledons, called strategy I, and another strategy relying on chelation, which is limited to monocotyledons, called strategy II [10]. Strategy-II plants produce plant iron carriers capable of chelating Fe^{3+}, which are then absorbed by specific epidermal root cell plasma membrane transporters [11]. In tomatoes, on the one hand, as Strategy-I species, acidification of the root mesenchyme by plasma membrane H^+-ATPase activity occurs to dissolve Fe^{3+}, and reduction to Fe^{2+} by Fe^{3+}-chelating reductase (FCR) activity increases iron solubility. Afterward, translocation of the resulting Fe^{2+} to the root cell via a specific Fe transporter (*IRT1*) takes place to accomplish iron acquisition in plants [12]. On the other hand, nitric oxide (NO) accumulates in the roots and promotes the expression of *FER/FIT*, as well as *IRT* and *FRO* genes, thus participating in the response to iron-deficiency stress in plants [13].

Polyamines are highly bioactive, low-molecular-weight aliphatic amines that occur as ubiquitous secondary metabolites in plants. Polyamines can bind to phospholipids and other biomolecules with negatively charged groups of nucleic acids and proteins through their ionic and hydrogen bonds, which widely affect the biological activity of plants [14]. In previous research reports, polyamines have been shown to perform an extremely important role in alleviating plant stress. Among the three widely distributed major polyamines, spermidine (Spd) plays a crucial role in abiotic stress tolerance. Due to its multivalent cation property, its physiological function is stronger and more associated with stress tolerance in plants [15]. Spd is a common polyamine in plants and is involved in adaptations to salinity [16], drought [17], cold [18] and heavy metals [19]. Some studies have shown that Spd modulates antioxidant enzyme activity and the expression of related genes in tomato seedlings exposed to high temperatures [20]. Exogenous Spd has been found to play an important role in remediating the effects of environmental stress on plants [21]. However, to date, few studies have reported on the Spd-mediated tolerance to iron stress, particularly in tomato plants.

In the present research, using 'Micro-tom' tomato as the object of study, we explored the effect of exogenously sprayed Spd on the growth, physiology and metabolism of tomato seedlings under low-Fe stress. The physiological analysis, combined with transcriptomic analyses, shed new light on the mechanism of Spd-mediated low-Fe tolerance in tomato seedlings from both physiological and molecular perspectives, which provides a theoretical basis for improving the uptake and utilization of Fe in protected cultivation.

2. Materials and Methods

2.1. Plant Materials, Growth Conditions and Experimental Treatments

Tomato (*Solanum lycopersicum* L.) cv 'Micro-Tom' seeds were purchased from the Ball Horticulture Company (West Chicago, IL, USA). Healthy seeds were selected and sown on a petri dish with distilled water. The germinated seedlings were transferred to 72-well trays and cultured under artificial climate chamber conditions: temperature 28 °C/22 °C (14 h day/10 h night), humidity 80% and light intensity 600 μmol $m^{-2}s^{-1}$. When the plants had four fully expanded leaves, uniformly grown tomato seedlings were planted in a hydroponic tank filled with half-strength Japanese Yamazaki tomato formula nutrient solution [22]. After five days of seedling culture in hydroponics, the following treatments were applied: (1) CK (control), Yamazaki formula nutrient solution (Fe concentration was 100 μM); (2) LF, low-Fe nutrient solution (Fe concentration was 10 μM); (3) Spd, Yamazaki formula nutrient solution (100 μM Fe) + 0.25 mM Spd foliar spray; (4) LF + Spd, low-Fe (10 μM) nutrient solution + 0.25 mM Spd foliar spray. The Spd was purchased from the Beijing Solarbio Technology Company. Both sides of the tomato leaves were sprayed with freshly prepared Spd solution (approximately 10 mL per plant). Low-Fe stress was imposed 1d after the Spd treatment. Foliar-spraying of Spd was repeated every two days. The control tomato plants were foliar-sprayed with an equal volume of distilled water. The nutrient solution was changed every three days, the pH value was adjusted to 6.0 ± 0.2 and an intermittent supply of oxygen was provided using an aeration pump. On the 10th day of treatments, unless otherwise stated, samples were collected/used for various analyses such as photosynthetic fluorescence indicators, osmoregulatory substance content, organic acid and polyamine contents and RNA sequencing. Biomass measurements were performed on day 15 of treatment.

2.2. RNA-Seq Analysis and Quantitative Real-Time PCR Analysis

Transcriptome sequencing was performed on samples from four treatments—CK, Spd, LF and LF + Spd—collected on day 10 of treatments by Hangzhou Lianchuan Biological Technology Co., Ltd. RNA-seq was performed with three biological replicates for each treatment. All raw sequencing data from the current study were deposited into the NCBI database under the accession number "PRJNA834903" (https://www.ncbi.nlm.nih.gov/sra/PRJNA834903), (accessed on 4 May 2022). Analysis of significant differences between samples was performed using R packages edgeR or DESeq2. Genes with differential fold FC > twofold or FC < 0.5-fold and a p-value < 0.05 were defined as differentially expressed genes [23]. GO (Gene Ontology) enrichment and KEGG (Kyoto Encyclopedia of Genes and Genomes) pathway enrichment were analyzed using the clusterProfiler R package. GO functional enrichment and KEGG pathway analysis were performed by Goatools (https://github.com/tanghaibao/Goatools), (accessed on 6 June 2022) and KOBAS (http://kobas.cbi.pku.edu.cn/home.do), (accessed on 6 June 2022). The qRT-PCR test reaction system and primers used for qRT-PCR are shown in Supplementary Tables S1 and S2, respectively. Samples were added to a 96-well plate and then reacted in an Applied Biosystems Quant Studio three real-time fluorescence quantitative PCR system (QuantStudio 3, ThermoFisher Scientific™, Waltham, MA, USA). The qRT-PCR amplification procedure consisted of Stage 1: pre-denaturation, one cycle 95 °C, 30 s; Stage 2: PCR reaction, 40 cycles of 95 °C for 10 s, 60 °C for 30 s, 72 °C for 40 s. Relative gene expression was estimated using the $2^{-\Delta\Delta Ct}$ method [24]. qRT-PCR experiments were performed in biological triplicates.

2.3. Determination of Biomass and Root Morphology, Root Vigor and Root Fe^{3+} Reductase Activity

After 15 days of treatment, five seedlings were randomly picked from each treatment, and the selected plants were cut from the same part, divided into above-ground and below-ground parts, any water on the plant surface was dried with absorbent paper and the fresh weight was measured. The samples were then placed in an electric thermostatic drying oven (Heratherm™ General Protocol Ovens, 51028148, Thermo Scientific™, Waltham, MA, USA) set to 105 °C for 30 min. After adjusting the temperature to 80 °C, the material was dried to a constant weight before measuring the dry weight. For root morphology measurements, the whole root system of a plant was scanned with a root system scanner (Epson Perfection V800 Photo, B11B223201, Epson America, Inc., Los Alamitos, CA, USA). The analysis was done using a root scanner (WinRhizo PRO, version 2017, Regent Instruments Inc., Quebec City, QC, Canada), and parameters such as the total root length, total surface area, total volume and average diameter were read [25]. Root vigor was determined by the triphenyl tetrazolium chloride (TTC) method [26]. Fe^{3+} reductase activity was determined according to the method of Ekmekcioglu C [27]. Three biological replicates for each treatment were set in treatments.

2.4. Determination of Photosynthetic Pigment Content and Photosynthetic Index

On the 10th day, chlorophylls (Chl) such as Chla, Chlb and carotenoids were measured in the third fully-expanded leaf [28]. About 0.1 g of leaf tissue was placed in a tube containing 96% ethanol in the dark for about 24 h until the leaves turned completely white. The absorbance values of chlorophyll extracts at 470 nm, 649 nm and 665 nm were measured with a spectrophotometer (UV-2450, Shimadzu, Kyoto Prefecture, Japan), and chlorophyll a, chlorophyll b and carotenoid contents were calculated.

The photosynthetic indexes such as the net photosynthetic rate (Pn), transpiration rate (Tr), intercellular CO_2 concentration (Ci) and stomatal conductance (Gs) were measured using a portable photosynthetic apparatus (LI-6800, Li-COR Inc., Lincoln, NE, USA) on a clear day at around 10 a.m. The parameters were set to a flow rate of 500 $\mu mol \cdot s^{-1}$, leaf temperature of 28 °C and CO_2 concentration of 400 $\mu mol \cdot mol^{-1}$; a CO_2 cylinder was used to stabilize the CO_2 environment [22].

Following 24 h of darkness, the seedling leaves were sampled to test the maximum photochemical efficiency, i.e., Fv/Fm [29]. In addition, the actual photochemical efficiency of PSII (ΦPSII), photosynthetic electron transfer rate (ETR), photochemical quenching coefficient (qP) and non-photochemical quenching coefficient (NPQ) were measured after 30 min of plant exposure to natural light conditions [30].

2.5. Determination of Antioxidant Properties and Osmoregulatory Substances

A fresh-leaf or root sample (0.3 g) was placed in a pre-cooled pestle and mortar and ground to a fine frozen powder under liquid nitrogen, followed by homogenization in 3 mL 50 mM phosphate buffer (pH 7.8) in an ice bath. Then, homogenate centrifugation was done at 12,000× *g* for 15 min at 4 °C. The supernatant was used to determine the peroxidase (POD) [31], catalase (CAT) [32] and superoxide dismutase (SOD) [33] activity. Activity analyses of POD, CAT and SOD were performed as described previously [34]. Three biological replicates for each treatment were performed. The lipid peroxidation level was measured by estimating the malondialdehyde (MDA) content in roots using thiobarbituric acid (TBA) [35]. Electrolyte leakage (%) was estimated by measuring ion leakage from roots according to the method of Shou [36]. The roots (which weighed 0.1 g) were placed in centrifuge tubes, then each tube was filled with 20 mL of distilled water. The conductivity (A1) was first measured after shaking the tube well, then the conductivity (A2) was again measured after shaking the tube in the shaker for 2 h. Finally, the sample was boiled and cooled to room temperature to measure the conductivity (A3). Relative electrolyte leakage was measured as follows: Relative conductivity = (A2−A1)/(A3−A1). The content of H_2O_2 in leaves and roots was determined by the method of Willekens [37]. The content of $O^{2\bullet-}$ in leaves and roots was analyzed by the method previously described by Li et al. [38]. Proline and soluble protein contents were determined by the methods of Bates [39] and Bradford [40], respectively. Meanwhile, the free amino acids and soluble sugar contents were determined by the method of Zhang et al. [41]. Each treatment was repeated three times to ensure the reliability of the results. The organic acid content was determined by high-performance liquid chromatography [42]. Parameter settings were as follows: a ZORBAX Eclipse XDB-C18 column (4.6 × 250 mm, 5 mm) was used; the mobile phase was set at 0.04 $mol \cdot mL^{-1}$, pH 2.4, KH_2PO_4-H_3PO_4 buffer solution; the flow rate was 0.8 $mL \cdot min^{-1}$; the column temperature was 30 °C, the detection wavelength was 210 nm and the injection volume was 10 μL.

2.6. Determination of Sucrose Content and Metabolism-Related Enzyme Activities

The sucrose content was determined by the hydrochloric acid-resorcinol method previously described by Zhang et al. [43]. We accurately weighed 0.1 g of leaves and roots and took 0.2 and 0.4 mL of supernatants, respectively. After adding 200 μL NaOH, the solution was boiled for 5 min at 100 °C, then cooled, before 2.8 mL 30% HCL and 0.8 mL 0.1% resorcinol were added, with the contents shaken well. Then, they were placed in a water bath at 80 °C for 10 min for the reaction to occur, and after cooling, the OD value was measured at 480 nm. Three replicates of each treatment were performed. Standard curves with different concentration gradients of sucrose were prepared with the standard solution and used to calculate the actual sucrose content in leaves and roots. To analyze the activities of sugar metabolism-related enzymes, frozen samples of leaves were weighed to 0.1 g. Sucrose synthase (SS), sucrose phosphate synthase (SPS), acid convertase (AI) and neutral convertase (NI) activities were determined using the corresponding enzyme activity assay kits (Beijing Solarbio Science & Technology Co., Ltd., Beijing, China).

2.7. Determination of Polyamine Content

Polyamines extraction from tomato seedlings was performed using the methods described by Flores and Galston [44]. The content of polyamines was determined by HPLC (high-performance liquid chromatograph UltiMate3000, ThermoFisher Scientific™, Waltham, MA, USA). The instrumentation and settings for endogenous polyamine analyses were as follows: a ZORBAX Eclipse XDB-C18 column (4.6 × 250 mm, 5 mm) and mobile phase (methanol: acetonitrile: water = 58:2.5:39.5) were used with a detection wavelength of 230 nm, flow rate of 1 mL·min^{-1}, column temperature of 30 °C and injection volume of 10 μL. The organic solvents used above were of chromatographic-grade purity and the water was ultrapure. The mobile phase was configured for use after ultrasonic sonication beforehand.

2.8. Statistical Analysis

All data were subjected to analysis of variance (ANOVA), analyzed with SPSS 21.0 statistical software and plotted with Microsoft Excel 2016. For multiple mean comparisons, differences between treatment means were separated by Duncan's multiple range test at $p < 0.05$.

3. Results

3.1. Overview of Sequencing Data-Quality Control

In this experiment, the leaves and roots of the Control (CK), Low Fe (LF), Spermidine (Spd) and Low Fe + Spd (LFS) were sequenced, and each treatment was replicated three times. The results showed that 99.98% of the nucleotides in the transcriptome sequencing data reached Q20, and 97.25% of the nucleotides exceeded Q30 (Table 1).

Table 1. Statistics of transcriptome sequencing data. Sample, sample name; Raw Read, the number of reads in total; Valid Read, the number of valid reads after de-junctioning, de-low quality, etc.; Valid Ratio, the proportion of valid reads; Mapped Reads, the number of reads that can be compared to the genome; Unique Mapped Reads, can only uniquely match to one position in the genome; Q20%, the percentage of bases with Q20% quality value ≥ 20 (sequencing error rate less than 0.01); Q30%, the percentage of bases with Q30% quality value ≥ 30 (sequencing error rate less than 0.001).

Sample	Raw Read	Valid Read	Valid Ratio (Reads)	Mapped Reads	Unique Mapped Reads	Q20%	Q30%
CK_L1	51,425,238	47,618,720	92.60	45,445,100 (95.44%)	38,907,284 (81.71%)	99.99	97.68
CK_L2	41,647,086	39,867,334	95.73	38,103,955 (95.58%)	32,333,283 (81.10%)	99.99	97.78
CK_L3	36,237,924	34,971,030	96.50	33,505,416 (95.81%)	28,442,929 (81.33%)	99.99	97.59
CK_R1	41,884,204	40,904,806	97.66	35,303,817 (86.31%)	30,123,337 (73.64%)	99.99	97.38

Table 1. *Cont.*

Sample	Raw Read	Valid Read	Valid Ratio (Reads)	Mapped Reads	Unique Mapped Reads	Q20%	Q30%
CK_R2	47,810,190	46,772,676	97.83	42,676,467 (91.24%)	36,960,152 (79.02%)	99.98	98.37
CK_R3	51,562,904	50,495,438	97.93	44,954,365 (89.03%)	38,867,119 (76.97%)	99.98	98.30
LF_L1	45,353,480	42,425,970	93.55	40,385,907 (95.19%)	34,478,301 (81.27%)	99.99	97.79
LF_L2	45,112,774	43,318,042	96.02	41,436,986 (95.66%)	35,189,847 (81.24%)	99.99	97.62
LF_L3	47,262,530	45,751,138	96.80	43,773,386 (95.68%)	37,222,647 (81.36%)	99.99	97.73
LF_R1	43,262,284	42,427,286	98.07	39,214,339 (92.43%)	33,844,982 (79.77%)	99.99	98.39
LF_R2	52,854,702	51,760,038	97.93	47,780,225 (92.31%)	41,266,840 (79.73%)	99.99	98.51
LF_R3	52,525,326	51,551,094	98.15	45,715,483 (88.68%)	39,276,860 (76.19%)	99.99	98.45
LFS_L1	46,825,170	44,333,134	94.68	42,409,857 (95.66%)	36,246,131 (81.76%)	99.99	97.50
LFS_L2	35,688,744	34,009,554	95.29	32,356,247 (95.14%)	27,660,644 (81.33%)	99.99	97.25
LFS_L3	41,306,858	39,800,370	96.35	38,018,606 (95.52%)	32,451,594 (81.54%)	99.99	97.60
LFS_R1	53,734,354	52,663,752	98.01	47,503,814 (90.20%)	40,935,810 (77.73%)	99.99	98.41
LFS_R2	52,372,096	51,298,494	97.95	45,352,741 (88.41%)	39,262,642 (76.54%)	99.98	98.37
LFS_R3	54,358,180	53,210,976	97.89	48,848,829 (91.80%)	42,040,996 (79.01%)	99.99	98.43
Spd_L1	50,060,896	45,020,762	89.93	43,045,480 (95.61%)	36,727,980 (81.58%)	99.99	97.52
Spd_L2	35,827,860	34,695,530	96.84	33,367,320 (96.17%)	28,459,478 (82.03%)	99.99	97.74
Spd_L3	50,682,352	48,290,700	95.28	46,229,329 (95.73%)	39,485,340 (81.77%)	99.99	97.53
Spd_R1	52,910,480	51,911,130	98.11	46,563,164 (89.70%)	40,136,000 (77.32%)	99.98	98.42
Spd_R2	51,487,988	50,453,420	97.99	45,578,298 (90.34%)	39,515,003 (78.32%)	99.98	98.44
Spd_R3	53,683,760	52,593,000	97.97	46,823,515 (89.03%)	40,562,480 (77.13%)	99.98	98.48

3.2. Analysis of Differentially Expressed Genes

To get a closer look at the differentially expressed genes, we mapped volcanoes (Figure 1). In the volcano maps, red represents significantly upregulated differently expressed genes, blue represents significantly downregulated differently expressed genes and gray represents non-significant differently expressed genes.

FPKM (fragments per kilobase of exon model per million mapped fragments) was used to count the expression abundance of known genes in different samples. In this experiment, we used the difference multiplier FC ≥ 2 or FC ≤ 0.5 (i.e., the absolute value of log2FC ≥ 1) as the threshold of change and a *p*-value <0.05 as the criterion for screening differential genes. The number of differentially expressed genes in each comparison group was counted, and a bar chart (Figure 2) was used to visualize the number of significantly differentially expressed genes in different comparison groups, as well as the specific changes (up- and downregulation). Compared to the control, 227 genes were upregulated and 201 genes were downregulated in the low-iron-treated leaves (LF), whereas the number of differentially expressed genes was higher in the root system, where 933 genes were upregulated and 1199 genes were downregulated, which indicated that the low-iron treatment had a more profound effect on transcription in the root system than in the leaves. Again, compared to the LF treatment, 606 genes were upregulated in the LF + Spd-treated leaves, and 302 genes were downregulated, while 422 genes were upregulated and 619 genes were downregulated in the root sample.

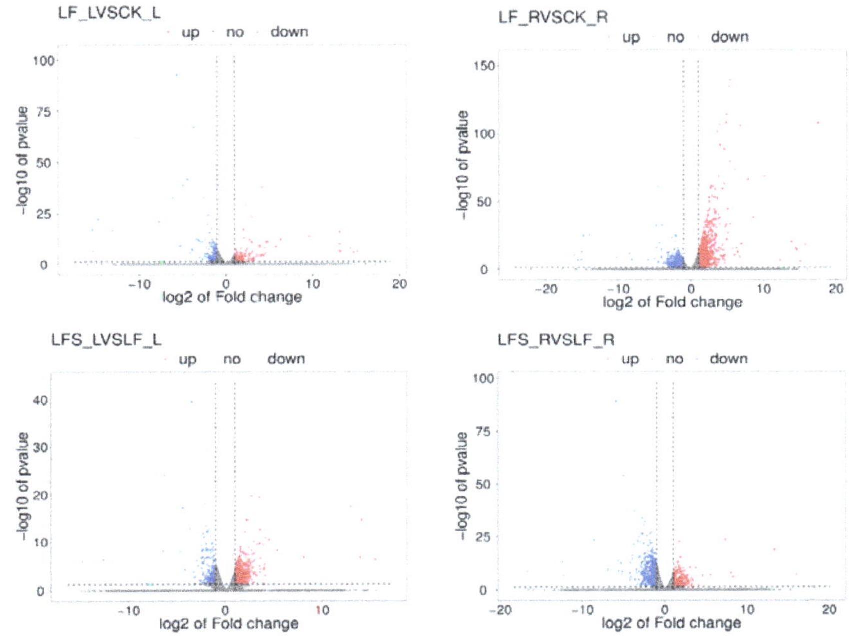

Figure 1. Volcano maps of expression differences. The horizontal coordinate represents the different expression fold changes of the gene in different samples, and the vertical coordinate represents the statistical significance of the difference in the gene expression change. LF_L vs. CK_L, Low Fe_Leaf sample vs. Control_Leaf sample; LF_R vs. CK_R, Low Fe_Root sample vs. Control_Root sample; LFS_L vs. LF_L, Low Fe + Spd_Leaf sample vs. Low Fe_Leaf sample; LFS_R vs. LF_R, Low Fe + Spd_Root sample vs. Low Fe_Root sample.

Then, we performed Venn diagram analysis for Control vs. Low Fe (CK vs. LF) and Low Fe vs. Low Fe + Spd (LF vs. LFS), gene ontology (GO) enrichment for Low Fe+ Spd_Leaf sample vs. Low Fe_Leaf sample (LFSL vs. LFL) and Low Fe + Spd_Root sample vs. Low Fe_Root sample (LFSR vs. LFR) and KEGG (Kyoto Encyclopedia of Genes and Genomes) pathway enrichment analysis for differently expressed genes as influenced by Spd treatment.

The Venn diagram can visualize not only the number of differently expressed genes in the different treatment groups but also the number of genes that are differently expressed in each treatment group in total. As shown in Figure 3, a total of 104 differently expressed genes were co-expressed among 428 differentially expressed genes in the treatment group CK vs. LF, and there were 908 differentially expressed genes in the treatment group LFS vs. LF in the leaves. In the case of the root sample, a total of 677 differently expressed genes were co-expressed among 2132 differently expressed genes in the treatment group CK vs. LF, and there were 1041 differently expressed genes in the treatment group LFS vs. LF.

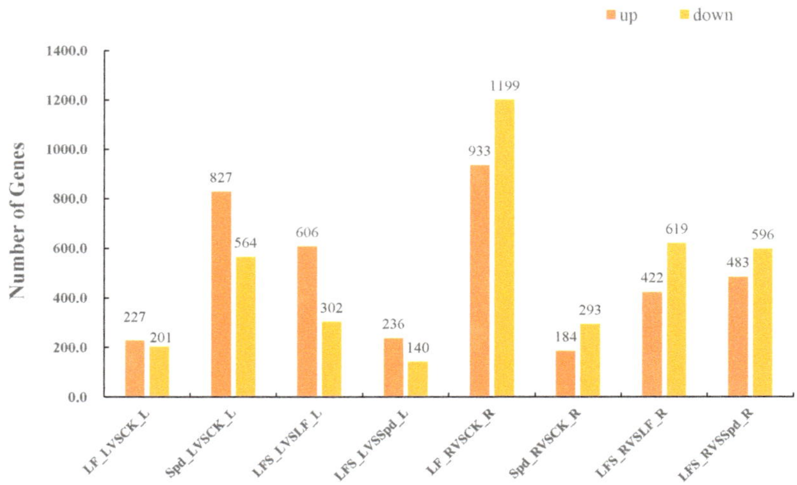

Figure 2. Number of significantly differentially expressed genes in different treatments. LF_L vs. CK_L, Low Fe_Leaf sample vs. Control_Leaf sample; Spd_L vs. CK_L, Spd_Leaf sample vs. Control_Leaf sample; LFS_L vs. LF_L, Low Fe + Spd_Leaf sample vs. Low Fe_Leaf sample; LFS_L vs. Spd_L, Low Fe + Spd_Leaf sample vs. Spd_Leaf sample; LF_R vs. CK_R, Low Fe_Root sample vs. Control_Root sample; Spd_R vs. CK_R, Spd_Root sample vs. Control_Root sample; LFS_R vs. LF_R, Low Fe + Spd_Root sample vs. Low Fe_Root sample; LFS_R vs. Spd_R, Low Fe + Spd vs. Spd_Root sample.

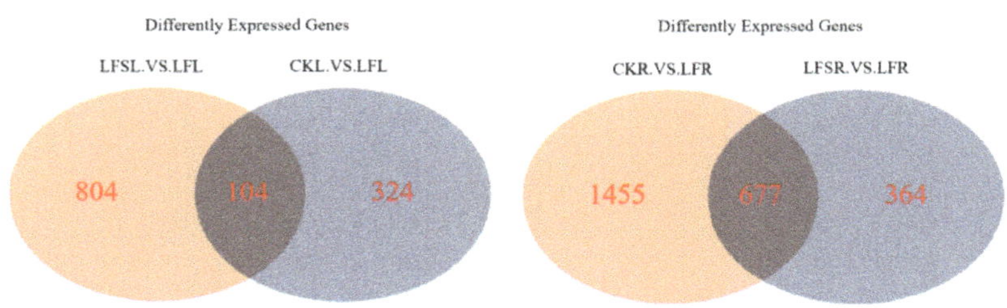

Figure 3. Venn diagram of significantly differentially expressed genes in different treatment comparisons. LFSL. vs. LFL, Low Fe+ Spd_Leaf sample vs. Low Fe_Leaf sample; CKL. vs. LFL, Control_Leaf sample vs. Low Fe_Leaf sample; CKR.VS.LFR, Control_Root sample vs. Low Fe_Root sample; LFSR. vs. LFR, Low Fe+ Spd_Root sample vs. Low Fe_Root sample.

The GO enrichment analysis of LFS vs. LF showed that enrichment in biological processes (BP) was mostly in functions such as transcriptional regulation with DNA as the template, protein phosphorylation, defense responses, redox processes, signal transduction processes, ethylene-activated signaling pathways, defense responses against fungi, and protein ubiquitination (Figure 4). In cellular components (CC), differentially expressed genes were involved in biological functions such as those of the nucleus, plasma membrane, membrane components, cytoplasm, chloroplast and extracellular regions. In molecular functions (MF), they were mainly enriched in sequence-specific protein-binding, specific DNA sequence-binding transcription factor activity, ATP-binding, DNA-binding, protein serine/threonine kinase activity, and metal ion-binding.

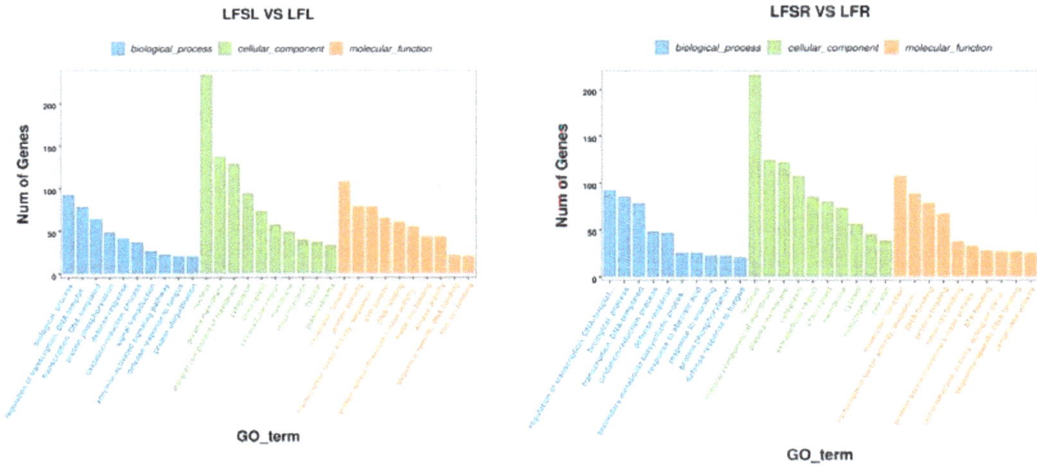

Figure 4. GO enrichment analysis of differently expressed genes. The abscissa represents different GO terms, blue represents biological processes, green represents cellular components, orange represents molecular functions and ordinate represents the number of differentially expressed genes. LFSL vs. LFL, Low Fe + Spd_Leaf sample vs. Low Fe_Leaf sample; LFSR vs. LFR, Low Fe + Spd_Root sample vs. Low Fe_Root sample.

To further explore the most important biochemical/metabolic pathways and signal transduction pathways involved in differentially expressed genes due to Spd treatment in low-iron-supplied tomato plants, the top-20 significantly enriched pathways were screened for KEGG enrichment analysis by the number of genes enriched in this pathway, and the enrichment results are presented in the form of bubble plots (Figure 5). KEGG enrichment analysis was performed on 908 differentially expressed genes in leaves and 1041 differentially expressed genes in roots, comparing the low-iron treatment and combined treatment of Spd and low iron.

The results showed that a total of 707 differently expressed genes in leaves were significantly enriched in 113 KEGG metabolic pathways, concentrated in metabolic pathways such as plant-pathogen interaction (79), phytohormone signaling (61), cytokinesis (30), amino and nucleotide sugar metabolism (29), phenyl propane biosynthesis (23) and starch and sucrose metabolism (23). A total of 867 differently expressed genes were significantly enriched in 117 KEGG metabolic pathways in the root system, concentrated in metabolic pathways such as plant-pathogen interaction (58), phytohormone signaling (56), benzyl propane biosynthesis (40), starch and sucrose metabolism (28), amino and nucleotide sugar metabolism (23) and carbon metabolism (22). It was found that the differentially expressed genes of iron-deficiency stress were mainly enriched in the pathways of phytohormone signaling, starch and sucrose metabolism and phenyl propane biosynthesis in both leaves and roots.

The results of GO enrichment analysis and KEGG pathway enrichment analysis showed that the differentially expressed genes in plant hormone signaling processes and starch and sucrose metabolism were significantly affected by low-Fe stress. Therefore, we performed a heat map analysis of differentially expressed genes in phytohormone signaling pathways and starch and sucrose metabolism. The results showed that a total of nine differently expressed genes in the phytohormone signal transduction pathway—*Solyc00g174330.3(PR1)*, *Solyc05g009610.1(GID1)*, *Solyc09g007010.1(PR1)*, *Solyc06g062460.3 (PIF3)*, *Solyc07g056000.2(TCH4)*, *Solyc09g089930.2(ERF1)*, *Solyc12g036470.2(PIF3)*, *Solyc01g1 07400.2(GH3)* and *Solyc03g093080.3(TCH4)*—were common to leaves in the comparison groups of LF vs. CK and LFS vs. LF (Figure 6). The expression of oleuropein sterol regulatory protein EBRU1 precursors *Solyc07g056000.2(TCH4)* and *Solyc03g093080.3(TCH4)* was downregulated under low-iron treatment, while all other genes were upregulated. In contrast, all nine differently expressed genes were upregulated in leaves after spraying with Spd under low-iron treatment. A total of 37 differently expressed genes were expressed in the root system, mostly concentrated in the growth hormone and ethylene metabolic pathways. The expression of *Solyc03g082510.1(SAUR)* and *Solyc10g076790.2(AUX1)* in the growth hormone metabolic pathway and *Solyc08g066660.1(ERF1)* and *Solyc03g114310.3(CTR1)* in the ethylene metabolic pathway were downregulated under low-Fe stress, while their expression with Spd treatment under low-iron stress was upregulated. It appears that differentially expressed genes related to hormone metabolism showed different trends in leaves and roots. For the upregulated genes, in leaves, Spd foliar-spray treatment could further upregulate gene expression, whereas, in roots, Spd foliar treatment downregulated genes to the control level. Of the 18 differentially expressed genes in the root system for starch and sucrose metabolic processes, seven differently expressed genes were downregulated and 11 differently expressed genes were upregulated, and the expression of genes related to hormone signaling was consistent with the Spd treatment; nonetheless, all of these were backregulated to the control level in the root sample.

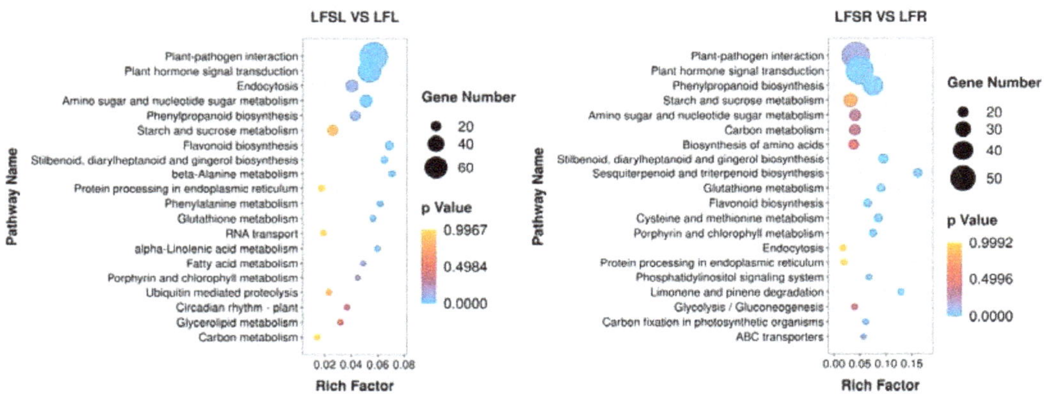

Figure 5. Enrichment analysis of differentially expressed genes in the KEGG pathway. The horizontal axis indicates the degree of enrichment (Rich factor), and the vertical axis indicates the enriched KEGG pathway; the size of the dots indicates the number of differentially expressed genes enriched in a KEGG pathway; the color of the dots indicates different *p* values; the Rich factor indicates the number of differentially expressed genes belonging to a KEGG pathway/the total number of genes belonging to this KEGG pathway. The larger the Rich factor, the higher the enrichment of the KEGG pathway. LFSL vs. LFL, Low Fe + Spd_Leaf sample vs. Low Fe_Leaf sample; LFSR vs. LFR, Low Fe + Spd_Root sample vs. Low Fe_Root sample.

Then, we analyzed the expression of genes related to hormone signaling pathways and sucrose metabolism, as well as differentially expressed genes of other metabolic pathways, as shown in Figure 7. In leaf blades, *Solyc01g008620.3(GN1-2-3)* expression was upregulated in the starch and sucrose metabolic pathways, which potentially accelerated glucose synthesis; *Solyc02g071620.3(CHLP)* and *Solyc07g064720.3(CHLP)* expression were upregulated in porphyrin and chlorophyll metabolism, which in turn, potentially functioned in the synthesis of chlorophyll a and chlorophyll b, respectively. *Solyc07g024000.3(NOL)* expression was downregulated, thus, perhaps, inhibiting the conversion of chlorophyll b to hydroxy-chlorophyll a. In the photosynthetic pathway, *Solyc11g006910.2(PetF)* iron oxytocin gene expression was upregulated during photosynthetic electron transfer; in the peroxisome pathway, i.e., the antioxidant enzyme system, *Solyc12g094620.2(CAT)* expression was upregulated in the antioxidant enzyme system and so on. In the root system, more genes are related to the expression of hormone metabolism, and among them, the expression of *Solyc10g076790.2(AUX1)* and *Solyc03g082510.1(SAUR)* was upregulated after Spd treatment under low iron, both of which are jointly involved in plant cell growth. However, *Solyc09g089610.3(ETR)*, *Solyc09g066360.1(ERF1)* and *Solyc04g071770.3(ERF2)* transcripts were downregulated, which potentially alleviated the effect of ethylene on cell senescence. Again, *Solyc12g038580.2(TPS)* expression was upregulated in the starch and sucrose metabolic pathways, which affected sugar synthesis, and *Solyc12g009300.3(SUS)* expression was downregulated, which might affect sucrose synthase activity. In the peroxisome pathway, the epoxidation process was promoted by upregulation of *Solyc01g066457.1(EPHX2)*. Upregulation of *Solyc01g058210.2(HMGCL)*, *Solyc10g007600.3(HAO)* and *Solyc12g099930.2 (AGXT)* contributed to amino acid metabolism, and upregulated expression of *Solyc12g094620.2 (CAT)* in hydrogen peroxide metabolism potentially increased redox levels. In addition to affecting the expression of related metabolic genes in each pathway, Spd-spraying under low iron upregulated the expression of *Solyc02g069200.3(IRT1)*, *Solyc01g094890.3((FRO2)* and *Solyc01g094910.3(FRO)*, which potentially improved the Fe uptake and transport capacity of the root system under low-iron stress.

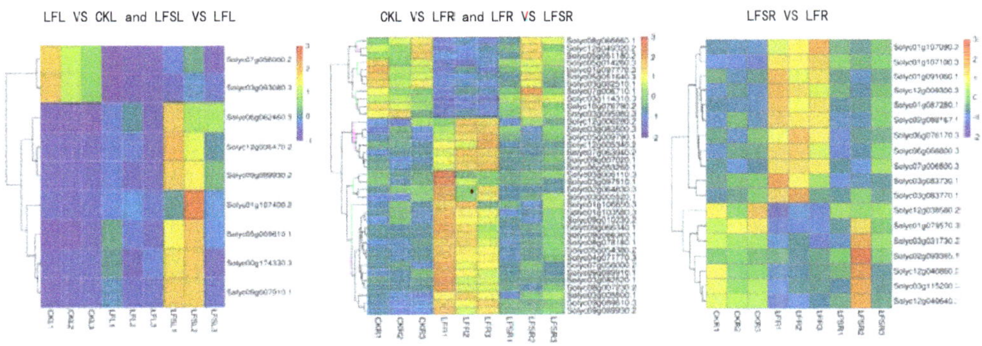

Figure 6. Heat map of differentially expressed genes related to plant hormone signal transduction and sucrose metabolism. Low Fe + Spd_Root sample vs. Low Fe_Root sample.

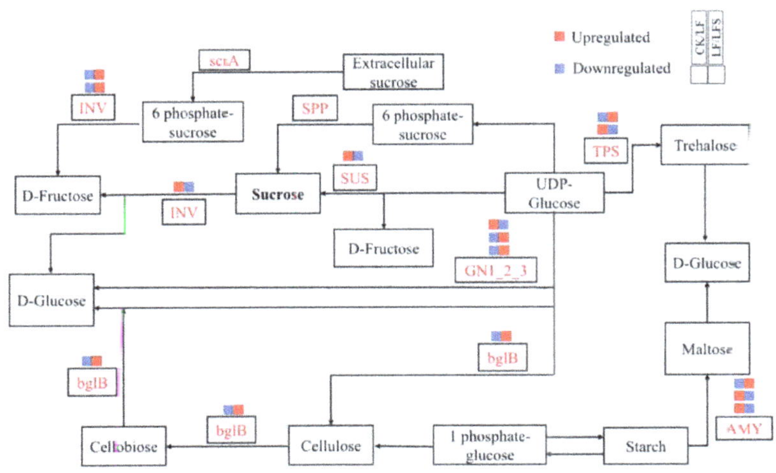

Figure 7. Diagram of plant sucrose metabolism pathway.

Finally, expression trends of six selected differentially-expressed genes related to iron transport or sucrose metabolism in the root were validated by qRT-PCR. The trends for the gene expression in qRT-PCR (Supplementary Figure S1) were approximately the same as the transcriptome sequencing results, indicating that the results were credible.

3.3. Exogenous Spd Improved the Growth and Photosynthetic Efficiency of Tomato Plants under Low-Iron Stress

The growth of tomato seedlings was significantly inhibited by low-Fe stress, along with significantly decreased dry and fresh weights by 28.57 and 27.91%, respectively. However, plant biomass was significantly increased by Spd foliar treatment under low-iron stress (Table 2). Likewise, root growth was significantly affected by low-iron stress, but Spd foliar treatment under low-Fe conditions increased the total root length, total root surface area and total root volume by 78.63, 41.35 and 40.91%, respectively, compared to the low-iron treatment (Table 2). It is evident that exogenous Spd-spraying has a mitigating effect on the growth of tomato seedlings under low-iron stress.

Table 2. Effects of exogenous Spd on tomato biomass under low-iron stress.

Treatments	Shoot Fresh Weight/g	Root Fresh Weight/g	Shoot Dry Weight/g	Root Dry Weight/g	Total Fresh Weight/g	Total Dry Weight/g
CK	5.43 ± 0.44 ab	1.45 ± 0.12 b	0.40 ± 0.01 b	0.09 ± 0.01 a	6.88 ± 0.43 b	0.49 ± 0.02 b
LF	3.80 ± 0.61 c	1.16 ± 0.13 c	0.29 ± 0.05 c	0.06 ± 0.01 b	4.96 ± 0.73 c	0.35 ± 0.03 c
Spd	6.19 ± 0.45 a	1.81 ± 0.12 a	0.48 ± 0.03 a	0.10 ± 0.01 a	7.99 ± 0.43 a	0.58 ± 0.03 a
LF + Spd	5.07 ± 0.24 b	1.42 ± 0.19 bc	0.36 ± 0.02 b	0.08 ± 0.01 a	6.49 ± 0.35 b	0.45 ± 0.01 b

CK, control; LF, Low Fe; Spd, spermidine; LF + Spd, Low Fe plus spermidine. Data are shown as mean ± SD. Within each column, entries followed by the same lowercase letters are not significantly different according to Duncan's test at $p \leq 0.05$.

Moreover, root vigor and Fe^{3+} reductase activity were significantly increased by either low-iron stress or Spd-spraying (Supplementary Figure S2). When compared to the low-Fe treatment, root vigor and Fe^{3+} reductase activity were further increased by 23.21 and 21.35%, respectively, after spraying with Spd under low-Fe stress.

The photosynthetic pigment content of tomato leaves was repressed by low-iron stress. However, the chlorophyll a, chlorophyll b and chlorophyll a + b contents were significantly increased by 23.58, 12.50 and 21.58%, respectively, in Spd treatment under low-Fe stress compared to low-iron stress only, though the carotenoid content was affected by Spd treatment under low-iron stress (Table 3).

Table 3. Effects of exogenous Spd on chlorophyll content in tomato leaves under low-iron stress.

Treatments	Chl a $mg{\cdot}g^{-1}$ FW	Chl b $mg{\cdot}g^{-1}$ FW	Carotenoid $mg{\cdot}g^{-1}$ FW	Chl a + b $mg{\cdot}g^{-1}$ FW
CK	1.47 ± 0.06 b	0.66 ± 0.02 b	0.24 ± 0.02 a	3.20 ± 0.08 b
LF	1.06 ± 0.03 d	0.56 ± 0.04 c	0.15 ± 0.01 b	2.41 ± 0.08 d
Spd	1.64 ± 0.02 a	0.70 ± 0.01 a	0.25 ± 0.02 a	3.52 ± 0.07 a
LF + Spd	1.31 ± 0.10 c	0.63 ± 0.02 b	0.18 ± 0.02 b	2.93 ± 0.16 c

CK, control; LF, Low Fe; Spd, spermidine; LF + Spd, Low Fe plus spermidine. Data are shown as mean ± SD. Within each column, entries followed by the same lowercase letters are not significantly different according to Duncan's test at $p \leq 0.05$.

In line with the photosynthetic pigment concentrations, the net photosynthetic rate was inhibited by 49.07% in leaves under low-Fe stress, and exogenous foliar-spraying of Spd alleviated the reduction of gas-exchange parameters in tomato leaves caused by low-Fe stress, and increased Pn, Tr, Gs and the intercellular CO_2 concentration (Ci) (Table 4).

Table 4. Effects of exogenous Spd on photosynthetic parameters in tomato leaves under low-iron stress.

Treatments	Pn/ (μmol·m^{-1}·s^{-1})	Gs/ (mmol·m^{-1}·s^{-1})	Ci/ (mmol·mol^{-1})	Tr/ (mmol·m^{-1}·s^{-1})
CK	7.56 ± 0.10 b	180.55 ± 1.68 a	130.91 ± 1.58 c	3.55 ± 0.00 a
LF	3.85 ± 0.08 d	92.73 ± 1.55 c	136.26 ± 2.55 b	2.04 ± 0.02 c
Spd	8.92 ± 0.03 a	178.94 ± 1.28 a	122.71 ± 0.80 d	3.54 ± 0.02 a
LF + Spd	4.73 ± 0.12 c	139.87 ± 0.43 b	153.42 ± 1.47 a	2.90 ± 0.05 b

CK, control; LF, Low Fe; Spd, spermidine; LF + Spd, Low Fe plus spermidine. Data are shown as mean ± SD. Within each column, entries followed by the same lowercase letters are not significantly different according to Duncan's test at $p \leq 0.05$.

Under low-Fe stress, chlorophyll fluorescence parameters such as the maximum photochemical efficiency of PSII (Fv/Fm), electron transfer efficiency (ETR), actual photochemical quantum yield of PSII (ΦPSII) and photochemical quenching coefficient (qP) of leaves significantly decreased by 7.47, 37.21, 37.32 and 35.47%, respectively, and the non-photochemical quenching coefficient (NFQ) increased by 85.94%. However, all these indicators, except for qP, increased significantly after spraying with Spd, suggesting that exogenous foliar-spraying with Spd under low-iron stress had a strong ameliorative effect on leaf chlorophyll fluorescence characteristics (Table 5).

Table 5. Effects of exogenous Spd on fluorescence parameters in tomato leaves under low-iron stress.

Treatments	Fv/Fm	ETR	ΦPSII	qP	NFQ
CK	0.763 ± 0.007 b	120.562 ± 0.95 b	0.276 ± 0.002 b	0.468 ± 0.026 b	1.380 ± 0.038 c
LF	0.706 ± 0.009 d	75.706 ± 3.14 d	0.173 ± 0.007 d	0.302 ± 0.012 d	2.566 ± 0.056 a
Spd	0.776 ± 0.008 a	146.958 ± 0.77 a	0.337 ± 0.002 a	0.518 ± 0.004 a	1.314 ± 0.012 c
LF + Spd	0.741 ± 0.002 c	106.407 ± 1.21 c	0.243 ± 0.003 c	0.416 ± 0.006 c	1.851 ± 0.022 b

CK, control; LF, Low Fe; Spd, spermidine; LF + Spd, Low Fe plus spermidine. Data are shown as mean ± SD. Within each column, entries followed by the same lowercase letters are not significantly different according to Duncan's test at $p \leq 0.05$.

3.4. Effect of Exogenous Spd on ROS Accumulation, Antioxidant System and Osmoregulatory Substances in Tomatoes under Low-Iron Stress

Low-iron stress increased the accumulation of intracellular O_2^- and H_2O_2, leading to increased membrane permeability, and disruption of plant cell membranes as evidenced by a significant increase in the relative electrolyte leakage and MDA content in the root sample. However, exogenous spraying of Spd decreased the O_2^- and H_2O_2 contents, which, in turn, reduced the levels of MDA and relative conductivity, thereby effectively alleviating the deleterious effects of low iron on the cell membrane (Figure 8, Supplementary Figure S3).

Next, we observed the ultrastructure of tomato leaves to reveal the effect of low-iron-induced oxidative stress on the plant cell structure. Figure 9 shows that under low-iron stress, the cell exhibited the phenomenon of plasma-wall separation, the cell membrane was damaged, the chloroplast and starch granules in the leaf were deformed, the chloroplast was irregularly spherical and the starch granule swelled obviously. However, with Spd foliar treatment under low-Fe stress, chloroplast deformity was recovered to some extent with elliptical bands, and the shape of starch grains was restored.

To study whether the alleviation of low-iron stress by exogenous Spd was related to the change in antioxidant enzyme activity in tomatoes, we analyzed the activities of SOD, POD and CAT in leaves and roots. The results, shown in Figure 10, revealed that the activities of SOD, POD and CAT decreased in leaves and roots under low-iron stress, which potentially indicated a weakened ROS scavenging ability. However, foliar-spraying with Spd increased the activities of SOD, POD and CAT to varying degrees, thereby effectively alleviating the ROS-induced damage to the cell membrane.

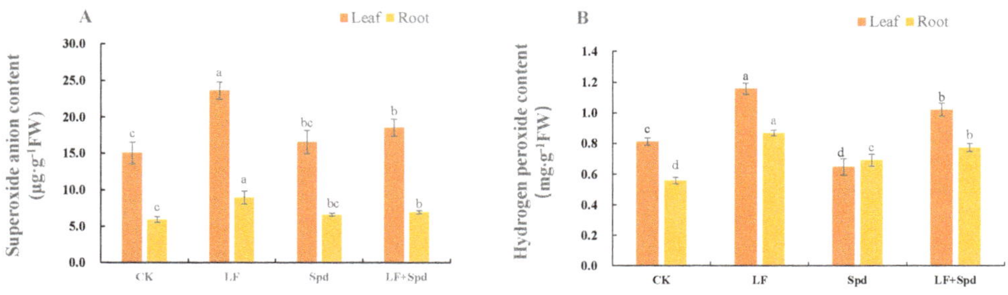

Figure 8. Effect of exogenous Spd on superoxide anion (O_2^-) content and hydrogen peroxide (H_2O_2) content under low-iron stress in tomato plants (**A**,**B**). Means denoted by the different lower case letters are significantly different according to Duncan's multiple range test ($p \leq 0.05$); the mean represents the average of three replicates and the vertical bar indicates $\pm$ standard deviation (SD).

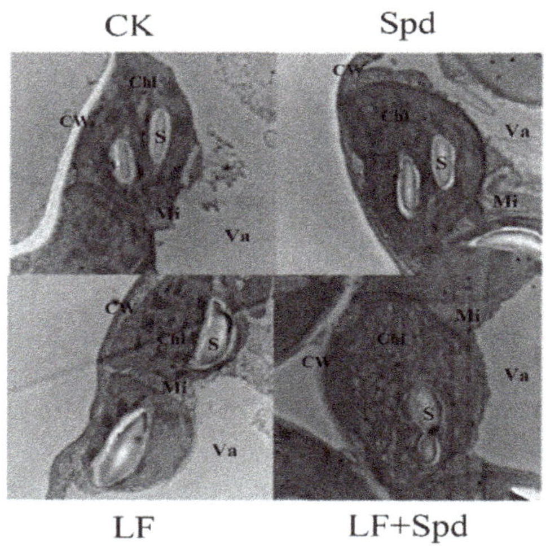

Figure 9. Ultra-structure of leaf cells revealed by transmission electron microscopy. CW, cell wall; Va, vacuole; Chl, chloroplast; Mi, mitochondria; S, starch grain.

We also analyzed the levels of osmoregulatory substances such as proline, sugars and proteins, which are vital for osmotic regulation under stressful conditions in plants. The proline contents in both leaves and roots significantly increased by 40.91 and 32.05%, respectively, and the free amino acid content significantly decreased by 31.20 and 14.79%, respectively, under low-Fe stress when compared with the control. Interestingly, the proline and free amino acid contents in leaves and roots increased with Spd foliar treatment under low-Fe stress compared to low-Fe stress only. The soluble protein content decreased in leaves and roots under low-Fe stress; however, it increased by 13.45% in leaves and 31.16% in roots after Spd foliar treatment under low-Fe stress. The soluble sugar content in leaves significantly decreased by 38.62% under low-Fe stress, while there was no significant change in this in roots. However, compared to low-Fe stress alone, treatment with Spd and low-Fe stress increased the soluble sugar content significantly in both leaves and roots (Figure 11).

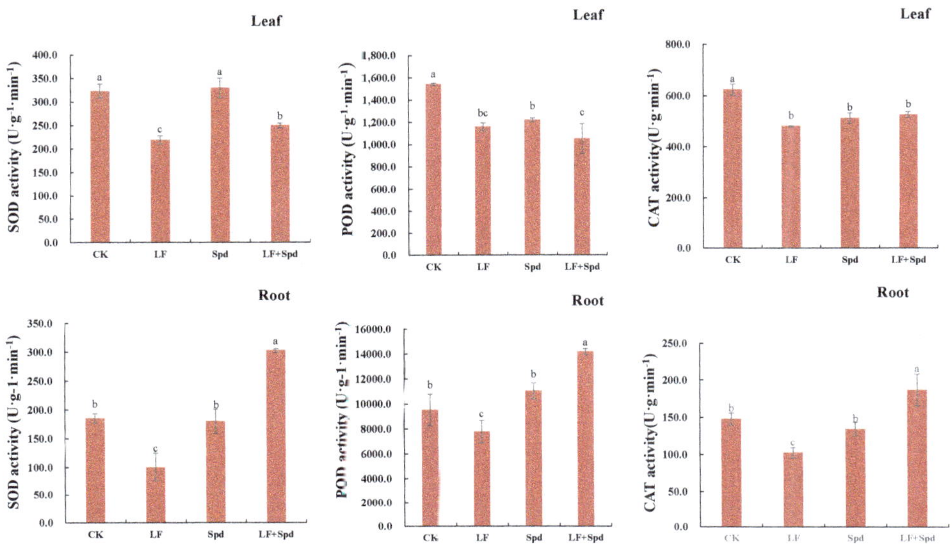

Figure 10. Effect of exogenous Spd on antioxidant enzyme activity under low-iron stress in tomatoes. The first row represents the leaves and the second row represents the roots. Means denoted by the different lower case letters are significantly different according to Duncan's multiple range test ($p \leq 0.05$); the mean represents the average of three replicates and the vertical bar indicates ± standard deviation (SD).

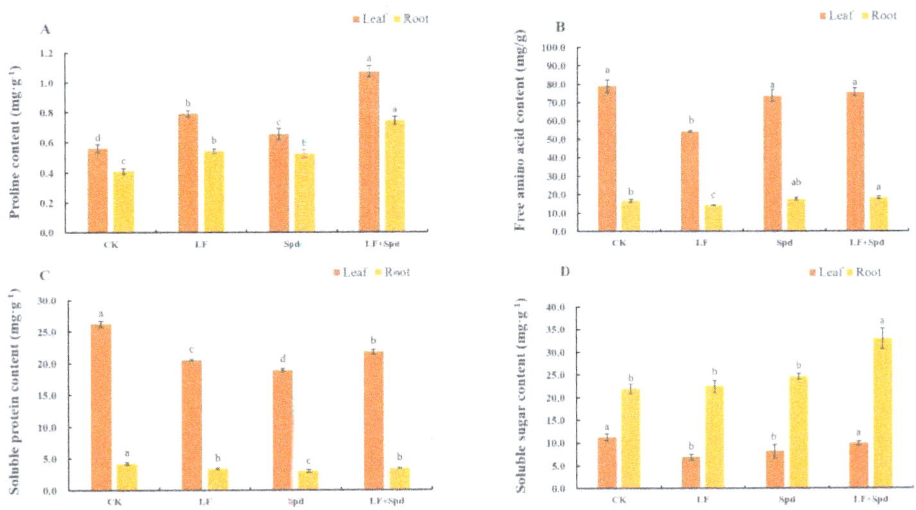

Figure 11. Effect of exogenous Spd on proline content, free amino acid content, soluble protein content and soluble sugar content under low-iron stress in tomato plants (**A–D**). Means denoted by the different lower case letters are significantly different according to Duncan's multiple range test ($p \leq 0.05$); the mean represents the average of three replicates and the vertical bar indicates ± standard deviation (SD).

3.5. Effect of Exogenous Spd on the Organic Acid Content in Roots and the Polyamine Content in Leaves under Low-Iron Stress in Tomato Plants

Oxalic, malic, acetic and citric acids in the root system responded differently to low-Fe stress (Table 6). The oxalic acid level was not significantly altered by low-Fe stress compared to the control; however, the citric and malic acid contents increased by 78.12 and 69.58%, respectively, and the acetic acid content decreased by 49.76% in tomato roots under low-Fe stress. Spd treatment under low-Fe stress significantly increased the contents of citric (49.15%), malic (172.76%) and acetic acids (310.88%) compared to low-Fe stress alone, suggesting that exogenous Spd treatment-induced increased secretion of organic acids from the root potentially enhanced the Fe transport capacity.

Table 6. Effects of exogenous Spd on organic acid content in tomato roots under low-iron stress.

Treatments	Oxalic Acid/(mg·g^{-1})	Malic Acid/(μg·g^{-1})	Citric Acid/(μg·g^{-1})	Acetic Acid/(μg·g^{-1})
CK	2.68 ± 0.13 a	357.77 ± 28.45 c	90.95 ± 4.85 c	114.52 ± 15.87 c
LF	2.73 ± 0.13 a	606.70 ± 97.39 bc	162.00 ± 23.73 b	57.54 ± 12.40 d
Spd	2.24 ± 0.09 b	684.47 ± 167.26 b	133.55 ± 9.60 bc	210.01 ± 11.88 b
LF + Spd	2.75 ± 0.35 a	1654.85 ± 218.26 a	241.63 ± 40.10 a	236.42 ± 15.10 a

CK, control; LF, Low Fe; Spd, spermidine; LF + Spd, Low Fe plus spermidine. Data are shown as mean ± SD. Entries within each column followed by the same lowercase letters are not significantly different according to Duncan's test at $p \leq 0.05$.

Meanwhile, under low-Fe stress, soluble and bound Put, Spd and Spm concentrations increased in the leaves, while free Put decreased and free Spd and Spm did not significantly change (Table 7). It is likely that free polyamines were converted to bound polyamines, which increased the bound polyamines under stress conditions. However, all three forms of polyamines, except for bound Spd, increased to different degrees after Spd foliar-spraying under low-Fe stress. This showed that exogenous Spd treatment could improve the biosynthesis and interconversion of endogenous polyamines to increase the plants' ability to withstand stress.

Table 7. Effects of exogenous Spd on the polyamine content in tomato leaves under low-iron stress.

Treatments	Free Polyamine (nmol·g^{-1})			Soluble Conjugated Polyamine (nmol·g^{-1})			Bound Polyamine (nmol·g^{-1})		
	Put	Spd	Spm	Put	Spd	Spm	Put	Spd	Spm
CK	662.64 ± 16.76 b	429.20 ± 57.33 b	142.69 ± 1.68 b	111.21 ± 39.69 c	30.44 ± 3.91 d	60.37 ± 2.54 c	828.56 ± 91.03 c	238.62 ± 27.29 c	451.48 ± 36.43 c
LF	545.76 ± 43.10 c	568.52 ± 43.73 b	160.10 ± 3.17 ab	229.51 ± 46.45 b	85.60 ± 4.36 b	96.29 ± 2.73 b	6909.74 ± 433.32 a	2064.22 ± 130.05 a	794.57 ± 43.54 b
Spd	848.99 ± 64.31 a	574.61 ± 30.55 b	160.37 ± 13.66 ab	129.79 ± 18.81 c	55.02 ± 4.26 c	63.07 ± 2.60 c	1060.74 ± 47.86 c	308.22 ± 14.31 c	496.38 ± 18.08 c
LF + Spd	702.82 ± 55.56 b	878.03 ± 140.96 a	170.73 ± 12.87 a	408.15 ± 78.32 a	103.22 ± 4.38 a	119.36 ± 9.19 a	4963.23 ± 340.45 b	1479.83 ± 102.24 b	1103.26 ± 55.84 a

CK, control; LF, Low Fe; Spd, spermidine; LF + Spd, Low Fe plus spermidine. Data are shown as mean ± SD. Entries within each column followed by the same lowercase letters are not significantly different according to Duncan's test at $p \leq 0.05$.

3.6. Effect of Exogenous Spd on Sugar Metabolism in Tomato Leaves under Low-Iron Stress

In addition to being a source of energy for plant metabolism, sucrose has also been identified as a signaling molecule involved in the regulation of Fe deficiency. The sucrose content in leaves increased by 41.52, 28.24 and 48.57% with time after low-Fe treatment, and was higher than the control. Spd-spraying under low-Fe stress significantly reduced the sucrose content in the leaves compared to the low-Fe treatment (Figure 12).

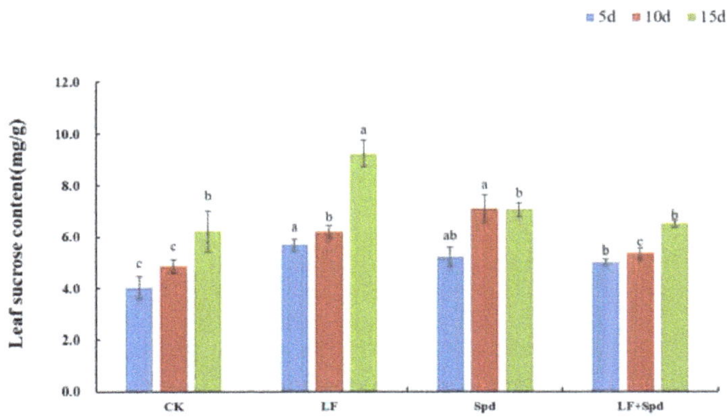

Figure 12. Effects of exogenous Spd on the sucrose content in tomato leaves under low-iron stress. Means denoted by the different lowercase letters on the same color bars are significantly different according to Duncan's multiple range test ($p \leq 0.05$); the mean represents the average of three replicates and the vertical bar indicates ± standard deviation (SD).

The results of the measurement of enzymes' activities related to sugar metabolism showed that under low-iron stress, the activities of both SS and SPS enzymes decreased, while the activities of two conversion enzymes, NI and AI, increased on day 10 after low-Fe treatment, although the sucrose content increased rather than decreased. This shows that the catabolic direction of SS and SPS enzyme activities was greater than the synthetic direction under low-Fe stress, and with the decrease in enzyme activities, the transport of photosynthetic products was blocked, causing the accumulation of sucrose in leaves, while the degradation and utilization of sucrose were weakened, which, in turn, stimulated the activities of two converting enzymes, NI and AI, and maintained the stability of sucrose anabolism. Exogenous Spd treatment significantly increased the SS activity and decreased the sucrose content, indicating that Spd promotes the degradation of sucrose, accelerates the consumption of sucrose transported from the leaves, promotes the transfer of photosynthetic products from the source to the reservoir and prevents the inhibitory effect of sucrose accumulation on photosynthetic efficiency (Supplementary Table S4).

4. Discussion

Iron is a vital element for the metabolism, growth and development of plants. Nevertheless, the lack of adaptive mechanisms to combat iron deficiency severely impairs plant biomass accumulation. Biomass is a direct manifestation of plant growth variation and can be an important basis for assessing the degree of plant injury due to stress [45]. Roots not only provide structural support to the above-ground parts of the plant but also provide nutrients and water. Therefore, the survival of a plant depends on its proper growth, development and root function [46]. Under low-iron stress, a decrease in above-ground and below-ground biomass (Table 2), and a suppressed total root length, total root surface area and total root volume of seedlings were observed (Supplementary Table S3). Morphological inhibition is one of the adverse effects caused by low-iron stress, and our results were consistent with earlier accounts of iron-deficiency effects on crop plants [47]. This is because adverse stress conditions inhibit both the division and growth of root cells, causing a significant decline in root biomass [48]. However, foliar-spraying with Spd increased not only above-ground and below-ground biomass but also the total root length, total root surface area, total root volume, root vigor and Fe^{3+} reductase activity, which potentially improved nutrient acquisition and alleviated low-iron stress in tomato seedlings.

Since photosynthesis is the most essential plant process, its efficiency has a significant influence on growth, yield and stress resistance in plants [49]. In this study, the photosynthetic pigment content of tomato leaves was significantly inhibited under low-iron stress and leaf photosynthetic activity was drastically reduced (Tables 3 and 4), which is consistent with the findings of Yao et al. [50]. This is because iron-deficiency stress hinders chlorophyll synthesis in tomato seedlings, leading to a reduction in chloroplast lamellae and disruption of the chloroplast structure. However, the chlorophyll contents in tomato leaves increased significantly after Spd foliar-spraying. Such effects support the hypothesis that the ability to capture and convert light energy was restored, and the exogenously sprayed Spd could safeguard chloroplasts and protect the photosynthetic mechanism from the adverse effects of environmental stress [51]. Moreover, chlorophyll fluorescence parameters such as Fv/Fm, PSII, ETR, etc., decreased significantly and NPQ increased under low-iron stress (Table 5), which was in agreement with the previous findings [52]. This is because damage to the photosystem II reaction centered on low-iron stress-inhibited PS II photochemical activity, reduced PS II primary light energy conversion efficiency and hindered the photosynthetic electron transfer process. Consistent with the previous reports in Sweet Corn [53], exogenous Spd increased the chlorophyll content and stabilized the photosynthetic system in tomato seedlings, thus alleviating the damage to the photosystem and enhancing or restoring photosynthetic efficiency. It can be inferred that exogenous Spd-spraying is crucial to improve the photosynthetic efficiency of tomato seedlings, leading to increased biomass and dry matter accumulation.

Polyamines protect plants from environmental stress by regulating the accumulation of sugar, proline and other osmotic substances [54]. Proline is an important osmotic adjustment substance in plants that functions in maintaining the membrane structure and is used as a physiological and biochemical indicator for the plant stress response [55]. Du [56] showed that the proline content in plants under stress increased, and was further increased by Spd treatment, which is in agreement with our results showing that proline content in leaves and roots of tomato seedlings under low-iron stress increased significantly compared to the control, and were further significantly increased after foliar-spraying of Spd under low-iron stress compared to LF treatment. The proteins synthesized and stored during plant growth are degraded to free amino acids for biosynthesis to maintain normal plant life activities [57]. When plants are subjected to stress, particularly osmotic stress, the soluble sugar content increases, which can improve the osmoregulatory capacity of leaves and provide carbon and nitrogen sources for plant organic matter synthesis [58]. The soluble sugar content in leaves and roots of tomato seedlings decreased under low-iron stress; however, exogenous Spd treatment increased the soluble sugar content in tomato seedlings under low-iron stress, suggesting that Spd improves the ability of plants to synthesize sugars [59]. To improve the plant tolerance to iron deficiency, roots can reduce the inter-root pH by secreting organic acids and increasing Fe^{3+} solubility [60]. Exogenous spraying with Spd significantly increased the content of citric and malic acids in the root system, which indicates that Spd potentially increases the secretion of organic acids in the root system, thus enhancing the iron transport in plants [61].

Plant performance under multiple abiotic stresses is linked to the accumulation of Put, Spd and Spm [62]. In this study, the content of all three forms of polyamines increased to different degrees after Spd-spraying, which is consistent with the results of Shan et al. [61]. It is highly likely that exogenous Spd treatment potentially improves the biosynthesis of endogenous polyamines and significantly enhances the ability of plants to withstand adversity. Moreover, the study also found that Spd treatment significantly increased SS enzyme activity, reduced sucrose content, promoted sucrose degradation, accelerated sucrose consumption, facilitated the transfer of photosynthetic products from source to sink and prevented the inhibitory effect of sucrose accumulation on photosynthetic efficiency in tomato seedlings [63], which is consistent with the results of our study.

Under stress conditions, reactive oxygen species (ROS) are profusely generated in plants, causing oxidative stress and damage to important molecules in plants [55,64]. The cell membrane is a barrier that maintains the relative stability of plant cells. Under stress conditions, the degree of membrane lipid peroxidation intensifies due to excessive accumulation of ROS, which changes the membrane permeability and affects the normal physiological and biochemical reactions [65]. In this study, low-iron stress reduced SOD, POD and CAT activities in tomato plants and weakened their ability to scavenge ROS, resulting in excessive intracellular O_2^- and H_2O_2 accumulation, increased membrane permeability and disruption of plant cell membranes (Figure 8). This relies on the fact that iron acts as a component of enzymes such as SOD, POD and CAT, and the three enzymes' activities were significantly inhibited when plants were subjected to a low-iron environment. After exogenous spraying of Spd treatment, the SOD, POD and CAT activities increased to different degrees and O_2^-, H_2O_2, MDA and the relative conductivity decreased, indicating that Spd effectively alleviated the extent of cell membrane disruption.

In iron-chelating reductase *FRO7* mutant plants, the iron content in chloroplasts and the activity of iron reductase are significantly lower than in wild-type plants, and the electron transport chain in the photosystem is interrupted, causing impaired photosynthesis [66]. Moreover, *FRO7* mutant plants show a severe yellowing phenotype, along with the occurrence of seedling lethality, indicating that the *FRO7* gene is important for maintaining iron homeostasis in chloroplasts and for the proper performance of photosynthesis in the plant [66]. In the present study, Spd treatment under low-iron stress upregulated the expression of the *FRO* gene and related Fe transporter genes *IRT1* and *IRT2* in the root, which is consistent with the results of a previous study in *Pyrus betulaefolia* [67].

Previous studies established that IAA plays an important role as a signaling molecule in the response to iron deficiency in plants, and that the local iron supply affects the plant lateral root growth and development by inducing the growth hormone AUX-1 transporter [68]. The strategy-I plants induce ethylene synthesis in response to iron-deficiency stress, and ethylene positively regulates the iron-deficiency response [69]. The ethylene response factor *ERF4/ERF72* is involved in iron-deficiency response in apple rootstocks, and interference with these two genes results in upregulated expression of iron-uptake genes in *Ziziphus jujube* roots, promoting iron uptake by the roots [70]. Accordingly, we also found that transcript levels of *ERF1* and *ERF2* genes were upregulated in the root system under low-iron stress, and exogenous spraying of Spd treatment further upregulated the expression of *ERF1* genes in the leaves, while it downregulated them in the root. Differential expression of these genes related to growth hormones and ethylene, together with the expression of downstream *FRO* and *IRT1* genes, potentially contributed to improved iron acquisition and transport under low-iron stress.

Meanwhile, sucrose accumulation in leaves increased under low-iron stress, which indicated that the translocation capacity of sucrose to the root system was possibly reduced; nonetheless, sucrose could act as a long-range signal to regulate the response of plants to Fe deficiency [71]. It is worth noting at this point that the expression of genes such as *COX15* in chlorophyll metabolism was downregulated under low-iron stress, indicating that the transport of sucrose to the lower part of the ground was inhibited [72]. The upregulated expression of genes such as *CHLP*, *PetF* and *CAT*, which are involved in chlorophyll synthesis and antioxidant enzyme activities, as well as significantly upregulated *SUS* and *TPS* gene expression and significantly increased sucrose synthase activity after Spd-spraying, indicated that Spd treatment also affected sugar metabolism to confer tolerance to low-iron stress in tomato plants.

5. Conclusions

Iron (Fe) deficiency severely limits agricultural crop yield due to its low availability, particularly in soils with a high pH. The success of iron fertilization largely depends on soil pH management, which is very challenging in field conditions. In this study, we showed that foliar application of exogenous plant growth regulator Spd could improve plant tolerance to low-iron stress. Briefly, the transcriptomic analysis revealed that exogenous Spd could regulate the plant response to low-iron stress by modulating the expression of genes involved in the processes of hormone metabolism, sucrose metabolism, antioxidant defense system, photosynthesis, chlorophyll metabolism and Fe uptake and transport. Besides this, biochemical and physiological analyses revealed that low-iron stress-induced suppression, in photosynthesis and growth of tomato seedlings, were significantly alleviated by exogenous Spd treatment, which was closely associated with differential modulation of photosynthetic pigment contents, gas exchange, chlorophyll fluorescence capacity, proline content, sucrose content, root vigor, citric and malic acid contents, ROS metabolism and polyamine synthesis and interconversion. Overall, this study reveals the critical mechanism of exogenous Spd-induced enhanced tolerance to low-iron stress in tomatoes and provides a novel characterization of the key traits associated with the adaptation of tomatoes to a low-iron environment. Traits associated with changes in low-iron-tolerance genes can potentially be used to improve yields of greenhouse tomatoes in low-iron environments. Nonetheless, large-scale experimentation is required to unveil and extend this knowledge, to develop better agricultural practices.

Supplementary Materials: The following supporting information can be downloaded at: https://www.mdpi.com/article/10.3390/antiox11071260/s1, Figure S1: The relative expression of selected differentially expressed genes verified by qRT-PCR; Figure S2: Effect of exogenous Spd on (A) root vigor and (B) Fe^{3+} reductase activity of tomato under low-iron stress; Figure S3: Effect of exogenous Spd on lipid peroxidation (A) and ion leakage (B) in tomato roots under low-iron stress; Table S1: qRT-PCR test reaction system; Table S2: Primers used for qRT-PCR; Table S3: Effects of exogenous Spd on tomato root morphological indexes under low-iron stress; Table S4: Effects of Spd on enzyme activities related to sucrose metabolism in tomato leaves under low-iron stress.

Author Contributions: Y.S., conceptualization, methodology, formal analysis, investigation and writing—original draft. Y.Z. (Yihong Zhao), formal analysis, investigation and writing—original draft. Q.Y., formal analysis and investigation. F.L., methodology and funding acquisition. X.L., formal analysis and investigation. X.J., formal analysis and investigation. Y.Z. (Yi Zhang), conceptualization, supervision, resources, writing—original draft, funding acquisition and project administration. G.J.A., conceptualization, writing—review and editing, funding acquisition and project administration. All authors have read and agreed to the published version of the manuscript.

Funding: This work was supported by the National Key R&D Program of China (2019YFD1000300), Shanxi Province Key R&D Plan (201903D211011), the Basic Research Program in Shanxi (20210302123366) and the National Natural Science Foundation of China (3195041055, 31501750, 31550110201).

Institutional Review Board Statement: Not applicable.

Informed Consent Statement: Not applicable.

Data Availability Statement: Data are contained within the article and Supplementary Materials.

Conflicts of Interest: The authors declare no conflict of interest.

References

1. Kaya, C.; Akram, N.; Ashraf, M. Influence of exogenously applied nitric oxide on strawberry (Fragaria × ananassa) plants grown under iron deficiency and/or saline stress. *Physiol. Plant.* **2018**, *165*, 247–263. [CrossRef] [PubMed]
2. Kaur, G.; Shukla, V.; Meena, V.; Kumar, A.; Singh, J.; Kandoth, K.; Mantri, S.; Raouched, H.; Pandey, A.K. Underpinning wheat physiological and molecular responses to co-occurring iron and phosphate deficiency stress. *HAL* **2020**, hal-02749803.
3. Hänsch, R.; Mendel, R.R. Physiological functions of mineral micronutrients (Cu, Zn, Mn, Fe, Ni, Mo, B, Cl). *Curr. Opin. Plant Biol.* **2009**, *12*, 259–266. [CrossRef]
4. Molassiotis, A.N.; Diamantidis, G.C.; Therios, I.N.; Tsirakoglou, V.; Dimassi, K.N. Oxidative stress, antioxidant activity and Fe (III)-chelate reductase activity of five Prunus rootstocks explants in response to Fe deficiency. *Plant Growth Regul.* **2005**, *46*, 69–78. [CrossRef]
5. Broadley, M.; Brown, P.; Cakmak, I.; Rengel, Z.; Zhao, F. Function of Nutrients: Micronutrients-ScienceDirect. In *Marschner's Mineral Nutrition of Higher Plants*, 3rd ed.; Academic Press: Cambridge, MA, USA, 2012; pp. 191–248.
6. Kumar, S.; Li, G.; Yang, J.; Huang, X.F.; Ji, Q.; Zhou, K.; Khan, S.; Ke, W.; Houet, H.W. Investigation of an Antioxidative System for Salinity Tolerance in Oenanthe javanica. *Antioxidants* **2020**, *9*, 940. [CrossRef] [PubMed]
7. Reimer, J.; Shaaban, B.; Drummen, N.; Sanjeev Ambady, S.; Genzel, F.; Poschet, G.; Wiese-Klinkenberg, A.; Usadel, B.; Wormit, A. Capsicum Leaves under Stress: Using Multi-Omics Analysis to Detect Abiotic Stress Network of Secondary Metabolism in Two Species. *Antioxidants* **2022**, *11*, 671. [CrossRef] [PubMed]

8. Song, Y.; Dong, Y.; Tian, X.; Wang, W.; He, Z. Mechanisms of exogenous nitric oxide and 24-Epibrassinolide alleviating chlorosis of peanut plants under iron deficiency. *Pedosphere* **2018**, *28*, 926–942. [CrossRef]
9. Chen, W.W.; Yang, L.J.; Cheng, J.; Mo, J.; Ye, T.; Zheng, S. Nitric oxide acts downstream of auxin to trigger root ferric-chelate reductase activity in response to iron deficiency in arabidopsis. *Plant Physiol.* **2010**, *154*, 810–819. [CrossRef]
10. Römheld, V.; Marschner, H. Iron defificiency stress induced morphological and physiological changes in root tips of sunflflower. *Physiol. Plant.* **1981**, *53*, 354–360. [CrossRef]
11. Kobayashi, T.; Nishizawa, N.K. Iron uptake, translocation, and regulation in higher plants. *Annu. Rev. Plant Biol.* **2012**, *63*, 131–152. [CrossRef]
12. Vert, G.; Grotz, N.; Dédaldéchamp, F.; Gaymard, F.; Guerinot, M.L.; Briat, J.F.; Curie, C. Correction: IRT1, an arabidopsis transporter essential for iron uptake from the soil and for plant growth. *Plant Cell* **2021**, *33*, 439–440. [CrossRef]
13. Lingam, S.; Mohrbacher, J.; Brumbarova, T.; Potuschak, T.; Fink-Straube, C.; Blondet, D.; Genschik, P.; Bauer, P. Interaction between the bHLH transcription factor fit and ethylene insensitive3/ethylene insensitive3-like1 reveals molecular linkage between the regulation of iron acquisition and ethylene signaling in arabidopsis. *Plant Cell* **2011**, *23*, 1815–1829. [CrossRef] [PubMed]
14. Aktar, F.; Islam, M.S.; Milon, M.A.-A.; Islam, N.; Islam, M.A. Polyamines: An essentially regulatory modulator of plants to abiotic stress tolerance: A review. *Asian J. Appl. Sci.* **2021**, *9*, 195–204. [CrossRef]
15. Duan, J.; Li, J.; Guo, S.; Kang, Y. Exogenous spermidine affects polyamine metabolism in salinity-stressed Cucumis sativus roots and enhances short-term salinity tolerance. *Plant Physiol.* **2008**, *165*, 1620–1635. [CrossRef] [PubMed]
16. ElSayed, A.I.; Rafudeen, M.S.; El-hamahmy, M.A.M.; Odero, D.C.; Hossain, M.S. Enhancing antioxidant systems by exogenous spermine and spermidine in wheat (*Triticum aestivum*) seedlings exposed to salt stress. *Funct. Plant Biol.* **2018**, *45*, 745–759. [CrossRef]
17. Amri, E.; Shahsavar, A.R. Response of lime seedlings (*Citrus aurantifolia* L.) to exogenous spermidine treatments under drought stress. *Aust. J. Basic Appl. Sci.* **2010**, *4*, 4483–4489.
18. Sun, X.; Xie, L.; Han, L. Effects of exogenous spermidine and spermine on antioxidant metabolism associated with cold-induced leaf senescence in Zoysiagrass (*Zoysia japonica* Steud.). *Hortic. Environ. Biotechnol.* **2019**, *60*, 295–302. [CrossRef]
19. Xu, X.; Shi, G.; Ding, C.; Xu, Y.; Zhao, J.; Yang, H.; Pan, Q. Regulation of exogenous spermidine on the reactive oxygen species level and polyamine metabolism in *Alternanthera philoxeroides* (Mart.) Griseb under copper stress. *Plant Growth Regul.* **2011**, *63*, 251–258. [CrossRef]
20. Sang, Q.; Shan, X.; An, Y.; Shu, S.; Sun, J.; Guo, S. Proteomic analysis reveals the positive effect of exogenous spermidine in tomato seedlings' response to high-temperature stress. *Front. Plant Sci.* **2017**, *8*, 120. [CrossRef]
21. Qian, R.; Ma, X.; Zhang, X.; Hu, Q.; Liu, H.; Zheng, J. Effect of exogenous spermidine on osmotic adjustment, antioxidant enzymes activity, and gene expression of gladiolus gandavensis seedlings under salt stress. *Plant Growth Regul.* **2020**, *40*, 1353–1367. [CrossRef]
22. Zhang, Y.; Liang, Y.; Zhao, X.; Jin, X.; Hou, L.; Shi, Y. Silicon compensates phosphorus deficit-Induced growth inhibition by improving photosynthetic capacity, antioxidant potential, and nutrient homeostasis in tomato. *Agronomy* **2019**, *9*, 733. [CrossRef]
23. Wang, N.; Qian, Z.; Luo, M.; Fan, S.; Zhang, X.; Zhang, L. Identification of salt stress responding genes using transcriptome analysis in green alga Chlamydomonas reinhardtii. *Int. J. Mol. Sci.* **2018**, *19*, 3359. [CrossRef] [PubMed]
24. Kenneth, J.L.; Thomas, D.S. Analysis of Relative Gene Expression Data Using Real-Time Quantitative PCR and the 2-ΔΔCт Method. *Methods* **2001**, *25*, 402–408.
25. Zhang, Y.; Li, S.; Liang, Y.; Zhao, H.; Hou, L.; Shi, Y.; Ahammed, G.J. Effects of exogenous spermidine and elevated CO_2 on physiological and biochemical changes in tomato plants under iso-osmotic salt stress. *Plant Growth Regul.* **2018**, *37*, 1222–1234. [CrossRef]
26. Mellor, R.S.; Pietro, A.S. Experimental plant physiology. *Bioence* **1974**, *24*, 418.
27. Ekmekcioglu, C.; Strauss-Blasche, G.; Marktl, W. The plasma membrane Fe3+-reductase activity of $CaCO_2$ cells is modulated during differentiation. *Biochem. Mol. Biol. Int.* **1998**, *46*, 951–961.
28. Lichtenthaler, H.K.; Wellburn, A.R. Determinations of total carotenoids and chlorophylls a and b of leafextracts in different solvents. *Biochem. Soc. Trans.* **1983**, *11*, 591–592. [CrossRef]
29. Schreiber, U.; Armond, P.A. Heat-induced changes of chlorophyll fluorescence in isolated chloroplasts and related heat-damage at the pigment level. *Biochim. Biophys. Acta* **1978**, *502*, 138–151. [CrossRef]
30. Cai, S.; Zhang, Y.; Xu, Y.; Qi, Z.; Li, M.; Ahammed, G.; Xia, X.; Shi, K.; Zhou, Y.; Retier, R.; et al. HsfA1a upregulates melatonin biosynthesis to confer cadmium tolerance in tomato plants. *Pineal. Res.* **2017**, *62*, e12387. [CrossRef]
31. Reuveni, R.; Shimoni, M.; Karchi, Z.; Kuc, J. Peroxidase Activity as a Biochemical Marker for Resistance of Muskmelon (*Cucumis melo*) to Pseudoperonospora cubensis. *Phytopathology* **1992**, *82*, 749–753. [CrossRef]
32. Garcia, C.; Fedrigo, K.; Gabriel, A.; Botelho, R.V.; Rodrigues, J.D.; Ono, E.O. Control of mildew in vines with cinnamon extract and catalase activity in organic production. *Res. Soc. Dev.* **2021**, *10*, e214101018885. [CrossRef]
33. Ukeda, H.; Kawana, D.; Maeda, S.; Sawamura, M. Spectrophotometric assay for superoxide dismutase based on the reduction of highly water-soluble tetrazolium salts by xanthine-xanthine oxidase. *Biosci. Biotechnol. Biochem.* **1999**, *63*, 485–488. [CrossRef] [PubMed]

34. Sheteiwy, M.; Shen, H.; Xu, J.; Guan, Y.; Song, W.; Hu, J. Seed polyamines metabolism induced by seed priming with spermidine and 5-aminolevulinic acid for chilling tolerance improvement in rice (*Oryza sativa* L.) seedlings. *Environ. Exp. Bot.* **2017**, *137*, 58–72. [CrossRef]
35. Li, S.; Jiang, H.; Wang, J.; Wang, Y.; Pan, S.; Tian, H.; Duan, M.; Wang, S.; Tang, X.; Mo, Z. Responses of plant growth, physiological, gas exchange parameters of super and non-super rice to rhizosphere temperature at the tillering stage. *Sci. Rep.-UK* **2019**, *9*, 10618. [CrossRef]
36. Shou, H.; Bordallo, P.; Fan, J.; Yeakley, J.M.; Bibikova, M.; Sheen, J.; Wang, K. Expression of an active tobacco mitogen-activated protein kinase kinase kinase enhances freezing tolerance in transgenic maize. *Proc. Natl. Acad. Sci. USA* **2004**, *101*, 3298–3303. [CrossRef] [PubMed]
37. Willekens, H.; Chamnongpol, S.; Davey, M.; Schraudner, M.; Langebartels, C.; Montagu, M.V.; Inzé, D.; Camp, W.V. Catalase is a sink for H_2O_2 and is indispensable for stress defence in C3 plants. *EMBO J.* **1997**, *16*, 4806–4816. [CrossRef]
38. Li, Z.; Gong, M. Improvement of measurement method for superoxide anion radical in plant. *Acta Bot. Yunnanica* **2005**, *27*, 211–216.
39. Bates, L.; Waldren, R.; Teare, I. Rapid determination of free proline for water-stress studies. *Plant Soil* **1973**, *39*, 205–207. [CrossRef]
40. Bradford, M.M. A rapid and sensitive method for the quantitation of microgram quantities of protein utilizing the principle of protein-dye binding. *Anal. Biochem.* **1976**, *72*, 248–254. [CrossRef]
41. Zhang, Y.; Hu, X.; Shi, Y.; Zou, Z.; Yan, F.; Zhao, Y.; Zhang, H.; Zhao, J. Beneficial role of exogenous spermidine on nitrogen metabolism in tomato seedlings exposed to saline-alkaline stress. *Am. Soc. Hortic. Sci.* **2013**, *138*, 38–49. [CrossRef]
42. Güler, A.; Candemir, A. Determination of physicochemical characteristics, organic acid and sugar profiles of Turkish grape juices. *Int. J. Agric. Environ. Food Sci.* **2020**, *4*, 149–156. [CrossRef]
43. Zhang, Y.; Zhou, Y.; Chen, S.; Liu, J.; Fan, K.; Li, Z.; Liu, Z.; Lin, W. Gibberellins play dual roles in response to phosphate starvation of tomato seedlings, negatively in shoots but positively in roots. *Plant Physiol.* **2019**, *234–235*, 145–153. [CrossRef] [PubMed]
44. Clarke, J.R.; Tyms, S. Rapid analysis of polyamines in cell culture by high performance liquid chromatography. *Med. Lab. Sci.* **1986**, *43*, 258–261. [PubMed]
45. Guo, L.; Born, M.L.; Niu, W.; Liu, F. Biochar amendment improves shoot biomass of tomato seedlings and sustains water relations and leaf gas exchange rates under different irrigation and nitrogen regimes. *Agric. Water Manag.* **2021**, *245*, 106580. [CrossRef]
46. Altaf, M.A.; Shahid, R.; Ren, M.; Naz, S.; Altaf, M.M.; Khan, L.; Tiwari, R.K.; Lal, M.K.; Shahid, M.A.; Kumar, R.; et al. Melatonin Improves Drought Stress Tolerance of Tomato by Modulating Plant Growth, Root Architecture, Photosynthesis, and Antioxidant Defense System. *Antioxidants* **2022**, *11*, 309. [CrossRef]
47. Carrasco-Gil, S.; Allende-Montalbán, R.L.; HernándezApaolaza, L.; Lucena, J.J. Application of seaweed organic components increases tolerance to Fe deficiency in tomato plants. *Agronomy* **2021**, *11*, 507. [CrossRef]
48. Lu, Y.; Ye, H.; Geng, S.; Huang, Z.; Liu, Z.; Long, X.; Liu, Z. Effects of NaCl stress on growth, leaf photosynthetic parameters and ion distribution of Helianthus tuberosus seedling. *Plant Resour. Environ.* **2010**, *19*, 86–91.
49. Roosta, H.; Estaji, A.; Niknam, F. Effect of iron, zinc and manganese shortage-induced change on photosynthetic pigments, some osmoregulators and chlorophyll fluorescence parameters in lettuce. *Photosynthetica* **2018**, *56*, 606–615. [CrossRef]
50. Yao, Y.; Jiang, C. Alteration of mineral elements and FTIR characterization of Hovenia orange live seedlings under iron deficiency stress. *China South Fruit Tree* **2016**, *45*, 10–14. (In Chinese)
51. Hussain, A.; Nazir, F.; Fariduddin, Q. Polyamines (spermidine and putrescine) mitigate the adverse effects of manganese induced toxicity through improved antioxidant system and photosynthetic attributes in Brassica juncea. *Chemosphere* **2019**, *236*, 124830. [CrossRef]
52. Kong, J.; Dong, Y.; Xu, L.; Liu, S.; Bai, X. Effects of foliar application of salicylic acid and nitric oxide in alleviating iron deficiency induced chlorosis of Arachis hypogaea L. *Bot. Stud.* **2014**, *55*, 9. [CrossRef] [PubMed]
53. Cai, S.; Wang, G.; Xu, H.; Liu, J.; Luo, J.; Shen, Y. Exogenous Spermidine Improves Chilling Tolerance in Sweet Corn Seedlings by Regulation on Abscisic Acid, ROS and Ca^{2+} Pathways. *Plant Biol.* **2021**, *64*, 487–499. [CrossRef]
54. Li, L.; Gu, W.; Li, C.; Li, W.; Li, C.; Li, J.; Wei, S. Exogenous spermidine improves drought tolerance in maize by enhancing the antioxidant defence system and regulating endogenous polyamine metabolism. *Crop Pasture Sci.* **2018**, *69*, 1076–1091. [CrossRef]
55. Zhang, T.; Wang, Y.; Ma, X.J.; Ouyang, Z.; Deng, L.; Shen, S.; Dong, X.; Du, N.; Dong, H.; Guo, Z.; et al. Melatonin Alleviates Copper Toxicity via Improving ROS Metabolism and Antioxidant Defense Response in Tomato Seedlings. *Antioxidants* **2022**, *11*, 758. [CrossRef]
56. Du, J.; Shu, S.; An, Y.; Zhou, H.; Guo, S.; Sun, J. Influence of exogenous spermidine on carbon–nitrogen metabolism under Ca $(NO_3)2$ stress in cucumber root. *Plant Growth Regul.* **2017**, *81*, 103–115. [CrossRef]
57. Fatima, A.; Adnan, A.S.; Arif, S.A.; Abdullah, Q.A.A. The effects of free amino acids profiles on seeds germination/dormancy and seedlings development of two genetically different cultivars of Yemeni Pomegranates. *Stress Physiol. Biochem.* **2012**, *8*, 114–137.
58. Long, W.; Li, Q.; Wan, N.; Feng, D.; Kong, F.; Zhou, Y.; Yuan, J. Root morphological and physiological characteristics in maize seedlings adapted to low iron stress. *PLoS ONE* **2020**, *15*, e0239075. [CrossRef]
59. Ahanger, M.A.; Qin, C.; Qi, M.; Dong, X.; Parvaiz, P.; FathiAbd_Allah, E.; Zhang, L. Spermine application alleviates salinity induced growth and photosynthetic inhibition in Solanum lycopersicum by modulating osmolyte and secondary metabolite accumulation and differentially regulating antioxidant metabolism. *Plant Physiol. Biochem.* **2019**, *144*, 1–13. [CrossRef]

60. Kabir, A.H.; Paltridge, N.G.; Able, A.J.; Paull, J.G.; Stangoulis, J. Natural variation for Fe-efficiency is associated with upregulation of Strategy I mechanisms and enhanced citrate and ethylene synthesis in *Pisum sativum* L. *Planta* **2012**, *235*, 1409–1419. [CrossRef]
61. Boquete, M.T.; Lang, I.; Weidinger, M.; Richards, C.L.; Alonso, C. Patterns and mechanisms of heavy metal accumulation and tolerance in two terrestrial moss species with contrasting habitat specialization. *Environ. Exp. Bot.* **2021**, *182*, 104336. [CrossRef]
62. Ma, Y.; Wang, P.; Gu, Z.; Tao, Y.; Shen, C.; Zhou, Y.; Han, Y.; Yang, R. Ca^{2+} involved in GABA signal transduction for phenolics accumulation in germinated hulless barley under NaCl stress. *Food Chem. X* **2019**, *2*, 100023. [CrossRef] [PubMed]
63. Shan, X.; Zhou, H.; Sang, T.; Shu, S.; Sun, J.; Guo, S. Effects of Exogenous Spermidine on Carbon and Nitrogen Metabolism in Tomato Seedlings under High Temperature. *Am. Soc. Hortic. Sci.* **2016**, *141*, 381–388. [CrossRef]
64. Lamhamdi, M.; Bakrim, A.; Aarab, A.; Lafont, R.; Sayah, F. A comparison of lead toxicity using physiological and enzymatic parameters on spinach (*Spinacia oleracea* L.) and wheat (*Triticum aestivum* L.) growth. *Moroc. J. Biol.* **2010**, *12*, 6–7.
65. Srivastava, R.K.; Pandey, P.; Rajpoot, R.; Rani, A.; Dubey, R.S. Cadmium and lead interactive effects on oxidative stress and antioxidative responses in rice seedlings. *Protoplasma* **2014**, *251*, 1047–1065. [CrossRef]
66. Wang, Y.Y. *Studies on the Role of bZIP44 Gene in Response to Iron Deficiency in Plants*; Hefei University of Technology: Hefei, China, 2019. (In Chinese)
67. Xie, C.; Jin, X.; Yan, L.; Shi, X.; Liu, H.; Xu, Y.; Dong, C. Effects of iron deficiency stress on physiology and gene expression of iron absorption and transpotation in Pyrus betulaefolia. *Nanjing Agric. Univ.* **2019**, *42*, 465–473. (In Chinese)
68. Giehl, R.H.; Lima, J.; von Wirén, N. Localized Iron Supply Triggers Lateral Root Elongation in Arabidopsis by Altering the AUX1-Mediated Auxin Distribution. *Plant Cell* **2012**, *24*, 33–49. [CrossRef]
69. Liu, W. *Functional Study of Ethylene Response Factor ERF4/ERF72 in Response to Iron Deficiency in Apple Rootstock*; China Agricultural University: Beijing, China, 2017. (In Chinese)
70. Bai, Z.Q.; Zu, H.H.; Wang, R.; Gao, X.X.; Zou, T.; Chen, G.L.; Wu, J.W. Molecular role of ethylene in fruit ripening of Ziziphus jujube Mill. *Plant Signal Behav.* **2020**, *15*, 1834749. [CrossRef]
71. Chen, P.F.; Chen, L.; Jiang, Z.R.; Wang, G.P.; Wang, S.H.; Ding, Y.F. Sucrose is involved in the regulation of iron deficiency responses in rice (*Oryza sativa* L.). *Plant Cell Rep.* **2018**, *37*, 789–798. [CrossRef]
72. Bashir, K.; Hanada, K.; Shimizu, M.; Seki, M.; Nakanishi, H.; Nishizawa, N.K. Transcriptomic analysis of rice in response to iron deficiency and excess. *Rice* **2014**, *7*, 18. [CrossRef]

antioxidants

MDPI

Article

Radio-Protective Effects of Stigmasterol on Wheat (*Triticum aestivum* L.) Plants

Hebat-Allah A. Hussein [1,2,*], Shifaa O. Alshammari [3], Fatma M. Elkady [4], Amany A. Ramadan [4], Sahar K. M. Kenawy [1] and Aisha M. Abdelkawy [1]

1 Botany and Microbiology Department, Faculty of Science (Girls Branch), Al-Azhar University, Cairo 11754, Egypt; saharkenawy.2052@azhar.edu.eg (S.K.M.K.); aisha.abdelgalil.5921@azhar.edu.eg (A.M.A.)

2 Biology Department, University College of Nairiyah, University of Hafr Al-Batin, Nairiyah 31991, Saudi Arabia

3 Biology Department, College of Science, University of Hafr Al-Batin, Hafr Al-Batin 39524, Saudi Arabia; dr.shifaa@uhb.edu.sa

4 Department of Botany, Agriculture, Biological Research Institute, National Research Centre, Dokki, Giza 12311, Egypt; fatma_elkady99@yahoo.com (F.M.E.); amanyramadan66@yahoo.com (A.A.R.)

* Correspondence: hebahussein@azhar.edu.eg or hebaali@uhb.edu.sa

Abstract: Ionizing radiation is abiotic stress limiting the growth and productivity of crop plants. Stigmasterol has positive effects on the plant growth of many crops. The role of stigmasterol in alleviating the effects of ionizing radiation on plant metabolism and development is still unclear. Therefore, the study aimed to investigate the effects of pretreatments with γ-radiation (0, 25, and 50 Gy), foliar application of stigmasterol (0, 100, and 200 ppm), and their interaction on the growth, and biochemical constituents of wheat (*Triticum aestivum* L., var. Sids 12) plants. Gamma radiation at 25 Gy showed no significant difference in plant height, root length, no. of leaves, shoot fresh weight, root fresh weight, Chl *a*, ABA, soluble phenols, and MDA compared to the control values. Gamma rays at 50 Gy inhibited shoot and root lengths, flag leaf area, shoot fresh and dry weights, photosynthetic pigments, total soluble sugars, proline, and peroxidase activity. However, it stimulated total phenols, catalase activity, and lipid peroxidation. On the other hand, stigmasterol at 100 ppm showed no significant effects on some of the physiological attributes compared to control plants. Stigmasterol at 200 ppm improved plant growth parameters, photosynthetic pigments, proline, phenols, antioxidant enzyme, gibberellic acid, and indole acetic acid. Correspondingly, it inhibited total soluble sugars, abscisic acid, and lipid peroxidation. Moreover, the application of stigmasterol caused the appearance of new polypeptides and the reappearance of those missed by gamma radiation. Overall, stigmasterol could alleviate the adverse effects of gamma radiation on wheat plants.

Keywords: antioxidants; biochemical metabolites; plant growth; stigmasterol; phytohormones; wheat

Citation: Hussein, H.-A.A.; Alshammari, S.O.; Elkady, F.M.; Ramadan, A.A.; Kenawy, S.K.M.; Abdelkawy, A.M. Radio-Protective Effects of Stigmasterol on Wheat (*Triticum aestivum* L.) Plants. *Antioxidants* **2022**, *11*, 1144. https://doi.org/10.3390/antiox11061144

Academic Editor: Nafees A. Khan

Received: 9 May 2022
Accepted: 6 June 2022
Published: 10 June 2022

1. Introduction

Wheat (*Triticum aestivum*, L.) is the main strategic cereal crop. It has high nutritional value as it is rich in carbohydrates, essential amino acids, fiber components, vitamins, and minerals [1]. Global wheat production amounted to around 778.6 million tons from 2020 to 2021 [2]. Wheat is used for making bread, starch, and wheat germ oil. The nutritional value of grains of bread-making quality depends mainly on the various protein constituents [3]. Wheat plants are exposed to numerous stressors, such as ionizing radiation stress, that seriously affect plant growth and productivity [4].

Ionizing radiation is a pollutant that could potentially lead to disturbances in ecosystems [5]. Gamma-radiation is an electromagnetic wave that results in ionizing radiation that impacts different biological macromolecules and induces variable biological effects [6,7]. The effects of gamma-rays on plant growth and development are diverse, ranging from stimulatory to inhibitory effects depending on the radiation dose, exposure duration, the response and sensitivity of different plant species and cultivars, and their interaction inside the cell, especially in water, to produce free radicals [8] that change the biochemical processes of plants [9]. Previous studies have investigated the effect of dose and exposure time of gamma-ray irradiation on seed germination and physiological parameters. Gamma rays at 100 Gy enhanced growth, yield characters, and certain biochemical constitutes of fenugreek [10] and wheat plants [11]. A recent study revealed that IS-Jarissa wheat varieties were able to live after exposure to radiation doses of 0 Gy, 100 Gy, 200 Gy, 300 Gy, and 400 Gy [12]. Additionally, gamma radiation at 200 Gy enhanced flavonoid compounds in 37 wheat lines [13], while ionizing radiation from 10 to 1000 Gy caused dramatic alterations in the composition of plant cells and induced cell death [14,15]. Mean germination time, root and shoot length, and seedling dry weight of wheat genotypes (Roshan and T-65-58-8) decreased with increasing radiation doses (100, 200, 300, and 400 Gy) [16]. Additionally, the radiation doses of 10, 15, 20, 25 and 30 kR caused different types of chromosomal anomalies in wheat plants, which increased with the increasing intensity of gamma radiation [17].

Plants produce mixtures of sterols, including stigmasterol, campesterol, and sitosterol. Stigmasterol belongs to the plant sterols and is a precursor of brassinosteroids, which act as growth regulators [18]. Additionally, Campesterol is the precursor of BR; the crucial role of BR in plant growth and development is well established [19], while sitosterol participates in cellulose synthesis [20]. Stigmasterol is implicated in the structure of phospholipid constituents, maintaining plasma membrane fluidity and permeability [19,21,22]. Moreover, foliar application of stigmasterol improved the growth characters, yield, anatomical structures, and percentage/composition of essential oil of basil plants [23]. Stigmasterol may be involved in gravitropism and tolerance to abiotic stress [19]. Indeed, stigmasterol application of germinating seeds enhanced the salt tolerance of faba beans and flax plants [24,25]. It is, however, unclear whether stigmasterol can overcome the harmful effects of ionizing radiation stress. As the cost of stigmasterol is brought down to affordable levels [24], the present research theme may contribute greatly to the usage of stigmasterol in agriculture production as well as to overcome the threat of ionizing radiation on crop plants around the world. Therefore, our purpose is to examine the role of stigmasterol in alleviating the adverse effects of γ-radiation on wheat plants and to understand the direct role of stigmasterol in plant growth and stress responses.

2. Materials and Methods

2.1. Experimental Design

A greenhouse experiment was conducted during the winter season of 2017/2018 at the Faculty of Science (Girls Branch), Al-Azhar University (latitude 30°03′22.3″ N longitude 31°19′25.4″ E), Nasr City, Cairo, Egypt. The study aimed to test the effects of pre-treatment with γ-radiation (0, 25, and 50 Gy) and foliar application of stigmasterol (0, 100, and 200 ppm) on growth, biochemical constituents, and yield characters of wheat plants (*Triticum aestivum*, L.). Wheat grains (Cultivar Sids 12) were received from the Agriculture Research Center (latitude 30°01′13.4″ N longitude 31°12′24.2″ E), Giza, Egypt. Stigmasterol, purity $\geq$ 95, soluble in chloroform, 50 mg/mL, MP Biomedicals, LLC, France, was used. The seeds were irradiated by gamma ^{60}Co at different doses including 0 (non-irradiated), 25, and 50 Gray (Gy) at the Egyptian Atomic Energy Authority (latitude 30°02′41.8″ N longitude 31°20′41.0″ E), Cairo, Egypt. Wheat grains were sown on November 21 in earthenware pots (no. 50) filled with sandy soil with six replicates for each treatment. The soil texture was sandy, field capacity 11.5%, pH 8.7, EC 0.35 dSm^{-1}, Cl^- 1.7, HCO_3^- 1.10, Na^+ 1.2, K^+ 0.25, Ca^{2+} 1.27%, and Mg^{2+} 0.58 meq L^{-1}. Phosphorus fertilizer was added before sowing at a rate of 6.0 g per pot of calcium superphosphate (15.5% P_2O_5). Nitrogen fertilizer was applied in two equal portions at a rate of 0.60 g/pot for each in the form of ammonium nitrate (33.5%N) at 30 and 60 days after planting. Potassium fertilizer was applied as a soil application at the rate of 2 g/pot in the form of potassium sulfate (48–52% K_2O) 45 days after planting. The foliar spray of stigmasterol treatments was applied at the vegetative stage (45 days after sowing).

2.2. Growth Parameters

At 65 days after sowing, three representative samples were collected from each treatment for detecting the tested growth traits (shoot length (cm), root length (cm), leaves (no. per plant, flag leaf area), as well as fresh and dry weights of shoot and root per plant).

2.3. Photosynthetic Pigments

According to the procedure in [26], photosynthetic pigments—chlorophyll-a (Chl *a*), chlorophyll-b (Chl *b*)—and total carotenoids in samples of wheat fresh leaves tissues using 85% acetone with 0.1 g fresh weight (FW) of wheat leaves after grinding in the solvent was measured. The homogenized samples were centrifuged at 3000 rpm, and the filtrate was topped up to 10 mL with acetone (85%). The absorbance was recorded at 663, 644, and 452 nm using a spectrophotometer (VEB, Carl-Zeiss-Promenade, Jena, Germany) using acetone as a blank. The concentration of the pigment fractions (Chl *a*, Chl *b*, and carotenoids) was accounted for as μg/mL using the following equations:

$$\text{Chl } a = [(10.3 \times \text{E663}) - (0.918 \times \text{E644})] = \mu\text{g mL}^{-1} \quad (1)$$

$$\text{Chl } b = [(19.7 \times \text{E644}) - (3.870 \times \text{E663})] = \mu\text{g mL}^{-1} \quad (2)$$

$$\text{Carotenoids} = (4.2 \times \text{E452}) - [(0.0264 \times \text{Chl } a) + (0.426 \times \text{Chl } b)] = \mu\text{g mL}^{-1} \quad (3)$$

The concentrations of chlorophylls and carotenoids were expressed as mg g^{-1} fresh weight (FW) of plant material. Pigment contents are represented as mg g^{-1} FW.

2.4. Identification of Endogenous Hormones

A total of 10 g of fresh tissue per sample was homogenized with 80% (v/v) ethanol and stirred overnight at 4 °C. The extract was filtered through a Whatman filter and the methanol was evaporated under a vacuum. The aqueous phase was adjusted to pH 2.5 with 1 N HCl, then partitioned with ethyl acetate (3 times), and finally passed through anhydrous sodium sulfate. After that, the ethyl acetate phase was evaporated under a vacuum. The dry residue containing acidic hormones (fraction I) was dissolved in 2.0 mL of methanol and stored in vials at 4 °C. The phytohormones (auxins, gibberellins, and abscisic acid) were determined by high-performance liquid chromatography (Shimadzu, Tokyo, Japan), isocratic UV analysis, and a reverse-phase C18 column (RP-C18 μ Bondapak, Waters). The column used included octadecylsilane (ODS) ultra-sphere particles (5 μm), and the mobile phases used were acetonitrile/water (26:74 v/v) at pH 4.00. The flow rate was 0.8 mL min^{-1}, and detection was UV 208 nm. The standard solutions of the individual acids (auxins, gibberellins, and abscisic acid (Sigma, St. Louis, MO, 63178, USA)) were prepared in the mobile phase and chromatographed. All solvents were purchased from Aldrich (Munich, Germany).

2.5. Total Soluble Sugars

Total soluble sugars (TSS) were estimated in dry flag leaves of the wheat plant by the anthrone technique [27]. The TSS was analyzed by reacting 0.1 mL of ethanol extract with 3.0 mL freshly prepared anthrone (150 mg anthrone (Aldrich Chemical Company Inc., Milwaukee, WI 53233, USA) + 100 mL 72% H_2SO_4) in a boiling water bath for 10 min. The cooled samples were read at 625 nm using a spectrophotometer (VEB, Carl-Zeiss-Promenade, Jena, Germany). Total soluble sugar was calculated using a standard curve of glucose.

2.6. Proline

Proline content was determined according to [28]. A total of 0.5 g of fresh leaves was extracted in 10 mL of aqueous sulfosalicylic acid (3%). A total of 2 mL of the extract was taken and mixed with 2 mL of acid ninhydrin reagent (Alpha, Mumbai, 400 002, India) and 2 mL of glacial acetic acid for one h at 100 °C. After cooling, 4 mL of toluene was added to the reaction mixture to extract the proline content. The absorbance was recorded using a spectrophotometer (VEB, Carl-Zeiss-Promenade, Jena, Germany) at 520 nm using toluene as a blank. Free proline was determined from the standard curve of *L*-proline.

2.7. Total Phenolics

Total phenolic content in dry leaves was estimated by the method described by [29,30]. One milliliter of the extract was added to ten drops of concentrated HCl in a boiling water bath for ten min and cooled. Then 1 mL of Folin–Ciocalteau reagent and 1.5 mL of 14% sodium carbonate were added to the extract. The mixture was completed with distilled water up to 5 mL, shaken well, and then kept in a boiling water bath for five minutes. The absorbance at 650 nm was noted, and the data were represented as mg g^{-1} FW using a pyrogallol standard curve.

2.8. Assay of Antioxidant Enzymes

The crude enzyme was obtained according to the assay of antioxidant enzyme activities. A fresh flag leaf (2 g) of wheat plants was extracted in 10 mL of 100 mM phosphate buffer (pH 6.8) and kept at 4 °C overnight. The extract was centrifuged at 5000 rpm for ten minutes and reserved to assay the activities of enzymes [31].

2.8.1. Peroxidase (POX) Assay

POX activity was assayed according to [32]. A total of 0.2 mL of crude extract was reacted with 5.8 mL of phosphate buffer (50 mM, pH 7.0), 2.0 mL pyrogallol (20 mM), and 2.0 mL hydrogen peroxide (20 mM). The increase in absorbance was determined within 60 s against a reagent without enzyme at 470 nm using a spectrophotometer. The amount of crude enzyme that converts one micromole of hydrogen peroxide in one minute at room temperature equals one unit of enzyme activity [33].

2.8.2. Catalase (CAT) Assay

The CAT activity was assayed [34] by mixing 40 µL of enzyme extract and 9.96 mL phosphate buffer (pH 7.0) containing H_2O_2 (0.16 mL of 30% H_2O_2 in 100 mL of 50 mM phosphate buffer). CAT activity was determined by measuring the rate of H_2O_2 absorbance change in one minute against a buffer blank at 250 nm using a spectrophotometer. One unit of enzyme activity is equivalent to the amount of enzyme that reduced 50% of the hydrogen peroxide in one minute at room temperature.

2.9. Lipid Peroxidation

The level of lipid peroxidation (malondialdehyde; MDA) was measured according to [35]. A total of 200 mg of fresh flag leaf of wheat plants was ground in 10 mL of 5% trichloroacetic acid (TCA) and centrifuged at 15,000 rpm for 10 min. Then, 2.0 mL of the extract was added to 4.0 mL of 0.5% thiobarbituric acid (Mallinckrodt. Inc., Paris, KY 40361, USA) in 20% TCA, heated in a boiling water bath for a half-hour, immediately cooled, and centrifuged at 10,000 rpm for ten minutes. The reading was noted at 532 and 600 nm using a spectrophotometer. By subtracting the absorption value at 600 nm, the absorption coefficient of 155 nmol cm^{-1} was used to assess the MDA content as nmol g^{-1} FW.

2.10. Protein Profile

The rapid freeze-dried leaf samples (0.2 g) were extracted with 1 mL of protein buffer and kept in the freezer overnight and then vortexed for 15 s and centrifuged at 5000 rpm at 4 °C for 15 min. Then, sodium dodecyl sulfate-polyacrylamide gel electrophoresis (SDS–PAGE) was performed [36]. The molecular weight of the isolated proteins was estimated using standard molecular weight markers (standard protein markers, 11–180 kDa; Sigma, St. Louis, MO, USA). The protein bands were stained with Coomassie Brilliant Blue G-250 (Sigma, St. Louis, MO, USA).

2.11. Statistical Analysis

The experiment was statistically analyzed as a split-plot design according to [37]. The significant differences were statistically evaluated by Duncan's test and one-way analysis of variance (ANOVA) using SPSS, version 18.0 (Statistical Package for Social Science, Copyright 2010, Chicago, IL, USA) to discriminate significance (defined as $p \leq 0.05$). The least significant differences (LSD) at 5% were calculated for means comparisons.

3. Results

3.1. Growth Parameters

Gamma radiation at 25 Gy significantly decreased flag leaf area, shoot dry weight, and root dry weight while showing no significant difference in plant height, root length, no. of leaves, shoot fresh weight, and root fresh weight. Moreover, except for root fresh weight, gamma radiation at 50 Gy significantly decreased ($p < 0.05$) the shoot and root lengths, no. of leaves, flag leaf area, shoot fresh and dry weights, and root dry weight of wheat plants compared with the control group (Table 1).

Table 1. Effect of stigmasterol treatments on the growth parameters of wheat plants grown from irradiated grains. The different letters (a–f) show statistical significance at $p < 0.05$. Differences are statistically significant at $p < 0.05$; vertical bars indicate ±SD.

Radiation Dose (Gy)	Stigmasterol (ppm)	Plant Height (cm)	Root Length (cm)	No. of Leaves	Flag Leaf Area (cm^2)	Shoot Fresh Weight (g)	Shoot Dry Weight (g)	Root Fresh Weight (g)	Root Dry Weight (g)
0	0	60.00 ^{b}	10.67 ^{ab}	6.00 ^{c}	20.10 ^{b}	6.08 ^{ab}	0.75 ^{bc}	1.50 ^{cd}	0.17 ^{c}
	100	63.00 ^{a}	11.00 ^{a}	6.00 ^{c}	25.77 ^{a}	6.12 ^{ab}	0.76 ^{bc}	1.26 ^{de}	0.24 ^{b}
	200	68.00 ^{a}	11.00 ^{a}	8.33 ^{a}	22.67 ^{ab}	6.21 ^{ab}	0.88 ^{a}	1.61 ^{bc}	0.28 ^{a}
25	0	58.67 ^{b}	9.67 ^{bc}	5.67 ^{c}	16.37 ^{c}	5.64 ^{bc}	0.48 ^{e}	1.45 ^{cd}	0.12 ^{e}
	100	52.00 ^{c}	10.00 ^{bc}	6.33 ^{c}	14.17 ^{c}	4.70 ^{a}	0.66 ^{cd}	1.34 ^{cd}	0.16 ^{c}
	200	68.00 ^{a}	12.00 ^{a}	7.33 ^{b}	23.90 ^{a}	6.70 ^{a}	0.83 ^{ab}	1.51 ^{e}	0.14 ^{de}
50	0	46.00 ^{d}	7.50 ^{d}	4.67 ^{d}	15.33 ^{c}	4.77 ^{d}	0.38 ^{f}	1.22 ^{de}	0.09 ^{f}
	100	59.00 ^{b}	9.10 ^{c}	5.67 ^{c}	20.90 ^{b}	5.00 ^{cd}	0.58 ^{d}	1.95 ^{a}	0.14 ^{de}
	200	66.00 ^{a}	10.00 ^{bc}	6.33 ^{c}	22.60 ^{ab}	6.63 ^{a}	0.87 ^{a}	1.83 ^{ab}	0.15 ^{cd}
LSD (0.05) for radiation or stigmasterol		2.388	0.844	0.504	1.611	0.389	0.052	0.192	0.015
LSD 0.05 for the interaction		4.136	1.463	NS	2.791	0.6743	0.091	0.332	0.025

Foliar application of stigmasterol at 100 ppm showed a significant effect on flag leaf area, root length, and root dry weight but showed no changes in plant height, no. of leaves, shoot fresh and dry weights, and root fresh weight of the non-irradiated. Stigmasterol at 200 ppm increased significantly plant height, root length, no. of leaves, shoot dry weight, and root dry weight while showing no changes in flag leaf, shoot fresh weight, and root fresh weight compared to the control plants.

The treated plants with 25 Gy + stigmasterol at 100 ppm caused significant increases in shoot and root dry weights, decreases in plant height and shoot fresh weight, and no significant effects on root length, flag leaf area, and root fresh weight compared to the corresponding control. Regarding the irradiated plants with 25 Gy, stigmasterol at 200 ppm significantly ($p < 0.05$) improved the shoot length, root length, number of leaves/plant, flag leaf area, shoot fresh and dry weights, and root dry weight over the control values.

The irradiated plants with 50 Gy and stigmasterol at 100 ppm, except for shoot fresh weight, showed significant increases in the following growth parameters: shoot length, root length, number of leaves/plant, flag leaf area, shoot dry weight, and root fresh and dry weights (28.26%, 21.33%, 21.41%, 36.33%, 52.63%, 59.84%, and 55.56%, respectively) of wheat plants over the corresponding control. Regarding the irradiated plants with 50 Gy, stigmasterol at 200 ppm significantly ($p < 0.05$) improved the shoot length, root length, number of leaves/plant, flag leaf area, fresh and dry weight of shoot, and root by 43.5%, 33.3%, 42.8%, 47.4%, 39.1%, 129.0%, 95.0%, and 66.7%, respectively, over the control values.

The effect of stigmasterol wheat plant growth depends on the applied concentration. However, stigmasterol at 100 ppm showed no significant effects on some of the physiological attributes compared to control plants. Foliar application of stigmasterol at 200 ppm could improve the growth parameters in non-irradiated and 25 Gy-irradiated plants. Moreover, 200 ppm could counteract the effects of 50 Gy gamma radiation on wheat plant growth.

3.2. Photosynthetic Pigments

Irradiated grains had a significant ($p < 0.05$) influence on the photosynthetic pigments, Chl *a*, Chl *b*, and carotenoids of wheat plants (Figure 1). Radiation at 25 Gy showed no significant effect on Chl *a*, while it decreased significantly Chl *b* and carotenoids compared to the control plants. Radiation at 50 Gy produced progressive damage in the photosynthetic pigments compared to the control plants. On the other hand, the application of stigmasterol at 100 and 200 ppm showed significant ($p < 0.05$) increases in Chl *a* and Chl *b*, and no significant effect in carotenoids in non-irradiated plants compared to the control plants. Moreover, the application of stigmasterol at 100 and 200 ppm significantly ($p < 0.05$) increased Chl *a*-, Chl *b*-, and carotenoid-irradiated plants compared with corresponding control values. Stigmasterol at 200 ppm was the most effective treatment for increasing the photosynthetic pigments.

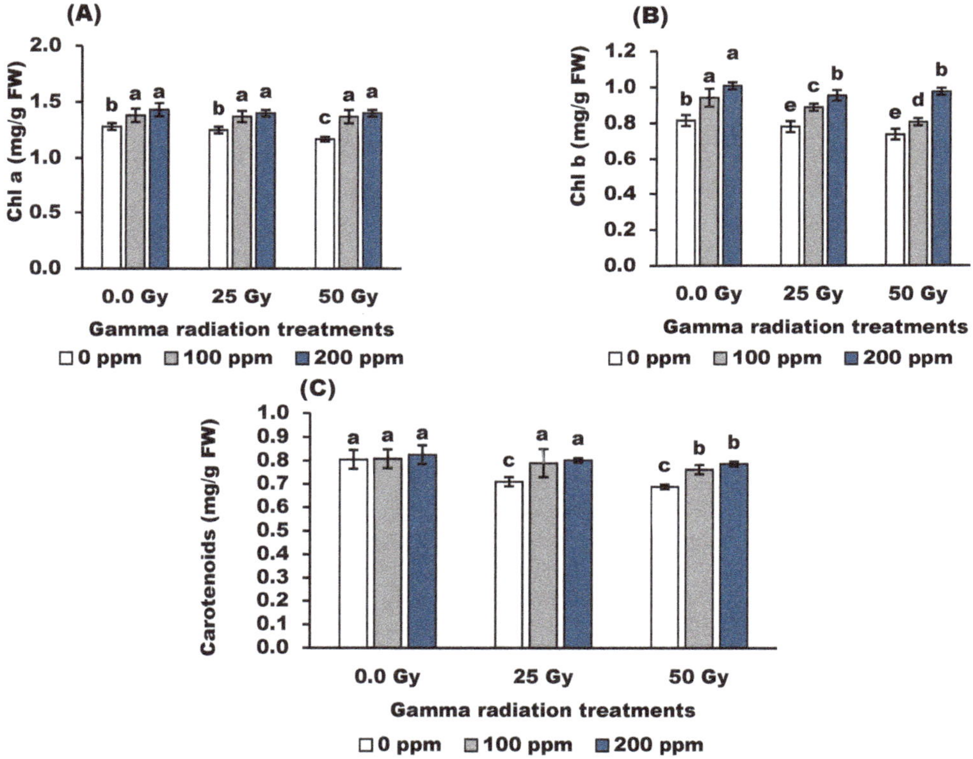

Figure 1. Effect of stigmasterol treatments on photosynthetic pigment; (**A**) Chl *a*, (**B**) Chl *b*, and (**C**) carotenoids contents of wheat plants grown from grain irradiated with gamma rays. The different letters (a–e) show statistical significance at $p < 0.05$; vertical bars indicate ± SD.

3.3. Endogenous Phytohormones

Data in Figure 2 revealed that gamma radiation (25 and 50 Gy) significantly decreased in GA and IAA compared to the control values. Moreover, 25 Gy showed no significant effect on ABA content, while 50 Gy caused a significant increase compared to the control value. In terms of stigmasterol, 200 ppm seemed to be more effective than the control, while 100 ppm treatments were not analyzed. The results showed that stigmasterol at 200 ppm markedly increased the growth promoter (GA3 and IAA) while decreasing the growth inhibitor (ABA) compared with control (non-irradiated plants).

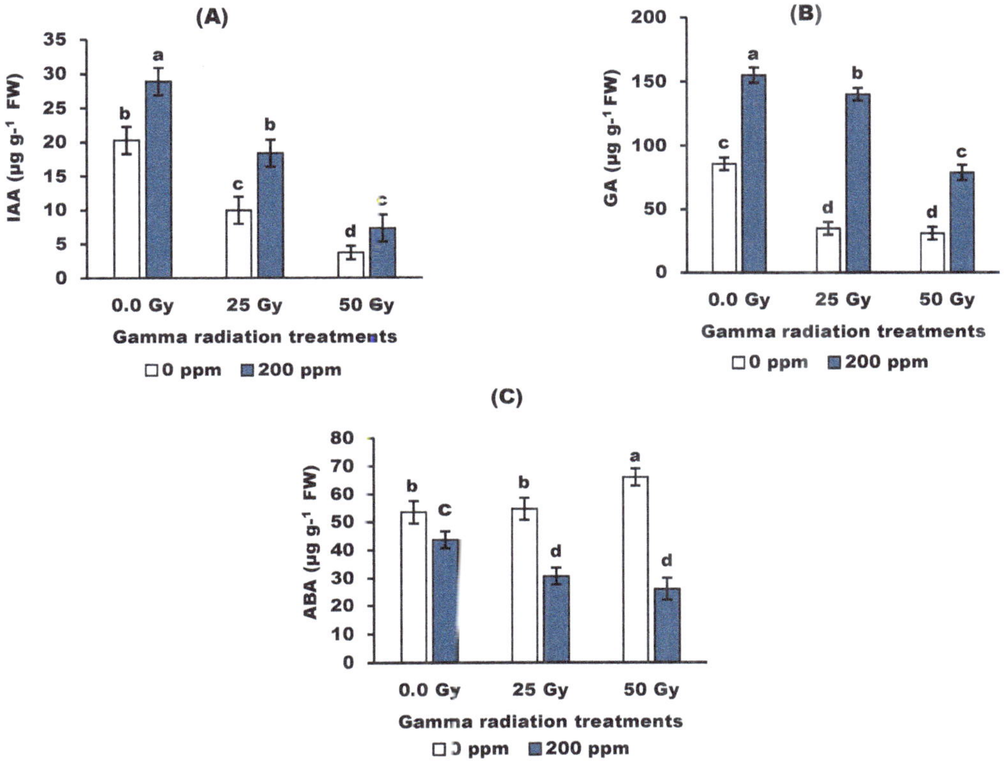

Figure 2. Effect of stigmasterol treatments on endogenous phytohormones; (**A**) IAA, (**B**) GA, and (**C**) ABA of wheat plants grown from grain irradiated with gamma rays. The different letters (a–d) show statistical significance at $p < 0.05$; vertical bars indicate ±SD.

3.4. Compatible Solutes

Concerning TSS content, the pretreatments with gamma radiation resulted in dramatic decrements ($p < 0.05$) in total soluble sugar (TSS) in wheat plants. Similarly, TSS decreased with increasing concentrations of stigmasterol in non-irradiated and irradiated plants. The control plants had the highest values of TSS compared to the other treatments.

Total soluble phenols showed no significant response to gamma radiation at 25 but were significantly stimulated ($p < 0.05$) with 50 Gy compared with the control value (Figure 3). In addition, stigmasterol at 100 and 200 ppm increased significantly the soluble phenol content in non-irradiated and 25 Gy-irradiated wheat plants over the control values. However, 100 ppm stigmasterol showed no significant effect on total phenols in 50 Gy-irradiated plants compared to the corresponding control. The interaction between 50 Gy and 200 ppm stigmasterol achieved the highest value of total phenols compared to the other treatments.

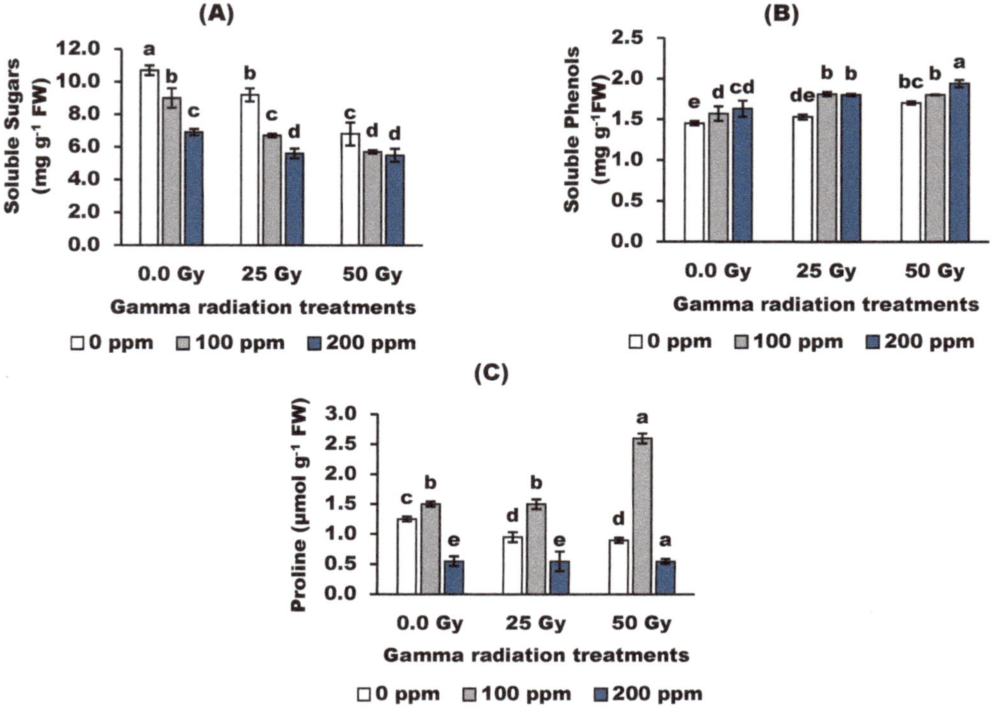

Figure 3. Effect of stigmasterol treatments on compatible solutes; (**A**) soluble sugars, (**B**) soluble phenols, and (**C**) proline of wheat plants grown from grain irradiated with gamma rays. The different letters (a–e) show statistical significance at $p < 0.05$; vertical bars indicate ±SD.

Gamma radiation dramatically decreased proline content compared to the control value. The effect of foliar application of stigmasterol on the proline content depended on the applied concentration. Stigmasterol at 100 ppm significantly increased the proline content in wheat plants over the corresponding control values. However, the maximum increase (108%) in proline content was in 50 Gy-irradiated plants treated with 100 ppm stigmasterol compared to the control value. Stigmasterol at 200 ppm shifted the proline content to the minimum value below the control value.

3.5. Antioxidant Enzymes

The data in Figure 4 indicated that irradiated grain with gamma radiation at 25 Gy and 50 Gy significantly ($p < 0.05$) increased the activity of the catalase (CAT) enzyme. Stigmasterol at 100 ppm and 200 ppm significantly ($p < 0.05$) increased the activity of CAT enzyme in wheat plants originating from irradiated and un-irradiated grains. Irradiated plants with 50 Gy and treated with 200 ppm stigmasterol increased the CAT activity by 13.0 (U/g) over the control 3.6 (U/g).

Peroxidase (POX) induced a significant decrease ($p < 0.05$) in response to the tested dose of gamma radiation in wheat plants. Radiation at 25 Gy and 50 Gy significantly decreased POX activity by 25.0% and 50.7%, respectively, compared to control values. On the other hand, stigmasterol increased POX activity compared to untreated wheat plants. Stigmasterol at 100 ppm resulted in the highest increase percentage, reaching 49.73, 84.43, and 173.0% in the 0 Gy-, 25 Gy-, and 50 Gy-irradiated plants, respectively, compared to the corresponding untreated plants.

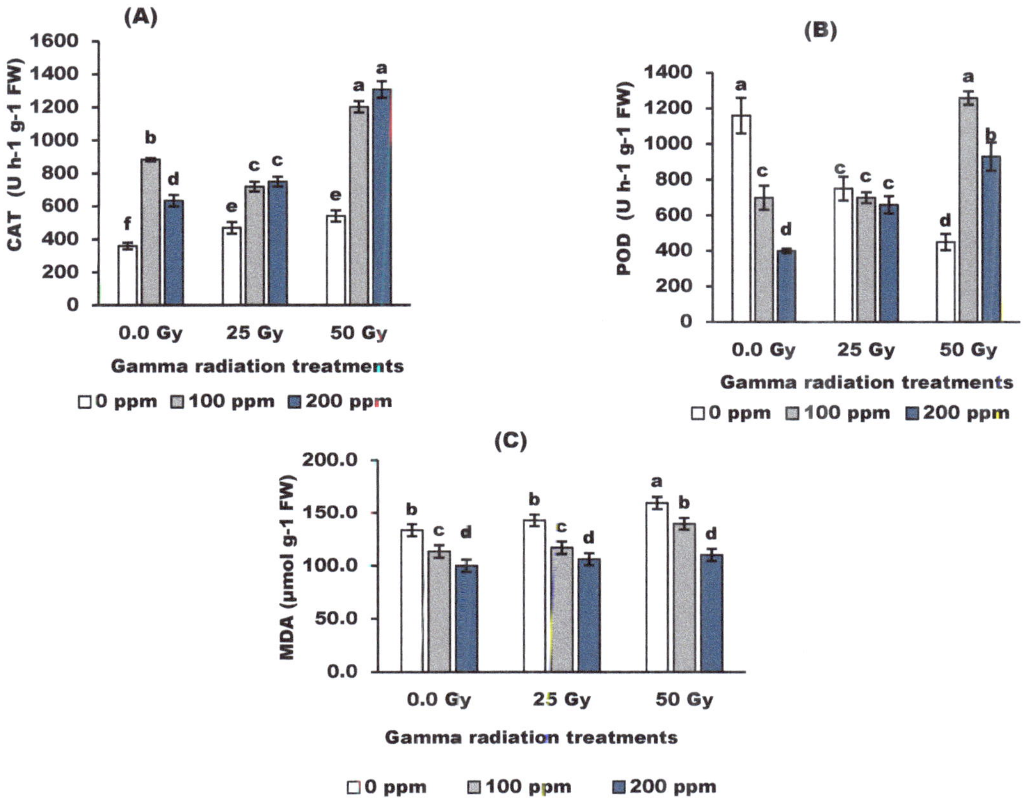

Figure 4. Effect of stigmasterol treatments on the activities of antioxidant enzymes; (**A**) CAT and (**B**) POD, and (**C**) lipid peroxidation of wheat plants grown from grain irradiated with gamma rays. The different letters (a–f) show statistical significance at $p < 0.05$; vertical bars indicate ±SD.

3.6. Lipid Peroxidation

Gamma radiation at 25 Gy showed a non-significant effect on lipid peroxidation MDA), while 50 Gy significantly increased ($p < 0.05$) MDA compared with the control value. Moreover, foliar application of stigmasterol lowered ($p < 0.05$) the MDA content in treated plants below the control. Stigmasterol at 200 ppm decreased MDA by 25%, 25.2%, and 31.25%, respectively, in the control plants and the 25 Gy- and 50 Gy-irradiated plants.

3.7. Protein Profile

Irradiated grain and application of stigmasterol induced a synthesis or disappearance of different protein bands compared to control plants (Table 2 and Figure 5). There were 23 polypeptide bands ranging from 11 to 185 kDa in the wheat proteins profile. Control plants had six protein bands with molecular weights of 11, 23, 30, 35, 43, and 48 KDa. The results revealed that 25 Gy induced the new polypeptides with 26 and 51 kDa. In addition, gamma radiation at 50 Gy induced polypeptides with 51 and 96 kDa. Moreover, foliar application of 200 ppm stigmasterol caused the appearance of new polypeptide bands with molecular weights of 185, 150, 141, 137, 89, 74, and 37 kDa.

Table 2. Changes in protein profile in leaves of wheat plants originated from irradiated grains and treated with stigmasterol. L1: Control, L2: 25 Gy, L3: 50 Gy, L4: 100 ppm stigmasterol, L5: 25 Gy + 100 ppm, L6: 25 Gy + 200 ppm, L7: 200 ppm stigmasterol, L8: 50 Gy + 100 ppm, L9: 50 Gy + 200 ppm.

No.	MW	L1	L2	L3	L4	L5	L6	L7	L8	L9
1	180	-	-	-	-	-	-	+	-	-
2	157	-	-	-	-	-	+	-	-	-
3	150	-	-	-	-	-	-	+	-	-
4	141	-	-	-	-	-	-	+	-	-
5	137	-	-	-	-	-	+	+	+	+
6	134	-	-	-	-	-	-	-	-	+
7	105	-	-	-	-	-	-	-	+	-
8	104	-	-	-	-	-	+	-	-	-
9	96	-	-	+	-	-	-	-	-	-
10	89	-	-	-	-	-	+	+	+	+
11	79	-	-	-	-	-	+	-	-	-
12	74	-	-	-	-	-	-	-	-	+
13	64	-	-	-	-	-	-	+	-	-
14	51	-	+	+	-	-	-	-	-	-
15	48	+	+	+	+	+	+	+	+	+
16	43	+	+	+	+	+	+	+	+	+
17	38	-	-	-	-	-	-	-	+	+
18	37	-	-	-	-	-	-	+	-	-
19	35	+	+	+	+	+	+	+	+	+
20	30	+	+	-	-	-	+	-	+	+
21	26	-	+	-	-	-	-	-	-	-
22	23	+	+	+	+	+	+	+	+	+
23	11	+	+	+	+	+	+	+	+	+
No. of Bands		6	8	8	6	8	13	11	9	12

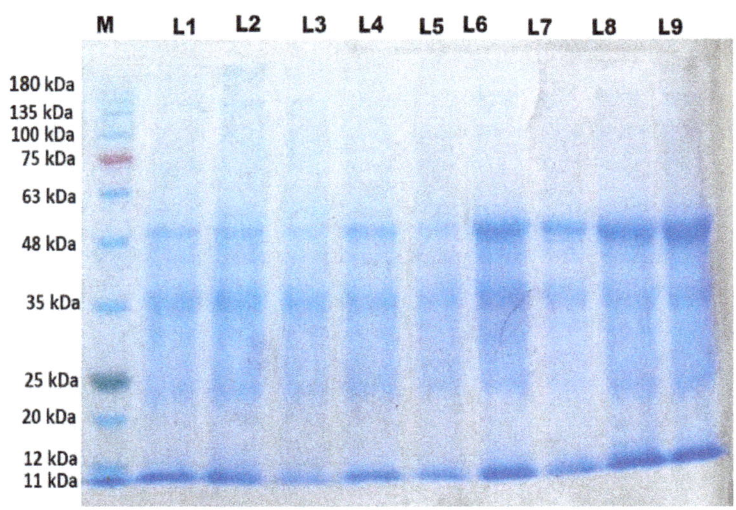

Figure 5. Changes in protein profile in leaves of wheat plants originated from irradiated grains and treated with stigmasterol. L1: Control, L2: 25 Gy, L3: 50 Gy, L4: 100 ppm stigmasterol, L5: 25 Gy + 100 ppm, L6: 25 Gy + 200 ppm, L7: 200 ppm stigmasterol, L8: 50 Gy + 100 ppm, L9: 50 Gy + 200 ppm.

The combination of 25 Gy and 200 ppm stigmasterol induced the formation of new polymorphic polypeptides with molecular weights of 157, 137, 104, 89, and 79 kDa and unique bands at 137, 105, 89, and 30 KDa. Furthermore, the interaction of 50 Gy and 200 ppm stigmasterol resulted in the appearance of the protein bands at 137, 134, 89, 74, and 38 KDa. The treatment of 50 Gy, stigmasterol, and the interaction (25 Gy and 100 ppm stigmasterol) led to the disappearance of 30 kDa compared to the control plants.

4. Discussion

The results indicated no significant difference in plant height, root length, no. of leaves, shoot fresh weight, or root fresh weight between 0 Gy and 25 Gy treatments, while the vegetative growth of wheat plants decreased with radiation dose (50 Gy). The changes in the plant growth exposed to gamma radiation are due to the conversation of metabolic energy and metabolites to ameliorate the oxidative stress effects imposed by ionizing radiation [8]. The growth inhibition of wheat plants may be due to gamma radiation inhibition and the photosynthetic pigments, IAA, GA3, and the antioxidant capacity while inducing ABA and MDA, as mentioned in this study. The negative response of wheat plants is dependent on the nature and extent of the disturbance of cellular metabolism and finally cell damage. For the stigmasterol applied treatments after gamma pretreatment, some of the growth parameters did not differ between 0 ppm and 100 ppm of stigmasterol treatments (Table 1). On the other hand, the application of stigmasterol at 200 ppm improved the growth parameters of both irradiated and un-irradiated wheat plants (Table 1). Stigmasterol plays a vital role in regulating plant growth and development [21,23]. Furthermore, they added that these improvements may be due to increasing the efficiency of water uptake and utilization, enhancing cell division, and/or cell enlargement.

Leaf chlorophylls are indicators of chloroplast structure and are considered the central part of photosynthetic systems. Data in Figure 1 indicated that Chl *a*, Chl *b*, and carotenoids decreased significantly with 50 Gy gamma radiation doses. These results agree with the previous study [38] on barley plants. The reduction in photosynthetic pigments may be related to the increasing chlorophyll photo-oxidation and damage to the photosynthetic apparatus. Interestingly, the application of stigmasterol can alleviate the damaging effects of gamma radiation on photosynthetic pigment content. The increase in photosynthetic pigment in response to stigmasterol could be attributed to increased photosynthetic apparatus and antioxidant enzyme activities [5].

Interestingly, gamma radiation modulates and alters the endogenous hormones of wheat plants represented by the inhibition of GA and IAA, associated with the accumulation of ABA. These results were confirmed by the findings recorded by [39] on barley plants using gamma radiation (50 Gy). Moreover, foliar spraying of the irradiated plants with 200 ppm stigmasterol showed a marked increase in growth promoter levels (GA3, IAA), while the growth inhibitor (ABA) was decreased as compared with control plants. Similarly, the authors in [40] found that 150 or 200 ppm of stigmasterol resulted in the highest GA3 and IAA and the lowest ABA in both cultivars when compared to untreated plants. In addition, the authors in [41,42] stated that stigmasterol increases phytohormones, which can act as messengers and regulators of plant growth and development. The changes in the endogenous hormones were correlated with the changes in the activity of antioxidant enzymes, MDA, and proline, suggesting that the application of stigmasterol plays a central role in counteracting the injurious effects of gamma radiation.

In addition, changes in total soluble sugars in wheat plants may indicate that gamma radiation has induced oxidative stress [43]. The accumulation of TSS by stigmasterol may be due to scavenging ROS to maintain them at the optimum level, hence protecting cell metabolism from free radicals. This explanation was supported by the fact that stigmasterol may act as a photosynthesis activator in wheat plants (Figure 1).

Phenolics are important constituents with scavenging ability due to their hydroxyl groups, which hence may contribute directly to their antioxidant properties. Total phenol content was increased in 50 Gy-irradiated wheat plants. The alterations in the effect of gamma radiation on phenols may occur because irradiation can break the chemical bonds of bioactive compounds, releasing soluble phenolics with low molecular weight and increasing the antioxidant potential of these compounds [44]. On the contrary, stigmasterol increased total phenol content compared to untreated plants. The significant positive correlation between stigmasterol and total phenol content may be induced by protective mechanisms against cell damage resulting from oxidative stress. These findings are consistent with those of [45], who demonstrated that stigmasterol promoted the antioxidant defense mechanism to counteract the negative effects of gamma radiation on faba bean plants.

Regarding proline, gamma radiation leads to a marked decrease in proline content. These results may be due to the radiosensitivity of proline or its oxidation using free radicals generated by gamma irritation. Moreover, stigmasterol stimulated proline content in irradiated and un-irradiated wheat plants. Similar results were obtained by [42] on flax plants. The results may be due to the role of stigmasterol, which stimulates proline content with high antioxidant properties for use in repair mechanisms against radiation effects on plant cell metabolism.

The application of gamma radiation (25 and 50 Gy) decreased the activity of POX enzymes while increasing the CAT activity. Ionizing radiation can trigger the production of ROS by interacting with atoms or molecules in the cell, which is called water radiolysis [38]. The results suggest that the increased activity of CAT induced by gamma irradiation can scavenge excess ROS, especially with the inhibition of peroxidase activity, consequently leading to the enhancement of the antioxidant capacity to overcome oxidative stress induced by water radiolysis. Moreover, treatment with stigmasterol increased POX and CAT activities (Figure 4). It appears that the increase in antioxidant enzymes may be due to the mediated role of stigmasterol in detoxification mechanisms against radio-oxidative stress.

Lipid peroxidation was markedly increased in the 50 Gy-irradiated wheat plants while decreasing with exogenous stigmasterol treatment. The increase in lipid peroxidation due to gamma radiation treatments was confirmed by [46] on the Black gram (*Vigna mungo* L.), who explained that ROS can react with nearly all cell constituents, which triggers free radical chain reactions that eventually cause membrane lipid peroxidation. They added that the membranes lose their stability, and their permeability is enhanced, leading to damage to the cell structure and disturbances of normal physiological functions as a result of free radical reactions. Moreover, treated wheat plants with stigmasterol improved stress tolerance by decreasing membrane lipid peroxidation in comparison to the corresponding control. This indicates that stigmasterol has a protective role in counteracting the damage induced by gamma radiation, resulting in the induction of antioxidant enzymes (POX and CAT) and non-enzymatic compounds (phenols and proline) associated with lower lipid peroxidation of wheat plants.

Gamma radiation induced the new polypeptides with molecular weights (Table 2 and Figure 5). These results agree with [47] on fenugreek and [38] on barley plants. They indicated that the formation of new bands (unique) is frequently caused by various DNA structural changes (e.g., breaks, transpositions, and deletions), which cause changes in amino acids and, as a result, the protein generated. Furthermore, proteins may play a role in signal transduction, anti-oxidative defense, and osmolyte production, all of which are crucial to the function and growth of plants [48]. In this regard, the authors in [49] suggested that a band with a molecular weight of 51 kDa could be linked to the Rubisco activase enzyme. Ribulose-1, 5-bisphosphate carboxylase activase is a key enzyme that initiates both photosynthetic and photo-respiratory carbon metabolism. Moreover, according to [45], a protein with a molecular weight of 26 KDa appears to be osmotically expressed in flax and sunflower plants under salinity stress to aid survival in stressed environments.

5. Conclusions

Gamma radiation at 25 Gy showed no significant difference in some growth parameters, Chl *a*, ABA, soluble phenols, and MDA compared to the control values. Gamma radiation at 50 Gy caused growth inhibition and lowered the photosynthetic pigments, promoting hormone and antioxidant capacity while inducing ABA hormone and lipid peroxidation. The foliar application of stigmasterol, especially at 200 ppm, on wheat plants originating from gamma-irradiated grains, improved the photosynthetic pigments, induced an accumulation of osmolytes, phenols, the activity of antioxidant enzymes, and new polypeptides, as well as resynthesized the missed bands by radiation. It is also noteworthy that the stimulatory effects persisted in plants pretreated with stigmasterol throughout, promoting hormones while lowering the ABA hormone. Overall, the results proved the effectiveness of stigmasterol at 200 ppm application in alleviating the adverse effects of radiation stress on wheat plants, as indicated by their protective and stimulatory effects on growth attributes and biochemical constituents.

Author Contributions: Conceptualization, H.-A.A.H.; methodology, H.-A.A.H., S.K.M.K., F.M.E. and A.M.A.; software, H.-A.A.H., S.O.A. and S.K.M.K.; validation, H.-A.A.H. and A.A.R.; formal analysis, H.-A.A.H., F.M.E. and A.A.R.; investigation, H.-A.A.H. and F.M.E.; resources, H.-A.A.H., S.O.A., S.K.M.K. and A.M.A.; data curation, H.-A.A.H., S.O.A., F.M.E. and A.A.R.; writing—original draft preparation, H.-A.A.H. and A.A.R.; writing—review and editing, H.-A.A.H. and S.O.A.; visualization, H.-A.A.H. and S.O.A.; supervision, H.-A.A.H., S.O.A., S.K.M.K. and A.M.A.; project administration, H.-A.A.H. and S.O.A.; funding acquisition, H.-A.A.H., S.O.A., S.K.M.K., F.M.E. and A.M.A. All authors have read and agreed to the published version of the manuscript.

Funding: This research received no external funding.

Institutional Review Board Statement: Not applicable.

Informed Consent Statement: Not applicable.

Data Availability Statement: The data presented in this study are available in the manuscript.

Acknowledgments: The authors are thankful to the Botany and Microbiology Department, Faculty of Science (Girls Branch), Al-Azhar University, and the Biology Department, College of Science, University of Hafr Al-Batin. The authors also would like to thank the Botany Department, Agriculture and Biological Research Institute, National Research Centre.

Conflicts of Interest: The authors declare no conflict of interest.

References

1. Ouda, S.; Noreldin, T.; Alarcón, J.J.; Ragab, R.; Caruso, G.; Sekara, A.; Abdelhamid, M.T. Response of Spring Wheat (*Triticum aestivum*) to Deficit Irrigation Management under the Semi-Arid Environment of Egypt: Field and Modeling Study. *Agriculture* **2021**, *11*, 90. [CrossRef]
2. OECD-FAO. *OECD-FAO Agricultural Outlook 2021–2030*; OECD-FAO Agricultural Outlook; OECD: Paris, France, 2021; ISBN 978-92-5-134608-2.
3. Grote, U.; Fasse, A.; Nguyen, T.T.; Erenstein, O. Food Security and the Dynamics of Wheat and Maize Value Chains in Africa and Asia. *Front. Sustain. Food Syst.* **2021**, *4*, 317. [CrossRef]
4. Di Pane, F.J.; Lopez, S.C.; Cantamutto, M.Á.; Domenech, M.B.; Castro-Franco, M. Effect of Different Gamma Radiation Doses on the Germination and Seedling Growth of Wheat and Triticale Cultivars. *Aust. J. Crop Sci.* **2018**, *12*, 1921–1926. [CrossRef]
5. Amirikhah, R.; Etemadi, N.; Sabzalian, M.R.; Nikbakht, A.; Eskandari, A. Gamma Radiation Negatively Impacted Seed Germination, Seedling Growth and Antioxidant Enzymes Activities in Tall Fescue Infected with Epichloë Endophyte. *Ecotoxicol. Environ. Saf.* **2021**, *216*, 112169. [CrossRef]
6. Reisz, J.A.; Bansal, N.; Qian, J.; Zhao, W.; Furdui, C.M. Effects of Ionizing Radiation on Biological Molecules—Mechanisms of Damage and Emerging Methods of Detection. *Antioxid. Redox Signal.* **2014**, *21*, 260–292. [CrossRef]
7. Alshammari, F.H.; Hussein, H.-A.A. Sterilization of Paper during Crisis. *AMB Express* **2022**, *12*, 13. [CrossRef]
8. Abdulhafiz, F.; Kayat, F.; Zakaria, S. Effect of Gamma Irradiation on the Morphological and Physiological Variation from In Vitro Individual Shoot of Banana Cv. Tanduk (*Musa* Spp.). *J. Plant Biotechnol.* **2018**, *45*, 140–145. [CrossRef]
9. Hussein, H.A.A.; Darwesh, O.M.; Mekki, B.B. Environmentally Friendly Nano-Selenium to Improve Antioxidant System and Growth of Groundnut Cultivars under Sandy Soil Conditions. *Biocatal. Agric. Biotechnol.* **2019**, *18*, 101080. [CrossRef]
10. Hanafy, R.S.; Akladious, S.A. Physiological and Molecular Studies on the Effect of Gamma Radiation in Fenugreek (*Trigonella foenum-graecum* L.) Plants. *J. Genet. Eng. Biotechnol.* **2018**, *16*, 683–692. [CrossRef] [PubMed]
11. Al-Rumaih, M.M.; Al-Rumaih, M.M. Influence of Ionizing Radiation on Antioxidant Enzymes in Three Species of Trigonella. *Am. J. Environ. Sci.* **2008**, *4*, 151–156. [CrossRef]
12. Dwinanda, P.; Syukur, S.; Suliansyah, I.; Suliansyah, I. Induction of Mutations with Gamma Ray Radiation to Improve the Characteristics of Wheat [*Triticum aestivum* L.] Genotype IS-Jarissa. *IOP Conf. Ser. Earth Environ. Sci.* **2020**, *497*, 012013. [CrossRef]

13. Han, A.R.; Hong, M.J.; Nam, B.; Kim, B.R.; Park, H.H.; Baek, I.; Kil, Y.S.; Nam, J.W.; Jin, C.H.; Kim, J.B. Comparison of Flavonoid Profiles in Sprouts of Radiation Breeding Wheat Lines (*Triticum aestivum*). *Agronomy* **2020**, *10*, 1489. [CrossRef]
14. United Nations Scientific Committee on the Effects of Atomic Radiation (UNSCEAR). *Sources and Effects of Ionizing Radiation, United Nations Scientific Committee on the Effects of Atomic Radiation UNSCEAR 2000 Report to the General Assembly, with Scientific Annexes*; United Nations Scientific Committee on the Effects of Atomic Radiation (UNSCEAR) Reports; United Nations: New York, NY, USA, 2000; Volume I.
15. Caplin, N.; Willey, N. Ionizing Radiation, Higher Plants, and Radioprotection: From Acute High Doses to Chronic Low Doses. *Front. Plant Sci.* **2018**, *9*, 847. [CrossRef] [PubMed]
16. Borzouei, A.; Kafi, M.; Khazaei, H.; Naseriyan, B.; Majdabadi, A. Effects of Gamma Radiation on Germination and Physiological Aspects of Wheat (*Triticum aestivum* L.) Seedlings. *Pak. J. Bot.* **2010**, *42*, 2281–2290.
17. Verma, R.C.; Khah, M.A. Assessment of Gamma Rays Induced Cytotoxicity in Common Wheat (*Triticum aestivum* L.). *Cytologia* **2016**, *81*, 41–45. [CrossRef]
18. Valitova, J.N.; Sulkarnayeva, A.G.; Minibayeva, F.V. Plant Sterols: Diversity, Biosynthesis, and Physiological Functions. *Biochemistry* **2016**, *81*, 819–834. [CrossRef]
19. Aboobucker, S.I.; Suza, W.P. Why Do Plants Convert Sitosterol to Stigmasterol? *Front. Plant Sci.* **2019**, *10*, 354. [CrossRef]
20. Peng, L.; Kawagoe, Y.; Hogan, P.; Delmer, D. Sitosterol-β-Glucoside as Primer for Cellulose Synthesis in Plants. *Science* **2002**, *295*, 147–150. [CrossRef]
21. Rogowska, A.; Szakiel, A. The Role of Sterols in Plant Response to Abiotic Stress. *Phytochem. Rev.* **2020**, *19*, 1525–1538. [CrossRef]
22. Shahzad, R.; Ewas, M.; Harlina, P.W.; Khan, S.U.; Zhenyuan, P.; Nie, X.; Nishawy, E. β-Sitosterol Differentially Regulates Key Metabolites for Growth Improvement and Stress Tolerance in Rice Plants during Prolonged UV-B Stress. *J. Genet. Eng. Biotechnol.* **2021**, *19*, 79. [CrossRef]
23. El-Tantawy, A.A.; Azoz, S.N. Enhancement of Growth and Increased Productivity of Fresh Herb and Aromatic Oil in Basil Plant by Foliar Application with Stigmasterol. *J. Med. Bot.* **2019**, *3*, 7–13. [CrossRef]
24. Hassanein, R.A.; Hashem, H.A.; Khalil, R.R. Stigmasterol Treatment Increases Salt Stress Tolerance of Faba Bean Plants by Enhancing Antioxidant Systems. *Plant Omics* **2012**, *5*, 476–485.
25. Hashem, H.A.; Bassuony, F.M.; Hassanein, R.A.; Baraka, D.M.; Khalil, R.R. Stigmasterol Seed Treatment Alleviates the Drastic Effect of NaCl and Improves Quality and Yield in Flax Plants. *Aust. J. Crop Sci.* **2011**, *5*, 1858–1867.
26. Metzner, H.; Rau, H.; Senger, H. Untersuchungen Zur Synchronisierbarkeit Einzelner Pigmentmangel-Mutanten von Chlorella. *Planta* **1965**, *65*, 186–194. [CrossRef]
27. Cerning, B.J. A Note on Sugar Determination by the Anthrone Method. *Cereal Chem.* **1975**, *52*, 857–860.
28. Bates, L.S.; Waldren, R.P.; Teare, I.D. Rapid Determination of Free Proline for Water-Stress Studies. *Plant Soil* **1973**, *39*, 205–207. [CrossRef]
29. Savitree, M.; Isara, P.; Nittaya, S.L.; Worapan, S. Radical Scavenging Activity and Total Phenolic Content of Medicinal Plants Used in Primary Health Care. *J. Pharm. Sci.* **2004**, *9*, 32–35.
30. Pourmorad, F.; Hosseinimehr, S.J.; Shahabimajd, N. Antioxidant Activity, Phenol and Flavonoid Contents of Some Selected Iranian Medicinal Plants. *Afr. J. Biotechnol.* **2006**, *5*, 1142–1145. [CrossRef]
31. Mukherjee, S.P.; Choudhuri, M.A. Implications of Water Stress-induced Changes in the Levels of Endogenous Ascorbic Acid and Hydrogen Peroxide in Vigna Seedlings. *Physiol. Plant.* **1983**, *58*, 166–170. [CrossRef]
32. Bergmayer, H.U. *Methods of Enzymatic Analysis*; Academic Press: New York, NY, USA; London, UK, 1974; pp. 1205–1214.
33. Kong, F.X.; Hu, W.; Chao, S.Y.; Sang, W.L.; Wang, L.S. Physiological Responses of the Lichen Xanthoparmelia Mexicana to Oxidative Stress of SO_2. *Environ. Exp. Bot.* **1999**, *42*, 201–209. [CrossRef]
34. Chen, C.; Yu, R.; Owuor, E.D.; Tony Kong, A.N. Activation of Antioxidant-Response Element (ARE), Mitogen-Activated Protein Kinases (MAPKs) and Caspases by Major Green Tea Polyphenol Components during Cell Survival and Death. *Arch. Pharm. Res.* **2000**, *23*, 605–612. [CrossRef] [PubMed]
35. Hodges, D.M.; DeLong, J.M.; Forney, C.F.; Prange, R.K. Improving the Thiobarbituric Acid-Reactive-Substances Assay for Estimating Lipid Peroxidation in Plant Tissues Containing Anthocyanin and Other Interfering Compounds. *Planta* **1999**, *207*, 604–611. [CrossRef]
36. Laemmli, U.K. Cleavage of Structural Proteins during the Assembly of the Head of Bacteriophage T4. *Nature* **1970**, *227*, 680–685. [CrossRef] [PubMed]
37. Snedecor, G.W.; Cochran., W.G. *Statistical Methods*; The Iowa State University Press: Ames, IA, USA, 1989.
38. Hussein, H.-A.A. Influence of Radio-Grain Priming on Growth, Antioxidant Capacity, and Yield of Barley Plants. *Biotechnol. Rep.* **2022**, *34*, e00724. [CrossRef]
39. Bitarishvili, S.V.; Volkova, P.Y.; Geras'kin, S.A. γ-Irradiation of Barley Seeds and Its Effect on the Phytohormonal Status of Seedlings. *Russ. J. Plant Physiol.* **2018**, *65*, 446–454. [CrossRef]
40. El-Greedly, N.H.; Mekki, B.B. Growth, Yield and Endogenous Hormones of Two Sesame (*Sesamum indicum* L.) Cultivars as Influenced by Stigmasterol. *J. Appl. Sci. Res.* **2005**, *1*, 63–66.
41. Bassuony, F.; Hashem, H.; Hassanein, R.; Baraka, D.; Khalil, R. Ameliorative Effect of Stigmasterol on the Productivity of *Vicia faba* Plants Grown under Salt Stress. *J. Plant Prod.* **2011**, *2*, 597–615. [CrossRef]

42. Bakry, A.B.; Sadak, M.S.; Younis, A.S.M. Some Agro-Physiological Studies of Stigmasrterol on Growth and Productivity of Some Flax Cultivars under Sandy Soil Conditions. *Am J. Agron.* **2019**, *12*, 50–63.
43. Golz, A.L.; Bradshaw, C. Gamma Radiation Induced Changes in the Biochemical Composition of Aquatic Primary Producers and Their Effect on Grazers. *Front. Environ. Sci.* **2019**, *7*, 100. [CrossRef]
44. Ito, V.C.; Zielinski, A.A.F.; Demiate, I.M.; Spoto, M.; Nogueira, A.; Lacerda, L.G. Effects of Gamma Radiation on the Stability and Degradation Kinetics of Phenolic Compounds and Antioxidant Activity during Storage of (*Oryza sativa* L.) Black Rice Flour. *Braz. Arch. Biol. Technol.* **2019**, *62*, 19180470. [CrossRef]
45. Bassuony, F.; Hashem, H.A.; Hassanein, R.; Baraka, D.; Khalil, R.R. The Impact of Stigmasterol on Growth, Productivity and Biochemical Response of *Vicia faba* L. Plants Grown under Salt Stress. *Egypt. J. Bot.* **2014**, *54*, 203–218. [CrossRef]
46. Yasmin, K.; Arulbalachandran, D.; Soundarya, V.; Vanmathi, S. Effects of Gamma Radiation (γ) on Biochemical and Antioxidant Properties in Black Gram (*Vigna mungo* L. Hepper). *Int. J. Radiat. Biol.* **2019**, *95*, 1135–1143. [CrossRef] [PubMed]
47. Khater, M.; El-Awadi, M.; Elashtokhy, M.; Abdel-Baky, Y.; Shalaby, M. Physiological and Molecular Changes in Fenugreek (*Trigonella foenum-graecum* L.) as a Response to Gamma Rays. *Int. J. PharmTech Res.* **2016**, *9*, 306–316.
48. Hussein, H.-A.A.; Alshammari, S.O.; Kenawy, S.K.M.; Elkady, F.M.; Badawy, A.A.; Hussein, H.-A.A.; Alshammari, S.O.; Kenawy, S.K.M.; Elkady, F.M.; Badawy, A.A. Grain-Priming with L-Arginine Improves the Growth Performance of Wheat (*Triticum aestivum* L.) Plants under Drought Stress. *Plants* **2022**, *11*, 1219. [CrossRef]
49. Abedi, T.; Alemzadeh, A.; Kazemeini, S.A. Wheat Yield and Grain Protein Response to Nitrogen Amount and Timing. *Aust. J. Crop Sci.* **2011**, *5*, 330–336.

antioxidants

MDPI

Communication

Flooding Tolerance in Sweet Potato (*Ipomoea batatas* (L.) Lam) Is Mediated by Reactive Oxygen Species and Nitric Oxide

Sul-U Park [1,2,†], Chan-Ju Lee [1,†], Sung-Chul Park [3], Ki Jung Nam [4], Kang-Lok Lee [4], Sang-Soo Kwak [1,2], Ho Soo Kim [1,*] and Yun-Hee Kim [4,*]

1 Plant Systems Engineering Research Center, Korea Research Institute of Bioscience and Biotechnology (KRIBB), 125 Gwahak-ro, Daejeon 34141, Korea; sulu0849@kribb.re.kr (S.-U.P.); moda22@kribb.re.kr (C.-J.L.); sskwak@kribb.re.kr (S.-S.K.)

2 Department of Environmental Biotechnology, KRIBB School of Biotechnology, University of Science and Technology (UST), 217 Gajeong-ro, Daejeon 34113, Korea

3 Biological Resource Center, Korea Research Institute of Bioscience and Biotechnology (KRIBB), Jeongeup 56212, Korea; heypsc@kribb.re.kr

4 Department of Biology Education, IALS, Gyeongsang National University, Jinju 52828, Korea; prin225@gnu.ac.kr (K.J.N.) leekl@gnu.ac.kr (K.-L.L.)

* Correspondence: hskim@kribb.re.kr (H.S.K.); cefle@gnu.ac.kr (Y.-H.K.); Tel.: +82-42-860-4464 (H.S.K.) +82-55-772-2237 (Y.-H.K.); Fax: +82-42-860-4608 (H.S.K.); +82-55-772-2230 (Y.-H.K.)

† These authors contributed equally to this work.

Abstract: Flooding is harmful to almost all higher plants, including crop species. Most cultivars of the root crop sweet potato are able to tolerate environmental stresses such as drought, high temperature, and high salinity. They are, however, relatively sensitive to flooding stress, which greatly reduces yield and commercial value. Previous transcriptomic analysis of flood-sensitive and flood-resistant sweet potato cultivars identified genes that were likely to contribute to protection against flooding stress, including genes related to ethylene (ET), reactive oxygen species (ROS), and nitric oxide (NO) metabolism. Although each sweet potato cultivar can be classified as either tolerant or sensitive to flooding stress, the molecular mechanisms of flooding resistance in ET, ROS, and NO regulation-mediated responses have not yet been reported. Therefore, this study characterized the regulation of ET, ROS, and NO metabolism in two sweet potato cultivars—one flood-tolerant cultivar and one flood-sensitive cultivar—under early flooding treatment conditions. The expression of *ERFVII* genes, which are involved in low oxygen signaling, was upregulated in leaves during flooding stress treatments. In addition, levels of respiratory burst oxidase homologs and metallothionein-mediated ROS scavenging were greatly increased in the early stage of flooding in the flood-tolerant sweet potato cultivar compared with the flood-sensitive cultivar. The expression of genes involved in NO biosynthesis and scavenging was also upregulated in the tolerant cultivar. Finally, NO scavenging-related *MDHAR* expressions and enzymatic activity were higher in the flood-tolerant cultivar than in the flood-sensitive cultivar. These results indicate that, in sweet potato, genes involved in ET, ROS, and NO regulation play an important part in response mechanisms against flooding stress.

Keywords: ethylene response factor; flooding stress; metallothionein; monodehydroascorbic acid reductase; resistant cultivar; respiratory burst oxidase; sensitive cultivar

Citation: Park, S.-U.; Lee, C.-J.; Park, S.-C.; Nam, K.J.; Lee, K.-L.; Kwak, S.-S.; Kim, H.S.; Kim, Y.-H. Flooding Tolerance in Sweet Potato (*Ipomoea batatas* (L.) Lam) Is Mediated by Reactive Oxygen Species and Nitric Oxide. *Antioxidants* **2022**, *11*, 878. https://doi.org/10.3390/antiox11050878

Academic Editor: Stanley Omaye

Received: 22 March 2022
Accepted: 27 April 2022
Published: 29 April 2022

1. Introduction

Flooding is a major environmental stress that causes loss of crop yield in various agricultural settings. The global risk of flooding has increased dramatically in recent decades due to climate change [1]. Over the past 50 years, increasingly frequent and severe flood events have negatively impacted the lifespan of terrestrial plants. When flooding occurs, gas exchange between the plant and its environment is severely restricted, causing several internal changes in the plant. Plant organs that are submerged in water lack O_2

and/or CO_2 and may accumulate high levels of the gaseous hormone ethylene (ET). In addition, there are changes in the concentrations of oxygen-derived free radicals, reactive oxygen species (ROS), and nitric oxide (NO) [2]. The production, kinetics, concentration, and precise balance of these substances in flooded cells depend on the plant organ and the type of flooding. Some crop species can withstand soil inundation for only a few hours, while other flood-tolerant crops can cope with partial or complete flooding for days or months [2].

Flood-tolerant crops have anatomical, morphological, or metabolic adaptations. One of the responses to flooding is the induction of ET [3]. ET production is enhanced in crops such as rice and barley. Only flood-tolerant rice varieties, however, can withstand prolonged soil submersion or submergence phases [2]. ET is usually produced in a two-step reaction involving 1-aminocyclopropane-1 carboxylate (ACC) synthetase and ACC oxidase. ET is not only produced in response to flooding stresses but is also a modulator of stress-related morphological responses during plant growth [4,5]. In *Arabidopsis thaliana*, ET is sensed by ET receptors (ETRs), which trigger a downstream signaling cascade of transcription factors (TFs) including *APETALA2/ETHYLENE RESPONSE FACTOR* (*AP2/ERF*) [6,7].

ROS are among the key molecules mediating plant responses to flood stress [8]. Apoplastic ROS are usually generated via the respiratory burst oxidase homolog (RBOH) protein located in the plasma membrane. RBOH is a homolog of the mammalian NADPH oxidase subunit gp91phox that produces a superoxide anion (O_2^-) [9]. The short-lived and highly reactive toxic chemical O_2^- is converted to the nonradical hydrogen peroxide (H_2O_2), either spontaneously or by catalysis by superoxide dismutase (SOD) [10]. ROS might also be formed by mitochondria and/or the photosynthetic electron transport chain [8,10]. Nonenzymatic ROS-scavenging proteins, including cysteine-rich metallothionein (MT) and antioxidant enzymes such as catalase and ascorbate-glutathione cycle-related enzymes, are essential for ROS homeostasis. During flooding stress, the ROS balance of cells is disturbed due to either enhanced ROS production or a decrease in ROS scavenging capacity [8].

NO is a short-lived, highly reactive molecule that regulates several plant developmental and stress responses [11,12]. The main method of NO production is through the enzymatic and nonenzymatic reduction of nitrites [13,14]. Nitrite production is highly dependent on nitrate reductase (NR) activity. The dependence of NO production on nitrite availability means that NR plays the major role in NO production [14,15]. There are limited and contrasting data on the temporal and spatial dynamics of NO in flooded plants [15–17]. NO emissions from unsubmerged atmospheric plant tissues decreased in Arabidopsis and cotton under waterlogging, whereas NO emissions increased in three submerged deciduous tree species [15–17]. In addition, cellular and exogenous NO levels increased in various plant species during hypoxia [13,18,19]. During hypoxic conditions, this NO burst is thought to reflect an increase in NO production due to increased NR activity and nitrite accumulation. The spatiotemporal dynamics of NO and its effects may vary, therefore, depending on the flood conditions. Studying the changes in ET, ROS, and NO levels is thus important for fully understanding the mechanisms of plant flooding responses and tolerance.

We previously studied 33 sweet potato cultivars, using phenotypic and biochemical characteristics to identify flood-tolerant cultivars [20]. In addition, we recently used comparative transcriptome profiling to compare a flood-tolerant sweet potato cultivar, Yeonjami (YJM), with a flood-sensitive cultivar, Jeonmi (JM) [21]. The expression levels of several candidate genes thought to be involved in flooding tolerance correlated with the comparative transcriptomic data. However, although each sweet potato cultivar can be classified as either tolerant or sensitive to flooding stress, the molecular mechanisms of flooding resistance in ET, ROS, and NO regulation-mediated responses have not yet been reported. In this study, therefore, we conducted a transcriptome-based expression analysis of genes involved in ET, ROS, and NO regulation in two sweet potato cultivars during flooding stress.

2. Materials and Methods

2.1. Plant Materials

This study involved two previously characterized cultivars of sweet potato (*Ipomoea batatas* (L.) Lam), namely, Yeonjami (YJM) and Jeonmi (JM). YJM is highly tolerant to flooding stress, whereas JM is highly sensitive [20,21]. Both cultivars were obtained from the Bioenergy Crop Research Center, National Crop Research Institute (RDA, Muan, Jeonnam, Korea). Sweet potato plants were cultivated and subjected to flooding stress as described by Park et al. [21]. Plants were grown in a growth chamber at 25 °C under a 16 h/8 h light/dark photocycle and 50% relative humidity. Experimental plants were grown for 2 weeks under normal conditions and then subjected to flooding stress. To induce flooding stress, water was added to the pots until approximately 65% of the aboveground tissue was submerged. Leaves were sampled 0 (control), 0.5, and 3 days after the beginning of flooding treatment. For leaf samples used in the experiment, the 3rd and 4th leaves from the top were used in four plants. The harvested leaves were above the water surface. Leaf samples were frozen immediately in liquid nitrogen and stored at −70 °C until further analysis.

2.2. Analysis of H_2O_2 Levels

The H_2O_2 content of sweet potato leaves was assessed using the xylenol orange method [22]. A total of 0.1 g of leaf tissue was ground and homogenized in a solution of 50 mM potassium phosphate buffer (pH 6.5) and 1 mM hydroxylamine. The homogenate was centrifuged at 12,000× *g* for 15 min at 4 °C. A total of 100 μL of the supernatant was added to a reaction mixture containing ferric ammonium sulfate ($FeNH_4[SO_4]$), 25 mM sulphuric acid, 250 μM xylenol orange, and 100 mM sorbitol. After 30 min in the dark at room temperature for incubation, the absorbance of the samples was determined at 560 nm. H_2O_2 measurements were expressed as relative values by control and treatment.

2.3. Analysis of Nitrite Content

The nitrite (NO_2^-) content was analyzed using a Promega Griess Reagent System assay kit (Promega, USA). Sweet potato leaves were homogenized on ice using a pestle and mortar in 25 mM HEPES buffer (pH 7.7), incubated for 20 min at 4 °C, and centrifuged for 10 min at 12,000 rpm. The NO_2^- concentration was determined using the Griess Reagent System, according to the manufacturer's instructions. Briefly, 40 μL of each sample was added to 50 μL sulfanilamide solution in a 96-well microtiter plate. The microtiter plate was covered and incubated at room temperature for 10 min before 40 μL NED (N-(1-naphthyl) ethylenediamine) solution was added to each well. The plate was then incubated at room temperature for a further 10 min. The absorbance of nitrite was read at 540 nm using a plate reader.

2.4. RNA Isolation and Analysis of Gene Expression

Total RNA was isolated from leaves using TRIzol reagent (Invitrogen, Carlsbad, CA, USA) and treated with RNase-free DNase I to avoid contamination with genomic DNA. Quantitative real-time PCR analysis was performed using a Bio-Rad CFX96 thermal cycler (Bio-Rad) with EvaGreen fluorescent dye, according to the manufacturer's instructions. Linear data were normalized to the mean threshold cycle (Ct) of the reference gene *ADPRI-BOSYLATION FACTOR* (*ARF*) [23]. Gene-specific PCR primers are listed in Table S1.

2.5. Monodehydroascorbate Reductase Assay

Monodehydroascorbate reductase (MDHAR) activity was measured in sweet potato leaves according to the methodology of Truffault et al. [24]. The MDHAR enzymatic activity assay is based on NADH oxidation. Extractions were performed on ground sweet potato leaf powder in 50 mM Tris-HCl at pH 7.8. The soluble extract was mixed with 1 mM ascorbate, 0.2 mM NADH, and ascorbate oxidase to give a linear production of the monodehydroascorbate (MDHA) radical. Measurements were performed in triplicate at 340 nm.

2.6. Statistical Analyses

Data were analyzed by two-way analysis of variance (ANOVA). The subsequent multiple comparisons were examined using the least significant difference (LSD) multiple range test. All statistical analyses were performed using Statistical Package for the Social Sciences software (SPSS 12). Statistical significance levels were set at $p < 0.05$.

3. Results

3.1. Expression of ERFVII during Early Flooding Treatment

To analyze the responses of sweet potato plants to flood stress, the tolerance of whole plants grown in soil was measured during early flooding treatment. Approximately 65% of the aboveground tissue of 2-month-old plants was submerged by adding excess water to the pot in a growth chamber at 25 °C under a 16 h/8 h light/dark photocycle. When the susceptible cultivar JM and the resistant cultivar YJM were exposed to flooding for 0, 0.5, and 3 days, the above atmospheric leaves of the JM plants showed slight wilting and curling, whereas the leaves of the YJM plants were less damaged (Figure 1A).

ET biosynthesis and signaling pathways are usually activated during flood treatment in plants [25]. TFs belonging to the *ERFVII* group, including *ERF71/HER2*, *ERF72/RAP2.3*, *ERF74/RAP2.12*, and *ERF75/RAP2.2*, are important representatives of genes that function in the ET signaling pathway activated during flood treatment at the transcriptional and post-translational levels [26,27]. Although expression analysis at the post-translational level is more important in the *ERFVIIs* [27], in this study, the expression of *ERFVII* group genes was investigated during early flood treatment at the transcriptional level (Figure 1B). The expression patterns of *ERF72/RAP2.3* (g6122), *ERF74/RAP2.12* (g55154), and *ERF74/RAP2.12* (g60549) were investigated after flood treatment using our previous transcriptomic data [21]. *ERF72/RAP2.3* was strongly induced from 0.5 d after flooding treatment in the flood-tolerant cultivar, YJM, whereas its expression decreased in the flood-sensitive JM. In contrast, *ERF74/RAP2.12* showed a weak increase in expression level in YJM, and increased expression after 3 d of flooding treatment. In JM, there was no clear pattern of changes in expression under flooding conditions, as levels increased or decreased at different times.

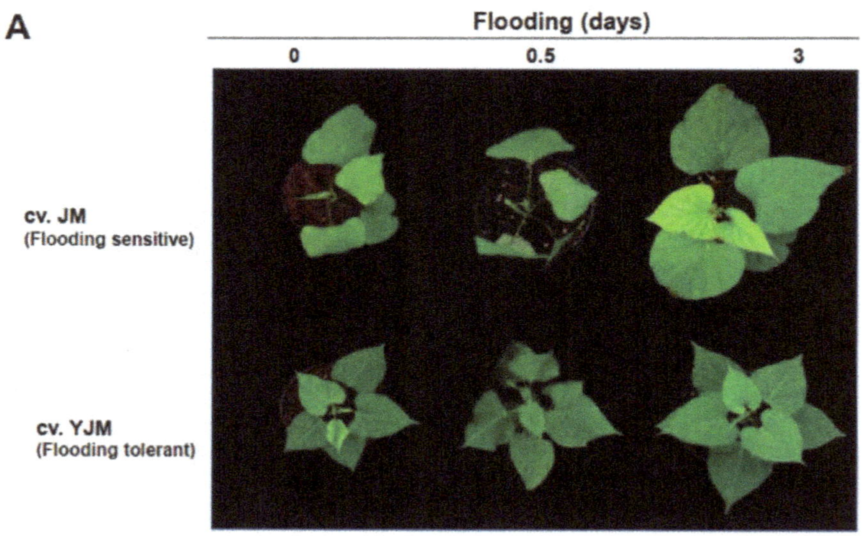

Figure 1. *Cont.*

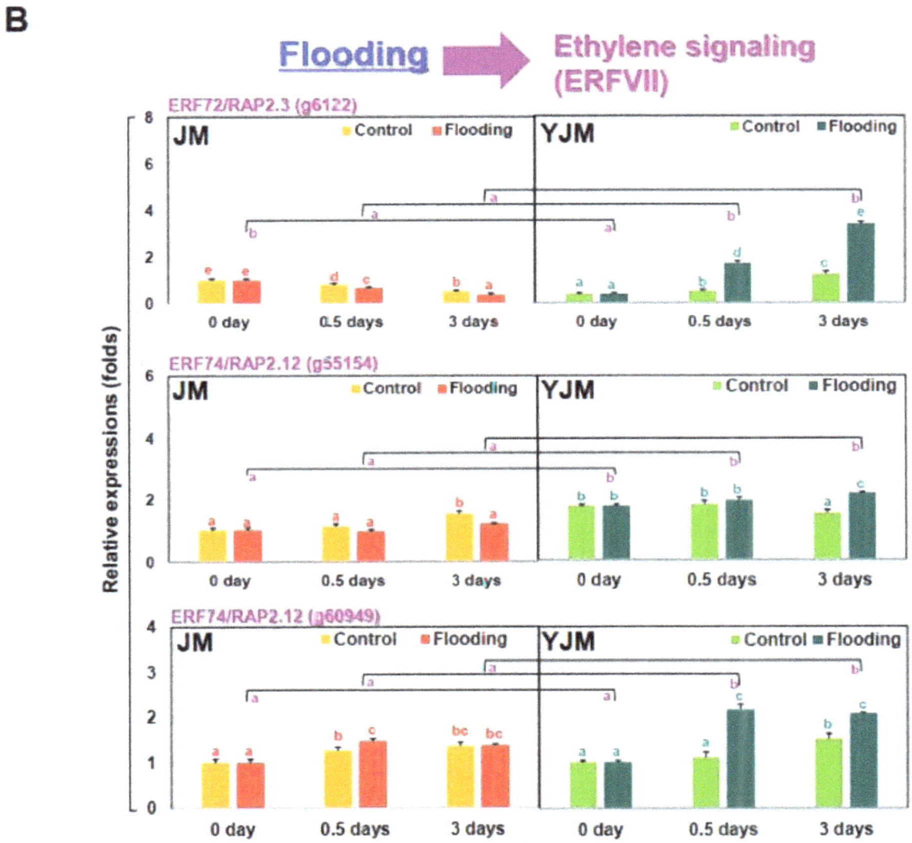

Figure 1. Effects of flooding stress on the sweet potato cultivars Yeonjami (YJM) and Jeonmi (JM). (**A**) Phenotypes of aboveground portions of YJM and JM after 3 d of flooding treatment. (**B**) Expression of *ERFVII* transcription factors in each cultivar after 0, 0.5, or 3 d of flooding treatment. Different letters represent statistically significant differences between control and flooding treatment, and between flooding-treated JM and YJM, determined using two-way ANOVA with the LSD post hoc test; $p < 0.05$.

3.2. The Responses of ROS during Early Flooding Treatment

The levels of free radicals and ROS in plants usually increase during flood treatment [25]. Increases in ET biosynthesis and signaling during flood treatment modulate the amount of ROS and free radical signaling. Therefore, changes in H_2O_2, a representative ROS, were investigated during the early flood treatment process (Figure 2A). On day 3 of flood treatment, the H_2O_2 content of the flooding-tolerant cultivar, YJM, was 38% lower than that of the flooding-sensitive cultivar, JM.

We also investigated the expression levels of *RBOH*, a gene which is involved in early ROS production, during flood treatment (Figure 2B). Expression levels of *RBOHA* (g19623), *D* (g51882), and *E* (g55049) were higher in the resistant cultivar YJM than in the susceptible JM under normal untreated conditions; expression increased to higher levels in YJM than in JM within 0.5 days of flood treatment. On the other hand, *RBOHC* (g19835) showed similar changes in expression levels in both cultivars under the normal control and the early flooding treatment. After 3 days of flood treatment, the expression levels of all the

RBOH genes measured in the experiment were higher in JM than in YJM, due to decreases in expression in the resistant YJM cultivar between 0.5 and 3 days of flood treatment.

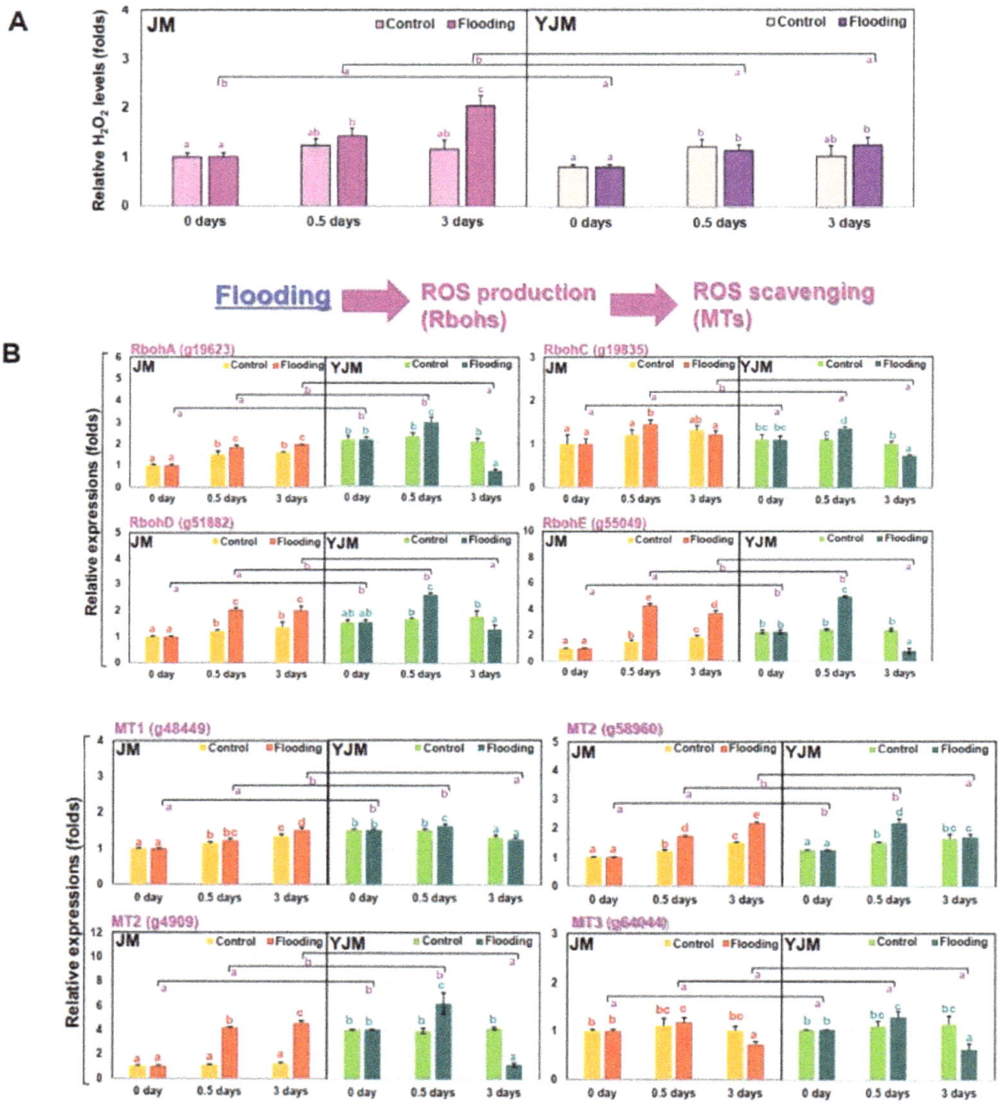

Figure 2. Expression of genes related to hydrogen peroxide (H_2O_2) and ROS pathways in sweet potato cultivars treated with flooding stress for 3 d. (**A**) Relative levels of H_2O_2 in each cultivar after 0, 0.5, or 3 d of flooding treatment. (**B**) Expression of *RBOH* genes in each cultivar after 0, 0.5, or 3 d of flooding treatment. (**C**) Expression of *MT* genes in each cultivar after 0, 0.5, or 3 d of flooding treatment. Different letters represent significant differences between control and flooding treatment, and between flooding-treated JM and YJM, determined using two-way ANOVA with the LSD post hoc test; $p < 0.05$.

Next, we measured the expression of *METALLOTHIONEIN* (*MT*) genes, which encode proteins that remove reactive oxygen species during flood treatment (Figure 2C). Expression levels of all the *MT* genes were higher in the flood-resistant YJM than in the flood-sensitive JM under normal untreated conditions. After 0.5 days of flood treatment, expression levels of *MT* genes remained higher in YJM than in JM; in particular, *MT2* (g58960 and g4909) showed a greater increase in expression in YJM than in JM. In contrast, *MT3* (g64044) showed similar expression levels in both cultivars under both control conditions and early flooding treatment. By the third day of flood treatment, however, the expression levels of *MT1* and *MT2* decreased in YJM and increased in JM.

3.3. The Responses of NO during Early Flooding Treatment

An increase in NO, together with rises in ET and ROS levels, is seen in plants at the initial stage of flood treatment; this increase in NO controls the increase in ET during flood treatment [26]. Plants usually regulate endogenous NO levels via the control of biosynthesis and scavenging. The main source of NO production is via the enzymatic and nonenzymatic reduction of nitrite (NO_2^-) [13,14]. NO_2^- production is highly dependent on nitrate reductase (NR). The dependency of NO production on nitrite availability makes NR the major player in NO production [14,15]. The NO generation during hypoxia is also thought to result from enhanced NR activity and NO_2^- accumulation, providing a substrate for NO production [13,19].

We therefore tested the changes in nitrite (NO_2^-) levels in sweet potato plants during early flood treatment. In contrast to the changes in H_2O_2 levels (Figure 2A), the increase in NO_2^- was 31% higher in YJM than in JM (Figure 3A).

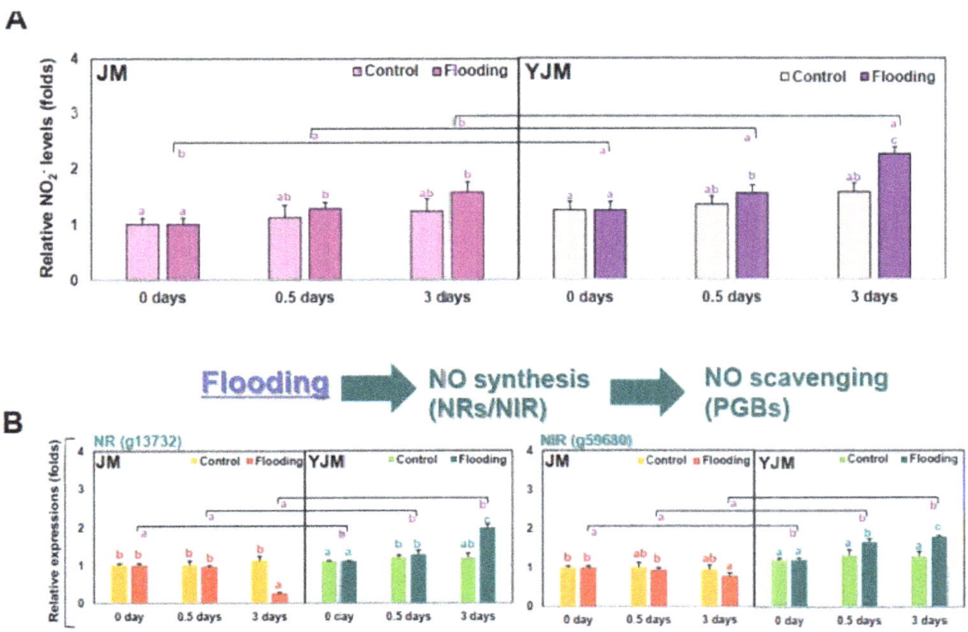

Figure 3. *Cont.*

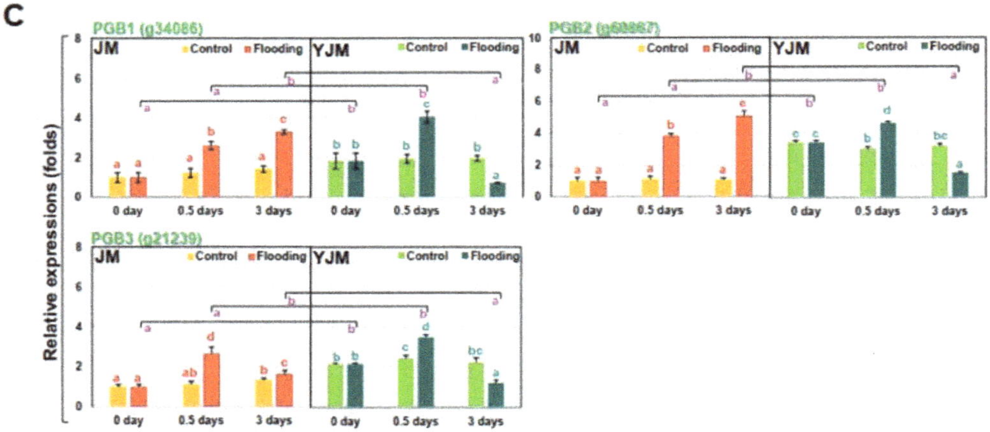

Figure 3. Levels of nitrite (NO_2^-) and expression of NO-related genes in sweet potato cultivars treated with flooding stress for 3 d. (**A**) Relative levels of NO_2^- in each cultivar after 0, 0.5, or 3 d of flooding treatment. (**B**) Expression of the NO biosynthesis genes *NR* and *NIR* in each cultivar after 0, 0.5, or 3 d of flooding treatment. (**C**) Expression of *PGB* genes in each cultivar after 0, 0.5, or 3 d of flooding treatment. Different letters represent statistically significant differences between control and flooding treatment, and between flooding-treated JM and YJM, determined using two-way ANOVA with the LSD post hoc test; $p < 0.05$.

Next, we investigated changes in the expression of the NO biosynthesis genes *NITRATE REDUCTASE* (*NR*) and *NITRITE REDUCTASE* (*NIR*) (Figure 3B). The expression of *NR* (g1372) and *NIR* (g59680), both genes involved in early NO biosynthesis, did not differ significantly between YJM and JM under control conditions. After flood treatment, their expression gradually increased in YJM but gradually decreased in JM. We also examined the expression of the NO-scavenging *PHYTOGLOBIN* (*PGB*) genes during early flooding treatment (Figure 3C). *PGB1* (g34086), *2* (g60867), and *3* (g21239) showed higher levels of expression in YJM than in JM under control conditions and during early flood treatment (0.5 days). Their expression decreased, however, in YJM by 3 days after flooding so that, by this timepoint, the expression of *PGB*s was higher in JM than in YJM.

3.4. The Responses of MDHAR during Early Flooding Treatment

Previous studies reported that ascorbic acid (AsA) and PGB are involved in NO metabolism in plants. In vitro enzymatic assays have shown that NO scavenging is catalyzed by monodehydroascorbic acid reductase (MDHAR), which mediates binding reactions involving iron reduction in PGB in the presence of AsA and NADH [28]. AsA supports NO scavenging, directly reducing PGB in plants to produce nitrate and monodehydroascorbate (MDA). The final product of this reaction is MDA, which is efficiently reduced back to AsA in the presence of MDHAR and NADH. NO scavenging is also promoted by MDHAR [29].

We therefore investigated *MDHAR* gene expression and MDHAR-associated enzymatic activity (Figure 4). It was confirmed that the expression of different *MDHAR* genes (cytosolic, g25138; peroximal, g17489; and mitochondrial, g35177) and MDHAR enzymatic activity were higher in the flood-tolerant YJM than in the flood-sensitive JM. Changes in MDHAR expression and activity in tolerant sweet potato during flood treatment may therefore play an important role in NO regulation through interactions with PGB as well as via ROS regulation.

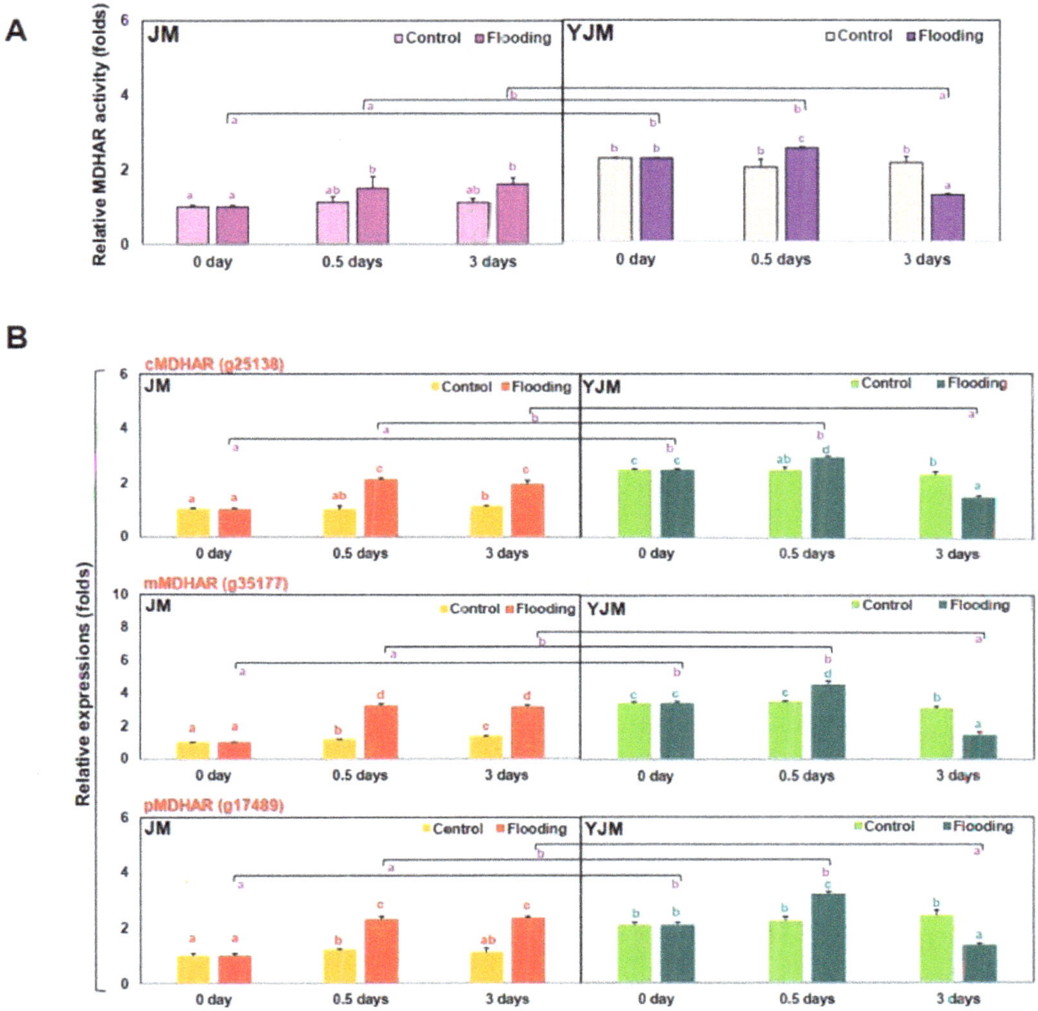

Figure 4. Changes in MDHAR activity and *MDHAR* gene expression in sweet potato cultivars treated with flooding stress for 3 d. (**A**) Relative MDHAR activity in each cultivar after 0, 0.5, or 3 d of flooding treatment. (**B**) Expression of *MDHAR* genes in each cultivar after 0, 0.5, or 3 d of flooding treatment. c*MDHAR*: cytosolic *MDHAR*; p*MDHAR*: peroxisomal *MDHAR*; m*MDHAR*: mitochondrial *MDHAR*. Different letters represent statistically significant differences between control and flooding treatment, and between flooding-treated JM and YJM, determined using two-way ANOVA with the LSD post hoc test; $p < 0.05$.

4. Discussion

Plants that exhibit resistance to flooding use two general survival strategies: low-O_2 escape syndrome (LOES) and low-O_2 quiescence syndrome (LOQS) [2,30]. Regulated anaerobic metabolism in both LOES and LOQS is equivalent to survival in low-O_2 conditions. At the heart of this mechanism is an evolutionarily conserved group of TFs, ethylene response factor VIIs (ERFVIIs), which activate the genes required for anaerobic metabolism in Arabidopsis [31]. A reduction in O_2 levels in plants stabilizes ERFVII TFs such as

RAP2.12 and RAP2.3, which play an important role in activating anaerobic responses at the transcriptional level [32]. Giuntoli et al. [33] also reported that the expression of a set of genes involved in the oxidative stress response is induced in flood-exposed Arabidopsis. Thus, ERFVII is a positive regulator of oxidative stress-related genes and genes involved in fermentative metabolism. The *ERFVII* TFs *RAP2.2*, *RAP2.3*, and *RAP2.12* mediate responses to oxidative stress and may act redundantly [34]. Overexpression of *RAP*-type *ERFVII* TFs also confers resistance to oxidative stress after H_2O_2 application. *RAP*-type *ERFVII* genes, which are involved in ROS scavenging and signaling, are themselves positively regulated by ERFVII TFs [33,34].

We investigated the expression of the representative *ERFVII* TFs *RAP2.3* and *RAP 2.12* in the sweet potato cultivars YJM and JM, which were identified as flood-resistant and flood-sensitive, respectively, in a previous study [21] (Figure 1). The expression of *ERF72/RAP2.3* and *ERF74/RAP2.12* was higher in YJM than in JM during flood stress; expression levels of these genes were increased, especially during the early period of flooding. We also measured the expression of four genes belonging to other groups within the ERF transcription factor family, *ERF1* (g31279), *ERF2* (g25395), *ERF4* (g54463), and *ERF5* (g20475). These showed either similar increases in expression in the two cultivars after flood treatment or slightly higher expression levels in JM (data not shown). A recent transcriptomic analysis of the sweet potato cultivars YJM and JM revealed changes in the expression of *ETR*, *EIN*, and *ERF*, which are all genes related to signal transduction involved in ET signaling pathways [21]. As indicated by previous studies in Arabidopsis and rice, it is highly likely that *RAP2.3* and *RAP2.12*, which are ERFVII TFs, are an important part of the mechanism enabling flood resistance in sweet potato.

The onset of anaerobic conditions triggers a burst of ROS in Arabidopsis due to changes in NADPH oxidase activity in membranes and an imbalance in the mitochondrial electron transfer system [35]. Arabidopsis *RBOHD* mutants are less tolerant of anaerobic conditions and show negative effects on *ALCOHOL DEHYDROGENASE1* (*ADH1*) expression, compared to wild-type seedlings immersed in water [35–37]. This suggests that ROS generated by *RBOHD* under these conditions may be a positive signal necessary for plant tolerance of flooding-mediated hypoxia. Yamauchi et al. [38] observed high levels of induction of *RBOH* expression in parallel with repression of *MT*, which encodes an ROS scavenger, in maize roots that did not show tissue reduction following treatment with the NADPH oxidase inhibitor diphenyleneiodonium (DPI). High expression of *RBOH* can therefore induce cell death in plants during flood treatment through oxidative burst. High *RBOH* expression appears to act either by suppressing the expression of ROS-scavenging genes, including the *MT*s, or, during flooding, through signal transduction pathways that increase the expression of a downstream gene that causes ROS to be removed and activates an appropriate defense mechanism. We observed that the expression of *RBOHA*, *D*, and *E* increased during the initial flood treatment in the flood-resistant cultivar YJM, and the expression of *MT2* significantly increased in the flooding treatment after 12 h (Figure 2). Therefore, flood tolerance in sweet potato appears to involve activation of the pathway involving the ROS scavenger protein MT.

Like ROS, NO is detrimental to plant cells, but it is also a key component of plant response-related signaling pathways [39]. In plants, NO is involved in the degradation of transcriptional regulators, which leads to the activation of key hypoxia-responsive genes [40]. Indeed, in anaerobic plants, NO availability negatively modulates the activation of responses induced by ERFVII TFs. Arabidopsis NR mutants such as *nitrate reductase 1* (*nia1*) and *nia2* exhibit changes in anaerobic gene transcription due to the impaired production of NO, which alters the activation of genes downstream of *ERFVII* in reaction to anaerobic conditions [40]. Therefore, NO destabilizes *ERFVII* and inhibits downstream signaling pathways. Hartman et al. [41] reported that early ET capture from flooded Arabidopsis increases transcription of the NO scavenger nonsymbiotic *PGB1*, thereby reducing the solubility of NO and promoting the stability of ERFVII. This phenomenon occurs prior to severe hypoxia in plants and acts as a priming event, enhancing the plant's tolerance to upcoming stressful conditions. In the current study, a greater increase in NO_2^- levels during flood treatment was observed in the flood-resistant cultivar YJM than in JM, a flood-sensitive cultivar (Figure 3A); although there was no significant change during the initial flood treatment, a difference was observed after 3 days of treatment. During the flood treatment, the expression of NR and NIR, genes involved in NO_2^- levels, gradually increased in YJM but decreased gradually in JM (Figure 3B). It therefore appeared that exposure to flooding activated response mechanisms regulated by NO_2^- generation in YJM, a flood-resistant sweet potato cultivar.

PGB, which scavenges NO, potentially promotes the interaction between NO and ET. Levels of *PGB* mRNA increase during submergence and hypoxic conditions in several plant species [42,43]. NO also increases the levels of *PGB* mRNA in rice, cotton, and spinach, indicating a feedback mechanism [44–46]. Interactions and feedback regulation between NO and ET have also been shown at the level of ERFVII, as ERFVII stability and action are highly dependent on NO and oxygen levels [40,47]. When NO or oxygen concentrations are reduced, ERFVII TFs accumulate, thereby promoting transcription of downstream target genes. Interestingly, several genes regulating ERFVII TFs contain hypoxia-responsive promoter elements, including several ET signaling genes and the NO-scavenging gene *PGB* [31]. Therefore, ERFVII TFs are upregulated by NO, oxygen, and ET, whereas ERFVII action can, in turn, induce ET biosynthesis and NO scavenging, creating a positive feedback loop [31]. We observed that the expression of *PGB*s increased strongly within 0.5 days of flood treatment in the flood-resistant cultivar YJM, i.e., during the initial period of flooding (Figure 3); in contrast, the expression of *PGB*s increased slowly and gradually in JM. In this study, therefore, the expression of NO_2^- generation-related genes gradually increased in response to flooding, but the expression of *PGBs* increased immediately, suggesting that the overall level of NO_2^- increased gradually in the flooding-tolerant YJM. It is possible that NO_2^- generated through the nonenzymatic pathway via mitochondria may have influenced the increase in NO_2^- levels, in addition to NO_2^- generated by the enzymatic pathway acting through NR and NIR. These results suggest that increases in NO_2^- generation and elimination affected the ROS signaling mechanism through the expression of *ERFVII*, thereby activating the flood resistance mechanism.

We previously reported comparative transcriptome profiling to compare a flood-tolerant sweet potato cultivar with a flood-sensitive cultivar [21]. A higher number of some of the ROS signaling-related genes such as mitogen-activated protein kinase (MAPK) and ET signaling-related genes were upregulated in the tolerant cultivars than in the susceptible cultivar. Recently, another study also reported comparative transcriptome profiling using different soybean cultivars [48]. A higher number of some of the ROS-related genes such as glutathione S-transferase and lipoxygenase were upregulated in the tolerant cultivars than in the susceptible cultivar. The number of some phytohormone ABA-related TFs of the basic leucine zipper domain was also higher in the tolerant cultivars than in the susceptible cultivar. Similar to our previous study, the expression levels of several candidate genes-related to ROS and phytohormones thought to be involved in flooding tolerance correlated with the comparative transcriptomic data.

This study addressed the transmission of the plant signaling molecules ET, ROS, and NO and showed how these signals were correlated in flood-tolerant and flood-sensitive cultivars of sweet potato (Figure 5). Consistent with previous research, a signaling pathway acting through ERFVII was activated in flood-resistant plants. In addition, our results suggest that the expression of *RBOH* and *MT* genes, which encode components of an ROS signaling pathway involving ERFVII, played an important role in enabling the flood resistance of sweet potato. This study also suggests that activation of the NO biosynthesis and scavenging cycle and an increase in MDHAR activity likely regulated ERFVII expression and levels of ROS via other pathways.

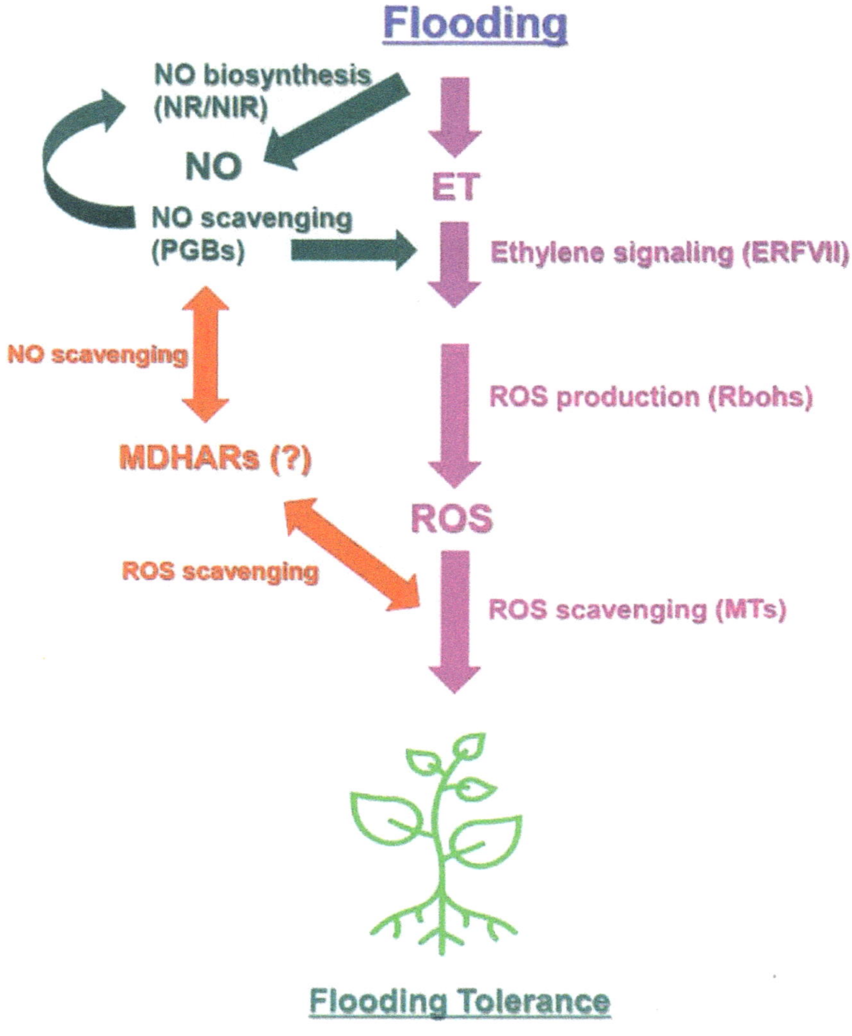

Figure 5. A suggested model of the ET-, ROS-, and NO-related biological processes and genes involved in influencing the flooding response that occurs in sweet potato leaves under early flooding. Flooding conditions result in the induction of ERFVII transcription factors and the expression of RBOHs, which produce O_2^- radicals from oxygen in the apoplast. RBOH expression may enhance RBOH activity and increase ROS levels in the apoplast as well as the cytosol. Expression of genes encoding ROS scavengers, such as MTs, increases, leading to an increase in ROS signaling and scavenging responses and inducing the response in flooding-tolerant sweet potato. The NO synthesis-scavenging cycle regulates ERFVII-mediated signaling. MDHARs are also potentially involved in NO/ROS-mediated responses in sweet potato under early flooding conditions. ERF: ethylene response factor; ET: ethylene; ROS: reactive oxygen species; RBOH: respiratory burst oxidase homolog; MT: metallothionein; NO: nitric oxide; NR: nitrate reductase; NIR: nitrite reductase; PGB: phytoglobulin; MDHAR: monodehydroascorbate reductase.

5. Conclusions

In conclusion, we characterized changes in the expression of flooding response-related genes in flood-tolerant and flood-sensitive sweet potato cultivars that were associated with response mechanisms mediated by ET, ROS, and NO. This approach identified candidate sweet potato genes whose expression was associated with changes in specific responses involved in metabolic and signaling pathways related to ET, ROS, and NO. The use of marker-assisted selection to identify genes linked to flood tolerance offers several advantages over flooding control in an integrated management system. Further investigation, based on large-scale genomic and transcriptomic analysis, is required to elucidate the exact role played by each candidate gene in regulating the signaling pathways involved in flooding tolerance responses of sweet potato during flooding stress. Transgenic plants with an enhanced or reduced expression of candidate genes will be generated to determine their role in the mechanisms conferring flooding tolerance. Overall, our results provide valuable information for enabling the development of crops with enhanced tolerance to hypoxic stress induced by flooding.

Supplementary Materials: The following are available online at https://www.mdpi.com/xxx/s1. Table S1: List of primers used for quantitative real-time PCR (qRT-PCR) analysis of sweet potato genes.

Author Contributions: Conceptualization, S.-U.P., C.-J.L., H.S.K. and Y.-H.K.; data curation, S.-C.P.; formal analysis, K.J.N., K.-L.L. and S.-S.K.; funding acquisition, S.-S.K.; methodology, S.-U.P., C.-J.L., H.S.K. and Y.-H.K.; resources, S.-S.K., H.S.K. and Y.-H.K.; validation, K.J.N., K.-L.L. and S.-S.K.; visualization, K.J.N., K.-L.L. and S.-S.K.; writing—original draft, S.-U.P., C.-J.L., H.S.K. and Y.-H.K.; writing—review and editing, S.-U.P., C.-J.L., H.S.K. and Y.-H.K. All authors have read and agreed to the published version of the manuscript.

Funding: This research was supported by the Basic Science Research Program through the National Research Foundation of Korea (NRF) funded by the Ministry of Science, ICT, and Future Planning (2020R1A2C1004560, 2021R1A2C400188711) and the Korea Research Institute of Bioscience and Biotechnology (KRIBB) Research Initiative Program (KGM5372113).

Institutional Review Board Statement: Not applicable.

Informed Consent Statement: Not applicable.

Data Availability Statement: Data are contained within the article.

Conflicts of Interest: The authors declare no conflict of interest.

Abbreviations

ACC	1-aminocyclopropane-1 carboxylate
ADH	alcohol dehydrogenase
ANOVA	analysis of variance
AP2/ERF	Apetala2/ethylene response factor
AsA	ascorbic acid
DPI	diphenyleneiodonium
ET	ethylene
ETR	ET receptor
Hb	hemoglobin
JM	Jeonmi
LOES	low-O_2 escape syndrome
LOQS	low-O_2 quiescence syndrome
LSD	least significant difference
MAPK	mitogen-activated protein kinase
MDHA	monodehydroascorbate
MDHAR	monodehydroascorbate reductase
MT	metallothionein

NED	N-(1-naphthyl) ethylenediamine
NIR	nitrite reductase
NO	nitric oxide
NR	nitrate reductase
PGB	phytoglobin
RBOH	respiratory burst oxidase homolog
ROS	reactive oxygen species
SOD	superoxide dismutase
TF	transcription factor
YJM	Yeonjami

References

1. Hirabayashi, Y.; Mahendran, R.; Koirala, S.; Konoshima, L.; Yamazaki, D.; Watanabe, S.; Kim, H.; Kanae, S. Global flood risk under climate change. *Nat. Clim. Change* **2013**, *3*, 816–821. [CrossRef]
2. Bailey-Serres, J.; Voesenek, L.A.C.J. Flooding stress: Acclimations and genetic diversity. *Annu. Rev. Plant Biol.* **2008**, *59*, 313–339. [CrossRef] [PubMed]
3. Sasidharan, R.; Voesenek, L.A.C.J. Ethylene-mediated acclimations to flooding stress. *Plant Physiol.* **2015**, *169*, 3–12. [CrossRef]
4. Parlanti, S.; Kudahettige, N.P.; Lombardi, L.; Mensuali-Sodi, A.; Alpi, A.; Perata, P.; Pucciariello, C. Distinct mechanisms for aerenchyma formation in leaf sheaths of rice genotypes displaying a quiescence or escape strategy for flooding tolerance. *Ann. Bot.* **2011**, *107*, 1335–1343. [CrossRef] [PubMed]
5. Steffens, B.; Geske, T.; Sauter, M. Aerenchyma formation in the rice stem and its promotion by H_2O_2. *New Phytol.* **2011**, *190*, 369–378. [CrossRef] [PubMed]
6. Chen, Y.F.; Randlett, M.D.; Findell, J.L.; Schaller, E.G. Localization of the ethylene receptor ETR1 to the endoplasmic reticulum in Arabidopsis. *J. Biol. Chem.* **2002**, *277*, 19861–19866. [CrossRef]
7. Jung, K.H.; Seo, Y.S.; Walia, H.; Cao, P.; Fukao, T.; Canlas, P.E.; Amonpant, F.; Bailey-Serres, J.; Ronald, P.C. The submergence tolerance regulator Sub1A mediates stress-responsive expression of AP2/ERF transcription factors. *Plant Physiol.* **2010**, *152*, 1674–1692. [CrossRef]
8. Steffens, B.; Steffen-Heins, A.; Sauter, M. Reactive oxygen species mediate growth and death in submerged plants. *Front. Plant Sci.* **2013**, *4*, 179. [CrossRef]
9. Torres, M.A.; Onouchi, H.; Hamada, S.; Machida, C.; Hammond-Kosack, K.E.; Jones, J.D. Six *Arabidopsis thaliana* homologues of the human respiratory burst oxidase (gp91phox). *Plant J.* **1998**, *14*, 365–370. [CrossRef]
10. Mittler, R.; Vanderauwera, S.; Gollery, M.; Van Breusegem, F. Reactive oxygen gene network of plants. *Trends Plant Sci.* **2004**, *9*, 490–498. [CrossRef]
11. Manjunatha, G.; Gupta, K.J.; Lokesh, V.; Mur, L.A.; Neelwarne, B. Nitric oxide counters ethylene effects on ripening fruits. *Plant Signal. Behav.* **2012**, *7*, 476–483. [CrossRef] [PubMed]
12. Farnese, F.S.; Menezes-Silva, P.E.; Gusman, G.S.; Oliveira, J.A. When bad guys become good ones: The key role of reactive oxygen species and nitric oxide in the plant responses to abiotic stress. *Front. Plant Sci.* **2016**, *7*, 471. [CrossRef] [PubMed]
13. Planchet, E.; Jagadis Gupta, K.; Sonoda, M.; Kaiser, W.M. Nitric oxide emission from tobacco leaves and cell suspensions: Rate limiting factors and evidence for the involvement of mitochondrial electron transport. *Plant J.* **2005**, *41*, 732–743. [CrossRef] [PubMed]
14. Chamizo-Ampudia, A.; Sanz-Luque, E.; Llamas, A.; Galvan, A.; Fernandez, E. Nitrate reductase regulates plant nitric oxide homeostasis. *Trends Plant Sci.* **2017**, *22*, 163–174. [CrossRef] [PubMed]
15. Magalhaes, J.R.; Monte, D.C.; Durzan, D. Nitric oxide and ethylene emission in *Arabidopsis thaliana*. *Physiol. Mol. Biol. Plant* **2000**, *6*, 117–127.
16. Copolovici, L.; Niinemets, U. Flooding induced emissions of volatile signalling compounds in three tree species with differing waterlogging tolerance. *Plant Cell Environ.* **2010**, *33*, 1582–1594. [CrossRef]
17. Zhang, Y.; Kong, X.; Dai, J.; Luo, Z.; Li, Z.; Lu, H.; Xu, S.; Tang, W.; Zhang, D.; Li, W.; et al. Global gene expression in cotton (*Gossypium hirsutum* L.) leaves to waterlogging stress. *PLoS ONE* **2017**, *12*, e0185075. [CrossRef]
18. Mugnai, S.; Azzarello, E.; Baluška, F.; Mancuso, S. Local root apex hypoxia induces NO-mediated hypoxic acclimation of the entire root. *Plant Cell Physiol.* **2012**, *53*, 912–920. [CrossRef]
19. Gupta, K.J.; Igamberdiev, A.U. Reactive nitrogen species in mitochondria and their implications in plant energy status and hypoxic stress tolerance. *Front. Plant Sci.* **2016**, *7*, 369. [CrossRef]
20. Park, S.U.; Lee, C.J.; Kim, S.E.; Lim, Y.H.; Lee, H.U.; Nam, S.S.; Kim, H.S.; Kwak, S.S. Selection of flooding stress tolerant sweetpotato cultivars based on biochemical and phenotypic characterization. *Plant Physiol. Biochem.* **2020**, *155*, 243–251. [CrossRef]
21. Park, S.U.; Kim, Y.H.; Lee, C.J.; Kim, S.E.; Lim, Y.H.; Yoon, U.H.; Kim, H.S.; Kwak, S.S. Comparative transcriptome profiling of two sweetpotato cultivars with contrasting flooding stress tolerance levels. *Plant Biotechnol. Rep.* **2020**, *14*, 743–756. [CrossRef]
22. Bindschedler, L.V.; Minibayeva, F.; Gardner, S.L.; Gerrish, C.; Davies, D.R.; Bolwell, G.P. Early signaling events in the apoplastic oxidative burst in suspension cultured French bean cells involved cAMP and Ca^{2+}. *New Phytol.* **2001**, *151*, 185–194. [CrossRef] [PubMed]

23. Park, S.C.; Kim, Y.H.; Ji, C.Y.; Park, S.; Jeong, J.C.; Lee, H.S.; Kwak, S.S. Stable internal reference genes for the normalization of real-time PCR in different sweetpotato cultivars subjected to abiotic stress conditions. *PLoS ONE* **2012**, *7*, e51502. [CrossRef] [PubMed]
24. Truffault, V.; Riqueau, G.; Garchery, C.; Gautier, H.; Stevens, R. Is monodehydroascorbate reductase activity in leaf tissue critical for the maintenance of yield in tomato? *J. Plant Physiol.* **2018**, *222*, 1–28. [CrossRef]
25. Sasidharan, R.; Hartman, S.; Liu, Z.; Martopawiro Sajeev, N.; van Veen, H.; Yeung, E.; Voeseneka, L.A.C.J. Signal dynamics and interactions during flooding stress. *Plant Physiol.* **2018**, *176*, 1106–1117. [CrossRef]
26. Kolbert, Z.; Feigl, G.; Freschi, L.; Poor, P. Gasotransmitters in action: Nitric oxide-ethylene crosstalk during plant growth and abiotic stress responses. *Antioxidants* **2019**, *8*, 167. [CrossRef]
27. Giuntoli, B.; Perata, P. Group VII ethylene response factors in arabidopsis: Regulation and physiological roles. *Plant Physiol.* **2018**, *176*, 1143–1155. [CrossRef]
28. Wang, X.; Hargrove, M.S. Nitric oxide in plants: The roles of ascorbate and hemoglobin. *PLoS ONE* **2013**, *8*, e82611. [CrossRef]
29. Igamberdiev, A.U.; Bykova, N.V.; Hill, R.D. Nitric oxide scavenging by barley hemoglobin is facilitated by a monodehydroascorbate reductase-mediated ascorbate reduction of methemoglobin. *Planta* **2006**, *223*, 1033–1040. [CrossRef]
30. Voesenek, L.A.C.V.; Bailey-Serres, J. Flood adaptive traits and processes: An overview. *New Phytol.* **2015**, *206*, 57–73. [CrossRef]
31. Gasch, P.; Fundinger, M.; Müller, J.T.; Lee, T.; Bailey-Serres, J.; Mustroph, A. Redundant ERF-VII transcription factors bind an evolutionarily conserved cis-motif to regulate hypoxia-responsive gene expression in *Arabidopsis*. *Plant Cell* **2016**, *28*, 160–180. [CrossRef] [PubMed]
32. Bui, L.T.; Giuntoli, B.; Kosmacz, M.; Parlanti, S.; Licausi, F. Constitutively expressed ERF-VII transcription factors redundantly activate the core anaerobic response in *Arabidopsis thaliana*. *Plant Sci.* **2015**, *236*, 37–43. [CrossRef] [PubMed]
33. Giuntoli, B.; Shukla, V.; Maggiorelli, F.; Giorgi, F.M.; Lombardi, L.; Perata, P.; Licausi, F. Age-dependent regulation of ERF-VII transcription factor activity in *Arabidopsis thaliana*. *Plant Cell Environ.* **2017**, *40*, 2333–2346. [CrossRef] [PubMed]
34. Papdi, C.; Pérez-Salamó, I.; Joseph, M.P.; Giuntoli, B.; Bögre, L.; Koncz, C.; Szabados, L. The low oxygen, oxidative and osmotic stress responses synergistically act through the ethylene response factor VII genes *RAP2.12*, *RAP2.2* and *RAP2.3*. *Plant J.* **2015**, *82*, 772–784. [CrossRef] [PubMed]
35. Pucciariello, C.; Parlanti, S.; Banti, V.; Novi, G.; Perata, P. Reactive oxygen species-driven transcription in *Arabidopsis* under oxygen deprivation. *Plant Physiol.* **2012**, *159*, 184–196. [CrossRef] [PubMed]
36. Liu, B.; Sun, L.; Ma, L.; Hao, F.S. Both *AtrbohD* and *AtrbohF* are essential for mediating responses to oxygen deficiency in *Arabidopsis*. *Plant Cell Rep.* **2017**, *36*, 947–957. [CrossRef]
37. Sun, L.; Ma, L.; He, S.; Hao, F. *AtrbohD* functions downstream of ROP2 and positively regulates waterlogging response in *Arabidopsis*. *Plant Signal Behav.* **2018**, *13*, e1513300. [CrossRef]
38. Yamauchi, T.; Rajhi, I.; Nakazono, M. Lysigenous aerenchyma formation in maize root is confined to cortical cells by regulation of genes related to generation and scavenging of reactive oxygen species. *Plant Signal. Behav.* **2011**, *6*, 759–761. [CrossRef]
39. Umbreen, S.; Lubega, J.; Cui, B.; Pan, Q.; Jiang, J.; Loake, G.J. Specificity in nitric oxide signalling. *J. Exp. Bot.* **2018**, *69*, 3439–3448. [CrossRef]
40. Gibbs, D.J.; Md Isa, N.; Movahedi, M.; Lozano-Juste, J.; Mendiondo, G.M.; Berckhan, S.; Marín-dela Rosa, N.; Vicente Conde, J.; Sousa Correia, C.; Pearce, S.P.; et al. Nitric oxide sensing in plants is mediated by proteolytic control of group VII ERF transcription factors. *Mol. Cell* **2014**, *53*, 369–379. [CrossRef]
41. Hartman, S.; Liu, Z.; van Veen, H.; Vicente, J.; Reinen, E.; Martopawiro, S.; Zhang, H.; van Dongen, N.; Bosman, F.; Bassel, G.W.; et al. Ethylene-mediated nitric oxide depletion pre-adapts plants to hypoxia stress. *Nat. Commun.* **2019**, *10*, 4020. [CrossRef]
42. Hebelstrup, K.H.; van Zanten, M.; Mandon, J.; Voesenek, L.A.; Harren, F.J.; Cristescu, S.M.; Møller, I.M.; Mur, L.A. Haemoglobin modulates NO emission and hyponasty under hypoxia-related stress in *Arabidopsis thaliana*. *J. Exp. Bot.* **2012**, *63*, 5581–5591. [CrossRef] [PubMed]
43. Mira, M.M.; Hill, R.D.; Stasolla, C. Phytoglobins improve hypoxic root growth by alleviating apical meristem cell death. *Plant Physiol.* **2016**, *172*, 2044–2056. [CrossRef] [PubMed]
44. Ohwaki, Y.; Kawagishi-Kobayashi, M.; Wakasa, K.; Fujihara, S.; Yoneyama, T. Induction of class-1 non-symbiotic hemoglobin genes by nitrate, nitrite and nitric oxide in cultured rice cells. *Plant Cell Physiol.* **2005**, *46*, 324–331. [CrossRef] [PubMed]
45. Qu, Z.L.; Zhong, N.Q.; Wang, H.Y.; Chen, A.P.; Jian, G.L.; Xia, G.X. Ectopic expression of the cotton non-symbiotic hemoglobin gene GhHbd1 triggers defense responses and increases disease tolerance in *Arabidopsis*. *Plant Cell Physiol.* **2006**, *47*, 1058–1068. [CrossRef] [PubMed]
46. Bai, X.; Long, J.; He, X.; Yan, J.; Chen, X.; Tan, Y.; Li, K.; Chen, L.; Xu, H. Overexpression of spinach non-symbiotic hemoglobin in *Arabidopsis* resulted in decreased NO content and lowered nitrate and other abiotic stresses tolerance. *Sci. Rep.* **2016**, *6*, 26400. [CrossRef]
47. Licausi, F.; Kosmacz, M.; Weits, D.A.; Giuntoli, B.; Giorgi, F.M.; Voesenek, L.A.C.J.; Perata, P.; Van Dongen, J.T. Oxygen sensing in plants is mediated by an N-end rule pathway for protein destabilization. *Nature* **2011**, *479*, 419–422. [CrossRef]
48. Dhungana, S.K.; Kim, H.S.; Kang, B.K.; Seo, J.H.; Kim, H.T.; Oh, J.H.; Shin, S.O.; Baek, I.Y. Analysis of differentially expressed genes in soybean leaf tissue of tolerant and susceptible cultivars under flooding stress revealed by RNA sequencing. *J. Crop Sci. Biotechnol.* **2021**, *24*, 83–91. [CrossRef]

antioxidants

MDPI

Article

Marker-Free Rice (*Oryza sativa* L. cv. IR 64) Overexpressing *PDH45* Gene Confers Salinity Tolerance by Maintaining Photosynthesis and Antioxidant Machinery

Ranjan Kumar Sahoo [1,2], Renu Tuteja [2], Ritu Gill [3], Juan Francisco Jiménez Bremont [4,*], Sarvajeet Singh Gill [3,*] and Narendra Tuteja [2,*]

1 Department of Biotechnology School of Engineering and Technology, Centurion University of Technology and Management, Bhubaneswar 752050, Odisha, India; ranjan.sahoo@cutm.ac.in

2 International Centre for Genetic Engineering and Biotechnology, Aruna Asaf Ali Marg, New Delhi 110067, India; renu@icgeb.res.in

3 Stress Physiology & Molecular Biology Lab, Centre for Biotechnology, Maharshi Dayanand University, Rohtak 124001, Haryana, India; ritu_gill@mdurohtak.ac.in

4 Laboratorio de Biotecnología Molecular Plantas, Division de Biología Molecular, Instituto Potosino de Investigacion Científicay Tecnologica AC, San Luis Potosí 78395, Mexico

* Correspondence: jbremont@ipicyt.edu.mx (J.F.J.B.); ssgill14@mdurohtak.ac.in (S.S.G.); ntuteja@icgeb.res.in (N.T.); Tel.: +91-11-2674-1358 (N.T.); Fax: +91-11-2674-2316 (N.T.)

Abstract: Helicases function as key enzymes in salinity stress tolerance, and the role and function of *PDH45* (pea DNA helicase 45) in stress tolerance have been reported in different crops with selectable markers, raising public and regulatory concerns. In the present study, we developed five lines of marker-free *PDH45*-overexpressing transgenic lines of rice (*Oryza sativa* L. cv. IR64). The overexpression of *PDH45* driven by CaMV35S promoter in transgenic rice conferred high salinity (200 mM NaCl) tolerance in the T_1 generation. Molecular attributes such as PCR, RT-PCR, and Southern and Western blot analyses confirmed stable integration and expression of the *PDH45* gene in the *PDH45*-overexpressing lines. We observed higher endogenous levels of sugars (glucose and fructose) and hormones (GA, zeatin, and IAA) in the transgenic lines in comparison to control plants (empty vector (VC) and wild type (WT)) under salt treatments. Furthermore, photosynthetic characteristics such as net photosynthetic rate (Pn), stomatal conductance (gs), intercellular CO_2 (Ci), and chlorophyll (Chl) content were significantly higher in transgenic lines under salinity stress as compared to control plants. However, the maximum primary photochemical efficiency of PSII, as an estimated from variable to maximum chlorophyll a fluorescence (Fv/Fm), was identical in the transgenics to that in the control plants. The activities of antioxidant enzymes, such as catalase (CAT), ascorbate peroxidase (APX), glutathione reductase (GR), and guaiacol peroxidase (GPX), were significantly higher in transgenic lines in comparison to control plants, which helped in keeping the oxidative stress burden (MDA and H_2O_2) lesser on transgenic lines, thus protecting the growth and photosynthetic efficiency of the plants. Overall, the present research reports the development of marker-free *PDH45*-overexpressing transgenic lines for salt tolerance that can potentially avoid public and biosafety concerns and facilitate the commercialization of genetically engineered crop plants.

Keywords: antioxidants; reactive oxygen species; oxidative stress; marker-free transgenic rice; mature seed-derived calli; pea DNA helicase 45; photosynthesis; salinity stress tolerance

Citation: Sahoo, R.K.; Tuteja, R.; Gill, R.; Jiménez Bremont J.F.; Gill, S.S.; Tuteja, N. Marker-Free Rice (*Oryza sativa* L. cv. IR 64) Overexpressing *PDH45* Gene Confers Salinity Tolerance by Maintaining Photosynthesis and Antioxidant Machinery. *Antioxidants* **2022**, *11*, 770. https://doi.org/10.3390/antiox11040770

Academic Editor: Nafees A. Khan

Received: 23 March 2022
Accepted: 5 April 2022
Published: 12 April 2022

1. Introduction

Rice (*Oryza sativa* L., family Gramineae (Poaceae)) is an important staple food crop that is produced (518 million tonnes, milled), cultivated, and consumed globally in >122 countries (excluding Antarctica), being susceptible to salt amongst cereal crops [1–3]. Major abiotic stresses (salinity, drought, extreme temperatures, heavy metal, etc.) are a significant limitation in rice cultivation globally [1]. Soil salinity is a major problem that reduces productivity of crops in irrigated as well as in tropical fields, where the deterioration of agricultural lands occur due to salinity [4–6]. It brings series of changes at the physiological, biochemical, and molecular levels by affecting the photosynthetic machinery (partial stomatal closure and hampered photosystem II (PSII), reactive oxygen species (ROS)-led molecular injury, restricted water/nutrient availability, and disturbed sodium (Na^+)/potassium ion (K^+) homeostasis), which ultimately poses serious yield penalty [7–12]. Due to rapidly growing global population and urbanization, it is impossible to increase the cultivated land area, and therefore to fulfill the demand of rice consumers, it becomes imperative to discover new techniques for developing salinity-tolerant crop plants by protecting the photosynthetic machinery (net photosynthetic rate, stomatal conductance, chlorophyll content), efficient ROS scavenging, membrane integrity, Na^+ exclusion, etc. [13–15]. Robust antioxidant machinery consisting of enzymatic (SOD, CAT, APX, GPX, GR, etc.) and non-enzymatic antioxidants (glutathione (GSH) and ascorbic acid (AsA) is efficient enough to protect the photosynthetic machinery, cellular components, and membranes under various abiotic stresses [8]. Therefore, strong antioxidant machinery can be well correlated with salinity stress tolerance in crop plants [8]. Nidumukkala et al. [16] reviewed the fact that overexpression of helicases in different model and crop plants provides salt tolerance though increased antioxidant capacity, photosynthetic efficiency, and ion homeostasis, as well as by regulating the expression of various stress responsive genes. Therefore, introduction of a stress-tolerant gene in rice is one of the effective ways to develop stress-tolerant cultivars without yield penalty. The presence of selectable marker genes (SMGs, antibiotic or herbicide resistance genes) in genetically engineered crops may arouse public and regulatory concerns due to biosafety issues because the weeds or pathogenic microorganisms present in soil may become resistant to herbicides or antibiotics and can harm public health [17]. The problem of transgene expression arises due to the sexual crossing, which can lead to homology-dependent gene silencing in the genome [17]. Due to consumer, environmental, and biosafety concerns, the regulatory bodies also encourage the development of marker-free transgenic crops with an array of different transformation strategies such as homologous recombination, site-specific recombination, co-transformation, transposon-mediated transgene reintegration system, and CRISPR/Cas9 system [17–20]. The tissue culture methods are generally used to understand the mechanisms underlying salt tolerance of transgenic lines [21,22]. Several techniques have been developed to improve Agrobacterium-mediated transformation of indica rice [23–25]. The development of an efficient large-scale transformation system requires a large number of transformants for successful gene transfer [24]. Previously, many researchers developed a transformation protocol for marker-free transgenic rice plants using anther culture [26,27], but the unavailability of explants (anther) throughout the year is a major limitation of this method and it is very laborious to screen the transgenic plants by a PCR-based method.

In the present study, we report that overexpression of *PDH45* gene in an elite indica rice variety IR64 (*Oryza sativa* L., cv. IR64) showed tolerance against salinity stress as well as improved growth, photosynthesis, and better antioxidant machinery in the transgenic rice. We exploited the potential of transgenic technologies for crop improvement through developing marker-free transgenic *PDH45* rice. Thus, we also successfully developed a screening technique using 200 mM NaCl salt to screen marker-free *PDH45* transgenic rice plants. Development of rice transgenic lines overexpressing the *PDH45* gene without the antibiotic marker gene for stable expression of the stress-tolerant trait in a predictable manner avoids the transfer of undesirable transgenic material to non-transgenic crops and related species.

2. Materials and Methods

2.1. Cloning and Transformation of PDH45 Gene in IR64 Rice

PDH45 gene (accession number: Y17186) was used to establish the tissue culture technique. The coding region of *PDH45* gene (1.2 kb) was cloned in reporter gene-free plant transformation vector pCAMBIA1300 in place of hygromycin to generate complete reporter and antibiotic marker-free plasmid pCAMBIA1300-*PDH45*. An empty vector (pCAMBIA1300) construct, called vector control (VC), was used to compare the function of the gene, and the VC construct comprises all components except the *PDH45* gene. The above two constructs (pCAMBIA1300-*PDH45* and pCAMBIA1300) were used for the *Agrobacterium tumefaciens* (LBA4404)-mediated transformation method [28]. The same conditions were used to generate all the plants.

2.2. Development of Selection Technique for Marker-Free Transgenic Plants

A new selection technique was developed by adding 200 mM NaCl in selection media, shoot induction, and root induction media for the selection of *PDH45* marker-free transgenic plants during the plant induction stage. We modified the media described by Sahoo and Tuteja [28]. Here, we used 200 mM NaCl in place of hygromycin as the gene *PDH45* has already been reported as being responsible for salinity tolerance in different plants [29–34]. The other compositions of media were the same as described earlier [28].

2.3. Molecular Analysis (PCR, Southern Blot, qRT-PCR, and Western Blot) of T1 Transgenic PDH45 Plants

The genomic DNA was extracted from the healthy leaves of marker-free *PDH45* transgenic plants and used to check the integration of the gene by PCR and Southern blot analysis. About 25 μg of genomic DNA was used for Southern blot analysis. First, the genomic DNA was digested with XbaI and resolved on 0.8% agarose gel followed by transfer to nylon membrane (Hybond-N, Amersham, Inc., Amersham, UK) as previously described [35]. The probe was radiolabelled by the gene amplification method using α–[32P] dCTP. Hybridization with the probe was conducted using the method described [35]. The qRT-PCR experiment was performed to check the transcript levels of the gene using gene-specific primers such as forward 5′-ATGGCGACAACTTCTGTG-3′ and reverse 5′-TATATAAGATCACCAATATTCATTGG-3′. For Western blot analysis, the crude plant extract was denatured and separated by SDS PAGE and transferred onto a polyvinylidene fluoride (PVDF) membrane using the method described [36]. Polyclonal antibodies (1:1000 dilutions) from rabbit were used as a probe against the *PDH45* gene.

2.4. Leaf Disk Senescence Assay and Chlorophyll Content

The chlorophyll content after leaf disk senescence assay was measured using the method described earlier [37].

2.5. Biochemical Analysis of Antioxidant Activities of Marker-Free PDH45 Transgenic Lines

The seeds of *PDH45* transgenic, WT, and VC plants were kept in hydroponics for germination, and 21-d-old plants were dipped in 200 mM NaCl for 24 h. The experiments were conducted in green houses of the International Centre for Genetic Engineering and Biotechnology (ICGEB), New Delhi, where 16 h light photoperiod at 25 °C temperature was maintained. Similar stress treatment and stress conditions as described were also used in the present study [12]. After 24 h salt stress, the plant tissues were used for biochemical analysis such as catalase (CAT), ascorbate peroxidase (APX), glutathione reductase (GR), proline, hydrogen peroxide (H_2O_2), lipid peroxidation, relative water content (RWC), and electrolytic leakage. All the parameters were measured using the methods described earlier [38].

2.6. Measurement of Photosynthetic Activities and Agronomic Characteristics of PDH45 Transgenic Plants

The different photosynthetic measurements such as photosynthetic yield, rate, intercellular CO_2 concentration, CO_2 release, stomatal conductance, and transpiration rate were recorded using an infrared gas analyzer (IRGA; LI-COR, http://www.licor.com, (accessed on 2 November 2021), on a sunny day between 11:00 and 12:00 noon. The plants were grown under 200 mM NaCl stress in a large tank, and all the parameters were measured using the expanded leaves of mature plants (60 d old). After 12 d of salt stress, different agronomic characteristics were measured using the method described earlier [12].

2.7. Chlorophyll a Fluorescence Measurements

Plants were grown in green houses of the International Centre for Genetic Engineering and Biotechnology (ICGEB), New Delhi, where 16 h light [photosynthetically active radiation (750 μmol m^{-2} s^{-1})] photoperiod at 25 °C temperature was maintained. Minimal fluorescence (Fo), maximal fluorescence (Fm), maximal variable fluorescence (Fv), and Fv/Fm ratio were included, where Fv = Fm − Fo.

Chlorophyll a (Chl a) fluorescence from the leaves of 25-day-old WT, VC, and transgenic rice seedlings was measured with a PAM-2100 fluorometer (Walz, Germany). Before each measurement, the leaf sample was kept in the dark for 20 min [39]. Optimum quantum efficiency (uPSII, also referred to as Y) of Photosystem II (PSII) was inferred from Fv/Fm = (Fm − Fo)/Fm [40].

2.8. Estimation of Sugar, Hormones (GA, Zeatin and IAA), and Ion Contents

Shoots and roots from mature (60 d old) *PDH45* T1 transgenic, VC, and WT plants after 12 d of salt stress were used in this study. The sugar content was estimated as described earlier [41]. The endogenous plant hormones (GA, zeatin and IAA) were estimated as described earlier [42]. The flame ionization photometer was used for the estimation of potassium, as described by Chapman and Pratt [43]. The sodium content was estimated as described by Munns et al. [44].

2.9. Salinity Tolerance of Transgenic Plant under 200 mM NaCl Stress

The *PDH45* transgenic lines (L4, L7, L8, L11 and L13) and VC and WT rice plants (60 d old) were grown in one large metal pot filled with soil and dipped in 200 mM NaCl. The plants were allowed to grow up to maturity (harvest), and the phenotypic conditions of these plants were recorded.

2.10. Statistical Analysis

The experimental data were statistically analyzed, and standard error was calculated from three independent observations. Analysis of variance (ANOVA) was performed on the data using SPSS (10.0 Inc., Chicago, IL, USA) to determine the least significant difference (LSD) for the significant data to identify the differences between means and presented as mean ± SE. The means were separated by Duncan's multiple range tests. Different letters indicate significant difference at $p < 0.05$.

3. Results

3.1. Molecular Analysis of Marker-Free PDH45 Transgenic Lines

The marker-free *PDH45* transgenic IR64 rice plants were developed using the pCAMBIA1300-*PDH45* gene construct (Figure 1a). Phenotypically, the transgenic rice plants were not significantly different from WT and VC plants (Figure 1b). The desired *PDH45* gene (1.2 kb) fragment was detected by PCR (Figure 1c). The Southern blot results confirmed the integration of a single-copy *PDH45* gene in transgenic rice plants in all the five transgenic lines (L4, L7, L8, L11, and L13) (Figure 1d). The real-time PCR (qRT-PCR) provided ≈8-fold induction in the transcript level of *PDH45* in transgenic lines (L4, L7, L8, L11 and L13) (Figure 1e). The Western blot results showed that *PDH45* protein was expressed to almost similar levels in all the transgenic lines (L4, L7, L8, L11 and L13) as compared to WT and VC plants (Figure 1f).

3.2. PDH45 Transgenic Lines Showed Salinity Tolerance

The damage caused in the leaf pieces by salt stress was observed in all the plants after 72 h; however, the *PDH45*-overexpressing lines displayed darker green leaves, in contrast to the yellowish leaves of the WT and VC plants (Figure 2a). In this sense, the reduction of chlorophyll content in leaf tissues was lesser in transgenic lines as compared to WT and VC plants under salt stress (Figure 2b). The lesser chlorophyll content in the leaf tissues of WT and VC plants as compared to transgenic lines provided strong evidence towards tolerance against salinity stress (Figure 2a,b). The transgenic lines (L4, L7, L8, L11 and L13) along with WT and VC plants were allowed to grow up to maturity in a metal tank filled with 200 mM NaCl. After 3d, WT and VC plants showed dropping characteristics, whereas *PDH45*-overexpressing transgenic lines L4, L7, L8, L11 and L13 grew well and produced viable seeds (Figure 2c,d).

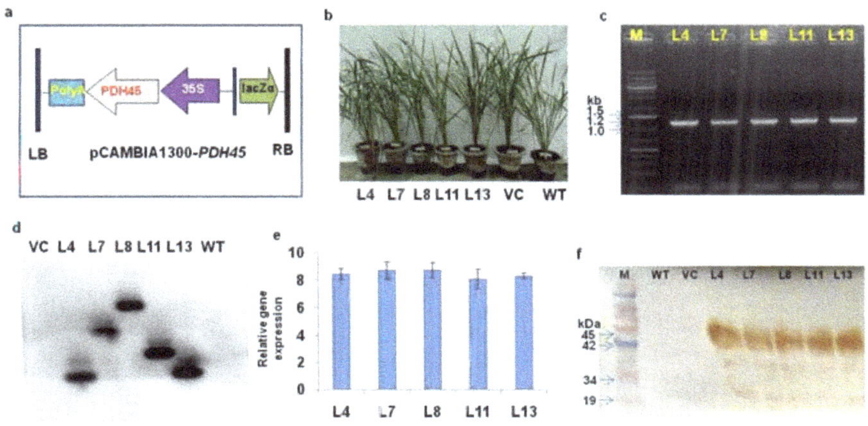

Figure 1. Screening and analysis of *PDH45* marker-free transgenic lines. (**a**) T-DNA construct of pCAMBIA 1300-*PDH45*. (**b**) Transgenic lines (L4, L7, L8, L11, L13, VC, and WT). (**c**) PCR conformation of the *PDH45*-overexpressing transgenic (T_1) lines showed the amplification of 1.2 Kb fragment. (**d**) Southern blot analysis showing the integration and copy number of the *PDH45* gene. (**e**) Relative gene expression of *PDH45* transgenic lines. (**f**) Western blot analysis showing the *PDH45* protein (≈45 kDa).

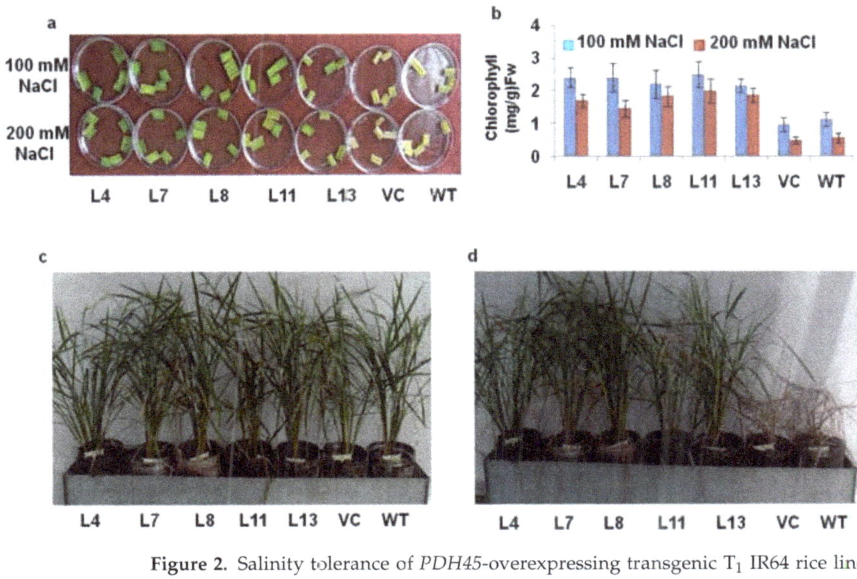

Figure 2. Salinity tolerance of *PDH45*-overexpressing transgenic T_1 IR64 rice lines. (**a**) Leaf disk senescence assay under 100 and 200 mM NaCl treatment. (**b**) Chlorophyll content (mg/g fw) in *PDH45* transgenic lines after salt treatment. (**c**) Third day in 200 mM NaCl treatment. (**d**) After 15 days of NaCl treatment.

3.3. Agronomic Performance of Marker-Free PDH45 Transgenic Plants under Stress

The agronomic performance of T_1 transgenic lines under 200 mM NaCl treatment was compared with WT and VC without NaCl treatment. Better agronomic characteristics were observed in *PDH45* transgenic plants as compared to WT and VC plants. Several phenotypic characteristics of transgenic plants were recorded and found to be almost similar to the WT and VC plants grown in 0 mM NaCl. However, under 200 mM NaCl treatment, the WT and VC plants did not survive until flowering stage (Figure 2d).

3.4. Photosynthetic Characteristics and Endogenous Ion Content of Marker-Free PDH45 T1 Transgenic Plants under Stress

The photosynthetic characteristics of transgenic plants were observed and compared to WT and VC plants after 12 d of induction of 200 mM NaCl salt treatment. The photosynthetic rate declined by 33% in WT and 35% in VC plants as compared to *PDH45* marker-free transgenic lines. The net photosynthetic rate, stomatal conductance, intracellular CO_2, CO_2 release, and transpiration rate were also higher in transgenic lines as compared to the WT and VC plants (Figure 3a–e).

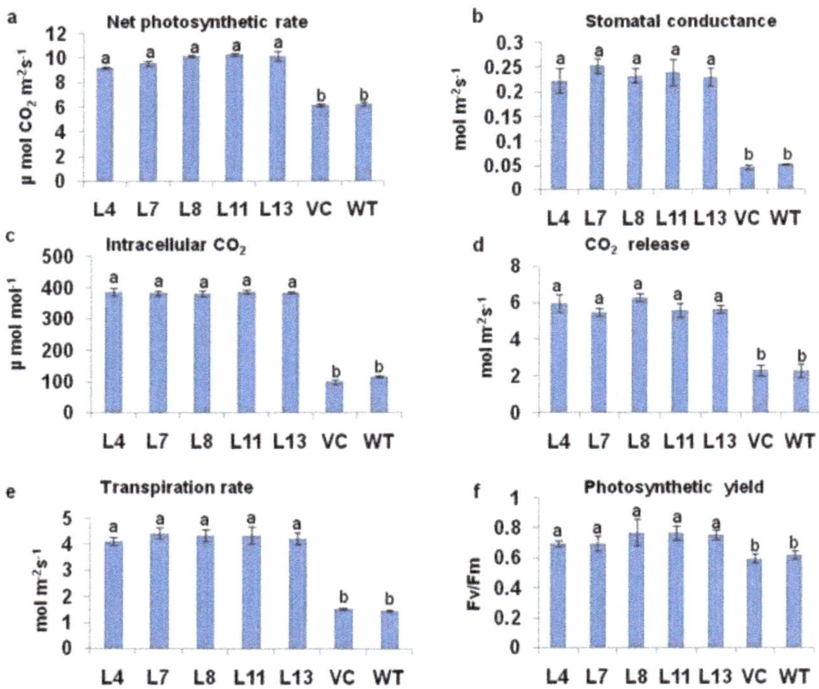

Figure 3. Measurement of photosynthetic characteristics and chlorophyll a fluorescence of WT, VC, and *PDH45* marker-free transgenic lines (L4, L7, L8, L11 and L13) under 200 mM NaCl treatment. (**a**) Photosynthetic rate. (**b**) Stomatal conductance. (**c**) Intracellular CO_2. (**d**) CO_2 release. (**e**) Transpiration rate. (**f**) Photosynthetic yield (Fv/Fm). Values are mean ± SE (n = 3). Different letters on the top of bars indicate significant differences at $p \leq 0.05$ level as determined by Duncan's multiple range test (DMRT).

3.5. Chlorophyll a Fluorescence

The chlorophyll fluorescence rose from a low minimum level ("O" level or Fo) to a higher maximum level ("P" level or Fm) when exposed from dark to light. The maximum primary photochemical efficiency of PSII, estimated from Fv/Fm, was almost identical in the transgenics to that in the VC and WT (Figure 3f).

3.6. Analysis of MDA, H_2O_2, Ion Leakage, and Antioxidant Response in Marker-Free PDH45 T1 Transgenic Plants

The salt-induced changes in the ion leakage, H_2O_2, proline content, accumulation of MDA, RWC, and antioxidant machineries in T_1 *PDH45* transgenic lines (L4, L7, L8, L11 and L13) were compared with WT and VC rice seedlings. We observed reduced levels of MDA, H_2O_2, and ion leakage, alongside an increase in proline content in *PDH45* transgenic lines in comparison to the WT and VC plants under salt stress at 200 mM NaCl (Figure 4a–d). The activities of CAT, APX, GPX, GR and RWC were increased in *PDH45* transgenic plants as compared to WT and VC plants (Figure 4e–i).

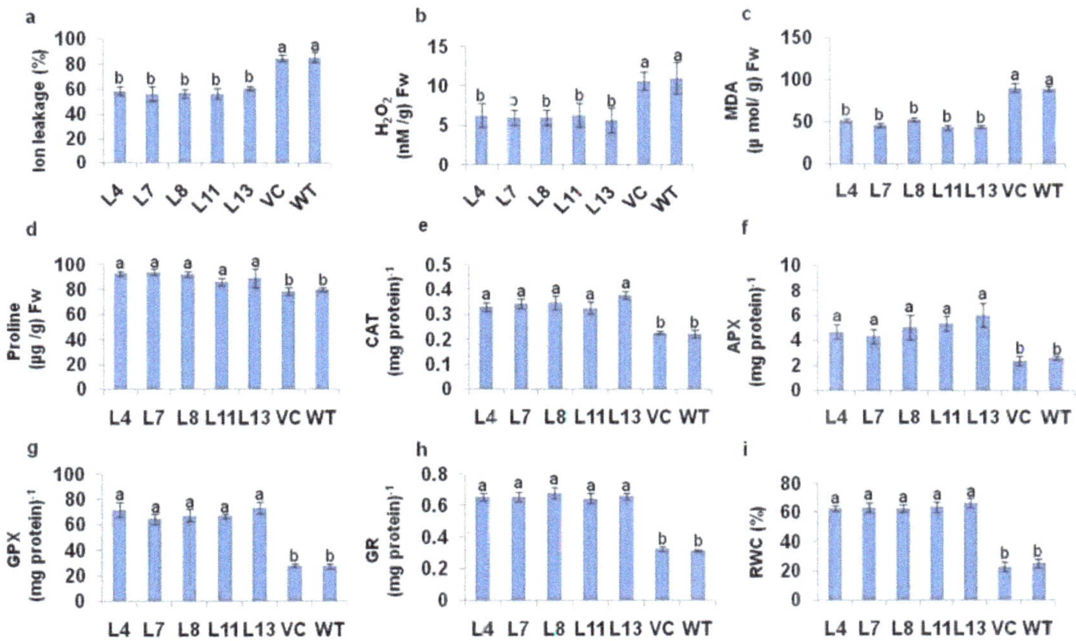

Figure 4. Biochemical analysis of *PDH45*-overexpressing T_1 transgenic lines (L4, L7, L8, L11, L13, VC) and WT plants exposed to 24 h at 200 mM NaCl treatment. (**a**) Ion leakage. (**b**) Hydrogen peroxide (H_2O_2) content. (**c**) Lipid peroxidation expressed in terms of MDA content. (**d**) Level of proline accumulation. (**e**) Catalase (CAT) activity; one unit of enzyme activity defined as 1 μmol H_2O_2 oxidized min^{-1}. (**f**) Ascorbate peroxidase (APX) activity; one unit of enzyme activity defined as 1 μmol of ascorbate oxidized min^{-1}. (**g**) Guaiacol peroxidase (GPX) activity. (**h**) Glutathione reductase (GR) activity; one unit of enzyme activity is defined as 1 μmol of GS-TNB formed min^{-1} due to reduction of DTNB. (i) Percent relative water content (RWC). Values are mean ± SE (n = 3). Different letters on the top of bars indicate significant differences at $p \leq 0.05$ level as determined by Duncan's multiple range test (DMRT).

3.7. The Sugar and Hormone Content of Marker-Free PDH45 T1 Transgenic Plants

The *PDH45* L4, L7, L8, L11 and L13 transgenic lines showed higher endogenous sugar (glucose and fructose) content in roots as well as in shoots when compared with WT and VC plants (Figure 5a,b). The endogenous hormones such as GA, zeatin, and IAA content were also higher in roots and shoots of *PDH45* transgenics as compared to WT and VC plants (Figure 5c–e). The potassium content in transgenic plants was higher, whereas sodium content was lower in marker-free *PDH45* transgenic plant tissues as compared to WT and VC plants (Figure 5f).

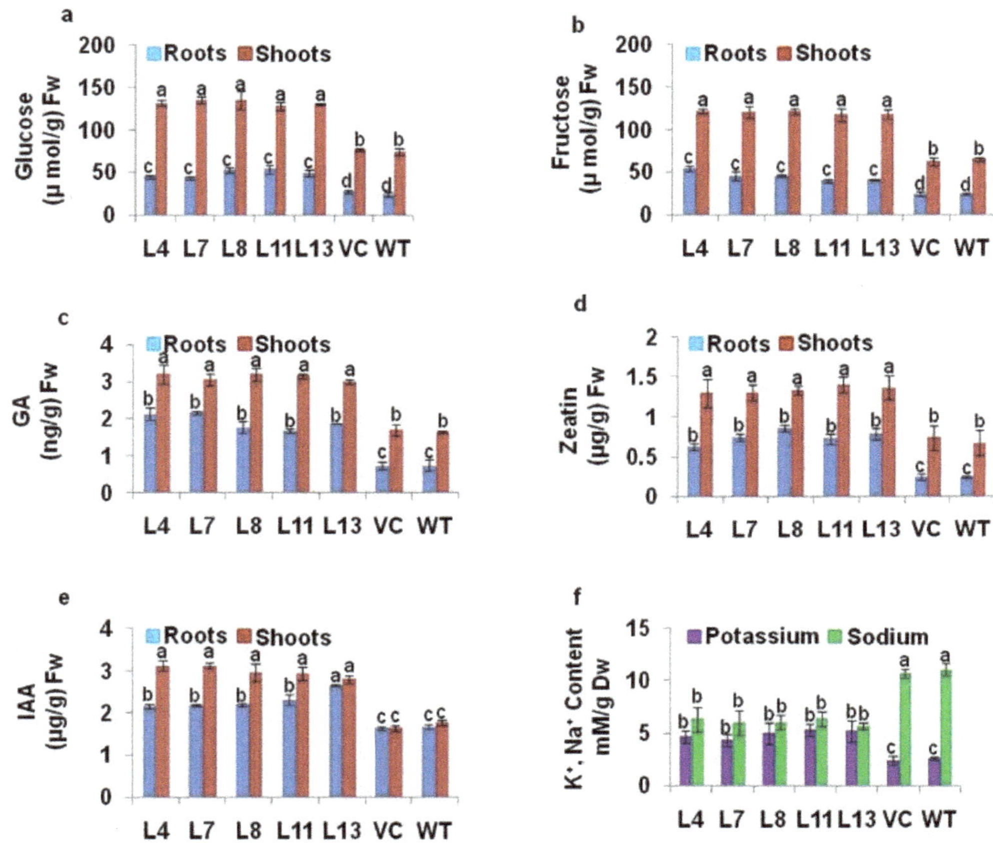

Figure 5. Soluble sugar, hormones, and K^+ and Na^+ content in the roots and shoots of *PDH45*-overexpressing marker-free transgenic lines (L4, L7, L8, L11 L13) as compared to WT and VC plants exposed to 24 h at 200 mM NaCl treatment. (**a**) Glucose content. (**b**) Fructose content. (**c**) Endogenous GA content. (**d**) Endogenous zeatin content. (**e**) Endogenous IAA content. (**f**) Endogenous potassium and sodium content. Values are mean ± SE (n = 3). Different letters on the top of bars indicate significant differences at $p \leq 0.05$ level as determined by Duncan's multiple range test (DMRT).

4. Discussion

In the era of frequently changing global climatic conditions, shortage of irrigation water, reduced agriculturally suitable cultivable land area, degradation and salinization of the agricultural soil, and unpredictable onset of abiotic stresses, agricultural productivity is severely affected, posing a serious threat to food security. Therefore, it is imperative to develop genetically engineered stress-tolerant crops with all the qualifications of global acceptance. It has been reported that overexpression of helicases (*PDH45*/*PDH47*) in different model and crop plants provides salt/cold tolerance through increased antioxidant capacity, photosynthetic efficiency, and ion homeostasis, as well as by regulating the expression of various stress responsive genes [16,29–32,37,45]. Genetically engineered transgenic crops with selectable markers (antibiotic or herbicide resistance) have public and regulatory concerns; therefore, development of marker-free transgenic plants is needed in order to avoid public and biosafety concerns and to facilitate the commercialization of genetically engineered crop plants [17].

We developed the method to select marker and reporter free transgenic lines using the previously published reports [46,47]. In this research, marker-free *PDH45* transgenic rice plants were raised using Agrobacterium-mediated transformation followed by screening with 200 mM NaCl in selection, shoot, and root regeneration media to select only the transformed calli for plant regeneration because *PDH45* is responsible for salinity tolerance [31–33,45]. The elevated stress tolerance in *PDH45*-expressing plants correlated with MH1 (*M. sativa* helicase 1) transgenic plants, showing that MH1 functions in abiotic stress tolerance by elevating reactive oxygen species (ROS) burden and through osmotic adjustment [48]. Five independent transgenic lines (L4, L7, L8, L11 and L13) along with empty VC and WT plants were used for functional validation under salt stress. These lines express almost similar levels of *PDH45* protein. Similar to previous reports, these *PDH45* transgenic rice plants also showed high salinity tolerance. This was indicated by the presence of higher chlorophyll content in the leaf disks of salinity-stressed T_1 transgenic plants, whereas VC and WT plant leaves became yellow. Moreover, the transgenic plants were able to grow in the continuous presence of 200 mM NaCl stress. These results indicate that the introduced trait is functional in transgenic plants and that it is also stable. The transgenic lines also maintained higher endogenous nutrient contents as compared to the VC and WT plants under salinity stress, which revealed the salinity tolerance potential of the transgenic lines. Similar findings have been reported earlier [12,32,33,45,49]. Higher concentration of potassium and lower concentration of sodium were found in leaves of *PDH45*-overexpressing transgenic lines as compared with VC and WT plants.

PDH45-overexpressing marker-free transgenic lines maintained higher endogenous nutrient contents under salinity stress as compared with WT and VC plants, which proved the salt stress tolerance potential of the marker-free *PDH45* transgenic lines, which is in agreement with the previous reports [31–33]. The higher potassium and lower sodium concentration in T_1 transgenic plants indicates that the lower Na+/K+ ratio in the transgenic lines might be responsible for imparting better stress tolerance to salinity stress in comparison to the VC and WT plants. The better photosynthetic activities such as net photosynthetic rate, stomatal conductance, intercellular CO_2 concentration, CO_2 release and transpiration rate, and photosynthetic yield (Fv/Fm) were observed in *PDH45* transgenic lines as compared to the VC and WT plants. The retention of chlorophyll content in transgenic lines indicates the better control over the photosynthetic apparatus under salt stress. Our data are in agreement with the earlier reports on *PDH45*-, SUV3-, and BAT1-overexpressing rice plants under stress [12,31,32,49].

Sugars such as glucose and fructose may play a key role in salt defense mechanisms through ROS detoxification [49–53]. The sugar content in *PDH45*-overexpressing marker-free transgenic lines was higher as compared to VC and WT plants. The *PDH45*-overexpressing transgenic rice plants showed significantly higher endogenous content of plant hormones in leaf, stem, and root, directing the molecular and biochemical mechanisms to confer increased stress tolerance [54]. A similar trend of endogenous plant hormone profile was also reflected in OsSUV3 and OsBAT1 transgenic rice under stress conditions [7,49] This is a very simple, reproducible, and improved protocol for selection of marker-free transgenic rice plants using Agrobacterium-mediated transformation of mature seed-derived callus tissues of indica rice variety, IR64.

5. Conclusions

The present study provides reporter and marker-free transgenic rice plants that has a scope for future commercialization and approval from regulatory agencies as they are focusing on the removal of reporter and marker genes from transgenic plants. We developed a unique successful salt screening method for screening transgenic lines during tissue culture and also utilized the unique function of *PDH45* helicase in providing salt tolerance in marker-free transgenic rice cv. IR64. It also provides a good example for the exploitation of helicases for enhanced agricultural production, while withstanding extreme climatic conditions, maintaining biosafety regulations, and ensuring food security.

Author Contributions: R.T. and N.T. designed the research; R.K.S. performed the experiments; R.K.S., R.T. and N.T. analyzed the data; R.K.S., R.G., S.S.G., J.F.J.B. and N.T. wrote the manuscript. All authors have read and agreed to the published version of the manuscript.

Funding: This research received no external funding. The APC was funded by JFJB.

Institutional Review Board Statement: Not applicable.

Informed Consent Statement: Not applicable.

Data Availability Statement: The data presented in this study are available in the article.

Acknowledgments: Work on plant abiotic stress tolerance in N.T.'s laboratory was partially supported by the Department of Science and Technology (DST), Government of India, and Department of Biotechnology (DBT), Government of India. S.S.G. & R.G. acknowledges partial support from DBT-BUILDER grant, Department of Biotechnology, Govt. of India (BT/INF/22/SP43043/2021). JFJB acknowledges funding support by CONACYT (Ciencia Básica A1-S-25233). The authors gratefully acknowledge the help of Govindjee for his critical review and suggestions for the improvement of the manuscript.

Conflicts of Interest: The authors declare no conflict of interest.

References

1. Bandumula, N. Rice Production in Asia: Key to Global Food Security. *Proc. Natl. Acad. Sci. India Sec. B Biol. Sci.* **2018**, *88*, 1323–1328. [CrossRef]
2. Zeigler, R.S.; Adams, A. The relevance of rice. *Rice* **2008**, *1*, 3–10. [CrossRef]
3. FAO. *Crop Prospects and Food Situation—Quarterly Global Report No. 4*; FAO: Rome, Italy, 2021. [CrossRef]
4. Mahajan, S.; Tuteja, N. Cold, salinity and drought stresses: An overview. *Arch. Biochem. Biophys.* **2005**, *444*, 139–158. [CrossRef] [PubMed]
5. Munns, R.; Tester, M. Mechanisms of salinity tolerance. *Ann. Rev. Plant. Biol.* **2008**, *59*, 651–681. [CrossRef]
6. Cushman, J.C.; Denby, K.; Mittler, R. Plant responses and adaptations to a changing climate. *Plant J.* **2022**, *109*, 319–322. [CrossRef]
7. Sahoo, R.K.; Ansari, M.W.; Tuteja, R.; Tuteja, N. OsSUV3 transgenic rice maintains higher endogenous levels of plant hormones that mitigates adverse effects of salinity and sustains crop productivity. *Rice* **2014**, *7*, 17. [CrossRef]
8. Gill, S.S.; Tuteja, N. Reactive oxygen species and antioxidant machinery in abiotic stress tolerance in crop plants. *Plant Physiol. Biochem.* **2010**, *48*, 909–930. [CrossRef]
9. Kerchev, P.I.; Van Breusegem, F. Improving oxidative stress resilience in plants. *Plant J.* **2022**, *109*, 359–372. [CrossRef]
10. Shahzad, B.; Yun, P.; Shabala, L.; Zhou, M.; Sellamuthu, G.; Venkataraman, G.; Chen, Z.-H.; Shabala, S. Unravelling the physiological basis of salinity stress tolerance in cultivated and wild rice species. *Funct. Plant Biol.* **2022**, *49*, 351–364. [CrossRef]
11. Liu, C.; Mao, B.; Yuan, D.; Chu, C.; Duan, M. Salt tolerance in rice: Physiological responses and molecular mechanisms. *Crop J.* **2022**, *10*, 13–25. [CrossRef]
12. Wu, H.; Hill, C.B.; Stefano, G.; Bose, J. Editorial: New Insights Into Salinity Sensing, Signaling and Adaptation in Plants. *Front. Plant Sci.* **2021**, *11*, 604139. [CrossRef] [PubMed]
13. Tuteja, N.; Sahoo, R.K.; Garg, B.; Tuteja, R. OsSUV3 dual helicase functions in salinity stress tolerance by maintaining photosynthesis and antioxidant machinery in rice (*Oryza sativa* L. cv. IR64). *Plant J.* **2013**, *76*, 115–127. [CrossRef] [PubMed]
14. Rivero, R.M.; Mittler, R.; Blumwald, E.; Zandalinas, S.I. Developing climate-resilient crops: Improving plant tolerance to stress combination. *Plant J.* **2022**, *109*, 373–389. [CrossRef] [PubMed]
15. Hannachi, S.; Steppe, K.; Eloudi, M.; Mechi, L.; Bahrini, I.; Van Labeke, M.-C. Salt Stress Induced Changes in Photosynthesis and Metabolic Profiles of One Tolerant ('Bonica') and One Sensitive ('Black Beauty') Eggplant Cultivars (*Solanum melongena* L.). *Plants* **2022**, *11*, 590. [CrossRef] [PubMed]
16. Nidumukkala, S.; Tayi, L.; Chittela, R.K.; Vudem, D.R.; Khareedu, V.R. DEAD box helicases as promising molecular tools for engineering abiotic stress tolerance in plants. *Crit. Rev. Biotechnol.* **2019**, *39*, 395–407. [CrossRef] [PubMed]
17. Tuteja, N.; Verma, S.; Sahoo, R.K.; Raveendar, S.; Reddy, I.N.B.L. Recent advances in development of marker-free transgenic plants: Regulation and biosafety concern. *J. Biosci.* **2012**, *37*, 167–197. [CrossRef]
18. Sarkar, S.; Roy, S.; Ghosh, S.K. Development of marker-free transgenic pigeon pea (*Cajanus cajan*) expressing a pod borer insecticidal protein. *Sci. Rep.* **2021**, *11*, 10543. [CrossRef]
19. Li, X.; Pan, L.; Bi, D.; Tian, X.; Li, L.; Xu, Z.; Wang, L.; Zou, X.; Gao, X.; Yang, H.; et al. Generation of Marker-Free Transgenic Rice Resistant to Rice Blast Disease Using Ac/Ds Transposon-Mediated Transgene Reintegration System. *Front. Plant Sci.* **2021**, *12*, 644437. [CrossRef]

20. Wang, J.J.; Wang, C.; Wang, K.J. Generation of marker-free transgenic rice using CRISPR/Cas9 system controlled by floral specific promoters. *J. Genet. Genom.* **2019**, *46*, 61–64. [CrossRef]
21. Venkataiah, P.; Christopher, T.; Subhash, K. Selection and characterization of sodium chloride and mannitol tolerant callus lines of red pepper (*Capsicum annuum* L.). *Plant Physiol.* **2012**, *9*, 158–163.
22. Pérez-Jiménez, M.; Olaya Pérez-Tornero, O. In Vitro Plant Evaluation Trial: Reliability Test of Salinity Assays in Citrus Plants. *Plants* **2020**, *9*, 1352. [CrossRef] [PubMed]
23. Xiang, Z.; Chen, Y.; Chen, Y.; Zhang, L.; Liu, M.; Mao, D.; Chen, L. Agrobacterium-Mediated High-Efficiency Genetic Transformation and Genome Editing of Chaling Common Wild Rice (*Oryza rufipogon* Griff.) Using Scutellum Tissue of Embryos in Mature Seeds. *Front. Plant Sci.* **2022**, *13*, 849666. [CrossRef]
24. Terada, R.; Asao, H.; Iida, S. A large-scale Agrobacterium-mediated transformation procedure with a strong positive-negative selection for gene targeting in rice (*Oryza sativa* L.). *Plant Cell Rep.* **2004**, *22*, 653–659. [CrossRef] [PubMed]
25. Banu, M.S.A.; Ahmed, B.; Parveen, S.; Rashid, M.H.U.; Huda, K.M.K. Agrobacterium-mediated Genetic Transformation of Rice var. BRRI Dhan 58. *Plant Tissue Cult. Biotech.* **2021**, *31*, 71–80. [CrossRef]
26. Zhu, L.; Pu, Y.-P.; Liu, W.-Z.; Hu, G.-C.; Si, H.-M.; Tang, K.-X.; Sun, Z.-X. Rapid Generation of Selectable Marker-Free Transgenic Rice with Three Target Genes by Co-Transformation and Anther Culture. *Rice Sci.* **2007**, *14*, 239–246. [CrossRef]
27. Woo, H.-J.; Qin, Y.; Park, S.-Y.; Park, S.K.; Cho, Y.-G.; Shin, K.S.; Lim, M.H.; Cho, H.-S. Development of Selectable Marker-Free Transgenic Rice Plants with Enhanced Seed Tocopherol Content through FLP/FRT-Mediated Spontaneous Auto-Excision. *PLoS ONE* **2015**, *10*, e0132667. [CrossRef]
28. Sahoo, R.K.; Tuteja, N. Development of Agrobacterium-mediated transformation technology for mature seed-derived callus tissues of indica rice cultivar IR64. *GM Crops Food* **2012**, *3*, 123–128.
29. Pham, X.H.; Reddy, M.K.; Ehtesham, N.Z.; Matta, B.; Tuteja, N. A DNA helicase from Pisum sativum is homologous to translation initiation factor and stimulates topoisomerase I activity. *Plant J.* **2000**, *24*, 219–229. [CrossRef]
30. Vashisht, A.A.; Tuteja, N. Stress responsive DEAD-box helicases: A new pathway to engineer plant stress tolerance. *J. Photochem. Photobiol. B* **2006**, *84*, 150–160. [CrossRef]
31. Sahoo, R.K.; Gill, S.S.; Tuteja, N. Pea DNA helicase 45 promotes salinity stress tolerance in IR64 rice with improved yield. *Plant Signal. Behav.* **2012**, *7*, 1037–1041. [CrossRef]
32. Gill, S.S.; Tajrishi, M.; Madan, M.; Tuteja, N. A DESD-box helicase functions in salinity stress tolerance by improving photosynthesis and antioxidant machinery in rice (*Oryza sativa* L. cv. PB1). *Plant Mol. Biol.* **2013**, *82*, 1–22. [CrossRef] [PubMed]
33. Garg, B.; Gill, S.S.; Biswas, D.K.; Sahoo, R.K.; Kunchge, N.S.; Tuteja, R.; Tuteja, N. Simultaneous Expression of *PDH45* with EPSPS Gene Improves Salinity and Herbicide Tolerance in Transgenic Tobacco Plants. *Front. Plant Sci.* **2017**, *8*, 364. [CrossRef] [PubMed]
34. Shivakumara, T.N.; Sreevathsa, R.; Dash, P.K.; Sheshshayee, M.S.; Papolu, P.K.; Tuteja, N.; UdayaKumar, M. Overexpression of Pea DNA Helicase 45 (*PDH45*) imparts tolerance to multiple abiotic stresses in chili (*Capsicum annuum* L.). *Sci. Rep.* **2017**, *7*, 2760. [CrossRef] [PubMed]
35. Sambrook, J.; Fritsch, E.F.; Maniatis, T. *Molecular Cloning: A Laboratory Manual*, 2nd ed.; Cold Spring Harbor Laboratory Press: Cold Spring Harbor, NY, USA, 1989.
36. Hurkman, W.; Tanaka, C. Solubilization of plant membrane proteins for analysis by two-dimensional gel electrophoresis. *Plant Physiol.* **1986**, *81*, 802–806. [CrossRef]
37. Sanan-Mishra, N.; Pham, X.H.; Sopory, S.K.; Tuteja, N. Pea DNA helicase 45 overexpression in tobacco confers high salinity tolerance without affecting yield. *Proc. Natl. Acad. Sci. USA* **2005**, *102*, 509–514. [CrossRef]
38. Garg, B.; Jaiswal, J.P.; Misra, S.; Tripathi, B.N.; Prasad, M.A. A comprehensive study on dehydration-induced antioxidative responses during germination of Indian bread wheat (*Triticum aestivum* L. Em Thell) cultivars collected from different agroclimatic zones. *Physiol. Mol. Biol. Plants* **2012**, *18*, 217–228. [CrossRef]
39. Demmig, B.; Winter, K.; Kruger, A.; Czygan, F.C. Photoinhibition and zeaxanthin formation in intact leaves: A possible role of the xanthophyll cycle in the dissipation of excess light energy. *Plant Physiol.* **1987**, *84*, 218–224. [CrossRef]
40. Schreiber, U.; Armond, P.A. Heat-induced changes of chlorophyll fluorescence in isolated chloroplasts and related heat-damage at the pigment level. *Biochim. Biophys. Acta.* **1978**, *502*, 138–151. [CrossRef]
41. Karkacier, M.; Erbas, M.; Uslu, M.K.; Aksu, M. Comparison of different extraction and detection methods for sugars using amino-bonded phase HPLC. *J. Chromatogr. Sci.* **2003**, *41*, 331–333. [CrossRef]
42. Chen, J.G.; Du, X.M.; Zhao, H.Y.; Zhou, X. Fluctuation in levels of endogenous plant hormones in ovules of normal and mutant cotton during flowering and their relation to fiber development. *J. Plant Growth Regul.* **1996**, *15*, 173–177. [CrossRef]
43. Chapman, H.D.; Pratt, P.F. *Method of Analysis of Soils, Plants and Waters*, 2nd ed.; California University Agricultural Division: Berkeley, CA, USA, 1982; p. 170.
44. Munns, R.; Wallace, P.A.; Teakle, N.L.; Colmer, T.D. Measuring soluble ion concentrations (Na^+, K^+, Cl^-) in salt-treated plants. In *Plant Stress Tolerance. Methods in Molecular Biology (Methods and Protocols)*; Sunkar, R., Ed.; Springer: New York, NY, USA, 2010; pp. 371–382.
45. Amin, M.; Elias, S.M.; Hossain, A.; Ferdousi, A.; Rahman, M.S.; Tuteja, N.; Seraj, Z.I. Overexpression of a DEAD box helicase, *PDH45*, confers both seedling and reproductive stage salinity tolerance to rice (*Oryza sativa* L.). *Mol. Breed.* **2012**, *30*, 345–354. [CrossRef]

46. Shanthi, P.; Jebaraj, S.; Geetha, S. In vitro screening for salt tolerance in Rice (*Oryza sativa*). *Electronic. J. Plant Breed.* **2010**, *1*, 1208–1212.
47. Zinnah, K.M.A.; Zobayer, N.; Sikdar, S.U.; Liza, L.N.; Chowdhury, M.A.N.; Ashrafuzzaman, M. In vitro regeneration and screening for salt tolerance in rice (*Oryza sativa* L.). *Int. Res. J. Biological Sci.* **2013**, *2*, 29–36.
48. Luo, Y.; Liu, Y.B.; Dong, Y.X.; Gao, X.-Q.; Zhang, X.S. Expression of a putative alfalfa helicase increases tolerance to abiotic stress in Arabidopsis by enhancing the capacities for ROS scavenging and osmotic adjustment. *J. Plant Physiol.* **2009**, *166*, 385–394. [CrossRef] [PubMed]
49. Tuteja, N.; Sahoo, R.K.; Huda, K.M.K.; Tula, S.; Tuteja, R. OsBAT1 augments salinity stress tolerance by enhancing detoxification of ROS and expression of stress-responsive genes in transgenic rice. *Plant Mol. Biol. Rep.* **2015**, *33*, 1192–1209. [CrossRef]
50. Bohnert, H.J.; Jensen, R.G. Strategies for engineering water stress tolerance in plants. *Trends Plant Sci.* **1996**, *14*, 89–97. [CrossRef]
51. Bentsink, L.; Alonso-Blanco, C.; Vreugdenhil, D.; Tesnier, K.; Groot, S.P.C.; Koornneef, M. Genetic analysis of seed soluble oligosaccharides in relation to seed storability of Arabidopsis. *Plant Physiol.* **2000**, *124*, 1595–1604. [CrossRef]
52. Roy, P.; Niyogi, K.; Sengupta, D.N.; Ghosh, B. Spermidine treatment to rice seedlings recovers salinity stress-induced damage of plasma membrane and PM-bound H^+-ATPase in salt-tolerant and salt sensitive rice cultivars. *Plant Sci.* **2005**, *168*, 583–591. [CrossRef]
53. Pattanagul, W.; Thitisaksakul, M. Effect of salinity stress on growth and carbohydrate metabolism in three rice (*Oryza sativa* L.) cultivars differing in salinity tolerance. *Indian. J. Exp. Biol.* **2008**, *46*, 736–742.
54. Osakabe, Y.; Arinaga, N.; Umezawa, T.; Katsura, S.; Nagamachi, K.; Tanaka, H.; Ohiraki, H.; Yamada, K.; Seo, S.U.; Abo, M.; et al. Osmotic stress responses and plant growth controlled by potassium transporters in Arabidopsis. *Plant Cell* **2013**, *25*, 609–624. [CrossRef]

antioxidants

MDPI

Article

Genome-Wide Identification of the Bcl-2 Associated Athanogene (BAG) Gene Family in *Solanum lycopersicum* and the Functional Role of *SlBAG9* in Response to Osmotic Stress

Hailong Jiang, Yurong Ji, Jiarong Sheng, Yan Wang, Xiaoya Liu, Peixiang Xiao and Haidong Ding *

Joint International Research Laboratory of Agriculture and Agri-Product Safety of Ministry of Education of China, College of Bioscience and Biotechnology, Yangzhou University, Yangzhou 225009, China; jhl19923574146@163.com (H.J.); jyr0806@163.com (Y.J.); myyjjfb@163.com (J.S.); wy1123406945@163.com (Y.W.); arielnxy@outlook.com (X.L.); xiaopeixiang_1998@163.com (P.X.)

* Correspondence: hdding@yzu.edu.cn

Abstract: The Bcl-2-associated athanogene (BAG) proteins are a family of multi-functional group of co-chaperones regulators, modulating diverse processes from plant growth and development to stress response. Here, 10 members of *SlBAG* gene family were identified based on the available tomato (*Solanum lycopersicum*) genomic information and named as *SlBAG1-10* according to their chromosomal location. All SlBAG proteins harbor a characteristic BAG domain, categorized into two groups, and SlBAG4, SlBAG7, and SlBAG9 of group I contain a plant-specific isoleucine glutamine (IQ) calmodulin-binding motif located in the N terminus. The quantitative real-time PCR expression analysis revealed that these *SlBAG* genes had organ-specific expression patterns and most *SlBAG* genes were differentially expressed in multiple abiotic stresses including drought, salt, high temperature, cold, and cadmium stress as well as abscisic acid and H_2O_2. In addition, heterologous overexpression of *SlBAG9* increased the sensitivity of Arabidopsis to drought, salt, and ABA during seed germination and seedling growth. The decreased tolerance may be due to the downregulation of stress-related genes expression and severe oxidative stress. The expression levels of some stress and ABA-related genes, such as *ABI3*, *RD29A*, *DREB2A*, and *P5CS1*, were significantly inhibited by *SlBAG9* overexpression under osmotic stress. Meanwhile, the overexpression of *SlBAG9* inhibited the expression of *FSD1* and *CAT1* under stress conditions and the decreased levels of superoxide dismutase and catalase enzyme activities were detected accompanying the trends in the expression of both genes, which resulted in H_2O_2 accumulation and lipid peroxidation. Taken together, these findings lay a foundation for the future study of the biological function of *SlBAG* genes in tomato.

Keywords: antioxidant defense; arabidopsis; bioinformatic analysis; *SlBAG* genes; *SlBAG9*; *Solanum lycopersicum*

Citation: Jiang, H.; Ji, Y.; Sheng, J.; Wang, Y.; Liu, X.; Xiao, P.; Ding, H. Genome-Wide Identification of the Bcl-2 Associated Athanogene (BAG) Gene Family in *Solanum lycopersicum* and the Functional Role of *SlBAG9* in Response to Osmotic Stress. *Antioxidants* **2022**, *11*, 598. https://doi.org/10.3390/antiox11030598

Academic Editor: Nafees A. Khan

Received: 16 February 2022
Accepted: 17 March 2022
Published: 21 March 2022

Publisher's Note: MDPI stays neutral with regard to jurisdictional claims in published maps and institutional affiliations.

1. Introduction

B cell lymphoma 2 (Bcl-2)-associated athanogene (BAG) protein is a relatively conservative protein in animals and plants. BAG proteins share a common conserved BAG domain (BD), which interacts with the ATPase domain of heat-shock protein 70 (Hsc70/Hsp70 in the C-terminal region but is generally different in the N-terminal region, which gives specificity to specific proteins and pathways [1]. In animals, BAGs are widely involved in many biological processes, such as tumor regulation, apoptosis and stress response [1,2]. In plants, seven BAG family genes were first discovered in Arabidopsis [3,4]. The family of plant BAG protein can be divided into two categories. The first category contains N-terminal ubiquitin-like (UBL) domain and BAG domain, including AtBAG1–4 proteins. They may be direct homologues of animal BAG1, with high similarity in structure and function. The second type (BAG5-7 proteins) contains an isoleucine glutamine (IQ) motif binding to Ca^{2+}-free calmodulin (CaM) near BAG domain, which is unique to plants [3]

Plant BAG proteins are involved in biological processes such as plant programmed cell death (PCD) and autophagy, and play an important role in plant response to abiotic stresses such as salt, drought, temperature, and pathogen infection. Some BAG family genes have been explored in other plants such as rice, tomato, and wheat [5–9], but presently, as a model plant Arabidospis has gained the primary research attention, while relatively little is known about the functional research on other species.

The Arabidopsis seedlings overexpressing *AtBAG1* challenged with salinity stress and decreased seedling growth [10]. The *atbag4* mutant plants were more sensitive to salt stress, while the tobacco plants overexpressing *AtBAG4* gene showed stronger tolerance to ultraviolet (UV), low temperature stress, oxidative stress, drought, and salt stress than the wild type [3]. AtBAG1-AtBAG4 bound to Hsc70 through the BAG domain. The *atbag5* mutant showed the delayed aging, while the overexpression transgenic lines showed the premature aging. As a signaling hub, AtBAG5 linked the Ca^{2+} signaling network with the Hsc70 chaperone system to regulate plant senescence [11,12]. AtBAG6 cleavage triggered autophagy and plant defense and AtBAG6 [13]. AtBAG7 was an endoplasmic reticulum (ER) localization protein that played a central regulatory role in the heat-induced unfolded protein response (UPR) pathway [14]. The double-faced role of AtBAG7 in plant–phytophthora interaction has been found recently [15]. However, it is gratifying that, nowadays, studies on other plants species are becoming available. Transgenic rice plants overexpressing *OsBAG4* showed tolerance to NaCl stress [16]. OsBAG4 is an active regulator of disease resistance and a rice E3 ubiquitin ligase EBR1 targeted OsBAG4 for proteasome degradation [17]. Soybean *GmBAG6a* gene overexpressed in Arabidopsis had the ability to resist nematode infection [18]. Overexpressing BAG family gene *HSG1* from grapes in Arabidopsis plants showed obvious resistance to high temperature [19]. Overexpression of wheat *TaBAG2* increased Arabidopsis heat tolerance [6]. *SlBAG2* and *SlBAG5b* mediated tomato leaf tolerance to dark stress and senescence [8].

Based on the publication of plant genome sequence databases, the BAG gene family have been proved to exist in different species, such as *Arabidopsis thaliana* [3], *Oryza sativa* [5,20], and *Physcomitrium patens* [21]. We have previously tried to explore the key high-temperature-responsive genes in tomato using integrative analysis of transcriptome and proteome [22] and *SlBAG9* with a high expression level under high temperature stress was screened at the transcriptional and protein levels, which aroused our interest in the BAG gene family. In this study, ten *SlBAG* genes were identified, the related characteristics of *SlBAG* gene and protein were studied, and the gene expression pattern under different stress conditions were explored. In addition, further research on the function of *SlBAG9* was carried out. Heterologous overexpression of *SlBAG9* in Arabidopsis increased the sensitivity of Arabidopsis to drought, salt, and abscisic acid (ABA), which was reflected in the decrease in seed germination rate and seedling growth, low expression levels of stress/ABA-responsive genes and activities of superoxide dismutase (SOD) and catalase (CAT), and aggravated oxidative damage. Taken together, these findings lay a foundation for the future study of the biological function of *SlBAG* genes in tomato.

2. Materials and Methods

2.1. Genome-Wide Identification and Analysis of the Tomato BAG Gene Family

To identify the tomato BAG gene family, the BAG domain (PF02179) was used as a query to perform an HMMER (v3.3.2) search against the tomato proteome downloaded from Tomato Genome Annotation ITAG4.0 database (https://solgenomics.net, accessed on 10 October 2021). At the same time, a comprehensive search was carried out by using the amino acid sequence of *Arabidopsis thaliana* according to a previous study [4], of which seven AtBAG proteins were obtained from the *Arabidopsis thaliana* Araport11. The reliability of *SlBAG* gene sequences and genome sequences was further confirmed by searching the NCBI database (https://www.ncbi.nlm.nih.gov, accessed on 20 October 2021). The genomic regions, transcripts, and products were from Tomato Genome version SL4.0 (https://solgenomics.net, accessed on 10 October 2021) and NCBI data comparison. The molecular weight (Mw) and isoelectric point (pI) were calculated by ProtParam on the ExPASy Server (https://web.expasy.org/protparam, accessed on 25 October 2021).

2.2. Phylogenetic Analysis

The BAG protein sequences were downloaded from database *Solanum lycopersicum* ITAG4.0, *Arabidopsis thaliana* Araport11, and *Oryza sativa* v7.0, respectively. Multiple sequence alignments of BAG proteins were carried out using MUSCLE and the tree was constructed by PhyML method (phylogeny.fr/simple_phylogeny.cgi, accessed on 5 November 2021).

2.3. Motif Analysis of SlBAG Proteins

The conserved motifs of tomato SlBAGs were predicted by the MEME tool (Version 5.4.1, http://meme-suite.org/tools/meme, accessed on 11 December 2021). The number of BAG conserved domain was validated using NCBI conserved domains (CDD) and shown using the HMMER web server (Biosequence analysis using profile hidden Markov Models | HMMER).

2.4. Gene Structure and Cis-Element Analysis of SlBAG Genes

The Gene Structure Display Server (http://gsds.cbi.pku.edu.cn, accessed on 15 December 2021) was used for the gene structure determination. The cis-regulatory elements of the SlBAG gene promoters were analyzed by the PLANTCARE database (http://bioinformatics.psb.ugent.be/webtools/plantcare/html, accessed on 20 December 2021) and the elements were displayed on the TBtool software (version v1.0986853, https://github.com/CJ-Chen/TBtools, accessed on 28 December 2021).

2.5. Expression Analysis of SlBAG Genes

Seeds of *Solanum lycopersicum* cv. Ailsa Craig were sterilized and sown in the solid Murashige and Skoog (MS) medium at 25 °C/20 °C, and 20-day-old seedlings were used for the following treatments. For *SlBAG* gene expression under stresses, tomato seedlings were cultured in liquid MS medium for 24 h, then exposed to salt, drought, high temperature, cold, cadmium, H_2O_2, and ABA treatment for 1, 3, 6, 12, 24, and 48 h. Tomato plants were transferred to liquid medium containing 100 mM NaCl, 20% polyethylene glycol-6000, 50 uM cadmium, 100 uM ABA, and 10 mM H_2O_2, respectively.

For HT and cold, the seedlings were exposed to HT (42 °C) or 4 °C conditions. All real leaves were collected, frozen, and preserved at −70 °C for RNA extraction. For organs-specific expression of *SlBAG* genes, organs from root, stem, leaf, flower, green fruit, and red fruit of different stages were collected according to the method of Ding et al. [23]. The transcription levels of *SlBAG* genes were determined by quantitative real-time PCR (qRT-PCR).

2.6. Phenotypic Analysis of SlBAG9 Overexpression Plants under Osmotic Stress

To analyze the phenotype of *SlBAG9*-overexpressing plants under osmotic stress (drought and salt), two developmental stages were used for processing. For seed germination assay, the seeds of *Arabidopsis thaliana* ecotype Columbia-0 (WT) and *SlBAG9* overexpression line 2–12 and 4–9 [22] were surface-sterilized and sowed on 1/2 MS plates supplemented with or without 200 and 300 mM mannitol (M1, M2), 175 and 200 mM NaCl (N1, N2), and 1.0 and 1.5 µM ABA (A1, A2), and placed under 4 °C for 3 days, then moved to a growth room under a 16 h-light/8 h-dark photoperiod in the greenhouse. The seed germination rate (root emergence) was evaluated every day, and the number of seedlings with significantly expanded and green cotyledons were also evaluated. For seedling growth assay, the seven-day-old seedlings were transferred to 1/2 MS medium supplemented with 0 or 300 mM mannitol, 175 mM NaCl, and 1.5 µM ABA and grown for different days. After 3 days, the rosette leaves were sampled for qRT-PCR and after 14 days the photographs were taken.

2.7. H_2O_2 Content, Lipid Peroxidation, and Antioxidant Enzymes

According to the method of Ding et al. [23], the lipid peroxidation of cell membrane was evaluated by the determination of malondialdehyde (MDA). The amount of MDA-TBA (thiobarbituric acid) complex was calculated from the coefficient of absorbance (155 mM^{-1} cm^{-1}). The content of H_2O_2 was estimated by monitoring the absorbance of the titanium peroxide complex at 415 nm. Next, 0.2 g of plant material were homogenized in 1.5 mL 50 mM phosphate buffer solution (PBS) (pH 7.5) containing 0.2 mM ethylenediamine tetra acetic acid (EDTA) and 2% polyvinylpyrrolidone (PVP). The homogenate was centrifuged in a refrigerated centrifuge at 13,000× *g* for 20 min and the supernatants were used for protein determination and enzyme assay. All the steps were carried out at 4 °C. superoxide dismutase (SOD) activity was measured by monitoring the inhibition of photochemical reduction of nitroblue tetrazolium (NBT) at 560 nm. The catalase (CAT) activity was determined by measuring the decrease in absorption of H_2O_2 at 240 nm.

2.8. RNA Isolation and qRT-PCR

Total RNA was extracted with an RNA Isolater Total RNA Extraction Reagent (Vazyme Biotech Co., Ltd., Nanjing, China). Reverse transcription was performed using the HiScript II Q RT SuperMix for qPCR (Vazyme, Nanjing, China). The primer pairs used in this study are all listed in Table S1. The qRT-PCR was performed on an ABI7500 instrument (Applied Biosystems, Foster, CA, USA) using SYBR Green qPCR kits (TaKaRa, Kusatsu, Japan). The PCR amplification conditions included an initial heat-denaturing step at 95 °C for 3 min, followed by 40 cycles of 30 s at 95 °C, 30 s at 58 °C, and 1 min at 72 °C. Relative mRNA levels were normalized to that of the internal reference genes. Each sample was divided into 3 biological replicates. The data were processed on the basis of the $2^{-\Delta\Delta CT}$ method.

3. Results

3.1. Genome-Wide Identification of BAG Gene Family in Tomato

The released whole-genome sequences of *Solanum lycopersicum* (ITAG2.4, 3.0, 4.0) were used in the present study. Firstly, we performed a keyword search using the BAG domain file (PF02179) as a query and 10, 14, and 12 genes were obtained from three databases, respectively. Secondly, we searched the genome protein sequence databases ITAG4.0 by BLASTP using seven Arabidopsis (AtBAG) proteins [4], and 11 gene-encoding proteins were identified. After InterProScan search and manual analysis to remove false positive and redundant genes, a total of 10 non-redundant, full-length BAG genes in *Solanum lycopersicum* were identified and designated as *SlBAG1-SlBAG10* based on their physical location. Interestingly, there were great differences in the information of the ten *SlBAG* genes in the three databases (Table S2). In order to further confirm the gene information, we cloned the full length of these genes to confirm the CDS sequence. Combining the sequence alignment through the NCBI database, the gene information was finalized (Table 1), which was partly different from the existing ITAG database and the reported characteristics of tomato *SlBAG* genes [24] to some extent. Compared with the BAG gene family reported by [24], there were differences in the genome sequences of eight genes and the coding protein sequences of five genes. In particular, the information of *SlBAG4* was completely different from the existing database. Therefore, we re-analyzed some characteristics of 10 *SlBAG* genes in the following sections.

These *SlBAG* genes were distributed on six chromosomes. *SlBAG1*, *SlBAG4*, and *SlBAG8* were located on chromosome 1, 4, and 10, respectively. Three *SlBAG* genes were on chromosome 6, and the rest were distributed on the other two chromosomes. The relevant characteristics of genes and proteins of SlBAGs were shown in Table 1. The range of coding sequence (CDS) was 513 bp to 3708 bp, and the range of protein was 170 to 1235 amino acids, with an average of 402 amino acids. The molecular weight range of SlBAG protein was 19.43–137.35 kDa and the isoelectric point range was 5.09–10.26.

Table 1. The sequence characteristics of SlBAG family members in tomato.

Name [a]	Gene Identifier [b]	Gene Symbol [c]	Gene [d]			Protein [e]		
			Chr	Intron	CDS (bp)	AA	MW (kDa)	pI
SlBAG1	Solyc01g095320	LOC101246665	1	2	3708	1235	137.35	5.30
SlBAG2	Solyc03g026220	LOC101264896	3	3	1026	341	38.25	9.52
SlBAG3	Solyc03g083970	LOC101246514	3	2	1188	395	45.28	9.43
SlBAG4	Solyc04g014740	LOC109119998	4	0	651	216	24.44	5.83
SlBAG5	Solyc06g007240	LOC101243790	6	3	837	278	31.53	6.28
SlBAG6	Solyc06g035720	LOC101267811	6	3	1002	333	37.56	9.45
SlBAG7	Solyc06g072430	LOC104648101	6	0	1122	373	42.50	5.66
SlBAG8	Solyc08g080320	LOC101246459	6	3	855	284	32.33	9.73
SlBAG9	Solyc10g084170	LOC101250069	10	0	513	170	19.43	10.26
SlBAG10	Solyc10g085290	LOC101258018	10	4	1185	394	41.91	5.09

[a]: Tomato BAG protein names given in this study; [b]: gene ID in the ITAG4.0 database; [c]: gene symbol in NCBI; [d]: gene characteristics; [e]: protein characteristics; CDS, coding DNA sequence; AA, amino acid; MW, molecular weight; pI, isoelectric point; kDa, kilodalton.

3.2. Phylogenetic Characterization of BAG Family Genes in Tomato

A comprehensive search was conducted for plant lineages, including 30 representative monocotyledons, eudicotyledons, basal angiosperms, ferns, bryophytes, and algae. As a result, 485 putative genes were identified from 30 plants with BAG as the keyword in the phytozome13 database, and 292 putative genes encoding BAG proteins were further identified by batch CD search, which were divided into five different groups (Figure 1). By now, some functions of the BAG family members have been studied in Arabidopsis and rice [7,25]. To predict the function of SlBAG proteins, the evolutionary relationship among *Arabidopsis thaliana* (7 genes), *Oryza sativa* (8 genes), and *Solanum lycopersicum* (10 genes) was constructed using the MUSCLE and PhyML method. The phylogenetic tree showed that these BAG proteins were divided into two groups (I, II) (Figure 2). Group I contained 14 members (with four, five, and five members of Arabidopsis, rice, and tomato, respectively). Cluster II included eleven members (with three, three, and five members of Arabidopsis, rice, and tomato, respectively). This classification is basically similar to the previous analysis of BAG proteins in Arabidopsis [3,21].

3.3. Exon–Intron Arrangement and Conserved Motifs Analysis

Gene structure diagrams of *SlBAG* genes, including the presence of exon-intron components, were constructed by the Gene Structure Display Server (GSDS) (Figure 3A). As a result, new gene structural sequences were presented in this study in addition to *SlBAG7*, and *SlBAG9*, which were reported [24]. *SlBAG7* and *SlBAG9* were predicted to exhibit no introns and two *SlBAG* genes (*SlBAG1*, *SlBAG4*) contained one intron, whereas five *SlBAG* genes in group II exhibited three introns. Besides, several conserved motifs among the SlBAG proteins were calculated using the MEME, a tool used to discover motifs in a group of related DNA or protein sequences (Figure 3A). The BAG domain was analyzed using the HMMER web server (Figure 3B). Besides the BAG domain, all members of group II (SlBAG2, 5, 6, 8, 10) contained ubiquitin-like domains similar to some animal counterparts. However, SlBAG4, SlBAG7, and SlBAG9 of group I contained an IQ calmodulin-binding motif located in the N terminus, which was unique to plants. The previous study showed that the SlBAG4 has no IQ calmodulin-binding motif [24]. However, here we proved that its gene and protein structure were completely different from that in the existing database.

3.4. Organ-Specific Expression Profiles of SlBAG Genes

The promoters of all ten *SlBAG* genes contained different cis elements for plant growth regulation, stress responses, hormone responses, and light responses (Figure S1; Table S3). To evaluate the function of *SlBAG* genes in the growth and development, the expression pattern of ten *SlBAG* genes in different organ (root, stem, leaves, flowers, green fruits, and red fruits) was determined with qRT-PCR. The expression of *SlBAG* genes, shown as a cluster heat map, was different in various organs and some genes were only expressed in specific organs (Figure 4). *SlBAG2, SlBAG3, SlBAG4,* and *SlBAG5* were highly expressed in roots. *SlBAG8* displayed the high expression in roots and leaves. For flowers and fruits, *SlBAG6* showed the highest transcript level. *SlBAG7* showed relatively higher expression in flowers and green fruits. However, *SlBAG1, SlBAG9,* and *SlBAG10* showed the highest level in red fruits. These results suggest that *SlBAG* genes might be involved in the growth and development of tomato.

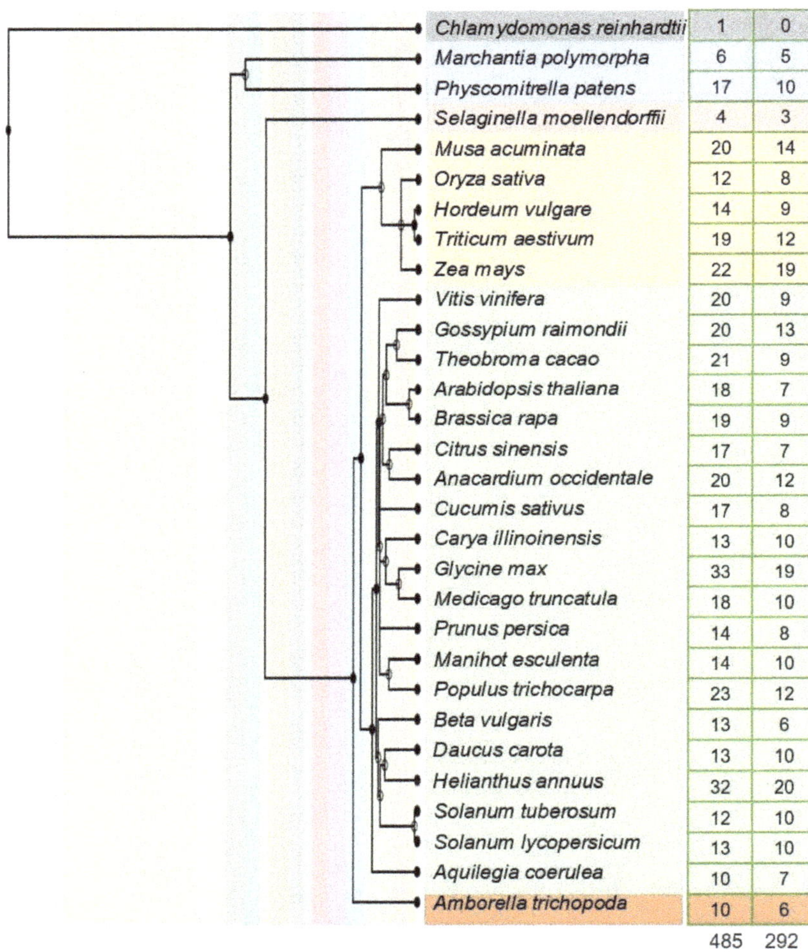

Figure 1. The number of BAG family in 30 plant species. The number of BAG proteins in 30 plant species and each group were shown on the right. The whole genomic data come from Phytozome 13. The number of BAG conserved domain analyzed by batch CD search were shown on the right. The species evolution tree was constructed by TimeTree 5 Beta (http://www.timetree.org, accessed on 10 January 2022).

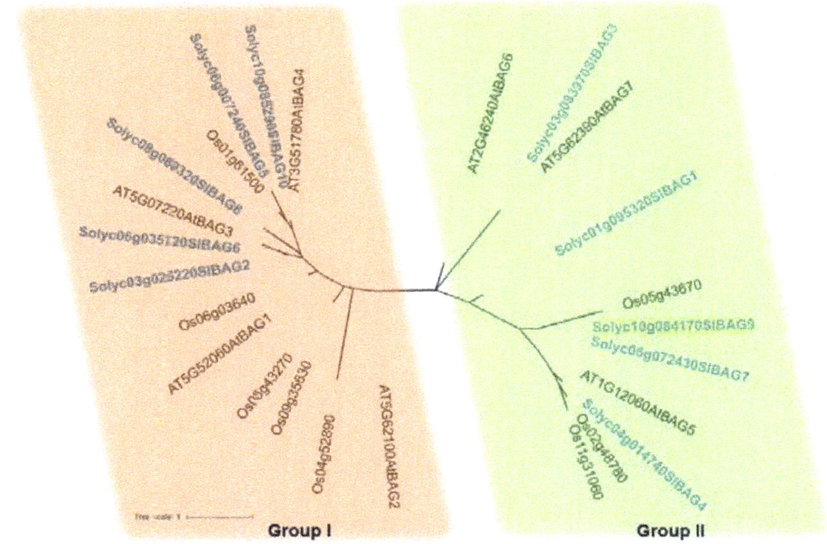

Figure 2. Phylogenetic tree of tomato SlBAG proteins. The 25 amino acid sequences of three plant species (Sl, *Solanum lycopersicum*; Os, *Oryza sativa*; At, *Arabidopsis thaliana*) were generated using the PhyML in MEGA 7.0. Based on the phylogenetic data, these proteins are divided into two distinct sub-groups, which are distinguished with different colors.

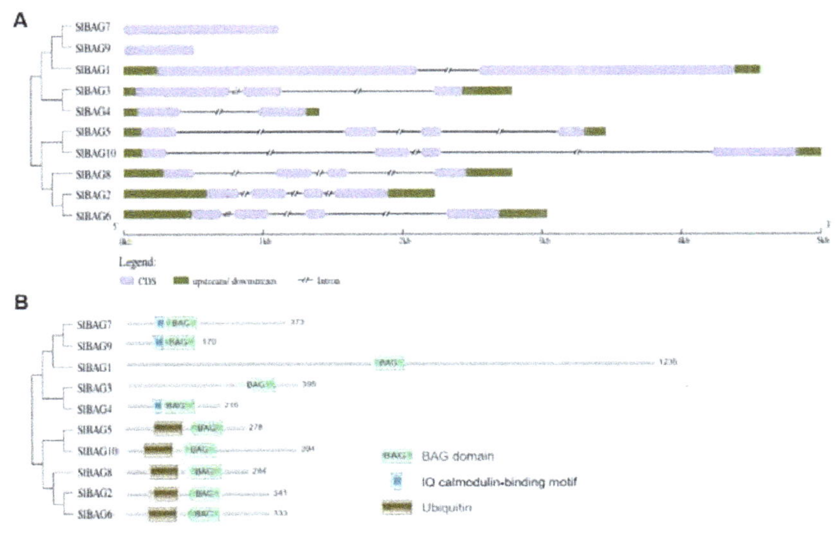

Figure 3. Gene structure and BAG domain of SlBAG family in tomato. (**A**) The gene structure was generated using a Gene Structure Display Server (GSDS version 2.0). The length of exons as well as introns and UTRs of each *SlBAG* gene are displayed proportionally. (**B**) BAG domains were generated through the HMMER web server (Biosequence analysis using profile hidden Markov Models | HMMER). The legends of BAG domain, Ubiquitin-like domain and IQ calmodulin-binding motif are listed on the right. The phylogenetic tree was constructed referring to Figure 2. SlBAG proteins are listed on the left.

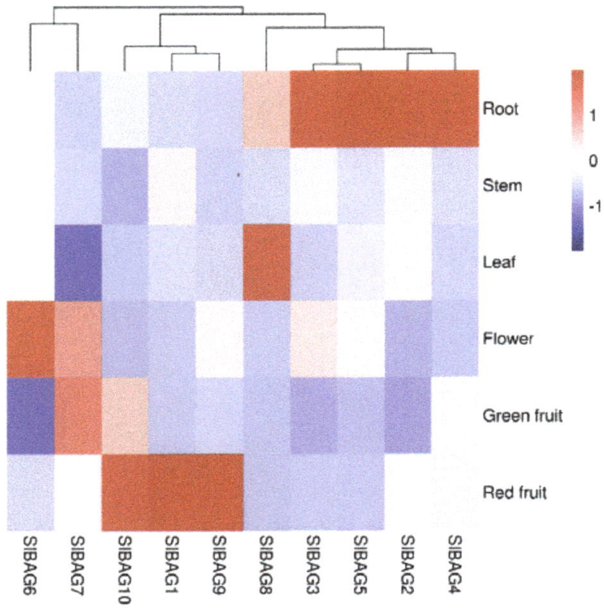

Figure 4. Expression profiles of *SlBAG* genes in different organs of tomato. The heatmap shows the expression change of ten *SlBAG* genes. Total RNA was extracted from different organs (root, stem, leaf, flower, green fruit, and red fruit) for qRT-PCR. Three independent biological repeats were performed (n = 3). The bar in the upper right corner represents the expression value of log2, and the change of expression level is represented by the change of color. Red indicates relatively high expression and blue indicates relatively low expression.

3.5. Expression Pattern of SlABG Genes under Abiotic Stress

The promoter analysis of *SlBAG* genes indicated that these genes might be related to the response of tomato to different stresses. Here, the *SlBAG* gene expression of tomato leaves after osmotic stress (drought and salt), temperature stress (high temperature and cold) and signal molecular (ABA and H_2O_2) was analyzed by qRT-PCR (Figure 5). Under all these stress conditions, almost all *SlBAG* genes showed different expression patterns. Under drought treatment, the expression of *SlBAG5* and *SlBAG6* generally increased with prolonged treatment time, while *SlBAG9* expression decreased with time under drought stress. Under salt stress, the expression of *SlBAG1*, *SlBAG2*, *SlBAG5*, and *SlBAG6* was upregulated. In response to high temperature, the expression level of *SlBAG1*, *SlBAG2*, *SlBAG3*, *SlBAG7*, *SlBAG8*, and *SlBAG9* in general were upregulated, while *SlBAG6* was downregulated. For cold stress, the expression levels of *SlBAG1* and *SlBAG2* in general were increased, while *SlBAG4*, *SlBAG6* and *SlBAG8* was decreased. As far as ABA is concerned, the expression of *SlBAG1*, *SlBAG5*, and *SlBAG6* generally showed an upward trend, while the expression of *SlBAG2*, *SlBAG8*, and *SlBAG10* showed a downward trend. For H_2O_2, the expression of *SlBAG1*, *SlBAG4*, and *SlBAG5* in general increased while expression of *SlBAG7*, *SlBAG9*, and *SlBAG10* decreased. The cluster heat map of all treatment showed that the expression of *SlBAG5* and *SlBAG6* induced by osmotic stress and signal molecules clustered in the same group and most of the *SlBAG* genes could be regulated by high temperature.

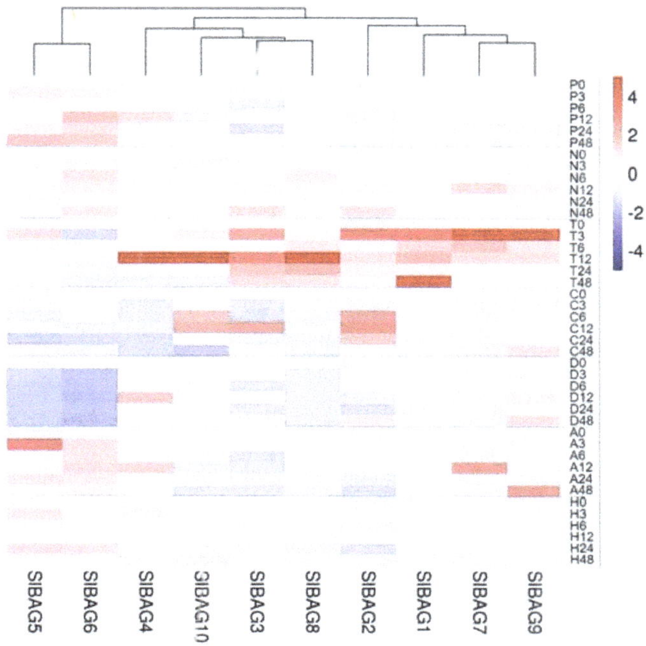

Figure 5. Expression profiles of *SlBAG* genes in tomato seedlings under different abiotic stresses. 20-day-old tomato plants were treated with PEG6000 (P, 20%), NaCl (N, 100 mM), high-temperature (T, 42 °C), cold (C, 4 °C), cadmium (D, 50 uM), ABA (A, 100 uM), or H_2O_2 (H, 10 mM) for 0, 3, 6, 12, 24, and 48 h. Total RNA from tomato leaves at indicated time was extracted from different organs (root, stem, leaf, flower, green fruit, and red fruit) for qRT-PCR. Three independent biological repeats were performed (n = 3). The bar in the upper right corner represents the expression value of log2, and the change of expression level is represented by the change in color. Red indicates relatively high expression and blue indicates relatively low expression.

3.6. Heterologous Expression of SlBAG9 Conferred Sensitivity to Drought, Salt, and ABA

Our previous studies showed that *SlBAG9* with a higher expression level at the transcriptional and protein levels in response to high temperature stress was involved in the negative regulation of thermotolerance [22,23]. The expression analysis of *SlBAG9* showed that it might be involved in the response to drought, salt, and ABA (Figure 5). To further investigate the biological functions of *SlBAG9*, heterologous overexpression lines of slbag9 in Arabidopsis (2–12, 4–9) were used here, which have been shown to be two homozygous lines [22]. We evaluated the seed germination and growth on 1/2 MS medium containing 200 and 300 mM mannitol, 175 and 200 mM NaCl, and 1.0 and 1.5 µM ABA treatments. No significant difference between the overexpression lines and WT was observed on 1/2 MS medium without treatment. However, the seed germination rate and seedling growth of *SlBAG9*-overexpressing lines were inhibited by mannitol, salt, and ABA compared to that of WT (Figures 6 and S2). For germination, after five days on MS medium containing 300 mM mannitol, about 85% of wild-type seeds, but less than 60% and 30% of the transgenic seeds (2–12, 4–9, respectively) germinated (Figure 6A). After ten days, more than 90% wild-type cotyledons, but less than 15% transgenic cotyledons turned green (Figure 6B), and the post-germination growth of transgenic materials was much worse than wild-type ones (Figure 6C). Two different treatments (200 and 300 mM mannitol) showed a certain concentration effect (Figure 6). For seedling growth, seven-day-old seedlings of wild-type, 2–12, and 4–9 were transferred to 1/2 MS containing 300 mM mannitol for

14 days and the growth pattern was observed. Under normal conditions, no significant difference was found in the aspects of seedlings growth of all the seedlings. Drought inhibited the growth of all seedlings, but the inhibition degree of 2–12, and 4–9 lines was greater than that of wild-type seedlings (Figure S2). A similar phenotype was also observed in NaCl treatment (Figure 6, Figure S2). These results demonstrated that *SlBAG9* conferred plant more sensitive to drought and salt stress, negatively regulating osmotic stress.

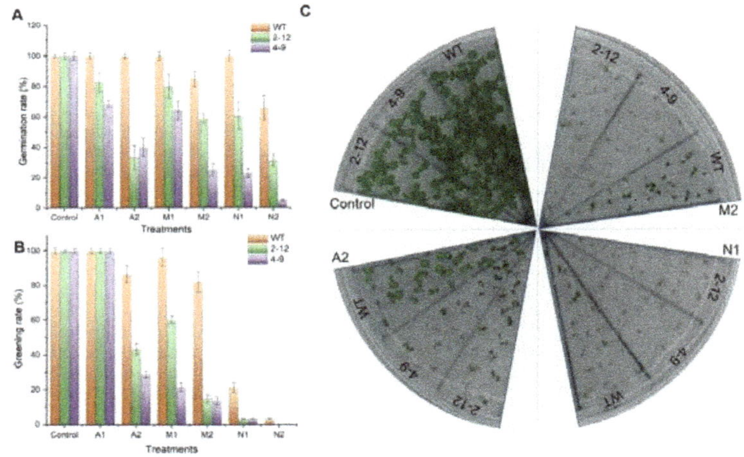

Figure 6. The increased sensitivity of *SlBAG9*-overexpressing Arabidopsis to mannitol, NaCl, and ABA stress. (**A**) The germination rate of *SlBAG9*-overexpressing lines (2–12 and 4–9) and wild type (WT) on 1/2 MS medium containing 200 and 300 mM mannitol (M1, M2), 175 and 200 mM NaCl (N1, N2), and 1.0 and 1.5 µM ABA (A1, A2) treatments for 5 days. (**B**) Cotyledon greening rate of *SlBAG9*-overexpressing lines and WT on 1/2 MS medium containing 200 and 300 mM mannitol (M1, M2), 175 and 200 mM NaCl (N1, N2), and 1.0 and 1.5 µM ABA (A1, A2) treatment for 10 days. Error bars indicate the SD of three replicated experiments. (**C**) The seedling growth of *SlBAG9*-overexpressing lines and WT Arabidopsis on 1/2 MS plates containing 300 mM mannitol (M2), 175 mM NaCl (N1), and 1.5 µM ABA (A2) treatments for 10 days.

ABA is an important mediator of drought and salt and senescence [26]. To further study whether the ABA was involved in *SlBAG9*-mediated response, we observed the phenotype of *SlBAG9*-overexpressing lines grown under ABA treatment. In total, 100% of wild-type seeds, but about 40% transgenic seeds germinated after five days on 1/2 MS medium supplemented with 1.5 µM ABA (Figure 6A). More than 80% cotyledons of wild-type cotyledons, but less than 50% of transgenic lines turned green as a result of ABA treatment for seven days (Figure 6B,C). For seedling growth, ABA treatment decreased leaf chlorophyll content, while the leaves of the WT exhibited an obvious stay-green phenotype (Figure S2).

3.7. *SlBAG9 Downregulated Stress/ABA-Responsive Genes*

To better understand the mechanism of *SlBAG9*-mediated osmotic stress, mRNA levels of two ABA-responsive genes *ABI3*, and *RD29A* and four stress-responsive genes *DREB2A*, *P5CS1*, *FSD1*, and *CAT1* were determined by qRT-PCR. The RNA transcript levels of the six genes were not significantly altered under normal growth conditions (Figure 7). However, the expression levels of these genes were upregulated in both WT and transgenic plants under osmotic stress or ABA treatment, but the increase was not as pronounced in transgenic plants as in WT plants (Figure 7). These results suggested that these genes in the *SlBAG9*-overexpressing plants are more sensitive to osmotic stress than WT plants, and that *SlBAG9* may be involved in the regulation of stress-regulated gene expression.

The expression levels of two key ROS scavenging enzyme gene *FSD1*, and *CAT1* were also regulated by *SlBAG9*, indicating that *SlBAG9* modulates the ROS scavenging system In view of this, it was speculated that the ROS scavenger-related enzyme activity was regulated by *SlBAG9*. Therefore, in the next section we further analyze the possible effect of *SlBAG9* on the oxidative damage and the activities of SOD and CAT.

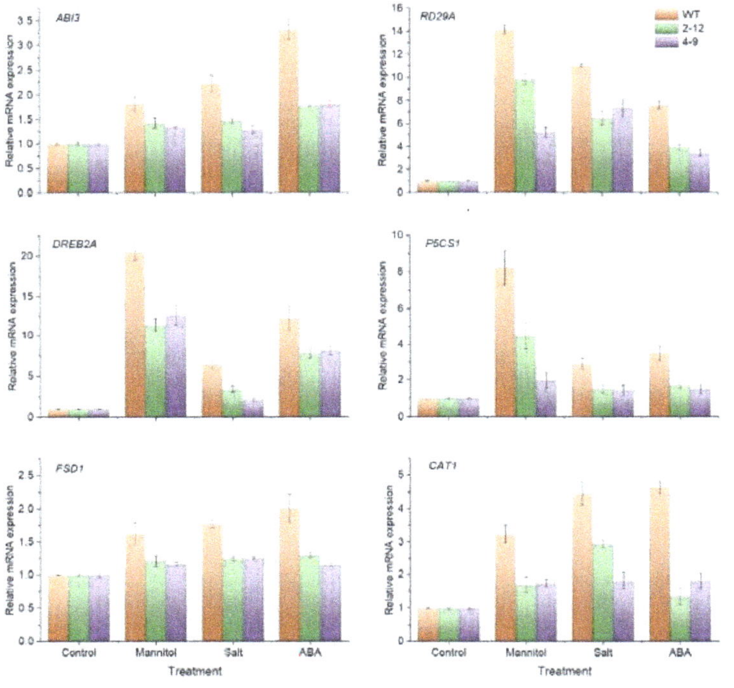

Figure 7. Relative expression of stress/ABA-responsive genes in *SlBAG9*-overexpressing Arabidopsis exposed to mannitol, NaCl, and ABA stress. The seven-day-old Arabidopsis plantlets of wild-type, 2–12, and 4–9 were transferred to 1/2 MS containing 0, 300 mM mannitol, 175 mM NaCl, and 1.5 µM ABA for three days. Total RNAs were extracted, and qRT-PCR analyses were performed. Error bars indicate the SD of three replicated experiments.

3.8. SlBAG9 Overexpression Aggravated Oxidative Damage under Drought, Salt, and ABA

Stress conditions trigger the accumulation of ROS molecules, resulting in oxidative damage to plant organelles [27]. To uncover whether *SlBAG9* was involved in drought, salt, and ABA-induced oxidative stress, the H_2O_2 level was estimated. Excessive accumulation of H_2O_2 was observed in the *SlBAG9*-overexpressing lines compared with the wild-type plants under drought, salt, and ABA conditions (Figure 8A). MDA is considered to be an effective marker of membrane damage caused by oxidative stress [28]. In accordance with the stressed phenotype, *SlBAG9*-overexpressing line 2–12 and 4–9 accumulated more MDA compared with the wild type under drought, salt, and ABA treatment (Figure 8B). These results indicated that *SlBAG9* can sensitively respond to drought, salt, and ABA to induce H_2O_2 generation and cause membrane lipid peroxidation, resulting in hypersensitivity responses. Since the above data suggested that *SlBAG9* mediates oxidative damage, and downregulate ROS scavenger-related gene *FSD1* and *CAT1*, the activities of SOD and POD were determined. There was little difference in SOD and CAT activities in WT, 2–12, and 4–9 genotypes under normal conditions. However, the SOD and CAT activities were lower in the 2–12 and 4–9 lines compared with the WT (Figure 8C,D). This result indicated that *SlBAG9* overexpression is implicated in ROS clearance and reduces tolerance to drought, salt, and ABA-induced oxidative stress by regulating antioxidant enzyme activity, which could support different change patterns of H_2O_2 and MDA (Figure 8).

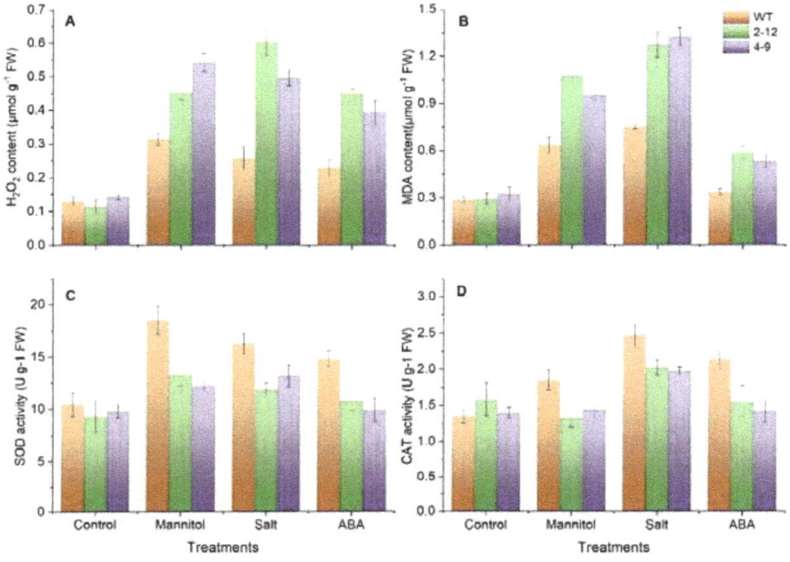

Figure 8. *SlBAG9* overexpression aggravated oxidative damage in Arabidopsis under mannitol, NaCl, and ABA treatment. The seven-day-old Arabidopsis plantlets of wild-type, 2–12, and 4–9 were transferred to 1/2 MS containing 0, 300 mM mannitol, 175 mM NaCl, and 1.5 μM ABA for three days for physiological change evaluation. The H_2O_2 content (**A**), MDA content (**B**), SOD activity (**C**), and CAT activity (**D**) were analyzed quantitatively in WT and SlBAG9-overexpressing lines 2–12 and 4–9. Error bars indicate the SD of three independent experiments.

4. Discussion

Cultivated tomato is easily affected by various environmental factors. Researchers have been studying potential resistance genes in plants for a long time. With the available tomato genome, the study of gene function is becoming more and more important [8]. Our previous research made us very interested in the tomato *SlBAG* gene family [22,29]. Many studies have shown that *BAG* gene plays an important role in plant growth, development, and stress response (reviewed by Thanthrige et al. [7]). Therefore, it is necessary to identify *SlBAG* genes and their biological function in tomato. Lately, Irfan et al. [24] identified 11 BAG genes using the tomato database (ITAG2.4). However, the database now includes three versions (2.4, 3.0, 4.0), in which different BAG gene information is presented. The present study aimed to a comprehensive genome-wide functional characterization of *SlBAG* genes and proteins in tomato. We obtained 10 BAG genes through sequence alignment and multi database comparison, cloned all ten CDS sequences, and finally determined the information of these genes (Table 1 and Table S2). Compared with the newly published genetic information obtained by Irfan et al. [24], *SlBAG6*, *SlBAG8*, *SlBAG10*, *SlBAG4*, and *SlBAG1* have new CDS sequences and eight gene structural sequences were new in addition to *SlBAG7*, and *SlBAG9*.

For the ten BAG proteins in this study, their protein structures and evolutionary relationships with BAG proteins in tomato, Arabidopsis and rice were analyzed. The proteins were divided into two groups, and the difference is mainly at their N-terminal. In addition to BAG domain, the first group has a UBL domain. A UBL domain can interact with 26S proteasome and is an indispensable part of BAG1 in stress response [30]. UBL domain in the group I suggests that they may also participate in the degradation of some proteins as molecular bridges. The second group has a specific CaM-binding domain IQ motif near the BD domain in plants, indicating that it may be involved in unique biological processes [31]. Irfan et al. [24] found that SlBAG7 and SlBAG9 had IQ domain. In the present study, three SlBAGs (SlBAG4, SlBAG7, SlBAG9) in the group II have IQ domain, indicating that their biological functions may be related to the Ca^{2+} signal. IQ motif can bind CaM and affect the formation of complex between CaM and targeted protein [22]. In vitro studies have shown that Ca^{2+} can affect the binding affinity of AtBAG6 and CaM and regulate the process of cell death mediated by AtBAG6. The CaM-binding motif and the BD are required for AtBAG6-mediated cell death [32]. As a signaling hub, AtBAG5 connects the Ca^{2+} signaling network with the Hsc70 chaperone system to regulate plant senescence. The IQ motif mutant retains the association between AtBAG5 with Hsc70 while disrupting the association of AtBAG5 with apo-CaM [11]. It is possible that the increase in Ca^{2+} in the mitochondrial matrix may protect mitochondria from senescence through a combination of Ca^{2+} and apo-CAM, so as to promote the release of Hsc70 from CaM/AtBAG5/Hsc70 signal complex and to inhibit ROS production [11,12].

The BAG family is widely distributed in the plant kingdom (Figure 1). The BAG family also exists in various plant tissues and organs. Results from many experiments have shown that the BAG gene family plays an important role in plant growth and development [3]. *AtBAG4* and *AtBAG6* expressions were detected in the root, stem, leaf, and flower of Arabidopsis. In the whole development process, *AtBAG2* and *AtBAG6* genes are expressed in various tissues in overlapping or specific expression pattern [33]. Arabidopsis knockout mutant *atbag4* or *atbag6* has the phenotype of early flowers and multi branching inflorescences, with a shortened life cycle and early aging. It was found that the rosette diameter of the 4-week-old *atbag2* mutant was larger than that of the wild type [34]. Arabidopsis plants overexpressing *AtBAG6* are shorter than wild-type plants [3]. The tissue-specific expression experiment of rice showed that *OsBAG1*, *OsBAG3*, and *OsBAG4* had the highest expression in roots, stems, and internodes, indicating that they may be involved in cell elongation and expansion. Rice OsBAG4 and EBR1 form a protein complex, which makes EBR1 control its protein stability level through ubiquitination of OsBAG4, inhibit growth and development, resulting in plant dwarfism [17]. Irfan et al. [24] showed that several *SlBAG* genes such as *SlBAG1*, *SlBAG3*, *SlBAG6*, and *SlBAG9* had differential expression during fruit development, which suggested that they might have a role in fruit development as well. In this study, many *SlBAGs* in tomato showed specific expression patterns in organs, indicating their important roles (Figure 4). Recently, He et al. [8] showed that *BAG2* and *BAG5b* were highly expressed in tomato stem and flower. Here, the corresponding name was *SlBAG7* and *SlBAG9*, respectively, which was not only highly expressed in flowers, but also in fruits (Figure 4).

The regulatory elements in plant promoters play an important regulatory role at the transcriptional level [35]. Here, it was found that in all *SlBAG* promoters there are a series of elements related to abiotic stress and hormones, such as MYC, MBS, DRE, HSE, W-box, and ABA and SA- responsive cis-elements (Figure S1). These results were consistent with the previous study that most of the same cis elements were found in 2000 bp upstream regions of *SlBAG* genes [24] and ~1000 bp upstream regions of all Arabidopsis BAG family genes, suggesting that they play a role in coping with different environmental stresses such as cold, drought and high salinity [4]. Take W-box as an example, AtBAG7 translocates from the ER to the nucleus, where it interacts with the transcription factor WRKY29, which then binds to the W-box in the promoter of *AtBAG7* to initiate the transcription of *AtBAG7* and other chaperones to promote stress tolerance [36]. The accumulated evidence shows that BAG expression can be regulated by various abiotic stresses [3,7,24,36]. Accordingly, here we monitored the transcriptional response of tomato *SlBAG* genes with the exposure to various stress conditions including drought, salt, HT, cold, as well as ABA and H_2O_2 signals (Figure 5). Plant *BAG* gene expression is related to its function to some extent.

Cold upregulated the expression of *AtBAG4* and *AtBAG4* overexpression increased tobacco plants tolerance to cold, salt, UV, and oxidative stress [3]. Heat stress significantly upregulated *AtBAG6* gene and protein level and the *atbag6* mutant is sensitive heat stress [4,33]. Transgenic rice plants overexpressing salt-induced *OsBAG4* showed tolerance to NaCl stress [25]. *AtBAG6* transcript levels were significantly upregulated by H_2O_2 [3]. The expression of several *SlBAG* genes was also induced by ACC, the precursor of ripening hormone ethylene and ABA, suggesting that *SlBAG* genes are potentially involved in the fruit ripening regulation and stress response [24]. In this study, some *SlBAG* genes also showed similar expression patterns. However, the involvement of H_2O_2 in the BAG-mediated biological function is unclear. The above results suggested that *SlBAG* family is involved in the response of tomato plants to abiotic stresses such as salt, drought, cold and HT, and ABA and H_2O_2 signals may be involved in these pathways.

SlBAG9 was noteworthy because in our previous studies it showed higher gene expression level and protein abundance under high temperature stress [22]. Overexpression of *SlBAG9* decreased the tolerance to HT [22,29]. Cis-elements and expression analysis indicated that *SlBAG9* may be involved in drought, salt, and ABA stress (Figures 5 and S1). This study further showed that *SlBAG9*-overexpressing Arabidopsis exhibited increased sensitivity to mannitol, salt, and ABA treatment (Figures 8 and S1). In terms of salt stress, many studies have shown that BAG can positively regulate plant salt tolerance. Arabidopsis with overexpression of *TaBAG* and *TaBAG2* showed significant enhancement of salt tolerance [6]. The *atbag4* mutant was more sensitive to salt stress [3]. Transgenic rice plants heterologously expressing *AtBAG4* showed higher salt tolerance than WT [37]. Recently, Wang et al. [25] reported that OsBAG4 functioned as a bridge between OsMYB106 and OsSUVH7 under salt stress to regulate OsHKT1;5 expression, so as to improve salt tolerance. Pan et al. [38] showed that salt suppressed *BAG6* and *BAG7* expression but addition of ACC (1-aminocyclo-propane-1-carboxylic acid) in the salt treatment could re-activate *BAG6* and *BAG7* expression, indicating that BAG genes are involved in the process of plant cell death induced by salt stress. The negative regulation mechanism of *SlBAG9* on salt tolerance remains to be studied. As far as drought and ABA are concerned, there are few studies on BAG gene function so far. Arabidopsis leaves with low-level *AtBAG4* overexpression appeared to be drought tolerant [3]. AtBAG4 interacted with potassium influx channel protein KAT1 in guard cells to regulate stomatal movement [39]. However, our observation was consistent with the latest data from Arabidopsis *AtBAG2* and *AtBAG6* [33]. Germination of *atbag2*, *atbag6*, and *atbag2atbag6* seeds was less sensitive to ABA compared to WT, whereas *AtBAG2* and *AtBAG6* overexpression lines showed the opposite results for ABA. The survival rate of *atbag2*, *atbag6*, and *atbag2atbag6* plants was higher than that of the WT under drought stress. In addition, these mutants showed differential expression of several stress and ABA-related genes and low ROS levels after drought and ABA treatment [33]. In this study, *ABI3*, *RD29A*, *DREB2A*, and *P5CS1* expression of transgenic plants was lower than that of WT plants under osmotic stress. In *Arabidopsis thaliana*, these genes had all been shown to be inducible by drought, salinity, or ABA. The higher expression of these stress genes was related to plant tolerance [40]. All these data indicated that the decreased transcription levels of these genes in *SlBAG9*-overexpressing lines lead to enhanced sensitivity to osmotic stress, which might be mediated by ABA. There are three points of view that support our hypothesis. First, the *SlBAG9* expression was induced by exogenous ABA (Figure 6). Second, *SlBAG9* overexpression in Arabidopsis resulted in hypersensitivity to ABA (Figures 7 and S2). Third, under ABA and osmotic stress, the *SlBAG9* overexpression significantly downregulated the expression levels of ABA signaling pathway genes such as *ABI3* and *RD29A* (Figure 8), which were reported to play positive regulators in ABA-associated abiotic stress [41]. The decrease in *ABI3* and *RD29A* expression of *SlBAG9* overexpression lines might be one of the reasons for the increased sensitivity to ABA and osmotic stress. The specific signaling pathway of ABA mediated by *SlBAG9* remains to be unveiled.

It is known that a variety of environmental stresses can promote ROS production and the senescence process will also increase the accumulation of ROS. Therefore, the regulation of ROS production plays a key role in senescence and stress response [42]. It has been reported that *atbag2*, *atbag6*, and *atbag2atbag6* seedlings lines accumulated lower H_2O_2 when treated with drought and ABA [33]. The transgenic plants overexpressing *AtBAG5* showed greater H_2O_2 accumulation than WT, indicating that *AtBAG5* is involved in leaf senescence by regulating the production of ROS. In the present study, overexpression of *SlBAG9* in Arabidopsis could induce the excess production of H_2O_2 in response to drought, salt, and ABA (Figure 8), suggesting that *AtBAG5* may be involved in regulating these stress responses though the proliferation of ROS. Excessive production of ROS will lead to oxidative stress and cell death in growing plants [43] H_2O_2 and MDA content is recorded with the extent of oxidative stress [28]. Consistent with the results of H_2O_2, *SlBAG9*-overexpressing seedlings exhibited significantly induced H_2O_2 and MDA accumulation compared to the WT seedlings after stress treatment. (Figure 8). These results corresponded well with the phenotype of these *SlBAG9* overexpression plants in response to drought, salt, and ABA, which suggested that *SlBAG9* accelerate H_2O_2 excess production, leading to a deeper degree of oxidative damage. Correspondingly, gene expression analyses showed *SlBAG9* overexpression downregulated the expression of key ROS scavenger-related genes *FSD1* and *CAT1* (Figure 7). *FSD1* encodes a chloroplast/nuclei/cytosol localized SOD that utilizes Fe as the cofactor (FeSOD) and FSD1 presents osmoprotection in Arabidopsis [44]. *CAT1* encodes a peroxisomal catalase (CAT1), which is implicated in the drought and salt stress responses [45]. SOD and CAT scavenge ROS by converting superoxide to H_2O_2 and H_2O_2 to oxygen and water, sequentially. There is a closely correlation between the expression of both two genes and enzyme activity. Li et al. [46] found that the expression of *CAT1* and *FSD1* were upregulated in *PpDHN*-overexpressing plants under drought and salt stress and elevated levels of SOD and CAT enzyme activities were detected accompanying the trends in expression of these genes. In the transgenic overexpressing *IpASR*, *CAT1* and *FSD1* expression and SOD and CAT enzyme activities showed the same trend under salt and drought stress [47]. In the current study, it was clearly shown that the expression of *FSD1* and *CAT1* (Figure 7) and the activities of SOD and CAT (Figure 8C,D) were not altered among wild-type and over-expressing *SlBAG9* lines; however, a difference was only detected under stress conditions or in response to exogenous ABA supply. Taken together all the above indicate that *SlBAG9* is involved in regulation of stress responses, which needs to be further explored. ABA signaling pathway is an important way for plants to deal with abiotic stress and ABA also regulates senescence [26]. Whether ABA was involved in SlBAG9-mediated antioxidant protection pathway still needed further study. Taken together, it was suggested that the increased sensitivity of *SlBAG9* overexpression to drought, salt, and ABA might be related to the oxidative damage regulated by antioxidant enzyme system such as SOD and CAT. This complex mechanism needs to be deeply investigated in the future.

5. Conclusions

In the present study, we identified and characterized ten *SlBAGs* at tomato genome level, which were supported by structural characteristics of the SlBAG proteins and genes, as well as by phylogenetic analysis. Large-scale expression of these *SlBAGs* by qRT-PCR further revealed tissue- and abiotic stress-specific expression patterns, which might be related to plant growth and abiotic stresses. Additionally, *SlBAG9* conferred sensitivity to osmotic stress when heterologously expressed in transgenic Arabidopsis. The enhanced sensitivity was ascribed to the decreased expression of several stress and ABA-responsive genes, the increased oxidative damage as well as the decreased ROS scavenging capability in the *SlBAG9*-overexpressing plants. Besides, it is worth noting that this study not only provide a foundation for understanding the functions of *SlBAG* gene family but also has important guiding significance for its final application in molecular resistance breeding.

Supplementary Materials: The following supporting information can be downloaded at: https://www.mdpi.com/article/10.3390/antiox11030598/s1, Figure S1: Cis-elements components in the promoters of tomato *SlBAG* genes; Figure S2: The increased sensitivity of *SlBAG9*-overexpressing Arabidopsis exposed to mannitol, NaCl, and ABA stress; Table S1: Primers used in this study for qRT-PCR analysis; Table S2: Information of 10 *SlBAG* genes from multiple data sources; Table S3: The cis-element of *SlBAG* gene promoters.

Author Contributions: H.D. and H.J., Conceptualization; H.J., Y.J., J.S. and Y.W., investigation; H.J., X.L. and P.X., formal analysis; H.D., writing. All authors have read and agreed to the published version of the manuscript.

Funding: This work was supported by the National Natural Science Foundation of China (grant no. 32172546) and the Natural Science Foundation of Jiangsu Province (BK20191437).

Institutional Review Board Statement: Not applicable.

Informed Consent Statement: Not applicable.

Data Availability Statement: The data presented in this study are available in the article and its Supplementary Materials.

Conflicts of Interest: The authors declare no conflict of interest.

References

1. Takayama, S.; Sato, T.; Krajewski, S.; Kochel, K.; Irie, S.; Millan, J.A.; Reed, J.C. Cloning and functional analysis of BAG-1: A novel Bcl-2-binding protein with anti-cell death activity. *Cell* **1995**, *80*, 279–284. [CrossRef]
2. Kabbage, M.; Dickman, M.B. The BAG proteins: A ubiquitous family of chaperone regulators. *Cell Mol. Life Sci.* **2008**, *65*, 1390–1402. [CrossRef] [PubMed]
3. Doukhanina, E.V.; Chen, S.; van der Zalm, E.; Godzik, A.; Reed, J.; Dickman, M.B. Identification and functional characterization of the BAG protein family in *Arabidopsis thaliana*. *J. Biol. Chem.* **2006**, *281*, 18793–18801. [CrossRef] [PubMed]
4. Nawkar, G.M.; Maibam, P.; Park, J.H.; Woo, S.G.; Kim, C.Y.; Lee, S.Y.; Kang, C.H. In silico study on *Arabidopsis BAG* gene expression in response to environmental stresses. *Protoplasma* **2017**, *254*, 409–421. [CrossRef] [PubMed]
5. Rana, R.M. Identification and characterization of the Bcl−2-associated athanogene (BAG) protein family in rice. *Afr. J. Biotechnol.* **2012**, *11*, 88–98.
6. Ge, S.; Kang, Z.; Li, Y.; Zhang, F.; Shen, Y.; Ge, R.; Huang, Z. Cloning and function analysis of BAG family genes in wheat. *Funct. Plant Biol.* **2016**, *43*, 393–402. [CrossRef]
7. Thanthrige, N.; Jain, S.; Bhowmik, S.D.; Ferguson, B.J.; Kabbage, M.; Mundree, S.; Williams, B. Centrality of BAGs in plant PCD, stress responses, and host defense. *Trends Plant Sci.* **2020**, *25*, 1131–1140. [CrossRef]
8. He, M.; Wang, Y.; Jahan, M.S.; Liu, W.; Raziq, A.; Sun, J.; Su, S.; Guo, S. Characterization of SlBAG genes from *Solanum lycopersicum* and its function in response to dark-induced leaf senescence. *Plants* **2021**, *10*, 947. [CrossRef]
9. Yan, J.; He, C.; Zhang, H. The BAG-family proteins in *Arabidopsis thaliana*. *Plant Sci.* **2003**, *165*, 1–7. [CrossRef]
10. Lee, D.W.; Kim, S.J.; Oh, Y.J.; Choi, B.; Lee, J.; Hwang, I. *Arabidopsis* BAG1 functions as a cofactor in Hsc70-mediated proteasomal degradation of unimported plastid proteins. *Mol. Plant* **2016**, *9*, 1428–1431. [CrossRef]
11. Li, L.; Xing, Y.; Chang, D.; Fang, S.; Cui, B.; Li, Q.; Wang, X.; Guo, S.; Yang, X.; Men, S.; et al. CaM/BAG5/Hsc70 signaling complex dynamically regulates leaf senescence. *Sci. Rep.* **2016**, *6*, 31889. [CrossRef] [PubMed]
12. Fu, S.; Li, L.; Kang, H.; Yang, X.; Men, S.; Shen, Y. Chronic mitochondrial calcium elevation suppresses leaf senescence. *Biochem. Biophys. Res. Commun.* **2017**, *487*, 672–677. [CrossRef] [PubMed]
13. Li, Y.; Kabbage, M.; Liu, W.; Dickman, M.B. Aspartyl protease-mediated cleavage of BAG6 is necessary for autophagy and fungal resistance in plants. *Plant Cell* **2016**, *28*, 233–247. [CrossRef] [PubMed]
14. Williams, B.; Kabbage, M.; Britt, R.; Dickman, M.B. *AtBAG7*, an *Arabidopsis* Bcl-2–associated athanogene, resides in the endoplasmic reticulum and is involved in the unfolded protein response. *Proc. Natl. Acad. Sci. USA* **2010**, *107*, 6088–6093. [CrossRef]
15. Zhou, Y.; Yang, K.; Cheng, M.; Cheng, Y.; Li, Y.; Ai, G.; Bai, T.; Xu, R.; Duan, W.; Peng, H.; et al. Double-faced role of Bcl-2-associated athanogene 7 in plant–Phytophthora interaction. *J. Exp. Bot.* **2021**, *72*, 5751–5765. [CrossRef]
16. Eckardt, N.A. Programmed cell death in plants: A role for mitochondrial associated hexokinases. *Plant Cell* **2006**, *18*, 2097–2099. [CrossRef]
17. You, Q.; Zhai, K.; Yang, D.; Yang, W.; Wu, J.; Liu, J.; Pan, W.; Wang, J.; Zhu, X.; Jian, Y.; et al. An E3 ubiquitin ligase-BAG protein module controls plant innate immunity and broad-spectrum disease resistance. *Cell Host Microbe* **2016**, *20*, 758–769. [CrossRef]
18. Yeckel, G.J. *Characterization of a Soybean BAG Gene and Its Potential Role in Nematode Resistance*; University of Missouri: Columbia, MO, USA, 2012.

19. Kobayashi, M.; Takato, H.; Fujita, K.; Suzuki, S. HSG1, a grape Bcl-2-associated athanogene, promotes floral transition by activating *CONSTANS* expression in transgenic *Arabidopsis* plant. *Mol. Biol. Rep.* **2012**, *39*, 4367–4374. [CrossRef]
20. Bansal, R.; Kumawat, S.; Dhiman, P.; Sudhakaran, S.; Rana, N.; Jaswal, R.; Singh, A.; Sonah, H.; Sharma, T.R.; Deshmukh, R. Evolution of Bcl-2 Anthogenes (BAG) as the regulators of cell death in wild and cultivated *Oryza* Species. *J. Plant Growth Regul.* **2022**. [CrossRef]
21. Castro, A.; Saavedra, L.L.; Ruibal, C.; Lascano, R.; Vidal, S. Genome-wide identification, characterization and expression analysis of the Bcl-2 associated athanogene (BAG) gene family in *Physcomitrium Patens*. *bioRxiv* **2020**. [CrossRef]
22. Ding, H.; Mo, S.; Qian, Y.; Yuan, G.; Wu, X.; Ge, C. Integrated proteome and transcriptome analyses revealed key factors involved in tomato (*Solanum lycopersicum*) under high-temperature stress. *Food Energy Secur.* **2020**, *9*, e239. [CrossRef]
23. Ding, H.; He, J.; Wu, Y.; Wu, X.X.; Ge, C.; Wang, Y.; Xu, W. The tomato mitogen-activated protein kinase SlMPK1 is as a negative regulator of the high temperature stress response. *Plant Physiol.* **2018**, *177*, 633–651. [CrossRef] [PubMed]
24. Irfan, M.; Kumar, P.; Ahmad, I.; Datta, A. Unraveling the role of tomato Bcl-2-associated athanogene (BAG) proteins during abiotic stress response and fruit ripening. *Sci. Rep.* **2021**, *11*, 21734. [CrossRef] [PubMed]
25. Wang, J.; Nan, N.; Li, N.; Liu, Y.; Wang, T.J.; Hwang, I.; Liu, B.; Xu, Z.Y. A DNA methylation reader-chaperone regulator transcription factor complex activates OsHKT1;5 expression during salinity stress. *Plant Cell* **2020**, *32*, 3535–3558. [CrossRef]
26. Gao, S.; Gao, J.; Zhu, X.; Song, Y.; Li, Z.; Ren, G.; Zhou, X.; Kuai, B. ABF2, ABF3, and ABF4 promote ABA-mediated chlorophyll degradation and leaf senescence by transcriptional activation of chlorophyll catabolic genes and senescence-associated genes in *Arabidopsis*. *Mol. Plant* **2016**, *9*, 1272–1285. [CrossRef]
27. Verslues, E.; Agarwal, M.; Katiyar-Agarwal, J.; Zhu, S.; Zhu, J.K. Methods and concepts in quantifying resistance to drought, salt and freezing, abiotic stresses that affect plant water status. *Plant J.* **2006**, *45*, 523–539. [CrossRef]
28. Jambunathan, N. Determination and detection of reactive oxygen species (ROS), lipid peroxidation, and electrolyte leakage in plants. *Methods Mol. Biol.* **2010**, *639*, 292–298. [PubMed]
29. Ding, H.; Qian, L.; Jiang, H.; Ji, Y.; Fang, Y.; Sheng, J.; Xu, X.; Ge, C. Overexpression of a Bcl-2-associated athanogene *SlBAG9* negatively regulates high-temperature response in tomato. *Int. J. Biol. Macromol.* **2022**, *194*, 695–705. [CrossRef] [PubMed]
30. Alberti, S.; Demand, J.; Esser, C.; Emmerich, N.; Schild, H.; Hohfeld, J. Ubiquitylation of BAG-1 suggests a novel regulatory mechanism during the sorting of chaperone substrates to the proteasome. *J. Biol. Chem.* **2002**, *277*, 45920–45927. [CrossRef] [PubMed]
31. Kabbage, M.; Kessens, R.; Dickman, M.B. A plant Bcl-2- associated athanogene is proteolytically activated to confer fungal resistance. *Microb. Cell* **2016**, *3*, 224–226. [CrossRef]
32. Kang, C.H.; Jung, W.Y.; Kang, Y.H.; Kim, J.Y.; Kim, D.G.; Jeong, J.C.; Baek, D.W.; Jin, J.B.; Lee, J.Y.; Kim, M.O.; et al. AtBAG6, a novel calmodulin-binding protein, induces programmed cell death in yeast and plants. *Cell Death Differ.* **2006**, *13*, 84–95. [CrossRef] [PubMed]
33. Arif, M.; Li, Z.; Luo, Q.; Li, L.; Shen, Y.; Men, S. The *BAG2* and *BAG6* genes are involved in multiple abiotic stress tolerances in *Arabidopsis thaliana*. *Int. J. Mol. Sci.* **2021**, *22*, 5856. [CrossRef] [PubMed]
34. Fang, S.; Li, L.; Cui, B.; Men, S.; Shen, Y.; Yang, X. Structural insight into plant programmed cell death mediated by BAG proteins in *Arabidopsis thaliana*. *Acta Crystallogr. D Biol. Crystallogr.* **2013**, *69*, 934–945. [CrossRef]
35. Danino, Y.M.; Even, D.; Ideses, D.; Juven-Gershon, T. The core promoter: At the heart of gene expression. *Biochim. Biophys. Acta* **2015**, *1849*, 1116–1131. [CrossRef]
36. Li, Y.; Williams, B.; Dickman, M. Arabidopsis B-cell lymphoma2 (Bcl-2)-associated athanogene 7 (BAG 7)-mediated heat tolerance requires translocation, sumoylation and binding to WRKY 29. *New Phytol.* **2017**, *214*, 695–705. [CrossRef]
37. Hoang, T.M.L.; Lalehvash, M.; Brett, W.; Harjeet, K.; James, D.; Mundree, S.G. Development of salinity tolerance in rice by constitutive overexpression of genes involved in the regulation of programmed cell death. *Front. Plant Sci.* **2015**, *6*, 175. [CrossRef] [PubMed]
38. Pan, Y.J.; Liu, L.; Lin, Y.C.; Zu, Y.G.; Li, L.P.; Tang, Z.H. Ethylene antagonizes saltinduced growth retardation and cell death process via transcriptional controlling of ethylene-, BAG- and senescence- associated genes in *Arabidopsis*. *Front. Plant Sci.* **2016**, *7*, 696. [CrossRef] [PubMed]
39. Locascio, A.; Marqués, M.C.; García-Martínez, G.; Corratgé-Faillie, C.; Andrés-Colás, N.; Rubio, L.; Fernández, J.A.; Véry, A.A.; Mulet, J.M.; Yenush, L. Bcl-2-associated athanogene regulates the KAT1 potassium channel and controls stomatal movement. *Plant Physiol.* **2019**, *181*, 1277–1294. [CrossRef]
40. Liu, L.; Hu, X.; Song, J.; Zong, X.; Li, D.; Li, D. Over-expression of a *Zea mays* L. protein phosphatase 2C gene (ZmPP2C) in *Arabidopsis thaliana* decreases tolerance to salt and drought. *J. Plant Physiol.* **2009**, *166*, 531–542. [CrossRef]
41. Bedi, S.; Sengupta, S.; Ray, A.; Chaudhuri, R.N. ABI3 mediates dehydration stress recovery response in *Arabidopsis thaliana* by regulating expression of downstream genes. *Plant Sci.* **2016**, *250*, 125–140. [CrossRef]
42. Zhang, Z.; Li, W.; Gao, X.; Xu, M.; Guo, Y. DEAR4, a member of DREB/CBF family, positively regulates leaf senescence and response to multiple stressors in *Arabidopsis thaliana*. *Front. Plant Sci.* **2020**, *11*, 367. [CrossRef] [PubMed]
43. Yang, Y.; Guo, Y. Unraveling salt stress signaling in plants. *J. Integ. Plant Biol.* **2018**, *60*, 796–804. [CrossRef] [PubMed]
44. Dvořák, P.; Krasylenko, Y.; Ovečka, M.; Basheer, J.; Zapletalová, V.; Šamaj, J.; Takáč, T. In vivo light-sheet microscopy resolves localisation patterns of FSD1, a superoxide dismutase with function in root development and osmoprotection. *Plant Cell Environ.* **2021**, *44*, 68–87. [CrossRef] [PubMed]

45. Piacentini, D.; Rovere, F.D.; Bertoldi, I.; Massimi, L.; Sofo, A.; Altamura, M.M.; Falasca, G. Peroxisomal PEX7 receptor affects cadmium-induced ROS and auxin homeostasis in Arabidopsis root system. *Antioxidants* **2021**, *10*, 1494. [CrossRef] [PubMed]
46. Li, Q.; Zhang, X.; Lv, Q.; Zhu, D.; Qiu, T.; Xu, Y.; Bao, F.; He, Y.; Hu, Y. Physcomitrella patens dehydrins (PpDHNA and PpDHNC) confer salinity and drought tolerance to transgenic Arabidopsis plants. *Front. Plant Sci.* **2017**, *8*, 1316. [CrossRef] [PubMed]
47. Zheng, J.X.; Zhang, H.; Su, H.X.; Xia, K.F.; Jian, S.G.; Zhang, M. *Ipomoea pes-caprae IpASR* improves salinity and drought tolerance in transgenic *Escherichia coli* and *Arabidopsis*. *Int. J. Mol. Sci.* **2018**, *19*, 2252. [CrossRef] [PubMed]

antioxidants

MDPI

Article

The E3 Ubiquitin Ligase Gene *Sl1* Is Critical for Cadmium Tolerance in *Solanum lycopersicum* L.

Chen-Xu Liu [1], Ting Yang [1], Hui Zhou [1], Golam Jalal Ahammed [2,*], Zhen-Yu Qi [3] and Jie Zhou [1,4,5,*]

1 Zhejiang Provincial Key Laboratory of Horticultural Plant Integrative Biology, Department of Horticulture, Zhejiang University, Yuhangtang Road 866, Hangzhou 310058, China; chenxuliu@zju.edu.cn (C.-X.L.); 22116163@zju.edu.cn (T.Y.); 3180100610@zju.edu.cn (H.Z.)
2 College of Horticulture and Plant Protection, Henan University of Science and Technology, Luoyang 471023, China
3 Agricultural Experiment Station, Zhejiang University, Hangzhou 310058, China; qizhenyu@zju.edu.cn
4 Key Laboratory of Horticultural Plants Growth, Development and Quality Improvement, Agricultural Ministry of China, Yuhangtang Road 866, Hangzhou 310058, China
5 Shandong (Linyi) Institute of Modern Agriculture, Zhejiang University, Linyi 276000, China
* Correspondence: ahammed@haust.edu.cn (G.J.A.); jie@zju.edu.cn (J.Z.)

Abstract: Heavy metal cadmium (Cd) at high concentrations severely disturbs plant growth and development. The E3 ubiquitin ligase involved in protein degradation is critical for plant tolerance to abiotic stress, but the role of E3 ubiquitin ligases in Cd tolerance is largely unknown in tomato. Here, we characterized an E3 ubiquitin ligase gene *Sl1*, which was highly expressed in roots under Cd stress in our previous study. The subcellular localization of Sl1 revealed that it was located in plasma membranes. In vitro ubiquitination assays confirmed that Sl1 had E3 ubiquitin ligase activity. Knockout of the *Sl1* gene by CRISPR/Cas9 genome editing technology reduced while its overexpression increased Cd tolerance as reflected by the changes in the actual quantum efficiency of PSII photochemistry (Φ_{PSII}) and hydrogen peroxide (H_2O_2) accumulation. Cd-induced increased activities of antioxidant enzymes including superoxide dismutase (SOD), catalase (CAT), ascorbate peroxidase (APX), and glutathione reductase (GR) were compromised in *sl1* mutants but were enhanced in *Sl1* overexpressing lines. Furthermore, the content of Cd in both shoots and roots increased in *sl1* mutants while reduced in *Sl1* overexpressing plants. Gene expression assays revealed that Sl1 regulated the transcript levels of heavy metal transport-related genes to inhibit Cd accumulation. These findings demonstrate that Sl1 plays a critical role in regulating Cd tolerance by relieving oxidative stress and resisting heavy metal transportation in tomato. The study provides a new understanding of the mechanism of plant tolerance to heavy metal stress.

Keywords: antioxidant enzymes; heavy metal stress; ubiquitination; protein degradation; tomato

Citation: Liu, C.-X.; Yang, T.; Zhou, H.; Ahammed, G.J.; Qi, Z.-Y.; Zhou, J. The E3 Ubiquitin Ligase Gene *Sl1* Is Critical for Cadmium Tolerance in *Solanum lycopersicum* L. *Antioxidants* **2022**, *11*, 456. https://doi.org/10.3390/antiox11030456

Academic Editor: Nafees A. Khan

Received: 22 January 2022
Accepted: 23 February 2022
Published: 25 February 2022

1. Introduction

The rapid development of industrialization and urbanization has resulted in severe environmental pollution [1]. Sewage or waste produced by industries and garbage generated by anthropogenic activities lead to the release of heavy metals into the environment, causing contamination of agricultural soil and water [2,3]. Thus, heavy metal pollution affects both human health as well as plant health, particularly plant growth and development [4].

Cadmium (Cd), a toxic heavy metal, severely inhibits plant growth and crop production when it occurs in high concentrations in soils or growth media [5]. The absorption and translocation of Cd in plants include distinct phases, such as the absorption of Cd in roots, the transportation of Cd into xylem and phloem, and the transportation of Cd into aboveground tissues [6]. The absorption and transportation of Cd in plants mainly depend on transport proteins such as heavy-metal-associated P-type ATPase family protein (HMA),

ATP-binding cassette transporters (ABC), natural resistance-associated macrophage protein (NRAMP), metal-tolerance protein (MTP), calcium exchanger protein (CAX), and zinc/iron-regulated transporter-like protein (ZRT/IRT) [6–12].

Owing to the interaction with the sulfhydryl group of proteins, Cd can affect the activity of multiple enzymes and the functions of proteins [13]. Hence, the accumulation of Cd in plant tissues disorders various growth, biochemical, and physiological processes, such as photosynthesis, antioxidant enzyme activity, cell structure, and plant morphology [13,14]. Specifically, Cd stress causes damage to chloroplast ultrastructure, inhibits pigment synthesis, and affects several photosynthesis-related protein complexes [15]. Cd stress also impairs the balance between reactive oxygen species (ROS) production and ROS scavenging [16]. In particular, Cd stress disturbs the electron transfer chain in mitochondria and chloroplasts and activates NADPH oxidases, which cause excessive ROS accumulation and associated oxidative stress in plants [17,18]. Moreover, Cd stress induces protein denaturation, which disturbs the balance of protein quality in plant cells [19].

Plants have also evolved very diverse and complex defensive mechanisms for resisting heavy metal stress [20]. The ascorbate–glutathione (AsA–GSH) cycle is an indispensable pathway for eliminating oxidative stress [21]. Moreover, antioxidant enzymes including superoxide dismutase (SOD), catalase (CAT), ascorbate peroxidase (APX), peroxidase (POD), and glutathione reductase (GR) are vital for scavenging ROS [22]. Antioxidant GSH not only functions on relieving oxidative stress, but also participates in the synthesis of phytochelatins (PCs) in plants, which can bind with heavy metal ions [20,23,24]. Moreover, the protein quality control system needs chaperone proteins to refold the denatured proteins or operates a protein degradation system to clear misfolded proteins under Cd stress [24,25].

Ubiquitin proteasome system (UPS) has been recognized as a critical process to control the abundance, quality, and function of protein in cells [26,27]. Ubiquitin is an indispensable component of UPS, which acts as identifying target protein for degradation [28]. Ubiquitin-dependent protein degradation requires reactions of multiple ubiquitin-related enzymes, including ubiquitin activating enzyme (E1), ubiquitin conjugating enzyme (E2), and ubiquitin ligase (E3) [29,30]. The ubiquitin molecule is activated by E1 and then transferred to E2. The E3 interacts with E2-ubiquitin and targets proteins for labeling substrate protein with ubiquitin. Finally, the 26S proteasome degrades the target proteins [31–33]. E3 ubiquitin ligases play critical roles in recognizing substrate proteins [29]. Among the ubiquitin-related enzymes, E3 ubiquitin ligases have the largest family in plants [19]. The E3 ligase has been divided into three types, as follows: really interesting new gene (RING)-type, homology to E6-associated carboxyl-terminus (HECT)-type, and U-box-type [31].

RING-type E3 ubiquitin ligases have a cysteine-rich domain that can bind with Zn ions [31]. Previous studies have revealed that RING-type E3 ubiquitin ligases are involved in mediating plant tolerance of heavy metal stress [19,25,34]. A RING-type E3 ubiquitin ligase HIR1 confers tolerance of arsenic (As) and Cd stress in rice [34]. Overexpression of *HIR1* increases root length in rice under As and Cd treatment. The E3 ubiquitin ligase protein HIR1 also interacts with tonoplast intrinsic protein TIP4;1 and regulates its abundance in rice, thereby alleviating heavy metal stress [34]. Tomato E3 ubiquitin ligase RING1 increases Cd tolerance by minimizing ROS levels due to enhanced antioxidant enzyme activities [19,25]. However, the roles of many tomato E3 ubiquitin ligases are largely unknown, particularly in Cd stress.

In a previous study, we found that the expression of the RING-type E3 ubiquitin ligase *Sl1* significantly increased in tomato roots when challenged with aluminum (Al) or Cd stress [25]. However, the precise role of *Sl1* remains elusive. We hypothesized that *Sl1* might play a crucial role in Cd tolerance in tomato plants. Here, we characterized the function of *Sl1* by generating *sl1* mutants and *Sl1* overexpressing lines in tomato. Our study unveils a novel role of *Sl1* in plant tolerance to Cd stress.

2. Materials and Methods

2.1. Plant Materials and Treatments

In this study, tomato (*Solanum lycopersicum* L., Tomato Genetics Resource Center, Davis, CA, USA, https://tgrc.ucdavis.edu (accessed on 4 March 2018)) cultivar "Ailsa Craig" was used to generate *sl1* mutants and *Sl1* overexpressing transgenic lines.

Tomato seedlings were raised in vermiculite and transferred to a hydroponic jar (40 cm × 25 cm × 15 cm) containing Hoagland nutrient solution, when two real leaves of seedlings unfolded. The growth conditions were temperature of 23/20 °C (day/night), 14 h photoperiod, 60% humidity, and photosynthetic photo flux density (PPFD) of 600 μmol m^{-2} s^{-1}. For Cd treatment, seedlings at the five-leaf stage were treated with 100 μM $CdCl_2$ and the hydroponic nutrient solution was changed every 5 days (d).

2.2. Generation of Transgenic Plants

The *sl1* mutant was generated by CRISPR/Cas9 technique. For CRISPR vector construction, the target sequence used as sgRNA on the only exon of *Sl1* was searched by the CRISPR-P program (http://cbi.hzau.edu.cn/cgi-bin/CRISPR (accessed on 10 May 2019)). For deletion of a large fragment, two targets were selected (sgRNA-*Sl1*-254: TCATTAAAGGGTCTTCAACA and sgRNA-*Sl1*-564: GTTGAACTTGGAGCAATGAT), primers were then generated by adding adapter sequence to two ends of target sequences (Figure S1A). The sense and antisense sgRNA primers were annealed and inserted into the *Bbs*I site of the AtU6-sgRNA-AtUBQ-Cas9 vector, then positive clones were confirmed by sequencing. The two clones were named sgRNA-*Sl1*-254 and sgRNA-*Sl1*-564. The fragment was amplified using the clone of sgRNA-*Sl1*-564 as a template, the fragment was then inserted into the backbone vector with Cas9 and sgRNA-254. pCAMBIA1301 was used as a binary expression vector, and the sgRNA-254-sgRNA-564-Cas9 was inserted into the *EcoR*I and *Hind*III site of pCAMBIA1301. Positive clones were transformed into *Agrobacterium tumefaciens* strain GV3101 for transgenic plant generation.

For the construction of *Sl1* overexpressing transgenic lines, the full-length coding sequences (CDS) of *Sl1* were amplified with primers (Forward primer 5′-TTACAATTACCATGGGGCGCGCCATGGATCTTGTTAGACTAAAGTATTTTGAA-3′, Reverse verse primer 5′-AACATCGTATGGGTAGGTACCTGACTCTAACTGAATAGGTAAAACTACATTTC-3′), then inserted the PCR products into the *Asc*I and *Kpn*I site of pFGC1008-HA vector. The positive vector was confirmed by sequencing and then transformed into *A. tumefaciens* strain GV3101.

The detailed method of generation of transgenic lines was described previously [35]. Two independent homozygous lines of the F2 generation of *Sl1* overexpressing plants were used in further experiments. Two homozygous lines of *sl1* mutants without CRISPR/Cas9 DNA were selected for further research. The special primers for mutant detection were designed as follows: Forward primer 5′-GCAGAGAGACAACATTCACCA- 3′, Reverse primer 5′-AAAGTTGTCGATCCGTCGCT-3′.

2.3. E3 Ubiquitin Ligase Activity Assay

The full length CDS of *Sl1* were amplified with the primers (Forward primer 5′-GAGGGAAGGATTTCAGAATTCATGGATCTTGTTAGACTAAAGTATTTTGAA-3′, Reverse primers 5′-CAGGTCGACTCTAGAGGATCCTGACTCTAACTGAATAGGTAAAACTACATTTC-3′). The PCR products were digested with restriction endonuclease *EcoR*I and *BamH*I and were then inserted into the pMAL-2c vector (New England Biolabs, Ipswich, MA, USA). The maltose-binding protein-empty vector (MBP-EV) and MBP-fused Sl1 protein were expressed in *Escherichia coli* strain BL21 (DE3) and purified with instructions of the manufacturer (New England Biolabs, Ipswich, MA, USA). The in vitro ubiquitination assay of Sl1 was performed by instructions described previously [36]. The reaction system was prepared as described previously [19].

After the reaction, a Western blot was used to detect whether the Sl1 protein has E3 ubiquitin ligase activity. Anti-His (A5C12; HUABIO, Hangzhou, China) and anti-MBP

(MBP61R; Thermo Fisher Scientific, Waltham, MA, USA) antibodies were used in Western blot assay.

2.4. Vector Construction and Subcellular Localization of *Sl1*

The full length CDS of *Sl1* were amplified with primers (Forward primer 5′-CTCTCGA-GCTTTCGCGAGCTCATGGATCTTGTTAGACTAAAGTATTTTGAA-3′, Reverse primer 5′-GCCCTTGCTCACCATGGATCCTGACTCTAACTGAATAGGTAAAACTACATTTC-3′), and the PCR products were inserted into *SacI* and *BamHI* site of pCAMBIA2300 with a GFP tag at the C terminus. The primers were designed by homologous recombination methods as described previously. After confirming positive clones by sequencing, the vector (35S-*Sl1*-GFP) and empty vector were transformed into *A. tumefaciens* strain GV3101. Transgenic *Nicotiana benthamiana* that expressed with nucleus-located signaling (RFP-H2B) was used for transient expression. The transient expression method was described previously [37]. For the fluorescence image acquisition, a Nikon A1 confocal microscope (Nikon, Tokyo, Japan) was used at 48 h post infiltration in tobacco leaves, and the excitation/emission wavelengths of GFP were 488 nm/480–520 nm, and the excitation/emission wavelengths of RFP were 561 nm/610–630 nm.

2.5. Measurement of Actual Quantum Efficiency of PSII Photochemistry

After 15 d treatment of Cd, plants were dark-adapted for 30 min before measurement. The actual quantum efficiency of PSII photochemistry (Φ_{PSII}) was determined by the Imaging PAM (IMAG-MAXI, Heinz Walz, Germany) in the fifth fully expanded leaves as described previously [38].

2.6. Measurements of Hydrogen Peroxide and Antioxidants Enzyme Activity

Three-tenths gram of tomato root samples were collected and ground in liquid nitrogen. The samples were combined with 3 mL precooled 1 M $HClO_4$ in a 10 mL centrifuge tube and mixed thoroughly. The pH of the sample was adjusted to 6–7 using 4 M KOH, 0.05 g activated carbon was then added to absorb pigments. The samples were centrifuged at 12,000× *rpm* for 10 min at 4 °C and supernatants were collected in 5 mL tubes for reaction. The reaction buffer was as follows: 1 mL supernatant or H_2O_2, 996 μL 1 mM ABTS (dissolved in 100 mM potassium acetate, pH 4.4), and 4 μL POD (P8375, Merck KGaA, Darmstadt, Germany). The absorbance of reaction buffer was detected at a wavelength of 412 nm by a SHIMADZU UV-2410PC spectrophotometer (Shimadzu Company, Kyoto, Japan) and the content of H_2O_2 was calculated by a standard curve [39].

Three-tenths gram fresh tomato roots were ground in liquid nitrogen and dissolved in a precooled buffer that included 50 mM phosphate buffered saline (pH 7.8), 0.2 mM EDTA, 2 mM ascorbic acid, and 2% (*w*/*v*) poly-vinylpolypyrrolidone. The samples were mixed thoroughly using vortex and then centrifuged at 12,000× *g* for 20 min at 4 °C. The supernatant was collected in new centrifuge tubes for detecting enzyme activity. The method of measuring enzyme activities of SOD, CAT, APX, and GR was described previously [40].

2.7. Measurement of Cd Content and Cd Localization

Tomato leaves and roots were washed and collected in liquid nitrogen after 10 d Cd treatment. Samples were put in the oven for 30 min at 115 °C then transferred to 60 °C for total dryness. The dried sample was ground and mixed with an acid solution that contained $HClO_4$ and HNO_3 (1:3, *v*:*v*). The samples were digested at 180 °C and the solution was then evaporated to 1-2 mL for dilution. The remained solution was diluted with deionized water to a final volume of 50 mL and inductively coupled plasma mass spectrometry (ICPMS-2030, Shimadzu Company, Kyoto, Japan) was used for determining Cd content [25].

For Cd localization, the Leadmium™ Green AM probe (Invitrogen, Carlsbad, CA, USA) was used for Cd staining according to the manufacturer's instructions. The root tips were stained by immersing in dye solution for 3 h in dark at room temperature; the sample was then washed three times with buffer (0.85% NaCl). A Nikon A1 confocal microscope (Nikon, Tokyo, Japan) was used to detect the Cd localization, and the excitation/emission wavelengths of GFP were 488 nm/510–530 nm [41]. The mean fluorescence intensity values of Cd stained root tips were detected by ImageJ 1.53 analysis software (National Institute of Health, Bellevue, WA, USA) and relative fluorescence intensity normalized to the intensity of the wild-type group under Cd stress.

2.8. Total RNA Isolation and qRT-PCR Analysis

One-tenth gram tomato samples were collected in a 2 mL centrifuge tube and frozen in liquid nitrogen. An RNAprep pure plant kit (Tiangen Biotech, Beijing, China) was used for total RNA extraction according to the manufacturer's instructions. Five-tenths milligram total RNA was used to reverse transcribe to cDNA template by HiScript II Q RT SuperMix for qPCR Kit (Vazyme, Nanjing, China). qRT-PCR was conducted with ChamQ Universal SYBR qPCR Master Mix (Vazyme, Nanjing, China) on a Light Cycle 480 II Real-Time PCR detection system (Roche, Basel, Switzerland). The total 20 μL reaction system was as follows: 10 μL SYBR qPCR Master Mix, 1 μL cDNA template, 10 μM forward and reverse primer, and deionized water. The PCR program was performed with 30 s at 95 °C, followed with 35–40 cycles of 10 s at 95 °C, 30 s at 58 °C, and 1 min at 72 °C. *ACTIN* was used for calculating the relative expression level of the target gene. The primers are listed in Table S1.

2.9. Immunoblotting Assays

The protein extraction and Western blot procedures were described previously [42,43]. The antibodies used in this assay included anti-HA (26183, Thermo Fisher Scientific, Waltham, MA, USA) and rabbit antimouse (ab6728, abcam, Cambridge, UK). The signals were visualized with FDbio-Femto ECL (FD8030, Fdbio science, Hangzhou, China).

2.10. Statistical Analysis

All experiments were conducted independently three times and three replications of each experiment were performed. All data were analyzed by SPSS 16.0 statistical software package and Tukey's test ($p < 0.05$) was used for significance analysis.

3. Results

3.1. Structure and Expression Analysis of Sl1

The *Sl1* (Solyc09g039890) gene that was previously reported as one of the highly expressed genes in tomato roots under Cd stress was selected as the target gene in the current study [25]. We examined the time course of the relative expression level of *Sl1* in roots exposed to Cd stress. After 7 d treatment, the expression of *Sl1* reached the peak, which was 4.9-fold of that in control (Figure 1). Moreover, we examined the relative expression of the *Sl1* gene in different tissues including leaf, stem, root, flower, and fruit. The results showed that the relative expression level of *Sl1* in roots was 6-fold of that in leaves and 1.6-fold of that in stems (Figure 1).

The *Sl1* gene locates on chromosome 9 with only one exon on coding sequences (CDS) that encodes 349 amino acids. We used InterPro (https://www.ebi.ac.uk/interpro/ (accessed on 20 October 2021)) and Expasy (https://web.expasy.org/compute_pi/ (accessed on 22 October 2021)) to analyze thC terminus structure of the Sl1 protein. The results show that the isoelectric point and molecular weight of Sl1 protein are 8.56 and 40.14 kDa, respectively, and Sl1 contains a RING domain of 127 to 170 amino acids (Figure S2A,B).

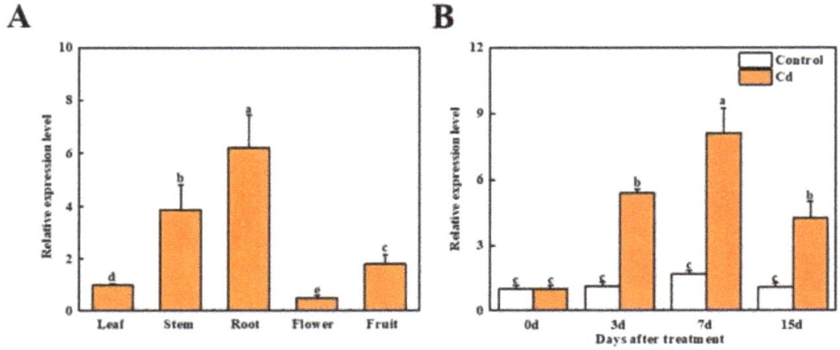

Figure 1. The relative expression level of *Sl1* in tomato plants. (**A**) The relative expression level of *Sl1* in different tissues of tomato. (**B**) Time course of the relative expression level of *Sl1* in tomato root with and without Cd stress. The data presented here are the average of three biological replicates (±SD). Different letters indicate a significant difference (p < 0.05, Tukey's test).

3.2. E3 Ubiquitin Ligase Activity and Subcellular Localization of Sl1

To verify the E3 ubiquitin ligase activity of Sl1 protein, we successfully purified the maltose-binding protein-Sl1 (MBP-Sl1) fusion protein and maltose-binding protein-empty vector (MBP-EV) fusion protein following the manufacturer instructions (Figure 2A). We performed the in vitro ubiquitination assay and showed that Sl1 protein had self-ubiquitination ability when present with E1, E2, and ubiquitin. However, lacking any of E1, E2, or ubiquitin in the reaction system undermined the self-ubiquitination ability of Sl1. Meanwhile, MBP-EV also did not show self-ubiquitination ability (Figure 2B).

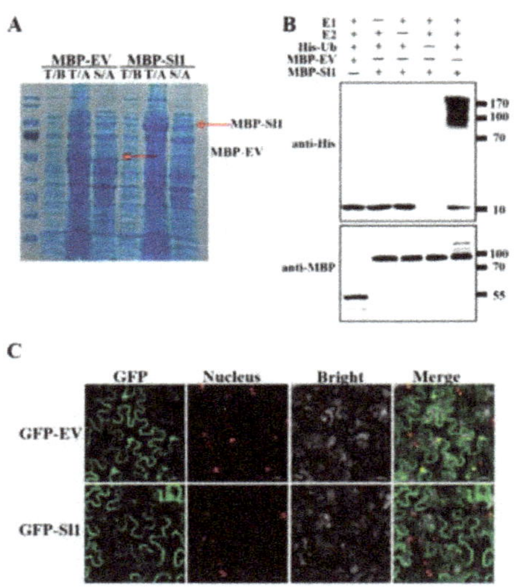

Figure 2. E3 ligase activity and subcellular localization of Sl1. (**A**) In vitro expression of MBP-Sl1 and MBP-EV. T: total protein; S: soluble protein; B: before expression; A: after expression. (**B**) In vitro E3 ligase activity of Sl1 protein. The reaction system included E1, E2, MBP-Sl1, and ubiquitin-His, the replacement of MBP-Sl1 with MBP-EV and the absence of E1, E2, and His-Ub as control. The Western blot was detected with anti-MBP and anti-His. (**C**) Subcellular localization of GFP-Sl1 and GFP-EV. The GFP-Sl1 was transiently expressed in *Nicotiana benthamiana* (tobacco with nucleus-located mCherry). Images were pictured by confocal microscope after 48 h infiltration. Bar = 25 μm.

To examine the subcellular localization of Sl1 protein, we transiently expressed GFP-Sl1 in transgenic tobacco (with nucleus signal). Images captured by a confocal microscope revealed that the Sl1 protein is located in plasma membranes rather than the nucleus (Figure 2C).

3.3. Sl1 Positively Regulates Cd Tolerance in Tomato

Since the expression of the *Sl1* gene was induced by Cd stress, we investigated the tolerance of *sl1* mutant lines, wild-type, and *Sl1* overexpressing lines to Cd stress. We generated the *sl1* mutant lines and *Sl1* overexpressing lines as described in the Methods and Materials section (Figure S1). Two *sl1* mutant lines were mutated at different sites that both induced early termination of translation. The *sl1-1* mutant line was found 311 bp deletion between sgRNA1 and sgRNA2 and the translation was terminated after 98 amino acids (Figure S1B). The *sl1-2* mutant line was deleted 1 bp after the protospacer adjacent motif (PAM) of sgRNA1 and the translation was terminated after 89 amino acids (Figure S1B). Two overexpressing lines of *Sl1* were checked with Western blot that both had bright bands near the predictive molecular weight (Figure S1C). The phenotypes of *sl1* mutants were similar to wild-type plants; however, *Sl1* overexpressing lines grew more slowly and showed smaller leaves compared with wild-type plants when they were grown under optimal (nonstress) environments (Figure 3A).

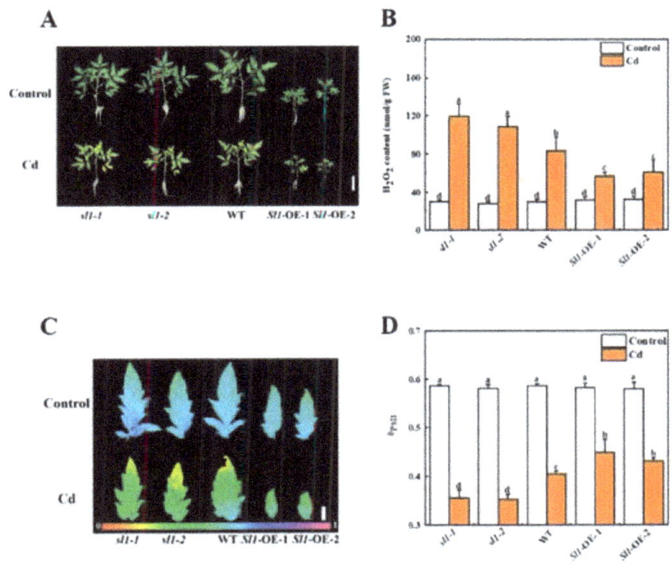

Figure 3. Sl1 positively regulates tomato Cd tolerance. (**A**) The phenotype of *Sl1* mutant lines (*sl1-1/2*), wild-type (WT), and *Sl1* overexpressing lines (*Sl1*-OE-1/2) under control and Cd stress after 15 d treatment. Bar = 10 cm. (**B**) The content of hydrogen peroxide (H_2O_2) in the roots of *Sl1* mutant lines (*sl1-1/2*), WT, and *Sl1* overexpressing lines (*Sl1*-OE-1/2) under control and Cd stress after 3 d treatment. (**C**,**D**) the image and level of actual quantum efficiency of PSII photochemistry (Φ_{PSII}) of *sl1-1/2*, WT, and *Sl1*-OE-1/2 plants with and without Cd treatment for 15 d. Bar = 1 cm. The data presented here are the average of three biological replicates (±SD). Different letters indicate a significant difference ($p < 0.05$, Tukey's test).

After 15 d of Cd treatment, *sl1* mutants showed sensitivity to Cd stress, while *Sl1*-OE showed enhanced tolerance to Cd stress (Figure 3). Compared to control conditions, Cd-stress-induced changes in leaf size, leaf color, level of Φ_{PSII}, and content of hydrogen peroxide (H_2O_2) in *sl1* mutants and wild-type plants. In particular, the leaves of *sl1* were more etiolated than those of wild-type, and the leaf sizes of mutants were smaller than that of wild-type. However, the leaf size and color in *Sl1* overexpressing plants were just slightly affected by Cd stress (Figure 3). Since H_2O_2 is a major ROS generated under stress conditions, we then detected the accumulation of H_2O_2 in roots under Cd stress. Importantly, the accumulation of H_2O_2 increased by 43.1% and 30.5% in *sl1-1* and *sl1-2* mutants, while it decreased by 31.7% and 26.5% in two lines of *Sl1* overexpressing plants compared with wild-type plants after Cd treatment, respectively (Figure 3B). Moreover, we detected the value of actual quantum efficiency of PSII photochemistry, Φ_{PSII}, which reflects the state of photosystem II as a reliable marker of plant health status. As shown in Figure 3C,D, the Φ_{PSII} value of *sl1* mutants, wild-type, and *Sl1* overexpressing lines exhibited no significant difference under control conditions. However, the Φ_{PSII} levels of *sl1-1* and *sl1-2* mutants decreased by 12.1% and 12.7% respectively, compared with wild-type under Cd stress (Figure 3D). The Φ_{PSII} levels of two lines of *Sl1* overexpressing plants were significantly greater than that in wild-type under Cd stress (Figure 3C,D). These results indicate that Sl1 is critical for alleviating Cd-induced H_2O_2 accumulation and damage to the photosynthetic system.

3.4. *Sl1 Promotes Antioxidant Enzyme Activity*

To understand whether *Sl1* influenced antioxidant enzyme activities in tomato under Cd stress, we examined the enzyme activities of superoxide dismutase (SOD), catalase (CAT), ascorbate peroxidase (APX), and glutathione reductase (GR). The results showed that Cd stress increased antioxidant enzyme activities in wild-type and overexpressing lines. However, in *sl1* mutant lines, there were no significant differences in antioxidant enzyme activities between control and Cd treatment (Figure 4). The activities of SOD, CAT, APX, and GR in two lines of *Sl1* overexpressing plants were all induced compared with wild-type plants under Cd stress (Figure 4). These results suggest that *Sl1* promotes the activities of antioxidant enzymes under Cd stress.

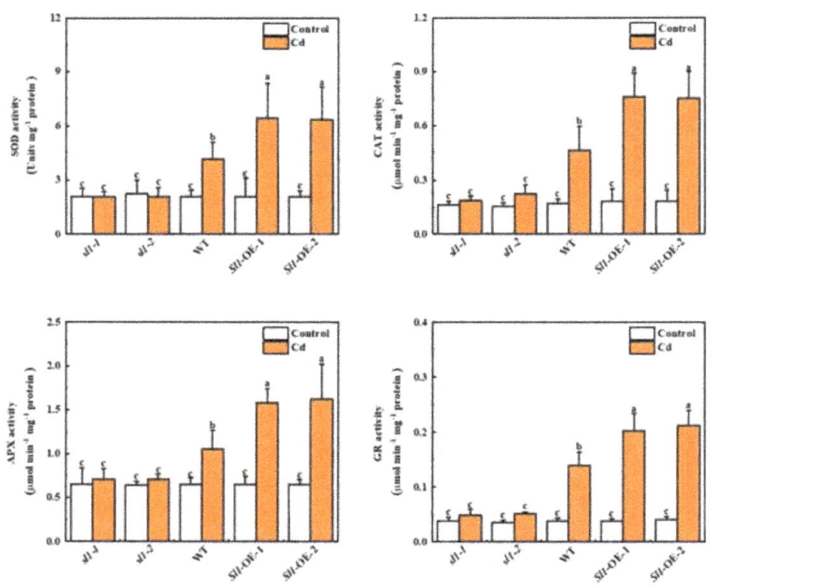

Figure 4. Sl1 promotes antioxidant enzyme activities in tomato plants under Cd stress. The activities of SOD, CAT, APX, and GR in the roots of *Sl1* mutant lines (*sl1-1/2*), wild-type (WT), and *Sl1* overexpressing lines (*Sl1*-OE-1/2) under Cd stress for 3 d. The data presented here are the average of three biological replicates (±SD). Different letters indicate a significant difference ($p < 0.05$, Tukey's test).

3.5. Sl1 Reduces Cd Accumulation and Transportation

To investigate whether *Sl1* is involved in Cd accumulation in tomato plants, we detected the content of Cd in shoots and roots under Cd stress. Results revealed that the Cd content in the roots was higher than that in the shoots (Figure 5A,B). Obviously, overexpression of *Sl1* decreased Cd content in both shoots and roots, while Cd content in *sl1* mutants significantly increased compared with that in wild-type (Figure 5A,B). The content of Cd in roots of two *Sl1* overexpressing lines both decreased by 25.6% compared with wild-type, while the Cd content increased by 34.7% and 41.6% in roots of *sl1-1* and *sl1-2* mutants compared with wild-type, respectively. Similarly, Cd accumulation was also higher in the shoots of *sl1* mutants than wild-type plants, while it was lower in the shoots of *Sl1* overexpressing lines.

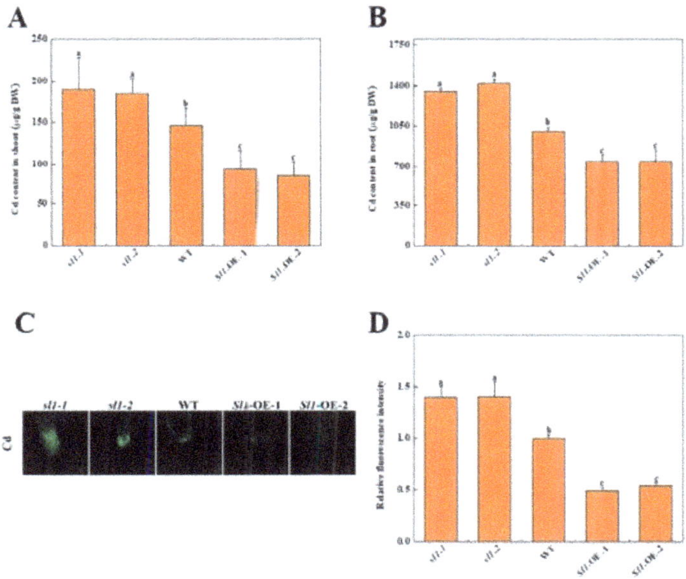

Figure 5. Sl1 decreases Cd content in tomato plants under Cd stress. The Cd content in the shoot (**A**) and root (**B**) of *Sl1* mutant lines (*sl1-1/2*), wild-type (WT), and *Sl1* overexpressing lines (*Sl1*-OE-1/2) under Cd stress for 10 d. (**C**) Cd accumulation in tomato root tips stained by the Cd-specific probe Leadmium™ Green AM. Bar = 25 μm. (**D**) Relative fluorescence intensity of Cd staining over tomato root tips of *sl1* mutants, wild-type, and *Sl1* overexpressing lines after 10 d Cd treatment. The relative fluorescence intensity is normalized to the intensity of wild-type in (**C**). The data presented here are the average of three biological replicates (±SD). Different letters indicate a significant difference ($p < 0.05$, Tukey's test).

To further investigate whether *Sl1* decreased Cd accumulation by altering Cd delivery, we used a Cd-specific probe to study the Cd distribution in the root tips. The Cd-specific probe stained signals were not detected in the root tips of all plants without Cd treatment (Figure S2E). However, Cd treatment induced the accumulation of Cd in the root tips as reflected by the increased fluorescence intensity. The relative fluorescence intensity of *sl1-1* and *sl1-2* mutants were 1.40-fold and 1.41-fold of that in wild-type plants, respectively, while the relative fluorescence intensity of *Sl1*-OE-1 and *Sl1*-OE-2 plants were only 49.2% and 54.7% of that in wild-type plants (Figure 5C,D).

To investigate how *Sl1* regulated Cd transportation, we examined the expression of heavy metal transportation-related genes (*CAX3*, *HMA-A*, *HMA-B*, and *IRT1*). There was no significant difference in the expression of these four genes between *sl1* mutants, wild-type, and *Sl1* overexpressing lines under control conditions. Although these four genes were highly expressed in *sl1* mutants under Cd stress, no significant difference was found between control and Cd treatment in two lines of *Sl1* overexpressing plants. We found that Cd stress dramatically increased the transcript level of *CAX3* in *sl1-1* and *sl1-2* mutants under Cd stress, which were 3.2-fold and 3.4-fold of that in wild-type, respectively (Figure 6). The heavy metal transport gene *HMA-A/B* was also upregulated by Cd stress in *sl1-1* and *sl1-2* plants by 76.5%/73.6% and 86.5%/79.9% compared with wild-type, respectively (Figure 6). Similarly, the expression of *IRT1*, which plays a prominent role in heavy metal transportation, was also increased by Cd stress in *sl1* mutants compared with wild-type plants (Figure 6). These results suggest *Sl1* potentially functions in resisting heavy metal transportation through repressing the transcription of heavy metal transportation-related genes.

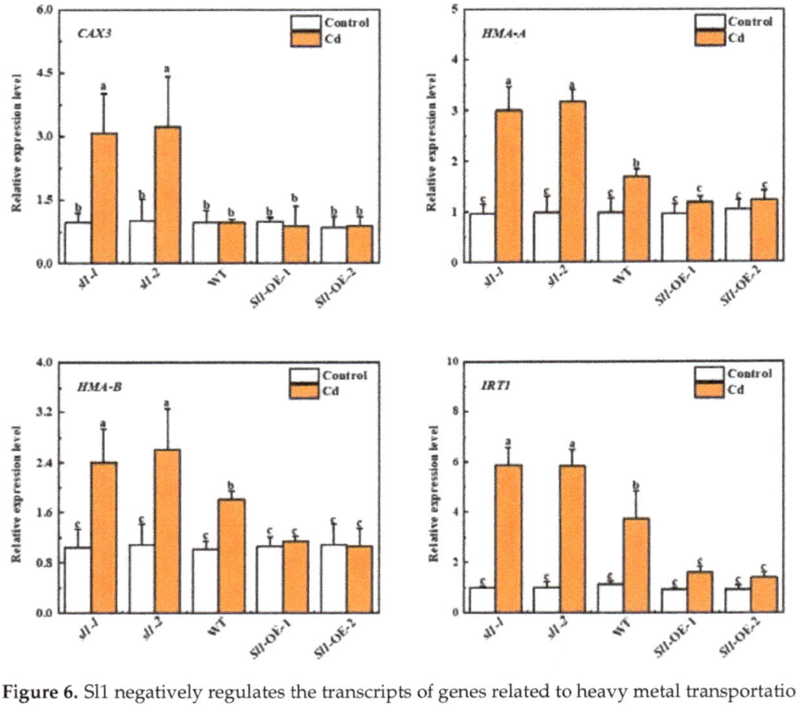

Figure 6. Sl1 negatively regulates the transcripts of genes related to heavy metal transportation. The relative expression of *CAX3*, *HMA-A*, *HMA-B*, and *IRT1* in the roots of *Sl1* mutant lines (*sl1-1/2*), wild-type (WT), and *Sl1* overexpressing lines (*Sl1*-OE-1/2) under Cd stress for 3 d. The data presented here are the average of three biological replicates (±SD). Different letters indicate a significant difference ($p < 0.05$, Tukey's test).

4. Discussion

As a significant component of the food chain, plants play a crucial role in the transportation and accumulation of toxic elements such as Cd in humans [44]. Nonetheless, plants also suffer from the stress induced by heavy metals and they address the stress by multiple pathways, including eliminating ROS, resisting heavy metal transportation, and maintaining protein quality [19,20,24,25]. The E3 ubiquitin ligase-mediated protein degradation plays an important role in plant stress tolerance [25,34,36,45,46]. Here, we characterized a RING-type E3 ubiquitin ligase Sl1, which conferred Cd tolerance in tomato. Our study advances the understanding of the mechanism of UPS-mediated heavy metal tolerance in plants.

The RING-type E3 ligase is closely associated with plant tolerance to various stress [47]. For example, a C3H2C3-type E3 ubiquitin ligase AtAIRP1 positively regulates ABA-dependent drought tolerance by mediating AtAIRP1 degradation [48,49]. The roles of E3 ubiquitin ligases in the positive regulation of heavy metal stress tolerance have been reported recently in different plant species [19,25,34,46]. Overexpression of *HIR1* increases the tolerance of rice to As and Cd stress [34]. *HIR1* that encodes an E3 ubiquitin ligase interacts with TIP4;1 for resisting heavy metal absorption in rice [34]. Moreover, a U-box type E3 ubiquitin ligase, *SlUPS*, is highly expressed under Cd stress in tomato. Heterologous expression of *SlUPS* in yeast increases the concentration of yeast bacterial fluid exposed to Cd and overexpression of *SlUPS* in *Arabidopsis* enhances Cd tolerance [46]. The RING-type E3 ligase AtIDF1 degrades IRT1 to modulate iron homeostasis in *Arabidopsis* [50]. Heterologous expression of a soybean RING-type E3 ubiquitin ligase gene *GmARI1* in *Arabidopsis* enhances Al tolerance [51]. Furthermore, tomato E3 ubiquitin ligase SlRING1 positively mediates Cd tolerance by enhancing antioxidant enzyme activities and inhibiting Cd accumulation [19,25]. Similar to *SlRING1*, *Sl1*, which is highly expressed under Cd stress in tomato roots, plays a pivotal role in Cd tolerance (Figures 1 and 3). We also found that Sl1 protein possessed the E3 ligase activity and overexpression of *Sl1* inhibited Cd accumulation in tomato. These results are consistent with previous studies [19,34], suggesting that RING-type E3 ubiquitin ligases play crucial roles in regulating metal ion transport.

Cadmium has a broad variety of negative impacts on plants, including oxidative stress, nutrient absorption disruption, and even plant mortality. Antioxidant enzymes such as SOD, POD, CAT, APX, and GR function in collaboration with nonenzymatic antioxidants such as AsA and GSH to prevent Cd-induced oxidative damage [52,53]. Furthermore, GSH directly participates in the synthesis of PCs [54,55]. PCs form complexes with Cd that can be compartmentalized into the vacuoles; thus, PCs and other thiols play an important role in determining the sensitivity or tolerance in contrasting genotypes of a plant species [56]. Moreover, studies on the semihalophytic plant *Mesembryanthemum crystallinum* L. and different Cd hyperaccumulators such as *Arabidopsis halleri*, *Thlaspi caerulescens*, *Solanum nigrum*, and *Sedum alfredii* species indicate that both antioxidative enzymes and nonenzymatic antioxidants play a vital role in Cd tolerance [57–59]. Previously, we found that an E3 ligase gene *SlRING1* positively regulates relative expression levels of *CAT*, *MDHAR*, *GSH1*, and *PCS*, while it decreases H_2O_2 content in tomato under Cd stress [19,25]. Consistent with those studies, *Sl1* overexpression increased the activities of SOD, CAT, APX, and GR, and decreased the content of H_2O_2 under Cd stress in *Sl1* overexpressing lines compared with those in wild-type and mutant lines (Figures 3 and 4).

Metal absorption and transportation are critical for plant tolerance to heavy metal stress [44]. The rice E3 ubiquitin ligase OsHIR1 targets TIP4;1 that functions as a heavy metal absorption protein, and thus OsHIR1-induced degradation of TIP4;1 increases rice tolerance to As and Cd stress [34]. Moreover, another RING-type E3 ubiquitin ligase in rice, OsAIR3, regulates protein degradation of molybdate transporter (OsMOT1;3) in rice to increase plant tolerance to arsenate stress [60]. In agreement with these studies, we found that overexpression of *Sl1* significantly attenuated the relative expression level of several genes related to heavy metal transportation, such as *CAX3*, *HMA-A*, *HMA-B*, and *IRT1*, along with decreased Cd content in *Sl1* overexpressing lines compared with the wild-type and mutant lines (Figures 5 and 6). *CAX* gene family is an important plant gene family involved in heavy metal transportation [11]. It is plausible that Sl1 functions as an E3 ubiquitin ligase for the degradation of proteins involved in heavy metal transporters or regulating the abundance of transcription factors upstream of those transporters. Interestingly, heavy-metal-induced stress also results in protein denaturation that aggravates the oxidative stress in plants [24]. Thus, further studies are essentially needed to investigate whether *Sl1* could degrade denatured proteins to relieve cell oxidative stress. Moreover, it will also be interesting to study whether *Sl1* can coordinate with autophagy to clear denatured proteins [61].

Plants do not have a Cd-selective transporter, therefore, Cd absorption happens through plasma membrane transporters that also take up other divalent cations [62]. Thus, it is indeed difficult to reduce plant Cd accumulation without compromising plant growth since the majority of well-known Cd transporters also transport various essential micronutrients such as Zn, Fe, Mn, and Cu. Despite the fact that CAXs are mostly Ca^{2+} specific transporters, AtCAX2 and AtCAX4 have been demonstrated to transport various other metals, such as Cd, Zn, and Mn in *Arabidopsis* [58]. Similarly, Fe-specific transporter OsIRT1 has been found to participate in Cd uptake in rice [63]. Thus, suppression of the metal transporter may lead to essential nutrient deficiency and compromised plant growth. In our study, overexpression of *Sl1* in tomato not only decreased Cd accumulation but also suppressed plant growth. Reduced Cd accumulation was associated with decreased expression of several metal transporters such as *CAX3*, *HMA-A*, *HMA-B*, and *IRT1*. Thus, it is possible that decreased plant growth in *Sl1* overexpressing lines could be a consequence of the suppression of the metal transporters under Cd stress. However, reduced growth of the *Sl1* overexpressing lines under control conditions could be attributed to some other reasons such as impaired hormone homeostasis since the expression levels of *CAX3*, *HMA-A*, *HMA-B*, and *IRT1* were not significantly different among *sl1* mutants, wild-type, and *Sl1* overexpressing lines under control conditions. Thus, it would be interesting to explore hormonal involvement in *Sl1*-regulated plant growth and stress tolerance in future studies.

5. Conclusions

In the present study, we characterized an E3 ubiquitin ligase Sl1 in tomato, which is located in plasma membranes and highly expressed in roots under Cd stress. For functional characterization of Sl1, we generated knockout lines and overexpressing lines of *Sl1* in tomato. The parameters of chlorophyll fluorescence and content of H_2O_2 demonstrated that *Sl1* overexpressing lines suffered less photosystem damage and oxidative stress compared with wild-type and mutants, suggesting that *Sl1* overexpressing lines are resistant to Cd stress, while mutant lines are sensitive to Cd stress. Moreover, Sl1 positively regulated antioxidant enzyme activities and negatively mediated gene expression associated with heavy metal transportation. Thus, the current study unveils a novel role of an E3 ubiquitin ligase Sl1 in tomato that may have potential implications in enhancing heavy metal tolerance in plants. However, identification of the substrate protein of Sl1 needs further study to precisely verify its association with heavy metal transportation.

Supplementary Materials: The following supporting information can be downloaded at: https://www.mdpi.com/article/10.3390/antiox11030456/s1, Figure S1: Generation of *Sl1* mutant lines and overexpression lines; Figure S2: Structure analysis of Sl1 protein and histochemical staining of Cd accumulation of tomato root tips; Table S1: The primers used for qRT-PCR. Supplementary material related to this article can be found in the online version.

Author Contributions: J.Z. and G.J.A. designed the research; C.-X.L., T.Y., H.Z. and Z.-Y.Q. performed the experiments; C.-X.L. and J.Z. analyzed the data; J.Z., C.-X.L. and G.J.A. wrote the manuscript. All authors have read and agreed to the published version of the manuscript.

Funding: This work was supported by the National Key Research and Development Program of China (2019YFD1000300), the National Natural Science Foundation of China (31922078 and 31872089), the Starry Night Science Fund of Zhejiang University Shanghai Institute for Advanced Study (SN-ZJU-SIAS-0011), the National Natural Science Foundation of China (31950410555), and the Ministry of Science and Technology of the People's Republic of China (QNJ20200226001, QNJ2021026001).

Institutional Review Board Statement: Not applicable.

Informed Consent Statement: Not applicable.

Data Availability Statement: Data is contained within the article and Supplementary Materials.

Conflicts of Interest: The authors declare no conflict of interest.

References

1. Haider, F.U.; Liqun, C.; Coulter, J.A.; Cheema, S.A.; Wu, J.; Zhang, R.; Wenjun, M.; Farooq, M. Cadmium toxicity in plants: Impacts and remediation strategies. *Ecotoxicol. Environ. Saf.* **2021**, *211*, 111887. [CrossRef] [PubMed]
2. Clemens, S.; Ma, J.F. Toxic heavy metal and metalloid accumulation in crop plants and foods. *Annu. Rev. Plant Biol.* **2016**, *67*, 489–512. [CrossRef] [PubMed]
3. Zhang, G.; Fukami, M.; Sekimoto, H. Genotypic differences in effects of cadmium on growth and nutrient compositions in wheat. *J. Plant Nutr.* **2000**, *23*, 1337–1350. [CrossRef]
4. Clemens, S.; Aarts, M.G.M.; Thomine, S.; Verbruggen, N. Plant science: The key to preventing slow cadmium poisoning. *Trends Plant Sci.* **2013**, *18*, 92–99. [CrossRef] [PubMed]
5. Rizwan, M.; Ali, S.; Adrees, M.; Ibrahim, M.; Tsang, D.C.W.; Zia-ur-Rehman, M.; Zahir, Z.A.; Rinklebe, J.; Tack, F.M.G.; Ok, Y.S. A critical review on effects, tolerance mechanisms and management of cadmium in vegetables. *Chemosphere* **2017**, *182*, 90–105. [CrossRef]
6. Song, Y.; Jin, L.; Wang, X. Cadmium absorption and transportation pathways in plants. *Int. J. Phytoremediation* **2017**, *19*, 133–141. [CrossRef]
7. Thomine, S.; Lelièvre, F.; Debarbieux, E.; Schroeder, J.I.; Barbier-Brygoo, H. AtNRAMP3, a multispecific vacuolar metal transporter involved in plant responses to iron deficiency. *Plant J.* **2003**, *34*, 685–695. [CrossRef]
8. Thomine, S.; Wang, R.; Ward, J.M.; Crawford, N.M.; Schroeder, J.I. Cadmium and iron transport by members of a plant metal transporter family in *Arabidopsis* with homology to Nramp genes. *Proc. Natl. Acad. Sci. USA* **2000**, *97*, 4991–4996. [CrossRef]
9. Hirschi, K.D. Expression of *Arabidopsis CAX1* in tobacco: Altered calcium homeostasis and increased stress sensitivity. *Plant Cell* **1999**, *11*, 2113–2122. [CrossRef]
10. Rea, P.A.; Li, Z.S.; Lu, Y.P.; Drozdowicz, Y.M.; Martinoia, E. From vacuolar GS-X pumps to multispecific ABC transporters. *Annu. Rev. Plant Physiol. Plant Mol. Biol.* **1998**, *49*, 727–760. [CrossRef]
11. Shigaki, T.; Pittman, J.K.; Hirschi, K.D. Manganese specificity determinants in the *Arabidopsis* metal/H^+ antiporter CAX2. *J. Biol. Chem.* **2003**, *278*, 6610–6617. [CrossRef] [PubMed]
12. Blaudez, D.; Kohler, A.; Martin, F.; Sanders, D.; Chalot, M. Poplar metal tolerance protein 1 confers zinc tolerance and is an oligomeric vacuolar zinc transporter with an essential leucine zipper motif. *Plant Cell* **2003**, *15*, 2911–2928. [CrossRef] [PubMed]
13. Zulfiqar, U.; Ayub, A.; Hussain, S.; Waraich, E.A.; El-Esawi, M.A.; Ishfaq, M.; Ahmad, M.; Ali, N.; Maqsood, M.F. Cadmium toxicity in plants: Recent progress on morpho-physiological effects and remediation strategies. *J. Soil Sci. Plant Nutr.* **2021**, *22*, 1–58. [CrossRef]
14. Choppala, G.; Saifullah; Bolan, N.; Bibi, S.; Iqbal, M.; Rengel, Z.; Kunhikrishnan, A.; Ashwath, N.; Ok, Y.S. Cellular mechanisms in higher plants governing tolerance to cadmium toxicity. *Crit. Rev. Plant Sci.* **2014**, *33*, 374–391. [CrossRef]
15. Alyemeni, M.N.; Ahanger, M.A.; Wijaya, L.; Alam, P.; Bhardwaj, R.; Ahmad, P. Selenium mitigates cadmium-induced oxidative stress in tomato (*Solanum lycopersicum* L.) plants by modulating chlorophyll fluorescence, osmolyte accumulation, and antioxidant system. *Protoplasma* **2018**, *255*, 459–469. [CrossRef]
16. Sharma, S.S.; Dietz, K.-J. The relationship between metal toxicity and cellular redox imbalance. *Trends Plant Sci.* **2009**, *14*, 43–50. [CrossRef]
17. Hernández, L.E.; Sobrino-Plata, J.; Montero-Palmero, M.B.; Carrasco-Gil, S.; Flores-Cáceres, M.L.; Ortega-Villasante, C.; Escobar, C. Contribution of glutathione to the control of cellular redox homeostasis under toxic metal and metalloid stress. *J. Exp. Bot.* **2015**, *66*, 2901–2911. [CrossRef]
18. Ahammed, G.J.; Wu, M.; Wang, Y.; Yan, Y.; Mao, Q.; Ren, J.; Ma, R.; Liu, A.; Chen, S. Melatonin alleviates iron stress by improving iron homeostasis, antioxidant defense and secondary metabolism in cucumber. *Sci. Hortic.* **2020**, *265*, 109205. [CrossRef]
19. Ahammed, G.J.; Li, C.-X.; Li, X.; Liu, A.; Chen, S.; Zhou, J. Overexpression of tomato RING E3 ubiquitin ligase gene *SlRING1* confers cadmium tolerance by attenuating cadmium accumulation and oxidative stress. *Physiol. Plant.* **2021**, *173*, 449–459. [CrossRef]
20. Hasan, M.K.; Ahammed, G.J.; Sun, S.; Li, M.; Yin, H.; Zhou, J. Melatonin inhibits cadmium translocation and enhances plant tolerance by regulating sulfur uptake and assimilation in *Solanum lycopersicum* L. *J. Agric. Food Chem.* **2019**, *67*, 10563–10576. [CrossRef]
21. Ahmad, P.; Jaleel, C.A.; Salem, M.A.; Nabi, G.; Sharma, S. Roles of enzymatic and nonenzymatic antioxidants in plants during abiotic stress. *Crit. Rev. Biotechnol.* **2010**, *30*, 161–175. [CrossRef] [PubMed]
22. Ahammed, G.J.; Xu, W.; Liu, A.; Chen, S. Endogenous melatonin deficiency aggravates high temperature-induced oxidative stress in *Solanum lycopersicum* L. *Environ. Exp. Bot.* **2019**, *161*, 303–311. [CrossRef]
23. Cobbett, C.; Goldsbrough, P. Phytochelatins and metallothioneins: Roles in heavy metal detoxification and homeostasis. *Annu. Rev. Plant Biol.* **2002**, *53*, 159–182. [CrossRef] [PubMed]
24. Hasan, M.K.; Cheng, Y.; Kanwar, M.K.; Chu, X.-Y.; Ahammed, G.J.; Qi, Z.-Y. Responses of plant proteins to heavy metal stress—A review. *Front. Plant Sci.* **2017**, *8*, 01492. [CrossRef] [PubMed]
25. Qi, Z.-Y.; Ahammed, G.J.; Jiang, C.-Y.; Li, C.-X.; Zhou, J. The E3 ubiquitin ligase gene *SlRING1* is essential for plant tolerance to cadmium stress in *Solanum lycopersicum*. *J. Biotechnol.* **2020**, *324*, 239–247. [CrossRef]
26. Xu, F.-Q.; Xue, H.-W. The ubiquitin-proteasome system in plant responses to environments. *Plant Cell Environ.* **2019**, *42*, 2931–2944. [CrossRef]

27. Hua, Z.; Vierstra, R.D. The cullin-RING ubiquitin-protein ligases. *Annu. Rev. Plant Biol.* **2011**, *62*, 299–334. [CrossRef]
28. Bosu, D.R.; Kipreos, E.T. Cullin-RING ubiquitin ligases: Global regulation and activation cycles. *Cell Div.* **2008**, *3*, 7. [CrossRef]
29. Stone, S.L. The role of ubiquitin and the 26S proteasome in plant abiotic stress signaling. *Front. Plant Sci.* **2014**, *5*, 00135. [CrossRef]
30. Guerra, D.; Mastrangelo, A.M.; Lopez-Torrejon, G.; Marzin, S.; Schweizer, P.; Stanca, A.M.; del Pozo, J.C.; Cattivelli, L.; Mazzucotelli, E. Identification of a protein network interacting with TdRF1, a wheat RING ubiquitin ligase with a protective role against cellular dehydration. *Plant Physiol.* **2011**, *158*, 777–789. [CrossRef]
31. Shu, K.; Yang, W. E3 ubiquitin ligases: Ubiquitous actors in plant development and abiotic stress responses. *Plant Cell Physiol.* **2017**, *58*, 1461–1476. [CrossRef] [PubMed]
32. Kraft, E.; Stone, S.L.; Ma, L.; Su, N.; Gao, Y.; Lau, O.-S.; Deng, X.-W.; Callis, J. Genome analysis and functional characterization of the E2 and RING-type E3 ligase ubiquitination enzymes of *Arabidopsis*. *Plant Physiol.* **2005**, *139*, 1597–1611. [CrossRef] [PubMed]
33. Hotton, S.K.; Callis, J. Regulation of cullin RING ligases. *Annu. Rev. Plant Biol.* **2008**, *59*, 467–489. [CrossRef]
34. Lim, S.D.; Hwang, J.G.; Han, A.R.; Park, Y.C.; Lee, C.; Ok, Y.S.; Jang, C.S. Positive regulation of rice RING E3 ligase *OsHIR1* in arsenic and cadmium uptakes. *Plant Mol. Biol.* **2014**, *85*, 365–379. [CrossRef] [PubMed]
35. Fillatti, J.J.; Kiser, J.; Rose, R.; Comai, L. Efficient transfer of a glyphosate tolerance gene into tomato using a binary *Agrobacterium Tumefaciens* vector. *Bio/technology* **1987**, *5*, 726–730. [CrossRef]
36. Zhang, Y.; Yang, C.; Li, Y.; Zheng, N.; Chen, H.; Zhao, Q.; Gao, T.; Guo, H.; Xie, Q. SDIR1 Is a RING finger E3 ligase that positively regulates stress-responsive abscisic acid signaling in *Arabidopsis*. *Plant Cell* **2007**, *19*, 1912–1929. [CrossRef]
37. Liao, Y.; Tian, M.; Zhang, H.; Li, X.; Wang, Y.; Xia, X.; Zhou, J.; Zhou, Y.; Yu, J.; Shi, K.; et al. Salicylic acid binding of mitochondrial alpha-ketoglutarate dehydrogenase E2 affects mitochondrial oxidative phosphorylation and electron transport chain components and plays a role in basal defense against *tobacco mosaic virus* in tomato. *New Phytol.* **2015**, *205*, 1296–1307. [CrossRef]
38. Chi, C.; Xu, X.; Wang, M.; Zhang, H.; Fang, P.; Zhou, J.; Xia, X.; Shi, K.; Zhou, Y.; Yu, J. Strigolactones positively regulate abscisic acid-dependent heat and cold tolerance in tomato. *Hortic. Res.* **2021**, *8*, 237. [CrossRef]
39. Willekens, H.; Chamnongpol, S.; Davey, M.; Schraudner, M.; Langebartels, C.; Van Montagu, M.; Inzé, D.; Van Camp, W. Catalase is a sink for H_2O_2 and is indispensable for stress defence in C3 plants. *EMBO J.* **1997**, *16*, 4806–4816. [CrossRef]
40. Xia, X.-J.; Wang, Y.-J.; Zhou, Y.-H.; Tao, Y.; Mao, W.-H.; Shi, K.; Asami, T.; Chen, Z.; Yu, J.-Q. Reactive oxygen species are involved in brassinosteroid-induced stress tolerance in cucumber. *Plant Physiol.* **2009**, *150*, 801–814. [CrossRef]
41. Li, M.-Q.; Hasan, M.K.; Li, C.-X.; Ahammed, G.J.; Xia, X.-J.; Shi, K.; Zhou, Y.-H.; Reiter, R.J.; Yu, J.-Q.; Xu, M.-X.; et al. Melatonin mediates selenium-induced tolerance to cadmium stress in tomato plants. *J. Pineal Res.* **2016**, *61*, 291–302. [CrossRef] [PubMed]
42. Wang, F.; Chen, X.; Dong, S.; Jiang, X.; Wang, L.; Yu, J.; Zhou, Y. Crosstalk of PIF4 and DELLA modulates CBF transcript and hormone homeostasis in cold response in tomato. *Plant Biotechnol. J.* **2020**, *18*, 1041–1055. [CrossRef] [PubMed]
43. Wang, Y.; Cao, J.-J.; Wang, K.-X.; Xia, X.-J.; Shi, K.; Zhou, Y.-H.; Yu, J.-Q.; Zhou, J. BZR1 mediates brassinosteroid-induced autophagy and nitrogen starvation in tomato. *Plant Physiol.* **2018**, *179*, 671–685. [CrossRef] [PubMed]
44. Jogawat, A.; Yadav, B.; Chhaya; Narayan, O.P. Metal transporters in organelles and their roles in heavy metal transportation and sequestration mechanisms in plants. *Physiol. Plant.* **2021**, *173*, 259–275. [CrossRef] [PubMed]
45. Zhang, H.; Cui, F.; Wu, Y.; Lou, L.; Liu, L.; Tian, M.; Ning, Y.; Shu, K.; Tang, S.; Xie, Q. The RING finger ubiquitin E3 ligase SDIR1 targets SDIR1-INTERACTING PROTEIN1 for degradation to modulate the salt stress response and ABA signaling in *Arabidopsis*. *Plant Cell* **2015**, *27*, 214–227. [CrossRef]
46. Dong, D.; Jing, C.; Li, Q.; Cao, Y.; Chen, C.; Yu, L.; Dong, Y.; Jin, X. *SlUPS*, a U-box type E3 ubiquitin ligase gene of *Lycopersicon esculentum* Mill., confers the capacity of cadmium stress resistance in *Arabidopsis*. *N. Zeal. J. Crop Hortic. Sci.* **2021**, *50*, 1–16. [CrossRef]
47. Lyzenga, W.J.; Stone, S.L. Abiotic stress tolerance mediated by protein ubiquitination. *J. Exp. Bot.* **2011**, *63*, 599–616. [CrossRef]
48. Oh, T.R.; Yu, S.G.; Yang, H.W.; Kim, J.H.; Kim, W.T. AtKPNB1, an *Arabidopsis* importin-β protein, is downstream of the RING E3 ubiquitin ligase AtAIRP1 in the ABA-mediated drought stress response. *Planta* **2020**, *252*, 93. [CrossRef]
49. Ryu, M.Y.; Cho, S.K.; Kim, W.T. The *Arabidopsis* C3H2C3-type RING E3 ubiquitin ligase AtAIRP1 is a positive regulator of an abscisic acid-dependent response to drought stress. *Plant Physiol.* **2010**, *154*, 1983–1997. [CrossRef]
50. Shin, L.-J.; Lo, J.-C.; Chen, G.-H.; Callis, J.; Fu, H.; Yeh, K.-C. IRT1 DEGRADATION FACTOR1, a RING E3 ubiquitin ligase, regulates the degradation of IRON-REGULATED TRANSPORTER1 in *Arabidopsis*. *Plant Cell* **2013**, *25*, 3039–3051. [CrossRef]
51. Zhang, X.; Wang, N.; Chen, P.; Gao, M.; Liu, J.; Wang, Y.; Zhao, T.; Li, Y.; Gai, J. Overexpression of a soybean ariadne-like ubiquitin ligase gene GmARI1 enhances aluminum tolerance in *Arabidopsis*. *PLoS ONE* **2014**, *9*, e111120. [CrossRef] [PubMed]
52. Hasanuzzaman, M.; Bhuyan, M.H.M.B.; Anee, T.I.; Parvin, K.; Nahar, K.; Mahmud, J.A.; Fujita, M. Regulation of ascorbate-glutathione pathway in mitigating oxidative damage in plants under abiotic stress. *Antioxidants* **2019**, *8*, 384. [CrossRef] [PubMed]
53. Dvořák, P.; Krasylenko, Y.; Zeiner, A.; Šamaj, J.; Takáč, T. Signaling toward reactive oxygen species-scavenging enzymes in plants. *Front. Plant Sci.* **2021**, *11*, 618835. [CrossRef] [PubMed]
54. Jozefczak, M.; Remans, T.; Vangronsveld, J.; Cuypers, A. Glutathione is a key player in metal-induced oxidative stress defenses. *Int. J. Mol. Sci.* **2012**, *13*, 3145–3175. [CrossRef]
55. Seth, C.S.; Remans, T.; Keunen, E.; Jozefczak, M.; Gielen, H.; Opdenakker, K.; Weyens, N.; Vangronsveld, J.; Cuyoers, A. Phytoextraction of toxic metals: A central role for glutathione. *Plant Cell Environ.* **2012**, *35*, 334–346. [CrossRef]

56. Maghrebi, M.; Baldoni, E.; Lucchini, G.; Vigani, G.; Valè, G.; Sacchi, G.A.; Nocito, F.F. Analysis of Cadmium Root Retention for Two Contrasting Rice Accessions Suggests an Important Role for *OsHMA2*. *Plants* **2021**, *10*, 806. [CrossRef]
57. Liu, H.; Zhao, H.; Wu, L.; Liu, A.; Zhao, F.-J.; Xu, W. Heavy metal ATPase 3 (HMA3) confers cadmium hypertolerance on the cadmium/zinc hyperaccumulator *Sedum plumbizincicola*. *New Phytol.* **2017**, *215*, 687–698. [CrossRef]
58. Huang, X.; Duan, S.; Wu, Q.; Yu, M.; Shabala, S. Reducing Cadmium Accumulation in Plants: Structure–Function Relations and Tissue-Specific Operation of Transporters in the Spotlight. *Plants* **2020**, *9*, 223. [CrossRef]
59. Śliwa-Cebula, M.; Kaszycki, P.; Kaczmarczyk, A.; Nosek, M.; Lis-Krzyścin, A.; Miszalski, Z. The Common Ice Plant (*Mesembryanthemum crystallinum* L.)–Phytoremediation Potential for Cadmium and Chromate-Contaminated Soils. *Plants* **2020**, *9*, 1230. [CrossRef]
60. Kim, J.H.; Lee, J.E.; Jang, C.S. Regulation of *Oryza sativa* molybdate transporter1;3 degradation via RING finger E3 ligase OsAIR3. *J. Plant Physiol.* **2021**, *264*, 153484. [CrossRef]
61. Zhou, X.-M.; Zhao, P.; Wang, W.; Zou, J.; Cheng, T.-H.; Peng, X.-B.; Sun, M.-X. A comprehensive, genome-wide analysis of autophagy-related genes identified in tobacco suggests a central role of autophagy in plant response to various environmental cues. *DNA Res.* **2015**, *22*, 245–257. [CrossRef] [PubMed]
62. Zakaria, Z.; Zulkafflee, N.S.; Mohd Redzuan, N.A.; Selamat, J.; Ismail, M.R.; Praveena, S.M.; Tóth, G.; Abdull Razis, A.F. Understanding Potential Heavy Metal Contamination, Absorption, Translocation and Accumulation in Rice and Human Health Risks. *Plants* **2021**, *10*, 1070. [CrossRef] [PubMed]
63. Takahashi, R.; Ito, M.; Kawamoto, T. The Road to Practical Application of Cadmium Phytoremediation Using Rice. *Plants* **2021**, *10*, 1926. [CrossRef] [PubMed]

antioxidants

MDPI

Article

Co-Application of 24-Epibrassinolide and Titanium Oxide Nanoparticles Promotes *Pleioblastus pygmaeus* Plant Tolerance to Cu and Cd Toxicity by Increasing Antioxidant Activity and Photosynthetic Capacity and Reducing Heavy Metal Accumulation and Translocation

Abolghassem Emamverdian [1,2,*], Yulong Ding [1,2], James Barker [3], Guohua Liu [1,2,*], Mirza Hasanuzzaman [4,*], Yang Li [5], Muthusamy Ramakrishnan [1,2] and Farzad Mokhberdoran [1]

1 Co-Innovation Center for Sustainable Forestry in Southern China, Nanjing Forestry University, Nanjing 210037, China; ylding@njfu.com.cn (Y.D.); ramky@njfu.edu.cn (M.R.); mfarzad649@hotmail.com (F.M.)

2 Bamboo Research Institute, Nanjing Forestry University, Nanjing 210037, China

3 School of Life Sciences, Pharmacy and Chemistry, Kingston University, Kingston-upon-Thames, Surrey KT1 2EE, UK; j.barker@kingston.ac.uk

4 Department of Agronomy, Faculty of Agriculture, Sher-e-Bangla Agricultural University, Dhaka 1207, Bangladesh

5 Department of Mathematical Sciences, Florida Atlantic University, Boca Raton, FL 33431, USA; yangli@fau.edu

* Correspondence: emamverdiyan@njfu.edu.cn (A.E.); ghliu@njfu.edu.cn (G.L.); mhzsauag@yahoo.com (M.H.); Tel.: +86-1585-0741-241 (A.E.)

Citation: Emamverdian, A.; Ding, Y.; Barker, J.; Liu, G.; Hasanuzzaman, M.; Li, Y.; Ramakrishnan, M.; Mokhberdoran, F. Co-Application of 24-Epibrassinolide and Titanium Oxide Nanoparticles Promotes *Pleioblastus pygmaeus* Plant Tolerance to Cu and Cd Toxicity by Increasing Antioxidant Activity and Photosynthetic Capacity and Reducing Heavy Metal Accumulation and Translocation. *Antioxidants* **2022**, *11*, 451. https://doi.org/10.3390/antiox11030451

Academic Editor: Nafees A. Khan

Received: 25 January 2022
Accepted: 20 February 2022
Published: 24 February 2022

Abstract: The integrated application of nanoparticles and phytohormones was explored in this study as a potentially eco-friendly remediation strategy to mitigate heavy metal toxicity in a bamboo species (*Pleioblastus pygmaeus*) by utilizing titanium oxide nanoparticles (TiO_2-NPs) and 24-epibrassinolide (EBL). Hence, an in vitro experiment was performed to evaluate the role of 100 µM TiO_2 NPs and 10^{-8} M 24-epibrassinolide individually and in combination under 100 µM Cu and Cd in a completely randomized design using four replicates. Whereas 100 µM of Cu and Cd reduced antioxidant activity, photosynthetic capacity, plant tolerance, and ultimately plant growth, the co-application of 100 µM TiO_2 NPs and 10^{-8} M EBL+ heavy metals (Cu and Cd) resulted in a significant increase in plant antioxidant activity (85%), nonenzymatic antioxidant activities (47%), photosynthetic pigments (43%), fluorescence parameters (68%), plant growth (39%), and plant tolerance (41%) and a significant reduction in the contents of malondialdehyde (45%), hydrogen peroxide (36%), superoxide radical (62%), and soluble protein (28%), as well as the percentage of electrolyte leakage (49%), relative to the control. Moreover, heavy metal accumulation and translocation were reduced by TiO_2 NPs and EBL individually and in combination, which could improve bamboo plant tolerance.

Keywords: toxic metals/metalloid; nanoparticles; phytohormones; phytoremediation; reactive oxygen species

1. Introduction

In recent decades, increasing anthropogenic activities have led to increases in greenhouse gases in the natural environment, and chemical fertilization has led to increases in heavy metal contamination in forestland and agricultural soils, deleteriously contributing to global climate change [1]. Many reports have shown that heavy metals pose a major threat to agricultural land, animals, and plants, which can influence the human food chain, leading to negative effects on human health [2,3]. Copper (Cu) and cadmium (Cd) have been mentioned as being the most abundant toxic metals in Chinese farmland soils [4].

While there is no evidence of the biological activity of Cd in the plant growth process, trace amounts of Cu could have a positive impact as a dietary nutrient on plant growth; however, extreme Cu levels induce plant toxicity [5]. Cu, as a trace element, can enhance photosynthetic efficiency, such as electron transport. Additionally, Cu regulates structural proteins involved in cell wall metabolism and can elevate mitochondrial respiration to produce energy [6]. Conversely, surplus concentrations of Cu in the form of Cu^{2+} are responsible for oxidative stress due to the generation of reactive oxygen species (ROS) compounds [7], which can result in a reduction in plant growth with an altered functionality of the cell membrane, limitation of enzyme activities, and depression of photosynthetic efficiency, ultimately leading to plant death [8]. Cd, a nonessential element with high toxicity, is known as the most dispersed element in soils and irrigation water, which has a destructive influence on plant and human life [9]. Excess Cd in soil and the absorption of cadmium by plants increase ROS production, such as free radicals, which are the main factors in the initiation of oxidative stress in plants [9]. Cadmium has a damaging impact on plant cell functions and the metabolic pathways involved in the production of lipids, proteins, and nucleic acids. Cd injures the cell membrane, which leads to lipoperoxidation and oxidative toxicity in plants. It has been demonstrated that cadmium reduces the plant defense system with a reduction in antioxidant activity capacity. This phenomenon finally reduces plant photosynthesis and inhibits plant growth and development [9].

Nanoparticles with unique structures and sizes (1 to 100 nm) [10] have been observed to increase plant nutrients and crop production [11]. The diverse surface-to-volume ratios of nanoparticles could differ from their bulk counterparts [12]. Recently, some studies have reported that titanium nanoparticles (TiO_2) have the ability to increase plant growth under metal stress [13–17]. Therefore, we suggest that titanium could be a good material to reduce plant stress. Brassinosteroids (BRs) are a new phytohormone and belong to the polyhydroxy steroidal group. There are 70 types of BRs in plants. Among them, 24-epibrassinolide (EBL) is known as the top bioactive BR that can promote plant growth under stressful conditions [18]. In addition, 24-epibrassinolide EBL induces antioxidant activity, plant photosynthesis, seed yield, and oxidative production under stressful conditions [19–21]. It has been indicated that the interaction of EBL with other cellular molecules can enhance signaling efficiency within the plant defense grid under stress conditions [22]. This phenomenon can boost antioxidant capacity in the face of multiple stressful factors, such as HMs [23]. This research study represents the individual and co-application of TiO_2-NPs and EBL, as well as the investigation of their role in the alleviation of Cu and Cd toxicity in bamboo plants with an emphasis on antioxidant, photosynthetic, and plant growth parameters.

Bamboo (*Bambusoideae*) species occupy the largest portion of Chinese farmland (6 million hectares) [24,25]. This fast-growing plant provides nutrient sources for local family livelihoods in southern and western China [26]. *Pleioblastus pygmaeus* is a suitable species for landscape purposes, with a characteristic height of 30–50 cm. *Pleioblastus pygmaeus* originated in Japan but was transferred to China in the early 20th century. A desirable condition of this plant for this experiment was its adaptation to basic (alkaline), acidic, and neutral soils [27]. Conversely, the excess of heavy metals (frequently Cu and Cd) caused by anthropogenic activities has become a major dilemma for agricultural and forestry soils in this area [4], which can influence bamboo plant growth and development. Hence, it is essential to find appropriate biologic materials to reduce soil toxicity and increase plant tolerance under heavy metal toxicity. Therefore, we selected two applications of TiO_2 NPs and 24-epibrassinolide, individually and in combination, against heavy metal toxicity, which could aid in understanding the involved mechanisms in the combined application of nanoparticles and phytohormones against heavy metal toxicity. To our knowledge, this is the first comprehensive study to investigate the combination of TiO_2 NPs and EBL in the amelioration of Cu and Cd toxicity in bamboo species. Therefore, in this paper, we aim to investigate the impact of TiO_2 NPs and EBL on enhancing plant tolerance under heavy metal toxicity with an emphasis on antioxidant and nonantioxidant enzyme capacity, ROS production, photosynthesis, and growth indices under Cu and Cd.

2. Materials and Methods

2.1. Plant Material and In Vitro Conditions

This research study was performed under in vitro conditions in a plant tissue culture laboratory using MS medium (Murashige and Skoog, 1962) [28] consisting of 6-benzylaminopurine (6-BA) (4 mL), micronutrients (10 mL), macronutrients (100 mL), kinetin (KT) (0.5 mL), sucrose (30 g) and agar (8 g) at pH 5.8 ± 0.1. For this purpose, a completely randomized design (CRD) was employed that contained 100 μM TiO_2 NPs and 10^{-8} M 24 epibrassinolide individually and in combination with 100 μM Cu as well as 100 μM Cd in four replications (Table 1). We adjusted the pH value in MS to 5.8 for two reasons: firstly, to optimize nutrient absorption, the availability of the nutrients to the plants was optimum at pH 5.8; and secondly, the preparation of the gelling of the agar-solidified medium should be completed at ca. pH 5.8. To proliferate bamboo roots, young shoots (10 mm long nodal explants) were planted in MS medium supplemented with pyridoxine (3 μM), nicotinic acid (4 μM), thiamine–HCl (1.2 μM), myo-inositol (0.6 mM), 30 g L^{-1} sucrose, and 0.1 mg L^{-1} indole-3-acetic acid (IAA) as a regulator hormone involved in plant growth. The appropriate amount of each treatment (100 μM TiO_2 NPs and 10^{-8} M 24-epibrassinolide) was mixed in 1 L MS medium, adjusted to pH 5.8 ± 0.1, and then applied to 8–10 g/L agar. The solution was placed in 60 mm diameter glass petri dishes containing 100 mL of culture, and sterilization of the intended MS medium was conducted in an autoclave (HiClave HVE-50, ZEALWAY-USA, Delaware, DE, USA) at the optimum temperature of 110 °C for 40 min. The dishes were transferred to an Air Tech incubation hood with ultraviolet sterilization with white fluorescent lamps (wavelength between 10 and 420 nm) at a temperature of 25 °C for 4 h. In the final step, the plantlet treatments were preserved as research materials in a controlled tissue culture chamber with fluorescent lamps (white) at a wavelength between 10 and 420 nm. In terms of temperature, the growth was performed at 17/22 °C in the dark periods and 30/25 °C in the light periods for three weeks.

Table 1. The treatment combinations of the experiment.

Treatments	Concentrations
Control	0
Cu	100 μM Cu
Cd	100 μM Cd
TiO_2	100 μM TiO_2
TiO_2 + Cu	100 μM TiO_2 + 100 μM Cu
TiO_2 + Cd	100 μM TiO_2 + 100 μM Cd
EBL	10^{-8} M EBL
EBL + Cu	10^{-8} M EBL + 100 μM Cu
EBL + Cd	10^{-8} M EBL + 100 μM Cd
TiO_2 + EBL	100 μM TiO_2 + 10^{-8} M EBL
TiO_2 + EBL + Cu	100 μM TiO_2 + 10^{-8} M EBL + 100 μM Cu
TiO_2 + EBL + Cd	100 μM TiO_2 + 10^{-8} M EBL + 100 μM Cd

Titanium nanoparticles were provided by Nanjing Jiancheng Company, Jiangsu Province, China, and consisted of a white powder with a purity of >99% nanotitanium and a diameter of 25 nm. The levels of Cu and Cd were chosen according to the previous studies, which displayed high and low levels of toxicity in bamboo plants [13,14]. Bamboo (*A. pygmaeus*) was selected from local species by the Bamboo Research Institute, which is located at Nanjing Forestry University.

In this research study, biomass and growth indices, including root and shoot dry weight (DW) and shoot length, were quantified. To investigate photosynthesis pigments, total chlorophyll (Chl). Chl a and b, and carotenoid contents were measured. To determine the fluorescence parameters, 5 parameters were recorded, including: (i) actual photochemical efficiency of PSII (φPSII), (ii) maximum photochemical efficiency of PSII (Fv/Fm), (iii) photochemical quenching coefficient (qP), (iv) effective photochemical efficiency of

PSII (Fv′/Fm′), and (v) nonphotochemical quenching (NPQ). Heavy metal accumulation and TiO_2 NP contents were measured in leaves, stems, and roots. Plant defense enzymes and nonenzymatic antioxidants were measured. To assay cell membrane injury, ROS compounds, electron leakage, and malondialdehyde (MDA) content were estimated. Finally, the translocation factor (TF), bioaccumulation factor (BAF), and tolerance index (TI) of the shoots and roots were calculated.

2.2. *Preparation of Samples*

Leaf samples were collected from the different treatments, and then 0.5 g samples were placed in a container and crushed into a powder. An appropriate amount of liquid nitrogen was added to the samples, and the obtained powder was dissolved in PBS (pH 7.2–7.4) at 2–8 °C. The solution was centrifuged at 2500–3500× *g* for 17 min to extract the supernatant, which was kept for use in antioxidant enzyme activity tests.

2.3. *Protective Enzymes*

Superoxide dismutase (SOD, EC: 1.15.1.1) was measured based on the results of photoreduction obtained by nitro blue tetrazolium (NBT), which was conducted using the Zhang method [29]. Peroxidase (POX, EC: 1.11.1.7) was estimated by using the protocol of Upadhyaya [30]. Catalase (CAT, EC: 1.11.1.6) was estimated based on the results of reactions analyzing H_2O_2 at an absorbance of 240 nm, which was estimated by the Aebi protocol [31]. Glutathione reductase (GR, EC: 1.6.4.2) was estimated using the protocol reported by Foyer and Halliwell [32] with some modifications. Ascorbate peroxidase (APX, EC: 1.11.1.11) was measured using the Nakano and Asada method [33]. APX antioxidant activity was obtained by recording the reduction in absorbance at 290 nm (coefficient of absorbance at 2.8 mM^{-1} cm^{-1}). Phenylalanine ammonia-lyase (PAL, EC: 4.3.1.5) activity was assessed using the Berner [34] protocol.

2.4. *Assessment of Nonenzymatic Antioxidant Activities (Flavonols, Tocopherols, and Total Phenolics)*

Methanolic Extract Preparation

For this test, 0.5 g of dry leaf sample was dissolved in 4 mL of methanol (80%) and then centrifuged at 7000× *g* for 15 min. The methanolic extract was used for the tests. The total phenolics were measured according to the protocol of Conde [35]. According to this protocol, a 0.1-m methanolic extra was added to 2.5 mL of 10% Folin–Ciocalteu reagent. Then, for neutralization of the obtained mixture of sodium bicarbonate, 7% was added. The final mixture was transferred to a spectrometer machine to measure the total phenolics at an absorbance of 765 nm. The content of flavonol was determined according to the Akkol method [36]. A 0.5-mL methanolic extract was homogenized with 0.4 mL of aluminum chloride (2%) and 1.5 mL of sodium acetate (5%). After preparation of the supernatant, it was kept at room temperature for 2.5 h. The flavonoid content was determined in the supernatant at an absorbance of 445 nm. The content of tocopherol was determined according to the protocol of Kayden [37]. For this purpose, 3 mL ethanol was mixed with 0.1 g of leaf samples, and the soluble solution was then centrifuged at 7000× *g* for 15 min. The obtained mixture was added to 0.1 mL ethanol extract, 0.2 mL bathophenanthroline at a concentration of 0.2%, 0.001 M of 0.2 mL ferric chloride, and 1 mM 0.2 mL phosphoric acid. The content of tocopherol was recorded by measuring the absorbance of the supernatant at 534 nm.

2.5. *Assay of Hydrogen Peroxide (H_2O_2), Malondialdehyde (MDA), Superoxide Radical ($O_2^{\bullet-}$), Soluble Proteins (SP), and Electrolyte Leakage (EL)*

Malondialdehyde is representative of lipid peroxidation, which was measured by the protocol described by Siddiqui [38]. In this experiment, 0.1% trichloroacetic acid (TCA) was used for the homogenization of fresh leaves, after which the sample was centrifuged at 8000× *g* for 25 min. The obtained amount of supernatant was mixed with TCA solution in the range of 20%, which contained 0.5% thiobarbituric acid. In the next process, the soluble solution was kept at 98 °C for 25 min. Then, the soluble solution was kept at room temperature. The final soluble solution was centrifuged a second time at 2000× *g* for 15 min at 5 °C. Finally, to estimate malondialdehyde, the absorbance was determined at 532 nm.

The levels of H_2O_2 were determined using the protocol reported by Patterson [39]. For this study, samples (leaves) in the specified amount of 0.5 g were mixed in a mortar and pestle by adding 10 mL cold acetone. The mixture was centrifuged at 4000× *g* for 25 min. In the next step, titanium chloride at a concentration of 20% in 2 mL of concentrated HCl and 2 mL of ammonia at the specified level of 17 M were added to the supernatant (1 mL). The supernatant was extracted with acetone, which was conducted by the addition of 2 N H_2SO_4 in 10 mL for proper absorbance. To remove immiscible inputs, the mixture was centrifuged again. The absorbance of the supernatant was recorded at 410 nm. The levels of H_2O_2 were determined based on a standard curve, which was created based on the known levels of H_2O_2 and formulated as μmole g^{-1} FM. The soluble protein (SP) levels were assigned according to the protocol of Bradford [40] and measured based on the effect of Coomassie Brilliant Blue (G25) on changes in protein levels. The final data were obtained using a spectrometer machine. The amount of superoxide radical ($O_2^{\bullet-}$) was determined according to the method of Li [41]. According to this protocol, 200 mg leaf tissue samples were mixed with phosphate buffer at pH 7.8 in the amount of 65 mM and then centrifuged at 4000× *g* for 20 min. The supernatant was incubated in 10 mM of hydroxylamine hydrochloride and 65 mM of phosphate buffer (pH = 7.8) for 15 min at 27 °C. In the next step, 7 mM α-naphthylamine plus 17 mM sulfanilamide was added to the mixture, preserved for 25 min and then recorded at an absorbance of 530 nm at 25 °C. Finally, to determine the final rate of $O_2^{\bullet-}$, nitrogen dioxide radicals (NO_2) were applied to generate a standard curve. Electrolyte leakage (EL) was calculated based on the protocol of Valentovic [42]. According to this protocol, 0.3 g of leaf samples were mixed with 15 mL of deionized water. Then, the mixture was kept at the optimum temperature (25 °C) for 2.5 h. In this stage, EC_1 was recorded as the primary electrical conductivity of the mixture. To obtain EC_2 as the secondary electrical conductivity, the samples were transferred to one autoclave and kept at 120 °C for 17 min. At the end of the test, EL was determined based on the following formula:

$$\text{EL (\%)} = EC_1/EC_2 \times 100 \tag{1}$$

2.6. Measurement of Photosynthetic Pigments and Fluorescence Parameters

Photosynthetic pigments, such as Chl a and b, and carotenoid levels were determined according to the protocol of Lichtenthaler and Buschmann [43]. For pre-experiment bamboo samples, ca. 0.5 g was provided, and then the samples were transferred to a mortar with liquid nitrogen. To prepare the liquid sample extract, the obtained powder was mixed with 20 mL of acetone at a specific concentration of 80% at 0 to 5 °C. Then, it was centrifuged at 5000× *g* for 15 min. At the end of the experiment, Chl a, Chl b, and carotenoid levels were determined at absorbances of 663, 645, and 470 nm, respectively. Finally, the levels of Chl and carotenoids were calculated based on the following formulae, which were set in units equal to mg/*g* fresh weight:

$$\text{Chlorophyll a} = 12.25\text{A}663 - 2.79\text{A}647 \tag{2}$$

$$\text{Chlorophyll b} = 21.50\text{A}647 - 5.10\text{A}663 \tag{3}$$

$$\text{Total Chl} = \text{Chl a} + \text{Chl b} \tag{4}$$

$$\text{Carotenoid} = 1000\text{A}470 - 1.82\text{Chl a} - 95.15\ \text{Chl b}/225 \tag{5}$$

A chlorophyll fluorescence imager (CFI) (England) was used to measure fluorescence characteristics, which was conducted under specific dark-adapted conditions for 35 min. To measure the light fluorescence parameters, the fluorescence characteristics were recorded at 700 micromoles m^{-2} s^{-0} in an illumination incubator for actinic light activation. In this study, the main fluorescence indices were: (i) actual photochemical efficiency of

PSll (φPSll); (ii) effective photochemical efficiency of PSll (Fv′/Fm′); (iii) photochemical quenching coefficient (qP); (4) maximum photochemical efficiency of PSll (Fv/Fm); and (5) non-photochemical quenching (NPQ).

2.7. Measurement of TiO_2 NPs and Metal Accumulation in Roots, Stems, and Leaves of Bamboo

Copper (Cu) and cadmium (Cd) contents and titanium accumulation in roots, stems, and leaves of bamboo species were investigated. For their measurement, the samples were cleaned and dried in an oven, and subsequently 70% nitric acid was added to the samples, which were preserved at an optimum temperature of 70 °C for 20 min. Then, the obtained solution was centrifuged at 9000× *g* for 20 min. The contents of Cu and Cd in the plant organs, such as roots, stems, and leaves, were assigned and analyzed using an atomic absorption spectrometry machine (AAS HITACHI, High-Tech Company Tokyo, Japan). In this process, a spectrometer equipped with a furnace of graphite and correction system of the Zeeman-effect background was performed (AAnalyst 800, Perkin Elmer, Norwalk, CT, USA). For determination of the element contents, the different characteristics of the instruments were adjusted. Standardization of the metal was performed using 2.5% nitric acid (spectra scan). For calibration, confirmation of the standard (Perkin Elmer), which contained all of the elements in one inorganic target analyst list (TAL), was performed at optimum intervals in one unattended automatic analysis run.

2.8. Biomass Determination and Shoot Length

Plant biomass was determined by measuring the dry weight of roots and shoots. Firstly, samples (root and shoot treatment) were cleaned and then placed in an oven to remove the water from their surfaces during the process. The samples were fixed at 115 °C for 25 min. To determine the plant dry weight, the samples were maintained at 70 °C for 10 h and then weighed. To determine the shoot length, bamboo plants were measured at the beginning and end of the study. The final data were defined as the difference in plant height.

2.9. Determination of the Tolerance Index (TI), Bioaccumulation Factor (BAF), and Translocation Factor (TF)

To determine the plant tolerance to Cu and Cd toxicity, the indices of the tolerance index (TI), bioaccumulation factor (BAF), and translocation factor (TF) were calculated. This result was obtained based on the method of Souri and Karimi [44] and is considered to denote the efficiency of photoextraction. The formula below is used to find the value of TI, which represents the tolerance index in the shoot and root; TF, which represents the translocation factor; and BAF, which represents the bioaccumulation factor of leaves, stem, and root.

TF (leaves/stem) = the concentration of the co-application of Ti-EBL/heavy metals (Cu, Cd) in the leaves/stem of plants (mg/kg)/the concentration of the co-application of Ti-EBL/heavy metals (Cu, Cd) in the roots of plants (6)

TI shoot/root = the dry weight of plant shoot/root from the co-application of Ti-EBL/heavy metal (Cu, Cd) treatment (g)/the dry weight of plant shoot/root from the control (g) (7)

BF (leaves/stem/root) = concentrations of heavy metals in the leaves or stem or root/concentrations of heavy metals in the medium (8)

2.10. Statistical Analysis

A completely randomized design (CRD) was used in this study, consisting of a 2-way factorial with four replicates. R software was used for ANOVA (analysis of variance), and the Tukey's test was used to determine mean differences, which were conducted at the $p < 0.05$ probability level.

3. Results

3.1. 24-Epibrasinolide and Titanium Oxide Nanoparticles Promote Antioxidant Capacity in Plants under Cu and Cd Toxicity

The data analysis revealed a significant difference between the various treatments of the co-application of EBL–TiO_2 NPs with Cu and Cd ($p < 0.001$). According to the obtained data, TiO_2 and EBL individually could increase antioxidant activity under stress conditions. However, the greatest stimulation of antioxidant activity was observed with the combination of TiO_2 and EBL. The combination of TiO_2 and EBL showed the highest capacity for antioxidant activity stimulation, with a 1.71-fold enhancement of SOD, 1.49-fold increase in POX, 1.76-fold increase in CAT, 1.52-fold increase in APX, 1.73-fold enhancement of GR, and 1.23-fold enhancement of PAL activity in comparison with their control treatments (Figure 1). Conversely, the lowest amount of antioxidant activity was observed with 100 μM Cu and 100 μM Cd, which resulted in 36% and 61% reductions in SOD, 24% and 40% reductions in POX, 39% and 58% reductions in CAT, 50% and 73% reductions in APX, 44% and 59% reductions in GR, and 28% and 36% reductions in PAL activities, respectively, compared with the control treatment. According to these results, we suggest that TiO_2 and EBL individually have the potential to reduce Cu and Cd toxicity, but the combination of TiO_2 and EBL has a larger impact on the amelioration of heavy metal toxicity.

3.2. 24-Epibrasinolide and Titanium Oxide Nanoparticles Reduce Malondialdehyde (MDA), Soluble Proteins (SP), Electrolyte Leakage (EL), Hydrogen Peroxide (H_2O_2), and Superoxide Radicals ($O_2^{\bullet-}$)

Our results showed that the TiO_2 NP and EBL concentrations had the ability to reduce ROS compounds and prevent cell membrane injury. The data analyses showed a significant difference in the co-application of TiO_2 NPs and EBL based on the concentration of Cu and Cd in the indices of MDA, H_2O_2, SP, EL, and $O_2^{\bullet-}$ ($p < 0.001$). The most positive effect of the treatments on heavy metals was related to the combination of TiO_2–EBL with Cu and TiO_2–EBL with Cd which demonstrated 49% and 42% reductions in MDA content, 38% and 33% reductions in H_2O_2 content, 38% and 36% reductions in $O_2^{\bullet-}$ content, 26% and 25% reductions in SP content, and 50% and 44% reductions in EL, respectively. Additionally, the levels of TiO_2 and EBL individually showed a positive role in the amelioration of oxidative stress, which occurred by restraining ROS production, which in turn resulted in protecting the cell membrane against oxidative free radicals. Conversely, the results showed that 100 μM Cu and 100 μM Cd increased the levels of MDA, H_2O_2, $O_2^{\bullet-}$, SP, and EL, with 48% and 60% increases in MDA content, 35% and 49% increases in H_2O_2, 28% and 41% increases in $O_2^{\bullet-}$, 25% and 29% increases in SP, and 50% and 63% increases in EL, respectively, compared with their control treatments (Figure 2).

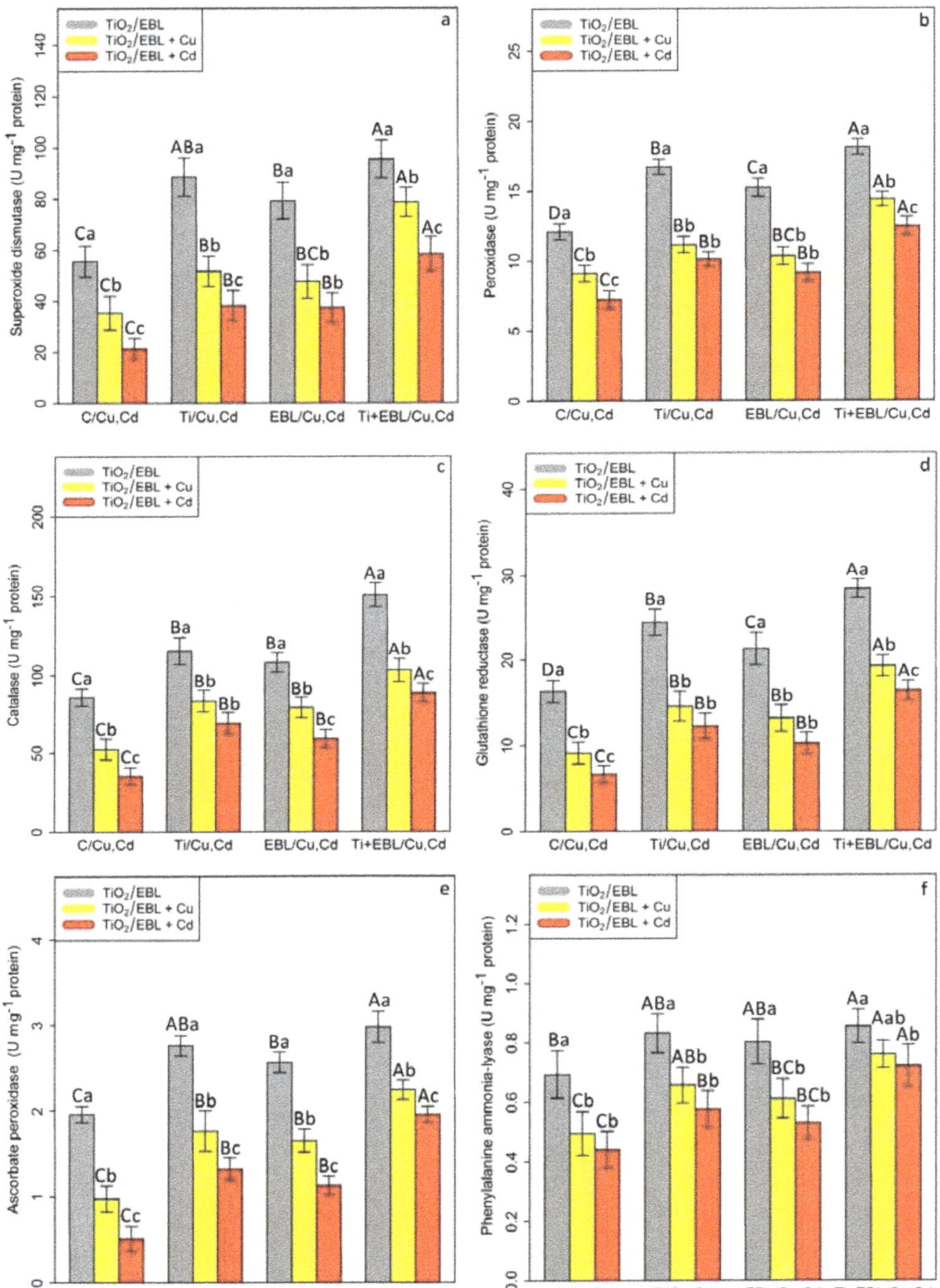

Figure 1. The impact of the co-application of 24-epibrassinolide and titanium oxide nanoparticles individually and combined on antioxidant enzyme activities (superoxide dismutase (SOD) (**a**), peroxidase

(POX) (**b**), catalase (CAT) (**c**), glutathione reductase (GR) (**d**), ascorbate peroxidase (APX) (**e**), and phenylalanine ammonia-lyase (PAL) (**f**)) in bamboo species (*Pleioblastus pygmaeus*) with 100 μM Cu and 100 μM Cd. In this study, 1-year-old branches of *P. pygmaeus* were used as plant treatments together with 100 μM TiO_2 NPs and 10^{-8} M 24-epibrassinolide, individually and in combination with 100 μM Cu and 100 μM Cd using four replications. Planting of the treated bamboo was performed in an Air Tech inoculation hood with fluorescent white lamps and ultraviolet light (wavelengths of 10–400 nm) at 15 °C and 30 °C. The bamboo plants were constantly exposed to excess heavy metals for three weeks. Sampling for the measurement of antioxidant enzyme activity (**a**–**f**) was conducted after three weeks of plant exposure to the co-application of 24-epibrassinolide and titanium oxide nanoparticles under 100 μM Cu and 100 μM Cd. The capital letters ($^{A–C}$) indicate significant differences between treatments of control (C), titanium (Ti), 24-epibrassinolide (EBL), and 24-epibrassinolide involving individual or combined application of titanium oxide nanoparticles (EBL–TiO_2 NPs) under 100 μM Cu and 100 μM Cd (the bars with similar colors), while the lowercase letters ($^{a–c}$) denote statistically significant differences at each concentration of the co-application of EBL and TiO_2 NPs, individually or in combination with 100 μM Cu and 100 μM Cd (the bars with various colors) based on Tukey's test ($p < 0.05$).

3.3. 24-Epibrasinolide and Titanium Oxide Nanoparticles Increase Nonenzymatic Antioxidant Activities (Flavonol, Tocopherol, and Total Phenolics) in Bamboo Species under Cu and Cd Toxicity

The effects of TiO_2 NPs and EBL concentrations on nonenzymatic activity (flavonol, tocopherol, and total phenolics) in the bamboo species revealed a significant difference between the co-application of 24-epibrassinolide and titanium oxide nanoparticles with Cu and Cd ($p < 0.001$). According to the results, the combination of TiO_2-HMs and EBL-HMs significantly increased nonenzymatic antioxidant activities in our bamboo species. However, the greatest increase in nonenzymatic activity under heavy metal stress was related to the combination of TiO_2 –EBL with Cu and TiO_2–EBL with Cd, with a 1.55-fold and 1.51-fold enhancement in flavonols, 1.53-fold and 1.51-fold enhancement in tocopherols, and 1.68-fold and 1.58-fold increase in total phenolics, respectively, in comparison with the control treatment (Figure 3). Conversely, the concentrations of 100 μM Cu and 100 μM Cd clearly reduced nonantioxidant activity, as demonstrated by a 21% and 23% reduction in flavonols, 12% and 24% reduction in tocopherols, and 34% and 28% reduction in total phenolics, respectively, in comparison with the control treatment. We suggest that the combination of TiO_2 and EBL has a positive impact on the reduction in heavy metal toxicity by stimulating nonenzymatic antioxidant activities (flavonols, tocopherols, and total phenolics).

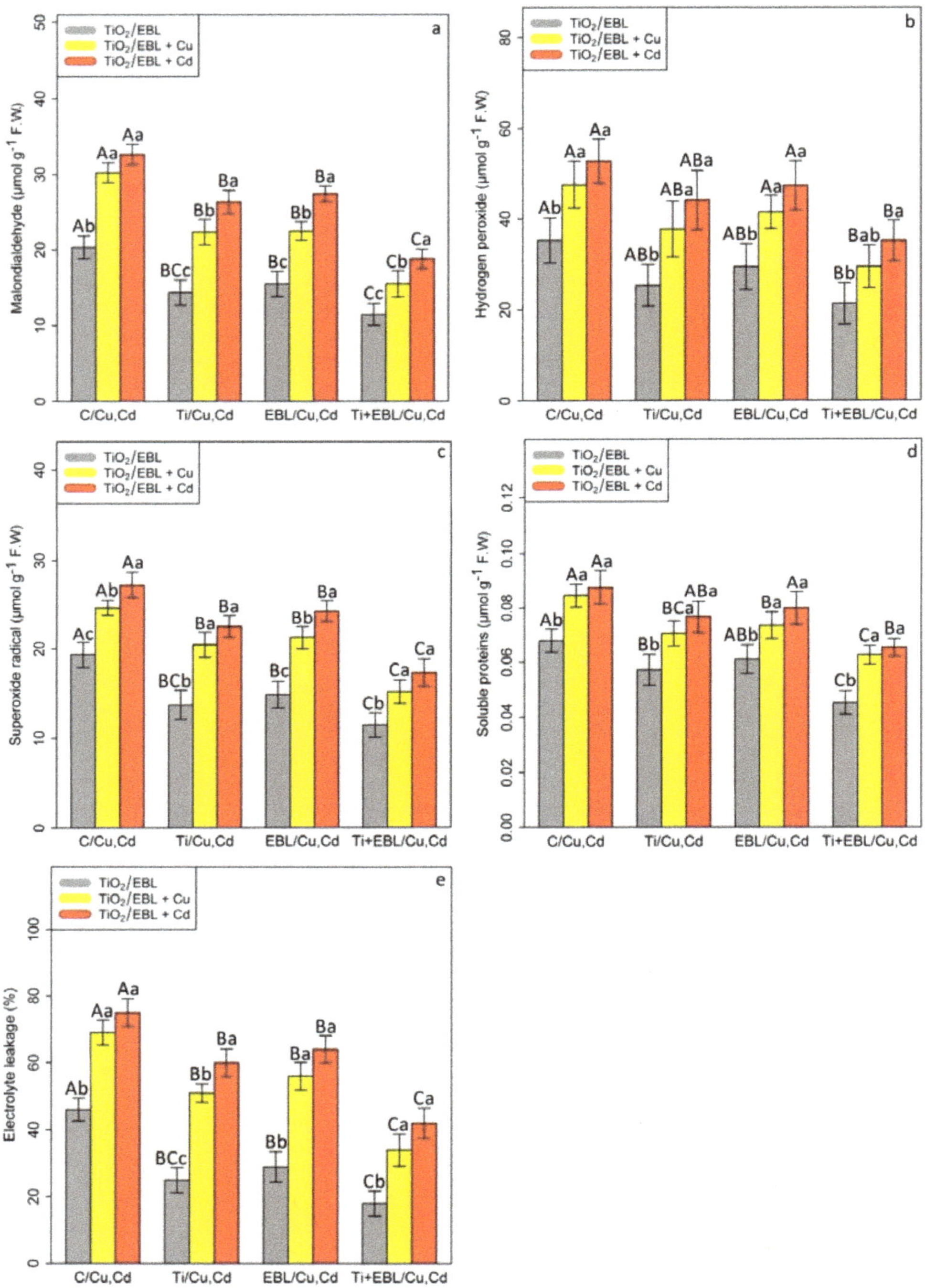

Figure 2. The impact of the co-application of 24-epibrassinolide and titanium oxide nanoparticles individually and combined on malondialdehyde content (MDA) (**a**), hydrogen peroxide (H_2O_2) (**b**), superoxide radical ($O_2^{\bullet-}$) (**c**), soluble proteins (SP) (**d**), and electrolyte leakage (EL) (**e**) in bamboo species (*Pleioblastus pygmaeus*) with 100 µM Cu and 100 µM Cd. In this study, 1-year-old branches of

P. pygmaeus were used as plant treatments together with 100 µM TiO_2 NPs and 10^{-8} M 24-epibrassinolide, individually and in combination with 100 µM Cu and 100 µM Cd using four replications. Planting of the treated bamboo was performed in an Air Tech inoculation hood with fluorescent white lamps and ultraviolet light (wavelengths of 10–400 nm) at 15 °C and 30 °C. The bamboo plants were constantly exposed to excess heavy metals for three weeks. Sampling for the measurement of MDA, H_2O_2, $O_2^{\bullet-}$, SP, and EL (**a**–**e**) was conducted after three weeks of plant exposure to the co-application of 24-epibrassinolide and titanium oxide nanoparticles under 100 µM Cu and 100 µM Cd. The capital letters ($^{A-C}$) indicate significant differences between treatments of control (C), titanium (Ti), 24-epibrassinolide (EBL), and 24-epibrassinolide involving individual or combined application of titanium oxide nanoparticles (EBL–TiO_2 NPs) under 100 µM Cu and 100 µM Cd (the bars with similar colors), while the lowercase letters ($^{a-c}$) denote statistically significant differences at each concentration of the co-application of EBL and TiO_2 NPs, individually or in combination with 100 µM Cu and 100 µM Cd (the bars with various colors) based on Tukey's test ($p < 0.05$).

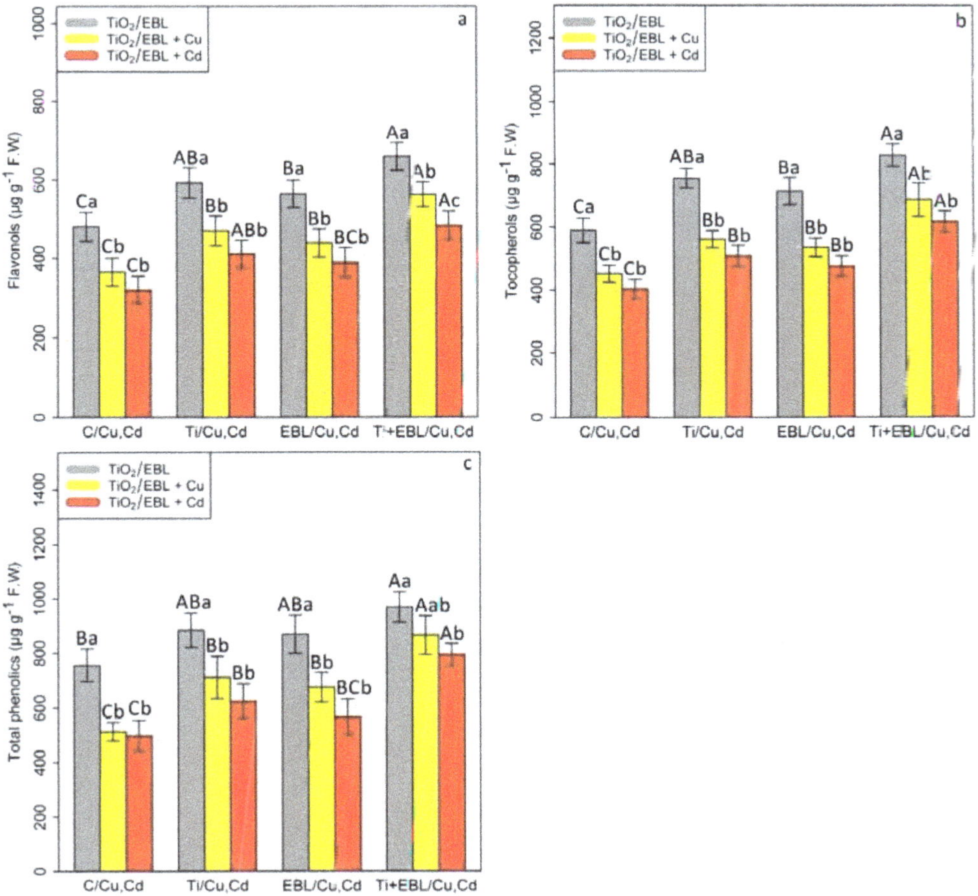

Figure 3. The impact of the co-application of 24-epibrassinolide and titanium oxide nanoparticles individually and combined on nonenzymatic antioxidant activities ((**a**) flavonols, (**b**) tocopherols, (**c**) total phenolics) in bamboo species (*Pleioblastus pygmaeus*) with 100 µM Cu and 100 µM Cd. In this study, 1-year-old branches of P. *pygmaeus* were used as plant treatments together with 100 µM TiO_2 NPs and 10^{-8} M 24-epibrassinolide, individually and in combination with 100 µM Cu and 100 µM Cd.

using four replications. Planting of the treated bamboo was performed in an Air Tech inoculation hood with fluorescent white lamps and ultraviolet light (wavelengths of 10–400 nm) at 15 °C and 30 °C. The bamboo plants were constantly exposed to excess heavy metals for three weeks. Sampling for the measurement of flavonols, tocopherols, and total phenolics (**a–c**) was conducted after three weeks of plant exposure to the co-application of 24-epibrassinolide and titanium oxide nanoparticles under 100 μM Cu and 100 μM Cd. The capital letters ($^{A–C}$) indicate significant differences between treatments of control (C), titanium (Ti), 24-epibrassinolide (EBL), and 24-epibrassinolide involving individual or combined application of titanium oxide nanoparticles (EBL–TiO_2 NPs) under 100 μM Cu and 100 μM Cd (the bars with similar colors), while the lowercase letters (^{a,b}) denote statistically significant differences at each concentration of the co-application of EBL and TiO_2 NPs, individually or in combination with 100 μM Cu and 100 μM Cd (the bars with various colors) based on Tukey's test ($p < 0.05$).

3.4. 24-Epibrassinolide and Titanium Oxide Nanoparticles Improve Photosynthetic Pigments and Fluorescence Parameters in Bamboo Species under Cu and Cd Toxicity

Photosynthetic pigments and fluorescence parameters are important indices in the evaluation of photosynthetic efficiency in different species of plants under stress conditions. The indicators of plant photosynthesis performance, including photosynthetic pigments (Chl a, Chl b, and total Chl, as well carotenoid contents), and fluorescence indices, including the maximum photochemical efficiency of PSII (Fv/Fm), photochemical quenching coefficient (qP), effective photochemical efficiency of PSII (Fv′/Fm′), actual photochemical efficiency of PSII (φPSII), and nonphotochemical quenching (NPQ), were measured. We found a significant difference between the co-application of 24-epibrassinolide and titanium oxide nanoparticles with Cu and Cd ($p < 0.001$). Based on the results, the levels of TiO_2NPs and EBL alone and in combination with heavy metals (Cu and Cd) could increase photosynthetic pigments in bamboo under Cu and Cd. However, the greatest increase in photosynthetic pigments under Cu and Cd was attributed to the co-application of TiO_2–EBL with Cu and the co-application of TiO_2–EBL with Cd, which resulted in 21% and 17% increases in Chl a, 85% and 83% increases in Chl b, 42% and 38% increases in total Chl, and 46% and 39% increases in carotenoid, respectively, in comparison with their control treatments (Table 2). Conversely, the measurement of the fluorescence parameters demonstrated a significant difference between the combination of TiO_2–EBL and Cu and Cd ($p < 0.001$). The data analysis revealed similar results, e.g., in the Chl and carotenoid contents and in the measurement of fluorescence parameters. Therefore, the greatest increase in fluorescence parameters was related to the combination of TiO_2 and EBL, which resulted in a 50% increase in the maximum photochemical efficiency of PSII (Fv/Fm), 41% increase in the photochemical quenching coefficient (qP), 54% increase in the effective photochemical efficiency of PSII (Fv′/Fm′), 56% increase in the actual photochemical efficiency of PSII (φPSII), and 58% increase in nonphotochemical quenching (NPQ) in comparison with their control treatments. We suggest that the combination of TiO_2 NPs and EBL has a strong ability to increase photosynthesis parameters in plants exposed to heavy metal stress (Cu and Cd) (Figure 4).

Table 2. The effect of the co-application of 24-epibrassinolide and titanium oxide nanoparticles individually and combined on photosynthetic pigments (Chl a, Chl b, and total Chl, as well as carotenoid contents) in bamboo species (*Pleioblastus pygmaeus*) with 100 μM Cu and 100 μM Cd.

Treatments	Chl a (mg g $^{-1}$ F.w.)	Chl b (mg g $^{-1}$ F.w.)	Chl a + b (mg g $^{-1}$ F.w.)	Caratenoids (mg g $^{-1}$ F.w.)
Control	9.64 ± 0.57 ^{Aa}	7.14 ± 1.24 ^{Ba}	16.78 ± 1.76 ^{Ba}	1.35 ± 0.23 ^{Ba}
100 μM Cu	8.51 ± 0.54 ^{Bab}	4.32 ± 1.75 ^{Bb}	12.83 ± 2.03 ^{Bb}	1.02 ± 0.11 ^{Bab}
100 μM Cd	8.35 ± 0.60 ^{Ab}	3.77 ± 1.16 ^{Bb}	12.12 ± 1.57 ^{Bb}	0.92 ± 0.12 ^{Ab}
100 μM TiO_2	10.71 ± 0.76 ^{Aa}	10.14 ± 1.57 ^{Aa}	22.35 ± 3.73 ^{Aa}	1.82 ± 0.11 ^{ABa}
100 μM TiO_2 + 100 μM Cu	9.52 ± 0.67 ^{ABab}	6.54 ± 0.95 ^{ABb}	16.06 ± 1.62 ^{Ab}	1.32 ± 0.33 ^{ABb}
100 μM TiO_2 + 100 μM Cd	9.17 ± 0.83 ^{Ab}	4.43 ± 1.27 ^{ABb}	13.60 ± 1.87 ^{Bb}	0.92 ± 0.21 ^{Ab}
10^{-8} M EBL	10.39 ± 0.54 ^{Aa}	9.50 ± 1.01 ^{ABa}	19.90 ± 1.35 ^{Aa}	1.50 ± 0.24 ^{Ba}
10^{-8} M EBL + 100 μM Cu	9.09 ± 0.77 ^{ABab}	5.80 ± 0.97 ^{ABb}	14.89 ± 1.66 ^{Ab}	1.09 ± 0.09 ^{ABa}
10^{-8} M EBL + 100 μM Cd	8.85 ± 0.67 ^{Ab}	4.33 ± 1.31 ^{ABb}	13.18 ± 1.97 ^{Ab}	1.09 ± 0.25 ^{Aa}
100 μM TiO_2 + 10^{-8} M EBL	11.05 ± 1.22 ^{ABa}	10.86 ± 0.82 ^{Aa}	21.92 ± 1.97 ^{Aa}	2.27 ± 0.27 ^{Aa}
100 μM TiO_2 + 10^{-8} M EBL + 100 μM Cu	10.23 ± 1.04 ^{Aa}	8.01 ± 1.68 ^{Ab}	18.24 ± 1.72 ^{Aab}	1.48 ± 0.08 ^{Ab}
100 μM TiO_2 + 10^{-8} M EBL + 100 μM Cd	9.74 ± 0.98 ^{ABa}	7.01 ± 1.44 ^{Ab}	16.75 ± 2.08 ^{Ab}	1.30 ± 0.12 ^{Ab}

In this study, 1-year-old branches of *P. pygmaeus* were used as plant treatments together with 100 μM TiO_2 NPs and 10^{-8} M 24-epibrassinolide individually and in combination with 100 μM Cu and 100 μM Cd using four replications. Planting of the treated bamboo was performed in an Air Tech inoculation hood with fluorescent white lamps and ultraviolet light (wavelengths of 10–400 nm) at 15 °C and 30 °C. The bamboo plants were constantly exposed to excess heavy metals for three weeks. Sampling for the measurement of photosynthesis pigments was conducted after three weeks of plant exposure to the co-application of 24-epibrassinolide and titanium oxide nanoparticles under 100 μM Cu and 100 μM Cd. The capital letters (^{A,B}) indicate significant differences between treatments of control (C), titanium (Ti), 24-epibrassinolide (EBL), and 24-epibrassinolide with titanium oxide nanoparticles (EBL–TiO_2 NPs) individually or in combination with 100 μM Cu as well as 100 μM Cd (the bars with similar colors), while the lowercase letters (^{a,b}) denote statistically significant differences at each concentration of the co-application of EBL and TiO_2 NPs individually or in combination with 100 μM Cu and 100 μM Cd (the bars with various colors) based on Tukey's test ($p < 0.05$).

3.5. 24-Epibrasinolide and Titanium Oxide Nanoparticles Reduce Heavy Metal Accumulation in Bamboo Leaves, Stems, and Roots

The decrease in metal accumulation in different types of plants is one of the main mechanisms responsible for the reduction of metal toxicity and increase in plant resistance when exposed to oxidative stress. Our results showed that TiO_2 and EBL had a positive impact on the reduction in heavy metal concentrations in bamboo species; therefore, TiO_2 and EBL alone or in combination significantly reduced heavy metal accumulation in leaves, stems, and roots (Table 3). This phenomenon is related to the role of TiO_2 as a physical barrier that leads to a reduction in metal translocation from roots to aerial parts. The roots in plants are typically the first contact points where exposure to heavy metals occurs. Therefore, root physical traits are extremely decisive in limiting metal entry into plants. The root-based cellular layers comprised of epiblema, endodermis, and exodermis form apoplastic barriers in the roots, which can restrict heavy metal uptake by plants. We hypothesized that TiO_2 NPs, through strengthening the apoplastic barriers in the roots and enhancing their impermeability, can significantly diminish the uptake of heavy metals. On the other hand, TiO_2 NPs with high adsorption capacity act as an efficient binder of metal ions. Hence, TiO_2 NPs can restrain the movement of heavy metals within the extracellular or intercellular parts of roots, thereby restricting heavy metal translocation from the root to shoot. TiO_2 NPs may also have the ability to influence the expansion of the epidermal layer in plants, preventing heavy metal accumulation in nonphotosynthetic tissues by providing additional physical resistance. Conversely, we indicated that EBL, a hormone that is involved in plant growth regulation alone and in combination with TiO_2, plays a positive role in the stimulation of antioxidant activity, which can scavenge ROS components in plant organs and inhibit plant oxidative stress caused by heavy metal toxicity. As shown in Table 3, the combination of TiO_2–EBL with heavy metals showed the greatest reduction in heavy metal accumulation in the leaves, stems, and roots of bamboo species (Table 3). We suggest that TiO_2 and EBL can reduce heavy metal contents in plant leaves, stems, and roots, thus demonstrating that the combination of TiO_2 and EBL has the greatest impact on the decrease in metal toxicity.

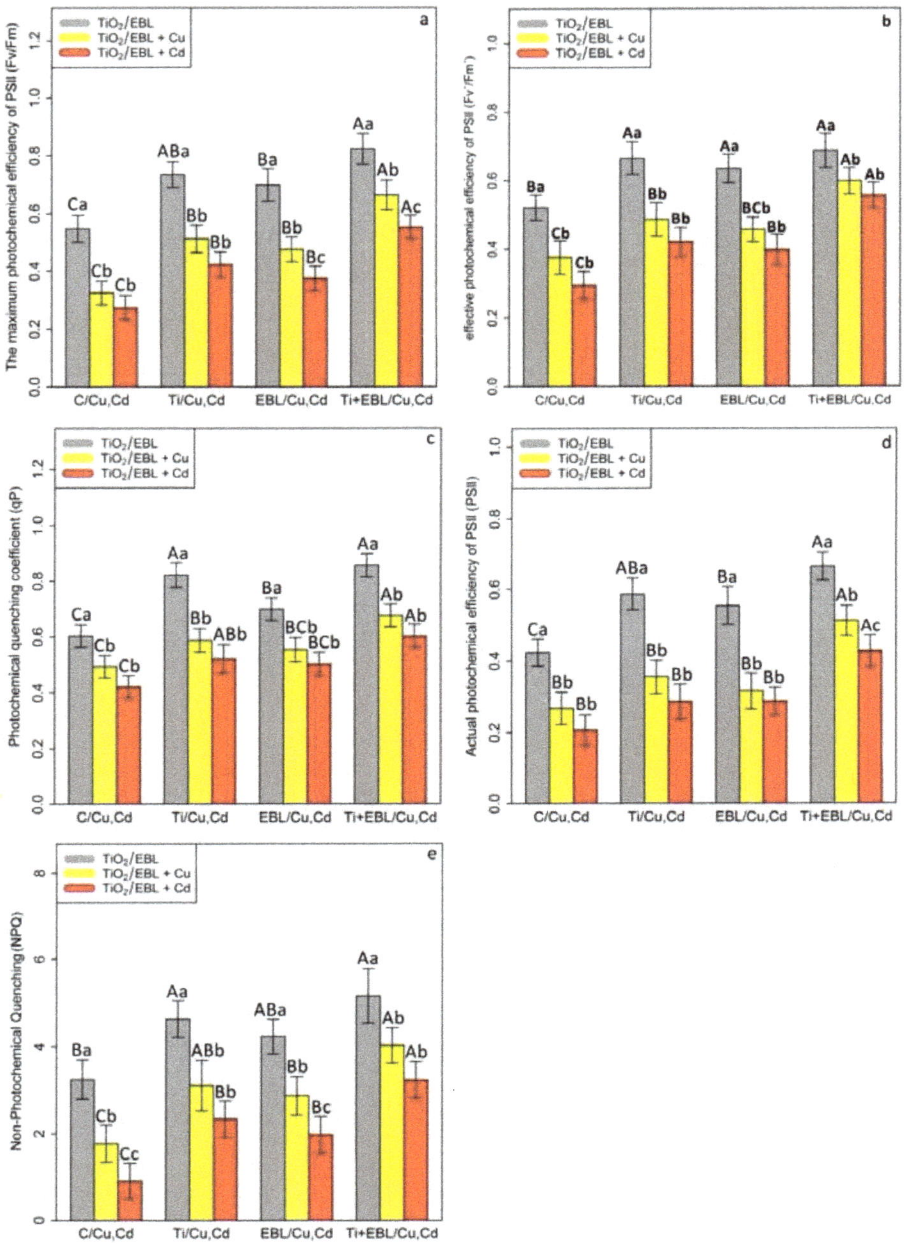

Figure 4. The effect of the co-application of 24-epibrassinolide and titanium oxide nanoparticles individually and combined on fluorescence parameters, including the maximum photochemical efficiency of PSll (Fv/Fm) (**a**), effective photochemical efficiency of PSll (Fv′/Fm′) (**b**), photochemical quenching coefficient (qP) (**c**), actual photochemical efficiency of PSll (φPSll) (**d**), and nonphotochemical quenching

(NPQ) (**e**) in bamboo species (*Pleioblastus pygmaeus*) with 100 μM Cu and 100 μM Cd. In this study, 1-year-old branches of *P. pygmaeus* were used as plant treatments together with 100 μM TiO_2 NPs and 10^{-8} M 24-epibrassinolide individually and in combination with 100 μM Cu and 100 μM Cd using four replications. Planting of the treated bamboo was performed in an Air Tech inoculation hood with fluorescent white lamps and ultraviolet light (wavelengths of 10–400 nm) at 15 °C and 30 °C. The bamboo plants were constantly exposed to excess heavy metals for three weeks. Sampling for the measurement of fluorescence parameters (**a–e**) was conducted after three weeks of plant exposure to the co-application of 24-epibrassinolide and titanium oxide nanoparticles under 100 μM Cu and 100 μM Cd. The capital letters ($^{A–C}$) indicate significant differences between treatments of control (C), titanium (Ti), 24-epibrassinolide (EBL), and 24-epibrassinolide involving individual or combined application of titanium oxide nanoparticles (EBL–TiO_2 NPs) under 100 μM Cu and 100 μM Cd (the bars with similar colors), while the lowercase letters ($^{a–c}$) denote statistically significant differences at each concentration of the co-application of EBL and TiO_2 NPs individually or in combination with 100 μM Cu and 100 μM Cd (the bars with various colors) based on Tukey's test ($p < 0.05$).

Table 3. The accumulation concentrations of titanium oxide nanoparticles and corresponding heavy metals (Cu and Cd) in bamboo shoots, stems, and roots.

Treatments	Heavy Metal in Leaves (μmol L^{-1})	TiO_2 NP in Leaves (μmol L^{-1})	Heavy Metal in Stem (μmol L^{-1})	TiO_2 NP in Stem (μmol L^{-1})	Heavy Metal in Root (μmol L^{-1})	TiO_2 NP in Root (μmol L^{-1})
Control	0	0	0	0	0	0
100 μM Cu	19.22 ± 1.10 ^{Ab}	0	25.3 ± 1.03 ^{Ab}	0	31.2 ± 1.00 ^{Ab}	0
100 μM Cd	24.20 ± 1.12 ^{Aa}	0	29.8 ± 1.02 ^{Aa}	0	36.3 ± 1.02 ^{Aa}	0
100 μM TiO_2	0	18.5 ± 0.90 ^{Aa}	0	24.5 ± 0.95 ^{Aa}	0	34.6 ± 1.04 ^{Aa}
100 μM TiO_2 + 100 μM Cu	13.42 ± 1.10 ^{Bb}	14.3 ± 0.79 ^{Ab}	16.6 ± 0.90 ^{Cb}	18.5 ± 0.86 ^{Ab}	19.1 ± 1.11 ^{Cb}	28.5 ± 0.85 ^{Ab}
100 μM TiO_2 + 100 μM Cd	16.80 ± 0.97 ^{Ba}	10.8 ± 0.87 ^{Ac}	21.2 ± 0.86 ^{Ca}	12.3 ± 0.86 ^{Ac}	25.4 ± 0.95 ^{Ca}	20.5 ± 0.90 ^{Ac}
10^{-8} M EBL	0	0	0	0	0	0
10^{-8} M EBL + 100 μM Cu	14.60 ± 1.07 ^{Bb}	0	18.8 ± 0.98 ^{Bb}	0	22.3 ± 0.94 ^{Bb}	0
10^{-8} M EBL +100 μM Cd	17.41 ± 0.97 ^{Ba}	0	23.4 ± 0.94 ^{Ba}	0	28.6 ± 0.90 ^{Ba}	0
100 μM TiO_2 + 10^{-8} M EBL	0	16.4 ± 0.77 ^{Ba}	0	21.3 ± 0.95 ^{Ba}	0	31.3 ± 0.94 ^{Ba}
100 μM TiO_2 + 10^{-8} M EBL +100 μM Cu	8.40 ± 1.07 ^{Cb}	11.2 ± 0.87 ^{Bb}	12.4 ± 1.02 ^{Db}	15.4 ± 0.95 ^{Bb}	15.4 ± 0.94 ^{Db}	24.5 ± 1.02 ^{Bb}
100 μM TiO_2 + 10^{-8} M EBL +100 μM Cd	10.36 ± 0.99 ^{Ca}	8.3 ± 0.90 ^{Bc}	15.4 ± 0.97 ^{Da}	10.2 ± 0.94 ^{Bc}	17.7 ± 0.94 ^{Da}	16.4 ± 0.86 ^{Bc}

In this study, 1-year-old branches of *P. pygmaeus* were used as plant treatments together with 100 μM TiO_2 NPs and 10^{-8} M 24-epibrassinolide individually and combined with 100 μM Cu and 100 μM Cd using four replications. The capital letters ($^{A–D}$) indicate significant differences between treatments of control (C), titanium (Ti), 24-epibrassinolide (EBL), and 24-epibrassinolide with titanium oxide nanoparticles (EBL–TiO_2 NPs) individually or in combination with 100 μM Cu as well as 100 μM Cd (the bars with similar colors), while the lowercase letters ($^{a–c}$) denote statistically significant differences at each concentration of the co-application of EBL and TiO_2 NPs individually or in combination with 100 μM Cu and 100 μM Cd (the bars with various colors) based on Tukey's test ($p < 0.05$). 2–6 24-Epibrasinolide and titanium oxide nanoparticles increase plant biomass indices (root and shoot dry weight) and plant growth (length of shoot) in bamboo species under Cu and Cd toxicity.

To evaluate the plant growth rate under Cu and Cd toxicity, the plant biomass, including root and shoot dry weight, as well as the length of shoots, were measured. A significant difference was found for the co-application of TiO_2–EBL and heavy metals ($p < 0.001$) (Figure 5). Therefore, TiO_2 and EBL individually and in combination significantly increased plant growth and biomass under stressful conditions. Based on this result, the greatest increase in plant biomass and growth under heavy metal exposure was related to the combination of TiO_2 and EBL, which resulted in a 21% increase in the dry weight of shoots, a 23% increase in the dry weight of roots, and a 19% increase in the length of shoots in comparison with their control treatments (Table 4). Conversely, the measurements showed that the lowest plant growth was recorded under 100 μM Cu and 100 μM Cd, which resulted in 0.5 g and 0.46 g dry weights of shoots, 0.59 g and 0.53 g dry weights of roots, and 10.04 cm and 9.21 cm shoot lengths, respectively (Table 4).

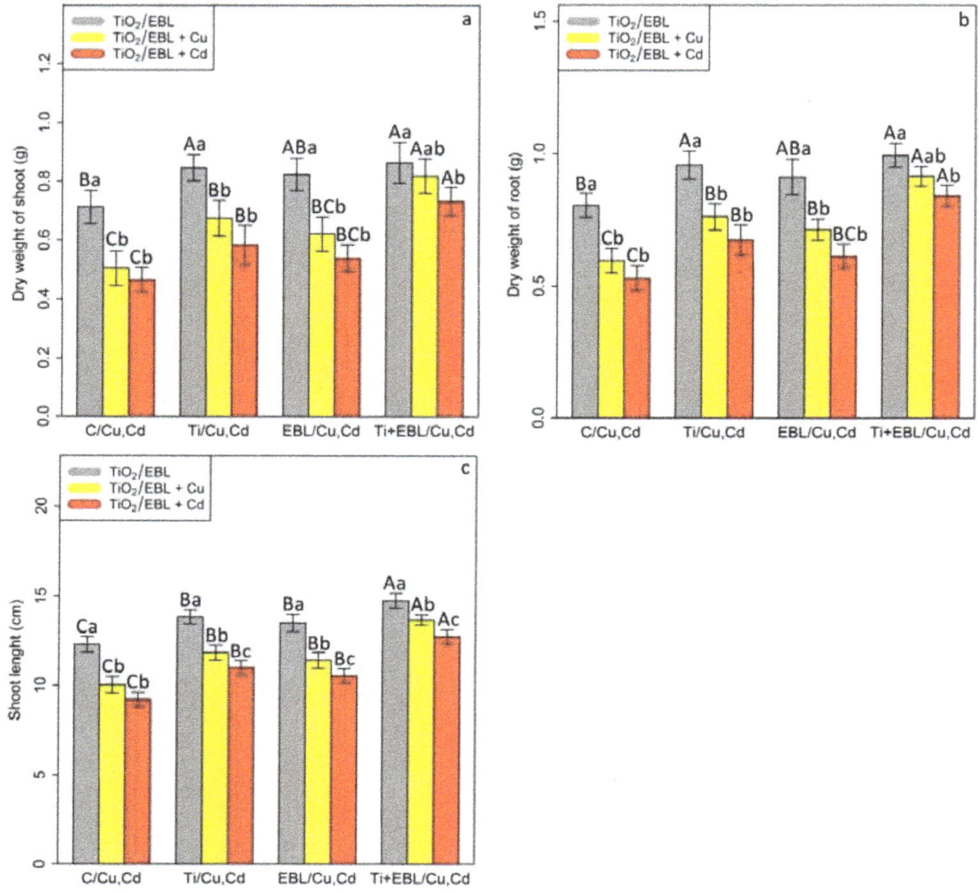

Figure 5. The impact of the co-application of 24-epibrassinolide and titanium oxide nanoparticles individually and combined on the dry weight of shoots (**a**), dry weight of roots (**b**), and shoot length (**c**) of bamboo species (*Pleioblastus pygmaeus*) with 100 µM Cu and 100 µM Cd. In this study, 1-year-old branches of *P. pygmaeus* were used as plant treatments together with 100 µM TiO_2 NPs and 10^{-8} M 24-epibrassinolide individually and combined with 100 µM Cu and 100 µM Cd through four replications. Planting of the treated bamboo was performed in an Air Tech inoculation hood with fluorescent white lamps and ultraviolet light (wavelengths of 10–400 nm) at 15 °C and 30 °C. The bamboo plants were constantly exposed to excess heavy metals for three weeks. Sampling for the measurement of biomass indexes and plant growth (**a**–**c**) was conducted after three weeks of plant exposure to the co-application of 24-epibrassinolide and titanium oxide nanoparticles under 100 µM Cu and 100 µM Cd. The capital letters ($^{A–C}$) indicate significant differences between treatments of control (C), titanium (Ti), 24-epibrassinolide (EBL), and 24-epibrassinolide involving individual or combined application of titanium oxide nanoparticles (EBL–TiO_2 NPs) under 100 µM Cu and 100 µM Cd (the bars with similar colors), while the lowercase letters ($^{a–c}$) denote statistically significant differences at each concentration of the co-application of EBL and TiO_2 NPs individually or in combination with 100 µM Cu and 100 µM Cd (the bars with various colors) based on Tukey's test ($p < 0.05$).

Table 4. The changes in bamboo biomass in root and shoot dry weight as well as shoot length with 24-epibrassinolide and titanium oxide nanoparticles individually or combined with 100 µM Cu and 100 µM Cd in comparison with the control treatment.

Treatments	Dry Weight of Shoot (%)	Dry Weight of Root (%)	Shoot Length (%)
100 µM Cu	29%↓	26%↓	18%↓
100 µM Cd	35%↓	34%↓	25%↓
100 µM TiO_2	18%↑	19%↑	12%↑
100 µM TiO_2 + 100 µM Cu	5%↓	5%↓	3%↓
100 µM TiO_2 + 100 µM Cd	17%↓	16%↓	10%↓
10^{-8} M EBL	15%↑	13%↑	10%↑
10^{-8} M EBL + 100 µM Cu	12%↓	11%↓	7%↓
10^{-8} M EBL + 100 µM Cd	24%↓	23%↓	14%↓
100 µM TiO_2 + 10^{-8} M EBL	21%↑	23%↑	19%↑
100 µM TiO_2 + 10^{-8} M EBL +100 µM Cu	15%↑	13%↑	11%↑
100 µM TiO_2 + 10^{-8} M EBL +100 µM Cd	3%↑	4%↑	3%↑

3–7 24-Epibrasinolide and titanium oxide nanoparticles reduce the translocation factor (TF) and bioaccumulation factor (BAF) and improve the tolerance index (TI) in roots and shoots of bamboo species. ↑ indicates increases and ↓ indicates decreases.

The translocation factor (TF) is one of the main mechanisms used to evaluate the remediation efficiency of heavy metals in plant organs and the reduction of toxicity in plants under heavy metal stress. Therefore, it is calculated according to differences in the accumulation of Cu and Cd in shoots and roots, and it serves as an important factor in increasing plant tolerance to toxicity. In the present study, the addition of 24-epibrassinolide and titanium oxide nanoparticles significantly reduced Cu and Cd translocation from roots to shoots, which led to a reduction in toxicity by limiting metal accumulation in the plant aerial organs. Therefore, according to the results, the co-application of 24-epibrassinolide and titanium oxide nanoparticles in combination with heavy metals (Cu and Cd) resulted in a low level of translocation factor, which could reduce metal toxicity in the aerial parts of bamboo plants (Table 5). Additionally, the result showed that the co-application of EBL and TiO_2 NPs significantly reduced Cu and Cd concentration in the leaves ($p < 0.001$), implicating the positive role of EBL–TiO_2NPs in the reduction of heavy metal toxicity in the aerial parts of the bamboo plant (Table 5). Conversely, the calculation of the tolerance indices in shoots and roots revealed a significant difference between the co-application of TiO_2 NPs and EBL alone under Cu and Cd ($p < 0.001$). Therefore, the levels of TiO_2 NPs and EBL indicated an increase in shoot and root tolerance under heavy metal stress, which was obtained by the amelioration mechanism of the co-application of TiO_2 and EBL against heavy metal toxicity, such as the stimulation of antioxidant activity and the increase in plant biomass. We suggest that TiO_2 NP and EBL concentrations alone increase plant tolerance under metal stress; however, the most positive effect was more pronounced with the co-application of TiO_2 NPs and EBL under Cu and Cd toxicity (Table 5).

Table 5. Changes in the translocation factor and tolerance index of shoots and roots in response to 24-epibrassinolide and titanium oxide nanoparticles individually or in combination with 100 µM Cu and 100 µM Cd compared with the control treatment. Each data point is the mean ± SE of four replicates. The capital letters ($^{A–C}$) indicate significant differences between treatments of control (C), titanium (Ti), 24-epibrassinolide (EBL), and 24-epibrassinolide involving individual or combined application of titanium oxide nanoparticles (EBL–TiO_2 NPs) under 100 µM Cu and 100 µM Cd (the bars with similar colors), while the lowercase letters ($^{a–c}$) denote statistically significant differences at each concentration of the co-application of EBL and TiO_2 NPs individually or in combination with 100 µM Cu and 100 µM Cd (the bars with various colors) based on Tukey's test ($p < 0.05$).

Treatment	Translocation Factor (Leaves)	Tolerance Index (Shoot)	Tolerance Index (Root)	Bioaccumulation Factor (Leaves)
Control	0.00 ± 0.00 ^{Bb}	1.00 ± 0.00 ^{Ba}	1.00 ± 0.00 ^{Ba}	0.00 ± 0.00 ^{Aa}
100 µM Cu	0.65 ± 0.02 ^{Aa}	0.70 ± 0.06 ^{Cb}	0.74 ± 0.09 ^{Bb}	0.19 ± 0.01 ^{Ab}
100 µM Cd	0.66 ± 0.01 ^{Aa}	0.65 ± 0.05 ^{Cb}	0.66 ± 0.09 ^{Bb}	0.24 ± 0.01 ^{Ac}
100 µM TiO_2	0.52 ± 0.04 ^{Ab}	1.19 ± 0.13 ^{ABa}	1.23 ± 0.12 ^{Aa}	0.00 ± 0.00 ^{Aa}
100 µM TiO_2 + 100 µM Cu	0.58 ± 0.01 ^{BCa}	0.94 ± 0.07 ^{ABb}	0.95 ± 0.11 ^{ABb}	0.13 ± 0.01 ^{Bb}
100 µM TiO_2 + 100 µM Cd	0.60 ± 0.01 ^{ABa}	0.80 ± 0.05 ^{Bb}	0.84 ± 0.10 ^{Bb}	0.16 ± 0.01 ^{Bc}
10^{-8} M EBL	0.00 ± 0.00 ^{Bb}	1.15 ± 0.09 ^{ABa}	1.13 ± 0.13 ^{ABa}	0.00 ± 0.00 ^{Aa}
10^{-8} M EBL + 100 µM Cu	0.60 ± 0.02 ^{Ba}	0.87 ± 0.08 ^{BCb}	0.89 ± 0.09 ^{ABb}	0.14 ± 0.01 ^{Bb}
10^{-8} M EBL + 100 µM Cd	0.61 ± 0.05 ^{Aa}	0.75 ± 0.01 ^{Bb}	0.76 ± 0.09 ^{Bb}	0.17 ± 0.00 ^{Bc}
100 µM TiO_2 + 10^{-8} M EBL	0.49 ± 0.01 ^{Ab}	1.21 ± 0.08 ^{Aa}	1.24 ± 0.12 ^{Aa}	0.00 ± 0.00 ^{Aa}
100 µM TiO_2 + 10^{-8} M EBL + 100 µM Cu	0.53 ± 0.02 ^{Cab}	1.1 ± 0.09 ^{Aab}	1.09 ± 0.10 ^{Aa}	0.08 ± 0.01 ^{Cb}
100 µM TiO_2 + 10^{-8} M EBL + 100 µM Cd	0.54 ± 0.02 ^{Ba}	1.02 ± 0.04 ^{Ab}	1.04 ± 0.08 ^{Aa}	0.10 ± 0.01 ^{Bc}

4. Discussion

Titanium, as a form of TiO_2, has the ability to alter the bioavailability and behavior of metals in the environment [45]. The impact of TiO_2 NPs on increasing antioxidant activity and plant growth has been reported in several studies [13,14,46]. This increase can be attributed to the inductive role of TiO_2 NPs in enhancing signaling associated with the activation of antioxidant enzyme activity [13]. This finding is consistent with the reported results in the present study. Therefore, our results demonstrated that the individual levels of TiO_2 NPs could increase antioxidant and nonantioxidant activity in bamboo plants under certain Cu and Cd levels. Conversely, the level of EBL regulates plant stress by stimulating antioxidant activity [47]. The increasing capacity of antioxidant activity based on the levels of EBL in plants under stress has been reported in many studies [48–50]. EBL seems to play a main role in the activation of genes responsible for antioxidants by stimulating the expression of genes responsible for SOD, CAT, and APX in plants exposed to heavy metal stress [51]. Hence, it is interesting to note that EBL has the ability to ameliorate oxidative stress caused by metals, which has previously been reported for many plant species, such as *Brassica juncea* [20], *Cicer arietinum* [52], and *Raphanus sativa* [53]. The main reason can be attributed to the role of BR signaling kinase (BSK 1) in the stimulation of salicylic acid levels against oxidative damage [54]. In our studies, the application of TiO_2 and EBL individually and in combination enhanced antioxidant enzyme activities, including SOD, POD, CAT, GR, APX, and PAL. However, the combination of TiO_2 NP and EBL was more effective in increasing antioxidant levels than TiO_2 NPs and EBL alone. Phenolic compounds, as nonenzymatic antioxidant activities, alleviate the negative effect of reactive oxygen radicals and have a strong ability to chelate metals [55,56]. There seems to be a relationship between enhancing phenylalanine ammonia-lyase (PAL) and the total phenolic compound, and it has been reported that PAL is a key enzyme responsible for the activation of the synthesis of phenolic compounds under stress [57,58]. This phenomenon has been reported in some studies on the reduction of Cu and Cd toxicity [55,59]. Our results demonstrated that the application of TiO_2 NPs and EBL individually and in combination increased nonantioxidant activity (total phenolics, flavonols, and tocopherols) under Cu and Cd toxicity. This phenomenon could be related to PAL gene transcript levels as well as increasing PAL activity in response to EBL levels under heavy metal stress, which ultimately reduces ROS compounds by synthesizing phenolic compounds.

When antioxidant activity increases, the plant experiences cellular injuries (H_2O_2, MDA, and EL) [60]. It has been reported that TiO_2 NPs induce certain stress-combative mechanisms, such as an improvement of the defense mechanism against ROS accumulation in plant intercellular space, which has been shown to attenuate H_2O_2 induction [60]. An increase in MDA content has been described as the initial stage of plant injury, which shows the rate of membrane lipid peroxidation [61]. This reveals the extent to which plants face this serious problem. Based on the present findings, we suggest that TiO_2 NPs protect the plant cell membrane from ROS, a phenomenon that has been related to the role of TiO_2 NPs in boosting antioxidant activity. Conversely, EBL has the ability to positively alter the membrane structure and membrane stability in plants exposed to stresses, such as heavy metals, which leads to a reduction in membrane lipid peroxidation [62]. In one study, the level of EBL was observed to diminish the concentrations of H_2O_2 and MDA (20–60% reductions) in plants under Pb stress [63]. In another study, EBL diminished the oxidative toxicity in cowpea under Cd by reducing lipid peroxidation, MDA content, and electrolytes [64,65]. In the present study, the level of EBL improved the ROS content and reduced the plant cell membrane under heavy metal toxicity. It is interesting to note that in this study, the application of TiO_2 NPs and EBL individually and in combination decreased ROS and lipid peroxidation, including H_2O_2, $O_2^{\bullet-}$, MDA, SP, and EL, in bamboo plants exposed to metal stress. One of the mechanisms underlying the amelioration of lipid peroxidation and ROS by EBL can be attributed to the increase in endogenous plant hormones that regulate plant growth, such as salicylic acid and ethylene, and the cross-talk between them. These mechanisms can improve plant tolerance under metal toxicity [66]. In this study, the co-application of TiO_2 NPs and EBL was more efficient in reducing ROS compounds and ameliorating lipid peroxidation than TiO_2 NPs and EBL alone.

Studies have indicated that TiO_2 NPs enhance photosystem II in spinach by promoting oxygen evolution and energy transfer [67]. Additionally, TiO_2 reduces Chl degradation and stimulates Chl biosynthesis, which can promote photosynthesis by stabilizing chlorophylls and carotenoids [60]. Chlorophylls are the most abundant component of the chloroplast and play an efficient role in the rate of photosynthesis [59]. The role of EBL in enhancing the cell number and photosynthetic pigment content (Chl a, Chl b, and carotenoids) has been demonstrated in some studies [68,69]. It seems that the increase in photosynthesis and Chl pigments by EBL is related to the stimulation activity of ribulose 1,5-bisphosphate carboxylase oxygenase as well as the increase in the Calvin cycle enzymes [20]. Therefore, the level of EBL with increasing carotenoids ameliorates photodamage during photosynthesis [70]. According to the above mechanisms, the co-application of EBL and Ti can improve photosynthesis and Chl pigment levels via Chl biosynthesis and reduce photodamage via the activation of the Calvin cycle enzymes in plants under heavy metal stress. Additionally, the results revealed a positive impact of the co-application of EBL and TiO_2 NPs on fluorescence parameters, which showed an increase in the efficiency of fluorescence indices, including the maximum photochemical efficiency of PSll (Fv/Fm), photochemical quenching coefficient (qP), effective photochemical efficiency of PSll (Fv′/Fm′), actual photochemical efficiency of PSll (φPSll), and nonphotochemical quenching (NPQ). Therefore, we suggest that the co-application of EBL and TiO_2 NPs can increase photosynthetic properties in plants exposed to heavy metals (Cu and Cd), and this phenomenon could be related to the increase in antioxidant activity and the reduction in heavy metal accumulation under metal toxicity stress.

Titanium is known to be the most abundant transition element after iron, with a level of 1–578 mg kg^{-1} in different species of non-hyperaccumulator plants [71]. However, low mobility in the soil may impact its absorption by plants [72]. Our results showed that the root accumulation of TiO_2 NPs was higher than that of stems and shoots, indicating that bamboo roots prefer to be storage organs of titanium, which has been reported in another study [73]. Therefore, the accumulation of titanium NPs led to the adsorption of heavy metals in roots, which, as a physical barrier, reduces metal translocation from roots to shoots. Conversely, EBL has the potential to reduce metal accumulation in plants by enhancing phytochelatin synthesis (PC) [48–50]. In a study on sugar beet, the levels of EBL reduced the absorption of heavy metals by plants by 50% [74–76]. It has also been reported [69] to reduce Cd accumulation in the roots, stems, and shoots of pea seedlings. Conversely, EBL levels reduce the accumulation of Cd by preserving ion homeostasis through the acceleration of calcium absorption [77,78]. Therefore, EBL leads to increased absorption of K^+, mg^+, and K^+, which can be transported to the aerial parts of plants, such as leaves, and finally limit Cd and metal translocation from roots to shoots [79]. In the present study, the application of EBL and TiO_2 could individually and in combination diminish the accumulation of heavy metals. These results are related to the role of EBL in preserving ion homeostasis, which can limit heavy metal uptake by coperception, which is associated with the role of TiO_2 in the adsorption and absorption of heavy metals on the root surface. Similar to other transition metals, titanium, present in small fractions, absorbs to and accumulates in roots, as it is translocated through the xylem stream from roots to shoots [80]. The translocation of TiO_2 NPs to aerial plant parts has been demonstrated [81,82]. As shown in Table 5, the co-application of TiO_2 NPs and EBL significantly reduced metal translocation from roots to shoots, which is an important mechanism in increasing the tolerance of bamboo plants. However, the results showed that BAF in the roots was higher than in the stem and the leaves, which indicated that TiO_2 NPs–EBL could effectively reduce BAF in the aerial parts of the bamboo plant. This could be explained by the mechanisms involved in the absorbance of Cu and Cd in the root surface by the co-application of TiO_2 NPs and EBL. Therefore, we suggest that TiO_2 NPs–EBL has an important role in the reduction of adsorption and uptake as well as the translocation of Cu and Cd to the aerial parts (leaves and stem). Thus, the application of TiO_2 NPs–EBL can retain the heavy metals on the bamboo root surface. On the other hand, the results showed that the TF in the leaves was less than that in the stem, which was an indication that the heavy metals had been accumulated in the leaves less than in the stem (Figure 6).

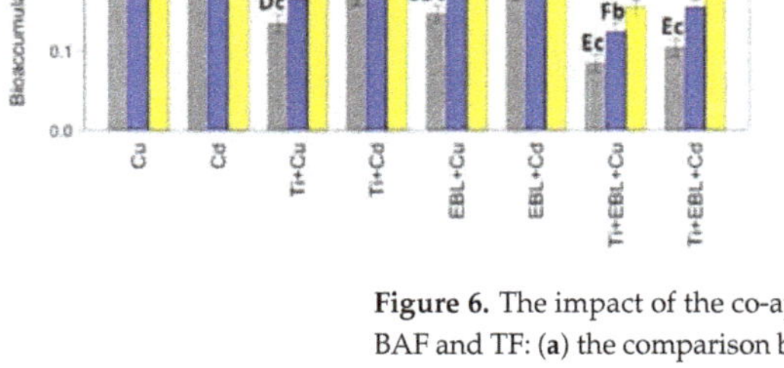

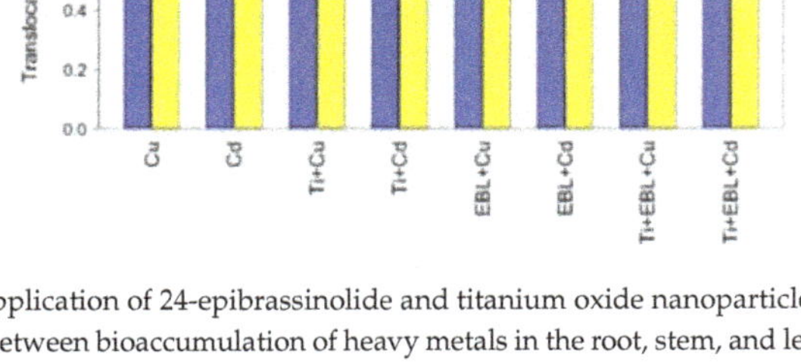

Figure 6. The impact of the co-application of 24-epibrassinolide and titanium oxide nanoparticles on BAF and TF: (**a**) the comparison between bioaccumulation of heavy metals in the root, stem, and leaves; (**b**) the comparison between translocation factor in the stem and the leaves. Bioaccumulation factor (BAF) is obtained by the difference between the concentrations of heavy metals in the leaf, stem, or root and concentrations of heavy metals in the medium, while the translocation factor (TF) is obtained by the difference between the concentration of heavy metals (Cu, Cd) in the leaves or stem of plants and the concentration of the heavy metals (Cu, Cd) in the roots of plants. The capital letters ($^{A–G}$) indicate significant differences between treatments of control (C), titanium (Ti), 24-epibrassinolide (EBL), and 24-epibrassinolide involving individual or combined application of titanium oxide nanoparticles (EBL

–TiO_2 NPs) under 100 μM Cu and 100 μM Cd (the bars with similar colors), while the lowercase letters ([a–c]) denote statistically significant differences between leaves, stem, and root at each treatment (the bars with various colors) based on Tukey's test ($p < 0.05$).

Titanium is a useful element that can aid plant growth by increasing plant photosynthesis and enzymatic activity as well as increasing plant uptake of other nutrients [72]. One study reported that TiO_2 can enhance the absorption rate of micro-and macronutrients, which could be the main factor in plant growth and biomass [83]. In our study, the role of TiO_2 in increasing the plant biomass was related to an increase in nutrient absorbance by the plants and reduced toxicity in response to an increase in antioxidant activity. Many studies have reported that EBL increases plant growth under heavy metal stress [21,84,85]. The reduction in plant growth by heavy metal toxicity is related to the number of intercellular metal ions bound to the surface of the cell [86,87]. Our results showed that EBL could increase the plant biomass and plant growth under heavy metal toxicity, which could be related to the role of EBL in the reduction in intercellular metal ions, a phenomenon that has been confirmed in a study on A. *obliquus* [63]. However, the role of EBL in plant growth regulation during the stimulation of plant defense mechanisms can also be considered. Therefore, we hypothesized that EBL increased the plant growth under heavy metal stress by promoting antioxidant capacity. The increase in Chl as well as carotenoid contents in response to EBL has been confirmed in many studies [88–91]. Conversely, EBL has the ability to control cell division and elongation by regulating xyloglucan endotransglucosylase [92,93], thus demonstrating the positive role of EBL on plant growth and development, especially under stressful conditions. In the present research, the role of the co-application of TiO_2 NPs and EBL in improving plant biomass and plant growth seemed to be related to the ameliorative mechanisms activated by the levels of both TiO_2 NPs and EBL. The application of small-sized nanoparticles in the range of 1–100 nm is a new strategy to maintain plant growth and development under heavy metals and other abiotic stresses. However, the build-up of TiO_2 NPs within the plant organs can have dual effects of growth promotion and suppression. Titanium dioxide nanoparticles (TiO_2 NPs) lead to several beneficial outcomes on the physiological, morphological, and biochemical traits of some plant species, which has been indicated in some studies [94,95] as well as our present study. Conversely, some researchers have reported the detrimental impacts of high levels of TiO_2 NPs on plants [96,97]. These implications might arise due to various environmental conditions, different plant species, and the applied levels [97,98]. Therefore, the safety/danger of TiO_2 NPs for plants depends on a myriad of factors including size, concentration, method of treatment application, plant type and growth pattern, uptake amount by plants, cellular chemical properties, translocation rate, and reactivity of TiO_2 NPs in various tissues, which determine NP interaction with a wide array of metabolic activities of the plants that can thus lead to their advantageous or toxic effects [99,100]. Additionally, the TiO_2 NP surface area, their predisposition for accumulation in tissues, and their inherent reactivity are the possible reasons for their toxicological repercussions [101]. Therefore, there is a great need to take the aforementioned variables into careful consideration while applying TiO_2 NPs in the agriculture and food industry, which can minimize health risk for humans. This finding revealed the effective role of the co-application of TiO_2 NPs and EBL in comparison with TiO_2 NPs and EBL individually.

5. Conclusions

Heavy metals are deemed a considerable environmental safety hazard with inhibitory impacts on plant growth due to the induction of excessive levels of ROS compounds, which causes oxidative stress in cells and tissues. The use of nanoparticles and phytohormones as two possible agents that mitigate the deleterious effects of heavy metals has been on the rise in the recent years. Thus, conducting extensive research using various plant species with distinct growth and morphological characteristics is needed. Based on our experimental results, the individual application of EBL as a phytohormone and TiO_2 NPs contributed to the amelioration of toxicity in bamboo plants under excess Cu and Cd. However, the co-application of EBL and TiO_2 demonstrated a greater effective influence on increasing plant tolerance under metal toxicity. Therefore, our results indicated that while Cu and Cd stress led to increased ROS production, causing injury to the plant membrane, by boosting oxidative activity, the co-application of EBL and TiO_2 significantly reduced the ROS content and oxidative stress in the plants, resulting in an increase in the photosynthetic properties and an enhancement in the plant growth and development. Conversely, the co-application of EBL and TiO_2 increased the plant tolerance under metal toxicity by reducing the heavy metal accumulation within the plant and restricting the metal translocation from the roots to the shoots. Overall, our study revealed the cellular-and tissue-level mechanisms involved in increasing bamboo plant tolerance to Cu and Cd toxicity through the integrated use of EBL and TiO_2. This result requires further investigation with different plant species.

Author Contributions: Conceptualization, A.E., Y.D. and J.B.; statistical analysis, A.E. and Y.L.; Investigation, A.E. and G.L.; Supervision, A.E., Y.D., and G.L.; Project Administration, A.E.; Funding Acquisition, A.E. and G.L.; writing—original draft and revised preparation, A.E., Y.D., J.B., F.M., M.R., and G.L.; writing—review and editing, J.B., F.M., and M.H.; visualization, M.H. All authors have read and agreed to the published version of the manuscript.

Funding: This work received financial support from Nanjing Forestry University (Start-Up Research Fund) and the Bamboo Research Institute. Special Funding for this work was provided by Jiangsu Agricultural Science and Technology Innovation Fund, No. CX (18) 2031.

Institutional Review Board Statement: The study was conducted according to the guidelines of the Declaration of Helsinki and approved by the Institutional Review Board of Nanjing Forestry University.

Informed Consent Statement: Not applicable.

Data Availability Statement: The data presented in this study are available in article.

Acknowledgments: We would like to extend our sincere gratitude and appreciation to Peijian Shi, Co-Innovation Center for Sustainable Forestry in Southern China, Nanjing Forestry University, Nanjing, Jiangsu, China, for helping in the statistical analysis of the manuscript.

Conflicts of Interest: The authors declare that there is no conflict of interests regarding the publication of this paper.

References

1. Bhat, J.A.; Shivaraj, S.M.; Singh, P.; Navadagi, D.B.; Tripathi, D.K.; Dash, P.K.; Solanke, A.U.; Sonah, H.; Deshmukh, R. Role of silicon in mitigation of heavy metal stresses in crop plants. *Plants* **2019**, *8*, 71. [CrossRef]
2. Kim, J.J.; Kim, Y.S.; Kumar, V. Heavy metal toxicity: An update of chelating therapeutic strategies. *J. Trace Elem. Med. Biol.* **2019**, *54*, 226–231. [CrossRef] [PubMed]
3. Qi, Z.Y.; Ahammed, G.J.; Jiang, C.Y.; Li, C.X.; Zhou, J. E3 ubiquitin ligase gene SlRING1 is essential for plant tolerance to cadmium stress in *Solanum lycopersicum*. *J. Biotechnol.* **2020**, *324*, 239–247. [CrossRef] [PubMed]
4. Zhang, X.; Zhong, T.; Liu, L.; Ouyang, X. Impact of soil heavy metal pollution on food Safety in China. *PLoS ONE* **2015**, *10*, e0135182. [CrossRef] [PubMed]
5. Adrees, M.; Ali, S.; Rizwan, M.; Zia-ur-Rehman, M.; Ibrahim, M.; Abbas, F.; Farid, M.; Qayyum, M.F.; Irshad, M.K. Mechanisms of silicon-mediated alleviation of heavy metal toxicity in plants: A review. *Ecotoxicol. Environ. Saf.* **2015**, *119*, 186–197. [CrossRef]
6. Yruela, I. Copper in plants. Braz. *J. Plant Physiol.* **2015**, *17*, 145–156. [CrossRef]
7. Bouazizi, H.; Jouili, H.; Geitmann, A.; El Ferjani, E. Copper toxicity in expanding leaves of *Phaseolus vulgaris* L.: Antioxidant enzyme response and nutrient element uptake. *Ecotoxicol. Environ. Saf.* **2010**, *73*, 1304–1308. [CrossRef]
8. Jouilia, H.; Bouazizia, H.; Rossignol, M.; Borderies, G.; Jamet, E.; El Ferjani, E. Partial purification and characterization of a copper-induced anionic peroxidase of sunflower roots. *Plant Physiol. Biochem.* **2008**, *46*, 760–767. [CrossRef]
9. Singh, S.; Susan, E.; D'Souza, S.F. Cadmium accumulation and its influence on lipid peroxidation and antioxidative system in an aquatic plant, *Bacopa monnieri* L. *Chemosphere* **2006**, *62*, 233–246. [CrossRef]
10. Li, M.; Ahammed, G.J.; Li, C.; Bao, X.; Yu, J.; Huang, C.; Yin, H.; Zhou, J. Brassinosteroid ameliorates zinc oxide nanoparticles-induced oxidative stress by improving antioxidant potential and redox homeostasis in tomato seedling. *Front. Plant Sci.* **2016**, *7*, 1–13. [CrossRef]
11. Dimkpa, C.O.; Bindraban, P.S. Nanofertilizers: New products for the industry? *J. Agric. Food Chem.* **2018**, *66*, 6462–6473. [CrossRef] [PubMed]

12. Nemec, S.; Kralj, S.; Wilhelm, C.; Abou-Hassan A.; Rols, M.P.; Kolosnjaj-Tabi, J. Comparison of iron oxide nanoparticles in photothermia and magnetic hyperthermia: Effects of clustering and silica encapsulation on nanoparticles' heating yield. *Appl. Sci.* **2020**, *10*, 7322. [CrossRef]
13. Emamverdian, A.; Ding, Y.; Mokhberdoran, F.; Ahmad, Z.; Xie, Y. The Investigation of TiO_2NPs Effect as a Wastewater Treatment to Mitigate Cd Negative Impact on Bamboo Growth. *Sustainability* **2021**, *13*, 3200. [CrossRef]
14. Emamverdian, A.; Ding, Y.; Mokhberdoran, F.; Ramakrishnan, M.; Ahmad, Z.; Xie, Y. Different Physiological and Biochemical Responses of Bamboo to the Addition of TiO_2 NPs under Heavy Metal Toxicity. *Forests* **2021**, *12*, 759. [CrossRef]
15. Rizwan, M.; Ali, S.; ur Rehman, M.Z.; Malik, S.; Adrees, M.; Qayyum, M.F.; Alamri, S.A.; Alyemeni, M.N.; Ahmad, P. Effect of foliar applications of silicon and titanium dioxide nanoparticles on growth, oxidative stress, and cadmium accumulation by rice (*Oryza sativa*). *Acta Physiol. Plant* **2019**, *41*, 35. [CrossRef]
16. Rehana Sardar, G.; Ahmed, S.; Ahmad Yasin, N. Titanium dioxide nanoparticles mitigate cadmium toxicity in *Coriandrum sativum* L. through modulating antioxidant system, stress markers and reducing cadmium uptake. *Environ. Pollut* **2022**, *292*, 118373. [CrossRef]
17. Faraji, J.; Sepehri, A. Titanium Dioxide Nanoparticles and Sodium Nitroprusside Alleviate the Adverse Effects of Cadmium Stress on Germination and Seedling Growth of Wheat (*Triticum aestivum* L.). *Univ. Sci.* **2018**, *23*, 61–87. [CrossRef]
18. Emamverdian, A.; Ding, Y.; Xie, Y. The Role of New Members of Phytohormones in Plant Amelioration under Abiotic Stress with an Emphasis on Heavy Metals. *Pol. J. Environ. Stud.* **2020**, *29*, 1009–1020. [CrossRef]
19. Zafari, M.; Ebadi, A.; Sedghi, M.; Jahanbakhsh, S.; Miransari, M.J. Alleviating effect of 24-epibrassinolide on seed oil content and fatty acid composition under drought stress in safflower. *J. Food Compost. Anal.* **2020**, *92*, 103544. [CrossRef]
20. Siddiqui, H.; Hayat, S.; Bajguz, A. Regulation of photosynthesis by brassinosteroids in plants. *Acta Physiol. Plant.* **2018**, *40*, 11738–12018. [CrossRef]
21. Hayat, S.; Hasan, S.A.; Yusuf, M.; Hayat, Q.; Ahmad, A. Effect of 28-homobrassinolide on photosynthesis, fluorescence and antioxidant system in the presence or absence of salinity and temperature in *Vigna radiata*. *Environ. Exp. Bot.* **2010**, *69*, 105–112. [CrossRef]
22. Ahanger, M.A.; Ashraf, M.; Bajguz, A.; Ahmad, P. Brassinosteroids regulate growth in plants under stressful environments and crosstalk with other potential phytohormones. *J. Plant Growth Regul.* **2018**, *37*, 1007–1024. [CrossRef]
23. Zhou, J.; Liu, D.; Wang, P.; Ma, X.; Lin, W.; Chen, S.; Mishev, K.; Lu, D.; Kumar, R.; Vanhoutte, I.; et al. Regulation of Arabidopsis brassinosteroid receptor BRI1 endocytosis and degradation by plant U-box PUB12/PUB13-mediaed ubiquitination. *Proc. Natl. Acad. Sci. USA* **2018**, *115*, 1906–1915. [CrossRef]
24. Li, Y.M.; Feng, P.F. Bamboo resources in china based on the ninth national forest inventory data. *World Bamboo Ratt.* **2019**, *17*, 45–48.
25. Kang, F.; Li, X.; Du, H.; Mao, F.; Zhou, G.; Xu, Y.; Huang, Z.; Ji, J.; Wang, J. Spatiotemporal Evolution of the Carbon Fluxes from Bamboo Forests and their Response to Climate Change Based on a BEPS Model in China. *Remote Sens.* **2022**, *14*, 366. [CrossRef]
26. Emamverdian, A.; Ding, Y.; Ranaei, F.; Ahmad, Z. Application of bamboo plants in nine aspects. *Sci. World J.* **2020**, *2020*, 7284203. [CrossRef]
27. Huang, W.; Olson, E.; Wang, S.H.; Shi, P. The growth and mortality of *Pleioblastus pygmaeus* under different light availability. *Glob. Ecol.* **2020**, *24*, e01262. [CrossRef]
28. Murashige, T.; Skoog, F. A Revised Medium for Rapid Growth and Bio Assays with Tobacco Tissue Cultures. *Physiol. Plant.* **1962**, *15*, 473–497. [CrossRef]
29. Zhang, X.Z. The Measurement and Mechanism of Lipid Peroxidation and SOD, POD and CAT Activities in Biological System. In *Research Methodology of Crop Physiology*; Agriculture Press: Beijing, China, 1992; pp. 208–211.
30. Upadhyaya, A.; Sankhla, D.; Davis, T.D.; Sankhla, N.; Smith, B.N. Effect of paclobutrazol on the activities of some enzymes of activated oxygen metabolism and lipid peroxidation in senescing soybean leaves. *J. Plant Physiol.* **1985**, *121*, 453–461. [CrossRef]
31. Aebi, H. Catalase in vitro. In *Methods in Enzymology*; Elsevier: Amsterdam, The Netherlands, 1984; Volume 105, pp. 121–126. [CrossRef]
32. Foyer, C.H.; Halliwell, B. The presence of glutathione and glutathione reductase in chloroplasts: A proposed role in ascorbic acid metabolism. *Planta* **1976**, *133*, 1–25. [CrossRef]
33. Nakano, Y.; Asada, K. Hydrogen peroxide is scavenged by ascorbate-specific peroxidase in spinach chloroplasts. *Plant Cell Physiol.* **1981**, *22*, 867–880. [CrossRef]
34. Berner, M.; Krug, D.; Bihlmaier, C.; Vente, A.; Muller, R.; Bechthold, A. Genes and enzymes involved in caffeic acid biosynthesis in Actinomycete *Saccharothrix espanaensis*. *J. Bacteriol.* **2006**, *188*, 2666–2673. [CrossRef]
35. Conde, E.; Cadahia, E.; Garcia-Vallejo, M. HPLC analysis of flavonoids and phenolic acids and aldehydes in *Eucalyptus* spp. *Chromatographia* **1995**, *41*, 657–660. [CrossRef]
36. Akkol, E.K.; Goger, F.; Koşar, M.; Başer, K.H.C. Phenolic composition and biological activities of Salvia halophila and Salvia virgata from Turkey. *Food Chem.* **2008**, *108*, 942–949. [CrossRef] [PubMed]
37. Kayden, H.J.; Chow, C.K.; Bjornson, L.K. Spectrophotometric method for determination of tocopherol in red blood cells. *J. Lipid Res.* **1973**, *14*, 533–540. [CrossRef]

38. Siddiqui, H.; Ahmad, K.B.M.; Hayat, S. Comparative effect of 28-homobrassinolide and 24-epibrassinolide on the performance of different components influencing the photosynthetic machinery in *Brassica juncea* L. *Plant Physiol. Biochem* **2018**, *129*, 198–212. [CrossRef] [PubMed]
39. Patterson, B.D.; MacRae, E.A.; Ferguson, I.B. Estimation of hydrogen peroxide in plant extracts using titanium (IV). *Anal. Biochem.* **1984**, *139*, 487–492. [CrossRef]
40. Bradford, M.M. A rapid sensitive method for the quantification of microgram quantities of protein utilising the principle of protein-Dye Binding. *Anal. Biochem.* **1976**, *72*, 248–254. [CrossRef]
41. Li, C.; Bai, T.; Ma, F.; Han, M. Hypoxia tolerance and adaptation of anaerobic respiration to hypoxia stress in two Malus species. *Sci. Hortic.* **2010**, *124*, 274–279. [CrossRef]
42. Valentovic, P.; Luxova, M.; Kolarovic, L.; Gasparikova, O. Effect of osmotic stress on compatible solutes content, membrane stability and water relations in two maize cultivars. *Plant Soil. Environ.* **2006**, *52*, 186–191. [CrossRef]
43. Lichtenthaler, H.K.; Buschmann, C. Chlorophylls and carotenoids: Measurement and characterization by UV–VIS specroscopy. In *Current Protocols in Food Analytical Chemistry*; John Wiley & Sons, Inc.: Hoboken, NJ, USA, 2001; p. 4.
44. Souri, Z.; Karimi, N. Enhanced phytoextraction by as hyperaccumulator *Isatis cappadocica* spiked with sodium nitroprusside. *Soil Sediment Contam. Int. J.* **2017**, *26*, 457–468. [CrossRef]
45. Wang, J.; Nie, Y.; Dai, H.; Wang, M.; Cheng, L.; Yang, Z.; Chen, S.; Zhao, G.; Wu, L.; Guang, S. Parental exposure to TiO_2 NPs promotes the multigenerational reproductive toxicity of Cd in Caenorhabditis elegans via bioaccumulation of Cd in germ cells. *Environ. Sci. Nano* **2019**, *6*, 1332–1342. [CrossRef]
46. Singh, D.; Kumar, S.; Singh, S.C.; Lal, B.; Singh, N.B. Applications of liquid assisted pulsed laser ablation synthesized TiO_2 nanoparticles on germination, growth and biochemical parameters of *Brassica oleracea* var. *Capitata. Sci. Adv. Mather.* **2012**, *4*, 522–531. [CrossRef]
47. Bajguz, A.; Hayat, S. Effects of brassinosteroids on the plant responses to environmental stresses. *Plant Physiol. Biochem.* **2009**, *47*, 1–8. [CrossRef] [PubMed]
48. Bajguz, A. Brassinosteroids and lead as stimulators of phytochelatins synthesis in *Chlorella vulgaris*. *J. Plant Physiol.* **2002**, *159*, 321–324. [CrossRef]
49. Sharma, I.; Pati, P.K.; Bhardwaj, R. Effect of 28-homobrassinolide on antioxidant defence system in *Raphanus sativus* L. under chromium toxicity. *Ecotoxicology* **2011**, *20*, 862–874. [CrossRef]
50. Arora, P.; Bhardwaj, R.; Kanwar, M.K. Effect of 24-epibrassinolide on growth, protein content and antioxidative defense system of *Brassica juncea* L. subjected to cobalt ion toxicity. *Acta Physiol. Plant* **2012**, *34*, 2007–2017. [CrossRef]
51. Kohli, S.K.; Handa, N.; Sharma, A.; Gautam, V.; Arora, S.; Bhardwaj, R.; Alyemeni, M.N.; Wijaya, L.; Ahmad, P. Combined effect of 24-epibrassinolide and salicylic acid mitigates lead (Pb) toxicity by modulating various metabolites in *Brassica juncea* L. seedlings. *Protoplasma* **2018**, *255*, 11–24. [CrossRef]
52. Ahmad, P.; Abdel Latef, A.A.; Abd_Allah, E.F.; Hashem, A.; Sarwat, M.; Anjum, N.A.; Gucel, S. Calcium and potassium supplementation enhanced growth, osmolyte secondary metabolite production, and enzymatic antioxidant machinery in cadmium-exposed chickpea (*Cicer arietinum* L.). *Front. Plant Sci.* **2016**, *7*, 513. [CrossRef]
53. Choudhary, S.P.; Kanwar, M.; Bhardwaj, R.; Yu, J.-Q.; Tran, L.-S.P. Chromium stress mitigation by polyamine-brassinosteroid application involves phytohormonal and physiological strategies in *Raphanus sativus* L. *PLoS ONE* **2012**, *7*, e33210. [CrossRef]
54. Deng, X.-G.; Zhu, T.; Peng, X.-J.; Xi, D.-H.; Guo, H.; Yin, Y.; Zhang, D.-W.; Lin, H.-H. Role of brassinosteroid signaling in modulating tobacco mosaic virus resistance in *Nicotiana benthamiana*. *Sci. Rep.* **2016**, *6*, 20579. [CrossRef] [PubMed]
55. Mierziak, J.; Kostyn, K.; Kulma, A. Flavonoids as important molecules of plant interactions with the environment. *Molecules* **2014**, *19*, 16240–16265. [CrossRef] [PubMed]
56. Kováčik, J.; Bačkor, M. Phenylalanine ammonia-lyase and phenolic compounds in chamomile tolerance to cadmium and copper excess. *Water Air Soil Pollut.* **2007**, *185*, 185–193. [CrossRef]
57. Asghari, M.; Zahedipour, P. 24-Epibrassinolide acts as a growth-promoting and resistance-mediating factor in strawberry plants. *J. Plant Growth Regul.* **2016**, *35*, 722–729. [CrossRef]
58. Manquián-Cerda, K.; Escudey, M.; Zúñiga, G.; Arancibia-Miranda, N.; Molina, M.; Cruces, E. Effect of cadmium on phenolic compounds, antioxidant enzyme activity and oxidative stress in blueberry (*Vaccinium corymbosum* L. plantlets grown in vitro. *Ecotoxicol. Environ. Saf.* **2016**, *133*, 316–326. [CrossRef] [PubMed]
59. Sanjari, S.; Keramat, B.; Nadernejad, N.; Mozafari, H. Ameliorative effects of 24-epibrassinolide and thiamine on excess cadmium-induced oxidative stress in canola (*Brassica napus* L.) plants. *J. Plant Interact.* **2019**, *14*, 359–368. [CrossRef]
60. Mohammadi, R.; MaaliAmiri, R.; Mantri, N.L. Effect of TiO_2 nanoparticles on oxidative damage and antioxidant defense systems in chickpea seedlings during cold stress. *Russ. J. Plant Physiol.* **2014**, *61*, 768–775. [CrossRef]
61. Demin, I.N.; Deryabin, A.N.; Sinkevich, M.S.; Trunova, T.I. Insertion of cyanobacterial *desA* gene coding for Δ12-acyl-lipid desaturase increases potato plant resistance to oxidative stress induced by hypothermia. *Russ. J. Plant Physiol.* **2008**, *55*, 639–648. [CrossRef]
62. Surgun, Y.; Çöl, B.; Bürün, B. 24-Epibrassinolide ameliorates the effects of boron toxicity on *Arabidopsis thaliana* (L.) Heynh by activating an antioxidant system and decreasing boron accumulation. *Acta Physiol. Plant* **2016**, *38*, 71. [CrossRef]
63. Talarek-Karwel, M.; Bajguz, A.; Piotrowska-Niczyporuk, A. 24-Epibrassinolide modulates primary metabolites, antioxidants, and phytochelatins in *Acutodesmus obliquus* exposed to lead stress. *J. Appl. Phycol.* **2020**, *32*, 263–276. [CrossRef]

64. Santos, L.R.; Batista, B.L.; Lobato, A.K.S. Brassinosteroids mitigate cadmium toxicity in cowpea plants. *Photosynthetica* **2018**, *56*, 591–605. [CrossRef]
65. Pereira, Y.C.; Rodrigues, W.S.; Lima, E.J.A.; Santos, L.R.; Silva, M.H.L.; Lobato, A.K.S. Brassinosteroids increase electron transport and photosynthesis in soybean plants under water deficit. *Photosynthetica* **2019**, *57*, 181–191. [CrossRef]
66. Fariduddin, Q.; Yusuf, M.; Ahmad, I.; Ahmad, A. Brassinosteroids and their role in response of plants to abiotic stresses. *Biol. Plant.* **2014**, *58*, 9–17. [CrossRef]
67. Lei, Z.; Minqyu, S.; Xiao, W.; Chao, L.; Chunxiang, Q.; Liang, C.; Hao, H.; Xiaoqing, L.; Fashui, H. Antioxidant stress is promoted by nano-anatase in spinach chloroplasts under UVB radiation. *Biol. Trace Elem. Res.* **2008**, *121*, 69–79. [CrossRef] [PubMed]
68. Talarek-Karwel, M.; Bajguz, A.; Piotrowska-Niczyporuk, A.; Rajewska, I. The effect of 24-epibrassinolide on the green alga *Acutodesmus obliquus* (Chlorophyceae). *Plant Physiol. Biochem.* **2018**, *124*, 175–183. [CrossRef] [PubMed]
69. Bajguz, A. Effect of brassinosteroids on nucleic acids and protein content in cultured cells of *Chlorella vulgaris*. *Plant Physiol. Biochem.* **2000**, *38*, 209–215. [CrossRef]
70. Jan, S.; Alyemeni, M.N.; Wijaya, L.; Alam, P.; Siddique, K.H.; Ahmad, P. Interactive effect of 24-epibrassinolide and silicon alleviates cadmium stress via the modulation of antioxidant defense and glyoxalase systems and macronutrient content in *Pisum sativum* L. seedlings. *BMC Plant Biol.* **2018**, *18*, 146. [CrossRef]
71. Lyu, S.; Wei, X.; Chen, J.; Wang, C.; Wang, X.; Pa, D. Titanium as a beneficial element for crop production. *Front. Plant Sci.* **2017**, *8*, 597. [CrossRef]
72. Bacilieri, F.S.; Pereira de Vasconcelos, A.C.; Quintao Lana, R.M.; Mageste, J.G.; Torres, J.L.R. Titanium (Ti) in plant nutrition—A review. *Aust. J. Crop Sci.* **2017**, *11*, 382–386. [CrossRef]
73. Daryabeigi Zand, A.; Mikaeili Tabrizi, A.; Vaezi Heir, A. Co-application of biochar and titanium dioxide nanoparticles to promote remediation of antimony from soil by Sorghum bicolor: Metal uptake and plant response. *Heliyon* **2020**, *6*, e04669. [CrossRef]
74. Bajguz, A. Suppression of Chlorella vulgaris growth by cadmium, lead, and copper stress and Its restoration by endogenous brassinolide. *Arch. Environ. Contam. Toxicol.* **2011**, *60*, 406–416. [CrossRef] [PubMed]
75. Kanwar, M.K.; Bhardwaj, R.; Arora, P.; Chowdhary, S.P.; Sharma, P.; Kumar, S. Plant steroid hormones produced under Ni stress are involved in the regulation of metal uptake and oxidative stress in *Brassica juncea* L. *Chemosphere* **2012**, *86*, 41–49. [CrossRef] [PubMed]
76. Kroutil, M.; Hejtmánková, A.; Lachman, J. Effect of spring wheat (*Triticum aestivum* L.) treatment with brassinosteroids on the content of cadmium and lead in plant aerial biomass and grain. *Plant Soil Environ.* **2010**, *56*, 43–50. [CrossRef]
77. Dong, Y.; Chen, W.; Bai, X.; Liu, F.; Wan, Y. Effects of Exogenous Nitric Oxide and 24-Epibrassinolide on the Physiological Characteristics of Peanut Seedlings Under Cadmium Stress. *Pedosphere* **2019**, *29*, 45–59. [CrossRef]
78. Wani, A.S.; Tahir, I.; Ahmad, S.S.; Dar, R.A.; Nisar, S. Efficacy of 24-epibrassinolide in improving the nitrogen metabolism and antioxidant system in chickpea cultivars under cadmium and/or NaCl stress. *Sci. Hortic.* **2017**, *225*, 48–55. [CrossRef]
79. Ahmad, P.; Ahanger, M.A.; Alyemeni, M.N.; Wijaya, L.; Alam, P. Exogenous application of nitric oxide modulates osmolyte metabolism, antioxidants, enzymes of ascorbate-glutathione cycle and promotes growth under cadmium stress in tomato. *Protoplasma* **2018**, *255*, 79–93. [CrossRef]
80. Kelemen, G.; Keresztes, A.; Bacsy, E.; Feher, M.; Fodor, P.; Pais, I. Distribution and intracellular localization of titanium in plants after titanium treatment. *Food Struct.* **1993**, *12*, 8.
81. Silva, S.; Craveiro, S.C.H.; Oliveira, A.J.; Calado, R.J.B.; Pinto, A.M.S.; Silva, C. Santos Wheat chronic exposure to TiO_2-nanoparticles: Cyto- and genotoxic approach. *Plant Physiol. Biochem.* **2017**, *121*, 89–98. [CrossRef]
82. Larue, C.; Laurette, J.; Herlin-Boime, N.; Khodja, H.; Fayard, B.; Flank, A.M.; Brisset, F.M. Carriere Accumulation, translocation and impact of TiO_2 nanoparticles in wheat (*Triticum aestivum* spp.): Influence of diameter and crystal phase. *Sci. Total Environ.* **2012**, *431*, 197–208. [CrossRef]
83. Rahneshan, Z.; Nasibi, F.; Moghadam, A.A. Effects of salinity stress on some growth, physiological, biochemical parameters and nutrients in two pistachio (*Pistacia vera* L.) rootstocks. *J. Plant. Interact.* **2018**, *13*, 73–82. [CrossRef]
84. Bajguz, A. An enhancing effect of exogenous brassinolide on the growth and antioxidant activity in Chlorella vulgaris cultures under heavy metals stress. *Environ. Exp. Bot.* **2010**, *68*, 175–179. [CrossRef]
85. Rasool, S.; Urwat, U.; Nazir, M.; Zargar, S.M.; Zargar, M.Y. Cross talk between phytohormone signaling pathways under abiotic stress conditions and their metabolic engineering for conferring abiotic stress tolerance. In *Abiotic Stress-Mediated Sensing and Signaling in Plants: An Omics Perspective*; Zargar, S.M., Zargar, M.Y., Eds.; Springer: New York, NY, USA, 2018; pp. 329–350.
86. Tripathi, B.N.; Mehta, S.K.; Amar, A.; Gaur, J.P. Oxidative stress in Scenedesmus sp. during short- and long-term exposure to Cu^{2+} and Zn^{2+}. *Chemosphere* **2006**, *62*, 538–544. [CrossRef] [PubMed]
87. Polonini, H.C.; Brandao, H.M.; Raposo, N.R.; Brandao, M.A.; Mouton, L.; Coute, A.; Yepremian, C.; Sivry, Y.; Brayner, R. Size-dependent ecotoxicity of barium titanate particles: The case of *Chlorella vulgaris* green algae. *Ecotoxicology* **2015**, *24*, 938–948. [CrossRef] [PubMed]
88. Guo, J.; Zhou, R.; Ren, X.; Jia, H.; Hua, L.; Xu, H.; Lv, X.; Zhao, J.; Wei, T. Effects of salicylic acid, epi-brassinolide and calcium on stress alleviation and Cd accumulation in tomato plants. *Ecotoxicol. Environ. Saf.* **2018**, *157*, 491–496. [CrossRef] [PubMed]
89. Yusuf, M.; Khan, T.A.; Fariduddin, Q. Interaction of epibrassinolide and selenium ameliorates the excess copper in *Brassica juncea* through altered proline metabolism and antioxidants. *Ecotoxicol. Environ. Saf.* **2016**, *129*, 25–34. [CrossRef]

90. Wu, C.; Li, F.; Xu, H.; Zeng, W.; Yu, R.; Wu, X.; Shen, L.; Liu, Y.; Li, J. The potential role of brassinosteroids (BRs) in alleviating antimony (Sb) stress in *Arabidopsis thaliana*. *Plant Physiol. Biochem.* **2019**, *141*, 51–59. [CrossRef]
91. Lima, M.D.R.; Barros, U.D.; Batista, B.L.; Lobato, A.K.S. Brassinosteroids mitigate iron deficiency improving nutritional status and photochemical efficiency in *Eucalyptus urophylla* plants. *Trees Struct. Funct.* **2018**, *32*, 1681–1694. [CrossRef]
92. Li, Y.; Song, Y.; Shi, G.; Wang, J.; Hou, X. Response of antioxidant activity to excess copper in two cultivars of Brassica campestris ssp. chinensis Makino. *Acta Physiol. Plant.* **2008**, *31*, 155–162. [CrossRef]
93. Saeidnejad, A.; Mardani, H.; Naghibolghora, M. Protective effects of salicylic acid on physiological parameters and antioxidants response in maize seedlings under salinity stress. *J. Appl. Environ. Biol. Sci.* **2012**, *2*, 364–373.
94. Singh, J.; Lee, B.K. Influence of nano-TiO_2 particles on the bioaccumulation of Cd in soybean plants (*Glycine max*): A possible mechanism for the removal of Cd from the contaminated soil. *J. Environ. Manag.* **2016**, *170*, 88–96. [CrossRef]
95. Abdel Latef, A.A.H.; Srivastava, A.K.; El-sadek, M.S.A.; Kordrostami, M.; Tran, L.S.P. Titanium Dioxide Nanoparticles Improve Growth and Enhance Tolerance of Broad Bean Plants under Saline Soil Conditions. *Land Degrad. Dev.* **2017**, *29*, 1065–1073. [CrossRef]
96. Li, F.M.; Zhao, W.; Li, Y.Y.; Tian, Z.J.; Wang, Z.Y. Toxic effects of nanoTiO_2 on *Gymnodinium breve*. *Environ. Sci.* **2012**, *33*, 233–238.
97. Feizi, H.; Moghaddam, P.R.; Shahtahmassebi, N.; Fotovat, A. Impact of bulk and nanosized titanium dioxide (TiO_2) on wheat seed germination and seedling growth. *Biol. Trace. Elem. Res.* **2012**, *146*, 101–106. [CrossRef] [PubMed]
98. Tan, W.; Peralta-Videa, J.R.; Gardea-Torresdey, J.L. Interaction of titanium dioxide nanoparticles with soil components and plants: Current knowledge and future research needs—A critical review. *Environ. Sci. Nano* **2018**, *5*, 257–278. [CrossRef]
99. Mattiello, A.; Filippi, A.; Pošćić, F.; Musetti, R.; Salvatici, M.C.; Giordano, C.; Vischi, M.; Bertolini, A.; Marchiol, L. Evidence of phytotoxicity and genotoxicity in *Hordeum vulgare* L. exposed to CeO_2 and TiO_2 nanoparticles. *Front. Plant. Sci.* **2015**, *6*, 1043. [CrossRef]
100. Rastogi, A.; Zivcak, M.; Sytar, O.; Kalaji, H.M.; He, X.; Mbarki, S.; Brestic, M. Impact of Metal and Metal Oxide Nanoparticles on Plant: A Critical Review. *Front. Chem.* **2017**, *5*, 78. [CrossRef]
101. Kobayashi, K.; Kubota, H.; Hojo, R.; Miyagawa, M. Effective dispersal of titanium dioxide nanoparticles for toxicity testing. *J. Toxicol. Sci.* **2019**, *44*, 515–521. [CrossRef]

MDPI
St. Alban-Anlage 66
4052 Basel
Switzerland
Tel. +41 61 683 77 34
Fax +41 61 302 89 18
www.mdpi.com

Antioxidants Editorial Office
E-mail: antioxidants@mdpi.com
www.mdpi.com/journal/antioxidants

www.ingramcontent.com/pod-product-compliance
Lightning Source LLC
Chambersburg PA
CBHW041936260926
51850CB00018B/197

* 9 7 8 3 0 3 6 5 7 7 8 2 1 *